Advances in Speech Signal Processing

ELECTRICAL ENGINEERING AND ELECTRONICS

A Series of Reference Books and Textbooks

1. Rational Fault Analysis, *edited by Richard Saeks and S. R. Liberty*
2. Nonparametric Methods in Communications, *edited by P. Papantoni-Kazakos and Dimitri Kazakos*
3. Interactive Pattern Recognition, *Yi-tzuu Chien*
4. Solid-State Electronics, *Lawrence E. Murr*
5. Electronic, Magnetic, and Thermal Properties of Solid Materials, *Klaus Schröder*
6. Magnetic-Bubble Memory Technology, *Hsu Chang*
7. Transformer and Inductor Design Handbook, *Colonel Wm. T. McLyman*
8. Electromagnetics: Classical and Modern Theory and Applications, *Samuel Seely and Alexander D. Poularikas*
9. One-Dimensional Digital Signal Processing, *Chi-Tsong Chen*
10. Interconnected Dynamical Systems, *Raymond A. DeCarlo and Richard Saeks*
11. Modern Digital Control Systems, *Raymond G. Jacquot*
12. Hybrid Circuit Design and Manufacture, *Roydn D. Jones*
13. Magnetic Core Selection for Transformers and Inductors: A User's Guide to Practice and Specification, *Colonel Wm. T. McLyman*
14. Static and Rotating Electromagnetic Devices, *Richard H. Engelmann*
15. Energy-Efficient Electric Motors: Selection and Application, *John C. Andreas*
16. Electromagnetic Compossibility, *Heinz M. Schlicke*
17. Electronics: Models, Analysis, and Systems, *James G. Gottling*
18. Digital Filter Design Handbook, *Fred J. Taylor*
19. Multivariable Control: An Introduction, *P. K. Sinha*
20. Flexible Circuits: Design and Applications, *Steve Gurley, with contributions by Carl A. Edstrom, Jr., Ray D. Greenway, and William P. Kelly*
21. Circuit Interruption: Theory and Techniques, *Thomas E. Browne, Jr.*
22. Switch Mode Power Conversion: Basic Theory and Design, *K. Kit Sum*
23. Pattern Recognition: Applications to Large Data-Set Problems, *Sing-Tze Bow*
24. Custom-Specific Integrated Circuits: Design and Fabrication, *Stanley L. Hurst*
25. Digital Circuits: Logic and Design, *Ronald C. Emery*
26. Large-Scale Control Systems: Theories and Techniques, *Magdi S. Mahmoud, Mohamed F. Hassan, and Mohamed G. Darwish*
27. Microprocessor Software Project Management, *Eli T. Fathi and Cedric V. W. Armstrong (Sponsored by Ontario Centre for Microelectronics)*
28. Low Frequency Electromagnetic Design, *Michael P. Perry*
29. Multidimensional Systems: Techniques and Applications, *edited by Spyros G. Tzafestas*
30. AC Motors for High-Performance Applications: Analysis and Control, *Sakae Yamamura*

31. Ceramic Materials for Electronics: Processing, Properties, and Applications, *edited by Relva C. Buchanan*
32. Microcomputer Bus Structures and Bus Interface Design, *Arthur L. Dexter*
33. End User's Guide to Innovative Flexible Circuit Packaging, *Jay J. Miniet*
34. Reliability Engineering for Electronic Design, *Norman B. Fuqua*
35. Design Fundamentals for Low-Voltage Distribution and Control, *Frank W. Kussy and Jack L. Warren*
36. Encapsulation of Electronic Devices and Components, *Edward R. Salmon*
37. Protective Relaying: Principles and Applications, *J. Lewis Blackburn*
38. Testing Active and Passive Electronic Components, *Richard F. Powell*
39. Adaptive Control Systems: Techniques and Applications, *V. V. Chalam*
40. Computer-Aided Analysis of Power Electronic Systems, *Venkatachari Rajagopalan*
41. Integrated Circuit Quality and Reliability, *Eugene R. Hnatek*
42. Systolic Signal Processing Systems, *edited by Earl E. Swartzlander, Jr.*
43. Adaptive Digital Filters and Signal Analysis, *Maurice G. Bellanger*
44. Electronic Ceramics: Properties, Configuration, and Applications, *edited by Lionel M. Levinson*
45. Computer Systems Engineering Management, *Robert S. Alford*
46. Systems Modeling and Computer Simulation, *edited by Naim A. Kheir*
47. Rigid-Flex Printed Wiring Design for Production Readiness, *Walter S. Rigling*
48. Analog Methods for Computer-Aided Circuit Analysis and Diagnosis, *edited by Takao Ozawa*
49. Transformer and Inductor Design Handbook, Second Edition, Revised and Expanded, *Colonel Wm. T. McLyman*
50. Power System Grounding and Transients: An Introduction, *A. P. Sakis Meliopoulos*
51. Signal Processing Handbook, *edited by C. H. Chen*
52. Electronic Product Design for Automated Manufacturing, *H. Richard Stillwell*
53. Dynamic Models and Discrete Event Simulation, *William Delaney and Erminia Vaccari*
54. FET Technology and Application: An Introduction, *Edwin S. Oxner*
55. Digital Speech Processing, Synthesis, and Recognition, *Sadaoki Furui*
56. VLSI RISC Architecture and Organization, *Stephen B. Furber*
57. Surface Mount and Related Technologies, *Gerald Ginsberg*
58. Uninterruptible Power Supplies: Power Conditioners for Critical Equipment, *David C. Griffith*
59. Polyphase Induction Motors: Analysis, Design, and Application, *Paul L. Cochran*

60. Battery Technology Handbook, *edited by H. A. Kiehne*
61. Network Modeling, Simulation, and Analysis, *edited by Ricardo F. Garzia and Mario R. Garzia*
62. Linear Circuits, Systems and Signal Processing: Advanced Theory and Applications, *edited by Nobuo Nagai*
63. High-Voltage Engineering: Theory and Practice, *edited by M. Khalifa*
64. Large-Scale Systems Control and Decision Making, *edited by Hiroyuki Tamura and Tsuneo Yoshikawa*
65. Industrial Power Distribution and Illuminating Systems, *Kao Chen*
66. Distributed Computer Control for Industrial Automation, *Dobrivoje Popovic and Vijay P. Bhatkar*
67. Computer-Aided Analysis of Active Circuits, *Adrian Ioinovici*
68. Designing with Analog Switches, *Steve Moore*
69. Contamination Effects on Electronic Products, *Carl J. Tautscher*
70. Computer-Operated Systems Control, *Magdi S. Mahmoud*
71. Integrated Microwave Circuits, *edited by Yoshihiro Konishi*
72. Ceramic Materials for Electronics: Processing, Properties, and Applications, Second Edition, Revised and Expanded, *edited by Relva C. Buchanan*
73. Electromagnetic Compatibility: Principles and Applications, *David A. Weston*
74. Intelligent Robotic Systems, *edited by Spyros G. Tzafestas*
75. Switching Phenomena in High-Voltage Circuit Breakers, *edited by Kunio Nakanishi*
76. Advances in Speech Signal Processing, *edited by Sadoaki Furui and M. Mohan Sondhi*

Additional Volumes in Preparation

Pattern Recognition and Image Preprocessing, *Sing-Tze Bow*
Energy-Efficient Electric Motors, *John C. Andreas*
Large-Scale Stochastic Systems: Control, Estimation, and Detection, *edited by Spyros G. Tzafestas and Keigo Watanabe*

Electrical Engineering-Electronics Software

1. Transformer and Inductor Design Software for the IBM PC, *Colonel Wm. T. McLyman*
2. Transformer and Inductor Design Software for the Macintosh, *Colonel Wm. T. McLyman*
3. Digital Filter Design Software for the IBM PC, *Fred J. Taylor and Thanos Stouraitis*

Advances in Speech Signal Processing

edited by

Sadaoki Furui
NTT Human Interface Laboratories
Tokyo, Japan

M. Mohan Sondhi
AT&T Bell Laboratories
Murray Hill, New Jersey

Marcel Dekker, Inc. New York • Basel • Hong Kong

Library of Congress Cataloging-in-Publication Data

Advances in speech signal processing / edited by Sadaoki Furui, M. Mohan Sondhi.
p. cm. – (Electrical engineering and electronics)
Includes bibliographical references and index.
ISBN 0-8247-8540-1 (alk. paper)
1. Speech processing systems. 2. Signal processing—Digital techniques. I. Furui, Sadaoki. II. Sondhi, M. Mohan. III. Series.
TK7882.S65A29 1991
621.382′2—dc20 91-26455
CIP

This book is printed on acid-free paper.

MARCEL DEKKER, INC.
270 Madison Avenue, New York, New York 10016

Current printing (last digit):
10 9 8 7 6 5 4 3 2

PRINTED IN THE UNITED STATES OF AMERICA

To James L. Flanagan, for all he has taught us and for the inspiration he has given us over the years.

S.F.
M.M.S.

Preface

This book originated in an invitation from Marcel Dekker, Inc., to put together a book of original articles on various aspects of speech signal processing. After discussing the possible scope of such a book with several of our colleagues, we decided that the chapters should stress the advances during the past five to ten years. The past decade has seen dramatic advances in all areas of speech research, and we believe that the ideas developed during this period will influence research for many years in the future.

We were fortunate that so many excellent authors accepted our invitations to contribute to our book. We can safely say that each of the 25 chapters was written by one or more leading authorities on the subject. We gave the authors considerable latitude concerning the organization and content of their chapters. Our only request was that they should emphasize the developments during the past five to ten years. Consequently, the book presents a wealth of material, in a variety of styles, which makes for interesting and informative reading.

One way in which we did have a strong influence on the content of the book is, of course, in the selection of topics. Our own interests lie mainly in the area of developing methods and algorithms for processing speech signals to perform various tasks such as coding and recognition. Neither of us has been involved with hardware implementation of such algorithms, and the chapters reflect this bias. However, the algorithms are described in a manner that enables one to implement them in hardware without undue difficulty.

The book is divided into four sections: Analysis and Coding, Enhancement and Quality Assessment, Recognition and Verification, and Text-to-Speech Synthesis. Together these sections cover almost all aspects of speech signal processing.

The section on analysis and coding has nine chapters. The first, by Hess, deals with the problem of accurately determining periods of voicing and the fundamental voice frequency. Several new methods have been developed recently, even for this long-standing problem. Chapter 2, by Gersho, Wang, and Zeger, deals with vector quantization—a topic that is of much more recent origin but that has already found application in many areas of signal processing. Efficient coding of speech signals is the subject matter of the next six chapters. Jayant gives a broadbrush overview of speech coding and applications, with emphasis on international standards for coding at various bit rates. Johnston and Brandenburg discuss coding of high-quality wideband (20 kHz) speech and music signals, with the aim of "transparent" coding, i.e., coding such that the original and coded versions are indistinguishable to a listener. The chapter by Kroon and Atal discusses linear predictive coding, which is perhaps the most widely used technique for the coding of speech signals. It has been highly successful for coding telephone speech at 8 kbits/s and, more recently, at bit rates as low as 4 kbits/s. McAulay and Quatieri discuss schemes for roughly the same transmission rates but in which the speech signal is represented as a sum of sinusoids with slowly varying amplitudes and frequencies. Coding of speech signals at bit rates lower than 4 kbits/s is of interest, for instance, in cellular radio telephony. However, the quality of speech coded at these low rates is not yet adequate for such applications. The chapter by Honda and Shiraki summarizes the present status of such low bit rate coders. The chapter by Schroeter and Sondhi also focuses on coding at very low bit rates. It discusses the representation of a speech signal in terms of parameters that specify not the speech signal but the vocal apparatus that produced it. Successful coders based on such parameters have not yet been realized. The final chapter in this section, by Ifukube, is concerned with speech coding of a very different sort. The aim here is to encode the speech into signals to be applied to electrodes implanted in the inner ear. Experimentation with such cochlear implants, as weapons against profound deafness, started almost 20 years ago. However, only in the past eight or ten years have they received widespread interest.

The next section of the book deals with enhancement and quality assessment. Of the four chapters in this section, two deal with enhancement and two with quality assessment. Boll's chapter deals with enhancement of a speech signal that has been degraded by additive noise. Such degradation can occur electrically (e.g., in amplifiers) or acoustically (e.g. , in a fighter cockpit). A very different type of degradation occurs when a speech signal is corrupted by the echo of another. The chapter by Sondhi and Kellermann deals with combating such echoes, which, again, might arise electrically or acoustically in a variety of communication situations. Assessment of speech quality is the subject of the next two chapters. Kitawaki discusses methods useful for assessing speech coders. In this application, the original and coded version of a signal are both available and may be used in arriving at an assessment. In the last chapter of this section, Pols discusses evaluation of the output of text-to-speech synthesizers, which is a significantly different problem. When the quality of a text-to-speech synthesizer is to be assessed, there is no "original" signal available, and the types of distortions are rather different from the ones encountered in speech coders.

The section on recognition and verification has nine chapters. The first, by Shikano and Itakura, describes one of the most basic problems in speech recognition, that is, the problem of spectral distance measures, with emphasis on improving phonetic unit recognition and realizing robust speech recognition. The chapter by Ghitza presents computational models that utilize detailed knowledge about the physiology of the auditory periphery. Such models have the potential of significantly improving future speech processing systems. The chapter by Sakoe summarizes the progress made in using dynamic programming (DP) algorithms for speech recognition in the past 10 years, including the incorporation of the DP algorithm into neural networks. The chapter by Juang and Rabiner reviews the statistical framework of hidden Markov models (HMM), which have become predominant in speech recognition during the past few years. It also discusses the promises and shortcomings of such models. The chapter by Waibel describes neural network approaches—the most recent development in new models for speech recognition. The chapter by Furui reviews speaker-independent and speaker-adaptive recognition techniques which have recently been actively investigated to cope with the problem of voice individuality. The chapter by Lee and Alleva summarizes the considerable progress made recently in continuous speech recognition based on HMMs. The chapter by Jelinek introduces principles and methods of language modeling, one of the most important issues in continuous speech recognition. The last chapter, by Rosenberg and Soong, summarizes recent advances in the automatic extraction of a talker's identity from measurements on the speech signal.

The section on text-to-speech (TTS) synthesis consists of three chapters. The chapter by Allen surveys text-to-speech systems which convert an input string of text characters into an output speech waveform. The overall TTS conversion process is characterized as text analysis followed by speech synthesis. The chapter by Liberman and Church summarizes techniques and problems in text analysis, including word pronunciation. The chapter by Sato describes methods of speech synthesis, focusing on formant-based and LPC-based techniques. Although currently available commercial systems can produce speech of acceptable intelligibility from unrestricted input text, the problem of providing high-quality TTS systems remains to be solved. These systems are expected to improve as future research brings deeper understanding of speech and natural language.

We thank all the authors for their contributions.

Sadaoki Furui
M. Mohan Sondhi

Contents

Preface v

Contributors xiii

I. Analysis and Coding

1. Pitch and Voicing Determination *Wolfgang J. Hess* 3
2. Vector Quantization Techniques in Speech Coding *Allen Gersho, Shihua Wang, and Kenneth Zeger* 49
3. High-Quality Coding of Telephone Speech and Wideband Audio *Nikil Jayant* 85
4. Wideband Coding—Perceptual Considerations for Speech and Music *James D. Johnston and Karlheinz Brandenburg* 109
5. Predictive Coding of Speech Using Analysis-by-Synthesis Techniques *Peter Kroon and Bishnu S. Atal* 141
6. Low-Rate Speech Coding Based on the Sinusoidal Model *Robert J. McAulay and Thomas F. Quatieri* 165

7. Very Low-Bit-Rate Speech Coding *Masaaki Honda and Yoshinao Shiraki* 209

8. Speech Coding Based on Physiological Models of Speech Production *Juergen Schroeter and M. Mohan Sondhi* 231

9. Signal Processing for Cochlear Implants *Tohru Ifukube* 269

II. Enhancement and Quality Assessment

10. Speech Enhancement in the 1980s: Noise Suppression with Pattern Matching *Steven F. Boll* 309

11. Adaptive Echo Cancellation for Speech Signals *M. Mohan Sondhi and Walter L. Kellermann* 327

12. Quality Assessment of Coded Speech *Nobuhiko Kitawaki* 357

13. Quality Assessment of Text-to-Speech Synthesis by Rule *Louis C. W. Pols* 387

III. Recognition and Verification

14. Spectrum Distance Measures for Speech Recognition *Kiyohiro Shikano and Fumitada Itakura* 419

15. Auditory Nerve Representation as a Basis for Speech Processing *Oded Ghitza* 453

16. Dynamic Programming–Based Speech Recognition Algorithms *Hiroaki Sakoe* 487

17. Issues in Using Hidden Markov Models for Speech Recognition *Biing-Hwang Juang and Lawrence R. Rabiner* 509

18. Neural Network Approaches for Speech Recognition *Alexander Waibel* 555

19. Speaker-Independent and Speaker-Adaptive Recognition Techniques *Sadaoki Furui* 597

20. Continuous Speech Recognition *Kai-Fu Lee and Fil Alleva* 623

21. Principles of Lexical Language Modeling for Speech Recognition *Frederick Jelinek, Robert L. Mercer, and Salim Roukos* 651

22. Recent Research in Automatic Speaker Recognition 701
Aaron E. Rosenberg and Frank K. Soong

IV. Text-to-Speech Synthesis

23 Overview of Text-to-Speech Systems *Jonathan Allen* 741

24. Text Analysis and Word Pronunciation in Text-to-Speech Synthesis *Mark Y. Liberman and Kenneth W. Church* 791

25. Speech Synthesis for Text-to-Speech Systems *Hirokazu Sato* 833

Index *855*

Contributors

Jonathan Allen Electrical Engineering and Computer Science Department, Massachusetts Institute of Technology, Cambridge, Massachusetts

Fil Alleva School of Computer Science, Carnegie-Mellon University, Pittsburgh, Pennsylvania

Bishnu S. Atal Speech Research Department, AT&T Bell Laboratories, Murray Hill, New Jersey

Steven F. Boll Research & Development, ITT Aerospace/Communications Division, San Diego, California

Karlheinz Brandenburg Lehrsruhl für Technische Electronik, Universität Erlangen-Nürnberg, Erlangen, Germany

Kenneth W. Church Linguistics Research Department, AT&T Bell Laboratories, Murray Hill, New Jersey

Sadaoki Furui Speech and Acoustics Laboratory, NTT Human Interface Laboratories, Tokyo, Japan

Allen Gersho Department of Electrical and Computer Engineering, University of California, Santa Barbara, California

Oded Ghitza Acoustics Research Department, AT&T Bell Laboratories, Murray Hill, New Jersey

Wolfgang J. Hess Institute for Communications Research and Phonetics, University of Bonn, Bonn, Germany

Masaaki Honda Information Science Laboratory, NTT Basic Research Laboratories, Tokyo, Japan

Tohru Ifukube Research Institute of Applied Electricity, Hokkaido University, Hokkaido, Japan

Fumitada Itakura Department of Electronics Engineering, Nagoya University, Aichi, Japan

Nikil Jayant Signal Processing Research Department, AT&T Bell Laboratories, Murray Hill, New Jersey

Frederick Jelinek Computer Sciences Department, Thomas J. Watson Research Center, IBM Research Division, Yorktown Heights, New York

James D. Johnston Signal Processing Research Department, AT&T Bell Laboratories, Murray Hill, New Jersey

Biing-Hwang Juang Speech Research Department, AT&T Bell Laboratories, Murray Hill, New Jersey

Walter L. Kellermann Acoustics Research Department, AT&T Bell Laboratories, Murray Hill, New Jersey

Nobuhiko Kitawaki Telecommunications Quality Laboratory, NTT Telecommunication Networks Laboratories, Tokyo, Japan

Peter Kroon Speech Research Department, AT&T Bell Laboratories, Murray Hill, New Jersey

Kai-Fu Lee Advanced Technology Group, Apple Computer, Inc., Cupertino, California

Mark Y. Liberman Department of Linguistics, University of Pennsylvania, Philadelphia, Pennsylvania

Robert J. McAulay Speech Systems Technology Group, Lincoln Laboratory, Massachusetts Institute of Technology, Lexington, Massachusetts

Robert L. Mercer Computer Sciences Department, Thomas J. Watson Research Center, IBM Research Division, Yorktown Heights, New York

Louis C. W. Pols Institute of Phonetic Sciences, University of Amsterdam, Amsterdam, The Netherlands

Thomas F. Quatieri Speech Systems Technology Group, Lincoln Laboratory, Massachusetts Institute of Technology, Lexington, Massachusetts

Lawrence R. Rabiner Information Principles Research, AT&T Bell Laboratories, Murray Hill, New Jersey

Aaron E. Rosenberg Speech Research Department, AT&T Bell Laboratories, Murray Hill, New Jersey

Salim Roukos Computer Sciences Department, Thomas J. Watson Research Center, IBM Research Division, Yorktown Heights, New York

Hiroaki Sakoe Department of Computer Science and Communication Engineering, Kyushu University, Fukuoka, Japan

Hirokazu Sato Speech and Acoustics Laboratory, NTT Human Interface Laboratories, Kanagawa, Japan

Juergen Schroeter Acoustics Research Department, AT&T Bell Laboratories, Murray Hill, New Jersey

Kiyohiro Shikano Speech and Acoustics Laboratory, NTT Human Interface Laboratories, Tokyo, Japan

Yoshinao Shiraki Information Science Laboratory, NTT Basic Research Laboratories, Tokyo, Japan

M. Mohan Sondhi Acoustics Research Department, AT&T Bell Laboratories, Murray Hill, New Jersey

Frank K. Soong Speech Research Department, AT&T Bell Laboratories, Murray Hill, New Jersey

Alexander Waibel School of Computer Science, Carnegie-Mellon University, Pittsburgh, Pennsylvania

Shihua Wang Department of Electrical and Computer Engineering, University of California, Santa Barbara, California

Kenneth Zeger Department of Electrical Engineering, University of Hawaii, Honolulu, Hawaii

Part I

Analysis and Coding

1

Pitch and Voicing Determination

WOLFGANG J. HESS University of Bonn, Bonn, Germany

Voicing determination and pitch determination are the two subproblems of *voice source analysis*. In voiced speech, the vocal cords vibrate in a quasi-periodic way. Speech segments with voiceless excitation are generated by turbulent air flow at a constriction or by the release of a closure in the vocal tract. The parameters we have to determine are the *manner of excitation*—i.e., the presence of a *voiced excitation* and the presence of a *voiceless excitation*, a problem we will refer to as voicing determination—and, for the segments of the speech signal in which a voiced excitation is present, the *rate of vocal cord vibration*, which is usually referred to as *pitch determination* or *fundamental frequency determination* in the literature.

Unlike the analysis of vocal tract parameters, where a number of independent and equivalent representations are possible (e.g., Rabiner and Schafer, 1978), there is no alternative to the parameters pitch and voicing, and the quality of synthesized signal critically depends on their reliable and accurate determination (Gold, 1977). This chapter presents a selection of the methods applied and pitch and voicing determination. Pitch determination algorithms and methods are presented in Section 1; Section 2 gives a brief overview of voicing determination methods. Section 3 finally discusses problems of error analysis and evaluation, accurate voicing and pitch determination using instruments such as the laryngograph, and various applications.

Especially pitch determination has a rather long history which goes back beyond the times of vocoding (e.g., Grützmacher and Lottermoser, 1937). The most important developments leading to today's state of the art were made in the 1960s and 1970s; most of these methods that are briefly reviewed in this chapter were extensively discussed in Hess (1983). Since then, few absolutely new principles have been invented; a number of methods, however, were improved and refined, whereas other solutions that required an amount of computational effort which appeared unrealistic at the time the algorithm was first developed were revived. With the widespread use of data bases containing many labeled and processed speech data, researchers nowadays tend toward thoroughly examining and checking the performance of their algorithms.

1. PITCH DETERMINATION

Pitch, i.e., *fundamental frequency* (or *rate of vocal fold vibration*) F_0, as well as *fundamental period* T_0, has a key position in the acoustic speech signal. The prosodic information of an utterance is predominantly determined by this parameter. The ear is by an order of magnitude more sensitive to changes of fundamental frequency than to changes of other speech signal parameters (Flanagan and Saslow, 1958; Klatt, 1973; Harris and Umeda, 1987). The quality of vocoder speech is essentially influenced by the quality and faultlessness of the pitch measurement (Gold, 1977). Hence the importance of this parameter for good and reliable measurement methods.

At first glance the task looks simple: one just has to detect the fundamental frequency of a quasi-periodic signal. When dealing with speech signals, however, the assumption of quasi-periodicity is often far from reality. Figure 1 shows an arbitrary (but typical) example of a speech signal. For a number of reasons, the task of pitch determination must be counted among the most difficult problems in speech analysis.

1. In principle, speech is a nonstationary process; the momentary position of the vocal tract may change abruptly at any time. This leads to drastic variations in the temporal structure of the signal, even between subsequent pitch periods.

2. Due to the flexibility of articulatory gestures and the wide variety of voices, there exist a multitude of possible temporal structures. Narrowband formants at low harmonics (especially at the second or third harmonic) are a particular source of trouble.

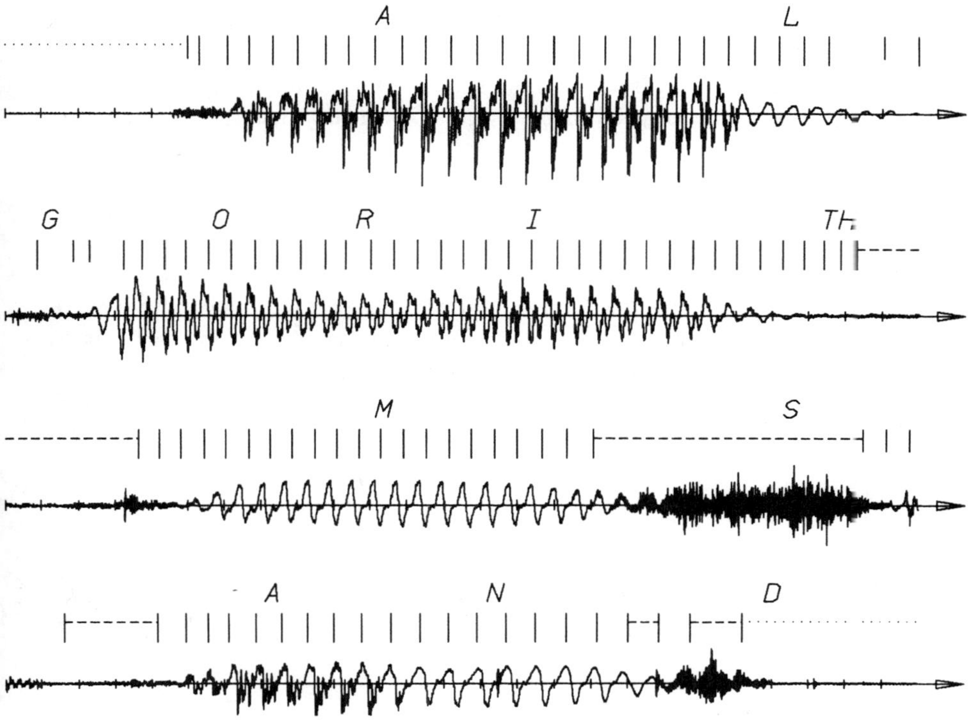

Figure 1 Example of a speech signal (after pitch determination). Beginning of the utterance "Algorithms and devices for pitch determination". Speaker: male; undistorted signal. Scale: 250 ms per line. The analysis was done using the algorithm by Hess (1979). (------) Voiceless; (······) pause; (||||||) pitch period boundaries ("markers"). Markers indicated by short lines were found irregular.

3. For an arbitrary speech signal uttered by an unknown speaker, the fundamental frequency can vary over a range of almost four octaves (50 to 800 Hz). Especially for female voices, F_0 thus often coincides with the first formant (the latter ranging from about 200 to 1400 Hz). This causes problems when inverse filtering techniques are applied.

4. The excitation signal itself is not always regular. Even under normal conditions, i.e., when the voice is neither hoarse nor pathologic, the glottal waveform exhibits occasional irregularities (Dolansky and Tjernlund, 1968; Fujimura, 1968; Lieberman, 1963). In addition, the voice may temporarily fall into vocal fry or creak (Hollien, 1974), which is a nonpathologic mode of

voice excitation with rather large and irregular intervals between subsequent glottal pulses (see Fig. 2).

5. Additional problems arise in speech communication systems, where the signal is often distorted or band limited (for instance, in the telephone channel).

Literally hundreds of methods for pitch determination have been developed. This section will give a survey of the prevailing principles and discuss some selected methods in more detail. First, we will deal with possible definitions of the parameter pitch itself, followed by a gross categorization of the various principles of its determination. After that we will go into a more detailed discussion of individual principles and individual solutions. As to the realization, we will not distinguish between a hardware device (whether analog or digital) and an algorithmic solution; they are all regarded as *pitch determination algorithms* (PDAs). In addition we will separate the problems

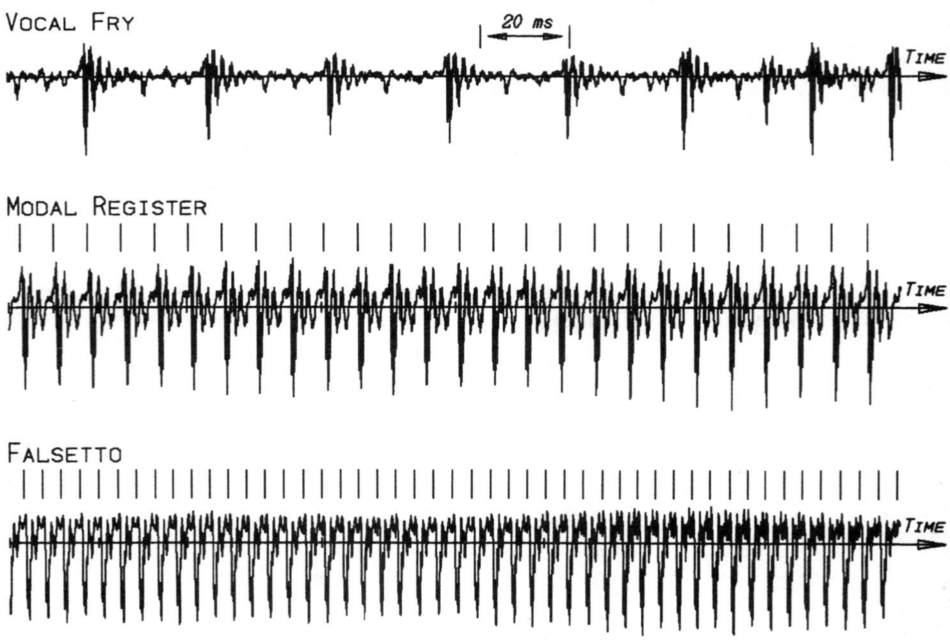

Figure 2 Speech signals in different phonation: vocal fry (upper line); modal register, i.e., normal speech (lower line, with pitch period delimiters). Signal: sustained vowel /ε/, speaker WGH (male).

of pitch determination and voicing determination, although the two are often realized within the same algorithm; we will assume in the remainder of this section that a voiced-unvoiced decision has been made already and that there are only voiced signals to be processed by the respective PDA.

1.1. Basic Definitions of Pitch

Pitch can be measured in many ways. If the signal is completely stationary and periodic, all these strategies—provided they operate correctly—lead to identical results. Since the speech signal is nonstationary and time variant, however, aspects of strategy such as the starting point of the measurement, the length of the measuring interval, the way of averaging (if any), or the operating domain (time, frequency, lag etc.) of an individual algorithm start to influence the results and may lead to estimates that differ from algorithm to algorithm even if all these results are "correct" and "accurate." Before entering a discussion of individual methods, we must therefore have a look at the parameter pitch and provide a clear definition of what should be measured and what is actually measured.

A word on terminology first. There are three points of view for looking at a speech processing problem (Zwicker et al., 1967): the *production*, the *signal processing*, and the *perception* points of view. In the actual case of pitch determination the production point of view is obviously oriented toward the generation of the excitation signal in the larynx; we will thus have to start from a time domain representation of the waveform as a train of laryngeal pulses. If an algorithm or device works in a speech-production oriented way, it measures individual *laryngeal excitation cycles* or, if some averaging is performed, it determines the *rate of vocal fold vibration*. The signal processing point of view can be characterized in such a way that (quasi-)periodicity is observed in the signal, wherever that signal comes from, and the task is just to extract the features that best represent this periodicity. The pertinent terms are *fundamental frequency* and *fundamental period*. If individual cycles are determined, we may (somewhat inconsistently) speak of *pitch periods* or simply of *periods*. The perception point of view leads to a frequency domain representation, since pitch sensation corresponds to a frequency and not to an average period or a sequence of periods (Goldstein, 1973; Terhardt, 1979; Plomp, 1976). This point of view is associated with the original meaning of the term *pitch*. However, the term pitch has consistently been used as some kind of "common denominator," i.e., as a general name for all the terms mentioned before, at least in the technical literature (Kohler, 1982). In addition, psychoacousticians have started to create new terms to describe the aspects of pitch perception, such as *spectral pitch* and *virtual pitch* (Terhardt, 1979), mostly because they feel it necessary to specify partial aspects of the complex phenomenon of pitch perception more precisely, but also in order to avoid confusion. In the following, we will therefore use the term pitch in this wider sense wherever a more restricted description is undesirable or impossible, and take the more precise terms otherwise.

Defining the different representations of pitch, it appears reasonable to proceed from production to perception. Going in that direction we will start at a local and detailed representation and arrive at a more global representation in the case of the perception-oriented view. The basic definitions could thus read as follows (Hess, 1983, p. 475; Hess and Indefrey, 1987):

> T_0 is defined as the elapsed time between two successive laryngeal pulses. Measurement starts at a well-specified point within the glottal cycle, preferably at the point of glottal closure or—if the glottis does not close completely—at the point where the glottal area reaches its minimum. (1)

PDAs that obey this definition will be able to locate the point of glottal closure and to delimit individual laryngeal excitation cycles. This task, which usually forms part of a glottal inverse filter, goes far beyond the scope of ordinary pitch determination; if the speech signal alone is available for the analysis, reliable results are to be expected only for selected algorithms and only if the signal is totally undistorted. With the aid of an instrument (see Section 3.2) this problem can be solved in a more general way.

> T_0 is defined as the elapsed time between two successive laryngeal pulses. Measurement starts at an arbitrary point within the glottal cycle. Which point that is depends on the individual method, but for a given PDA this point is always located at the same position within the glottal cycle. (2)

Ordinary time domain PDAs (Section 1.4) follow this definition. The reference point can be a significant extreme, a certain zero crossing, an excursion cycle, and so on. This is not necessarily the point of glottal closure itself. Usually, however, it is possible to derive the point of glottal closure from this reference point when the signal is undistorted. Yet the presence of phase distortions can even destroy this possibility. PDAs that follow this definition usually track the signal period by period in a synchronous way, and a commonly used term (although somewhat inconsistent with the definition of the term pitch as given above) for what is measured here is *individual pitch periods*. A well-known example of this kind of algorithm is the one by Reddy

(1967; see Section 1.4.1).

> T_0 is defined as the elapsed time between two successive laryngeal cycles. Measurement starts at an arbitrary instant which is fixed according to external conditions and ends when a complete cycle has elapsed. (3)

This is an *incremental* definition of T_0. T_0 is still defined as the length of an individual period but no longer from the speech production point of view, since the definition has nothing to do with the individual excitation cycle. The synchronous method of processing is maintained, but the phase relations between the laryngeal waveform and the markers, i.e., the pitch period delimiters at the output of the algorithm, are lost. Once a reference point in time has been established, it will be kept only as long as the measurement is correct and as long as voicing continues. If there is a measurement error, or if voicing ceases, the location of the reference point is lost, and the next reference point may be completely different with respect to its position within the excitation cycle.

> T_0 is defined as the average length of several periods, i.e., as the average elapsed time between a small number of successive laryngeal cycles. How the averaging is performed and how many periods are involved are matters of the individual method. (4a)

This is the standard definition of T_0 for any PDA that applies stationary short-term analysis, including the implementations of frequency domain PDAs. Well-known methods, such as cepstrum (Noll, 1967) and autocorrelation (Rabiner, 1977) methods, follow this definition. The corresponding frequency domain definition reads as follows.

> F_0 is defined as the fundamental frequency of an (approximately) harmonic pattern in the (short-term) spectral representation of the signal. It depends on the particular method whether F_0 is calculated as the frequency of a certain harmonic divided by the respective harmonic number m (including $m = 1$), as the frequency difference between adjacent spectral peaks, or as the greatest common divisor of the frequencies of the individual harmonics. (4b)

The perception point of view of the problem leads to a different definition of pitch. Pitch perception happens in the frequency domain. According to the existing theories (Plomp, 1976).

> F_0 is defined as the frequency of the sinusoid that evokes the same perceived pitch (residue pitch, virtual pitch, etc.) as the complex sound which represents the input speech signal. (5)

This definition is principally different from the previous ones. Above all, it is a *long-term* definition (Terhardt et al., 1982). The pitch perception theories were developed for stationary complex sounds and were extended toward short pulse trains with varying amplitude patterns and constant frequencies but not toward signals with varying fundamental frequencies. Except for some recent investigations which indicate that the difference limen for F_0 changes goes up by at least an order of magnitude when time-variant stimuli are involved (Harris and Umeda, 1987; Hart. 1981), the question of the behavior of the human ear with respect to *short-term* pitch perception is only partially answered, and our knowledge about what kind of short-term "analysis" is executed in the human ear and how it is executed is still incomplete. Hence even PDAs that claim to be perception oriented (e.g., Duifhuis et al., 1982) enter the frequency domain in a manner similar to that in definition (4b), i.e., by a standard short-term transformation such as the discrete Fourier transform (DFT) with previous windowing of the signal.

Since the results of individual algorithms may be different according to the definition they follow, and since definitions (1) through (5) are partly given in the time (or lag) domain and partly in the frequency domain, it is necessary to reestablish the relation between the time domain and frequency domain representations of pitch,

$$F_0 = \frac{1}{T_0} \tag{6}$$

in such a way that when a measurement is carried out in one of the domains, however T_0 or F_0 is defined there, the representation in the other domain will always be established by this equation.

1.2. Categorizing the Various Principles

We subdivide a PDA into three processing steps: (1) the preprocessor, (2) the basic extractor, and (3) the postprocessor (McKinney, 1965; Hess, 1983, p. 152). The basic extractor performs the main task: it converts the input signal into a series of pitch estimates. The task of the preprocessor is data reduction and enhancement in order to facilitate the operation of the basic extractor. The postprocessor operates in a more application-oriented way. Some of its typical tasks are error correction, smoothing the pitch contour, and graphic display.

The existing PDA principles can be split up into two gross categories when the input signal of the basic extractor is taken as a criterion. If this signal has the same time base as the original speech signal, the PDA operates in the time domain. It will thus measure T_0 according to one of definitions (1) through (3). In all other cases, somewhere in the preprocessor the time domain is left. Since the speech signal is time variant, this cannot be done other than by a short-term transformation; in this case we will usually determine T_0 or F_0 according to definition (4a), (4b), or (5); in some rare cases definition (3) may apply as well. Accordingly, we have the two PDA categories: *time-domain* PDAs and *short-term analysis* PDAs.

1.3. Short-Term Analysis PDAs

1.3.1. Principle of Short-Term Analysis

In any short-term analysis PDA a *short-term* (or *short-time*) *transformation* is performed in the preprocessor step. The speech signal is split up into a series of frames (Fig. 3); an individual frame is obtained by taking a limited number

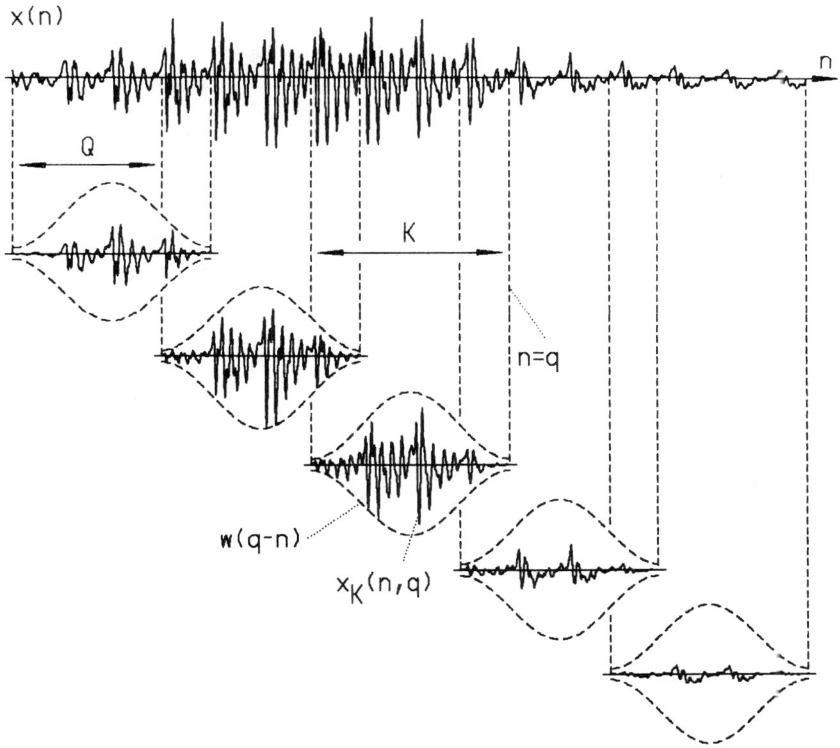

Figure 3 A principle of short-term analysis.

of consecutive samples of the signal $x(n)$ from the starting point, $n = q - K + 1$, to the ending point, $n = q$. The frame length, K, is chosen short enough so that the parameter(s) to be measured can be assumed approximately constant within the frame. On the other hand, K must be large enough to guarantee that the parameter remains measurable. For most short-term analysis PDAs a frame thus requires two or three complete periods at least. In extreme cases, when F_0 changes abruptly or when the signal is irregular, the contradiction of these two conditions can be a source of error (Fujisaki et al., 1986). The frame interval Q, i.e., the distance between consecutive frames (or its reciprocal, the frame rate), is determined in such a way that any significant parameter change is documented in the measurements.

The short-term transformation, so to speak, is intended to behave like a concave mirror which focuses all the scattered information on pitch, as far as it is available within the frame, into one single peak in the spectral domain. This peak is then determined by a peak detector (as the usual implementation of the basic extractor in this type of PDA). Hence the output signal of the basic extractor is a sequence of average pitch estimates. The short-term transform causes the phase relations between the spectral domain and the original signal to be lost. At the same time, however, the algorithm loses much of its sensitivity to phase distortions and signal degradation. Unfortunately, the increased reliability of the algorithm is accompanied by increased computing effort (which is at least one order of magnitude higher than for a time domain PDA). Much of this effort goes into the numeric calculation of the transform. Besides the search for reliability, the search for fast implementation has therefore been an important issue in the design of short-term analysis PDAs.

1.3.2. Overview

Not all the known spectral transforms show the desired focusing effect. Those ones which do are in some way related to the power spectrum: correlation techniques, frequency domain techniques, and a least-squares approach (Fig. 4). Among the correlation techniques we find the well-known autocorrelation function, which became successful in pitch determination of band-limited signals when it was combined with time domain center clipping (Sondhi, 1968; Rabiner, 1977). Its counterpart is given by applying a distance function, for instance, the average magnitude difference function (AMDF) (Sobolev and Baronin, 1968; Ross et al., 1974):

$$\mathrm{AMDF}(d) = \sum_n |x(n) - x(n+d)| \tag{7}$$

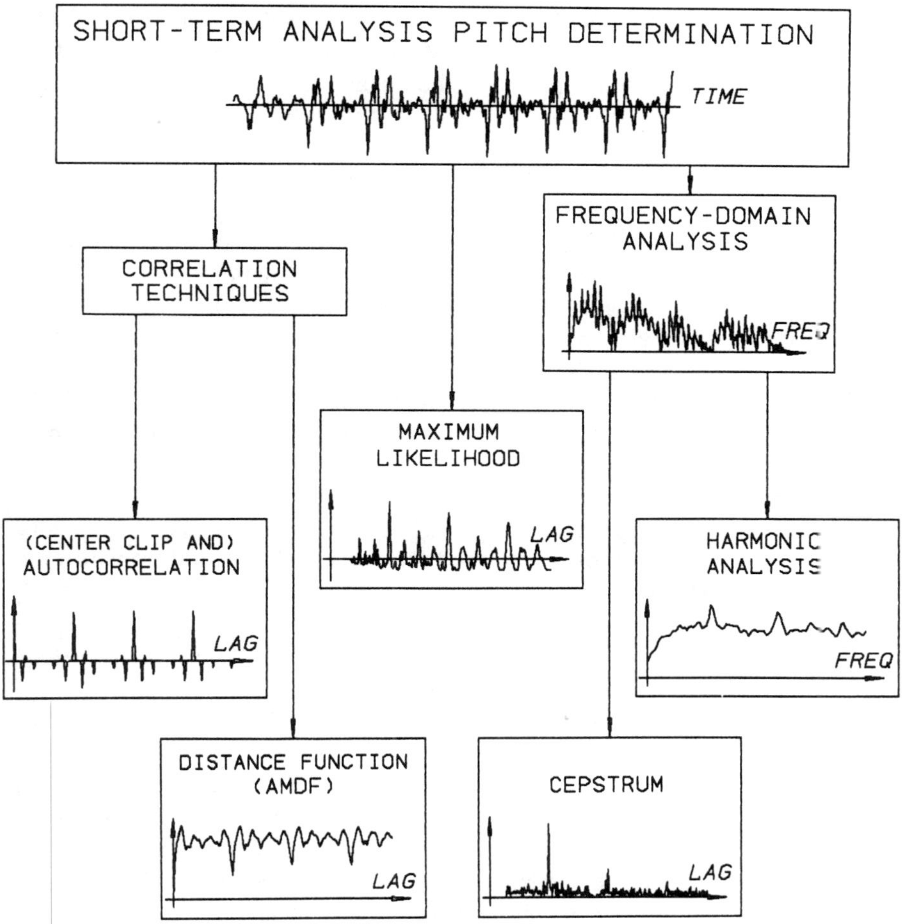

Figure 4 Methods of short-term analysis (short-time analysis) pitch determination.

If the signal were strictly periodic, the distance function would take on a value of zero at $d = T_0$. For the quasi-periodic speech signal there will be a strong minimum in the AMDF at this value of the lag (delay time) d. In contrast to all other short-term PDAs, where the estimate of T_0 or F_0 is indicated by a maximum whose position *and* value have to be determined, the minimum has an ideal target value of 0 so that we need only determine its position. For this reason, distance functions do not require (quasi-)stationarity within the measuring interval; they can cope with very short frames of one pitch period or even less. This principle thus represents the only short-term analysis PDA which is able to follow definition (3) (Moser and Kittel, 1977). The AMDF has also been successfully applied to the linear prediction residual (Un and Yang, 1977).

The frequency domain methods are also split up into two groups. Direct determination of F_0 as the location of the lowest peak in the power spectrum is unreliable and inaccurate. It is thus preferred to investigate the harmonic structure of the signal. One way to do this is spectral compression, which computes the fundamental frequency as the greatest common divider of all harmonics. The power spectrum is compressed along the frequency axis by a factor of two, three, etc. and then added to the original power spectrum. This operation gives a peak at F_0 resulting from the coherent additive contribution of the higher harmonics (Schroeder, 1968; Noll, 1970; Martin, 1981, 1987). Some of these PDAs stem from theories and functional models of pitch perception in the human ear (Terhardt, 1979; Terhardt et al., 1982; Duifhuis et al., 1982; Hermes, 1988). A corresponding (analog!) technique was developed in the time domain using a filter bank (R. L. Miller, 1970). The second frequency domain technique leads back into the time domain. Instead of transforming the power spectrum itself (which would lead to the autocorrelation function), however, the inverse transform is performed on the logarithmic power spectrum. This results in the well-known *cepstrum* (Noll, 1967), which shows a distinct peak at the "quefrency" (lag) $d = T_0$.

Some examples of these PDAs are discussed in more detail further below (Martin, 1981, 1987; Noll, 1967; Indefrey et al., 1985).

Finally, we have to mention the least-squares ("maximum likelihood") approach. This was originally a mathematical procedure for separating a periodic signal of unknown period T_0 (Noll, 1970) from Gaussian noise within a finite signal. Since the speech signal is not periodic and the background noise (plus the aperiodic components of the speech signal itself) cannot be expected to be Gaussian, the approach has to be modified slightly in order to work in a PDA (Wise et al., 1976; Friedman, 1977).

In summary, short-term analysis PDAs provide a sequence of average pitch estimates rather than a measurement of individual periods. They are not very sensitive to phase distortions and to absence of the fundamental harmonic. On the other hand, the computing effort is relatively high.

1.3.3. Examples

Harmonic Pattern Matching: The Frequency Domain Comb Filter Method
The PDA described by Martin (1981) (Fig. 5) measures F_0 in the frequency domain by harmonic analysis. Martin's algorithm modifies the well-known

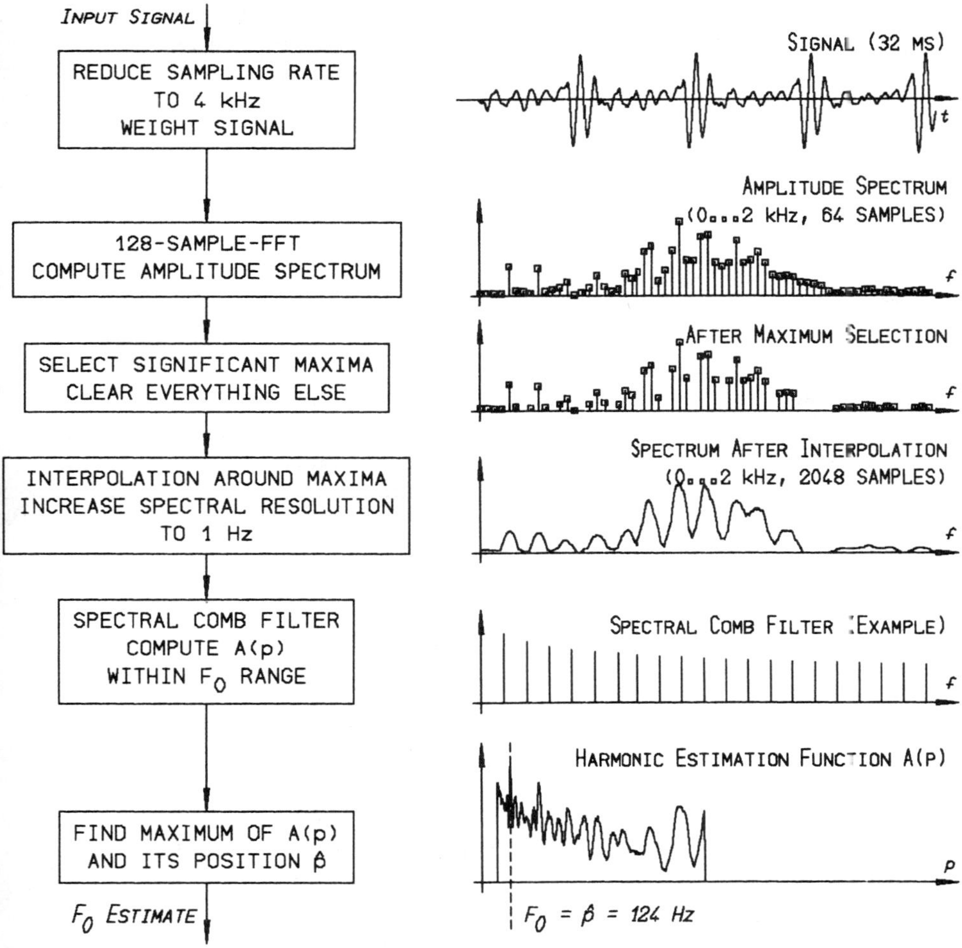

Figure 5 PDA by Martin (1981): example of performance. Signal: vowel [a], male voice, sampling frequency 16 kHz.

harmonic pattern-matching principle (Schroeder, 1968; Noll, 1970) in such a way that the computational effort for the spectral transform is minimized. The signal is first decimated to 4 kHz and then Fourier transformed by a 128-point fast Fourier transform (FFT). This yields a spectral resolution of about 30 Hz, which is sufficient to represent values of F_0 down to 60 Hz. The algorithm then enhances any spectral information which may pertain to a harmonic structure by preserving only those spectral samples which represent local maxima or their immediate neighbors and setting everything else to zero. To measure F_0 with sufficient accuracy, the spectral resolution is then increased to 1 Hz. A spectral comb filter, which is applied over the whole range of F_0, yields the harmonic estimation function $A(p)$; the value $\hat{p}$ at which this function reaches its maximum is taken as the estimate for F_0.

In a more recent version of this PDA, Martin (1987) applies a logarithmic frequency scale for computation of $A(p)$, which results in another substantial reduction of the computational effort. Charpentier (1986) uses the short-term phase spectrum at two successive samples to arrive at the frequencies of the individual harmonics without the need for spectral interpolation. Gong and Haton (1987) implement a comparable PDA which operates in the time domain and does not need a spectral transformation. Lahat et al. (1987) transfer the autocorrelation method into the frequency domain. The amplitude spectrum is passed through a bank of 16 spectral filters ("lifters") which cover the measuring range of F_0. At the output of each lifter a frequency domain autocorrelation function is calculated covering the respective range of each lifter. The estimate for F_0 is then determined as the location of the maximum of that function and refined by interpolation.

Double-Transform PDAs with Nonlinear Distortion in the Frequency Domain
The sensitivity against strong first formants, especially when they coincide with the second or third harmonic, is one of the big problems in pitch determination. This problem is suitably met by some procedure for *spectral flattening*.

Spectral flattening can be achieved in several ways. One of them is time domain nonlinear distortion, such as center clipping (Sondhi, 1968; Rabiner, 1977). A second way is linear spectral distortion by inverse filtering (Markel, 1972; Un and Yang, 1977). A third way is frequency domain amplitude compression by nonlinear distortion of the spectrum. This algorithm operates as follows: (1) short-term analysis and transformation into the frequency domain via a suitable discrete Fourier transform, (2) nonlinear distortion in the frequency domain, and (3) inverse Fourier transformation. The resulting domain is again equivalent to the time domain; to avoid confusion, we will henceforth call it the *lag domain*.

Two members of this group were already mentioned: the autocorrelation PDA (Rabiner, 1977) and the cepstrum PDA (Noll, 1967). It is well known that the autocorrelation function can be computed as the inverse Fourier transform of the power spectrum. Here, the distortion consists in taking the squared magnitude of the complex spectrum. The cepstrum, on the other hand, uses the logarithm of the spectrum. The two methods therefore differ only in the characteristics of the respective nonlinear distortions applied in the spectral domain. The cepstrum PDA is known to be rather insensitive to strong formants at higher harmonics but to develop a certain sensitivity to additive noise. The autocorrelation PDA, on the other hand, is insensitive to noise but rather sensitive to strong formants. Regarding the slope of the distortion characteristic, we observe the dynamic range of the spectrum being expanded by squaring the spectrum for the autocorrelation PDA, whereas the spectrum is substantially flattened by taking the logarithm. The two requirements—robustness against strong formants and robustness against additive (white) noise—are in some way contradictory. Expanding the dynamic range of the spectrum emphasizes strong individual components, such as formants, and suppresses wideband noise, whereas spectral flattening equalizes strong components and at the same time raises the level of low-energy regions in the spectrum, thus raising the level of the noise as well. Thus it is worth looking for other characteristics that perform spectral amplitude compression. Sreenivas (1981) proposes the fourth root of the power spectrum instead of the logarithm. For larger amplitudes this characteristic behaves very much like the logarithm; for small amplitudes, however, it has the advantage of going to zero and not to $-\infty$. Weiss et al. (1966) use the amplitude spectrum, i.e., the magnitude of the complex spectrum.

Indefrey et al. (1985) implemented these principles together with optional preprocessing to investigate systematically the performance of these PDAs. Four nonlinear spectral functions (power spectrum, amplitude spectrum, fourth root of power spectrum, and logarithm; see Fig. 6 for an example) were combined with three methods of time domain preprocessing (no preprocessing, center clipping, and frequency domain inverse filtering). This configuration was tested with four types of signals: high quality, telephone quality, and both high quality and telephone quality with added noise at various noise levels. Only gross pitch determination errors (see Section 3.1) were evaluated; it was not intended in this study to analyze the performance with respect to voicing.

The results for a test sentence (about 8 s) and three speakers are displayed in Fig. 7a and b. Without preprocessing, cepstrum and fourth root show best results for the clean signals, whereas amplitude spectrum and autocorrelation (power spectrum) are slightly worse (Fig. 7a). For the noisy signals, autocorrelation behaves best and cepstrum is significantly worse than the other three PDAs. For center clipping the tendency with respect to the algorithms is about the same; however, center clipping substantially degrades the performance of the PDAs as soon as additive noise is involved. A favorable effect of center clipping was found only in connection with the autocorrelation PDA and band-limited signals—the case it was designed for

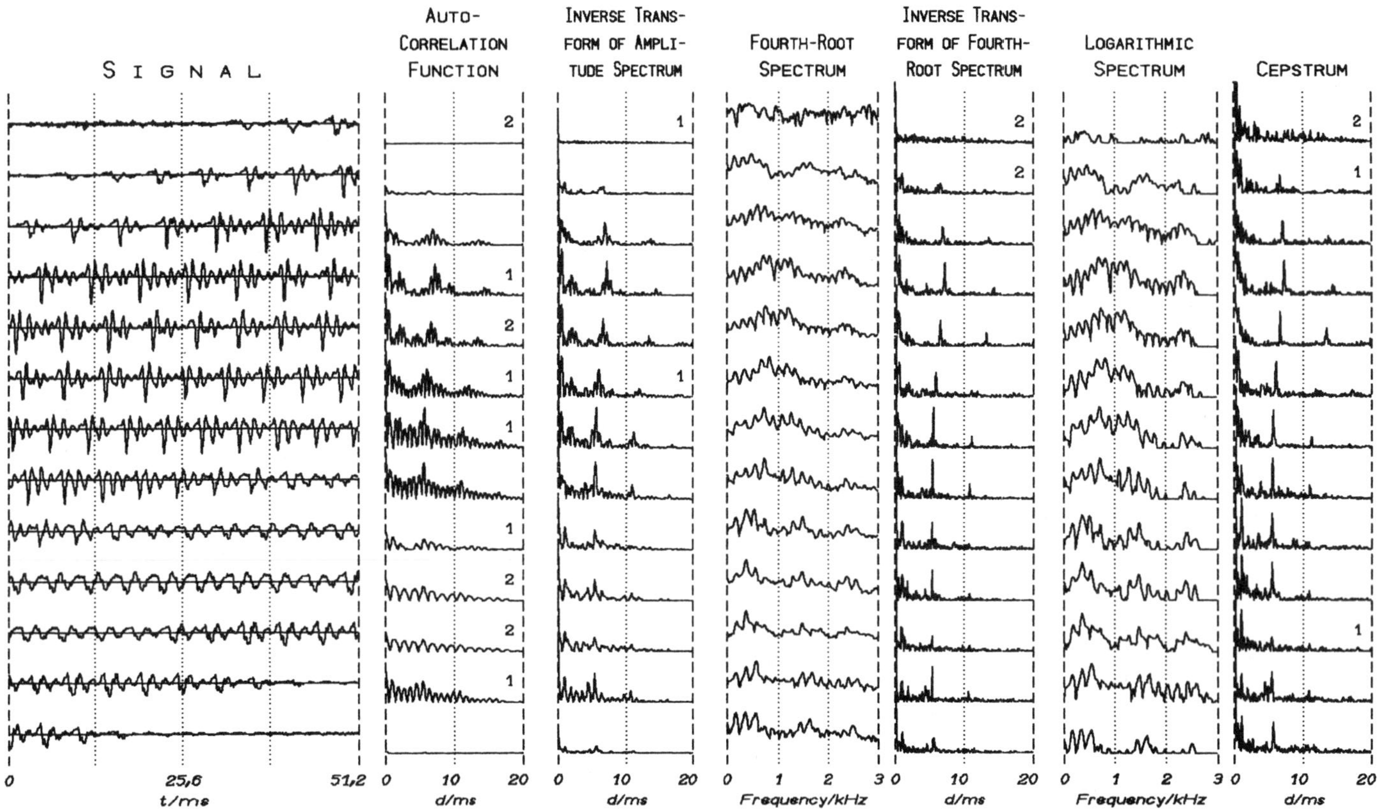

Figure 6 Pitch determination by double spectral transform and frequency domain nonlinear distortion (Indefrey et al., 1985): example of performance. Time runs from top to bottom; frame interval Q = 16 ms. Four nonlinear functions are displayed. The numbers at the right-hand side of a part of the inverse transforms indicate how many spurious peaks (exceeding 70% of the amplitude of the principal maximum) have been encountered within the measuring range (1 to 20 ms). Signal: VCV transition [ale] from the utterance "Digitale Verarbeitung . . . ," speaker WGH (male), good quality, no time domain preprocessing.

(Sondhi, 1968). The inverse filter, as a whole, did not have a favorable effect. Figure 7b shows the behavior for one selected configuration (high-quality signal with additive noise, fourth-root spectrum PDA, no time domain preprocessing) at various noise levels. Since the signal-to-noise ratio (SNR) is not uniformly defined in the literature, the PDA was evaluated using both constant-level noise with a global SNR and variable-level noise with a constant segmental ("local") SNR. The two SNR measures did not result in significant deviations of the performance. The PDA breaks down somewhere between −6 and −12 dB SNR. This value is consistent with data reported elsewhere in the literature for related PDAs (Schroeder, 1968; Noll, 1970; Wise et al., 1976) and shows that there exist a number of short-term PDAs that are extremely noise resistant.

According to these results, applying several methods of spectral flattening in the same configuration (e.g., time domain center clipping and taking the logarithm in the spectral domain) does not improve the performance of the PDA. Center clipping behaves well for band-limited signal but degrades the performance when the signal becomes noisy. Although all these PDAs are able to cope with telephone-quality speech, the absence of the first harmonic has an unfavorable effect which becomes detrimental when the signal is noisy.

Knowing that many errors arise from a mismatch during short-term analysis (which results in too few or too many pitch periods within a given frame), Fujisaki et al. (1986) investigated the influence of the relations between the error rate, the frame length, and the actual value of T_0 for an autocorrelation PDA which operates on the linear prediction residual. The optimum occurs when the frame contains about three pitch periods. Since this value is different for every individual voice, a fixed-frame PDA runs nonoptimally for most situations. For an exponential window, however, this optimum converges to a time constant of about 10 ms for all voices. For a number of PDAs, especially for the autocorrelation PDA, such a window permits recursive updating of the autocorrelation function, i.e., sample-by-sample pitch estimation without excessive computational effort.

"Active" Pitch Determination by High-Order Linear Prediction Linear prediction (LP) is usually applied to estimate the transfer function of the vocal tract. If a high-order LP filter is applied to a voiced speech signal, however, its poles will match the individual harmonics of the signal rather than the resonances of the vocal tract. A PDA based on this principle was designed by Atal (unpublished; see Rabiner et al., 1976, or Hess, 1983, p. 393). The algorithm operates as follows (Fig. 8).

1. After decimation to 2 kHz, the signal is analyzed with a 41th-order LP filter using the covariance method of linear prediction. The high

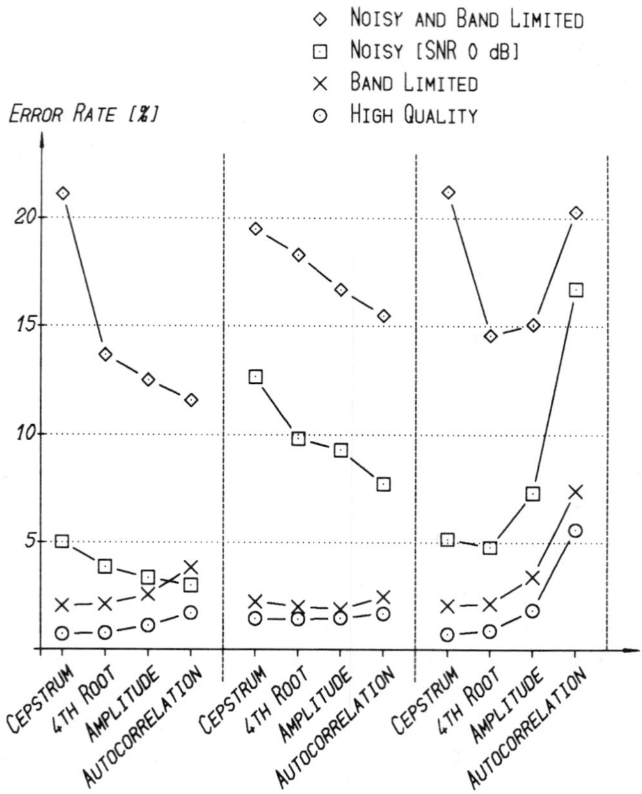

(a)

Figure 7 Performance evaluation of PDAs with double spectral transform. The rate of gross pitch determination errors is displayed. (a) Various configurations of signals, spectral distortions, and time domain preprocessing; for more details, see text. (b) Error rates for one configuration (fourth-root spectrum PDA, no time domain preprocessing, clean signal) and additive noise at various levels.

order guarantees that even at the low end of the F_0 range, i.e., at $F_0 = 50$ Hz, two complex poles are available for each harmonic.

2. Each complex pole pair represents an exponentially decaying (or rising) oscillation. To eliminate phase information, all residues at the pole locations in the z plane are set to unity. The pertinent compu-

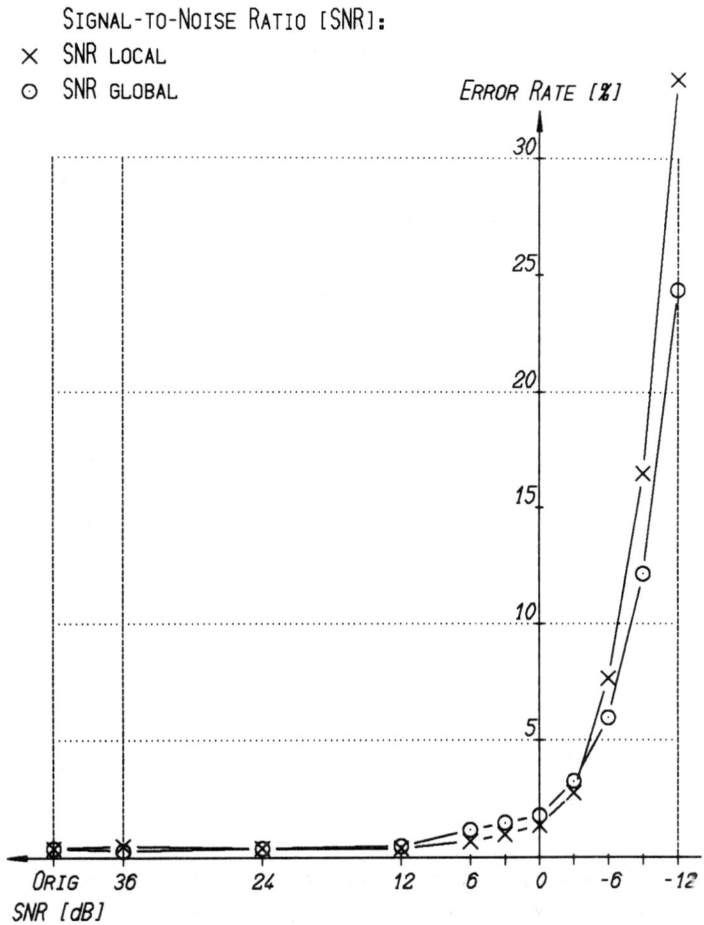

(b)

tation can be avoided when the locations of the poles are explicitly determined.

3. The impulse response of the filter now supplies a waveform for the estimation of T_0 (Fig. 8e and f). When the poles are explicitly available, it is sufficient to determine and to sum up the impulse responses of the individual second-order partial filters. This method has the advantage that the sampling frequency of the impulse response—and with it the measurement accuracy—can easily be increased. In addition, poles that are situated outside or far inside the unit circle can be modified or excluded from further processing.

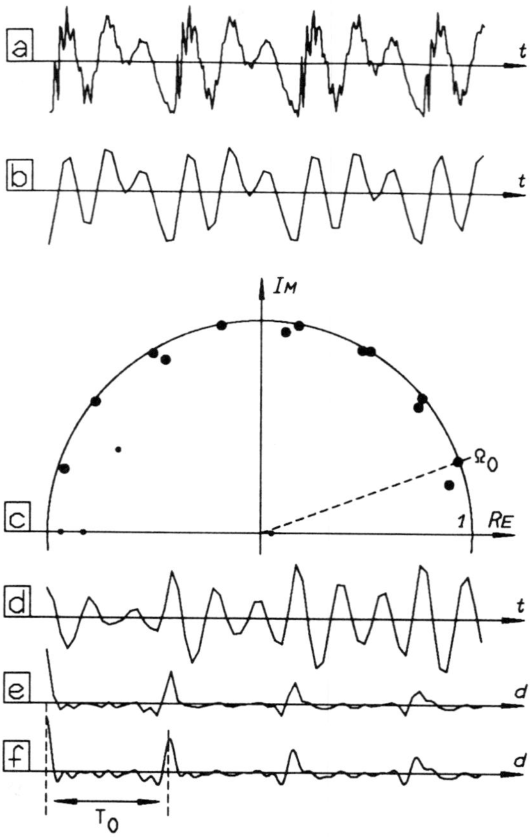

Figure 8 PDA by Atal (cited in Rabiner et al., 1976): example of performance. (a) Input signal (vowel [ɛ:], male voice, sampling frequency 8 kHz). (b) Same signal after decimation to 2 kHz. (c) Pole diagram of the 41st order LP filter in the z plane. (d) Impulse response of the LP filter (unstable!). (e) Impulse response of the filter after enforcing stability, setting the residues to unity, and eliminating all poles that were real or too far inside the unit circle [small filled circles in (c)]. (f) Same impulse response, sampled at 16 kHz. Ω_0, Angular frequency in the z plane corresponding to F_0.

1.4. Time Domain Pitch Determination Algorithms

This category of PDAs is less homogeneous than that of the short-term analysis methods. One way to split them up is according to how the burden of data reduction is distributed between the preprocessor and the basic extractor. Doing this, we find most time domain PDAs between two extremes (Fig. 9):

1. The burden is imposed on the preprocessor. In the extreme case, only the waveform of the first harmonic is offered to the basic extractor.
2. The burden is imposed on the basic extractor, which then has to cope with the whole complexity of the temporal signal structure. In the extreme case, the preprocessor is totally omitted.

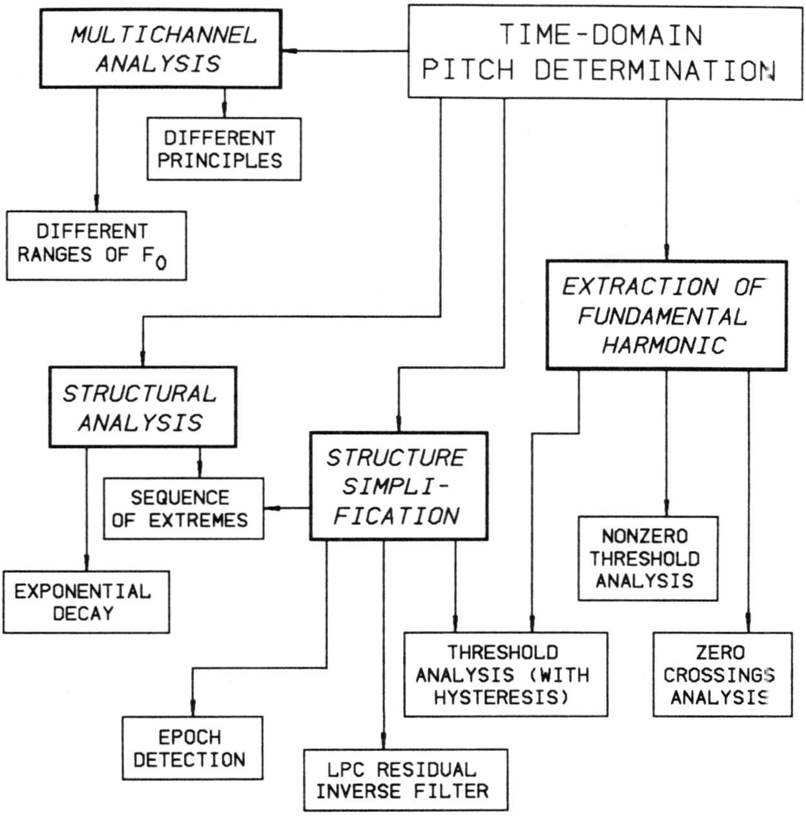

Figure 9 Principles of time domain PDAs.

Time domain PDAs are principally able to track the signal period by period. At the output of the basic extractor we find a sequence of period boundaries (pitch *markers*). Since the local information on pitch is taken from each period individually, time domain PDAs are more sensitive to local signal degradations and thus less reliable than the majority of their short-term analysis counterparts. On the other hand, time domain PDAs may still operate correctly even when the signal itself is irregular because of temporary voice perturbation or vocal fry.

1.4.1. Structural Analysis

A pitch period is the truncated response of the vocal tract to an individual glottal impulse. Since the vocal tract behaves like a lossy linear system, its impulse response consists of a sum of exponentially damped oscillations. It is therefore to be expected that the magnitude of the significant peaks in the signal is greater at the beginning of the period than at the end (Fig. 10). Appropriate investigation of the signal peaks (maxima and/or minima) leads to an indication of periodicity.

There are problems associated with this approach, however. First, the frequencies of the dominant damped waveforms are determined by the local formant pattern and may change abruptly. Second, the damping of the formants, particularly of a low first formant, is often quite weak and can be overrun by temporary changes of the signal level. Third, if the signal is phase distorted, different formants may be excited at different points in time. These problems are surmountable, but they lead to relatively complicated algorith-

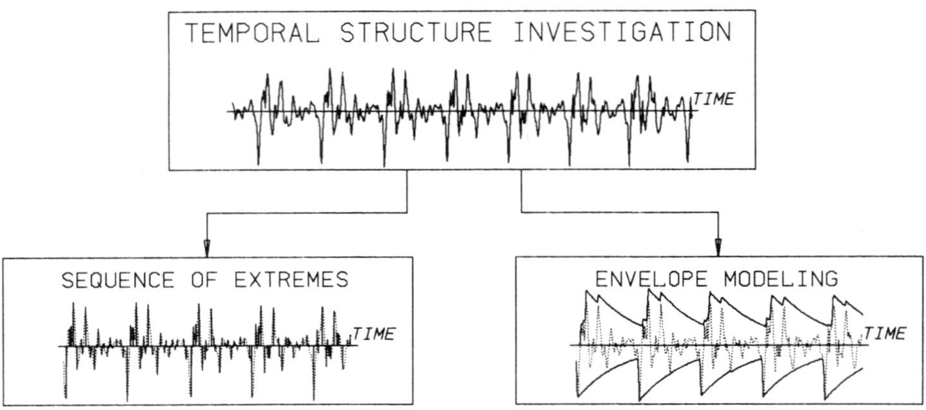

Figure 10 Time domain PDAs: temporal structure investigation.

mic solutions which have to consider a great variety of temporal structures. Since most of the program instructions are decisions, however, these PDAs run relatively fast. The usual way to carry out the analysis is as follows (Reddy, 1967; N. J. Miller, 1975; Tucker and Bates, 1978, D. M. Howard, 1989).

1. Do a moderate low-pass filtering to remove the influence of higher formants.
2. Determine all the local maxima and minima.
3. Exclude extremes which are found to be insignificant until one significant point per period is left.
4. Reject obviously incorrect points by local correction.

For restricted applications these PDAs may be remarkably simple (Gold and Rabiner, 1969; Tillmann, 1978).

Structural analysis, especially when many possible structures have to be reviewed, is a good application for self-organizing, nonlinear pattern recognition methods, e.g., for artificial neural networks. Such a PDA was introduced by I. S. Howard et al. (Howard and Huckvale, 1988; Howard and Walliker, 1989). The speech signal is first divided into nine subbands with subsequent half-wave rectification and second-order linear smoothing in each channel. The underlying idea is to obtain a representation similar to that in a wide-band spectrogram. The basic extractor consists of a four-layer perceptron structure, the input layer comprising 41 successive samples with 9 channels each. Two hidden layers with 10 units each and a fully connected network are followed by a 1-unit output layer which is intended to yield an impulse when the network encounters a signal structure associated with the instant of glottal closure. The network is trained using reference data obtained from the output signal of a laryngograph. Such a structure has the advantage that it can be based on several features that may occur at different instants during a pitch period. It was shown to outperform conventional devices of the same type, for instance, the peak-picking PDA (D. H. Howard, 1989; see next paragraph) which was evaluated for comparison.

A different solution originates from the analog domain (Dolansky, 1955; Filip, 1969; Winckel, 1964; Zurcher, 1977). The envelope of the period is modeled by a cascade of analog differentiators and diode-resistance-capacitance circuits with short rise time constants and comparatively long decay time constants. These circuits emphasize the principal peaks of the signal and suppress all the others. The performance, however, strongly depends on proper adjustment of the decay time constants. For that reason, this relatively simple device works well only for a restricted range of F_0 (about two octaves). A manual range switch or something similar is required if a wider range of F_0 is to be analyzed. Because of its simplicity, this principle has been revived in a recent application for cochlear prostheses (D. M. Howard, 1989). Using a logarithmic amplifier, Howard's PDA avoids many problems associated with the older devices, and his device compares favorably to a number of other PDAs tested for this special application.

1.4.2. Fundamental Harmonic Detection

F_0 can be detected in the signal via the waveform of the fundamental harmonic. If present in the signal, this harmonic is extracted from the signal by extensive low-pass filtering in the preprocessor. The basic extractor can then be relatively simple. Figure 11 shows the principle of three basic extractors: the zero-crossings analysis basic extractor as the simplest device,

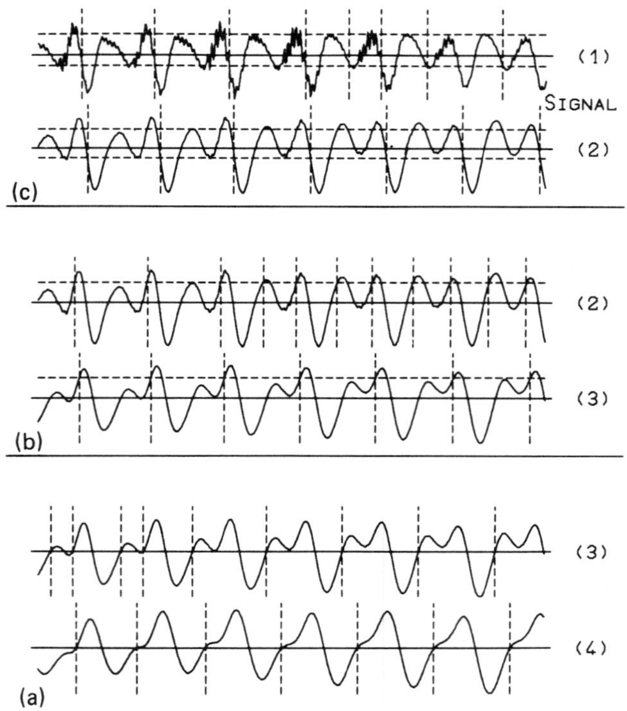

Figure 11 Example of the performance of basic extractors for fundamental harmonic extraction: (a) zero-crossings analysis basic extractor; (b) nonzero threshold analysis basic extractor; (c) threshold analysis basic extractor with hysteresis. Signals: (1) original, (2) low-pass filtered at 6 dB/octave, (3) low-pass filtered at 12 dB/octave, (4) low-pass filtered at 18 dB/octave.

the nonzero threshold basic extractor, and finally the threshold analysis basic extractor with hysteresis. The zero-crossings analysis basic extractor sets a marker whenever the zero axis is crossed with a defined polarity. This requires that the input waveform has two and only two zero crossings per period. The threshold analysis basic extractor sets a marker whenever a given nonzero threshold is exceeded. The threshold analysis basic extractor with hysteresis acts like the normal threshold analysis basic extractor except that the marker is not set before a second (lower) threshold is crossed in the opposite direction. This more elaborate device requires a lesser degree of low-pass filtering in the preprocessor.

The requirement of extensive low-pass filtering is one of two severe weak points of this otherwise fast and simple approach. For the zero-crossings analysis basic extractor, an attenuation of 18 dB/octave is necessary within the range of F_0 to be determined (McKinney, 1965). Accordingly, the amplitude of the signal at the basic extractor will vary by more than 50 dB due to the variations of F_0 alone. This dynamic range, increased by the intrinsic dynamic range of the signal (at least another 30 dB), is too much for the PDA to work correctly. The application of a zero-crossings analysis basic extractor thus limits the possible range of F_0. For the threshold analysis basic extractor the problem is not so acute, but the fact that the threshold must be adapted to the overall signal amplitude complicates the design of the PDA. The second weak point is that this principle is restricted a priori to environments in which the first harmonic is present in the signal. If such a PDA is nevertheless to be applied to processing band-limited signals, the first harmonic must be enhanced or reconstructed. One way to do this is nonlinear distortion. In that respect, many proposals have been made from the beginning (i.e., Grützmacher and Lottermoser, 1937; Risberg et al., 1960). No single nonlinear characteristic, however, is able to enhance the first harmonic of the signal in an optimal way for any situation, i.e., for any speaker or environmental condition (McKinney, 1965; Hess, 1979); some of them work well in a constrained environment (for instance only with band-limited signals or male voices) or in a realization in which several channels with different nonlinear functions are combined (Hess 1979; see Section 1.4.4).

1.4.3. Simplification of the Temporal Structure

Algorithms of this type take some intermediate position between the principles of structural analysis and fundamental harmonic extraction. The majority of these algorithms follow one of two principles: (1) inverse filtering and (2) epoch detection. Both of these principles deal with the fact that the laryngeal excitation function has a temporal structure which is much simpler and more regular than the temporal structure of the speech signal itself, and both methods, when they work, are able to follow definition (1) if the signal is not grossly phase distorted.

The inverse filter approach cancels the transfer function of the vocal tract and thus reconstructs the laryngeal excitation function. If one is interested in pitch only and not in the excitation function itself, a crude approximation of the inverse filter is sufficient. Such an approximation is realized, for instance, when the analysis is confined to the first formant (Hess, 1976). The inverse filter approach has one weak point which arises frequently with female voices. When F_0 is high, it may coincide with the first formant. If the inverse filter is not blocked, it then removes the fundamental harmonic (which is extremely strong in this case) from the signal and brings the PDA to failure.

The second principle, epoch extraction, is based on the fact that at the beginning of each laryngeal pulse there is a discontinuity in the second derivative of the excitation function. Usually this discontinuity cannot be reliably detected in the speech signal because of phase distortions which occur when the waveform passes the vocal tract. The signal is thus first phase shifted by 90° (applying a Hilbert transform). The squares of the original and the phase-shifted signals are then added and yield a new signal which shows a distinct peak at the time when the discontinuity in the excitation function occurs. The original method works only when the spectrum of the investigated signal is flat to some extent. To enforce spectral flatness, the analyzed signal can be band limited to high frequencies well above the narrow band lower formants (Ananthapadmanabha and Yegnanarayana, 1975). Another way is to analyze the linear predictive coding (LPC) residual (Ananthapadmanabha and Yegnanarayana, 1979) or to filter the signal into subbands (De Mori et al., 1977).

The epoch detection principle depends on the presence of a discontinuity in the second derivative of the laryngeal excitation function. This discontinuity is often weak, especially in back vowels like [u], when a formant exactly coincides with the first of a higher harmonic, or when speech is uttered with a soft or a falsetto voice. In two recent approaches (Di Francesco and Moulines, 1989; Cheng and O'Shaughnessy, 1989) this drawback was overcome by the finding that the global statistical properties of the waveform change with glottal opening and closing as well. These PDAs, which exploit different features of the signal and were developed independently, derive and apply a generalized maximum-likelihood measure that indicates the instant of glottal closure more precisely than previous epoch detection PDAs.

1.4.4. Multichannel Approaches

Except for the algorithmic investigation of the temporal structure and—nowadays—epoch detection, most simple time domain PDAs are restricted with respect to the range of F_0 or the type of signal to be processed. One way to increase the range or the reliability of these PDAs is to implement several of them in parallel and to make some decision as to which one has the "correct" output. The partial PDAs may be identical in design, and each of them may process a subrange of F_0 (McKinney, 1965; Léon and Martin, 1969). On the other hand, they may apply different principles without restriction of the frequency range. The PDAs of Risberg et al. (1960) or Hess (1979), for instance, use several nonlinear functions to enhance the first harmonic in different ways. Gold and Rabiner (1969) combined several simple peak detection basic extractors together with a pattern-matching procedure. The selection criteria for finding the most likely channel are defined by a certain channel hierarchy (Léon and Martin, 1970), by a regularity check applying a minimum-frequency selection principle (Risberg et al., 1960; Hess, 1979), by statistical measures (Bruno et al., 1982), or by syntactic rules (De Mori et al., 1977). The selection is continuously checked so that the PDA is able to change its choice at any time.

One problem with time domain multichannel PDAs is that the individual channels often mark period boundaries at different instants in time, for instance, when significant maxima and minima are exploited independently of each other (Gold and Rabiner, 1969). Unless there is a special synchronization routine (Hess, 1979), such PDAs are no longer able to synchronize themselves correctly with the signal and thus have to operate according to the incremental definition (3) or even the short-term definition (4) although they pertain to the time domain category.

1.5. Postprocessing and Smoothing

Error correction is one of the main postprocessing tasks in pitch determination. The most important routines in this domain are *list correction* and *pitch contour smoothing*, List correction routines correct only gross pitch determination errors (see Section 3.1); smoothing procedures are successfully applied to reduce measurement inaccuracies as well.

In contrast to the basic extractor, which processes every frame or every pitch period independently of its neighbors, the postprocessor regards the signal under more global aspects. For instance, it operates on a voiced segment as a whole. As the name *list correction* indicates, estimates or markers may be processed using a list without immediate reference to the signal. Of course, such a procedure delays the delivery of the results and is thus applicable only when permitted by the application.

The flow diagram of a prospective list correction algorithm for a time domain PDA is displayed in Fig. 12 (Reddy, 1967; Hess, 1983, p. 284). This figure needs some explanation. One of the advantages of a list correction algorithm is that the signal need not be processed sequentially. The routine may set its starting point where the pitch estimates are particularly reliable (for instance, within a vowel). Such an "island" is characterized by high signal amplitude and regularity of the estimates. From there the algorithm starts processing the list with and against time toward the margins of the voiced segment. For each period the algorithm determines an expected value, which depends on the adjacent regular periods, and computes the relative deviation between this expected value and the actual period length as given by the basic extractor. When the deviation exceeds a given threshold, the actual marker is regarded as erroneous and the algorithm performs a correction according to the size of the relative deviation as indicated in the figure.

Another procedure for reduction of pitch determination errors is *smoothing*. In contrast to list correction, smoothing is possible only when a pitch contour is given as a sequence of T_0 or F_0 estimates and not as delimiters of individual periods. Two smoothing methods are in use: linear smoothing using some kind of low-pass filter and (nonlinear) median smoothing (Rabiner et al., 1975). Linear smoothing reduces measurement inaccuracies but is unable to cope with the effect of gross pitch determination errors, which are reduced in size and distributed over a larger amount of time but are not really removed. Median smoothing, on the other hand, replaces each pitch estimate with the middle value of an ordered sequence of three, five, or seven adjacent estimates; gross outliers are removed, but measurement inaccuracies remain unchanged. Rabiner et al. (1975) combine these methods and propose a two-step smoothing procedure with median smoothing coming first, followed by a linear smoother.

Applying such a smoothing algorithm was shown to improve substantially the (objective and subjective) performance of any PDA to which it was applied (Rabiner et al., 1976; McGonegal et al., 1977). Secrest and Doddington (1982) used dynamic programming methods to find an optimal path through a list of pitch estimate candidates with the smoothness of the contour as the major criterion. Specker (1984) showed that post-processing is able to reduce the number of gross errors in a time domain PDA (in this case that of Tucker and Bates, 1978) by almost an order of magnitude.

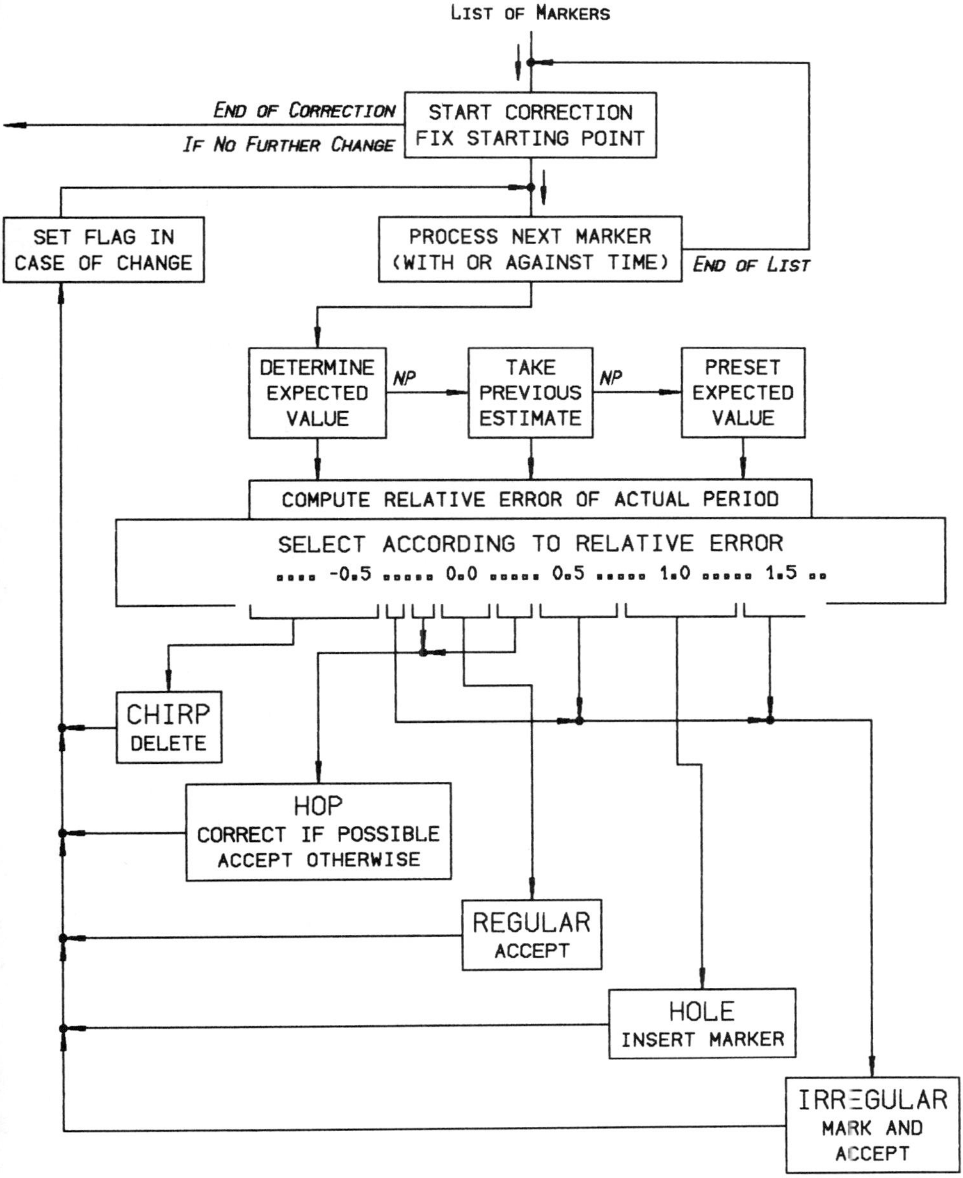

Figure 12 Flow diagram of a list correction algorithm for a time domain PDA. NP: Operation not possible.

2. VOICING DETERMINATION ALGORITHMS

The task of voicing determination may be split up into two subtasks: (1) a decision whether or not a *voiced* excitation is present and (2) a decision whether or not a *voiceless** excitation is present. If neither of these excitations is active, the current segment represents *pause* or *silence*; if both excitations are simultaneously active, we speak of *mixed excitation* (Siegel and Bessey, 1982). The two features *voiced* and *voiceless* are binary unless they occur simultaneously. In segments with mixed excitation the *degree of voicing*—for instance, the energy ratio of the voiced and voiceless excitations—may play a role, although this feature is rarely exploited.

Most voicing determination algorithms (VDAs) thus apply *decisions*. The following general schemes are frequently found.

1. *Two-step realization*. In the first step the algorithm checks whether or not there is a signal (*pause-speech separation*); the second step further investigates those segments classified as speech.
2. *Simplified one-step realization*. The algorithm distinguishes between the three categories *silence*, *voiced*, and *voiceless*. Mixed excitation is usually not taken into account; such segments are then mostly classified as voiced.

VDAs exploit almost any elementary speech signal parameter which may be computed independently of the type of input signal: energy, amplitude, short-term autocorrelation coefficients, zero-crossings count, ratio of signal amplitudes in different subbands or after different types of filtering, linear prediction error, or the salience of a pitch estimate. As to the method applied, the VDAs can be grouped into three major categories: (1) simple threshold analysis algorithms, which exploit only a few basic parameters (e.g., Hess, 1976); (2) more complex algorithms based on pattern recognition methods; and (3) integrated algorithms for both voicing and pitch determination. For reasons of space, only a few selected VDAs of the latter categories are briefly reviewed.

---

*With respect to the various algorithms, we distinguish between *voiceless* and *unvoiced* segments in this chapter. A signal segment is voiceless when a source of voiceless (i.e., turbulent) excitation is present. This excludes silent as well as purely voiced intervals but includes segments with mixed (i.e., simultaneous voiced and voiceless) excitation. On the other hand, any segment without voiced excitation is labeled unvoiced. This includes voiceless signals and silence but excludes segments with mixed excitation. In the literature the terms *voiceless* and *unvoiced* are widely regarded as synonyms.

2.1. Simultaneous Determination of Pitch and Voicing

A number of PDAs—usually pertaining to the short-term analysis category—permit estimation of the salience of their results without a need to know the actual value of the pitch estimate. This is always possible when the amplitude of the significant maximum or minimum at T_0 of F_0 in the basic extractor of the PDA can be compared to a reference value. As an example, the ratio of the values of the autocorrelation function at $d = T_0$ and at $d = 0$ (the latter equaling the signal energy) gives a direct measure of the degree of periodicity of the signal. It is dangerous, however, to rely on this feature alone. Of course, it is correct (and almost trivial) to state that pitch can exist only where the signal is voiced. But this statement cannot be simply reversed; i.e., we cannot say that a segment is unvoiced because pitch is not existent (or not measurable). The respective PDA may momentarily fail, or the signal may be voiced but irregular (Siegel and Bessey, 1982; Rabiner et al., 1976; see also Fig. 2 or Section 3.1). Amplitude changes and especially small intraframe changes of F_0 severely degrade the "salience" of the pitch estimate (Sreenivas, 1981). It is thus at least necessary to check adjacent frames before making a decision (Noll, 1967). In this respect, such a VDA behaves very much like a median smoother in pitch determination.

In principle, these VDAs do not make a voiced-unvoiced discrimination; rather they check the presence of a (sufficient but not absolutely necessary) condition for a voiced signal. It is thus likely that such an algorithm commits many more voiced-to-unvoiced errors than vice versa.

The VDA of Lobanov (1970) avoids this problem, although it is based on a similar principle. A voiceless segment of speech represents a stochastic signal which is continuously excited. In contrast, the excitation of a voiced signal is confined to a few instants per period; major parts of the pitch period are composed of exponentially decaying oscillations, and adjacent samples of the signal are highly correlated. This contrast of a stochastic versus a deterministic signal is preserved even when a voiced signal becomes irregular or aperiodic. Lobanov's VDA exploits this feature by computing the Hilbert transform of the speech signal, combining the original signal and its Hilbert transform to yield the complex analytic signal, and plotting the momentary amplitude and phase of the analytic signal in the so-called phase plane. For voiced frames the analytic signal will describe a closed curve in the phase plane (Fig. 13a and b). During unvoiced segments, where the signal and its Hilbert transform are much less correlated, the curve will touch almost any point in the phase plane within a short interval (Fig. 13c). In Lobanov's algorithm the phase plane is crudely quantized, and the algorithm simply counts the number of points which have been touched within a given frame. The

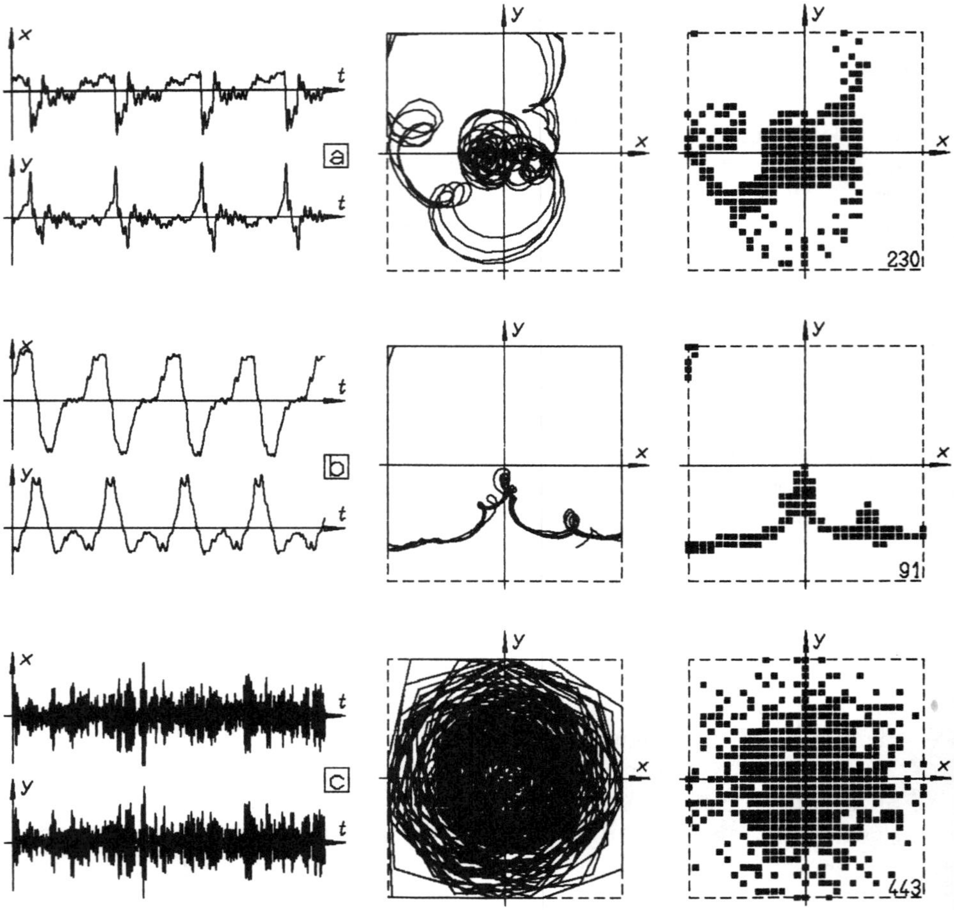

Figure 13 VDA by Lobanov (1970): example of performance. (a) Vowel [ε]; (b) nasal [m]; (c) fricative [s]. (Left) original signal frame $x(t)$ and signal $y(t)$ after Hilbert transform; frame length 25.6 ms. (Middle) x-y plane ("phase plane") without quantization. (Right) Phase plane after sampling and quantization (5 bits); sampling frequency 40 kHz (see text); the numbers in the corner indicate how many points were touched by the curve. The signal was normalized such that the peak amplitude exceeds the range of the phase plane by about 50%. According to Lobanov, this is necessary to ensure that a stochastic signal is likely to reach the most distant point of the phase plane within a given amount of time.

algorithm was originally designed to work with analog signals. To operate correctly in a digital environment, a high sampling rate or a special interpolation routine is needed.

2.2. Pattern Recognition VDAs

One of the motivations for applying a pattern recognition algorithm in a VDA was the wish to get away from the conjunction of voicing determination and pitch determination (Atal and Rabiner, 1976). Atal and Rabiner's VDA (the first of a series of VDAs developed at Bell Laboratories in the late 1970s) uses a statistical classifier and is based on five parameters: signal energy, zero-crossing rate, autocorrelation coefficient at a delay of one sample, first predictor coefficient of a 12-pole LP analysis, and energy of the normalized prediction error. For a given environmental condition the algorithm works well, but it is rather sensitive to environmental changes, e.g., from high-quality speech to a telephone channel (Rabiner et al., 1976).

The usual classification problems in speech recognition, where we have to cope with a large number of different classes, require that the input parameters form specific clusters in the parameter space, which are then separated by the classifier. In contrast, the voicing determination problem has at most four categories (silence, voiced, voiceless, and mixed) and the distribution of the patterns in the parameter space is rather diffuse. It is thus suitable to concentrate the VDA on patterns which are situated at or near the boundaries between the different categories in the parameter space. Such a VDA was developed by Siegel (1979) and extended by Siegel and Bessey (1982). For some applications, such as high-quality analysis-synthesis systems, incorporation of a mixed excitation source is desirable (1) for voiced fricatives and (2) for some high vowels which tend to become partly devoiced in connected speech (Fujimura, 1968). Siegel and Bessey further found that for the voiced-voiceless-mixed classification, the number of features used for a voiced-unvoiced classifier is insufficient. Their VDA is realized in two steps using a binary decision tree structure. The first step comprises a classifier which separates frames that are predominantly voiced from those that are predominantly unvoiced. It uses a minimum-distance statistical classifier exploiting seven features: normalized autocorrelation coefficient at unit sample delay, minimum normalized LP error, zero-crossings rate, signal energy, overall degree of periodicity (via AMDF), and degree of periodicity measured via cepstrum in two subbands. In both categories the mixed frames are split off during the second step. The voiced-mixed decision uses another six features, mostly cepstral and LP measures, whereas the voiceless-mixed decision is based on two features alone. The VDA is reported to work with 94% overall accuracy and 77% correct identification of the mixed frames.

3. EVALUATION AND APPLICATION

To evaluate the performance of a measuring device, one should have another instrument with at least the same accuracy. If this is not available, at least objective criteria—or data—are required to check and adjust the behavior of the new device. In pitch and voicing determination both these bases of comparison are tedious to generate. There is no VDA or PDA which operates without errors (Rabiner et al., 1976). There is no reference algorithm, even with instrumental support, that goes completely without manual inspection or control (Krishnamurthy and Childers, 1986; Hess and Indefrey, 1987). Only rather recently have speech data bases with reference pitch contours and voicing information become available (e.g., Carré et al., 1984; Picone et al., 1987), and only then did designers of new PDAs and VDAs start to provide detailed data on the performance of their algorithms (e.g. Fujisaki et al., 1986; Indefrey et al., 1985).

In the following, we will first discuss types of errors and error criteria in pitch and voicing determination (Section 3.1) and deal with the question of developing reference algorithms and data (Section 3.2). Section 3.3 will then provide some results of comparative evaluations in addition to those which were reported above (Section 1.3.3).

3.1. Error Analysis in Pitch Determination

According to the classical study by Rabiner et al. (1976), which established the guidelines for the performance evaluation of these algorithms, PDAs and VDAs commit four types of errors: (1) gross pitch determination errors; (2) fine pitch determination errors, i.e., measurement inaccuracies; (3) voiced-to-unvoiced errors; and (4) unvoiced-to-voiced errors. The latter two types represent errors of voicing determination, whereas the first ones refer to pitch determination.

Gross pitch determination errors are "drastic failures of a particular method or algorithm to determine pitch" (Rabiner et al., 1976). Usually an error is considered to be gross when the deviation between the correct value of T_0 or F_0 and the estimate of the PDA exceeds the maximum rate of change a voice can produce without becoming irregular [Rabiner et al. (1976): 1 ms; Hess and Indefrey (1987): 10%; Krubsack and Niederjohn (1989): 0.25 octave]. On the other hand, errorlike situations may also arise from "drastic failures of the voice to produce a regular excitation pattern," which is not very frequent in well-behaved speech (Dolansky and Tjernlund, 1968) but is nearly always the case when the voice temporarily falls into vocal fry (Fourcin, 1974; Hollien, 1974; Secrest and Doddington, 1982; see Fig. 2). Hence, gross errors arise mainly from three standard situations.

1. *Adverse signal conditions*: strong first formants, rapid change of the vocal tract position, band-limited or noisy recordings. Good algorithms reduce these errors to a great extent but cannot avoid them completely (Rabiner et al., 1976).
2. *Insufficient algorithm performance*, e.g., mismatch of F_0 and frame length (Fujisaki et al., 1986); temporary absence of the key feature in some algorithms.
3. "Errors" that arise from *irregular excitation of voiced signals*. Since most algorithms perform some averaging or regularity check, they can do nothing but fail when the source becomes irregular.

When a PDA is equipped with an error-detecting routine (and the majority of PDAs are even if no postprocessor is used), and when it detects that an individual estimate may be wrong, it is usually not able to decide reliably whether this situation is a true measurement error, which should be corrected or at least indicated, or a signal irregularity, where the estimate may be correct and should be preserved as it is. This inability of most PDAs to distinguish between the different sources of errorlike situations is one of the great unsolved problems in pitch determination.

Measurement inaccuracies cause noisiness of the obtained T_0 or F_0 contour. They are small deviations from the correct value but can nevertheless be annoying to the listener. Again there are three main causes.

1. *Inaccurate determination of the key feature.* This applies especially to algorithms that exploit the temporal structure of the signal, for instance, when the key feature is a principal maximum whose position within a pitch period depends on the formant F1.
2. *Intrinsic measurement inaccuracies*, such as the ones introduced by sampling in digital systems.
3. "Errors" from *small fluctuations of the voice*, which contribute to the perception of "naturalness" and should thus be preserved.

Voicing errors are misclassifications of the VDA. We have to distinguish between *voiced-to-unvoiced* errors, in which a frame is classified as unvoiced although it is in fact voiced, and *unvoiced-to-voiced* errors, with the opposite misclassification. This scheme, as established by Rabiner et al. (1976), does not take into account mixed excitation. Voiced-to-unvoiced errors and unvoiced-to-voiced errors must be regarded separately because they are perceptually not equivalent (Viswanathan and Russell, 1984), and the reasons for such errors in an actual implementation may be different and even contradictory.

3.2. Developing Reference PDAs with Instrumental Help

A number of evaluations, such as the one by El Mallawany (1977), compared the algorithm(s) to be tested to the results of a well-known algorithm such as the cepstrum PDA, whose performance was known to be good. Rabiner et al. (1976) used an interactive PDA (McGonegal et al., 1975) to generate reference data. This procedure proved reliable and accurate but required a great deal of human work. Dal Degan (1982) took the output signal of a vocoder, for which the pitch contour was exactly known, as the standard for his PDA evaluation. Bruno et al. (1982) evaluated the performance of a two-channel PDA using the output signal of a mechanic accelerometer which derives the information on pitch from the vibrations of the neck tissue at the larynx. The same device (Stevens et al., 1975) was used by Viswanathan and Russell (1984) for their evaluation of five PDAs. Indefrey et al. (1985) used a laryngograph to obtain the signal for generating a reference contour.

Among all the algorithms used for determining a reference pitch contour, the most promising methods appear to be those which make use of an instrument (such as a mechanic accelerometer or a laryngograph) that derives pitch directly from the laryngeal waveform. This type of algorithm avoids most errors pertinent to the problem of pitch determination from the speech signal, and it permits using natural speech for the evaluation of the performance of PDAs. Among the many instruments available [see (Hess, 1983, chapter 5) for a survey], the laryngograph (Fourcin and Abberton, 1971; Childers and Krishnamurthy, 1985) is especially well suited for this kind of application. It is robust and reliable, does not prevent the speaker from natural articulation, and gives a good estimate for the instant of glottal closure. A number of PDAs have been designed for this device (e.g., Krishnamurthy and Childers, 1986; Hess and Indefrey, 1987). In addition, Childers et al. (1989) propose a four-category VDA that exploits the speech signal and the laryngogram. In the following, one of these algorithms (Hess and Indefrey, 1987) is presented in some more detail.

The principle of the laryngograph is well known. A small high-frequency electric current is led through the larynx by a pair of electrodes which are pressed against the neck at the position of the larynx from both sides. The opening and closing of the glottis during each pitch period cause the laryngeal conductance to become time variant; thus the high-frequency current is amplitude modulated. In the receiver the current is demodulated and amplified. Finally, the resulting signal is high-pass filtered in order to remove unwanted low-frequency components due to vertical movement of the larynx.

Figure 14 shows an example of the laryngogram (the output signal of the laryngograph) together with the pertinent speech signal. In contrast to the speech signal, the laryngogram is hardly affected by the momentary position of the vocal tract, and the changes in shape or amplitude are comparatively small. Since every glottal cycle is represented by a single pulse, the use of the laryngograph reliably suppresses gross period determination errors. In addition, it supplies the basis for a good voiced-unvoiced discrimination since the laryngogram is almost zero during unvoiced segments, where the glottis is always open. Nonetheless, the laryngograph is not free from any problem: it may fail temporarily or permanently for some individual speakers, or it may miss the beginning or end of a voiced segment by a short interval—for instance, when the vocal folds, during the silent phase of a plosive, continue to oscillate without producing a signal, or when voicing is resumed after a plosive and the glottis does not completely close during the first periods (Lebrum, 1971; Childers and Krishnamurthy, 1985). For such reasons, visual inspection of the reference contour is necessary even with this configuration; these checks, however, can be confined to limited segments of the signal.

What key feature is best used for delimiting the individual periods? According to the theory of voice excitation (van den Berg, 1958; cf. also Stevens,

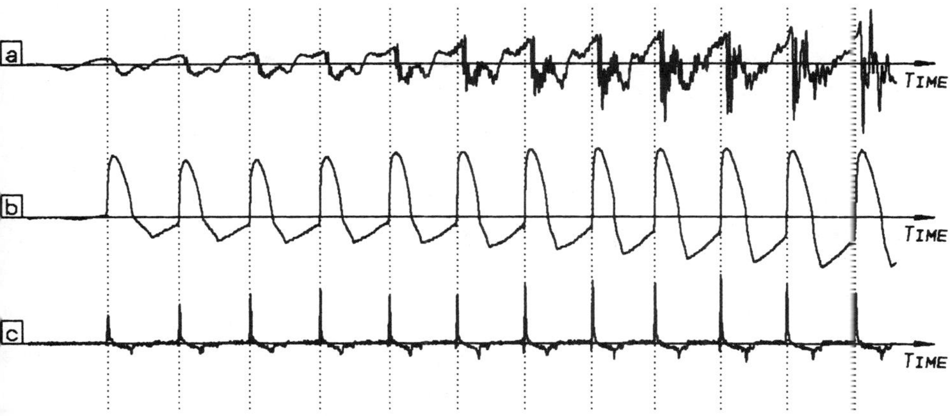

Figure 14 (a) Speech signal, (b) laryngogram, and (c) differentiated laryngogram. The markers delimiting the individual periods were derived from the maxima of (c). Signal: Transition /ja/; speaker WGH (male).

1977), the instant of glottal closure is the point of maximum vocal tract excitation and it is justifiable to define this instant as the beginning of a pitch period. In the laryngogram this feature is well documented. As long as the glottis is open, the conductance of the larynx is at a minimum and the laryngogram is low and almost flat. When the glottis closes, the laryngeal conductance goes up and the laryngogram shows a steep upward slope. The point of inflection during the steep rise of the laryngogram, i.e., the instant of maximum change of the laryngeal conductance, was found best suited to serve as the reference point for this event.

To press measurement inaccuracies due to signal sampling below the difference limen for perception of F_0 changes over the whole range of F_0, a temporal resolution corresponding to a sampling frequency of more than 100 kHz is required. The strategy of the algorithm is as follows.

1. The laryngogram, originally sampled at 16 kHz, is digitally differentiated by a first-order nonrecursive differentiator filter. The algorithm then determines the significant maxima or the differentiated laryngogram; spurious maxima due to noise are suppressed by simple threshold discrimination.
2. The locations of the maxima of the differentiated laryngogram represent the raw positions of the period delimiters ("markers"); around these points the sampling rate of the signal is increased by a factor of 8; after differentiating and interpolating the location of the maximum is determined with a temporal resolution of 7.8 μs.

With passband and stopband cutoff frequencies of 5 and 9 kHz, respectively, and a stopband attenuation of more than 72 dB (to keep aliasing distortions below the level of the quantizing noise of the laryngogram), a 144th-order linear-phase interpolator filter proved necessary. The first-order differentiator filter is sufficient to estimate the positions of the raw markers at the original sampling rate of 16 kHz. For the accurate measurement, however, a 7th-order differentiator filter (referring to the original sampling frequency of 16 kHz) gives a good approach to the ideal differentiator in the interesting frequency range below 5 kHz. To minimize roundoff errors, the two filters had to be combined to a 200th-order nonrecursive filter so that differentiation and interpolation finally are performed in one step at the increased sampling frequency of 128 kHz. As before, this filter is applied only in the immediate vicinity of the raw markers.

Although the use of the laryngograph reduces the number of gross errors to a minimum, there are still occasional failures of the algorithm in specific situations. Hence a simple error detection logic based on threshold analysis had to be incorporated. First of all, this logic suppresses weak markers that may occur because of noise in the laryngogram. If a rapid vertical movement of the larynx causes a marker to be set, this marker will occur in isolation, not embedded in a train of markers as in voiced speech. Hence, if a single marker or a sequence of not more than two markers is detected within an unvoiced interval of at least 200 ms on either side, the logic treats these markers as erroneous and removes them. In case this is not sufficient, an interactive routine for visual inspection has been provided that may be used to accept or reject manually individual markers that were not reliably accepted or rejected by the automatic procedure.

3.3. Comparative Evaluation of VDAs and PDAs—Some More Results

Because of the absence of reliable criteria and systematic guidelines, few publications on early PDAs included a quantitative evaluation of the algorithms presented. The classical studies by Rabiner et al. (1976) and McGonegal et al. (1977), without proposing a new algorithm, investigated seven PDAs (two time domain, five short-term analysis) with respect to pitch and voicing determination. A reference pitch contour was obtained interactively using the semiautomatic PDA by McGonegal et al. (1975). The main results read as follows.

1. None of the PDAs involved worked without errors, even under good recording conditions. Each PDA had its own "favorite" error; nevertheless, any error condition actually occurred for any of the PDAs.
2. Almost any gross error was perceptible; in addition, unnatural noisiness of a pitch contour was well perceived.
3. The subjective evaluation did not match the preference of the objective evaluation. In fact, none of the objective criteria (number of gross errors, noisiness of the pitch contour, voicing errors) correlated well with the subjective scale of preference.

Hence the question of which errors in pitch and voicing determination are the really annoying ones for the human ear remained open. This issue was further pursued by Viswanathan and Russell (1984), who developed objective evaluation methods that are more closely correlated to the subjective judgments. The individual error categories are weighted according to the consistency of the error, i.e., the number of consecutive erroneous frames, the momentary signal energy, the magnitude of the error, and the special context.

Indefrey et al. (1985), concentrating on the evaluation of PDAs only, investigated several short-term PDAs in various configurations. The results were discussed above (Section 1.3.3). In a sequel, Indefrey (1987) added several other PDAs to this evaluation. He showed that in many situations different short-term analysis PDAs behave in a complementary way so that combining them to a multichannel PDA could lead to better overall performance.

3.4. Aspects of Application

The area of speech communication systems is one of the important application areas of pitch determination. Other areas include (1) phonetics and linguistics (including musicology), i.e., the measurement of pitch contours as carriers of prosodic, phonetic, and musical information; (2) education, including training aids for the deaf or teaching aids for foreign languages; and (3) voice pathology and phoniatrics. Each of these applications has a different profile of requirements (Hess, 1983, p. 521). With respect to these requirements, the respective applications can be subdivided according to whether or not the human ear is the final "customer" of a measured pitch contour. If the human ear is at the end of the chain of which the PDA or VDA is a part, it is crucial to know whether a time delay for manual correction is permitted. There is no time in vocoder systems or in the recent application of speech-processing hearing prostheses, e.g., cochlear implants (D. M. Howard, 1989; Fourcin et al., 1983). On the other hand, there is time for manual correction in high-quality speech synthesis systems which concatenate original speech data in parametric or waveform-coded representation and need accurate pitch period determination to manipulate pitch and duration (e.g., Charpentier and Moulines, 1989). Even a laryngograph may be applied for such a purpose (Krishnamurthy and Childers, 1986). In the last few years the development of powerful waveform coding schemes which do not need a PDA at all or need only a very rudimentary one have made a vocoder unnecessary in many applications. The applications which will continue to require a PDA in speech communication systems, such as hearing prostheses or high-quality speech synthesis from stored data, are more fault tolerant than the vocoder. Future developments in the domain of pitch and voicing determination are thus likely to move away from the search for a new principle that is able "to solve everything" toward improved implementations of known algorithms that are cheap, fast, and robust at the same time.

REFERENCES

Ananthapadmanabha, T. V., and Yegnanarayana, B. (1975). Epoch extraction of voiced speech. *IEEE Trans. Acoust., Speech, Signal Process. ASSP-23*:562–569.

Ananthapadmanabha, T. V., and Yegnanarayana, B. (1979). Epoch extraction from linear prediction residual for identification of closed glottis interval. *IEEE Trans. Acoust., Speech, Signal Process. ASSP-27*:309–319.

Atal, B. S. (1975). Unpublished work on pitch determination, cited in Rabiner et al., 1976.

Atal, B. S., and Rabiner, L. R. (1976). A pattern recognition approach to voiced-unvoiced-silence classification with applications to speech recognition. *IEEE Trans. Acoust., Speech, Signal Process. ASSP-24*:201–212.

Bruno, G., Di Benedetto, M. G., and Gilio, A. (1982). A probabilistic pitch estimation method with combination of several techniques. In *Fortschritte der Akustik, FASE/DAGA '82 Göttingen*. VDE-Verlag, Berlin, pp. 975–978.

Carré, R., Descout, R., Eskénazi, M., Mariani, J., and Rossi, M. (1984). The French language database: Defining, planning, and recording a large database. *Proc. IEEE ICASSP-84*. IEEE, New York.

Charpentier, F. J. (1986). Pitch detection using the short-term phase spectrum. *Proc. IEEE ICASSP-86*, paper 3.9. IEEE, New York.

Charpentier, F. J., and Moulines, E. (1989). Pitch-synchronous waveform processing techniques for text-to-text speech synthesis using diphones. *Proc. EUROSPEECH-89*, Paris, CEP Consultants, Edinburgh. Vol. 2, pp. 13–19.

Cheng, Y. M., and O'Shaughnessy, D. (1989). Automatic and reliable estimation of glottal closure instant and period. *IEEE Trans. Acoust., Speech, Signal Process. ASSP-37*:1805–1815.

Childers, D. G., and Krishnamurthy, A. K. (1985). A critical review of electroglottography. *CRC Crit. Rev. Biomed. Eng. 12*:131–161.

Childers, D. G., Hahn, M., and Larar, J. N. (1989). Silent and voiced/unvoiced/mixed excitation (four-way) classification of speech. *IEEE Trans. Acoust., Speech, Signal Process. ASSP-37*(Corr.):1771–1774.

Dal Degan, N. (1982). Vocoder quality: an automatic procedure to measure the performance of pitch extractors. *Proc. Globecom'82*. IEEE, New York.

De Mori, R., Laface, P., Makhonine, V. A., Mezzalama, M. (1977). A syntactic procedure for the recognition of glottal pulses in continuous speech. *Pattern Recog. 9*:181–189.

Di Francesco, and Moulines, E. (1989). Detection of the glottal closure by jumps in the statistical properties of the signal. *Proc. EUROSPEECH-89*, Paris CEP Consultants, Edinburgh, pp. 39–42.

Dolansky, L. O. (1955). An instantaneous pitch-period indicator *J. Acoust. Soc. Am. 27*:67–72.

Dolansky, L. O., and Tjernlund, P. (1968). On certain irregularities of voiced speech waveforms. *IEEE Trans. AU-16*:51–56.

Duifhuis, H., Willems, L. F., and Sluyter, R. J. (1982). Measurement of pitch in speech: An implementation of Goldstein's theory of pitch perception *J. Acoust. Soc. Am. 71*:1568–1580.

El Mallawany, I. (1977). Evaluation comparative de mélographes numériques. *Recherches/Acoustique 4*:15–34 (CNET, Lannion, France)

Filip, M. (1969). Envelope periodicity detection. *J. Acoust. Soc. Am. 45*:719–732.

Flanagan, J. L., and Saslow, M. G. (1958). Pitch discrimination for synthetic vowels. *J. Acoust. Soc. Am. 30*:435–442.

Fourcin, A. J. (1974). Laryngographic examinations of vocal fold vibrations. In *Ventilatory and Phonatory Control*, F. Wyke (Ed.). Oxford University Press, London, pp. 315–333.

Fourcin, A. J., and Abberton, E. (1971). First applications of a new laryngograph. *Med. Biol. Illus. 21*:172–182.

Fourcin, A. J., Douek, E., Moore, B., Rosen, S., Walliker, J., Howard, D. M., Abberton, E., and Framton, S. (1983). Speech perception with promontory stimulation. *Ann. N.Y. Acad. Sci. 405*:280–294.

Friedman, D. H. (1977). Pseudo-maximum-likelihood speech pitch extraction. *IEEE Trans. Acoust., Speech, Signal Process. ASSP-25*:213–221.

Fujimura, O. (1968). An approximation to voice aperiodicity. *IEEE Trans. AU-16*: 68–72.

Fujisaki, H., Hirose, K., and Shimizu, K. (1986). A new system for reliable pitch extraction of speech. *Proc. IEEE ICASSP-86*, paper 34.16. IEEE, New York.

Gold, B. (1977). Digital speech networks. *Proc. IEEE 65*:1636–1658.

Gold, B., and Rabiner, L. R. (1969). Parallel processing techniques for estimating pitch periods of speech in the time domain. *J. Acoust. Soc. Am. 46*:442–448.

Goldstein, J. L. (1973). An optimum processor theory for the central formation of the pitch of complex tones. *J. Acoust. Soc. Am. 54*:1496–1516.

Gong, Y., and Haton, J. P. (1987). Time-domain harmonic matching pitch estimation using time-dependent speech modeling. *IEEE Trans. Acoust., Speech, Signal Process. ASSP-35*:1386–1400.

Grützmacher, M., and Lottermoser, W. (1937). Über ein Verfahren zur trägheitsfreien Aufzeichnung von Melodiekurven. *Akustische Z. 2*:242–248.

Harris, M. S., and Umeda, N. (1987). Difference limens for fundamental frequency contours in sentences. *J. Acoust. Soc. Am. 81*:1139–1145.

Hart, J. 't (1981). Differential sensitivity to pitch distance, particularly in speech. *J. Acoust. Soc. Am. 69*:811–822.

Hermes, D. J. (1988). Measurement of pitch by subharmonic summation. *J. Acoust. Soc. Am. 83*:257–264.

Hess, W. J. (1976). A pitch-synchronous digital feature extraction system for phonemic recognition of speech. *IEEE Trans. Acoust., Speech Signal Process. ASSP-24*: 14–25.

Hess, W. J. (1979). Time-domain pitch period extraction of speech signals using three nonlinear digital filters. *Proc. IEEE ICASSP-79*. IEEE, New York, pp. 773–776.

Hess, W. J. (1983). *Pitch Determination of Speech Signals—Algorithms and Devices*. Springer-Verlag, Berlin.

Hess, W. J., and Indefrey, H. (1984). Accurate pitch determination of speech signals by means of a laryngograph. *Proc. IEEE ICASSP-84*, paper 18B.1. IEEE, New York.

Hess, W. J., and Indefrey, H. (1987). Accurate time-domain pitch determination of speech signals by means of a laryngograph. *Speech Commun. 6*:55–68.

Hollien, H. (1974). On vocal registers. *J. Phonet. 2*:125–143.

Howard, D. M. (1989). Peak-picking fundamental period estimation for hearing prostheses. *J. Acoust. Soc. Am. 86*:902–910.

Howard, I. S., and Huckvale, M. A. (1988). Speech fundamental period estimation using a trainable pattern classifier. *Proc. Speech-88 (FASE)*, Edinburgh. CEP Consultants, Edinburgh.

Howard, I. S., and Walliker, J. R. (1989). The implementation of a portable real-time multilayer-perceptron speech fundamental period estimator *Proc. EUROSPEECH-89*, Paris. CEP Consultants, Edinburgh, pp. 206–209.

Indefrey, H. (1987). Vergleichende Untersuchungen zur Grundfrequenzbestimmung von Sprachsignalen mit dem Ausgangssignal eines Laryngographen als Referenzgröße. Dr.-Ing. dissertation, Technical University of Munich, Germany.

Indefrey, H., Hess, W. J., and Seeser, G. (1985). Design and evaluation of double-transform pitch determination algorithms with nonlinear distortion in the frequency domain. *Proc. IEEE ICASSP-85*, vol. 2, paper 11.12. IEEE, New York.

Klatt, D. (1973). Discrimination of fundamental frequency contours in synthetic speech: Implication for models of speech perception. *J. Acoust. Soc. Am. 53*:8–16.

Kohler, K. J. (1982). 25 Years of Phonetica. Preface to the special issue on pitch analysis. *Phonetica 39*(4–5):185–187.

Krishnamurthy, A. K., and Childers, D. G. (1986). Two-channel speech analysis. *IEEE Trans. Acoust., Speech, Signal Process. ASSP-34*:730–743.

Krubsack, D. A., and Niederjohn, R. J. (1989). A logarithmic approach to fundamental frequency error measurement in speech. *J. Acoust. Soc. Am. 85*: 1782–1784.

Lahat, M., Niederjohn, R. J., and Krubsack, D. A. (1987). A spectral autocorrelation method for measurement of the fundamental frequency of noise-corrupted speech. *IEEE Trans. Acoust., Speech Signal Process. ASSP-35*:741–750.

Lebrum, Y. (1971). On the so-called dissociations between electroglottogram and phonogram. *Folia Phoniatr. 23*:225–227.

Léon, P., and Martin, Ph. (1969). *Prolégomènes à l'étude des structures intonatives*. Studia Phonetica No. 2. Didier, Paris.

Lieberman, P. (1963). Some acoustic measures of the fundamental periodicity of normal and pathologic larynges. *J. Acoust. Soc. Am. 35*:344–353.

Lobanov, B. M. (1970). Automatic discrimination of noisy and quasi periodic speech sounds by the phase plane method. *Sov. Phys. Acoust. 16*:353–356. [Original (in Russian) in *Akust. Zh. 16*(1970):425–428]

Markel, J. D. (1972). The SIFT algorithm for fundamental frequency estimation. *IEEE Trans. AU-20*:367–377.

Martin, Ph. (1981). Détection de F_0 par intercorrélation avec une function peigne. *Journées d'Etude sur la Parole 12*:221–232. SFA/GALF, Lannion, France.

Martin, Ph. (1987). A logarithmic spectral comb method for fundamental frequency detection. *Proc. 11th Int. Congr. Phonetic Sciences*. Tallinn, USSR, paper 59.2. Estonian Academy of Sciences, Tallinn, USSR.

McGonegal, C. A., Rabiner, L. R., Rosenberg, A. E. (1975). A semiautomatic pitch detector (SAPD). *IEEE Trans. Acoust., Speech, Signal Process. ASSP-23*:570–574.

McGonegal, C. A., Rabiner, L. R., Rosenberg, A. E. (1977), A subjective evaluation of pitch detection methods using LPC synthesized speech. *IEEE Trans. Acoust., Speech, Signal Process. ASSP-25*:221–229.

McKinney, N. P. (1965). Laryngeal frequency analysis for linguistic research. Research Report 14, Communication Sciences Lab., University of Michigan; Ann Arbor.

Miller, N. J. (1975). Pitch detection by data reduction. *IEEE Trans. Acoust., Speech, Signal Process. ASSP-23*:72–79.

Miller, R. L. (1970). Performance characteristics of an experimental harmonic identification pitch extraction system (HIPEX). *J. Acoust. Soc. Am. 47*:1593–1601.

Moser, M., and Kittel, G. (1977). Rechnergestützte Tonhöhenbestimmung. *Folia Phoniatr. 29*:119–126.

Noll, A. M. (1967). Cepstrum pitch determination. *J. Acoust. Soc. Am. 41*:293–309.

Noll, A. M. (1970). Pitch determination of human speech by the harmonic product spectrum, the harmonic sum spectrum, and a maximum likelihood estimate. In *Symposium on Computer Processing in Communication*, Microwave Institute (Ed.), vol. 19. University of Brooklyn Press, New York, pp. 779–797.

Picone, J., Doddington, G. R., and Secrest, B. G. (1987). Robust pitch detection in a noisy telephone environment. *Proc. IEEE ICASSP-87*. IEEE, New York, pp. 1442–1445.

Plomp, R. (1976). *Aspects of Tone Sensation*. Academic Press, London.

Rabiner, L. R. (1977). On the use of autocorrelation analysis for pitch detection. *IEEE Trans. Acoust., Speech, Signal Process. ASSP-25*:24–33.

Rabiner, L. R., and Schafer, R. W. (1978). *Digital Processing of Speech Signals*. Prentice-Hall Signal Processing Series, A. V. Oppenheim (Ed.). Prentice-Hall, Englewood Cliffs, N.J.

Rabiner, L. R., Cheng, M. J., Rosenberg, A. E., and McGonegal, A. (1976). A comparative study of several pitch detection algorithms. *IEEE Trans. Acoust., Speech, Signal Process. ASSP-24*:399–413.

Rabiner, L. R., Sambur, M. R., and Schmidt, C. E. (1975). Applications of nonlinear smoothing algorithm to speech processing. *IEEE Trans. Acoust., Speech, Signal Process. ASSP-23*:552–557.

Reddy, D. R. (1967). Pitch period determination of speech sounds. *Commun. ACM 10*:343–348.

Risberg, A., Möller, A., and Fujisaki, H. (1960). Voice fundamental frequency tracking. STL-QPSR No. 1, 3–5, Royal Institute of Technology, Stockholm.

Ross, M. J., Shaffer, H. L., Cohen, A., Freudberg, R., and Manley, H. J. (1974). Average magnitude difference function pitch extractor. *IEEE Trans. Acoust., Speech, Signal Process. ASSP-22*:353–361.

Schroeder, M. R. (1968). Period histogram and product spectrum: New methods for fundamental-frequency measurement. *J. Acoust. Soc. Am. 43*:829–834.

Secrest, B. G., and Doddington, G. R. (1982). Postprocessing techniques for voice pitch trackers. *Proc. IEEE ICASSP-82*. IEEE, New York, pp. 172–175.

Siegel, L. J. (1979). A procedure for using pattern classification techniques to obtain a voiced/unvoiced classifier. *IEEE Trans. Acoust., Speech, Signal Process. ASSP-27*:83–89.

Siegel, L. J., and Bessey, A. C. (1982). Voiced/unvoiced/mixed excitation classification of speech. *IEEE Trans. Acoust., Speech. Signal Process. ASSP-30*: 451–461.

Sobolev, V. N., and Baronin, S. P. (1968). Investigation of the shift method for pitch determination. *Elektrosvyaz 12*:30–36 (in Russian)

Sondhi, M. M. (1968). New methods of pitch extraction. *IEEE Trans. AU-16*:262–266.

Specker, P. (1984). A powerful postprocessing algorithm for time-domain pitch trackers. *Proc. IEEE ICASSP-84*, paper 18B.2. IEEE, New York.

Sreenivas, T. V. (1981). Pitch estimation of aperiodic and noisy speech signals. Dissertation, Department of Electrical Engineering, Indian Institute of Technology, Bombay.

Stevens, K. N. (1977). Physics of laryngeal behavior and larynx modes. *Phonetica 34*: 264–279.

Stevens, K. N., Kalikow, D. N., and Willemain, T. R. (1975). A miniature accelerometer for detecting glottal waveforms and nasalization. *J. Speech Hear. Res. 18*: 594–599.

Terhardt, E. (1979). Calculating virtual pitch. *Hear. Res. 1*:155–182.

Terhardt, E., Stoll, H., and Seewann, M. (1982). Algorithm for extraction of pitch and pitch salience from complex tonal signals, *J. Acoust. Soc. Am. 71*:679–688.

Tillmann, H. G. (1978). Bestimmung der Stimmperiode in der Zeitfunktion des digitalen Sprachschallsignals. Forschungsberichte des Instituts für Phonetik und sprachliche Kommunikation *9*, pp. 207–213. University of Munich, Germany

Tucker, W. H., and Bates, R. T. H. (1978). A pitch estimation algorithm for speech and music. *IEEE Trans. Acoust., Speech, Signal Process. ASSP-26*:597–604.

Un, C. K., and Yang, S. C. (1977). A pitch extraction algorithm based on LPC inverse filtering and AMDF. *IEEE Trans. Acoust., Speech, Signal Process. ASSP-25*:565–572.

Van den Berg, J. (1958). Myoelastic-aerodynamic theory of voice production. *J. Speech Hear. Res. 1*:227–244.

Viswanathan, V. R., and Russell, W. H. (1984). Subjective and objective evaluation of pitch extractors for LPC and harmonic-deviations vocoders. BBN Report 5726, Bolt Beranek and Newman, Cambridge, Mass.

Weiss, M. R., Vogel, R. P., and Harris, C. M. (1966). Implementation of a pitch extractor of the double spectrum analysis type. *J. Acoust. Soc. Am. 40*:657–662.

Winckel, F. (1964). Tonhöhenextraktor für Sprache mit Gleichstromanzeige. *Phonetica 11*:248–256.

Wise, J. D., Caprio, J. R., and Parks, T. W. (1976). Maximum likelihood pitch estimation. *IEEE Trans. Acoust., Speech, Signal Process. ASSP-24*:418–423.

Zurcher, J. F. (1977). La mesure du fondamental par la détection de crêtes. Techniques employés, résultats. *Journées d'Etude sur la Parole 8*:119–126. SFA/GALF, Lannion, France.

Zwicker, E., Hess, W., and Terhardt, E. (1967). Erkennung gesprochener Zahlworte mit Funktionsmodell und Rechenanlage. *Kybernetik 3*:267–272.

2

Vector Quantization Techniques in Speech Coding

ALLEN GERSHO and SHIHUA WANG University of California, Santa Barbara, California

KENNETH ZEGER University of Hawaii, Honolulu, Hawaii

1. VECTOR QUANTIZATION

It has become widely recognized in the past decade that a vector, i.e., an ordered set of signal samples or parameters, can be efficiently coded by matching the input vector to a similar pattern or *code vector* in a codebook. For a given input vector, the encoder simply identifies the address, or index, of the best matching code vector. The index, as a binary word, is then transmitted and the decoder replicates the corresponding code vector by a table lookup from a copy of the same codebook. Since the codebook contains a finite set of entries, the matching code vector is one of a finite set of possible approximations. In this sense, analog-to-digital conversion has been performed. The index is a digital code that allows the receiver to reproduce an approximation to the original. Hence, the operation of quantization has taken place directly on a vector. The vector components are not coded individually as in scalar quantization, but rather all at once. Considerable efficiency is achieved, fractional bit rates (bits per vector component) become possible, and the average distortion (e.g., average squared error per component) for a given bit rate is generally much reduced in comparison to scalar quantization of the components. Figure 1 illustrates the basic idea of vector

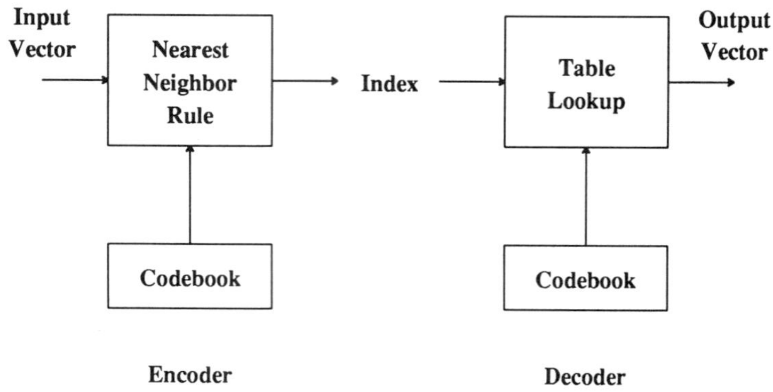

Figure 1 Vector quantization.

quantization (VQ). From this description, it is evident that VQ is in essence a pattern-matching algorithm; such an approach to pattern recognition was in use long before the age of digital electronics.

Perhaps the first application of pattern matching to speech coding dates back to Dudley (1958). The first major application of VQ to speech coding was reported by Buzo et al. (1980), who substantially reduced the bit rate of a linear predictive coding (LPC) vocoder by applying VQ to the LPC parameters. Subsequently VQ found its way into waveform coding as well and a generalization of DPCM using vector prediction together with VQ was developed. Today VQ is a well established and widely used technique. It has been applied to the efficient coding of LPC parameters, pitch predictor filter parameters, gain parameters, block waveform coding and coding of the excitation or residual signal in analysis-by-synthesis predictive coding techniques, such as VXC, CELP, and VAPC.

In this chapter we first review the fundamentals of vector quantization, including optimality and design considerations. Current applications of VQ to speech coding are then discussed, including waveform coding, analysis-by-synthesis predictive coding, and parameter coding. Important complexity issues inherent in VQ are examined, and finally, a discussion is presented of various recent techniques that are incorporated in VQ to combat some of the detrimental effects of channel noise.

1.1. Vector Quantization Fundamentals

Quantization plays a fundamental role in the coding of speech signals. Vector quantization is a generalization of scalar quantization to the quantization of a vector, an ordered set of real numbers. (For comprehensive discussions of VQ and its applications to speech coding, see Gray, 1984; Gersho, 1982; Abut et al., 1982; Gray et al., 1980, 1981; Jayant and Noll, 1984.) The jump from one dimension to multiple dimensions is a major step and allows a wealth of new ideas, concepts, techniques, and applications to arise that often have no counterpart in scalar quantization. While scalar quantization is primarily associated with analog-to-digital conversion, VQ deals with sophisticated digital signal processing in which the relevant input signals already have some form of digital representation and VQ is usually, but not exclusively, used for the purpose of data compression. Nevertheless, there are interesting parallels with scalar quantization and many of the structural models and analytical techniques used in VQ are natural generalizations of the scalar case.

A vector can be used to describe almost any type of *pattern*, such as a segment of a speech waveform, by forming a vector of samples from the waveform. An important example in speech coding arises when a set of parameters (forming a vector) is used to represent the spectral envelope of a speech sound. Vector quantization can be viewed as a form of pattern recognition in which an input pattern is "approximated" by one of a predetermined set of standard patterns; or in other words, the input pattern is matched with one of a stored set of templates or code words.

Vector quantization can also be regarded as a building block for a variety of complicated signal processing tasks, including classification and linear transformation. In such applications VQ can be viewed as a complexity-reducing technique because the reduction in bits can simplify the subsequent computations, sometimes permitting complicated digital signal processing to be replaced by simple table lookups. Thus VQ is far more than a generalization of scalar quantization. Its scope and implications are vast. In the last few years it has become an important technique in speech recognition as well as in speech compression. Recently, it has been used in speech enhancement and other speech processing applications. The importance of VQ continues to grow and the range of applications continues to expand.

A *vector quantizer* Q of dimension k and size N is formally a mapping from k-dimensional Euclidean space $\mathbf{R}^k$ into a finite set $\mathbf{C}$, containing N output or reproduction vectors from $\mathbf{R}^k$. Thus,

$$Q : \mathbf{R}^k \longrightarrow \mathbf{C}$$

where $\mathbf{C} = \{\mathbf{y}_0, \mathbf{y}_1, \ldots, \mathbf{y}_{N-1}\}$ and $\mathbf{y}_i \in \mathbf{R}^k$ for each $i \in \mathbf{J} = \{0, 1, \ldots, N-1\}$. The set $\mathbf{C}$ is called the *codebook* and has N elements, each a distinct vector in $\mathbf{R}^k$. The codebook is typically implemented as a table in a digital memory. The *resolution* or *rate* of the quantizer is $r = (\log_2 N)/k$, which measures the number of bits per vector component used to represent the input vector (and thus the degree of accuracy). It is important to recognize that for a fixed dimension k the resolution is determined by the size N of the codebook and not by the number of bits used to describe the code vectors stored in the codebook.

Associated with every N point vector quantizer is a *partition* of $\mathbf{R}^k$ into N regions or *cells*, R_i for $i \in \mathbf{J}$. The ith cell is defined by

$$R_i = \left\{ \mathbf{x} \in \mathbf{R}^k : Q(\mathbf{x}) = \mathbf{y}_i \right\}$$

or in terms of inverse image notation as $R_i = Q^{-1}(\mathbf{y}_i)$. From the definition of the cells it follows that the R_i are pairwise disjoint and their union covers all of $\mathbf{R}^k$.

A vector quantizer can be decomposed into two component operations, the vector *encoder* and the vector *decoder*. The encoder $\mathbf{E}$ is a mapping from $\mathbf{R}^k$ to the index set $\mathbf{J}$, and the decoder $\mathbf{D}$ maps the index set $\mathbf{J}$ into the reproduction set $\mathbf{C}$. Thus,

$$\mathbf{E} : \mathbf{R}^k \longrightarrow \mathbf{J} \quad \text{and} \quad \mathbf{D} : \mathbf{J} \longrightarrow \mathbf{R}^k$$

and the overall operation of VQ can be regarded as the cascade or composition of two operations:

$$Q(\mathbf{x}) = \mathbf{D}(\mathbf{E}(\mathbf{x}))$$

In the context of a speech coding system, the encoder of a vector quantizer performs the task of selecting an appropriately matching code vector $\mathbf{y}_i$ to approximate, or in some sense to describe or represent, an input vector $\mathbf{x}$. The vector $\mathbf{x}$ may represent a number of different possible speech coding parameters including, among others, consecutive speech samples, LPC coefficients, and prediction residual samples. The index i of the selected code vector is transmitted (as a binary word) to the receiver, where the decoder performs a table lookup procedure and generates the reproduction $\mathbf{y}_i$, the quantized approximation of the original input vector.

To conveniently assess the performance of a particular vector quantizer, we use a *distortion function* $d(\mathbf{x}, \mathbf{y})$ that indicates the distortion between any two vectors $\mathbf{x}$ and $\mathbf{y}$ in $\mathbf{R}^k$ where $\mathbf{x}$ is the original and $\mathbf{y}$ the reproduced vector. The "overall" measure of performance for a given quantizer is given by the average distortion between an input vector and the quantized output vector, $E[d(\mathbf{x}, Q(\mathbf{x}))]$, where E represents the expectation operator. As an example, the mth power distortion measure is defined by

$$d(\mathbf{x}, \mathbf{y}) = \|\mathbf{x} - \mathbf{y}\|^m$$

The most common and perhaps natural distortion measure is the squared error function, defined by setting $m = 2$ above. In this case the average quantizer distortion is the *mean squared error*, $E\|\mathbf{x}-Q(\mathbf{x})\|^2$. Another useful (and more general) distortion function is the *weighted squared error* function given by

$$d(\mathbf{x},\mathbf{y}) = (\mathbf{x}-\mathbf{y})^t\mathbf{W}(\mathbf{x}-\mathbf{y})$$

where $\mathbf{W}$ is a symmetric nonnegative definite matrix. In some cases $\mathbf{W}$ may be a function of the original vector, $\mathbf{x}$, to be quantized.

The *Itakura–Saito* distance function (Gray et al., 1980; O'Shaughnessy, 1987) computes a distortion between two input random vectors by using their spectral densities. It is defined by the following L_1 norm:

$$d(\mathbf{x},\mathbf{y}) = \left\| \frac{f_x}{f_y} - \ln\left(\frac{f_x}{f_y}\right) - 1 \right\|_1$$

where f_x and f_y are the spectral densities of the input random vectors $\mathbf{x}$ and $\mathbf{y}$, respectively. This measure can also be conveniently computed using autocorrelation functions.

An important issue for speech coding with vector quantization is the design and implementation of quantizers to meet performance objectives. That is, for a given source, distortion function, and constraint on the bit rate, what are the best vector quantizer encoding and decoding rules? In general there is no known method for explicitly specifying the optimal quantizer, although certain optimality conditions do exist.

For a general class of distortion functions, it is known that the best vector quantizer encoder for a given decoder satisfies the *nearest-neighbor rule* that is, each input vector $\mathbf{x}$ is quantized to an output vector $Q(\mathbf{x})$ which is at least as close to $\mathbf{x}$ as any other vector in the codebook. If there happen to be two or more code vectors which are at least as close to $\mathbf{x}$ as any other code vector, then the value of $Q(\mathbf{x})$ can be chosen arbitrarily from among the set of nearest code vectors to $\mathbf{x}$.

If the squared-error distortion function is assumed, then for a given quantizer encoder, an optimal decoder must satisfy the *centroid condition*; that is, each code vector $\mathbf{y}$ must be the conditional mean, or centroid, of its corresponding cell R_i.

The problem of designing a vector quantizer for a particular source is of fundamental importance. If one assumes the nearest neighbor and centroid rules, then the design problem reduces to determining the best codebook for a given source (the nearest-neighbor rule is implicitly assumed). The generalized Lloyd algorithm (GLA) (Linde et al., 1980) utilizes these two necessary rules to perform an iterative descent procedure based on a finite training set representation of the source. During each iteration in the GLA each vector in the training set is compared with every vector in the current codebook to find the closest code vector, thus partitioning the training set into nearest-neighbor cells. The centroid of each partition cell is then computed and serves as a code vector for the next iteration, and the process is repeated.

For a fixed transmission rate r (bits per vector component) the size of a VQ codebook is 2^{kr}, where k is the dimension of the vector. Hence the complexity of a full codebook search increases exponentially as the vector dimension grows. This is one major drawback of using the GLA for VQ design and for using a full-search VQ encoder in general.

The GLA yields locally optimal codebooks that perform well in practice but may not perform as well as is theoretically possible. Some probabilistic methods have been developed that can find near–globally optimal VQ codebooks in a reasonable amount of computation time (Vaisey and Gersho, 1988; Cetin and Weerackody, 1988; Flanagan et al., 1989; Zeger and Gersho, 1989). One of the most successful of these probabilistic search techniques is *simulated annealing*, which iteratively "perturbs" a given codebook in a random manner. Each perturbed codebook is "accepted" if the quantizer improves or is probabilistically accepted if the quantizer performance degrades. The probability of accepting degraded codebooks is gradually reduced as the algorithm progresses, which eventually settles the codebook into a good stationary point, avoiding many poor local minima.

Other VQ design techniques have been developed with the purpose of reducing the computational complexity of the GLA and achieving better local minima. The splitting method (Gray, 1984) begins by constructing a codebook containing one code vector. It then "splits" the code vector into two new code vectors and designs a new codebook of size two with the GLA, using these two vectors as initial code vectors. At each stage in the algorithm, all of the previous code vectors are split and a new codebook is designed which is twice the size of the previous codebook. The design terminates when the desired power-of-two codebook size is reached. Other VQ design techniques include using a gradient descent (Chang and Gray, 1986), a conjugate gradient descent (Yair et al., 1990), and a learning algorithm related to neural networks (Yair et al., 1991; Nasrabadi and Feng, 1988).

Because of the large search complexity of ordinary VQ, there have been several research efforts to develop fast algorithms for the nearest-neighbor search process. Algorithms exploiting the constrained geometry of the encoding regions find nearest-neighbor code vectors within a time whose expected value is a logarithmic function of the codebook size (Friedman et al., 1977; Cheng and Gersho, 1986). Other fast search techniques are given in Cheng et al. (1984), Fukunaga and Narendra (1975), Soleymani and Morgerai (1987), and Vidal (1986).

For a given codebook size N, the transmission rate of a VQ system is $\log_2 N$ bits per vector. The output indexes of a given VQ encoder constitute a finite ensemble and have a certain associated probability distribution. In general, this distribution is nonuniform and hence entropy coding can be exploited. If a VQ encoder is followed by a variable-rate entropy coder, such as a Huffman coder, the output average bit rate of the encoder can be reduced while maintaining the same average distortion. This increases the performance of the system in the rate-distortion sense. An algorithm for optimal design of a vector quantizer with a constraint on the entropy of the VQ indexes is given in Chou et al. (1989a). The algorithm, known as *entropy-constrained VQ (ECVQ)*, closely resembles the GLA and assumes that the quantizer is followed by an ideal entropy coder, which in practice works well.

In principle, if the probability distribution $f_{\mathbf{X}}(\mathbf{x})$ of the input vector $\mathbf{X}$ is known, it is possible to evaluate the average distortion achieved for a particular vector quantizer. In practice, it is sufficient to estimate this figure from the training set as part of the codebook design process. On the other hand, it is of great interest to know the minimum average distortion attainable for an optimal vector quantizer for a statistically specified input and a given codebook size. Unfortunately, no general theory is available to provide an answer. However, there exists a useful and relatively simple formula for the squared Euclidean distance measure in the asymptotic case of *high resolution*, that is, as the codebook size N approaches infinity. It turns out that this formula is reasonably accurate for moderately high-resolution applications, and it provides considerable insight into the gain that VQ offers compared to scalar quantization. Specifically, we have:

$$D \approx N^{-2/k} C_k H_k(f_{\mathbf{x}})$$

where C_k is the coefficient of quantization and depends only on the vector dimension and

$$H_k(f_{\mathbf{X}}) = \|f_{\mathbf{X}}(\mathbf{x})\|_{k/(k+2)} = \left[\int_{-\infty}^{\infty} (f_{\mathbf{X}}(\mathbf{x}))^{k/(k+2)}\, d\mathbf{x}\right]^{(k+2)/k}$$

The constant C_k drops relatively little as the dimension increases from one to infinity. The equation clearly shows how the minimum distortion decreases with increasing codebook size. The dependence on the joint probability density of the input vector is not so easy to see, but it can be shown that as the statistical interdependence of the vector component grows, the distortion decreases as expected. In particular, in the case of a jointly Gaussian random vector, we get

$$H_k(f_{\mathbf{X}}) = 2\pi \left(\frac{k+2}{k}\right)^{(k+2)/2} (\det R_{\mathbf{x}})^{1/k}$$

where $R_{\mathbf{x}}$ is the covariance matrix of $\mathbf{X}$. It is well known that the kth root of the determinant of the covariance matrix is a useful measure of the degree of correlation between the vector components. When the vector is a block of consecutive scalar samples of a stationary random process, $(\det R_{\mathbf{x}})^{1/k}$ approaches the minimum mean squared prediction error for one-step prediction with infinite memory.

Perhaps the simplest form of VQ for speech coding is block waveform coding or vector PCM (VPCM), as it is sometimes called. The digitized speech waveform is divided into consecutive blocks (or vectors) of length k, and each vector is encoded to the index of one of the k-dimensional vectors in a codebook. The speech waveform is thus converted into a sequence of binary data (code vector indexes) which can be transmitted or stored. The performance of this simple VQ technique is generally much lower than can be achieved when VQ is embedded in a more sophisticated system.

If the digitized speech waveform is immediately followed by a waveform predictor, then the output residual waveform can be vector quantized and an increase in performance generally results. The most common predictor is one employing scalar linear predictive coding but another possibility is vector linear prediction. A generalized version of scalar predictive quantization (as in DPCM) combines the use of vector prediction and VQ (Gersho and Cuperman, 1982; Gersho and Cuperman, 1985). A pth-order vector linear prediction $\hat{\mathbf{x}}_n$ of a k-dimensional vector $\mathbf{x}_n$, at time instant n, is of the form

$$\hat{\mathbf{x}}_n = \sum_{i=1}^{p} \mathbf{A}_i \mathbf{x}_{n-i}$$

where each $\mathbf{A}_i$ is a constant $k \times k$ matrix. The *prediction gain ratio* (in dB) in this case is defined as

$$10 \log_{10} \frac{E\|\mathbf{x}_n\|^2}{E\|\mathbf{x}_n - \hat{\mathbf{x}}_n\|^2}$$

where a zero-mean input is assumed. Vector predictive coding has also found applications to interframe coding of LPC parameters and to image coding.

2. GAIN-ADAPTIVE VECTOR QUANTIZATION

The wide dynamic range of a speech waveform contributes to the difficulty of directly applying VQ to blocks of samples. Vectors formed in this way will require an excessively large codebook to accommodate not only the wide variety of shapes of the waveform segment represented by the vector but also the wide range of amplitude levels that can arise. In scalar quantization methods for speech, the technique of gain-adaptive quantization has become widely used to reduce the inefficiency of directly coding samples that have a wide dynamic range. Its use in ADPCM is widely known.

The same concept has been extended to the vector case and has proved to be an effective means of combating the dynamic range problem for waveform coding (Chen and Gersho, 1987). In various forms, it is applicable not only for direct waveform coding of speech but also for coding subband signals generated by an analysis filter bank as in subband coding (Gersho et al., 1984).

In a typical application of gain-adaptive VQ, a codebook for gain-normalized speech vectors is first designed by a codebook design algorithm optimized specifically for this application. Input vectors are scaled in amplitude by a gain-normalizing factor which is adaptively generated by either forward or backward adaptation. Each normalized vector is then vector encoded and the received index is decoded to produce a corresponding code vector which is then scaled back to the normal level by using the same adaptive gain value as in the encoder. This effectively allows a much greater dynamic range for signal vectors than would be possible with conventional VQ with a reasonable codebook size.

If the gain is produced via forward adaptation, the input speech is buffered and an average gain value is measured, quantized, and then used for an entire frame consisting of several signal vectors. Side information is transmitted to the receiver to identify the needed gain value. Because one gain value serves several successive vectors, a relatively small bit rate is needed to specify the gain.

In the more interesting case of backward adaptation, no side information is needed and the gain control is generated as an *estimate* of the norm of the current input vector by operating on the past of the reconstructed vectors. The gain estimation operation can be viewed as a linear or nonlinear prediction operation and, in particular, LPC techniques can be applied to the random process consisting of the sequence of vector norms. Often it is convenient to work on the logarithm of the vector norm, the *log-norm*, in which case linear prediction methods can be applied to predict the current log-norm value from the past of the sequence of log-norms of the reconstructed vectors. One particularly important application of backward gain adaptive VQ is in low-delay speech coding (Cuperman et al., 1989; Chen, 1989), which will be discussed later in this chapter. It is expected that a forthcoming CCITT standard for 16 kb/s speech coding will incorporate a form of gain-adaptive VQ.

3. STRUCTURALLY CONSTRAINED VECTOR QUANTIZATION

Although it is not possible to quantize a vector better than by using an optimal full-search codebook, there are serious drawbacks to this owing to computational complexity and storage requirements. For a fixed transmission rate, both the storage and the search complexity of a vector quantizer grow exponentially with the vector dimension. For example, in simple VPCM coding of 8-kHz sampled speech at a rate of 1 bit/s, the size of the codebook is 2^k, where k is the vector dimension. The amount of time available to search the codebook grows linearly in proportion to k, whereas the codebook size grows exponentially in k. If we choose $k = 20$ then at most 2.5 ms would be available to search over one-million 20-dimensional vectors for a best match, requiring over 8 billion floating-point arithmetic operations per second, an infeasible task with existing signal processor chips.

There are, however, several suboptimal VQ coding schemes which provide reduced-complexity alternatives to full-search optimal VQ. Each such technique to be described in this section introduces some structural constraint on the VQ system. This means that the set of code vectors cannot have arbitrary locations as points in k-dimensional space but are distributed in a restricted manner that allows lower-complexity search methods for nearest neighbors. Some of the structurally constrained techniques do not even find the true nearest-neighbor code vectors. Hence the performance of these VQ codes are suboptimal due to both the constrained locations of the code vectors and the suboptimal search procedure. The constrained VQ techniques discussed in this section include tree-structured VQ, multistage VQ, and gain/shape VQ. All of these are applicable to speech compression and have been used in various speech coding schemes in the past.

3.1. Gain/Shape VQ

Gain/Shape VQ (*GSVQ*) is another structurally constrained VQ technique that can help reduce the inherent computational complexity of VQ. The basic premise of GSVQ is that many different input vectors will have similar "shapes" but will have varying "norms." This leads to the idea of storing in a codebook a finite set of typical shapes and storing in a scalar codebook a finite set of typical "gains." GSVQ was introduced in Buzo et al. (1980) and later optimized in Sabin and Gray (1984).

In the most basic form of GSVQ, the norm (or *gain*) $g = \|\mathbf{X}\|$ of each input vector $\mathbf{X}$ is extracted and encoded into $\hat{g}$ by a scalar quantizer, and the resulting unit norm *shape* vector $\mathbf{S} = \mathbf{X}/g$ is encoded into $\hat{\mathbf{S}}$ by a vector quantizer. The decoder generates the product $\hat{g}\hat{\mathbf{S}}$ as its approximation to the input vector **X**. One method of quantizing the gain and shape is to find independently nearest-neighbor matches for them in their codebooks. It turns out that this encoding technique is not optimal for GSVQ. The optimal encoding scheme consists of first finding which shape vector $\hat{\mathbf{S}}$ in a unit-norm shape codebook maximizes the scalar product $\mathbf{X}^t\mathbf{S}$, where the superscript t denotes the transpose, and then finding which gain $\hat{g}$ in the scalar codebook best approximates $\mathbf{X}^t\hat{\mathbf{S}}$. This solution, however, assumes a shape codebook with unit-norm code vectors.

In the more general setting, where the shape code vectors need not have unit norm, slightly better performance can be attained but the optimal encoding strategy involves a full search of all possible products of gain and shape code vectors, thus eliminating the complexity reduction advantage.

3.2. Multistage VQ

Another complexity reduction VQ technique that can be used when codebook storage size must also be reduced is *multistage VQ (MSVQ)*, In MSVQ an input vector **X** is successively encoded by a series of quantizers. The input to the first quantizer is **X** and the output is $\hat{\mathbf{X}}$. The residual vector $\mathbf{X} - \hat{\mathbf{X}}$ is the input to the second quantizer. At each stage, the input to a quantizer is the residual vector from the previous quantizer. The decoder output is a reconstructed vector obtained as the sum of $\hat{\mathbf{X}}$ plus all of the residual code vectors from each stage. The basic structure of MSVQ is illustrated in Fig. 2.

If the ith stage has a codebook with N_i code vectors, then both the search complexity for each input vector and the storage requirement are equal to the sum of the N_i's, whereas the size of the equivalent codebook (i.e., size of a full search) is equal to the product of the N_i's. Thus the complexity and storage requirements are greatly reduced using MSVQ.

Generally, the components of the quantization error vector at each stage of MSVQ tend to be less statistically dependent on one another than those of the input vector at the same stage. Thus, the coding gain of the quantizers

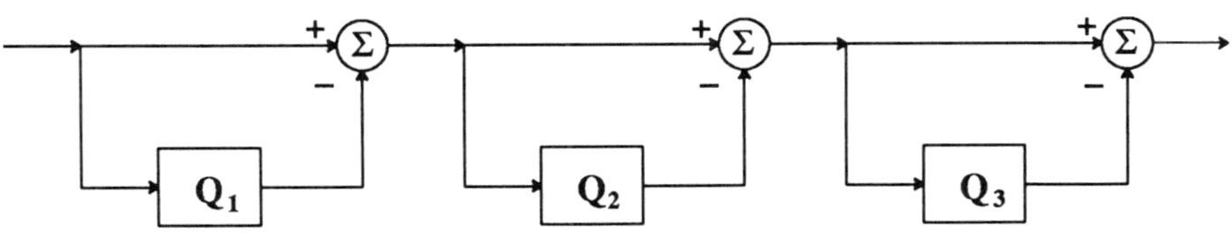

Figure 2 Multistage VQ.

tends to diminish in the later stages. In practice, usually only two or three stages are used in MSVQ because of diminishing returns.

3.3. Tree-Structured VQ

One of the most effective complexity reduction techniques for VQ, called tree-structured VQ (TSVQ), is based on a specially designed codebook that allows a successive approximation search procedure with a tree structure. The search for the best code vector is performed in stages; each stage eliminates a substantial number of candidate code vectors from consideration. The outcome of each stage determines how the search will proceed in the next stage.

An m-ary tree search begins at the root node of the tree and continues along a sequence of branches until a terminal node or *leaf* is reached. At each nonterminal node, the input vector is compared with m predesigned test vectors, each of which is associated with a particular child node connected by an edge from the parent node. A unique path through the tree is determined by following the edge at each stage associated with the test vector that is nearest to the input vector. The number of remaining candidate code vectors at each stage is equal to $1/m$ of those remaining at the previous stage.

An m-ary (balanced) tree of length d corresponds to a codebook of length $N = m^d$. In encoding an input vector, a particular path down the tree is determined that ends at a leaf. Each leaf node is associated with a particular code vector, which also serves as a test vector for the final stage of the search process. An index identifying the selected code vector is transmitted by the encoder to the decoder. Note that an m-ary tree with length $d = 1$ is equivalent to a full-search vector quantizer of size m.

The number of vector comparisons performed for a tree search is d, which equals $\log_m N$, a significant reduction from the N comparisons necessary in full-search VQ. The total search complexity for an m-ary tree is md because each of the d stages requires m vector comparisons. This is a substantial reduction from m^d, the total number of comparisons required in full-search VQ. In contrast, all of the test vectors must be stored in addition to the code vectors, increasing the storage requirement over conventional VQ by a factor of $m/(m - 1)$.

The decoder for TSVQ operates in the same manner as in conventional VQ and stores a copy of the codebook but does not need to store the test vectors. Alternatively, for *progressive* transmission of an encoded vector, digital information (an m-ary symbol) is sent as each edge in the path through the tree is determined by the encoder. This allows the decoder to make successively improved approximations to the ultimate code vector by using the test vectors associated with the traversed edges as successive approximations to the input vector. In this case, the decoder must also store the test vectors and utilize the tree structure.

A standard method of designing tree-structured vector quantizers is somewhat analogous to the "splitting method" for designing ordinary vector quantizers. Initially, the entire training set is used in the generalized Lloyd algorithm to design a codebook with m code vectors. These m code vectors are the test vectors for the first level of the tree being constructed. The subset of the training set lying in the nearest-neighbor region of each test vector is then used by the GLA to design an m-vector codebook for the subregions. The m new code vectors at each of the m branches of the root node become the test vectors for the second stage of the tree. As this process is repeated, the tree grows and the quantization error of the resulting codebook decreases. TSVQ has been successfully used in speech coding applications in Makhoul et al. (1985), Buzo et al. (1980), and Gray and Abut (1982).

Another successful approach in tree-structured VQ is to use unbalanced trees, in which some paths from the root node to a leaf node are longer than others. Unbalanced trees offer a natural way to obtain variable-rate transmission codes, which can be useful in some applications. For example, in a binary tree one bit can be allocated for the choice between the two edges at any given node. Any path through the tree is then uniquely specified by a binary word with length in bits equal to the number of edges traversed to reach the leaf. The rate-distortion performance of an unbalanced tree-structured VQ can be significantly better than that of a balanced tree. The most common technique for designing unbalanced TSVQ coders is to grow a large balanced tree and then prune away edges. An efficient algorithm that prunes a tree in an optimal rate-distortion sense has been given by Chou et al. (1989b). An improved technique is given in Riskin and Gray (1991), in which an unbalanced tree is grown and then pruned back.

4. WAVEFORM VECTOR QUANTIZATION

The most direct and primitive application of VQ to speech coding is to perform block waveform coding, or VPCM, on the speech signal vector with a codebook of size N and dimension k (Abut et al., 1982; Gersho and Cuperman, 1983). VPCM is illustrated in Fig. 3, where the box labeled S/P performs serial-to-parallel conversion on each group of k samples and the box P/S performs the corresponding parallel-to-serial conversion.

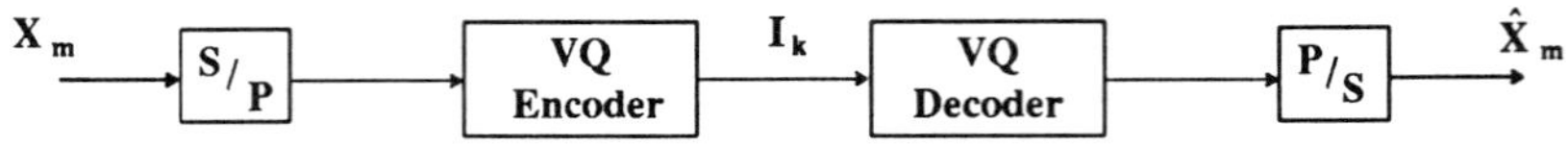

Figure 3 Vector PCM.

Although VPCM indeed offers a substantial coding gain over PCM (which does not exploit redundancy at all), it is nevertheless by itself a primitive coding technique and not of practical significance. For example, a bit rate of 8 kb/s based on $k = 8$ and $N = 256$ results in speech with a signal-to noise ratio in the neighborhood of 10 dB and with a substantial level of granular noise. The quality is roughly similar to that of delta modulation coding at 16 kb/s. This poor quality (by contemporary standards) is due largely to the limited vector dimension, implying that only local correlation among a small number of samples within a block is considered and no correlation between adjacent blocks is being exploited.

From distortion-rate theory we know that increasing the block size (vector dimension) for the same rate will asymptotically approach the ultimate performance limits. If we use dimension $k = 40$ and a rate of 1 bit per sample, we would indeed be exploiting substantial correlation and could expect a significant performance improvement. Indeed, it is reasonable to speculate that this would provide a very high-quality speech coding scheme. However, this implies the use of a VQ codebook of size N_2^{40}, which has an impractically high search complexity as well as storage requirement.

How do we circumvent this complexity barrier? Constrained codebook schemes such as tree structures or multiple stages of VQ can certainly reduce the complexity, but the price in performance to obtain the drastic reduction in complexity that is needed in the above example can be severe and may be overshadowed by alternative coding methods that do not use VQ.

Alternatively, we can greatly simplify the problem by making use of parametric modeling of speech (specifically linear prediction methods) as a redundancy removal technique. In particular, we can improve on direct waveform coding by performing *residual coding*, where the unpredictable residual is coded as a waveform and the prediction parameters are separately coded as side information. This idea is described below.

5. RESIDUAL CODING WITH VECTOR QUANTIZATION

When we subtract from the current speech sample a prediction of this sample based on past values of the speech signal, the prediction error waveform or *residual* is much less correlated and has much lower power. It also has a statistical distribution that is relatively invariant to changes in speech statistics. The residual signal is therefore an easier waveform to quantize with VPCM than is the original speech signal.

There are two primary ways to use VQ for coding the residual. One way is to extend the DPCM concept from sample-by-sample processing to process vectors of samples at a time. This approach leads to *vector DPCM* or *vector predictive coding*, where vector prediction is combined with VQ to code successive blocks of prediction residuals with the usual feedback structure corresponding to scalar DPCM (Gersho and Cuperman, 1985). An alternative approach is to use an open-loop configuration as shown in Fig. 4. Although both methods offer enhanced performance over direct VQ of the speech waveform, we focus here only on the second, which can be viewed as an evolutionary step toward analysis-by-synthesis predictive coding methods of particular importance in speech coding today.

In Fig. 4, the buffer stores a frame of speech samples so that analysis can be performed to determine the predictor parameters. The quantized parameters are used to control the time-varying prediction error filter A and, after transmission to the decoder, the corresponding synthesis filter B, which is the inverse of A. These filters may be based on long-term or pitch-based correlation as well as on short-term correlation. It is reasonable to expect that a codebook of dimension 40 with size $N = 1024$ might be adequate for coding this residual with reasonable accuracy. Thus, we may expect that 10 bits (rather than 40 bits) per vector will be sufficient for high-quality speech coding.

Although this coding method improves on VPCM of the speech waveform itself, there is one significant weakness to the residual encoding scheme of Fig. 4. The VQ encoder, as used here, performs a nearest-neighbor search to minimize the distortion between the input residual vector and the code

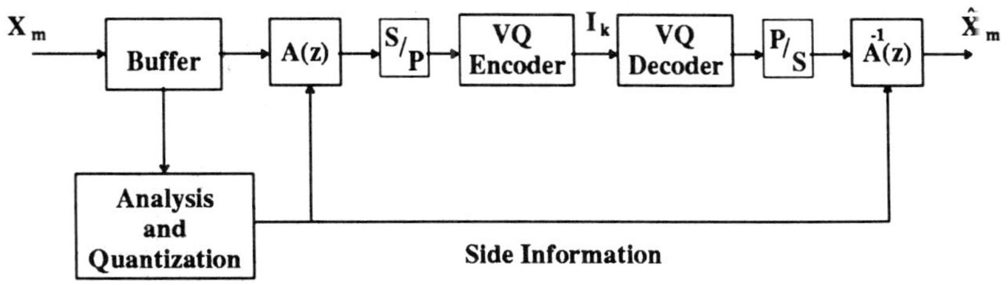

Figure 4 Residual coding structure with VQ.

vector. Regardless of the particular distortion measure used, there is something fundamentally wrong with this method: reducing the quantization error in the residual does not necessarily result in a corresponding reduction in the error between the reconstructed speech and the original speech. In fact, the quantization error produced by the VQ will be filtered by the synthesis filter so that the final synthesized output speech will have a spectrally altered quantization noise and may have a substantially higher power level than the error produced by the quantizer itself. We next describe a better way to search for the best code vector.

6. ANALYSIS-BY-SYNTHESIS PREDICTIVE CODING

Vector excitation coding (VXC) and code-excited linear prediction (CELP) are different names for a powerful and effective family of speech coding algorithms that are increasingly being adopted for a variety of applications at bit-rates ranging from 4.8 up to 16 kb/s. The term CELP is becoming associated with the specific coding algorithm adopted by the US government as a standard for 4.8 kb/s (Campbell et al., 1990). Henceforth we refer to the generic family of algorithms that perform predictive coding using analysis-by-synthesis techniques as VXC. Since this family of algorithms is described elsewhere in this book (see Chapter 5), in this section we focus only on the role of VQ in VXC algorithms and provide some insight into the operation of VXC by giving an evolutionary interpretation of the approach starting with basic waveform VQ, or vector PCM.

We now reexamine the residual coding structure and take an important conceptual leap that leads from the residual coding structure of Fig. 4 to VXC. Suppose that the codebook used in the scheme of Fig. 4 has an adequate set of candidate code vectors. In other words, if the encoder magically knew the right indexes to transmit to identify the sequence of code vectors to be used, the codebook would be adequate to provide the desired speech quality.

We wish to find the best sequence of code vectors that minimizes the distortion between the reproduced and original speech. Therefore, for each code vector search, we examine the corresponding block of *input* speech samples viewed as a vector and search for a code vector that leads to the best matching block of output (reconstructed speech) samples, also regarded as a vector. Thus, we directly examine the output speech vector which is produced by filtering the samples of the code vector through the synthesis filter. Of course, the output vector also depends on the filter memory due to prior inputs and this must also be taken into account. For each candidate code vector, the output speech vector is compared to the original input speech vector. The search process then identifies the best possible code vector.

It is striking to note that with this approach the residual VQ encoder and the residual signal itself are no longer needed! The new encoding system requires a replica of the decoder in order to compute what output speech vector would be produced for any given candidate code vector and given state (memory) of the synthesis filter. The resulting search process is called analysis by synthesis and the overall coding scheme is VXC.

In VXC it is helpful first to examine the operation of the VXC decoder, since it determines how the speech is synthesized (reconstructed) from transmitted data. The encoder is in a sense a servant of the decoder, since its job is to examine the original speech and determine what data must be delivered to the decoder. The VXC decoder structure remains the same as the decoder already shown in Fig. 4 for the residual VQ scheme. The VXC decoder receives and demultiplexes the data needed to specify the synthesis filter parameters and the excitation code vector. Each frame is divided into subframes of k samples corresponding to the duration of one excitation vector.

For each subframe, the decoder receives an index consisting of c excitation code bits which identifies the index that specifies one of 2^c excitation code vectors by means of a table-lookup procedure. This vector is serialized as k successive samples and is applied to the synthesis filter. The filter is clocked for k samples, feeding out the next k samples of the synthesized speech; then the filter is "frozen" until the next scaled excitation vector is available as the next input segment to the synthesis filter.

In many applications an adaptive postfilter is added to the decoder as a final postprocessing stage to enhance the quality of the recovered speech. This filter is usually adapted to correspond to the short-term spectrum of the speech (Chen and Gersho, 1986).

6.1. The VXC Encoder

The VXC encoder structure is shown in Fig. 5. We describe its operation in the simplest way, ignoring the many shortcuts and tricks which greatly reduce the complexity involved in the search process. For simplicity, the S/P and P/S blocks are omitted from Fig. 5; they are easily determined from the context. Also omitted are the frame buffer and the analysis and quantization of the synthesis filter parameters.

The encoder forms input speech vectors $\mathbf{V}(n)$ from successive blocks of k input speech samples. The task of the encoder is to determine the next c bits of data to be transmitted to the decoder so that the decoder will then be able

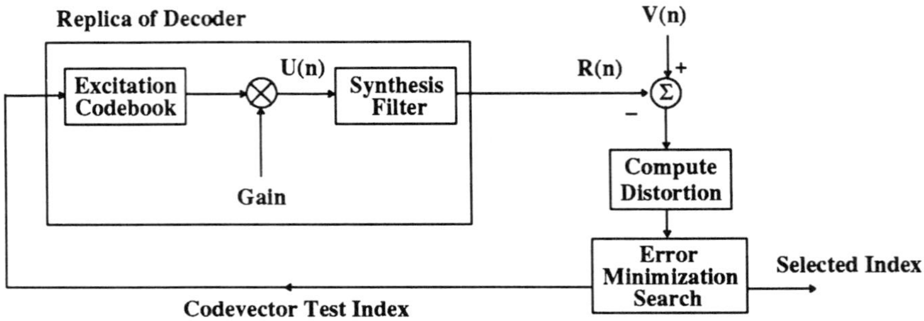

Figure 5 VXC encoder.

to synthesize a reconstructed output speech vector that closely approximates the original input speech vector.

This implies that the decoder must embody a replica of the decoder, as shown in Fig. 5, which can locally generate each of the $N = 2^c$ possible speech vector candidates that the decoder would produce for the same transmitted data values.

The error minimization search module sequentially generates new test indexes corresponding to particular code vectors. Each test index is fed to the replica of the decoder, which generates a synthesized speech vector that would be produced by the actual decoder if this index were actually transmitted. The replica decoder is initialized by setting the synthesis filter memory to the initial conditions that were determined after the prior search process was completed. Then the test index is applied to the excitation codebook, yielding an excitation vector. The excitation vector is then applied to the synthesis filter to produce the output vector $\mathbf{R}(n)$. The vector $\mathbf{R}(n)$ is then subtracted from the input speech vector $\mathbf{V}(n)$, and the distortion between these two vectors, i.e., the sum of the squares of the components of the difference vector, is computed by the distortion computation module. This error value is applied to the search module, which stores the distortion value, compares it with the lowest distortion value obtained so far in the current search process, and, if appropriate, updates the lowest distortion value and the corresponding vector index.

Speech samples emerging from the synthesis filter are configured into corresponding vectors of k contiguous samples.

Because the replica decoder is operating repeatedly in the search process, we must ensure that each candidate output speech vector, corresponding to a candidate index being tested, is produced under the conditions that will be present when the actual decoder generates the next output vector. After each test of a candidate index, the memory state of the replica decoder has changed and is no longer at the correct initial condition for the next test. Therefore, before generating each of these candidates, the memory in the replica decoder must also be reset to the correct initial conditions.

Several important modifications to this basic scheme are important to practical implementations but have not been described here. Among them is the use of a perceptual weighting filter to compute a more meaningful measure of distortion. Another is the use of GSVQ, with separate gain and shape codebooks instead of a single standard VQ codebook.

6.2. Vector Sum Excitation Codebooks

A question of practical importance is how the quality of a given VXC coder can be improved if more bits are made available and to which components of the coder these bits should be assigned for the maximum benefit. Generally, the best performance gain comes from increasing the codebook size. However, adding just one bit per code vector doubles the codebook size and the corresponding search complexity. Thus, computational and storage constraints quickly force one to limit the codebook size and lead to alternative designs in which the vector dimension is reduced and more bits are given to synthesis filter parameters. On the other hand, the use of specially constrained codebook structures offers the possibility of larger codebooks and significant performance improvements while maintaining tolerable complexity.

A novel and powerful technique for reducing the complexity of the excitation codebook search procedure was introduced by Gerson and Jasiuk (1990). Rather than have each of M code vectors be independently generated either randomly or by a design procedure, they design b "basis" vectors and then generate the $M = 2^b$ code vectors by taking binary linear combinations of the basis vectors. The resulting coding algorithm, a derivative of VXC, is called vector sum excited linear prediction (VSELP), and an 8-kb/s version of this algorithm has been adopted as a standard for the U.S. cellular mobile telephone industry. We next explain the basic idea of this technique for fast codebook search.

Let $\mathbf{v}_i$ denote the ith basis vector of a given set of b basis vectors. The code vectors are then formed as

$$\mathbf{u}_\theta = \sum_{i=1}^{b} \theta_i \mathbf{v}_i$$

by taking all possible linear combinations where $\theta_i = \pm 1$ for each i. Thus each binary-valued vector θ determines a particular code vector $\mathbf{u}_\theta$. Naturally, the b-bit binary word transmitted over the channel can simply correspond to a mapping of θ values with $+1$ being a binary 1 and -1 being a binary 0. Since the code vectors are so simply generated, only b basis vectors, rather than an entire codebook of M code vectors, need be stored.

This special codebook structure can be searched very efficiently. Instead of finding the vector output of the weighted synthesis filter for each of the M code vectors, only the filtered output of the b basis vectors needs to be determined because from these any synthesized output can be readily obtained by addition. Furthermore, the search for the optimal code vector is computationally simplified by noting that the mean squared error between the weighted input vector and a filtered code vector depends in a simple manner on the values of θ_i. By ordering the b-bit binary words in a Gray code, only 1 bit changes from one word to the next. This means that only a simple change is needed to compute the mean squared error for the next candidate code vector from the previous candidate code vector.

The basis vectors are designed from a training set of speech data by solving for each component of each basis vector in a large set of simultaneous equations. These equations are obtained by minimization of the total weighted squared error, which is a function of the basis vector components. The design is iterated in a closed-loop manner, starting with Gaussian random numbers for the components (Gerson and Jasiuk, 1990).

The vector sum approach can be augmented by combining it with MSVQ. MSVQ was applied to VXC in Davidson and Gersho (1988). It becomes more effective when each stage consists of a vector sum codebook and joint optimization of the gains for each stage is performed. The joint optimization becomes easy to implement by an orthogonalization procedure (Gerson and Jasiuk, 1990)

6.3. Low-Delay VXC

Vector excitation coding (VXC) combines techniques such as vector quantization, analysis-by-synthesis codebook searching, perceptual weighting, and linear predictive coding to successfully achieve good speech quality at low bit rates. However, one important aspect of coding has been ignored in the development of VXC or other conventional low-bit-rate excitation coding schemes; that is the *coding delay*. In fact, most existing speech coders with rates at or below 16 kb/s require high delays in their operation and cause various problems when they are applied to practical communication systems. In

VXC, a large net coding delay, excluding computational delays, results from the use of buffering needed to perform the LPC and open-loop pitch analysis. New methods have been proposed to adapt synthesis filters without the high coding delay mentioned above while maintaining the quality of encoded speech.

With the conventional VXC scheme described above, the synthesis filter is adaptively updated every frame by what is sometimes called forward adaptation, the process of recomputing and updating the desired filter parameters from the input speech. The use of forward adaptation has two disadvantages: it requires transmission of side information to the receiver to specify the filter parameters and it leads to a large encoding delay of at least one analysis frame due to the buffering of input speech samples. The input buffering and other processing typically result in a one-way codec delay of 50 to 60 ms. In certain applications in the telecommunications network environment, coding delays as low as 2 ms per codec are required. Recently, the CCITT adopted a performance requirement of less than 5 ms delay with a desired objective of less than 2 ms for candidate 16-kb/s speech coding algorithms to be considered for a new standard intended to achieve essentially the same quality as the 32-kb/s ADPCM standard, G.721. Such a low delay is not feasible with the established coders that are based on forward adaptive prediction coding systems. Although ADPCM satisfies the low-delay requirement, it cannot give acceptable quality when the bit rate is reduced to 16 kb/s.

An alternative solution is based on proposed backward adaptation configurations (Iyengar and Kabal, 1988; Watts and Cuperman, 1988; Cuperman et al., 1989). In a backward adaptive analysis-by-synthesis configuration, the parameters of the synthesis filter are not derived from the original speech signal but are computed by backward adaptation, extracting information only from the sequence of transmitted codebook indexes. Since both the encoder and decoder have access to the past reconstructed signal, side information is no longer needed for synthesis filters and the low-delay requirement can be met with a suitable choice of vector dimension. The remarkable feature of these low-delay coders is that only the indexes that specify excitation code vectors need to be transmitted. All side information usually associated with VXC-based algorithms is completely eliminated.

A particular version of low-delay coding with backward adaptation, known as low-delay CELP (LD-CELP), is under consideration for the 16-kb/s CCITT standardization (Chen 1989). This method uses *block* rather than *recursive* adaptation for updating a 50th-order LPC-based synthesis filter.

7. VECTOR QUANTIZATION OF LINEAR PREDICTOR COEFFICIENTS

Linear prediction is perhaps the most widely used technique in speech coding. It can remove near-sample or distant-sample correlations in a speech waveform. The former, which we usually call short-term or "spectral" prediction, effectively matches the spectral envelope of the signal. The latter, often referred to as "pitch" prediction, removes distant-sample redundancy in quasi-periodic speech segments, typical in voiced sounds.

In high-bit-rate waveform coding schemes, such as ADPCM and APC, short-term linear prediction results in an error signal of significantly lower power than the original waveform, and it can thus be encoded more accurately. In contrast, in low-bit-rate vocoding systems the short-term predictor is the basis of a time-varying filter modeling the vocal tract characteristics. In vector excitation systems this is augmented by long-term prediction to more fully remove redundancies in the error signal.

Since speech is a nonstationary process, it is necessary to update the prediction coefficients periodically. With forward adaptation, the predictor coefficients need to be transmitted to the decoder for synthesis. Therefore, the efficient quantization of these parameters becomes an important task in any adaptive prediction coding system.

LPC parameters have traditionally been individually (scalar) quantized. To obtain an acceptable quantization distortion with scalar quantization, usually 30 to 40 bits are needed to quantize a 10th order predictor (Tremain, 1982).

Since linear prediction can be viewed as a spectral matching process, distortion measures used for LPC quantization can be as effectively defined in the frequency domain as in the time domain. A number of spectral distortion measures have been discussed by Gray and Markel (1976). One of the most meaningful distortion measures, which has the property of being consistent with the residual energy minimization concept of the LPC analysis process, is the likelihood ratio:

$$d_{LR}\left(\frac{1}{A_M}, \frac{1}{A}\right) = \frac{r_x(0)}{\alpha_M} r_a(0) + 2\sum_{i=1}^{M} \frac{r_x(i)}{\alpha_M} r_a(i) - 1$$

where $r_x(i)$ and $r_a(i)$ are the autocorrelation sequences of the input frame and of the polynomial coefficients of $A(z)$, respectively, where $1/A(z)$ is any Mth order all-pole filter and $1/A_M(z)$ is the optimal LPC model of input frame $X(z)$. The term α_M is the residual energy resulting from inverse filtering

$X(z)$ with $A_M(z)$. The likelihood ratio is closely related to the Itakura–Saito distortion measure defined earlier.

Studies of this distortion measure (Juang et al., 1982) illustrate that the amount of parameter deviation allowed for a specified level of distortion is highly dependent on the entire input vector. This coupling effect is called *nonlinear dependence.* This property cannot be fully incorporated into the sensitivity approach for scalar quantization, in which the distortion is measured approximately by a sum of incremental distortions due to individual LPC parameters, without accounting for the coupling effect.

Vector quantization can remove redundancy in a vector effectively by making use of four interrelated properties of vector parameters: linear dependence (correlation), nonlinear dependence, the shape of the probability density function, and vector dimensionality itself. It is the existence of such significant nonlinear dependencies in the space of spectral parameters for speech which makes vector quantization truly attractive in this application. Kang and Coulter (1976) made the first effort in introducing the technique of vector quantization into LPC parameter quantization. Their work was mainly based on intuition, rather than on a well-defined mathematical distortion and methodology for designing an optimal codebook. Buzo et al. (1980) developed a mathematical framework and procedure for vector quantization of LPC parameters. Their experimental results showed that vector quantization permits a significant bit-rate savings over scalar quantization at the same distortion level, making efficient utilization of the statistics of coefficient vectors (clustering property) and of the coupling effect between coefficients. Wong et al. (1982) applied this framework to develop an 800-b/s vector-quantized LPC vocoder. Subjective evaluation showed that, overall, the 800-b/s vocoder approached a quality quite close to that of a 2400-b/s LPC vocoder. Their work has demonstrated that vector quantization appears to preserve more continuity from frame to frame than scalar quantization. Kang and Fransen (1984) also applied VQ to quantize the line spectral pairs in an 800-b/s vocoder which achieved a DRT (diagnostic rhyme test) score only 1.4 below that of a 2400-b/s LPC vocoder.

Most applications of vector quantization in speech coding have until recently focused on very low bit-rate coding, where the quantization resolution of LPC parameters with 10–12-bit vector quantizers are adequate to maintain acceptable speech intelligibility and quality. For some medium- and high-bit-rate coding systems, high-resolution LPC quantization, say 30 to 40 bits per frame, is demanded. If we improve the quantization resolution with straightforward VQ, the codebook search complexity and codebook storage will increase exponentially. For example, in a 10 bit/frame VQ for an Mth order linear prediction with the d_{LR} measure, $2^{10} \times (M + 1)$ multiply-adds are needed to quantize one input vector. If we use a 30 bit/frame LPC VQ, the computation for quantizing one input vector would be $2^{30} \times (M + 1)$ multiply-adds which is 10^6 times as many as the computations in the 10 bit/frame VQ. The complexity in both computation and storage has thwarted the direct application of vector quantization to high-resolution LPC parameter quantization.

The simplest solution for overcoming the difficulties in complexity is to use TSVQ. The major advantage of TSVQ is the substantial decrease in computational cost compared to full search with a relatively small decrease in performance. However, the storage requirements are increased and may be prohibitive. On the other hand, MSVQ reduces the storage requirement as well as search complexity but leads to a somewhat greater degradation in performance than does TSVQ, particularly when more than two stages are used. Nevertheless, MSVQ has been found to be a useful way to quantize log-area ratio (LAR) parameters.

An alternative to MSVQ is a product-code approach to VQ. In this type of VQ, the input space is partitioned into two or more subspaces; the vector dimension in each subspace is usually reduced and often considerably reduced. In fact, GSVQ is a special case of a product code. The subvector for each subspace is then quantized separately with a relatively small codebook designed for that subspace.

There are several ways to partition LPC parameter vectors, including vector splitting and band splitting. In the vector-splitting method, a vector is split into two or more subvectors and each subvector is quantized independently. For example, a vector with 14 LARs is divided into two vectors and two codebooks are used: a 6-dimensional, 10-bit codebook for parameters 1 to 4 and an 8-dimensional, 5-bit codebook for parameters 5 to 14. With a band-splitting method, proposed in Copperi and Sereno (1984), the spectral vector is split into two frequency bands and separate LPC model parameters for each band are vector quantized separately. Besides the advantage of reducing complexity in these methods, we gain the flexibility of controlling the bit allocation to the subvectors according to their perceptual importance in speech coding. For example, in the band-splitting method, we can make use of the fact that the human ear is generally less sensitive to high-frequency distortions than to low-frequency distortions and assign more bits for the lower-order vector.

In another VQ technique the LPC polynomial is represented as a cascade of two lower-order polynomials, a separate codebook is used for each polynomial, and quantization of the two components is jointly optimized using a likelihood-ratio distortion measure (Shoham, 1989). The LPC all-pole filter is represented as a cascade of two lower-order all-pole filters. The quantizer uses two codebooks to quantize each of the lower-order filters. Splitting the LPC filter into two filters reduces the coding complexity, while the efficiency of VQ is largely preserved.

7.1. Interframe Coding of LPC Parameters

So far, we have discussed only memoryless VQ for LPC parameter quantization, in which each input vector is quantized independently of any other input vector. Because of the nonstationary characteristics of speech, the LPC parameter set used for modeling the vocal tract transfer function needs to be updated during successive speech frames, with a frame size of 20 to 30 ms. However, for some segments of speech, especially for sustained vowels, the speech spectral envelope is actually a slow time-varying process, and spectra of adjacent frames are highly correlated. Figure 6 illustrates a typical evolution of log-magnitude speech spectral envelopes for successive frames, spaced 20 ms apart. This time dependence inherent in the LPC parameter sets suggests that any quantization method which exploits interframe redundancy will further reduce the coding bits for a given average distortion.

One way to exploit interframe correlation is to use vector linear prediction (VLP), introduced in Gersho and Cuperman (1985). Vector prediction of a frame's parameter set from previous frames is essentially equivalent to predicting the spectrum of the frame from the evolving pattern of spectra of the prior frames. By applying VLP, each spectral parameter in the

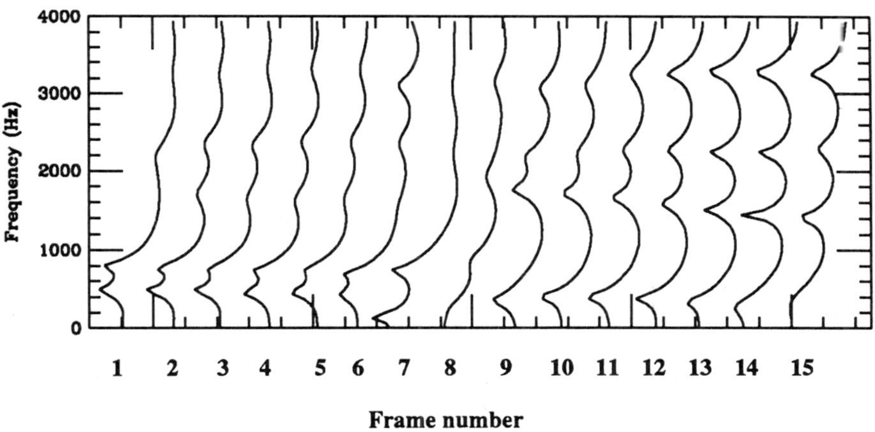

Figure 6 Log-spectral envelopes versus frames.

current frame is predicted not only from corresponding parameters of previous frames but also from the other spectral parameters of previous frames. In this way, correlation between different speech frames can be maximally exploited. Works which combine VLP with VQ for LPC parameter quantization (Davidson et al., 1987; Yong et al., 1988; Shoham, 1987) have shown that the coding performance is greatly enhanced by exploiting interframe correlation.

In Yong et al. (1988), a switched-adaptive interframe vector prediction (SIVP) scheme tracks statistical changes in the time-varying spectral parameters while maintaining reasonably low complexity. The line spectral frequencies (LSFs) were chosen as the spectral parameter set in this study, and it was found that very little gain is offered with higher than first-order prediction. Thus, a single predictor matrix is used to predict the LSF parameter vector $\mathbf{x}_n$ in the current frame from the corresponding vector $\mathbf{x}_{n-1}$ for the prior frame, according to $\hat{\mathbf{x}}_n = \mathbf{A}\mathbf{x}_{n-1}$. With switched adaptation, the predictor matrix $\mathbf{A}$ is updated vector by vector. For each input vector to be predicted, a predictor matrix is selected from a fixed set of such matrices via a statistical classification of the input vector. An index identifying the selected predictor matrix is transmitted to the receiver as side information. For a statistically specified source of vectors, the optimal predictor matrix is given by $\mathbf{A} = \mathbf{C}_{01}\mathbf{C}_{11}^{-1}$, where the covariance matrices are defined by $\mathbf{C}_{ij} = E\{\mathbf{x}_{n-i}\mathbf{x}_{n-j}^T\}$. To design the predictor matrices, a training sequence of LSF vectors is used to compute the covariance matrices for each class. Thus, if the current vector $\mathbf{x}_m$ is assigned to class i, the running estimate of C_{0j} is updated with the product $\mathbf{x}_m\mathbf{x}_{m-j}^T$. After passing through the training sequence in this way, the prediction matrix for each class is then computed from the corresponding covariance matrices.

The block diagram of a switched-adaptive predictive coding system is shown in Fig. 7 for the simple case with only two predictor matrices so that only 1 bit per frame of side information is required. In the figure, x_n represents an input vector and $\hat{x}_n$ represents its prediction. The prediction error vector e_n obtained by subtracting $\hat{x}_n$ from x_n is vector quantized and sent to the decoder. The positions of the switches at the encoder and the decoder are synchronized by a flag signal (index). The reconstructed signal $\tilde{x}_n$ is obtained by adding the quantized prediction error to the prediction of the signal.

Another approach to removing interframe redundancy from the spectral parameter coding is to apply vector quantization to quantize jointly the ordered set of spectral parameter vectors in a superframe (a group of consecutive speech frames). When the number of frames in a superframe is fixed, it is called matrix quantization (Tsao and Gray, 1985). If the number of frames is variable (according to a dynamic segmentation algorithm), the term segment quantization (Roucos et al., 1983; Wong et al., 1983) is used.

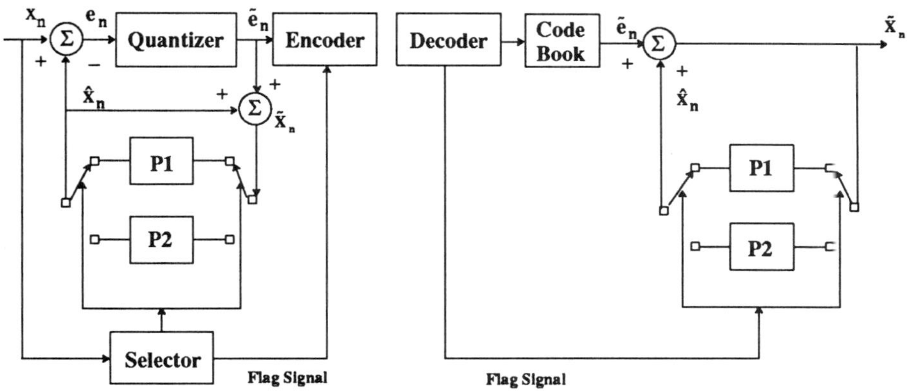

Figure 7 Switched-adaptive predictive coding system.

Motivated by rate-distortion theory, matrix quantization extends VQ techniques to matrices of n spectral parameter vectors. Matrix quantizers, designed with the GLA and a summed Itakura–Saito distortion measure, are locally optimal quantizers. For a given quality of reproduction, they generally require substantially lower coding rates than standard VQ. A matrix quantizer operating at 150 b/s with a 9-bit codebook and $N = 3$ is claimed to be comparable to a standard VQ operating at 350 b/s (Tsao and Gray, 1985).

In segment quantization, a segment which contains a number of LPC parameter sets corresponding to a spectral steady-state region in the input speech is quantized as a single entity to the nearest template in a codebook. Assuming the segmentation to the input is given, a segment $X = [x_1 x_2 \cdots x_l]$ is defined as a variable-length sequence of k-dimensional LPC vectors x_i. In order to simplify the codebook design and distortion measure in the codebook search, a time-warping transformation is performed on X prior to the quantization, which converts a variable-length segment to a fixed-length segment. The converted segment is denoted as $Y = [y_1 y_2 \cdots y_m]$ in the transformed space, where m is the segment length after the time-warping transformation. The segment quantizer uses a Euclidian distance to search for the nearest template in a codebook. The codebook given in Roucos et al. (1983) has 8192 segment templates where each template is a $k \times M$ matrix.

This type of quantization indeed removes most of the intra- and inter-frame redundancy from the LPC parameters but usually requires a large time delay and high complexity. Therefore, it is more attractive for very low bit-rate vocoding applications.

Vector quantization can also be applied to long-term (pitch) predictor coefficients. It not only codes pitch predictor parameters efficiently for multitap pitch predictors but also guarantees the stability of the pitch synthesis filter. Since the stability of a pitch synthesis filter depends on all the filter coefficients, if the parameters are quantized individually, small deviations in one coefficient might render the filter unstable. With VQ, a codebook can be designed with each code vector representing a stable synthesis filter.

8. VECTOR QUANTIZATION FOR NONLINEAR PREDICTION

Linear prediction of speech is the one most important technique used in speech coding. Yet speech is non-Gaussian, so linear prediction is not the optimal way to estimate a sample from prior values of the waveform. Nonlinear prediction of speech in principle offers a way to enhance the redundancy-removing feature of linear prediction and provide a superior model of the speech production mechanism. Recently, a method for performing nonlinear prediction was proposed based on the use of VQ (Wang et al., 1990). The method is based on a general theory of optimal nonlinear interpolation from vector-quantized observables, which includes nonlinear prediction as a special case (Gersho, 1990). The nonlinear prediction method, simply stated, is to vector quantize the set of past samples and then obtain the prediction of the new sample from a lookup table of prestored values. The table is addressed by the index of the quantized vector that approximates the past history.

In theory, the optimal predictor is the conditional expectation of the new value given the set of prior values. With this method, two approximations are made. One is to replace the past with a finite-resolution quantization of the past values using VQ. The accuracy of this approximation grows with the size of the VQ codebook. The second approximation is to compute the conditional expectation of the new value given the quantized past by using a training set of speech data rather than the "true" analytical model of the speech statistics.

The approach has been validated by testing it on a moving-average random process for which the optimal nonlinear predictor is known. When it is applied to open-loop prediction of speech, the results show that rather large codebook sizes are needed for representation of the past and very large training sets are necessary for designing the predictor table. Typically, the segmental prediction gain in this way exceeds that of linear prediction by only a very small margin. Nevertheless, the most striking indication of the effectiveness of nonlinear prediction is that the prediction residual is much less intelligible than the corresponding linear prediction residual. This result was observed with a fixed nonlinear predictor which was designed using speech training data different from the speech on which the effect was tested.

Figure 8 shows the spectrum of one frame of the residual for both linear and nonlinear prediction with a fourth-order predictor in each case. The

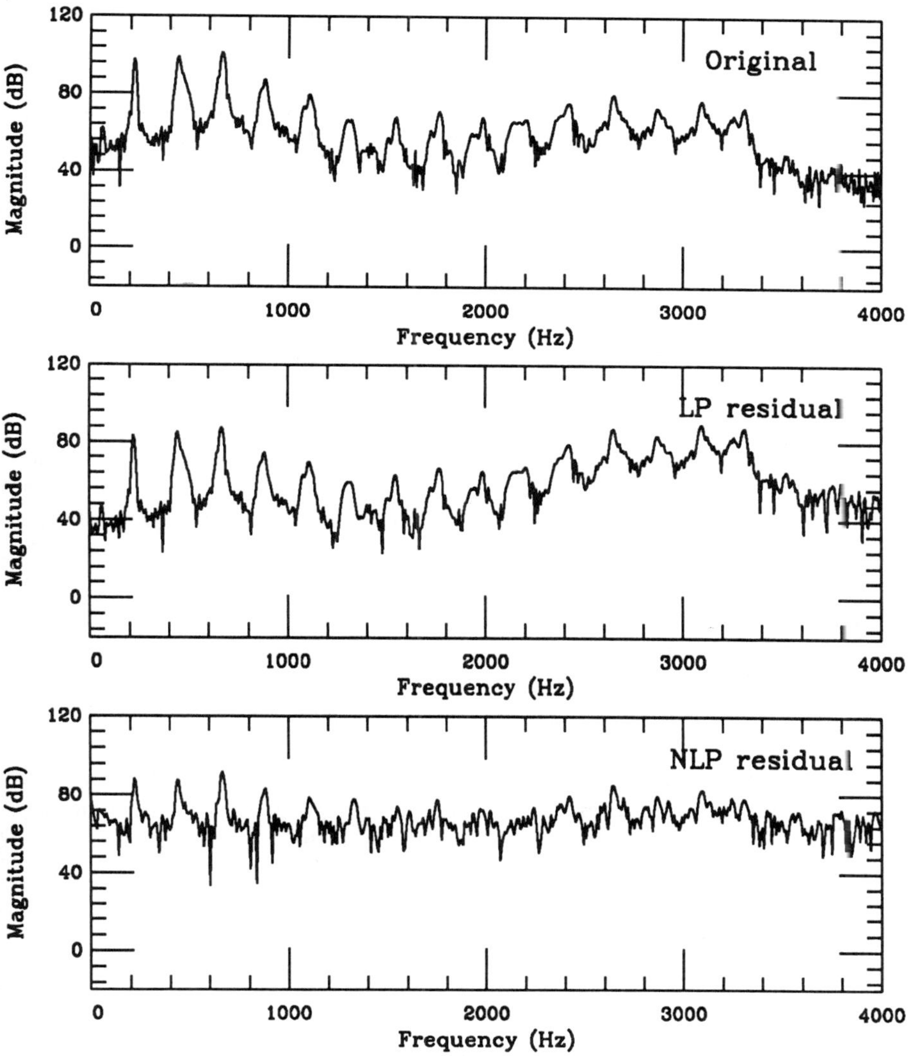

Figure 8 Spectra of original speech and linear and nonlinear prediction residuals.

spectrum of the original speech frame is also shown. It is remarkable that the spectral envelope has been notably flattened and the higher pitch harmonics have almost disappeared.

These results motivate further study of nonlinear prediction, but practical application to speech coding may be premature. The only speech coding algorithm tested so far with nonlinear prediction was (nonadaptive) DPCM. With a fifth-order predictor, DPCM with nonlinear prediction gives a somewhat higher coding gain than with linear prediction at rates of 3 or 4 bits per sample. There are several nontrivial problems with nonlinear prediction that must be solved before it can become a useful technique for analysis-by-synthesis coding systems. Nevertheless, it is an interesting new direction in speech coding that is worthy of further study.

9. COMBINED SOURCE AND CHANNEL CODING

Of particular importance in a voice communication system is the effect of transmission errors on the performance of the recovered speech. This issue is relatively minor for optical fiber or other guided-media channels, but it must receive substantial consideration in other speech coding applications such as in radio channels (because of transmitted power constraints and propagation conditions). In mobile satellite and cellular radio systems, bit error rates as high as 0.001 or 0.01 must be considered and, typically, reasonable speech quality must be maintained at the 0.001 error rate and tolerable quality at the 0.01 rate.

Traditional channel coding techniques, such as error correction or detection, can be employed to protect transmitted VQ information from the effect of channel noise. Channel coding normally requires an increase in the rate of the transmitted bit stream, as well as additional processing delay incurred in the process of calculating the redundant bits and decoding the information bits. For speech communication systems on narrow-bandwidth channels, every bit allocated for channel coding implies sacrificing a bit that would otherwise help to improve the intrinsic speech quality of the coder. Furthermore, error correction codes do not take into account information about the specific character of the data source; they merely attempt to reduce the likelihood that binary channel words are incorrectly received. Recent research has focused on methods for parsimoniously and efficiently alleviating the effect of channel noise on the output speech quality. Two different approaches have been investigated for tackling this problem: selective protection and joint source channel coding.

The first approach consists of determining which transmitted bits are the most important in terms of speech quality and selectively protecting them against channel noise using known channel coding methods. A typical data frame of encoded speech may contain several binary words representing VQ indexes for different vectors of parameters, and the more sensitive indexes can be selectively protected. By understanding the relative sensitivity of different speech coding parameters to bit errors, techniques for maintaining robustness to such errors can be developed. This form of selective channel coding can protect the transmitted speech information in a limited way without greatly increasing the overall bit rate (Cox et al., 1988).

The other approach is one that incorporates combined source and channel coding techniques with VQ. Combined source and channel coding can significantly improve the performance of the reconstructed speech signal on poor channels.

It is also possible to reduce the sensitivity of the bit stream to errors without adding redundant bits, simply by carefully allocating codewords to the signal parameter values. In particular, *redundancy-free* error protection codes for VQ have been introduced (Zeger and Gersho, 1991, 1987; Chen et al., 1987; De Marca and Jayant, 1987; Farvardin, 1988) and effective design algorithms for this purpose have been developed. An important feature of these techniques is that no added delay occurs in the coding system. The only cost is some fairly intensive off-line computation during the design process.

The basic idea of the technique, known as pseudo-Gray coding, is to determine carefully which channel words should be associated with which VQ code vector indexes, by way of an *index assignment function*. In the past, source code designers using VQ generally made the assignment arbitrary. Intuitively, vectors that are "close" to each other should be assigned indexes which differ in as few bit positions as possible. In this way, channel errors cause an index to be decoded as a vector which approximates the code vector that was supposed to be correctly decoded (i.e., without channel noise).

Another approach to joint source-channel coding with vector quantization is to design the quantizer's encoder and decoder taking into account the statistics of the channel (Kumazawa et al., 1984; Zeger and Gersho, 1988). For a given noisy channel, the problem is to find the best encoder for a fixed decoder and then find the best decoder for a fixed encoder. Equivalently, for a fixed partition of the input space, the goal is to find the best codebook, and vice versa. It has been shown that the optimal encoder and decoder of a quantizer of a noisy channel must satisfy two necessary conditions, known as the *weighted nearest neighbor* and *weighted centroid* rules, which generalize the nearest neighbor and centroid conditions that are necessary for quantizer optimally on a noiseless channel.

10. CONCLUDING REMARKS

The concept of vector quantization is clearly a general and powerful one with many specialized techniques that have emerged for combating the complexity obstacle. This chapter has given only a brief overview of the use of VQ in speech coding. There remain many coding applications where VQ techniques are still too costly in complexity or storage to be incorporated into hardware products. Research in VQ techniques is continuing with the aim of expanding its utility and applicability for the next generation of speech coding algorithms and implementations.

REFERENCES

Abut, H., Gray, R. M., and Rebolledo, G. (1982). Vector quantization of speech and speech-like waveforms, *IEEE Trans. Acoust., Speech, Signal Process. ASSP-30*: 423–435.

Buzo, A., Gray, A. H., Gray, R. M., and Markel, J. D. (1980). Speech coding based upon vector quantization. *IEEE Trans. Acoust., Speech, Signal Process. ASSP-28*: 562–574.

Campbell, J., Tremain, T., and Welch, V. (1990). The DoD 4.8 kbps standard (proposed federal standard 1016). In *Advances in Speech Coding*, B. S. Atal, V. Cuperman, and A. Gersho (Eds.). Kluwer Academic Publishers, Norwell.

Cetin, A. E., and Weerackody, B. (1988). Design vector quantizers using simulated annealing, *IEEE Trans. Circuits Syst. CAS-35*:1550.

Chang, P. C., and Gray, R. M. (1986). Gradient algorithms for designing predictive vector quantizers. *IEEE Trans. Acoust., Speech, Signal Process. ASSP-34*:679–690.

Chen, J. H. (1989). A robust low-delay CELP speech coder at 16 kb/s. *Proc. IEEE Global Communications Conf.*, November, Dallas.

Chen, J. H., and Gersho, A. (1986). Vector adaptive predictive coding of speech at 9.6 kb/s. *Proc. IEEE Int. Conf. on Acoustics, Speech, and Signal Processing*, vol. 3, pp. 1693–1696, Tokyo.

Chen, J. H., and Gersho, A. (1987). Gain-adaptive vector quantization with application to speech coding. *IEEE Trans. Commun. COM-35*:918–930.

Chen, J. H., Davidson, G., Gersho, A., and Zeger, K. (1987). Speech coding for the mobile satellite experiment. *Proc. IEEE Int. Conf. on Communications*, June, Seattle.

Cheng, D. Y. and Gersho, A. (1986). A fast codebook search algorithm for nearest-neighbor pattern matching. *Proc. IEEE Int. Conf. on Acoustics, Speech, and Signal Processing* vol. 1, pp. 265–268, Tokyo.

Cheng, D. Y., Gersho, A., Ramamurthi, B., and Shoham, Y. (1984). Fast search algorithms for vector quantization and pattern matching. *Proc. IEEE Int. Conf. on Acoustics, Speech, and Signal Processing*, pp. 911.1–911.4, March, San Diego.

Chou, P. A., Lookabaugh, T., and Gray, R. M. (1989a). Entropy-constrained vector quantization. *IEEE Trans. Acoust., Speech, Signal Process. ASSP-37*:31–42.

Chou, P. A., Lookabaugh, T., and Gray, R. M. (1989b). Optimal pruning with applications to tree-structured source coding and modeling. *IEEE Trans. Inform. Theory, IT-35*:299–315.

Copperi, M., and Sereno, D. (1984). 9.6 Kb/s piecewise LPC residual excited coder using multiple-stage vector quantization. *Proc. IEEE Int. Conf. on Acoustics, Speech, and Signal Processing*, 10.5.1–10.5.4, March, San Diego.

Cox, R. V., Hagenauer, J., Seshadri, N., and Sundberg, C.-E. (1988). A sub-bandcoder designed for combined source and channel coding. *Proc. IEEE Int. Conf. on Acoustics, Speech, and Signal Processing*, pp. 235–238, New York.

Cuperman, V., Gersho, A., Pettigrew, R., Shynk, J. S., and Yao, J.-H. (1989). Backward adaptation techniques for low delay vector excitation coding of speech. *Conf. Record IEEE Global Communications Conference*, pp. 1242–1246, November, Dallas.

Davidson, G., and Gersho, A. (1988). Multiple-stage vector excitation coding of speech waveforms. *Proc. IEEE Int. Conf. on Acoustics, Speech, and Signal Processing*, pp. 163–166, April, New York.

Davidson, G., Yong, M., and Gersho, A. (1987). Real-time vector excitation coding of speech at 4800 bps. *Proc. IEEE Int. Conf. on Acoustics, Speech, and Signal Processing*, pp. 2189–2192, April, Dallas.

Dudley, H. (1958). Phonetic pattern recognition vocoder for narrow-band speech transmission. *J. Acoust. Soc. Am.* 30(8):733–739.

Farvardin, N. (1988). Optimal Binary Code Word Assignment for Vector Quantization over a Noisy Channel—An Application of Simulated Annealing. *IEEE Int. Symp. on Information Theory*. June, Kobe, Japan.

Flanagan, J., Morrell, D., Frost, R., Read, C., and Nelson, B. (1989). Vector quantization codebook generation using simulated annealing. *Proc. ICASSP'89*, P32.M7.8, May, Glasgow.

Friedman, J. H., Bentley, J. L., and Finkel, R. A. (1977). An algorithm for finding best matches in logarithmic expected time. *ACM Trans. Math. Software*, *3*(3):209–226.

Fukunaga, K., and Narendra, P. M. (1975). A branch and bound algorithm for computing k-nearest neighbors. *IEEE Trans. Comput. C-24*:750–753.

Gersho, A. (1982). On the structure of vector quantizers. *IEEE Trans. Inform. Theory IT-28*:157–166.

Gersho, A. (1990). Optimal nonlinear interpolative vector quantization. *IEEE Trans. Commun. COM-38*:1285–1288.

Gersho, A., and Cuperman, V. (1982). Adaptive differential vector coding of speech. *Conf. Record GlobeCom 82*, December, pp. 1092–1096.

Gersho, A., and Cuperman, V. (1983). Vector quantization: A pattern matching technique for speech coding, *IEEE Commun. Mag.*, December, pp. 15–21.

Gersho, A., and Cuperman, V. (1985). Vector predictive coding of speech at 16 b/s. *IEEE Trans. Commun. COM-33*:685–696.

Gerson, I., and Jasiuk, M. (1990). Vector sum excited linear prediction (VSELP) speech coding at 8 kb/s. *Proc. IEEE Conf. Acoustics, Speech, and Signal Processing*, pp. 461–464, April, Albuquerque.

Gersho, A., Ramstad, T., and Versvik, I. (1984). Fully vector-quantized subband coding with adaptive codebook allocation, *Proc. Int. Conf. on Acoustics, Speech, and Signal Processing*, p. 10.7, March, San Diego.

Gray, A. H., Jr., and Markel, J. D. (1976). Distance measures for speech processing. *IEEE Trans. Acoust., Speech, Signal Process. ASSP-24*:380–391.

Gray, R. M. (1984). Vector quantization. *IEEE ASSP Mag.* 1:4–29.

Gray, R. M., and Abut, H. (1982). Full search and tree searched vector quantization of speech waveforms. *Proc. IEEE Int. Conf. on Acoustics, Speech, and Signal Processing*, pp. 593–596, May, Paris.

Gray, F. M., Buzo, A., Gray, A. H., Jr., and Matsuyama, Y. (1980). Distortion measures for speech processing. *IEEE Trans. Acoust., Speech, Signal Process. ASSP-28*:367–376.

Gray, R. M., Gray, A. H., Jr., Rebolledo, G., and Shore, J. E. (1981). Rate distortion speech coding with a minimum discrimination information distortion measure. *IEEE Trans. Inform. Theory IT-27*:708–721.

Iyengar, V., and Kabal, P. (1988). A low delay 16 kbits/sec speech coder. *Proc. IEEE Int. Conf. on Acoustics, Speech, and Signal Processing*, pp. 243–246, April, New York.

Jayant, N. S., and Noll, P. (1984). *Digital Coding of Waveforms*. Prentice-Hall, Englewood Cliffs, N. J.

Kang, G. S., and Coulter, D. C. (1976). 600 bps voice digitizer. *Proc. IEEE Int. Conf. on Acoustics, Speech, and Signal Processing*, pp. 91–94, April, Philadelphia.

Kang, G. S., and Fransen, L. J. (November 1984). Low-bit-rate speech encoders based on line-spectrum frequencies (LSFs). Naval Research Laboratory Report No. 8857.

Kumazawa, H., Kasahara, M., and Namekawa, T. (1984). A construction of vector quantizers for noisy channels. *Electron. Eng. Jpn.* 67-B(4):39–47.

Linde, Y., Buzo, A., and Gray, R. M. (1980). An algorithm for vector quantizer design, *IEEE Trans. Commun. COM-28*:84–95.

Makhoul, J., Roucos, S., and Gish, H. (1985). Vector quantization in speech coding. *Proc. IEEE 73*:1551–1588.

Marca, J. De, and Jayant, N. (1987). An algorithm for assigning binary indices to the codevectors of a multi-dimensional quantizer. *Proc. IEEE Int. Conf. on Communications*. June, Seattle.

Nasrabadi, N. M., and Feng, Y. (1988). Vector quantization of images based upon the Kohonen self-organization feature maps. *Proc. 2nd ICNN Conf.*, vol. I, pp. 101–105.

O'Shaughnessy, D. (1987). *Speech Communication*. Addison-Wesley, Reading, Mass.

Riskin, E. A., and Gray, R. M. (1991). A greedy tree growing algorithm for the design of variable rate vector quantizer. *IEEE Trans. Acoust., Speech, Signal Process.*, in press.

Roucos, S., Schwartz, R. M., and Makhoul, J. (1983). A segment vocoder at 150 b/s. *Proc. IEEE Int. Conf. on Acoustics, Speech, and Signal Processing*, pp. 61–64, April, Boston.

Sabin, M. J., and Gray, R. M. (1984). Product code vector quantizers for waveform and voice coding. *IEEE Trans. Acoust., Speech, Signal Process. ASSP-32*:474–488.

Shoham, Y. (1987). Vector predictive quantization of the spectral parameters for low rate speech coding. *Proc. IEEE Int. Conf. on Acoustics, Speech, and Signal Processing*, vol. 4, pp. 2181–2184, April, Dallas.

Shoham, Y. (1989). Cascaded likelihood vector coding of the LPC information. *Proc. IEEE Int. Conf. on Acoustics, Speech, and Signal Processing*, pp. 160–163, May, Glasgow.

Soleymani, M. R., and Morgerai, S. D. (1987). A high-speed search algorithm for vector quantization, *Proc. IEEE Int. Conf. on Acoustics, Speech, and Signal Processing*, pp. 45.6.1–3, April, Dallas.

Tremain, T. E. (1982). The government standard linear predictive coding algorithm: LPC-10. *Speech Technol.* April, pp. 40–49.

Tsao, C., and Gray, R. M. (1985). Matrix quantizer design for LPC speech using the generalized Lloyd algorithm, *IEEE Trans. Acoust., Speech, Signal Process. ASSP-33*:537–545.

Vaisey, J., and Gersho, A. (1988). Simulated annealing and codebook design. *Proc. ICASSP'88*, pp. 1176–1179, April, New York.

Vidal, E. (1986). An algorithm for finding nearest neighbours in (approximately) constant average time, *Pattern Recogn. Lett. 4*:145–157.

Wang, S., Paksoy, E., and Gersho, A. (1990). Performance of nonlinear prediction of speech. *Proc. Int. Conf. on Spoken Language Processing*, November, Kobe

Watt, L., and Cuperman, V. (1988). A vector ADPCM analysis-by-synthesis configuration for 16 kbit/s speech coding, *Conf. Record IEEE Global Communications Conference*, November, pp. 275–259.

Wong, D. Y., and Gray, A. H. (1982). Distortion performance of vector quantization for LPC voice coding. *IEEE Trans. Acoust., Speech, Signal Process. ASSP-30*:294–303.

Wong, D. Y., Juang, B. H., and Gray, A. H. (1982). An 800 b/s vector quantization LPC vocoder. *IEEE Trans. Acoust., Speech, Signal Process. ASSP-30*: no. 5, pp. 770–780.

Wong, D. Y., Juang, B. H., and Cheng, D. Y. (1983). Very low data rate speech compression with LPC vector and matrix quantization. *Proc. IEEE Int. Conf. on Acoustics, Speech, and Signal Processing*, pp. 65–68, April, Boston.

Yair, E., Zeger, K., and Gersho, A. (1992). Competitive learning and soft competition for vector quantizer design, *IEEE Trans. Signal Process.*, in press.

Yair, E., Zeger, K., and Gersho, A. (1990). Conjugate gradient methods for designing vector quantizers. *Proc. IEEE Int. Conf. on Acoustics, Speech, Signal Processing*, April, Albuquerque.

Yong, M., Davidson, G., and Gersho, A. (1988). Encoding of LPC spectral parameters using switched-adaptive interframe vector prediction. *Proc. IEEE Int. Conf. on Acoustics, Speech, and Signal Processing*, vol. 1, pp. 402–405, April, New York.

Zeger, K., and Gersho, A. (1987). Zero redundancy channel coding in vector quantisation. *Electron. Lett. 23*:654–656.

Zeger, K., and Gersho, A. (1988). Vector quantization design for memoryless noisy channels, *Proc. IEEE Int. Conf. on Communications*, pp. 1593–1597, June, Philadelphia.

Zeger, K., and Gersho, A. (1989). A stochastic relaxation algorithm for improved vector quantizer design, *Electron. Lett. 25*(14):896–898.

Zeger, K., and Gersho, A. (1990). Pseudo-Gray coding. *IEEE Trans. Commun., COM-38*:2147–2158.

3

High-Quality Coding of Telephone Speech and Wideband Audio

NIKIL JAYANT AT&T Bell Laboratories, Murray Hill, New Jersey

1. INTRODUCTION

This chapter is a review of digital speech technology, with emphasis on applications demanding high-quality reproduction of the speech signal. Examples of such applications are network telephony, ISDN terminals for audio teleconferencing, and systems for the storage of audio signals that include the important subclass of wideband speech. Depending on the application, the bandwidth of input speech can vary from about 3 to nearly 20 kHz. The bit rate in the digital representation could vary from 2 to 128 kilobits per second (kbps), depending on the application and the user's expectations of signal quality.

Section 2 is devoted to traditional 3.2-kHz telephony and Section 3 to wideband speech and audio.

The parameters describing the performance of a digital coding system are several. Examples are processing delay, tolerance of transmission errors and multiple stages of coding and decoding, and ability to handle nonvoice signals such as voiceband modem waveforms. The single most important descriptor of coder performance is the quality of digitized speech and the way it diminishes with decreasing bit rate.

1.1. Measurement of Speech Quality

Measurement of speech quality is a difficult and long-standing problem. For the purposes of this chapter, a commonly used subjective rating scale of 1 to 5 is used, wherever possible, to quantify the level of digital speech quality. This is the so-called mean opinion score (MOS) scale (Daumer, 1982; Jayant and Nell, 1984; Kitawaki and Nagabuchi, 1988). MOS scores of 4.0 or higher will be used to signify *high-quality* or *near-transparent* coding. The term *network quality* will imply high-quality coding as a necessary but not sufficient condition; sufficiency will require further capabilities of the speech coder, as demanded by the communication network environment. An MOS score of 3.5 to 4.0 will denote *communication quality*. This level of quality is characterized by a level of speech degradation that is easily detectable but not bad enough to impede natural telephone communication. Finally, *synthetic quality* speech will imply a signal characterized perhaps by high intelligibility but by an inadequate level of naturalness and speaker recognizability. These deficiencies are usually reflected by an MOS score not exceeding 3.0. In the rest of this chapter, adjectives used to describe speech quality will derive meaning partly from the numerical calibration defined in this paragraph and partly from context.

The five-point MOS scale will also be used in our discussion of coding algorithms for 7-kHz speech. The diagnostic rhyme test (DRT) is a word-intelligibility measure, while the diagnostic acceptability measure (DAM) reflects acceptability for speech communication in a broader multidimensional sense (Voiers, 1977a, 1977b).

1.2. Overview of Digital Telephony

Figure 1 is a description of the current state of telephone speech coding in terms of standards activity, bit rate, typical application, and decoded speech quality. In this summary, the bandwidth of speech is assumed to be 3.2 kHz, and quality is measured on the mean subjective rating (MOS) scale of 1 to 5. As mentioned earlier, MOS values of 4.0 and 3.5 can be regarded as minimum requirements for near-transparent quality and robust communications quality, respectively. Current goals in speech coding include the achievement of near-transparent or transparent quality at 8 kbps and the achievement of robust communications quality at 4.8 kbps and lower.

Figure 2 is a more quantitative description of speech quality as a function of bit rate. The solid characteristics in the figure refer to generic examples of coding algorithms. The historical progression is from right to left, and the

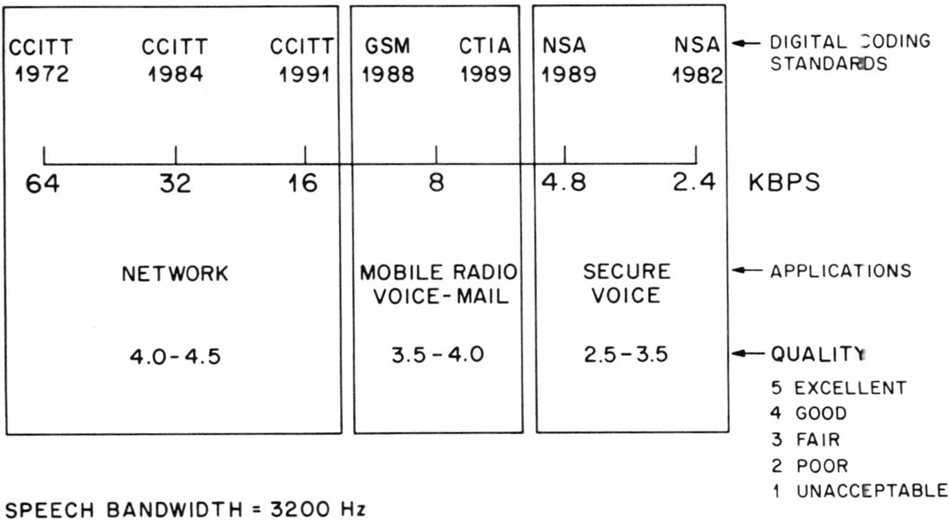

Figure 1 Digital telephony—standards, typical applications, and ranges of speech quality. CCITT, International Telegraph and Telephone Consultative Committee; CTIA, Cellular Technology Industry Association (USA); GSM, Group Spéciale Mobile (Europe); NSA, National Security Agency (USA). The frequency range of telephone speech is 200 to 3400 Hz.

broken curve is a description of a research goal that is thought to be achievable in the sense of not violating fundamental limits on coding capability. The PCM curve is that of the memoryless algorithm of pulse code modulation. The waveform coder curve is that of a high-complexity algorithm such as adaptive predictive coding, which uses redundancy-removing operations to reduce signal energy at the input to the amplitude quantizer, resulting in a lower bit rate for a specified level of output speech quality. The vocoder point at 2.4 kbps is that of an algorithm that produces intelligible but synthetic-sounding speech at very low transmission rates by means of a highly compact excitation-modulation model (Fig. 3a). The synthetic quality in this system is accepted in applications in which digital encryption and low transmission rate are of critical importance.

The hybrid coder curve describes the performance of a class of algorithms that combine the high quality potential of waveform coding with the compression efficiency of a model-based vocoder. The idea is to use a time-varying excitation model much more sophisticated than that of a traditional vocoder

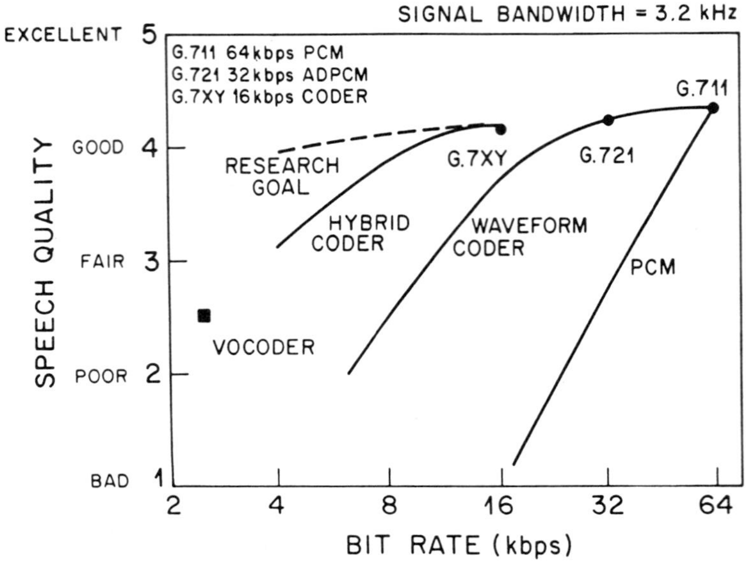

Figure 2 Quality of telephone-bandwidth speech as a function of transmission rate.

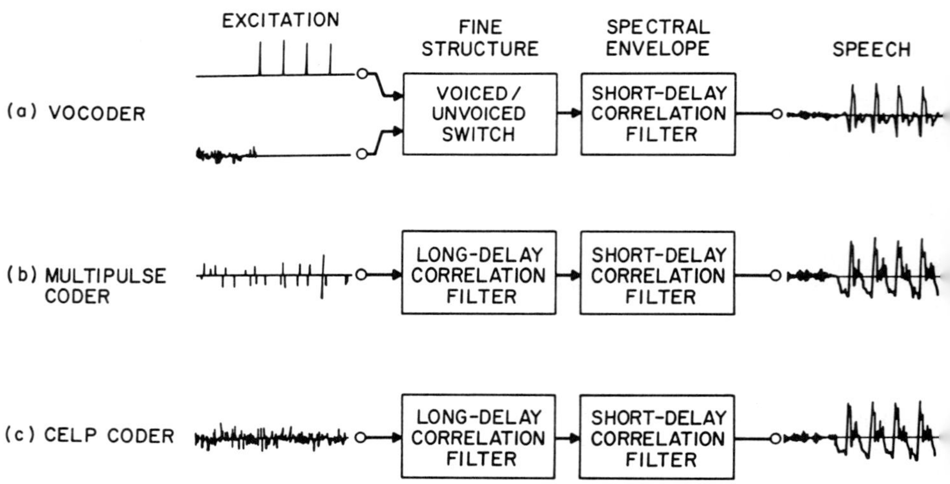

Figure 3 Models of speech excitation in (a) linear prediction (LPC) vocoder and (b)(c) hybrid coders; (b) multipulse LPC, and (c) codebook-excited LPC (Atal, 1986).

and to use waveform-coder principles to compute a distortion-minimizing excitation for every frame (say, 16 ms) of input speech (Fig. 3b and c). Hybrid coders using code-excited linear prediction (CELP) are discussed in Section 2 for applications at 4.8, 8, and 16 kbps.

The solid dots in Fig. 2 refer to coding algorithms that provide high quality at 64, 32, and 16 kbps. Pulse code modulation (PCM) at 64 kbps and adaptive differential PCM (ADPCM) at 32 kbps are CCITT standards and are respectively called G.711 and G.721. These algorithms provide network-quality coding, a necessary condition for which is high-quality speech encoding with an MOS rating of at least 4.0. Currently, the CCITT is considering the definition of a high-quality speech standard at 16 kbps. The technique is likely to be a hybrid coding algorithm.

Digital coding of telephone speech is elaborated in Section 2 with focus on network-quality coding at 16 kbps.

1.3. Overview of Digital Audio

Figures 1 and 2 have referred specifically to telephone-band speech. In Figure 2 the achievement of higher quality at a given bit rate implies reduced speech distortion without any change of bandwidth. We will next discuss the effects of changing signal bandwidth on speech quality and intelligibility. Figure 4 defines four commonly understood grades of audio bandwidth. If the audio signal is speech, rather than music, perceived gains in quality are perhaps greatest when one progresses from the telephone level to the commentary, or AM radio, level. The gains are in terms of increased intelligibility, naturalness, and speaker recognition. Low-frequency enhancement (50 to 200 Hz) contributes to increased naturalness and presence, and high-frequency enhancement (3400 to 7000 Hz) provides greater intelligibility and fricative differentiation (for example, *s* versus *f*).

We will briefly discuss speech intelligibility as a function of input frequency content. The articulation index score (Jayant and Noll, 1984) is a specific intelligibility measure that shows this dependence. The 20 nonoverlapping speech bands with equal (5%) contributions to the articulation index are shown in Table 1. Note that the 0 to 200 Hz band has a zero contribution to the articulation index measure of intelligibility. However, as discussed earlier, frequencies below 200 Hz contribute to enhanced naturalness.

High-quality compression of wideband audio and ISDN applications of digital audio are discussed in Section 3. We also discuss a 64-kbps CCITT standard for 7-kHz audio and a 128-kbps standard for 20-kHz audio being

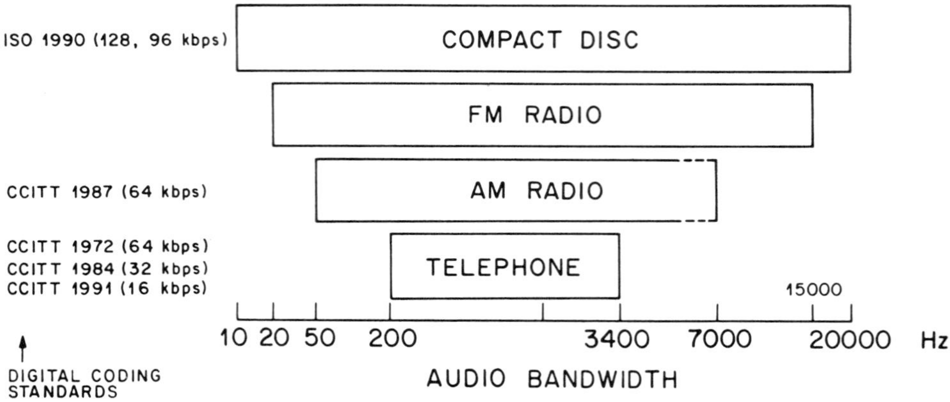

Figure 4 Four grades of audio signal bandwidth and corresponding standards for digital coding. ISO, International Standards Organization.

considered by the International Standards Organization (ISO). Audio coding standards are also being planned by national standards committees and by multinational organizations such as the European Broadcasters Union (EBU).

Table 1 Frequency Bands Contributing Equally to the Articulation Index

Number	Limits	Mean	Number	Limits	Mean
1	200 to 330	270	11	1660 to 1830	1740
2	330 to 430	380	12	1830 to 2020	1920
3	430 to 560	490	13	2020 to 2240	2130
4	560 to 700	630	14	2240 to 2500	2370
5	700 to 840	770	15	2500 to 2820	2660
6	840 to 1000	920	16	2820 to 3200	3000
7	1000 to 1150	1070	17	3200 to 3650	3400
8	1150 to 1310	1230	18	3650 to 4250	3950
9	1310 to 1480	1400	19	4250 to 5050	4650
10	1480 to 1660	1570	20	5050 to 6100	5600

2. DIGITAL CODING OF TELEPHONE SPEECH

The importance of low-bit rate telephony has been indicated in Fig. 1. Figure 2 has illustrated that our current understanding of coding has resulted in high-quality speech at bit rates down to about 8 kbps. Research is now focused on the important neighborhood of 4 kbps.

2.1. Digital Telephony at 4 kbps

High-quality coding at 4 kbps is a primary focus in speech research (Atal, 1986; Jayant and Chen, 1989; Kleijn et al., 1988; Kroon and Atal, 1990; McAulay and Quatieri, 1986; Shoham, 1989). It is of importance for enhancing secure telephony in government and military applications (Kemp et al., 1989) and also for providing the central capability of future band-efficient systems for digital cellular radio—for example, cellular channels with a user bandwidth of 5 kHz. These are wireless access channels for moving vehicles in rural, suburban, or urban environments served by terrestrial base stations. Speech coding at bit rates on the order of 4 kbps is also needed in mobile satellite communication (MSAT) applications (Kleijn et al., 1988) for providing wireless access to moving vehicles in remote areas and in International Maritime Satellite (INMARSAT) applications with a 6.4-kbps target for total transmission rate (speech coder bit rate plus overhead for channel error protection). Finally, speech coding at 4 kbps has an attractive implication for storage—1 hour of spoken material on a single 16-Mbit memory chip.

Recent advances in CELP coding have produced significant improvements relative to 2.4-kbps vocoding, and this is reflected in 4.8-kbps algorithms with very high levels of intelligibility and acceptability, as described by DRT and DAM scores. However, the subjective quality of 4.8-kbps speech, as noted in Fig. 2, is below the high quality level of 4.0 on the MOS scale.

Table 2 summarizes DRT, DAM, and MOS performances of various speech coders in the range of 64 to 2.4 kbps (Daumer, 1982; Jayant and Noll, 1984; Kemp et al., 1989). Coding algorithms discussed in the table include PCM, ADPCM, linear predictive (vo)coding (LPC), CELP, and low-delay CELP (LD-CELP). The CELP algorithms typify the hybrid coders of Fig. 2.

High-quality coding at 4 kbps is an important focus for speech research as well as communication technology. In the specific context of CELP coding, improvements in low-bit-rate CELP have been attained by means of adaptive control of the stochastic component of speech excitation (Shoham, 1989), by means of time-varying bit allocations to excitation and spectrum parameters (Jayant and Chen, 1989), and by the use of fractional-sample accuracy in pitch analysis (Kroon and Atal, 1990). In a broader sense, new research could

Table 2 DRT, DAM, and MOS Scores for Standard Speech Coders[a,b]

Coder	DRT	DAM	MOS
64-kbps PCM	95	73	4.3
32-kbps ADPCM	94	68	4.1
16-kbps LD-CELP	94*	70*	4.0
8-kbps CELP	93+	68*	3.7
4.8-kbps CELP	93−	67−	3.0+
2.4 kbps LPC	90	54	2.5*

[a]PCM, Pulse code modulation; ADPCM, adaptive differential PCM; CELP, code-excited linear predictive coding; LD-CELP, low-delay CELP; LPC, linear predictive vocoder.
[b]*, Estimates; −, upper bounds; +, lower bounds.

provide a better unification of CELP and alternative methodologies (such as sinusoidal coding) and could define more powerful ways of incorporating auditory noise masking in frequency and time.

2.2. Digital Telephony at 8 kbps

Systems for first-generation digital cellular radio use bandwidths greater than 5 kHz and hence can afford higher bit rates such as 8 kbps for speech coding. For example, the proposal in the United States is to use a user bandwidth of 10 kHz and a total transmission rate of about 13 kbps for speech coding and channel error protection (Chen et al., 1989; Cox et al., 1989; Gerson and Jasiuk, 1989). The bit rate for speech coding alone will be on the order of 8 kbps. Such a system will replace the current practice of analog FM speech with a 30-kHz user bandwidth. The digital system provides greater robustness to channel noise and fading, as well as better reuse of individual carrier frequencies. As a result, the improvement in terms of number of cellular users will exceed the factor of 3 implied by the changeover from 30 to 10 kHz. The net gain is expected to be a factor on the order of 5 to 7.

In the United States, cellular telephony is will be based on a hybrid coding (CELP) algorithm. Although speech quality at 8 kbps currently falls slightly below the high-quality threshold (MOS = 4.0), it provides a communication quality that is adequate to provide improvements over analog FM telephony, especially at low levels of radio channel quality—say, a signal-to-channel-noise ratio (C/N) below 15 to 18 dB. This is depicted in the impressionistic curves of Fig. 5. The communication delay due to the speech coder is expected to be on the order of 40 ms, a value considered acceptable for the cellular telephone application.

For digital cellular radio in Europe, the recommendation of Group Spéciale Mobile (GSM) is also a hybrid coder, a regular pulse excitation algorithm with a bit rate of 13.2 kbps (out of a total transmission rate of 22.8 kbps) and a coder delay of 40 ms (Kroon et al., 1986; Vary et al., 1988). The coding technique is algorithmically similar to a 9.6-kbps multipulse excitation coder for the Skyphone airline application.

2.3. Network-Quality Speech Coding

For ubiquitous application in networks, a speech coding algorithm has to satisfy several performance criteria. These include a level of speech quality high enough to withstand multiple stages of coding and decoding, a processing delay low enough to withstand echoes and additional delay components in the network, and the ability to handle nonspeech signals in the telephone band.

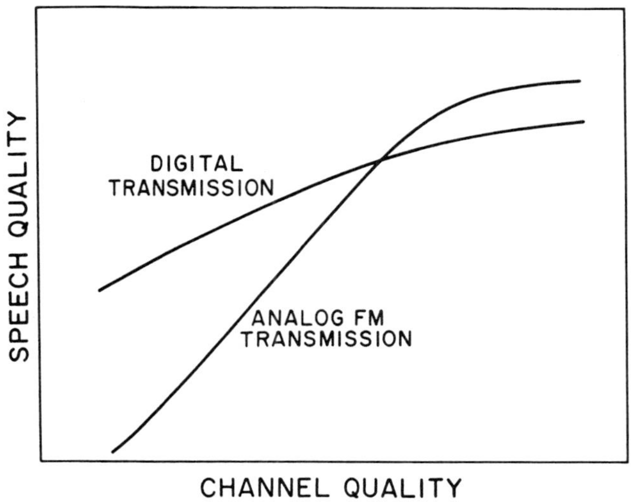

Figure 5 Speech quality versus channel quality in cellular telephony.

2.3.1. PCM (64 kbps) and Variable-Rate ADPCM (40, 32, 24, and 16 kbps)

Algorithms at 64 kbps (G.711, PCM) and 32 kbps (G.721, ADPCM) satisfy network requirements and are international CCITT standards (CCITT 1984; 1990; 1990; Rosenberger, 1988; Taka and Maitu, 1986). These standards are in ubiquitous use in both public and private speech telecommunications. The 32-kbps standard is relatively recent (1986). An important application of this codec is in digital circuit multiplication equipment (DCME). Here, the combination of 2 : 1 compression (64 to 32 kbps) with exploitation of the so-called 2.5 : 1 TASI or DSI gains (effect of silences in speech) provides effective circuit expansions of 5 : 1 over traditional telephone systems. The 32-kbps algorithm has been extended to 24 to 40 kbps in G.723. Also, embedded ADPCM is a draft CCITT standard G.727 at the rates of 40, 32, 24, and 16 kbps. G.727 can be used with G.764 wideband packet network applications. Lower transmission rates such as 24 and 16 kbps permit accommodation of occasional bursts of heavy telephone traffic. The lower speech quality of the 24- and 16-kbps systems can be optionally enhanced at the receiver by the use of adaptive algorithms for postfiltering (Ramamoorthy et al., 1988; Chen and Gersho, 1987). The higher ADPCM rate of 40 kbps provides the capability of transmitting 9.6-kbps modem waveforms. The simplicity of G.721 and related algorithms also makes them very attractive for wireless access applications requiring very low transmitter power; for example, terminals within or near buildings provided with indoor wireless communication (Cox et al., 1986; Cox, 1989; Steele and Prabhu, 1985; Steele, 1989).

The 32-kbps ADPCM algorithm is also more robust than 64-kbps PCM to digital errors in transmission. At a bit error rate of 1 in 1000 the degradation in speech quality is graceful, and at an error rate of 1 in 100 speech intelligibility is good although the quality is poor.

2.3.2. Low-Delay Speech Coding at 16 kbps

Currently, the CCITT is considering defining a low-delay network-quality speech standard at 16 kbps (Fig. 1). Possible applications include DCME, ISDN transmission, packetized speech, cordless telephone, and speech for videophone service. Table 3 lists the intended applications for such a standard, and Table 4 lists some of the performance requirements desired of the proposed speech encoding algorithm.

An important challenge for the proposed coding algorithm is the combination of high quality with low delay at the relatively low quantization rate of 16 kbps. The delay requirement means that estimation of the time-varying model for the speech spectral envelope has to be performed in a backward-adaptive mode, using a history of already quantized speech, rather than by buffering tens of milliseconds of input speech for forward spectral estimation

Table 3 CCITT, Appendix 3, Annex 1 to Q21/XV (As Amended, March 13–22, 1989)

List of possible applications for 16-kbit/s voice coding:

1. Videophone service using transmission rates 64, 2 × 64, 128 kbit/s.
2. Cordless telephone.
3. Low C/N digital satellite systems. This includes maritime, thin-route, and single channel per carrier satellite systems.
4. DCME. In this equipment 16-kbit/s speech coding is generally combined with DSI techniques. The equipment may be used for long terrestrial connections and for digital satellite links generally characterized by high C/N ratios.
5. PSTN.[a] This application covers the encoding of voice telephone channels in trunks, junction, or distribution networks (e.g., transcoder equipment). However, indiscriminate use of 16-kbps coding in trunk applications and junction national networks is not recommended.
6. ISDN.[b] Distinguishing features with respect to PSTN for setting 16-kbit/s speech coding requirements are the availability of unrestricted end-to-end digital bearer capabilities, absence of electrical echo control devices, and availability of SSN 7.
7. Digital leased lines. Two possibilities may be envisaged in this case; one is where the end-to-end digital leased circuits include only one encoding/decoding, the other is where the end-to-end digital leased circuits are connected into the public network and they may include digital transcodings.
8. Store and forward systems.
9. Voice messages for recorded announcements.
10. Land digital mobile radio (DMR) systems.
11. Packetized speech.
12. Audio channel for low-bit-rate one-way video service (e.g., surveillance).

[a]ISDN, Integrated Services Digital Network.
[b]PSTN, Public Switched Telephone Network

(AT&T, 1989; Chen and Cox, 1990; Gersho and Cuperman, 1989; Iyengar and Kabal, 1988).

Figure 6 shows the block diagram of a backward-adaptive coder proposed for 16 kbps (Chen and Cox, 1990). The coding algorithm is codebook-excited linear predictive coding. The only source of encoding delay in the system is the forward adaption of the excitation signal in the form of selecting the best out of a rich codebook of possible excitation vectors, each of length 5 samples (0.625 ms). The adaption of spectral information is backward adaptive, and an important requirement here is the attainment of significant spectral

Table 4 Desired Characteristics of 16-kbps Coder

Parameter	CCITT requirement	CCITT objective
Coding delay	≤ 5 ms	≤ 2 ms
Speech quality, bit error rate = 0	distortion ≤ 4 qdu[a]	
at bit error rate = 10^{-3}	not worse than G.721[a]	
at bit error rate = 10^{-2}	not worse than G.721	
Tandeming for speech	3 async. tandems with distortion ≤ 14 qdu	sync. tandems without distortion accumulation
Transmit signaling tones	DTMF[*]	
Transmit music		no annoying effects
Operate at lower rates		graceful degradation
Complexity		as low as possible

[a]G.721: CCITT standard for 32-kbps ADPCM speech coding [3.5 qdu where qdu = quantization distortion unit (distortion due to a single stage of 64-kbps PCM coding with an "average" codec)]. To calibrate codecs on the qdu scale the CCITT uses the MNRU (a reference unit with a known qdu characteristic). The MNRU is specified in a CCITT Recommendation. [*]DTMF, Dual-Tone Multifrequency

estimation (prediction gain) in spite of the presence of quantization noise in the past speech samples used for spectral analysis.

In one version of the low-delay CELP coder (Chen and Cox, 1990), the linear predictor and the gain of the excitation vector are backward adaptive, being based on analyses of previously quantized speech and previously used excitation. The LPC analysis uses a very high order, 50, and a nonsymmetrical (non-Hamming) window to provide high prediction gain (16 to 18 dB) in spite of backward adaption and a slow adaption rate (once every 20 samples). The noise-shaping filter is also updated once every 20 samples, but based on analysis of unquantized speech, because the shaping operation needs to be performed only at the encoder. The excitation gain is updated for every new 5-sample vector, based on a tenth-order adaptive linear gain predictor in the log-gain domain. The predictor for the gain is itself updated less frequently, once every 20 samples, based on the log-gain sequence of previously quantized and scaled excitation vectors. The excitation is vector quantized using a 10-bit codebook, with 3 bits for gain and 7 bits for shape (pattern of gain-normalized vector). The shape codebook is nonstochastic in that it is iteratively trained using an adequately extensive speech data base.

The backward-adaptive low-delay CELP algorithm (LD-CELP) described in Chen and Cox (1990) is quite complex, and the codebook search

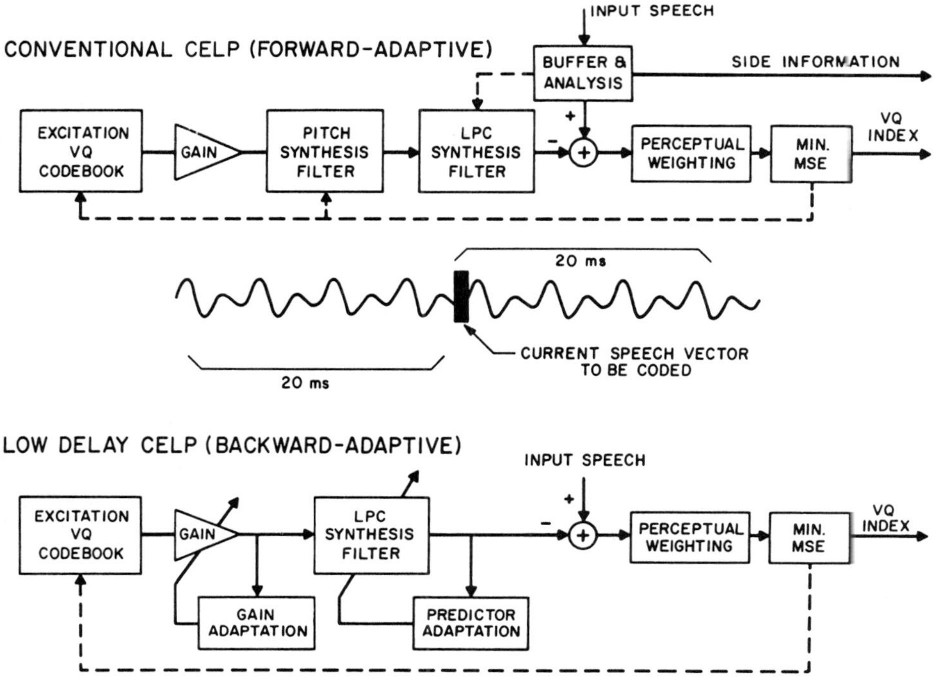

Figure 6 Backward-adaptive system for CELP.

is the single most demanding component of the algorithm. The availability of a 25-Mflop processor with advanced memory capability (the AT&T WE DSP32C) permits a single-chip (half-duplex) implementation. A full-duplex coder currently requires a two-chip implementation, but prospects for a single-chip implementation with nearly equal speech quality are good. Also of interest is the possibility of extending the low-delay property to lower bit rates such as 8 kbps, while maintaining the speech quality offered by traditional high-delay coders at those bit rates.

High-quality coders at 16 and 8 kbps are prime candidates for transmission applications as well as store-and-forward systems for voice mail.

3. DIGITAL CODING OF WIDEBAND SPEECH AND AUDIO

The naturalness of wideband speech is a significant feature for extended telecommunication processes such as audio conferencing and program broadcasting. Basic rate ISDN provides a natural framework for a 64-kbps algorithm to encode wideband audio for such applications. The digital connectivity afforded by ISDN has prompted a worldwide revising of audio transmission quality. End-to-end digital connectivity has particularly made possible the inclusion of low frequencies down to 50 Hz in the transmitted audio band (Mermelstein, 1988).

3.1. Coding of 7 kHz Audio

The CCITT standard for 7 kHz audio (G.722) is a 64 kbps algorithm developed primarily for ISDN teleconferencing and loudspeaker telephony. Because of the 64 kbps capability, a single "voice-grade" channel, on a digital or analog PSTN, can transport a commentary-quality sound program over any attainable distance and yield a broadcast-grade voice program at the receiving end.

The G.722 coding algorithm is based on a two-band subband coder, with ADPCM coding of each of the subbands (Fig. 7) (CCITT, 1986; Mermelstein, 1988; Modena et al., 1986; Soumagne et al., 1986). The bandwidth of the lower subband is slightly greater than that of conventional telephony. High-quality coding is provided by an unequal bit allocation procedure in

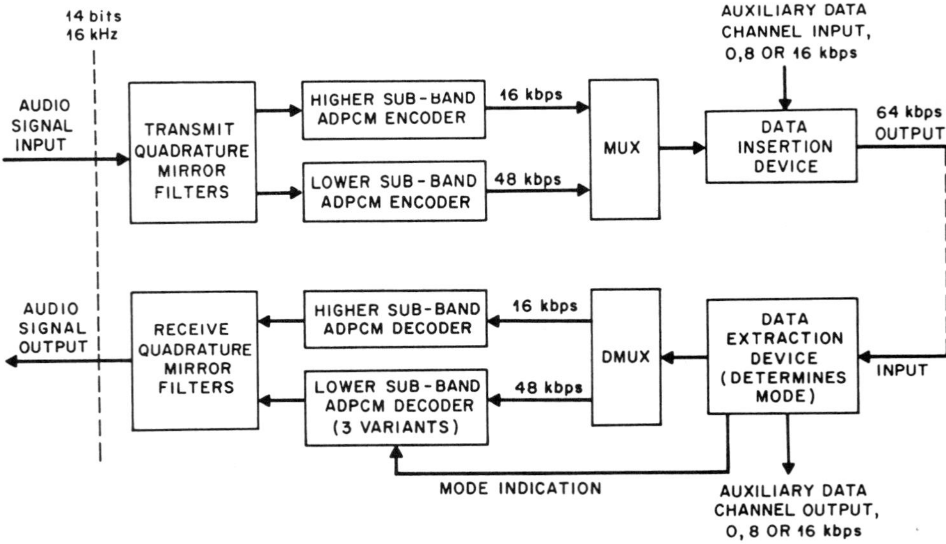

Figure 7 Block diagram of a two-band subband coder for 64 kbps coding of 7-kHz audio (Mermelstein, 1988).

which the low- and high-frequency subbands are quantized using 6 and 2 bits per sample, respectively. Since the 64-kbps standard is addressed primarily to wideband speech and only secondarily to music signals, a simple fixed bit allocation strategy is adequate. The two-band filter banks used for analysis and synthesis produce a communication delay of about 3 ms. This turns out to be a very desirable feature because of expected interconnections of G.722 with narrowband links where uncanceled echoes could pose a problem if compounded by codec delay. (In isolation, digital wideband links do not have two-wire/four-wire hybrids and resulting uncanceled echoes). The 64-kbps algorithm is robust to random error rates on the order of 1 in 10000 and to four tandem stages of repeated encoding and decoding. The simplicity of the quantizing, predicting, and filtering (24-tap) algorithms permits implementation on a single-chip fixed-point processor.

ISDN applications suggest operating the audio coding algorithms at slightly lower bit rates. Use of an embedded coding technique for ADPCM permits the operation of the low-frequency subband at one of three quantizing rates—6, 5, and 4 bits per sample, with graceful degradation of quality. Corresponding audio rates are 64, 56, and 48 kbps; in the last two instances, capacities of 8 and 16 kbps are available for simultaneous data transmission over the 64-kbps basic rate channel (Table 5).

Figure 8 shows audio quality on the MOS scale for speech and music material at the rates of 64, 56, and 48 kbps. Also shown for comparison is the performance of 240-kbps linear PCM (15-bit audio input sampled at 16 kHz). In another comparison involving G.722, G.721, and 128-kbps PCM (16 kHz × 8 bits per sample), it has been shown that the G.722 algorithm at 64 kbps has an equivalent SNR gain of 13 dB over G.721, of which 6 dB is attributable to increased input bandwidth (Mermelstein, 1988).

Figure 8 also shows the current research goal for the coding of the 7-kHz audio. This goal is believed to be quite realistic. One of its implications is the possibility of coding 7-kHz audio at 32 kbps with high quality (MOS = 4.0). This will enable the transmission of two stereo wideband channels at 64 kbps.

Table 5 CCITT Coder for 7-kHz Audio. Bit Allocations (kbps)

Lower subband	Higher subband	Data
48	16	0
40	16	8
32	16	16

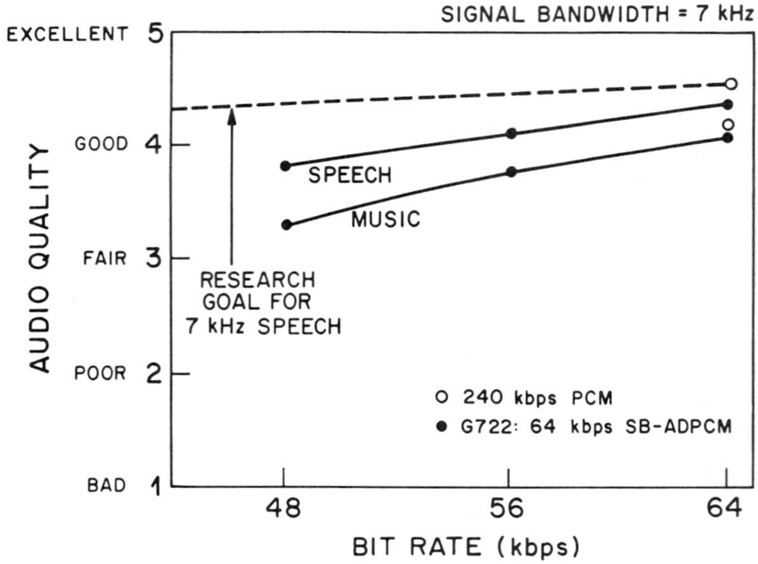

Figure 8 Quality of 7-kHz digital audio as a function of bit rate in the G.722 algorithm (Mermelstein, 1988).

The utilization of cross-channel correlations can provide a further increase of capability—for example, accommodation of a bandwidth greater than 7 kHz in the 64-kbps system. There are at least two reasonable approaches to the problem of high-quality coding of audio at 32 kbps—the linear prediction approach exemplified by CELP and the frequency domain technique of transform or subband coding. In both cases, the attainment of high audio quality will depend on the use of perpetual tuning of the algorithm to provide effective shaping of the quantization noise. The CELP technique offers the additional possibility of low-delay coding by the use of backward adaptation, as in Fig. 6.

3.2. Coding of 20-kHz Audio

Bit rates of 300 to 400 kbps have been used for audio broadcasting, usually with a bandwidth of about 15 kHz (and a sampling rate of 32 kHz). Specifically, NICAM-PCM (PCM with nearly instantaneous compounding) at 384 kbps and adaptive delta modulation at 330 kbps have been formally compared with 448-kbps (14-bit) linear PCM and shown to be competitive for the Direct Broadcast Satellite application.

The ISO is committed to the standardization of a low-bit-rate coding algorithm for 20-kHz audio in the year 1991. Applications for the low-bit-rate wideband speech include electronic publishing, travel and guidance, teleteaching, multilocation games, multimedia memo, and data base storage. Table 6 lists possible applications for the proposed ISO standard for 20-kHz audio. Seen in the table are several applications for digitally represented wideband speech. Another major application for 20-kHz digital audio is in advanced television systems such as high-definition television (HDTV).

The ISO standard is aimed at storing both video and audio on storage media such as an (audio) compact disc, with a total capacity on the order of 1.5 Mbps. The need to include high-quality video in this capacity implies that the audio signal should occupy a very small fraction of 1.5 Mbps. The role of the low-rate audio coder is therefore very challenging. In the field of entertainment audio, special applications such as studio processing and archiving call for transparent coding, implying MOS values well in excess of 4.0.

On a compact disc (CD), audio is stored at a bit rate of about 700 kbps per sound channel (16-bit PCM coding of 44.1 kHz-sampled signals). The emerging ISO standard calls for a CD-audio bandwidth of 20 kHz and a single-channel bit rate of 96 or 128 kbps. Production of high-quality audio at these very low rates calls for a new generation of coding algorithms in which the gains due to redundancy removal need to be augmented by the liberties offered by the human auditory process (Brandenburg, 1987; Johnston, 1988; Krahe, 1986; Krasner, 1979; Johnston and Brandenburg, Chapter 5, this volume).

A powerful way to incorporate a human auditory model in the process is to frequency-analyze the input audio signal and postulate a just-noticeable quantization noise threshold as a function of frequency for the given audio signal spectrum (Fig. 9). Quantization errors that are equal to, or lower than, this threshold are masked by the audio spectrum in the frequency vicinity of the errors. Given the noise threshold, a variable bit allocation procedure can be utilized to provide an actual noise spectrum close to the threshold characteristic. Extremely high coding efficiency can be realized because the sophisticated noise threshold typically permits very coarse, or even zero-bit, quantization of significant fractions of the input spectrum. The objective signal-to-noise ratio of these systems can be fairly low (for example, as low as 20 dB or less), while the subjective signal-to-noise ratio is at a level high enough to signify transparency or near-transparency.

The framework for frequency analysis and variable bit allocation can be either subband coding or transform coding (Fig. 10). In either case, the need

Table 6 MPEG* Applications

1. Electronic publishing (end user)
   Integrated and interactive text, graphics, sound, and video. (Example: "Hypercard with video and/or audio windows.")
   Electronic newspaper, including text, sound, video. (Example: "the New York Times on CD, perhaps with sound.")
2. Electronic publishing (production)
   Digital audio editing of resource materials for inclusion in electronic texts.
3. Travel guidance
   Preview of city views.
   Flight and driving simulation.
   Interactive video in public spaces.
   Active guidance.
   Surrogate travel.
4. Educational applications
   Electronic books.
   Remote classroom instruction.
5. Business applications
   Electronic books.
   Multimedia E-mail.
6. Games
   Standalone games, with audio and video from a storage medium.
   Multilocation games.
7. Entertainment applications.
   MTV audio on CD.
   TV program and movie storage and distribution.
   Broadband stereo distribution.
   Previewing systems.
8. Data base applications
   Audio data base applications.
   Audio + Video, with either onsite or offsite data bases.
9. Multimedia memo and bulletin board
   Bulletin board.
   Computer conferencing with multimedia memos including audio.
   Face-to-face teleconferencing.
10. Broadcast sound
    Production.
    Transmission of "contribution quality" audio.
    Transmission for distribution.
    Digital audio broadcasting.
11. Multimedia art
12. Medical applications
    Education

*MPEG, Moving Pictures Experts Group, ISO

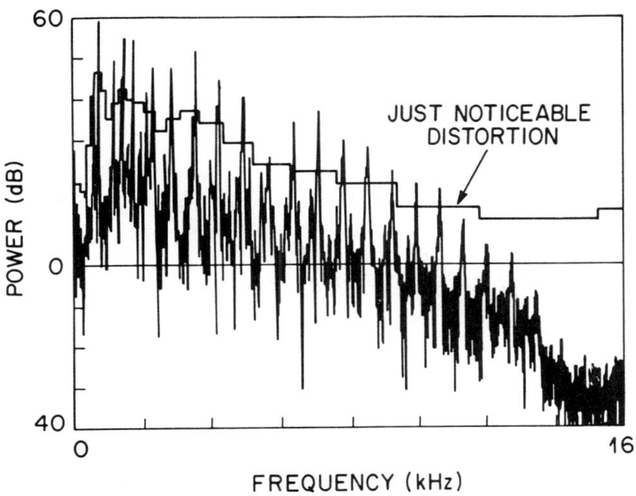

Figure 9 Masking threshold as a function of frequency for an illustrative audio signal (trumpet) (Johnston, 1988).

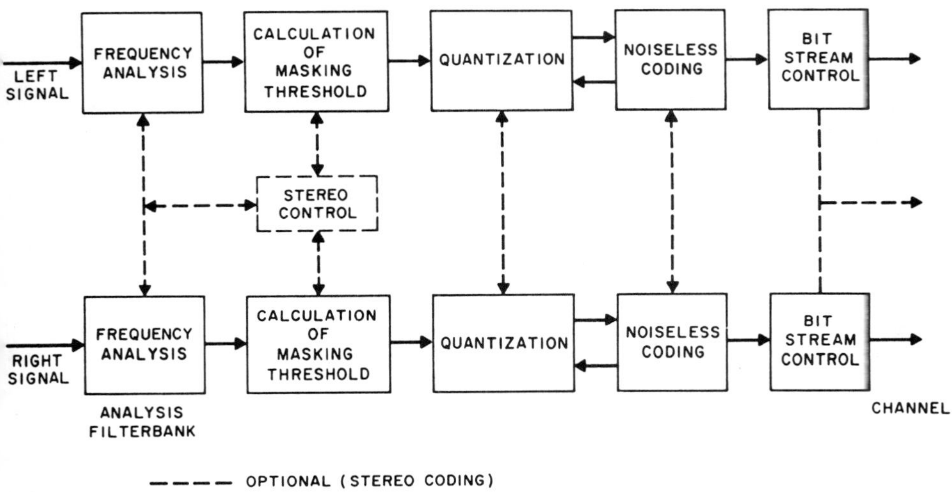

Figure 10 Perceptual coding of wideband audio—block diagram of a perceptual frequency-domain coder (Johnston, 1988; Johnston and Brandenburg, Chapter 5, this volume). Dashed lines indicate the option of utilizing left-right channel correlations for increasing efficiency in stereo coding.

for high quality at a low average bit rate means that the noise threshold should ideally be computed in a time-varying fashion, and this, together with noiseless entropy coding of the quantizer output, leads to encoding algorithms of very high complexity. Decoding is typically simpler in comparison.

Notice that the noise threshold in Fig. 9 is described with a very fine resolution in frequency. It is not clear whether this description can be implemented with satisfactory fidelity in the framework of LPC coding. Even if the noise shaping is thus implemented, possibly by LPC analysis of an extremely high order, the nonstationarity in certain types of audio (for example, music in all its generality) pose a problem for LPC-based coding.

4. FUTURE TRENDS

We conclude with a discussion of future directions in the field, in terms of anticipated technology applications and the algorithms needed to support those technologies.

4.1. Applications

Speech bandwidths in the range of 5 to 20 kHz are likely to see various subclasses of applications. The notion of quality as a function of speech bandwidth will become more pervasive, and subjective testing will lead to better quantification of the quality-bandwidth function. Sophisticated algorithms for coding will lead to transmission techniques in which quality is not limited by quantization noise. Coding systems in the range 8 to 64 kbps will provide graceful flexibilities in terms of selected bandwidth and special features such as stereo separation in teleconferencing. CD-like audio at 64 kbps will become commonplace. Advances in coding will be supported by new technologies for wideband transducers, noise-canceling systems for audio pickup, and autodirective microphone arrays (Flanagan et al., 1985; Kaneda and Ohga, 1986; Miyoshi and Kaneda, 1988; Sondhi and Elko, 1986).

4.2. Algorithms

An important gap in our understanding of audio compression is in the comparative assessment of linear predictive coding and linear transform (or subband) coding. Linear prediction has been extremely successful in the coding of telephone speech. Its application to higher-bandwidth audio needs further research, demonstration, and evaluation. Transform coding and subband coding have been very successful for fine-grained frequency analysis of very wideband audio signals and for the invocation of complex spectral masking. On the other hand, it is not clear if transform or subband methods have a fundamental advantage (or disadvantage) relative to linear predictive methods in the coding of intermediate-bandwidth (5 to 10 kHz) audio, especially speech.

Another gap in our understanding is in the design of asymmetric algorithms, systems in which the decoder is significantly simpler than the encoder. Such systems can provide very great levels of efficiency in signal storage. The encoder can even be non-real-time; decoding delay can be much shorter than the encoding delay; and for one-way applications such as broadcast and playback of stored data, constraints on decoding delay can also be very lenient.

As coding algorithms become increasingly efficient and approach fundamental capabilities, perception research will play a significant role by enhancing our understanding of noise masking, especially in the time domain.

Advances in signal processor technology will continue to support increasingly complex algorithms for coding and decoding. Applications that will benefit the most are those that are asymmetric and decode intensive. In these applications, the synergistic working of coding theory, perception science, and signal processing will bring to the human listener sophisticated speech technology in affordable forms.

ACKNOWLEDGMENT

I thank the following colleagues at AT&T Bell Laboratories for reviewing an earlier version of this chapter: B. S. Atal, K. H. Brandenburg, J.-H. Chen, R. V. Cox, J. D. Johnston, W. B. Kleijn, P. Noll, M. Sherif, and Y. Shoham.

REFERENCES

Aliphas, A., and Feldman, J. (1987). The versatility of digital signal processing chips. *IEEE Spectrum*, June, pp. 40–45.

Atal, B. S. (1986). High-quality speech at low bit rates: Multi-pulse and stochastically excited linear predictive coders. *Proc. ICASSP*, pp. 1681–1684.

Atal, B. S., and Schroeder, M. R. (1984). Stochastic coding of speech at very low bit rates. *Proc. ICC*, Amsterdam, May, pp. 1610–1613.

AT&T. (1989). Detailed description of AT&T's LD-CELP algorithm—contribution to CCITT Study Group XV, November 1989. Also AT&T low delay code-excited linear predictive coder—contribution to working party T1Y1.2, doc. no. T1Y1.2/89-35, January, 1989.

Brandenburg, K. H. (1987). OCF—a new coding algorithm for high quality sound signals. *Proc. ICASSP*, April, pp. 141–144.

CCITT Study Group XVIII. (1984). 32 kbits/s Adaptive Differential Pulse Code Modulation (ADPCM). Working Party 8, Draft Revision of Recommendation G.721, Temporary Document No. D.723/XVIII. U.S.A., Geneva.

CCITT. (1990). Recommendation G.727—5, 4, 3, 2 bit per Sample Embedded ADPCM. Study Group XV.

CCITT. (1990). Recommendation G.764—Packet Voice. Study Group XVIII.

CCITT Study Group XVIII. (1986). 7 kHz audio coding within 64 kbits/s. CCITT Draft Recommendation G.722, Report of Working Party XVIII/8, July.

Chen, J.-H., and Cox, R. V. (1990). LD-CELP: A high quality speech coder with low delay. *Proc. ICASSP*.

Chen, J.-H., and Gersho, A. (1987). Real-time vector APC speech coding at 4800 bps with adaptive postfiltering. *Proc. ICASSP*, April, pp. 2185–2188.

Chien, E. S. K., Goodman, D. J., and Russell, J. E. (1987). Cellular access digital network (CADN): Wireless access to networks of the future. *IEEE Commun. Mag. 25*(6).

Cox, D. C. (1989). Portable digital radio communications—an approach to tetherless access. *IEEE Commun. Mag.* July, pp. 30–40.

Cox, D. C., Arnold, H. W., and Porter, P. T. (1986). Universal portable digital communications. *Telephony*, October 13, pp. 80–87.

Cox, R. V., Kleijn, W. B., and Kroon, P. (1989). Robust CELP coders for noisy backgrounds and noisy channels. *Proc. ICASSP*, May.

Daumer, W. R. (1982). Subjective evaluation of several efficient speech coders. *IEEE Trans. Commun.*, April, pp. 655–662.

De Marca, J. R. B., Farvardin, N., Jayant, N. S., and Shoham, Y. (1988). Robust vector quantization for noisy channels. *Proc. M-SAT Conference*, Jet Propulsion Laboratory, Pasadena, Calif., May.

Flanagan, J. L., Johnston, J. D., Zahn, R., and Elko, G. (1985). Computer-steered microphone arrays for sound transduction in large rooms. *J. Acoust. Soc. Am. 78*: 1508–1518.

Gersho, A., and Cuperman, V. (1989). A high-level description of the Consortium for Speech Coding's low delay vector excitation coder (LC-VXC). Contribution to CCITT Study Group XV, March.

Gerson, I. A., and Jasiuk, M. A. (1989). Vector sum excited linear prediction (VSELP). IEEE Workshop on Speech Coding for Telecommunications. September 5–8.

Hoyle, R. D., and Falconer, D. D. (1987). A comparison of digital speech coding methods for mobile radio systems. *IEEE J. Selected Areas Commun.*, June, pp. 915–920.

Irmer, T. (1986). An idea turns into a reality—CCITT activities on the way to ISDN. *IEEE J. Selected Areas Commun.*, May, pp. 316–319.

Iyengar, V., and Kabal, P. (1988). A low delay 16 kbits/sec speech coder. *Proc. ICASSP*, April, pp. 243–246.

Jayant, N. S., and Chen, J.-H. (1989). Speech coding with time-varying bit allocations to excitation and LPC parameters. *Proc. ICASSP*, May.

Jayant, N. S., and Noll, P. (1984). *Digital Coding of Waveforms*. Prentice-Hall, Englewood Cliffs, N.J.

Johnston, J. D. (1988). Transform coding of audio signals using perceptual noise criteria. *IEEE J. Selected Areas Commun.*, February, pp. 314–323.

Kaneda, Y., and Ohga, J. (1986). Adaptive microphone array system for noise reduction. *IEEE Trans. Acoust. Speech Signal Process-ASSP*:1391–1400.

Kemp, D. P., Sueda, R. A., and Tremain, T. E. (1989). An evaluation of 4800 bps voice coders. *Proc. ICASSP*, May.

Kitawaki, N., and Nagabuchi, H. (1988). Quality Assessment of Speech Coding and Speech Synthesis Systems. *IEEE Commun. Mag.*, October, pp. 36–44.

Kleijn, W. B., Krasinski, D. J., and Ketchum, R. H. (1988). Analysis and improvement of the vector quantization in CELP. *Proc. MSAT Conf.*, Jet Propulsion Laboratory, Pasadena, Calif., May.

Krahe, D. (1986). Ein Verfahren zur Datenreduktion bei digitalen Audiosignalen unter Ausnutzung psychoakustischer Phänomene. *Rundfunktech. Mitt.*, Jahrg. 30.

Krasner, M. A. (1979). Digital encoding of speech and audio signals based on the perceptual requirements of the auditory system. Technical Report 535, Lincoln Laboratories, MIT.

Kroon, P., and Atal, B. S. (1990). Pitch predictors with high temporal resolution. *Proc. ICASSP*,

Kroon, P., Deprettere, E. F., and Sluyter, R. J. (1986). Regular-pulse excitation—a novel approach to effective and efficient multipulse coding of speech. *IEEE Trans. Acoust., Speech, Signal Process. ASSP-34*:1054–1063.

McAulay, R. H., and Quatieri, T. F. (1986). Speech analysis and synthesis based on a sinusoidal model. *IEEE Trans. Acoust., Speech, Signal Process. ASSP*:744–754.

Mermelstein, P. (1988). A new CCITT coding standard for digital transmission of wideband audio signals. *IEEE Commun. Mag.*, January, pp. 8–15.

Miyoshi, M., and Kaneda, Y. (1988). Inverse filtering of room acoustics. *IEEE Trans. Acoust. Speech Signal Process. ASSP*:145–152.

Modena, G., Coleman, A., Usai, P., and Coverdale, P. (1986). Subjective performance evaluation of the 7 kHz Audio Coder. *Proc. GLOBECOM*. December.

Nishitani, T., Kuroda, I., Satoh, M., Katoh, T., and Aoki, Y. (1987). A CCITT standard of 32 kbits/s ADPCM LSI coder. *IEEE Trans. Acoust. Speech Signal Process. ASSP*:219–225.

Ramamoorthy, V., Jayant, N. S., Cox, R. V., and Sondhi, M. M. (1988). Enhancement of ADPCM speech coding with backward-adaptive algorithms for postfiltering and noise feedback. *IEEE-JSAC*, February, pp. 364–382.

Rosenberger, J. (1988). Definition of qdu and use of MNRU. Study Group XII Experts Groups on Speech Quality, Contribution SQ-15.88, July 20.

Shoham, Y. (1989). Constrained excitation CELP coding. IEEE Workshop on Speech Coding for Telecommunications, Vancouver, September.

Sondhi, M. M., and Elko, G. W. (1986). Adaptive optimization of microphone arrays under a nonlinear constraint. *Proc. ICASSP*, April, pp. 19.9.1–19.9.4.

Soong, F. K., Cox, R. V., and Jayant, N. S. (1985). Subband coding of speech using backward adaptive prediction and bit allocation. *Proc. ICASSP*, April.

Soong, F. K., and Juang, B. H. (1988). Optimal quantization of LSP parameters. *Proc. ICASSP*, pp. 394–397.

Soumagne, J., Mabilleau, P., Morissette, S., Chouinard, G., and Bennett, D. (1986). A comparative study of the proposed high quality coding schemes for digital music. *Proc. ICASSP*, April.

Steele, R. (1989). The cellular environment of lightweight handheld portables. *IEEE Commun. Mag.*, July, pp. 20–29.

Steele, R., and Prabhu, V. K. (1985). High-user-density digital cellular mobile radio systems. *IEEE Proc.* Part F, No. 5, August, pp. 396–404.

Taka, M., and Maitre, X. (1986). CCITT standardizing activities on speech coding. *Proc. ICASSP*, April.

Talbot, D. B., and Cheney, J. F. (1986). A five-band companded technique for converting telephone quality to broadcast quality using two voice-grade phone lines. *J. Audio Eng. Soc.*, October, pp. 804–811.

Taniguchi, T., et al. (1986). A high-efficiency speech coding algorithm based on ADPCM with multi-quantizer. *Proc. ICASSP*, April, pp. 1721–1724.

Theile, G., Stoll, G., and Link, M. (1988). Low bit-rate coding of high-quality audio signals. *EBU Tech. Rev.*, No. 230, August, pp. 71–94.

Vary, P., Hellwig, K., Hofmann, R., Sluyter, R. J., Galand, C., and Rosso, M. (1988). Speech Codec for the European mobile radio system. *Proc. ICASSP '88*, April, pp. 227–230.

Voiers, W. D. (1977a). Diagnostic evaluation of speech intelligibility. In *Speech Intelligibility and Speaker Recognition*. M. Hawley (ed.), Dowden Hutchinson Ross, Stroudsburg, Pa.

Voiers, W. D. (1977b). Diagnostic acceptability measure for speech communication systems. *Proc. ICASSP*, May, pp. 204–207.

Yokoyama, Y. (1987). ADPCM equipment for 9.6 kbps data. *Telecommunications*, September, pp. 102–106.

4

Wideband Coding—Perceptual Considerations for Speech and Music

JAMES D. JOHNSTON AT&T Bell Laboratories, Murray Hill, New Jersey

KARLHEINZ BRANDENBURG Universität Erlangen-Nürnberg, Erlangen, Germany

1. INTRODUCTION

In the past year several schemes for high-quality audio coding at low bit rates have been developed. For these wideband coding schemes, the goal of the research effort is quite different from that of traditional speech coding. Rather than naturalness or "toll quality," complete transparency, when the coded signal cannot be discerned from the original of the coded speech or music, is desired. Newly developed algorithms claim transparent or nearly transparent quality at bit rates between 1.5 and 4 bits/sample at sampling rates from 32 to 48 kHz. This chapter will describe the basic ideas behind these "perceptually transparent" coders and give some examples from this rapidly evolving field. Additional examples of perceptual coders are listed in the references.

2. A COMPARISON WITH THE NARROWBAND SOURCE CODING PROBLEM

In the case of narrowband (200 to 3200 Hz or 50 to 7000 Hz passbands) coding of speech, advanced algorithms such as that of Chen (1989) take advantage of several kinds of redundancies generated by the source. In particular, for speech coding at telephone bandwidth the coding algorithms make a great deal of use of vocal tract models, knowledge of how fast speech models can change, limited bandwidth, and sampling rate. Linear predictive coding (LPC) gain is a good example of the source-coding gain that can be realized by using simple source statistics, and the spectral flatness measure (SFM) (Jayant and Noll, 1984) is a measure of the absolute predictability of a signal in the LPC sense.

Table 1 shows some typical LPC gains for telephone speech, the 256-sample LPC distance, and the SFM. The 256-sample LPC distance shows the misfit between an LPC predictor calculated 256 samples in advance (past time) of the signal to be predicted and the predictor calculated for the current signals. It shows a measure of the nonstationarity of the signal as well as information on how fast predictors must adjust in order to avoid problems due to signal nonstationarity.

As the signal passband increases to 50 to 7000 Hz, typical signals provide gains as shown in Table 2. At this bandwidth, both speech and music signals are included. Note that since the time represented by 256 samples is decreased, we expect a reduction of the LPC distance statistics. As is clear from Table 2, the increase in prediction gain is not very large. Going on to compact disc (CD) signals, sampled with 16 bits at 44/1 kHz, we obtain Table 3. Even at wide bandwidths, some signals do not have a substantial prediction gain, and the maximum distance is still substantial despite the reduced time interval that 256 samples represents.

This implies that doing source coding of either speech or music will not generally result in a substantial compression at wide bandwidths. For instance, the narrowband noise signal mentioned in Table 3 will require a total signal-to-noise ratio (SNR) of about 70 dB in order to be approach transparency unless a sophisticated noise-shaping strategy is used. Even given the mean LPC gain of 40.5 dB, about 5 bits/sample will be required for that SNR. Clearly, in order to provide good-quality coding at low bit rates, some sort of

Table 1 LPC Statistics for 16 Poles, 200–3200 Hz Passband

Speaker	Mean gain (dB)	Sigma (dB)	Max (dB)	Min (dB)	Max distance (dB)	Mean distance (dB)	Mean SFM (dB)
Male 1	17.7	5.3	26.4	2.8	14.2	2.4	21.8
Female	14.6	6.3	28.9	3.5	14.9	3.1	20.7
Male 2	11.5	5.3	23.6	.72	12.7	3.45	26.6

Table 2 LPC Statistics for 16 Poles, 50–7000 Hz Passband

Signal	Mean gain (dB)	Sigma (dB)	Max (dB)	Min (dB)	Max distance (dB)	Mean distance (dB)	Mean SFM (dB)
Female voice	26.5	7.57	43.4	6.4	12.1	2.8	30.3
Male voice	26.3	7.3	41.7	9.9	16.5	2.4	30.2
Female singing	20.7	3.34	27.4	5.5	5.8	.87	26.1
Acoustic guitar	18.5	2.8	27.0	9.9	5.9	.85	25.8
Male singing	21.6	3.85	37.6	6.01	8.7	.99	26.4
Pipe organ	11.8	2.06	17.4	7.4	9.5	1.1	22.2
Solo piano	23.9	3.77	34.6	15.5	8.5	1.4	30.0
Solo violin	15.4	3.36	25.0	7.5	3.2	.93	25.1

Table 3 LPC Statistics for 16 Poles, CD Passband

Signal	Mean gain (dB)	Sigma (dB)	Max (dB)	Min (dB)	Max distance (dB)	Mean distance (dB)	Mean SFM (dB)
Narrowband noise	40.5	5.2	59.2	25.0	19.1	1.0	43.5
Bassoon	40.1	3.8	23.8	28.8	10.9	.68	45.6
Viola	30.7	3.7	43.5	20.9	11.9	.92	34.1
Castenets	19.9	4.5	38.8	4.9	17.2	1.68	22.9
Solo piano	36.9	3.9	51.1	23.3	12.1	1.03	39.6
Cabasa	5.4	.5	7.6	3.7	1.6	.5	8.7
Female operatic	32.6	4.3	44.6	14.5	10.7	.77	36.9
Trumpet concerto	36.6	2.8	47.1	28.2	6.7	.64	39.7
Zarathustra	33.6	4.9	50.6	12.7	10.6	.65	37.7
Metal percussion	23.1	4.2	35.1	8.2	26.0	.84	29.2
Suzzane Vega	23.4	5.7	36.4	3.7	18.0	1.4	27.3

intelligent noise-shaping approach must be applied. In order to determine what sort of noise shaping would be advantageous, it helps to look at the sensitivities of the human ear to signals and changes in signals.

3. PSYCHOACOUSTICS

The term "masking" describes the effect by which a fainter but distinctly audible signal becomes inaudible when a correspondingly louder signal occurs simultaneously. The masking depends both on the spectral composition of the masker and the signal to be masked as well as their variation with time. Accordingly, one distinguishes between masking effects in the frequency domain and masking effects in the time domain.

In the case of high-quality audio coding the masker is the original input signal and the signal to be masked is the noise signal, defined as the difference between the original input signal and the output signal of the decoder (after compensation for the time delay of the coder-decoder chain). By proper manipulation the coding noise can be shaped to be less audible than a noise signal at the same or less noise energy but without noise shaping. As the foregoing male vocal signal can demonstrate, noise shaping can provide high coding quality without providing a high bit rate or high overall SNR. This signal, which consists of a male a-capella chorus, is perceptually transparent at an overall SNR of 13 dB when the noise is inserted according to psychoacoustic principles but is nontransparent to additive white noise even at an SNR of 60 dB and to locally adjusted white noise at an SNR of 45 dB.

In the past, many noise-shaping methods have been suggested (Jayant and Noll, 1984), including both implicit and explicit schemes based on various time and/or short-term spectral statistics. A particularly interesting noise-shaping scheme was described in Schroeder et al. (1979), in which a noise threshold generated from psychoacoustical criteria was suggested. Here we extend the idea of perceptual noise shaping from that of Schroeder et al. (1979) to a more general perceptual coding scheme.

In general, results for psychoacoustic masking show that while large total SNR may be necessary for signals with a disadvantageously shaped noise spectrum such as "white noise," low SNR may suffice for signals where the noise is cleverly injected. As an example, Scharf (1970) shows a set of results for tone-masking noise and Schroeder et al. (1979) show one way in which thresholds at one frequency are related to signals at another frequency. Clearly, as is shown by these results, a coding system with a flat noise floor will require a high SNR for sine waves, but a coding system with highly adaptive noise shaping may require a much lower SNR, for instance 25 dB rather than 90+ dB for the case of a 1-kHz tone.

4. THE HEARING PROCESS

Many people (Scharf, 1970) have examined the hearing process. Here, we will not attempt to make a thorough study of the hearing process; rather we will try to present a "composite" response of the hearing system and show how a coding system can be adapted to use a hearing model.

The ear is primarily a frequency analysis instrument. In general, a signal is separated by a number of processes, both mechanical and neurological, into frequency components. The scale that the ear appears to use for this separation, called the critical band, was introduced by Fletcher (1940) and verified by many individuals using different sets of tests (Scharf, 1970; Zwicker, 1956; Greenwood, 1961; Zwisloski, 1965). Hellman (1972), in particular, showed that the critical bands found by Scharf were also valid for other signals, and Greenwood (1961) showed that critical bands were mapped to a physical representation on the cochlea. In short, a critical band was originally defined by the following experiment:

1. A tone was presented at a fixed level, well above absolute threshold.
2. A narrowband noise source was injected into the sine wave, with the center of the noise band at the tone frequency.
3. The energy of the noise was adjusted until it reached the just noticeable difference (JND), i.e., the level at which the noise was perceptible.
4. The noise source was then adjusted to a slightly wider bandwidth by moving either the lower or upper cutoff a small distance away from the tone.
5. The level test was repeated.

At the end of these tests, it was found that there was a band of frequencies centered at each tone in which the noise energy remained nearly constant, even though the noise bandwidth and/or noise shape (in-band) was changed. This band of frequencies was called the "critical band".

Later, other researchers showed that noise maskers had the same sort of behavior (Hellman, 1972) and that the critical band corresponded to a physical measurement in the cochlea (Greenwood, 1961; Zwisloski, 1965). A tabulation of critical bands and bandwidths (Scharf, 1970) is shown in Table 4. Given this critical band scale, handling masking of steady-state tones and noise inside one critical band is relatively well described. Schroeder et al.

Table 4 Critical Band Center and Edge Frequencies

Band number (Hz)	Lower edge (Hz)	Center (Hz)	Upper edge (Hz)
1	0	50	100
2	100	150	200
3	200	250	300
4	300	350	400
5	400	450	510
6	510	570	630
7	630	700	770
8	770	840	920
9	920	1000	1080
10	1080	1170	1270
11	1270	1370	1480
12	1480	1600	1720
13	1720	1850	2000
14	2000	2150	2320
15	2320	2500	2700
16	2700	2900	3150
17	3150	3400	3700
18	3700	4000	4400
19	4400	4800	5300
20	5300	5800	6400
21	6400	7000	7700
22	7700	8500	9500
23	9500	10500	12000
24	12000	13500	15500
25	15500	19500	

Source: From Scharf (1970).

(1979) and others, in the course of investigating masking phenomena outside a critical band, referred to the "spreading function," which describes, for steady-state situations, the masking effect of a signal in one critical band on signals in another critical band. It is important to note that the shape of the spreading function varies with level, where the masking abilities of a signal spread farther from the base frequency as the level of the masker is increased. This spreading function is currently thought to be a by-product of the mechanical cochlear filtering mechanism.

Various researchers have used combinations of these classical results and new results (Delgutte, 1990) to generate a number of hypotheses about the masking properties of the arbitrary signal. The more complex of these methods take absolute level into account (Delgutte, 1990) and are thus difficult to apply in coding systems, where the user will want to control the listening level. Less complex versions of masking threshold calculation, making conservative assumptions (absolute threshold low, spreading function narrow), are useful in signal coding. The rest of this chapter will discuss such coding methods.

5. PERCEPTUAL CODING IN THE FREQUENCY DOMAIN

5.1. The Basics

As most of the masking properties of the human auditory system are normally described in the frequency domain, the quantization noise is most easily controlled there. Figure 1 shows a general block diagram that can be used to represent perceptual coders working in the frequency domain. It is interesting to note that similar coders using physiologically related filter banks have also been proposed for image compression (Stockham, 1989; Safranek-Johnston, 1989), **suggesting that a thorough study of perception is important to all coding algorithms.**

The following list describes Figure 1 in some detail:

- A sampled filter bank is used to decompose the input signal into subsampled spectral components. Different possibilities for generating

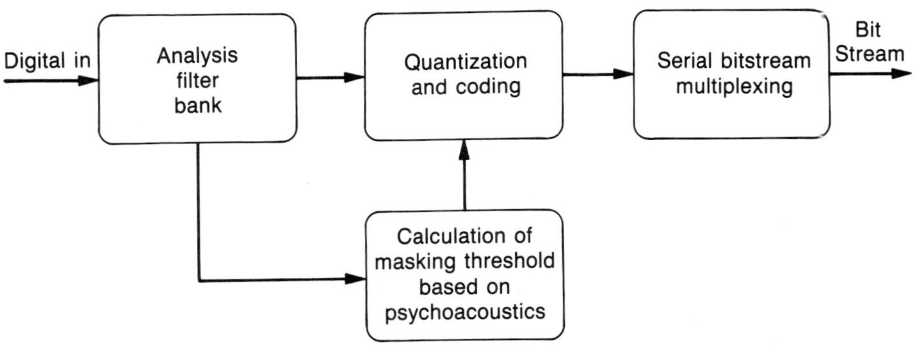

Figure 1 General block diagram of a perceptual coder working in the frequency domain.

an analysis-synthesis model which is invertible and computationally efficient will be discussed below.

- The subsampled spectral components, or separate calculation of frequency content, are used to calculate an estimate of the actual (time-dependent) masking threshold using rules derived from psychoacoustics.
- The spectral components are quantized and coded using the information about the masking threshold. Again there are different possible ways to do this quantization and coding, ranging from simple block companding of groups of values (in frequency or time) to sophisticated analysis-by-synthesis schemes. Either may include additional noiseless compression using variable-length coding. A rate loop is often included to provide fixed-rate coding at a cost of frequent overcoding and/or occasional undercoding.
- In the last step, the quantized and coded spectral components are multiplexed with additional side information to produce a serial bit stream. In some systems, additional error protection, time stamps, editing information, etc. are also added at this point.

5.2. Filter Banks Used in Perceptual Coding in the Frequency Domain

The sampled filter bank used to decompose the time domain input data into components with the desired spectral and time domain resolution is crucial for the overall performance of a perceptual coding system. Several competing design goals have to be met:

- The spectral decomposition and subsampling process should be invertible; i.e. the filter bank should be of the "perfect reconstruction" type. This has the advantage that all noise which is added in the coding/decoding process is generated in the quantizer. Because the noise source is known, it can be controlled so that it is masked by the signal. It is also desirable that the filter bank and its inverting process both maintain a high degree of frequency selectivity in order to make the application of the psychoacoustic threshold simple.
- The analysis system, consisting of the analysis filter bank and the subsampling, should be critically sampled or nearly critically sampled. The number of spectral components per time determines the number of values to be quantized and coded and should be as low as possible. The optimum in this respect is a critically sampled analysis-synthesis system.

- The bandwidth of the filter bank should be equal to or narrower than the width of a narrowest critical band. This makes it easier to control the perceptibility of the quantization noise. Simultaneously, the time window of the filter bank must be controlled to avoid introducing noise components over a time window wide enough that masking constraints are violated. In general, most uniformly spaced filter banks cannot meet both of these constraints because of the 40:1 variation in width of critical bands with frequency.
- Very high frequency resolution is desirable to take advantage of the "transform gain," which is the frequency domain equivalent of the prediction gain. The transform gain is highest for signals with low spectral flatness. These signals need to be coded very accurately because there is very little masking of the quantization noise.
- The time domain resolution of the filter bank should be low enough to mask the spreading in time of the error signal. In a subsampled system the lower bound for the time resolution is given by the Nyquist bound. The quantization noise in the frequency domain is spread in time according to the actual filter bank used. Figure 2 shows an example of this noise spreading. This effect is called pre-echo and will be discussed in more detail later.

5.2.1. Tree Structured Filter Banks

QMF Filter Banks Quadrature mirror filters (QMFs) (for a description see Jayant and Noll, 1984) are half-band filters with the property that alias terms introduced by critical sampling cancel each other in the reconstruction filter bank. In Krasner (1979) a perceptual coder based on a tree structure of QMF is proposed. The tree is designed to give a subband bandwidth which is roughly equivalent to the width of critical bands at a given frequency. The advantage of QMF filter banks is the near-perfect stopband rejection if, for example, a 64-tap prototype filter is used (see Fig. 3).

Generalized QMF, IIR Filter Banks As QMF filters have the disadvantage of only nearly perfect reconstruction and strictly two-band split, other filter banks with the aliasing cancellation property have been developed. For wideband coding of speech and music, generalized QMF banks and IIR filter banks have been used. More details can be found in Sauvagerd (1988), and Jayant and Noll (1984).

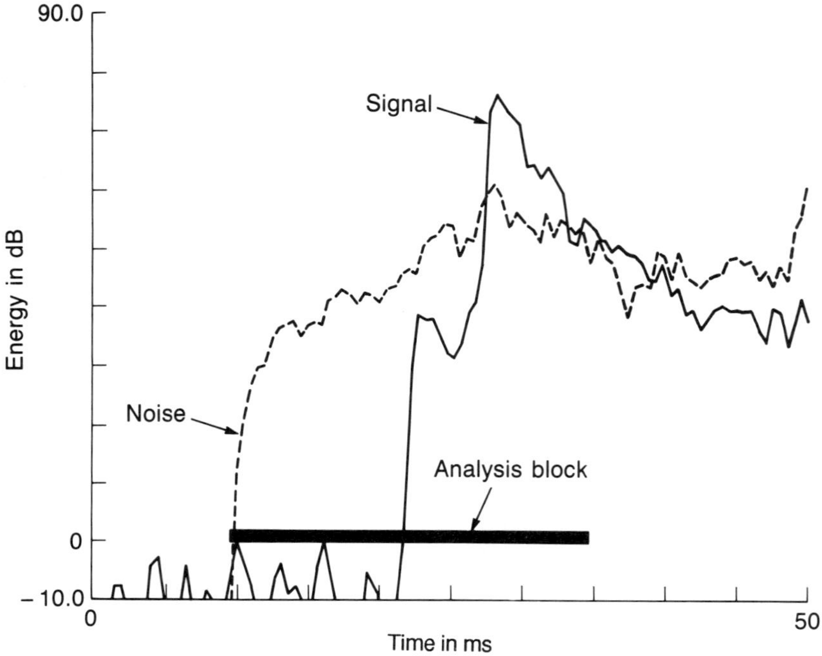

Figure 2 Example of spreading of noise in time. (a) Input signal (attack of a triangle). (b) Analysis window used by an example of a transform coder. (c) Noise signal produced by this transform coder.

5.2.2. *Equally Spaced Filter Banks*

Analysis-synthesis systems using a windowed and overlapped discrete Fourier transform (DFT) or related types of transforms are mathematically equivalent to special equally spaced and subsampled filter banks. The fast algorithms which are available to calculate these transforms (e.g., fast Fourier transform) make these filter banks especially attractive candidates for use in wideband audio coding. Discrete transforms have a double interpretation as equally spaced filter banks and as unitary transforms. The latter interpretation views a discrete transform as an approximation to the Karhunen–Loève transform, which has been shown to be optimum for coding with a mean squared error criterion (Zelinski et al., 1977). From this view, wideband audio coders using a windowed discrete transform are usually called "transform coders". We are more interested in the interpretation of the output values of the transform as spectral components given by a subsampled, equally spaced

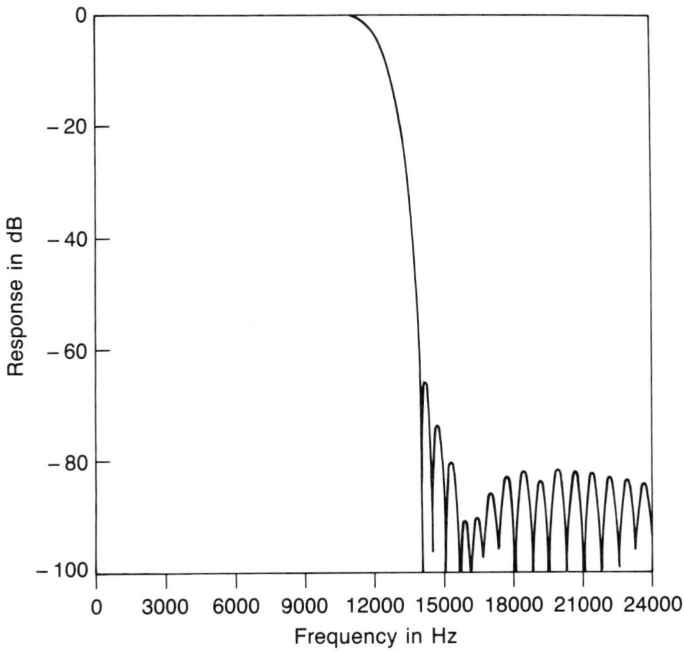

Figure 3 Frequency response of "64d" QMF (Johnston, 1979).

filter bank. The most important quality measure for such a transform scheme is the frequency response of the equivalent band selection filter.

Transform coding as used in speech coding (Zelinski, 1977) uses a DFT, calculated via an FFT, or a discrete cosine transform (DCT). The equivalent frequency response of a transform channel depends on the window used. Commonly used window functions are

$$\text{Rectangular} \qquad w(i) = 1 \quad \text{for all } i$$

$$\text{Sine taper} \qquad w(i) = \sin\left(\pi \frac{i}{N}\right)$$

and their frequency resolution are shown in Figs. 4 and 5. With a rectangular window as used in Brandenburg et al. (1982) the analysis-synthesis system is critically sampled, but the system suffers from poor frequency resolution and blocking noise, which can be regarded as poor filtering and aliasing of the noise introduced into the transformed coefficients.

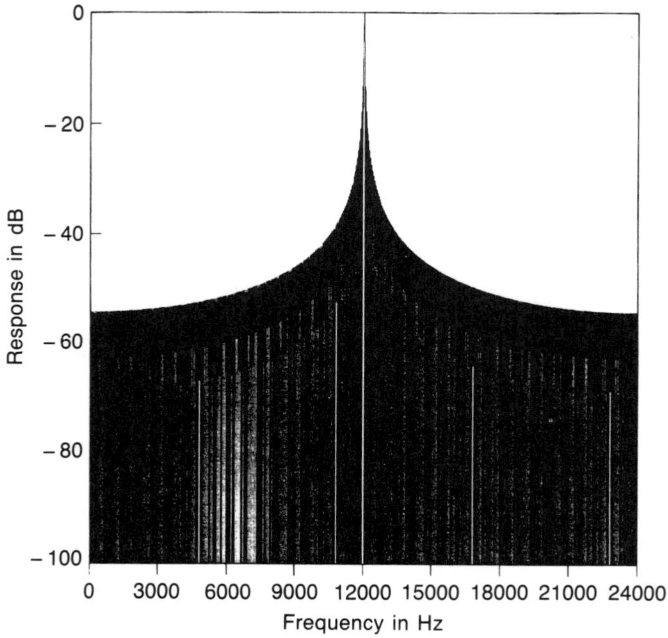

Figure 4 Frequency resolution of rectangular window.

Overlapped windows allow for better frequency response functions but carry the penalty of additional values in the frequency domain which have to be coded given a certain number of time domain samples. The system is not critically sampled. For use in coding, a compromise between large overlap giving better frequency resolution and low overhead due to overlap has to be found. The most widely used compromise is an overlap of 1/16 of the window length; (Johnston, 1988b, 1989).

Modified Discrete Cosine Transform In Princen et al. (1987) a "filterbank design based on time domain alias cancellation" was proposed. This design uses 50% window overlap and downsampling of the spectral values by a factor of 2, thus effectively implementing a critically sampled transform-based filter bank with the frequency response given by a sine window. The alias terms which are introduced by the subsampling in the frequency domain cancel each other in the synthesis stage in the same way that frequency domain alias terms cancel each other in QMFs. The scheme proposed by Princen et al. is known as an overlapping block transform (OBT) and as a modified discrete cosine

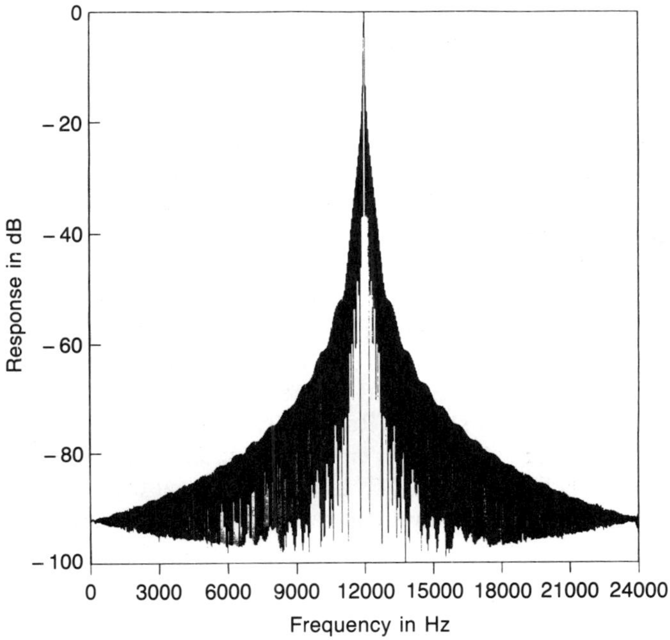

Figure 5 Frequency resolution of 1/16 overlap sine taper window.

transform (MDCT). We will use the name MDCT here because it describes the type of transform best.

The MDCT is basically a DCT with a modified transform kernel and subsampling of the values in the frequency domain. If the subsampling is included from the start in the transform formula, one obtains the following expression. The conversion to the frequency domain is given by

$$X_t(m) = \sum_{k=0}^{N-1} f(k)x_t(k) \cos\left[\frac{\pi}{2N}(2k+1+n)(2m+1)\right]$$

where N is the block length in time, $n = N/2$ is the block length in the frequency domain, $f(k)$, $k = 0, \ldots, N-1$, $x_t(k)$ is the samples of the tth block, and $X_t(m)$ are the frequency domain values. The inverse transform is given by

$$y_t(p) = \sum_{m=0}^{n-1} X_t(m) \cos\left[\frac{\pi}{2N}(2p+1+n)(2m+1)\right]$$

If a window of maximum length is used, a distinctly better selection curve of the filters in the frequency domain compared to an overlap of 1/16 is obtained. Figure 6 shows the frequency response of the equivalent filter at 12 kHz. Several fast algorithms for calculating an MDCT (e.g., via a standard DCT) are known. Because of these fast algorithms, the MDCT seems to be a good compromise between the freedom to choose filter design parameters, as in QMF and generalized QMF filter banks, and the computational efficiency of transform filter banks.

Polyphase Filter Banks Another compromise between unconstrained filter design and computational efficiency is the polyphase filter bank. In a polyphase filter band a transform (e.g., FFT) is used to modulate the input values, which have been multiplied by the coefficients of the low-pass prototype filter. It has been shown (Rothweiler, 1983) that with this structure multiband quadrature filter banks can be designed which have an alias cancellation property like that of QMF filters. The advantage of this technique is the combination of relatively small restrictions in the design of the proto-

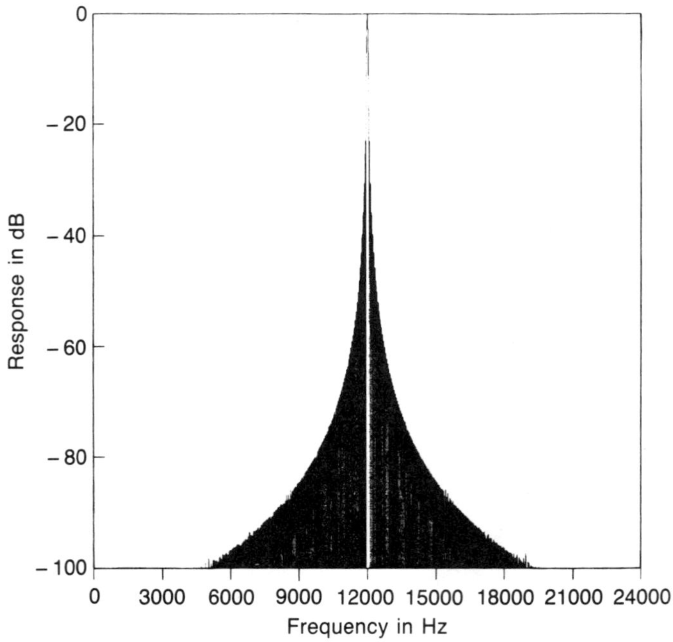

Figure 6 Frequency response of the prototype filter of the MDCT filter bank.

type filter (i.e., very steep filters can be used) and the computational savings of using an FFT to modulate the single bands into the right frequency range. One coder appears to use the window flexibility in order to implement a (fixed across frequency) partial control of noise time-aliasing (pre-echo).

5.2.3. Hybrid Filter Bank

Although equally spaced filter banks are used to get better computational efficiency while maintaining good frequency resolution, the tree-structured approach is closer to the characteristics of the human auditory system. To get the best of both worlds, Brandenburg and Johnston (1990) proposed an analysis-synthesis system that provides enhanced time-frequency resolution and a better fit of the analysis framework to the ear's filtering performance.

Figure 7 shows a block diagram of the basic analysis filter bank. The time domain values are first filtered by a conventional QMF-tree filter bank. This filter bank is used to get four channels with 3 to 12 kHz bandwidth (frequency resolution) and, accordingly, 2- to 8-sample time resolution. The output signals of the QMF-tree are filtered again, this time using a windowed transform to get a computationally efficient filter band with many channels. The window is a sine window, using 50% overlap of the analysis blocks. The combined filter bank has a frequency resolution of 23.4 Hz at low frequencies and 187.5 Hz at high frequencies, with a corresponding difference in time resolution.

5.3. Estimation of the Masking Threshold

Depending on the type of the analysis-synthesis filter bank, the masking threshold will be calculated either from the analysis output or by a separate process using a separate frequency analysis method. In either case, the output of the perceptual calculation process is a hypothetical masking threshold that is used to control the quantization of the spectral components. The output of the threshold calculation is either the threshold itself, used to control an analysis-by-synthesis quantization and coding scheme, a bit allocation which defines the number of bits allocated to each spectral component of each cluster on frequency lines, or a set of required SNR's.

The following subsections describe some options for the threshold estimation which have been used in perceptual coding algorithms.

5.3.1. Fixed Bit Allocation Derived from Worst-Case Considerations

From the bandwidth of the analysis bands and the masking curve as found in Zwicker (1956), worst-case values for the SNR needed in each critical band can be deduced. From this worst-case masking threshold the number of bits

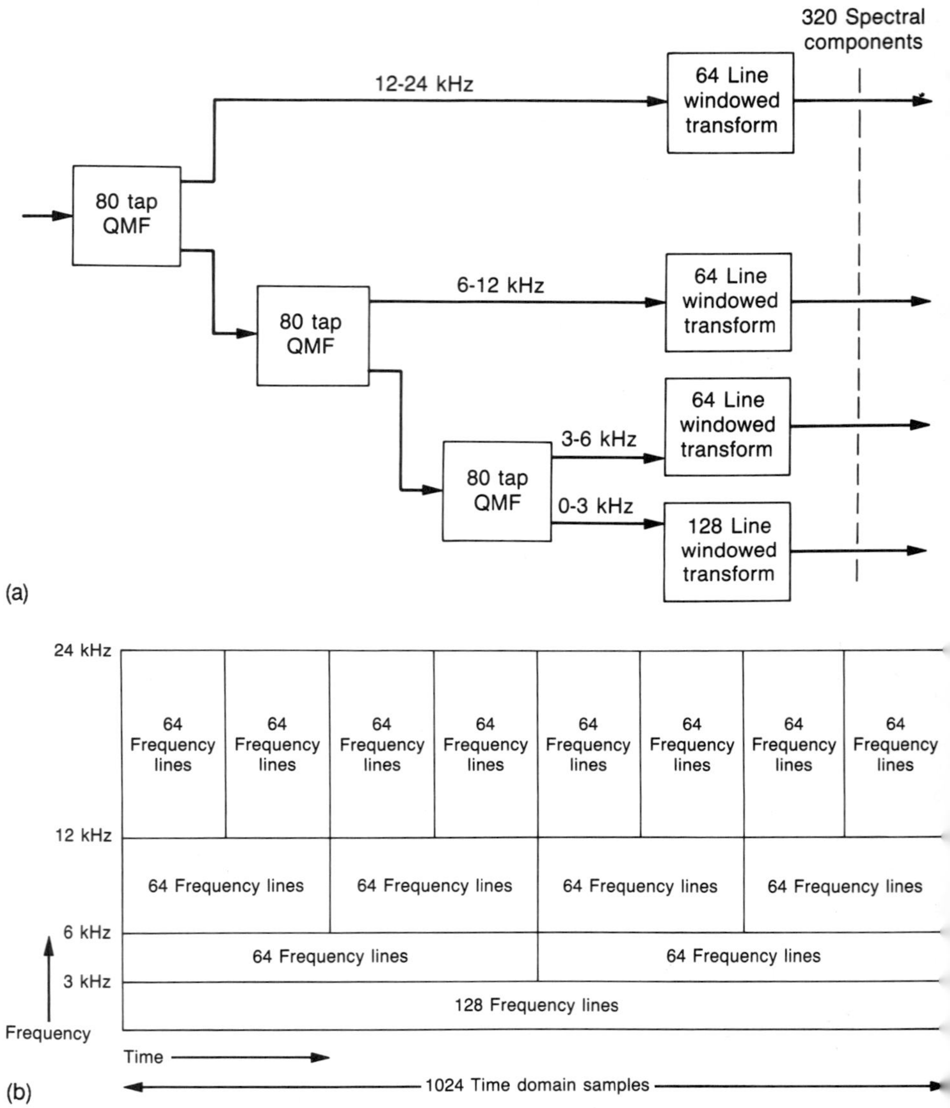

Figure 7 (a) Basic analysis system used in the hybrid coder. (b) Time-frequency breakdown of the hybrid analysis structure.

needed to do a uniform quantization can be estimated if differences due to different amplitude distributions are neglected. The number of bits needed to fulfill this worst-case consideration is low for high frequencies, where the masking curve is found to be very flat (in linear frequency), and is high for low frequencies. A fixed bit allocation works well for most speech and music signals. There are several disadvantages:

1. For many input signals too many bits are used at low frequencies.
2. The masking curve from Zwicker (1956) was measured for narrowband noise and does not show the worst case for tonal signals. This leads to audible quantization noise for tonal signals at high frequencies.
3. If the worst case including tonal signals is used for the fixed bit allocation, a much higher bit rate is needed compared to other bit allocation strategies.

5.3.2. Adaptive Bit Allocation Derived from the Spectral Envelope of the Signal (Low-Complexity ATC)

Another approach to bit allocation for a perceptual coder uses a technique similar to that in adaptive transform coding (ATC) (Zelinski, 1977). ATC uses a bit allocation proportional to the log of the power spectrum, which leads to the minimum noise power. Perceptually modified ATC allocates an minimum amount of bits to the spectral values to fulfill a worst-case criterion for narrowband noise similar to that in fixed bit allocation algorithms. A second part of the available bits is allocated dynamically as in the standard ATC. This has the result that tonal signals (with high energy at the spectral peaks) are coded more accurately, so noise is not audible. With this bit allocation strategy good coding quality can be achieved at bit rates of 3.5 bits/sample or more.

5.3.3. Indirect Application of a Calculated Noise Threshold

The two methods described above approximate the masking threshold without doing an explicit calculation of the masking threshold in each block of data. A closer fit to the correct threshold is possible by using rules derived from psychoacoustics directly on each actual data block, either from the output of the analysis filter bank or from a separate, more accurate estimation of the short-time power spectrum. The power spectrum is then used to estimate the actual masking threshold. To get this estimate, rules for masking of tones by noise and of noise by tones are used. Both the masking within critical bands and the masking from one frequency to spectral distances of more than one critical band are used. For a detailed description of the estimation of the masking threshold see Johnston (1988b). Noise allocation is then done using an estimate of the quantization error of the actual quantizer. For example, if a uniform quantizer is used, the noise allocation uses a standard $\delta^2/12$ estimate of the noise injected by the quantizer. The bit allocation is implicit rather than explicit, and some simple method of partial-bit allocation is used to avoid wasted quantizer levels.

5.3.4. *Direct Implementation of a Calculated Noise Threshold*

In the case of an analysis-by-synthesis technique, again no explicit bit allocation is needed. The analysis-by-synthesis technique is driven by the allowable noise power at each frequency, and bit allocation, as in the Johnston (1988b) example, is implicit rather than explicit. In this case it is not necessary to duplicate the bit allocation algorithm and with it the estimation of the masking threshold in the decoder. The calculation of the masking threshold is done much as in the case of psychoacoustically derived bit allocation. The advantage of direct use of the threshold in an analysis-by-synthesis system is greater flexibility.

5.4. Quantization and Coding

In the analysis-synthesis system, threshold calculation, and bit noise allocation system there are a number of possible ways to do the actual quantization and coding of the filter bank outputs. In principle, all time domain coding methods can be used to quantize and code spectral components; however, the following paragraphs describe some of the more easily applied methods, roughly in order of increasing complexity.

5.4.1. *Quantization with Known Estimate for the Power Spectrum*

In classical ATC (Zelinski and Noll, 1977) as well as in ATC versions for coding of high-quality digital audio (Brandenburg et al., 1982), an adaptive quantizer step size is derived from estimates of the power spectrum. The variance estimates are calculated in both the coder and the decoder from logarithmically quantized subband energies. A simple linear interpolation between the log subband energies is used to get the estimated variance for each spectral line.

5.4.2. *Block Companding with Transmission of Scale Factors*

The main problem of the classical ATC approach to estimating the quantizer step sizes is that quantizer overload is possible. A less efficient but safer method of calculating the quantizer step sizes is to use block companding (block floating point) for quantization and coding of short blocks of frequency components. The maximum of the data within a block is quantized using a logarithmic quantizer characteristic. This value is the exponent of the block floating representation. The values within the block are then normalized relative to the quantized maximum. The normalized values are quantized using a uniform quantizer, resulting in the mantissa values.

An additional advantage of block companding is the improved robustness of the coding scheme to transmission errors. If the exponents of the block floating values (often called "scale factors") are assumed to be transmitted without errors (e.g., by ECC), the errors in the single bands are bound by the maximum value in this band. In all of the frequency domain techniques, transmission errors affect the signal not throughout the spectrum but within a restricted band that is usually within the range of most effective masking. If the bit error rate is small enough, most errors are barely audible.

5.4.3. Entropy Coding Using a Rate Control Loop

Variable-length coding (also called Huffman coding, optimum coding, or entropy coding) is often used to get additional noiseless compression compared to direct storage of transmission of the quantizer outputs. It is well known in the literature (Jayant and Noll, 1984) and is also used in image coding. In coding of wideband digital audio, systems with Huffman coding are used (see Brandenburg, 1987b; Johnston, 1988a) to compensate for the disadvantageous distribution of quantizer outputs. A basic rate control iteration starts with some initial quantization step sizes, which can be derived from the total signal energy and a spectral flatness measure in the case of optimum coding in the frequency domain (OCF) or from the calculated masking threshold in the case of perceptual transform coding (PXFM). Using these step sizes, all the spectral components are quantized. The bits necessary to code the quantized values are counted, usually using table lookup. If the number of bits needed for coding an individual block exceeds the number of bits available on the average for a block, the quantizer step sizes are adjusted and the quantization and counting of necessary bits is done again. If the number of bits required is smaller than that which must be sent, the step sizes are decreased and the bits counted again.

5.4.4. Analysis-by-Synthesis System Using a Distortion Control Loop

As in speech coding, the best coding efficiency can be achieved using analysis-by-synthesis systems. The idea is not to estimate the quantization error from general rules and allocate the noise according to the masking calculation, but to compare the actual quantization noise to the masking threshold directly. This can be done using a distortion control loop similar to the rate control loop explained earlier.

The signal is first quantized using the rate control loop. Then the reconstructed spectral values are calculated. From these and the unquantized (input) spectral components, the actual quantization noise in computed by simply subtracting the reconstructed from the original values. It is assumed that the quantization process is phase preserving. The quantization noise is compared to the estimated masking threshold ("allowed noise"). If the actual noise exceeds this threshold for some part of the spectrum, the corresponding spectral components have to be quantized with a smaller quantization step size, giving less quantization error. This is done, for example, by scaling the frequency values in subblocks corresponding to critical bands (Brandenburg, 1987b). Every time this scaling has been done, the quantization has to be done again and will itself be adjusted to get the bit count below the available number of bits. Because finer quantization of some critical bands will lead to a somewhat bigger step size throughout the spectrum (under the constraint of a limited number of bits available for coding), this iteration loop will not necessarily converge. Special precautions are necessary to catch runaway conditions. On the other hand, after each loop through the distortion control, a valid combination of scale factors and quantized values is available. This combination is perhaps not optimal in the sense of best noise shaping, but it can be used as an approximation under the condition of limited computing time available.

A different example of distortion control is shown in PXFM, where the shape of the masking threshold is preserved and the entire masking threshold is moved up or down to meet rate requirements. This approach does not require a second loop for distortion control, but it may not provide the best choice of overcoding in regions of easily coded signal.

6. EXAMPLES OF ACTUAL CODERS

In recent years, numerous algorithms for high-quality digital audio coding systems have been proposed. The following descriptions are intended as examples, not as a thorough description of the state of the art. More examples of perceptual audio coders can be found in the references. The examples given below are ordered according to the complexity of the concepts used. This is not necessarily the order of increasing complexity of computation.

6.1. Subband ADPCM Coders

The first example of a wideband digital audio coder is an extension of the G722 7-kHz subband adaptive differentiated pulse code modulation (ADPCM) coder as standardized by CCITT. Subband ADPCM coders try to find a compromise between low-complexity time domain (full band) coder and higher-complexity perceptual frequency domain coders. Subband ADPCM coders for more than speech bandwidth date back to at least 1979; e.g., see Johnston (1979), where an algorithm anticipating G7222 is described.

In Smyth et al. (1988) a detailed description of two versions of a subband ADPCM coder for coding wide-bandwidth (15-kHz) signals can be found. The scheme proposed for higher quality uses 64-tap QMF filters as the analysis-synthesis filter banks. Three such filters are used in the standard tree structure to give a subband bandwidth of 4 kHz. Each band uses an 8-pole backward block adaptive predictor and a backward adaptive Laplacian quantizer. Because of the use of backward adaptation, no side information needs to be transmitted. The bit allocation between the four bands again is varied backward adaptively according to the short-term energy distribution of the prediction error across the four bands. According to Smith et al. the prediction gain can approach 50 dB for some signals. Very similar subband ADPCM coders have been proposed by Vörös (1988).

The subband ADPCM coders are somewhat different from the other examples because they do not use rules derived from psychoacoustics to shape the noise. They are not perceptual coders, but the combined effect of the subband partitioning, even if the bands are much broader than critical bands, and the use of prediction to enhance the redundancy removal for correlated (tonelike) signals does some noise shaping.

6.2. MUSICAM

MUSICAM (Masking-pattern Universal Sub-band Integrated Coding And Multiplexing), as described in Stoll and Dehery (1990), uses a 32-channel equal-bandwidth filter bank. The filter bank is implemented using a polyphase filter network (Rothweiler, 1983). Twelve successive samples in each band are quantized and coded jointly using block floating point (block companding). The exponent values for the blocks are called scale factors and are transmitted as side information. MUSICAM is a full perceptual coder in that the bit allocation for the frequency components is calculated using rules derived from psychoacoustics. This is done using a separate FFT which is carried out in parallel with the analysis filter bank. From the FFT output the masking threshold is calculated. For each of the 32 bands of the main filter bank, the minimum of the threshold within the band is used to determine the bit allocation for that band. The result is a varying bit allocation. According to Stoll and Dehery (1990), a bit rate in excess of 110 kb/s is seldom required by this bit allocation. The bit rate can be chosen not to exceed a predefined value. Current target bit rates are 64, 96 or 128, and 192 kb/s.

6.3. MSC

The coding scheme presented in Krahe (1986) under the name multiple spectral coding (MSC) is the first transform coder which is specifically adapted to the results from psychoacoustics. The analysis-synthesis filter bank used is an FFT filter bank with 1/16 overlap and a block length of 1024 samples. In contrast to the classical ATC and other transform coding schemes, the frequency components are processed further by magnitude and phase. The magnitude and, if the magnitude is not zero, phase value are initially quantized in a coarse quantization with a 2-bit quantizer. Maximum values in bands partitioned according to critical bands are used to control the quantizers. After the coarse quantization, a bit allocation is done according to rules derived from psychoacoustics. The quantization done according to this bit allocation is called fine quantization. According to Krahe (1986), a bit rate of 2.5 to 3 bits/sample is needed at a sampling frequency of 44.1 kHz to achieve a "very good" quality of the reconstructed signal. Preprocessing in the time domain was developed to limit pre-echo effects in this coder.

6.4. Low-Complexity ATC

The fourth example of a high-quality audio coder is the low-complexity ATC (LC-ATC) as described in Brandenburg et al. (1988a). This algorithm was developed from classical ATC as used for coding of speech signals (Zelinski and Noll, 1977). The algorithm was modified in such a way that adaptation to the characteristics of the human ear and a simplification were achieved at the same time. The price paid for these advantages was a slightly higher bit rate compared to previous versions of ATC. In terms of the terminology defined above, LC-ATC can be described as follows:

- An FFT or MDCT with 512 spectral components is used as the analysis and synthesis filter bank.
- The spectral components are grouped with a frequency resolution according to the critical bands at lower frequencies and better resolution at higher frequencies, totaling 46 bands.
- The bit allocation is derived from the logarithmically quantized maxima in each band. These maxima form an envelope of the estimated power spectrum and are transmitted to the receiver as side information. The bit allocation is done using a fixed minimum bit count and distributing additional bits proportional to the log signal energy.
- Block floating point (block companding) is used for quantization of the spectral components. As the maxima and the number of bits per band are derived from the estimated spectral envelope, no side infor-

mation in addition to the band energy maxima is necessary to decode the signal at the receiver.

6.5. OCF/PXFM

In Brandenburg (1987b, 1988, 1989) and Johnston (1988b, 1989) perceptual transform coders have been proposed under the names optimum coding in the frequency domain (OCF) and perceptual transform coding (PXFM). Because they are very similar, they will be described together.

OCF and PXFM use overlapped transforms as analysis-synthesis filter banks. PXFM uses 1/16 overlap FFT with a block length of 2048 samples. OCF in the newer form (Brandenburg, 1988) uses an MDCT as described earlier with a 512-sample block length and a 1024-sample sine window. The calculation of the masking threshold in both cases is done via rules derived from psychoacoustics. In PXFM, several steps are involved in calculating the masking threshold:

1. Critical band analysis of the signal, derived from the power spectrum
2. Tonality estimation, derived from the power spectrum or local-in-frequency measures
3. Application of the spreading function to the critical band spectrum
4. Calculation of the spread masking threshold using tonality information
5. Accounting for absolute thresholds
6. Relating the spread masking threshold to the critical band masking threshold

The quantization and coding in OCF look for the best possible coding for each block using two iteration loops. An outer iteration loop (distortion loop) exists in order to constrain the quantization noise to a level below the masking threshold for each critical band. The actual noise is measured by comparing original and quantized frequency domain data. If the noise exceeds the masking threshold in at least one critical band, the quantizer step size is reduced for the spectral values in this band and quantization and coding is repeated. This analysis-by-synthesis method is both efficient and flexible. It is flexible in the sense that the rules used to calculate the masking threshold or the workings of the distortion loop can be changed without any changes in the transmission format of the decoder.

The inner iteration loop (rate loop) can be found in both OCF and PXFM. The quantizer and variable-length code iteratively try to compress the frequency values to less than the "number of available bits." The spectral values are quantized. In OCF this is done using a nonuniform quantizer which by itself does some noise shaping. The quantized values are coded with a Huffman coder (sometimes called "optimum coder", hence the name OCF). As Huffman codes use a variable number of bits for each value, the number of bits needed to code a block of data is not known in advance. Buffer feedback quantizer control is used to keep the number of bits below the allowed number. If the number of bits needed to code a block of quantized spectral data exceeds the number of available bits, the quantizer step size is increased. Increased quantizer step size leads to fewer quantization intervals, which are coded with fewer bits. This is done iteratively until the number of bits needed is low enough. On the decoder side only the Huffman-coded data and the quantizer step size are needed to reconstruct the spectral values. In PXFM, the rate loop adjusts rate in both directions, as the noise shape is explicitly applied rather than generated by analysis by synthesis.

An additional step of bit rate reduction is done in the stereo version of PXFM as described in Johnston (1989). The channels of hi-fi stereo transmission usually do not have much correlation in the time domain. This does not hold for the frequency domain. Considerable correlation can be found when the magnitude values of the channels are compared. This correlation can be used to considerable advantage in the coding algorithm.

When loudspeakers are used, Jetzt (1979) suggests that in almost all cases one may use the sum of the power spectra to approximate the sound pressure level at each ear; hence one can combine the threshold calculation for the two channels as well as gain from the combination of the two channels' spectra.

Furthermore, by coding the L − R and L + R channels rather than L and R (left and right), one can gather additional coding advantage, partially because of the gain in the frequency domain correlation and partially because the noise inserted in the signal is localized either at both ears or in the center, depending on where most of the signal energy is localized, gaining some advantage in the case where masking level differences might cause additional perceived noise (Moore, 1982).

Typically, stereo coding provides enough gain to allow the PXFM coder to gain about half of one channel's bit rate without any noticeable loss of quality. Work on OCF stereo coding is proceeding.

7. TIME DOMAIN EFFECTS

As described above, forward masking and backward masking permit coding of audio signals using schemes in which the noise pertaining to acoustical events is distributed over a certain time interval. This applies to all frequency domain algorithms, regardless whether they are of the block-by-block signal processing (transform coder) or direct filtering (subband coder) type. Whereas forward masking permits the design to be restricted to simultaneous masking for all commonly used block lengths, the time constraints in the case of backward masking have to be observed closely.

If one fails to observe the constraints of backward masking, so-called pre-echoes are obtained. Figure 2 illustrates this effect. In the case of signals with very steeply rising amplitudes such as triangles or castanets, distinct pre-echoes may be audible in all frequency domain coders. They result, in general, from the use of analysis-synthesis systems that have "noncausual" noise spread. There are several methods for overcoming this effect:

- Design of the analysis-synthesis system

A trivial solution for complying with the backward masking time is to limit the time spread of the error signal by choosing a suitable block length of filter response. If a maximum forward masking time of 4 ms is assumed, the maximum block length is 176 samples for a sampling frequency of 44.1 kHz. The next power of 2 as used for transforms is 128. Compared to the currently used block length of 512 (OCF) to 2048 (PXFM), considerably poorer frequency domain resolution results, with a corresponding penalty in compression rate for stationary signals.

- Preprocessing in the time domain

Some authors propose increasing the noise margin of the signal in the time intervals in which pre-echoes can be expected. The basic idea is to control the level in subblocks in the time domain. If automatic gain control is used in individual blocks within a data block, the total noise energy is no longer uniform but is distributed according to the scalings.

- Adaptive control of the analysis-synthesis system

Analysis-synthesis systems have been described which allow "on-the-fly" switching between different block lengths. This technique—actually changing the time and frequency resolution according to the characteristics of the signal—can improve the total coding efficiency considerably.

- Inclusion of backward masking in the calculation of the masking threshold (pre-echo control)

In coders using an explicit calculation of the masking threshold the pre-echo problem can be solved by incorporation backward masking into the estimation of the masking threshold. This in done in the latest version of the combined result of PXFM and OCF (see Brandenburg, 1989). A pre-echo is inaudible if the level of the noise signal lies below the theoretical masking threshold of the block in the absence of the signal that causes the pre-echo (e.g., triangle attack). This theoretical threshold cannot be calculated from the current signal block, but it can be estimated from the threshold of the previous block. This means, for instance, that for every block the threshold of the previous block multiplied by a constant is used as an upper limit for the threshold of the current block.

In the case of sharp attacks this leads to much lower thresholds. To code the signal with higher accuracy more bits are needed. Therefore, this method can be used only in conjunction with coding algorithms which allow a locally varying bit rate.

8. LEVELS OF QUALITY, RESULTS

8.1. Evaluation of Quality

In order to develop, evaluate, and compare low-bit-rate coding schemes, one has to be able to measure the quality of the coded and reconstructed signal. A quantitative expression for the distortion of the original signal is desirable. The method most often used in speech coding is SNR. Methods based on listening tests, especially mean opinion score (MOS), are also used.

The results of SNR measures rarely correlate with the corresponding subjective results of listening tests. A signal can show a mean square distortion of 13 dB while maintaining transparent perceptual quality. On the other hand, quantizing noise with a measured SNR in excess of 60 dB may be agonizingly audible. When comparing coders, an algorithm which shows worse SNR measurements than another for all pieces of music tested may nevertheless have a much higher subjective rating. This discrepancy results from the fact that SNR does not take masking into effect. Methods other than absolute SNR must be used for testing of perceptually driven coders.

8.1.1. Listening Tests

A subjective evaluation by the largest possible number of listeners is regarded as an ultimate test for the quality of a data reduction scheme. Even in this case, an objective evaluation of the results is desired.

The human organism is subject to influence regarding sound perception; therefore blind tests with careful statistical evaluation are necessary when high-quality coding schemes are to be compared. The audible distortions even at 64 kb/s (for 48-kHz sampling frequency) are less than in the case of toll-quality speech coding. It is not sufficient to obtain quality ratings averaged over a number of test subjects. Because of the different grades of experience and hearing acuity between different test subjects and the need to fulfill the quality requirements of the most critical listeners, the testing must reveal problems which are audible only to a few "golden ears"; hence the tests must provide a statistically confident measure for each subject, preferably on each stimulus.

Subjective tests of high-quality digital audio coding systems are done using MOS (CCIR scales) or using AB or ABX double-blind tests and testing for nonrandom identification of A/B of A/B/X.

8.1.2. Objective Criteria

Several attempts have been made to develop an objective measurement system which incorporates models of hearing as known from psychoacoustics (Schroeder et al., 1979). The noise-to-mask ratio (NMR), as proposed in Brandenburg (1987a), has been used to compare several coding algorithms. The results have been reasonably close to the results of listening tests. However, one main obstacle to the widespread use of such measurement algorithms remains: as the models derived from psychoacoustics used in both advanced perceptual coders and measurement algorithms are far from complete, the measurement results are only as good as the model used. Naturally, if the same model is used in both a measurement system and a perceptual coder, as in the case for NMR and OCF, the coder will nearly always score best in tests done using the same measurement system. A universally accepted set of rules for deriving the masking threshold from the signal has to be found before tests based on measurement systems like NMR can be used to compare perceptual coders developed by different research groups.

8.1.3. Quality Levels

Even though there are no universally accepted comparison results for the different coders which have been introduced, we will reference some results about the quality achieved. It is useful to define levels of quality and give results about the bit rates needed to achieve them.

"Better Than FM" Quality The lowest level of quality claimed by wideband digital audio coders is sometimes called hi-fi quality (see Brandenburg, 1989b) or high quality. The definition used by the authors is "better than FM radio, but worse than CD quality." For some pieces of music there are audible differences between original and reconstructed signals, but on the average the quality is still much better than can be found on standard FM broadcast radio. For most pieces of music it is very near or equivalent to CD quality.

Transparent Quality The goal for "transparent" quality is that the original and coded and reconstructed signals are totally indistinguishable. Not even very subtle differences between original and reconstructed signals should be audible. This has to be true for all possible input signals (including artificial test signals) and all test subjects. The test for transparency of an algorithm is best done via AB or ABX double-blind listening tests.

Studio Quality This is required for systems in which the coded and reconstructed signals must be edited, filtered, or coded again. For this quality level a margin between the quantization noise and the noise threshold is needed. Special precautions have to be taken to prevent loss of quality in the case of editing, mixing, equalization, etc. Often some parts of the psychoacoustic model (i.e., spreading function) are turned off for this quality level.

8.2. Results for Bit Rate Versus Quality

The following claims of bit rate versus quality can be found for some of the algorithms described in this chapter. For OCF (Brandenburg, 1989), MUSICAM (Stoll and Dehery, 1990), and PXFM (Johnston, 1989), approximately the same claims are made for bit rate versus quality:

- At 128 to 192 kb/s per mono channel, studio quality is available.
- At 96 to 128 kb/s per mono channel, transparent coding for known sources is claimed.
- At 64 kb/s per mono channel or 128 kb/s for a stereo pair, better than FM quality is achieved. This bit rate is important because the transmission of a stereo signal over a 2B ISDN basic access channel is possible. For PXFM, the gain for a stereo pair over a dual-mono coder is said to be significant.

For simpler algorithms such as LC-ATC or subband ADPCM the quality level at 128 kb/s per mono channel is usually described as "high quality." At the moment, no independent test results for all the algorithms are available.

9. FUTURE RESEARCH

High-quality coding of digital audio is moving from a research topic to the first applications (Eberlein et al., 1990). This does not mean that there is no room for further improvement. The ultimate aim of wideband audio coding is transparent coding at the lowest possible bit rate, with robustness to source and channel noise. Transparent coding means no audible difference between the original signal and the coded and reconstructed signal for all input signals and for all possible listeners, including the so-called golden ears. Of course, achievement of this goal by definition cannot be proved. On the other hand, transparent quality to known sources has been reached. One research topic for the future is the search for ever new "critical signals." Once systems using high-quality bit rate reduction schemes are incorporated into consumer products, critical consumers will participate in the hunt for signals that might possibly break existing algorithms and necessitate new or improved algorithms.

In more technical terms, we believe that the optimum trade-off between time resolution and frequency resolution has not yet been found. Better analysis-synthesis systems, probably incorporating different frequency resolutions at different frequencies and dynamically varying systems, would be a possible step. A deeper understanding of the basic mathematics behind traditional filter banks using transforms, QMF and pseudo-QMF filters, and wavelet theory, and theoretical unification of the various filter banks, should provide some understanding of the issues involved.

We see further progress necessary not only in the signal processing aspects of wideband digital audio coding but also in psychoacoustics. Up to now, only relatively simple models from psychoacoustics have been implemented in coding algorithms. Some questions which are elementary for coding applications cannot yet be answered by common knowledge from psychoacoustics. Both implementation of advanced models from psychophysics and psychoacoustics into coding and new research in these fields triggered by questions arising from coding are necessary for advancement of the field of wideband coding of high-quality digital audio.

REFERENCES

Berger, T. (1971). *Rate Distortion Theory*. Prentice-Hall, Englewood Cliffs, N.J.

Brandenburg, K. (1987a). Evaluation of quality for audio encoding at low bit rates. 82nd AES Convention, Preprint 2433.

Brandenburg, K. (1987b). OCF—a new coding algorithm for high quality sound signals. *Proc. ICASSP 1987*, pp. 141–144.

Brandenburg, K. (1988a). High quality sound coding at 2.5 bit/sample. 84th AES Convention, Paris, Preprint 2582.

Brandenburg, K. (1989). Ein Beitrag zu den Verfahren und der Qualitaetsbeurteilung fuer hochwertige Musikcodierung, Ph.D. thesis, University of Erlangen, Erlangen, Federal Republic of Germany.

Brandenburg, K., and Johnston, J. D. (1990). Second generation perceptual audio coding: The hybrid coder. 88th AES Convention, Prepring 2937(H3).

Brandenburg, K., and Schramm, H. (1983). A 16 bit transform coder for real time processing of sound signals. In *Signal Processing*. II: *Theories and Applications*, H. W. Schuessler, ed. North-Holland, Amsterdam, pp. 359–362.

Brandenburg, K., and Seitzer, K. (1988b). OCF: Coding high quality audio with data rates of 64 kbit/sec. 85th AES Convention, Los Angeles, Preprint 2723.

Brandenburg, K., Langenbucher, G. G., Schramm, H., and Seitzer, D. (1982). A digital signal processor for real time adaptive transform coding of audio signals up to 20 kHz bandwidth. *Proc. IEEE Int. Conf. Circuits and Computers*, pp. 474–477.

Brandenburg, K., Kapust, R., Seitzer, D., Eberlein, E., Gerhaeuser, H., Popp, H., and Schott, H. (1988a). Fast signal processor encodes 48 kHz/16 bit audio into 3 bit in real time, *Proc. ICASSP 1988*, pp. 2528–2531.

Brandenburg, K., Kapust, R., Seitzer, D., Eberlein, E., Gerhaeuser, H., Grill, B., Kraegeloh, S., Popp, H., and Schott, H. (1988b). Low bit rate codecs for audio signals—implementation in real time. 85th AES Convention, Los Angeles, Preprint 2707.

Caine, C. R., English, A. R., and O'Cleary, J. B. H. (1980). NICAM 3: Near-instantaneously companded digital transmission system for high quality sound programmes. *Radio Electron. Eng. 50*:519.

Chen, J.-H. (1989). A robust low-delay CELP speech coder at 16 kbit/s. *Proc. IEEE Global Communications Conf.* November, pp. 1237–1241.

Crochiere, R. E., and Rabiner, L. R. (1983). *Multirate Digital Signal Processing.* Prentice-Hall, Englewood Cliffs, N.J.

Delgutte, B. (1990). *J. Acoust. Soc. Am.* 791–813.

Eberlein, E., Gerhaeuser, H., and Kraegeloh, S. (1990). Fraunhofer Arbeitsgruppe f Integierte Schal., *Wetterkrenz 13*:D-8520.

Fletcher, N. (1940). Auditory patterns, *Rev. Modern Phys.* 12:47–65

Greenwood, D. D. (1961). Critical bandwidth and frequency coordinates of the basilar membrane, *J. Acoust. Soc. Amer. 33*:1344–1356.

Hellman, R. P. (1972). Asymmetry of masking between noise and tone. *Percept. Psychophys.* 241–246.

Jakubowski, H., and Twietmeyer, H. (1985). The German DBS sound-broadcasting system: Audio coding methods. 77th AES Convention, Hamburg.

Jayant, N. and Noll, P. (1984). *Digital Coding of Waveforms*: *Principles and Applications to Speech and Video*. Prentice-Hall, Englewood Cliffs, N.J.

Jetzt, J. J. (1979). Critical distance measurements on rooms from the sound energy spectrum response. *J Acoust. Soc. Am. 65*:1204–1211.

Johnston, J. D. (1988a). Estimation of perceptual entropy using noise masking criteria. *Proc. ICASSP*, pp. 2524–2527.

Johnston, J. D. (1988b). Transform coding of audio signals using perceptual noise criteria. *IEEE J. Selected Areas Commun. 6*:314–323.

Johnston, J. D. (1989). Perceptual transform coding of wideband stereo signals. *ICASSP-89 Record*, pp. 1993–1996.

Johnston, J. D., and Crochiere, R. E. (1979). An all digital commentary grade subband coder. *J. Audio Eng. Soc. 27*:855–865.

Johnston, J. D., and Goodman, D. J. (1979a). Digital transmission of commentary grade (7 kHz) audio at 56 or 64 kb/s. *Proc. ICASSP*, pp. 442–444.

Krahe, D. (1986). Ein Verfahren zur Datenreduktion bei digitalen Audiosignalen unter Ausnutzung psychoakustischer Phaenomene. *Rundfunktech. Mitt. 30*:117 (in German).

Krasner, M. A. (1979). Digital encoding of speech and audio signals based on the perceptual requirements of the auditory system. Technical Report 535, MIT Lincoln Laboratory, Lexington, Mass.

Krasner, M. A. (1980). The critical band coder—digital encoding of speech signals based on the perceptual requirements of the auditory system. *Proc. ICASSP*, pp. 327–331.

Lovis, G. (1986) Untersuchungen zu datenreduzierenden Codierverfahren fuer diskrete Musiksignale. Ph.D. thesis

Moore, B. J. (1982). *An Introduction to the Psychology of Hearing*. Academic Press, London, pp. 168–170.

Nguygen, C. (1985). Implementation of digital audio systems for television multichannel sound. 78th AES Convention, Anaheim, Preprint 2226.

Nishiguchi, M., Akagiri, K., and Suzuki, T. (1986). A new audio bit rate reduction system for the CD-I format. 81st AES Convention, Los Angeles, Preprint 2375.

Plenge, G. (1985). The German DBS digital sound-broadcasting system: General description, 77th AES Convention, Hamburg, Preprint 2203.

Princen, J., Johnson, A., and Bradley, A. (1987). Subband transform coding using filter bank designs based on time domain aliasing cancellation. *Proc. ICASSP*, pp. 2161–2164.

Richardson, E. B., and Jayant, N. S. (1986). Subband coding with adaptive prediction for 56 kbits/s audio. *IEEE Trans. Acoust. Speech Signal Process. ASSP-34*:691–696.

Rothweiler, J. H. (1983). Polyphase quadrature filters-A new sub-band coding technique, Proc. ICASSP pp. 1280–1283.

Safranek, F. J., and Johnston, J. D. (1989). A Perceptually tuned sub-band image coder with image dependent quantization and post-quantization data compression. *ICASSP-1989 Record*, pp. 1945–1948.

Sauvagerd, U. (1988). Bit-rate reduction for high quality audio signals using floating-point wave digital filters. *Proc. ISCAS*, pp. 2031–2934.

Scharf, B. (1970). Critical bands. In *Foundations of Modern Auditory Theory*. (J. Tobias, ed.) Academic, New York, pp. 159–202.

Schroeder, E. F., and Voessing, W. (1986). High quality digital audio encoding with 3.0 bits/sample using adaptive transform coding. 80th AES Convention, Montreaux, Preprint 2321.

Schroeder, M. R., Atal, B. S., and Hall, J. L. (1979). Optimizing digital speech coders by exploiting masking properties of the human ear. *J. Acoust. Soc. Am. 66*:1647–1652.

Seitzer, D., Brandenburg, K., Kapust, R., Eberlein, E., Gerhaeuser, H., Popp, H., and Schott, H. (1988). Real-time implementation of low complexity adaptive transform coding. 84th AES Convention, Paris, Preprint 2581.

Smyth, S. M. F., and McCanny, J. V. (1988). 4 bit hi-fi: High quality music coding for ISDN and broadcasting applications. *Proc. ICASSP*, pp. 2532–2535.
Stautner, J. P., and Horowitz, D. M. (1986). Efficient data reduction for digital audio using a digital filter array. *Proc. ICASSP*, pp. 13–16.
Stautner, J. P., and Jayasimba, S. (1987). A digital signal processing algorithm for digital audio disk recording. 83rd AES Convention, New York, Preprint 2516.
Stockham, T. G., and Xie, Z. 1989. Optimal previsualized image vector quantization. *ICASSP-1989 Record*, pp. 1763–1966.
Stoll, G., and Dehery, F. (1990). High quality audio bit-rate reduction family for different applications. *Proc. IEEE Int. Conf. of Comm.*, pp. 937–941.
Stoll, G., Link, M., and Theile, G. (1988). Masking-pattern adapted subband coding: Use of the dynamic bit-rate margin. 84th AES Convention, Paris, Preprint 2585.
Theile, G., Stoll, G., and Link, M. (1987). Low bit-rate coding of high quality audio signals. 82nd AES Convention, Preprint 2432.
Theile, H., Stoll, G., and Link, M. (1988). Low bit-rate coding of high-quality audio signals—an introduction to the MASCAM system. EBU Review—Technical, No. 230.
Todd, C., and Gundry, K. (1984). A digital audio system for broadcast and prerecorded media. 75th AES Convention, Paris,
Tribolet, J. M., and Crochiere, R. E. (1979). Frequency domain coding of speech. *IEEE Trans. Acoust. Speech Signal process. ASSP-27*:512–530.
Vörös, P. (1988). High quality sound coding within 2 × 64 kb/s using spontaneous dynamic bit-allocation, Proc. *ICASSP*, pp. 2536–2539.
Zwicker, E. (1982). Psychoakustik. Springer-Verlag, New York (in German).
Zwicker, E., and Feldtkeller, R. (1956). Das Ohr als Nachrichtenempfaenger. Hirzel, Stuttgart (in German).
Zwisloski, J. (1965). Analysis of some auditory characteristics. *In* Handbook of mathematical psychology, (R. D. Luce, R. R. Bush and E. Galanter, eds.) Wiley, New York.

5

Predictive Coding of Speech Using Analysis-by-Synthesis Techniques

PETER KROON and BISHNU S. ATAL AT&T Bell Laboratories, Murray Hill, New Jersey

1. INTRODUCTION

The use of a digital representation of speech signals in communication systems has many advantages in terms of quality and reliability. Consequently, many existing analog networks, such as the telephone system, are being converted into digital networks. In a digital system, the analog speech signal is converted into a digital representation by an *encoder*, sent over a digital channel, and converted back into an analog signal by a *decoder*. A simple digital representation is *pulse code modulation* (PCM) (Jayant and Noll, 1984), which is obtained by bandlimiting and sampling the speech signal at its Nyquist frequency and representing the amplitudes of the samples by a finite number of values (*quantization*). For telephone applications, the analog speech signal is band limited to 4 kHz and sampled at 8 kHz. The resulting sample amplitudes are adequately represented with 13 bits/sample. By logarithmically compressing the input samples, this number is reduced to 8 bits/sample without compromising speech quality. This compression technique, referred to as μ-law or A-law PCM (Jayant and Noll, 1984), is in widespread use and results in a *bit rate* of 64 kb/s.

In many applications, such as mobile communications and voice storage applications, the available channel capacity is limited. Even for applications that do not have a hard limit on the available channel capacity, it is often economical to find signal representations that use the least number of bits. The bit rate can be reduced by representing the sample amplitudes with fewer bits. A coarse quantization increases the quantization error, since speech samples can take a wide range of possible values (that is, a large *dynamic range*). Speech samples, however, exhibit a strong correlation from sample to sample, and for voiced speech signals, such as vowels, there is a strong correlation between adjacent periods. Removal of these correlations reduces the dynamic range of the sample amplitudes, allowing more efficient quantization. *Linear predictive coding* (Atal, 1982; Jayant and Noll, 1984) is an efficient procedure for removing these correlations. The linear prediction method predicts the current input sample values based on previously reconstructed values. The difference between the current value and its predicted value is quantized and transmitted. This process is reversed at the decoder to reconstruct the speech signal. The procedure of generating the difference signal and using this signal, or its quantized version, to produce the reconstructed signal can be described as linear filtering with filters based on the predictors. The input is filtered through an *analysis* filter, and the difference signal forms the excitation for the *synthesis* filter. Since the speech signal characteristics vary over time, the prediction is made more efficient by adjusting the predictors periodically. This procedure is referred to as *adaptive prediction* and is, for example, used in 32 kb/s adaptive differential PCM (ADPCM) (Jayant and Noll, 1984).

At lower bit rates only a few bits per sample are available for encoding the excitation signal, and the key issue in designing coders for these rates is finding bit-efficient representations. Instead of trying to find a representation that matches the difference signal, it is more efficient to find an excitation sequence that for the given synthesis filter produces a signal that is close to the original input signal. Each sample of the excitation signal affects many samples of reconstructed speech because of the recursive structure of the synthesis filters. Therefore the choice of the excitation signal is made by measuring its effect on the reconstructed speech over more than one sample. In other words, the decision about the best quantized representation is not made instantaneously but is delayed for an interval that includes several samples. This approach is called *delayed decision coding*. Since this decision depends on the error between the original and reconstructed speech, it requires synthesis during analysis, and the procedure is generally referred to as *analysis-by-synthesis adaptive predictive coding*.

Section 2 of this chapter discusses the general structure of analysis-by-synthesis coders, followed by a description of commonly used excitation representations. Section 3 addresses issues related to the choice of an error criterion. Section 4 gives more details on the procedures used for finding the predictors and excitation signals. The effect of the various coder parameters on the quality of the reconstructed speech signal is discussed in Section 5. Techniques that are usually incorporated in analysis-by-synthesis coders are described in Section 6. Other coder issues besides speech quality, such as complexity, delay, and robustness, are discussed in Section 7. The coding techniques described in this chapter are applied in national and international standard coders, which are described in Section 8. Section 9 concludes this chapter with a brief summary and a short discussion of the advantages and disadvantages of analysis-by-synthesis coders.

2. ANALYSIS-BY-SYNTHESIS CODER STRUCTURE

Figure 1 shows an example of an analysis-by-synthesis coder. The reconstructed speech is produced by filtering the signal produced by the excitation generator through both a long-term predictor synthesis filter $1/P(z)$ and a short-term predictor synthesis filter $1/A(z)$. The excitation signal is found by minimizing the weighted mean-squared error over several samples, where the error signal is obtained by filtering the difference between the original and reconstructed signals through a weighting filter $W(z)$. Both short- and

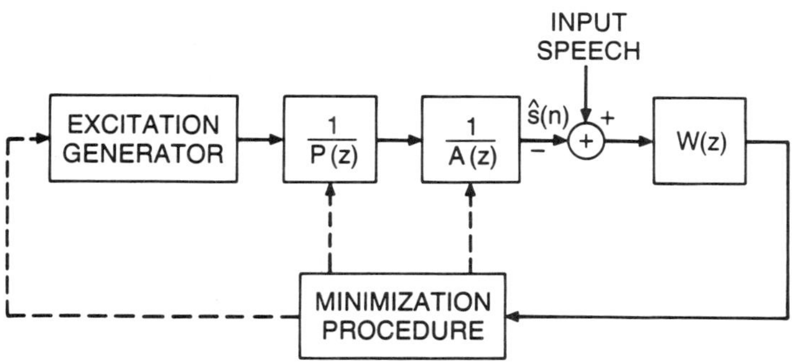

Figure 1 Example of an analysis-by-synthesis adaptive predictive coding scheme. The short- and long-term predictors are represented by $A(z)$ and $P(z)$. $W(z)$ is the error weighting filter.

long-term predictors are adapted over time, but the short-term predictor is usually adapted at a slower rate than the long-term predictor and excitation parameters. Note that since the analysis involves synthesis, the description of the analysis procedure completely describes the decoder. Although most discussions in this chapter are related to the configuration shown in Fig. 1, many variations are possible. For example, the long-term predictor can be omitted, or the locations of the long-term and short-term predictors can be interchanged.

The filter $1/A(z)$ models the short-term correlations (spectral envelope) in the speech signal and has the form

$$\frac{1}{A(z)} = \frac{1}{1 - \sum_{k=1}^{p} a_k z^{-k}} \tag{1}$$

where a_k are the short-term predictor coefficients and p is the order of the filter. Typical values of p are between 8 and 16. The predictor coefficients are determined from the speech signal using linear prediction techniques as described in Markel and Gray (1976) and Rabiner and Schafer (1978). The filter coefficients are adapted in time, with rates varying from 30 to as high as 500 times a second.

The filter $1/P(z)$ models the long-term correlations (spectral fine structure) in the speech signal and has the form

$$\frac{1}{P(z)} = \frac{1}{1 - \sum_{k=-(q-1)/2}^{(q-1)/2} b_k z^{-(M+k)}}, \qquad q = 1, 3, \ldots \tag{2}$$

where M is the delay in samples and b_k are the long-term prediction coefficients. The value of M corresponds to the number of samples of a delay in the range from 2 to 20 ms. For periodic signals, this delay would correspond to a pitch period (or possibly an integral number of pitch periods). The delay is random for nonperiodic signals. The delay M and the coefficients b_k are determined either from the speech signal or from the difference signal after removing the short-term correlations (Atal, 1982). Typically one to three predictor coefficients are used, and their values are adapted in time at rates varying from 50 to 200 times a second. Increasing the order of the predictor results in better prediction, but more bits are needed to encode the additional coefficients. For periodic segments, the main role of multiple coefficients is to provide interpolation between the samples if the period does not correspond to an integer number of samples. Instead of a higher-order predictor, one could use a first-order predictor with a noninteger delay, which allows more efficient quantization of the predictor parameters (Kroon and Atal, 1990).

To achieve a low overall bit rate, each frame of excitation samples has to be represented such that the average number of bits per sample is small. This can be achieved, for example, by keeping only a few non-zero-valued samples and using efficient quantization schemes for the many zero-valued samples. The zero samples could be obtained, for example, by center clipping the prediction error signal (Atal, 1982). More effective is to use an analysis-by-synthesis approach to find the nonzero excitation samples. The *multipulse excitation* coder (Atal and Remde, 1982) represents the excitation as a sequence of pulses located at nonuniformly spaced intervals. The excitation analysis procedure has to determine both amplitudes and positions of the

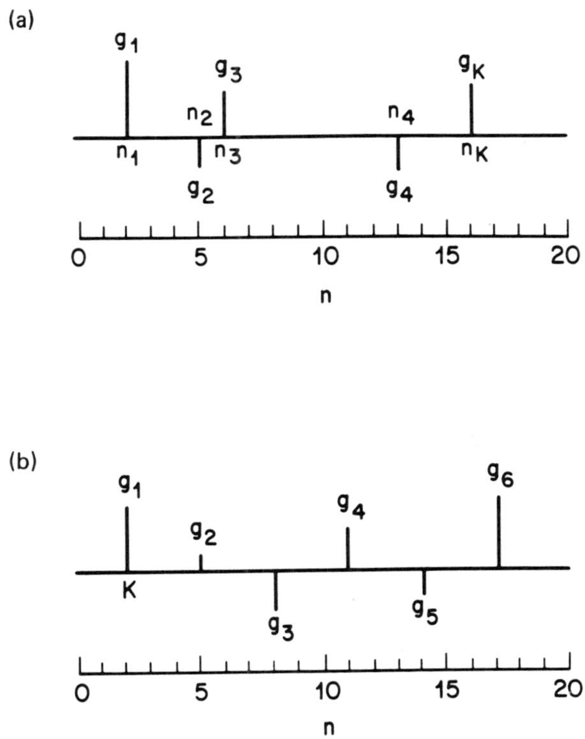

Figure 2 Examples of parametric representations using a few nonzero samples. (a) The multipulse excitation (MDE) is characterized by a set of pulses with positions n_i and amplitudes g_i. (b) The regular-pulse excitation (RPE) is characterized by a set of equidistance pulses with amplitudes g_i and position K of the first pulse.

pulses. Finding these parameters all at once is a difficult problem and simpler procedures such as determining locations and amplitudes one pulse at a time (Atal and Remde, 1982; Singhal and Atal, 1989), are used. The number of pulses required for an acceptable speech quality varies from 4 to 6 pulses per 5 ms. For each pulse both amplitude and location have to be transmitted, requiring about 7 to 8 bits per pulse (Kroon and Deprettere, 1988). In the *regular-pulse excitation* coder (Kroon et al., 1986) the excitation signal is represented by a set of uniformly spaced pulses (typically 10 pulses per 5 ms). The position of the first pulse within a frame and the amplitudes of the pulses are determined during the encoding procedure. For a given position of the first pulse, all other pulse positions are known and the amplitudes are found by solving a set of linear equations. For each excitation frame only the amplitudes and the position of the first pulse have to be transmitted, requiring about 3 to 4 bits per pulse. Figure 2 shows an example of the difference between multipulse (MPE) and regular-pulse excitation (RPE) sequences.

Code-excited or *vector excited* coders use another approach to reduce the number of bits per sample. Here, both encoder and decoder store a collection of C possible sequences of length L in a codebook. The excitation for each frame is described completely by the index to an appropriate vector in the codebook. In the configuration of Fig. 1, this index is found by an exhaustive search over all possible codebook vectors and selection of the one that produces the smallest error between the original and reconstructed signals. The transmission of the index requires $(\log_2 C)/L$ bits/sample, and typically values are in the range from 0.2 to 2 bits/sample.

3. SELECTION OF AN ERROR CRITERION

In the coder shown in Fig. 1, the coder parameters are selected such that the error between the original and reconstructed signals is minimized. A commonly used error criterion is the mean-squared error. Minimizing a mean squared error results in a quantization noise that has the same energy at all the frequencies of the input signal. Reducing the bit rate increases the noise energy and makes the noise more audible. The loudness of a low-level noise is strongly affected by the presence of a louder speech signal. This phenomenon is called *masking* (Moore, 1982), and frequency masking experiments as described in (Schroeder et al. 1979) showed that the hearing system has only a limited capability to detect noise in the frequency bands in which the speech signal has high energy (as happens in formant regions). To make use of the masking effect, the quantization noise has to be distributed in relation to the speech power over the different frequency bands. The desired distribution is illustrated in Fig. 3. Spectral shaping of the noise can be achieved by minimizing a weighted error in the block diagram of Fig. 1. Note that this weighting procedure does not affect the bit rate or the complexity of the synthesis procedure. It increases only the complexity of the encoder. The use of noise shaping procedures was thoroughly investigated for adaptive predictive coding (APC) schemes (Atal and Schroeder 1979, Makhoul and Berouti, 1979), and a suitable error weighting filter is

$$W(z) = \frac{F(z/\gamma_1)}{F(z/\gamma_2)} \tag{3}$$
$$= \frac{1 - \sum_{k=1}^{p} f_k \gamma_1^k z^{-k}}{1 - \sum_{k=1}^{p} f_k \gamma_2^k z^{-k}}, \qquad 0 \le \gamma_2 \le \gamma_1 \le 1$$

where $F(z)$ is a short-time predictor similar to (1). For most applications $F(z)$ is chosen to be identical to $A(z)$. The parameters γ_1 and γ_2 are fractions between 0 and 1 that control the energy of the error in the formant regions. Note that decreasing γ increases the bandwidth of the zeros of $F(z/\gamma)$. Noise

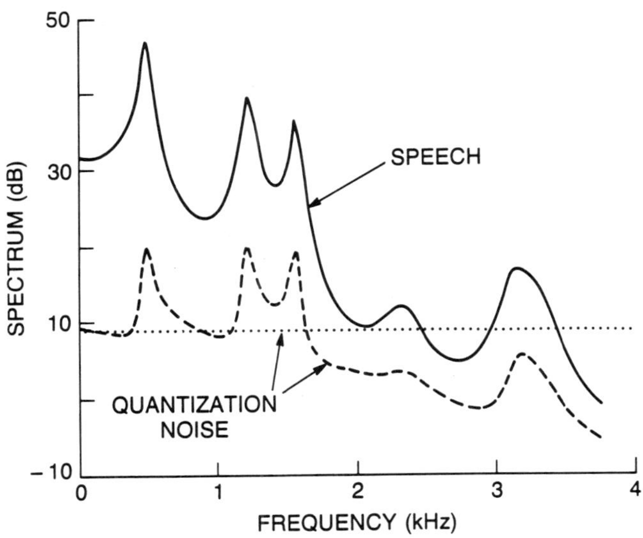

Figure 3 Illustration of the use of noise shaping to reduce the loudness of coding noise. The solid line shows the spectral envelope of the speech signal. The dotted straight line represents coding noise with a flat spectrum; the dashed line represents the same amount of noise shaped according to the speech spectrum. The shaped noise is less audible than the white noise.

shaping increases the mean squared error between the original and reconstructed speech, resulting in a reduction in segmental signal-to-noise ratio (SNR). The values of γ_1 and γ_2 must be determined by suitable listening tests (see Section 8).

At lower bit rates, where the quantization error can be large, the effect of error weighting is less noticeable. One reason is that the level of noise is so high that despite the shaping the noise remains audible. Moreover, the assumption that the quantization noise has a flat spectrum is no longer valid, which makes the results of the shaping procedure less predictable. The proposed noise weighting procedure is based on models of masking that were obtained from psychoacoustical experiments with simple stationary signals, such as single tones and white noise. A speech signal is a much more complex signal, with many harmonic components, whose relative amplitudes and phases vary with time. Therefore, the masking effects for speech signals will frequently be different from the results obtained by extrapolating the psychoacoustical data. Frequency masking is only one aspect of applying perceptual criteria. The proposed weighting does not take into account the spectral fine structure of the signal or temporal masking of one signal event by another (forward and backward masking). Besides lack of knowledge about the importance of these issues, there is, at this moment, no simple mechanism for including them in a distance criterion. A better understanding of the hearing mechanism may lead to more appropriate error criteria (Atal, 1989).

4. DETERMINING THE CODER PARAMETERS

4.1. Predictor Parameters

The predictor parameters are determined from a segment of samples from either the input signal or the reconstructed signal. For the short-term predictor a segment with a length between 10 and 30 ms is used. Hence, estimation of the predictor parameters from the input signal introduces a delay. This delay can be reduced by estimating the predictor from the previously reconstructed signal. This procedure, referred to as *backward-adaptive prediction*, has the additional advantage that it does not require explicit transmission of the predictor coefficients. *Forward-adaptive prediction* refers to the procedure in which the predictor coefficients are determined from the input speech and transmitted explicitly. Figure 4 shows the conceptual difference between forward and backward prediction.

For the configuration shown in Fig. 1 it is common first to determine the short-term predictor from either the input or the reconstructed speech samples and then to use the analysis-by-synthesis procedure to determine both

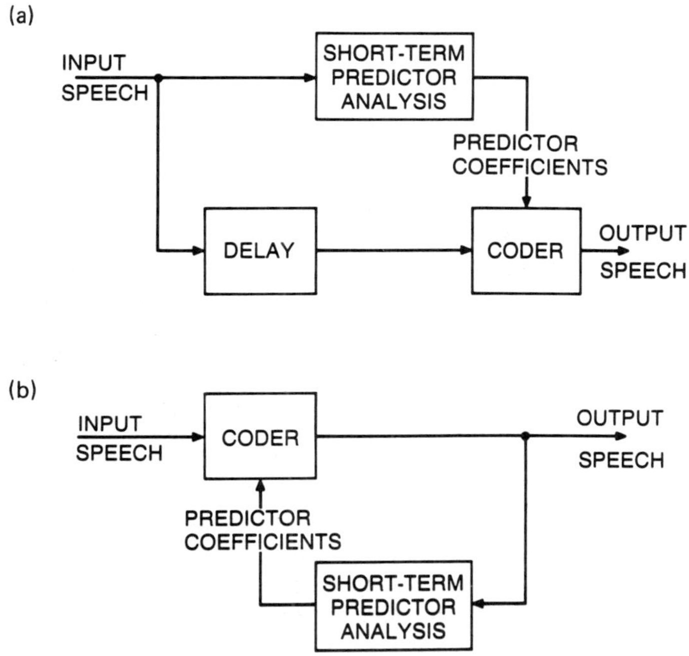

Figure 4 Illustration of the difference between (a) forward and (b) backward prediction. The coder for both cases is identical and is based on the block diagram of Fig. 1.

the long-term predictor and excitation signals. To make a distinction between parameters determined inside and those determined outside the analysis-by-synthesis loop, it is common to refer to these procedures as *closed loop* and *open-loop analysis*, respectively. The open-loop approach is not as effective as the closed-loop approach. For example, forward-adaptive predictors are determined from the input speech, whereas during synthesis the predictors use samples of reconstructed speech. This mismatch causes the performance of the predictors to deteriorate. It is possible to estimate or readjust the short-term predictor in an analysis-by-synthesis scheme, but this task is complicated. For long-term predictors this closed-loop procedure is feasible and often used.

The long-term predictor parameters are determined as follows. For a given delay M, the corresponding predictor coefficients are found by solving a system of linear equations. The best value of the delay M is found by doing an exhaustive search over all its possible values. For a long-term predictor with noninteger delay $M + l/D$ the predictor is expressed as

$$P(z) = 1 - b \sum_{k=0}^{q-1} p_l(k) z^{-(M-I+k)} \tag{4}$$

where p_l is the interpolation filter of length q for the fraction l/D and I represents the delay of the interpolation filter (Kroon and Atal, 1990). The advantage of the noninteger delay predictor is that it is easier to produce smooth delay contours during periodic speech segments. This allows for interpolation of the delay parameters and leads to more efficient encoding.

For closed-loop long-term predictor analysis, a problem arises when the delay is less than the excitation frame size, since the output of the predictor is now a function of the excitation signal that has not yet been determined. Since the predictor parameters b and M are quantized, they have a finite number of values, and the problem is avoided by doing an exhaustive search over all possible combinations of b, M, and the excitation parameters. A simpler approach is to redefine the structure of the long-term predictor and use an adaptive codebook (Kleijn et al., 1988). In this approach the long-term synthesis filter is replaced by a codebook that contains repeated versions of the previous excitation. The resulting vectors are searched and scaled in a way similar to that for a fixed codebook (see Section 4.2). The advantage of this procedure is that the closed-loop analysis is well defined and even allows for simultaneous optimization of adaptive and fixed codebook vectors. The difference between the traditional prediction filter formulation and the adaptive codebook representation is noticeable only for delays less than the frame duration. For segments in which the envelope of the periodic signal is rapidly expanding or decaying, the long-term predictor formulation is more accurate, especially when larger frame sizes are used.

4.2. Excitation Parameters

This section describes in more detail the procedures for finding the optimal excitation sequence. The parameters to be found depend on the excitation structure, such as pulse positions and amplitudes for the multipulse and regular-pulse coders or a codebook sequence for the code-excited coder. The search procedures for the different types of excitation are similar (Kroon and Deprettere, 1988), and this section describes only the codebook excitation search procedure.

The description is based on Fig. 5, which is a rearrangement of Fig. 1. It is assumed that the kth excitation vector $\mathbf{x}_k$ is scaled with a gain coefficient

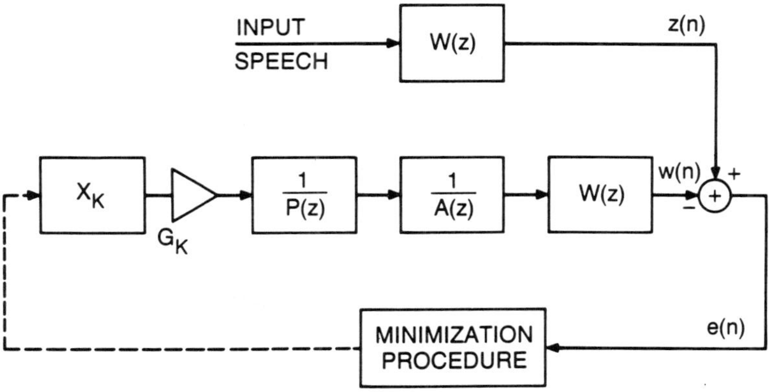

Figure 5 Alternative representation of Fig. 1 used for the determination of the excitation parameters. This arrangement reduces the complexity of the analysis, especially if $W(z) = A(z)/A(z/\gamma)$.

g_k. The signal $w(n)$ at the output of the cascade of the synthesis filters $1/P(z)$ and $1/A(z)$ and the weighting filter $W(z)$ is a function of both the input signal and the memory contents of the filters $1/A(z)$ and $W(z)$. Since the filters are linear, this output can be written as the sum of the memory response (zero input) and the scaled input vector convolved with the impulse response of the three filters. Using vector notation, this can be written as

$$\mathbf{w}_k = \mathbf{w}_0 + g_k \mathbf{y}_k \tag{5}$$

where $\mathbf{w}_0$ represents the memory response and $\mathbf{y}_k$ is the convolved excitation vector. Representing the weighted speech signal by $\mathbf{z}$, the error vector $\mathbf{e}_k$ corresponding to the excitation vector $\mathbf{x}_k$ is written as

$$\mathbf{e}_k = \mathbf{z} - \mathbf{w}_k = \mathbf{z}_0 - g_k \mathbf{y}_k \tag{6}$$

with

$$\mathbf{z}_0 = \mathbf{z} - \mathbf{w}_0 \tag{7}$$

The objective is to minimize the squared error:

$$E_k = \mathbf{e}_k' \mathbf{e}_k \tag{8}$$

where the prime denotes transpose. The gain coefficient g_k is derived from (6) and (8) by setting the derivative of (8) with respect to g equal to zero. Hence,

$$g_k = \frac{\mathbf{z}_0' \mathbf{y}_k}{\mathbf{y}_k' \mathbf{y}_k} \tag{9}$$

From (6), (8), and (9) the following expression for the error is obtained:

$$E_k = \mathbf{z}_0'\mathbf{z}_0 - \frac{(\mathbf{z}_0'\mathbf{y}_k)^2}{\mathbf{y}_k'\mathbf{y}_k} \tag{10}$$

Equation (10) is evaluated for each codebook vector $\mathbf{x}_k$, and the vector minimizing this expression is selected. Once the vector is known, the gain coefficient g_k is obtained from (9).

5. MORE ON PREDICTORS AND EXCITATION

This section discusses more specific aspects of the different coder parameters, such as their impact on the overall speech quality and their quantization properties. The coder performance is related to the number of excitation parameters and the rate at which they are adjusted. To achieve good coder performance, a compromise between the number of parameters and their update rate has to be found. If the parameters are updated frequently, memoryless encoding is inefficient since the temporal correlation between the parameters increases. Hence it is more efficient to apply backward prediction on these parameters and transmit the difference values. The choice of the various coder parameters is even more complicated because of the complex interaction between predictors and parameters. Although if enough bits are available the excitation can compensate for poor prediction, experience shows that it is better to do the best prediction possible and use the excitation to represent the nonpredictable elements of the speech signal.

5.1. Short-Term Predictor

The short-term predictor parameters are determined from either the input speech (forward estimation) or the reconstructed speech signal (backward estimation) using linear prediction analysis. Commonly used procedures are the autocorrelation and covariance techniques (Markel and Gray, 1976). The number of predictor coefficients is typically between 8 and 16. For backward-adaptive systems a higher order can be used because no bits are required to transmit the coefficients. The short-term predictor can be interpreted as modeling the vocal tract. Since the vocal tract is a physical system that cannot change instantaneously, a low update rate, between 50 and 100 times per second, is sufficient. For forward-adaptive predictive systems a typical choice is a 20-ms interval between updates and a 10th-order predictor.

If forward estimation is used, the predictor coefficients have to be transmitted and a low-distortion representation of the short-term predictor coefficients is crucial. As a rule, a log-spectral distortion below 1 dB is inaudible. Although many quantization procedures achieve an average distortion around 1 dB, it is important to reduce the peak values because they are the most perceptible. Efficient scalar quantization schemes require between 30 and 36 bits for a 10th-order predictor (Soong and Juang, 1990). By using differential coding this number can be reduced, but the coder becomes more sensitive to bit errors. Vector quantization schemes (see Chapter 2) will need about 24 to 28 bits to represent a 10th-order predictor (Paliwal and Atal, 1990) but are complex and sensitive to transmission errors.

The predictor coefficients are updated every 10 to 20 ms. This sudden change can introduce undesired transients in the reconstructed speech. To reduce these transients it is beneficial to interpolate the short-term filter parameters. To do this, the parameters must first be converted to a domain suitable for interpolation. A main requirement is that the interpolated coefficients produce stable synthesis filters. Linear interpolations in the log-area domain (Rabiner and Schafer, 1978) or line spectral frequency domain (Itakura, 1975) for every excitation subframe (4 to 8 ms) are good choices.

5.2. Long-Term Predictor

The long-term prediction filter models primarily the periodicity in the excitation signal. The multipulse and regular-pulse excitations are flexible enough to generate this periodicity without a long-term synthesis filter. For high-pitched voices, however, not enough pulses are available to model the excitation signal and the use of a long-term prediction filter improves the speech quality. The long-term predictor is more important in code-excited coders because a small codebook cannot easily generate the periodic components in the excitation signal. A suitable range of the delay is from 2.5 to 18 ms, which corresponds to a pitch range between 56 and 400 Hz and covers the variation in pitch for a wide range of speakers. The delay is expressed as a multiple of the sampling interval, which leads to a coarse quantization of shorter delays. A finer quantization, as used in noninteger delay predictors, leads to better performance (Kroon and Atal, 1990).

The value of M is quantized using 6 to 9 bits. The predictor coefficients b_k are treated as vectors and encoded by vector quantization techniques. Typically, about 2 bits per coefficient are needed (Atal, 1982). Since the long-term predictor coefficients are updated frequently, say every 5 ms, the resulting bit rate due to the long-term predictor is high. Reducing the update rate degrades the performance. Backward estimation of the long-term predictor is not commonly used because of its sensitivity to transmission errors. To reduce the number of bits needed to encode the delay M differential encoding can be used, but that reduces the flexibility in adaptation and results in degraded performance.

5.3. Excitation

The excitation parameters (pulses or codebook vectors) are determined for a frame of samples such that the (weighted) mean squared error over the same frame is minimized. The regular- and multipulse excitation representations can be considered special cases of the code-excited coder, and in the remainder of this chapter we will discuss only the latter.

The codebook can be generated with representative samples of the underlying signal or signals that have statistics similar to the signal to be coded, such as Gaussian noise for residual signals (Atal, 1982; Schroeder and Atal, 1985). Comparable results are obtained with other populations such as uniform and Laplace distributions (Kroon and Atal, 1987). Center clipping these codebooks (clip level $\pm$ 1.2 for unit-variance code vectors) (Davidson and Gersho, 1986) improves the performance and leads to fast search procedures (see Section 7). Codebooks can be trained with a modified generalized Lloyd procedure (Gersho and Zeger, 1990) to take into account a speech-dependent error criterion. The advantage of trained codebooks is that they improve the performance. The disadvantage is that for a mismatch between input data and training data the performance is worse than with a random codebook. Moreover, it is more difficult to incorporate fast search techniques because of lack of structure in the codebook.

All shape and energy variations in the excitation can be taken care of by using a larger codebook. A more efficient method is to normalize the codebook vectors to unit variance and apply a separate gain coefficient. The sign of this coefficient is considered to be part of the shape and is derived directly from (9). It effectively doubles the size of the codebook without increasing the complexity. The absolute value of the gain is quantized with a nonuniform quantizer and needs typically 3 to 5 bits. For short excitation frames the temporal correlation between absolute values of the gain is high, and it is more efficient to use backward-adaptive predictive encoding and quantize the difference signal. If the pitch predictor is represented as an adaptive codebook, the excitation is a linear combination of vectors from both codebooks (see Fig. 6). The corresponding gain coefficients of these codebooks can be efficiently encoded by using an overall gain coefficient and a value that specifies the relative contribution of each vector to the overall energy of the excitation signal (Gerson and Jasiuk, 1990).

The codebooks in Fig. 6 are searched sequentially; that is, the adaptive codebook is searched first, followed by a search of the fixed codebook. The

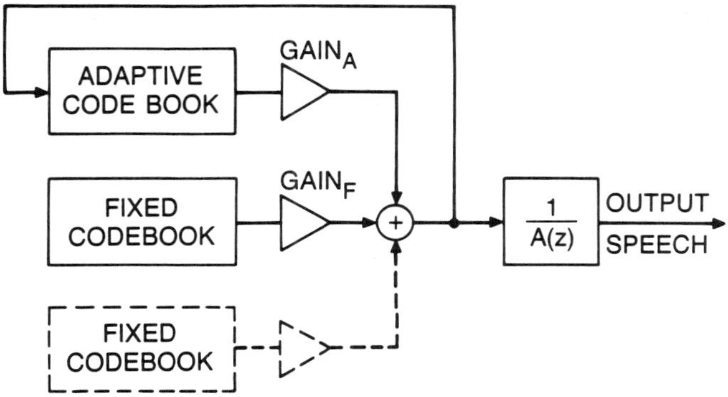

Figure 6 By using an adaptive codebook representation, the input to the short-term synthesis filter is a linear combination of both adaptive and fixed codebooks. Multiple fixed codebooks can be used to reduce the complexity.

use of multiple fixed codebooks reduces the complexity but is less efficient in terms of performance. Once the codebook indices have been found, the optimal gain coefficients are found by solving a set of linear equations. Although this procedure guarantees the lowest weighted mean squared error, it will not always produce the lowest perceived distortion. For example, the consistency introduced by the adaptive codebook during voiced segments is perceptually important. Hence, a sequential approach that starts with the adaptive codebook and solves sequentially for the remaining codebooks can produce results similar to or better than those of the simultaneous approach. Moreover, it is not always beneficial to select the best fixed codebook vector based on a weighted mean-squared error criterion (Kroon and Atal, 1988; Shoham, 1990), and it is sometimes better to make the contribution of the fixed codebooks a function of the performance of the adaptive codebook or long-term predictor.

The excitation is typically determined for short frames. Since the long-term predictor uses previous excitation signals, the selected excitation signal and long-term predictor coefficients will affect future choices. To take this into account, a delayed decision approach over multiple excitation frames can be used. For each frame a few best choices are maintained, and in the next frame, different excitation signals are selected given different initial conditions. Based on the combined results, the best path is maintained, and the process is repeated for the next frame (Mano and Moriya, 1990). This procedure increases both complexity and delay but leads to improved performance.

6. IMPROVING PERFORMANCE WITHOUT INCREASING BIT RATE

Analysis-by-synthesis coders are used to realize a variety of coders operating at different bit rates. The speech quality depends on the bit rate and varies from fair at 4 kb/s to excellent at 16 kb/s. This section describes techniques that are used either to further improve the performance at a given bit rate or to further reduce the bit rate without compromising speech quality.

For communication applications it is desirable to maintain a constant bit rate and the encoding is done frame by frame, while keeping the total number of bits per frame constant. For some applications, such as voice mail, a *variable bit rate* is allowed, resulting in much lower average bit rates (Francesco et al., 1990).

Speech enhancement techniques, such as *postfiltering* (Ramamoorthy et al., 1988), are used to reduce the loudness of the coding noise. The general idea of postfiltering is to emphasize the (perceptually more important) spectral peaks in the speech signal, thereby increasing the subjective signal-to-noise ratio, at the expense of signal distortion. Postfilters can be based on both short- and long-term predictors (Kroon and Deprettere, 1988). Figure 7 gives an example of a postfilter arrangement. The filter parameters are derived from the predictor parameters that were either explicitly transmitted or derived from the reconstructed speech signal. The global spectral postfilter has a form similar to (4) but with $\gamma_1 < \gamma_2$, usually combined with a filter that equalizes the spectral tilt (Chen and Gersho, 1987). The postfilter based on the long-term predictor operates directly on the excitation signal (Gerson and Jasiuk, 1990) to minimize the transient effects introduced by changing the filter parameters. The described postfiltering approach changes the energy of the filtered signal, and an energy normalization procedure is necessary to restore the energy envelope of the input signal. A disadvantage of postfiltering is that it degrades the performance of tandemed coders (Cox et al.,

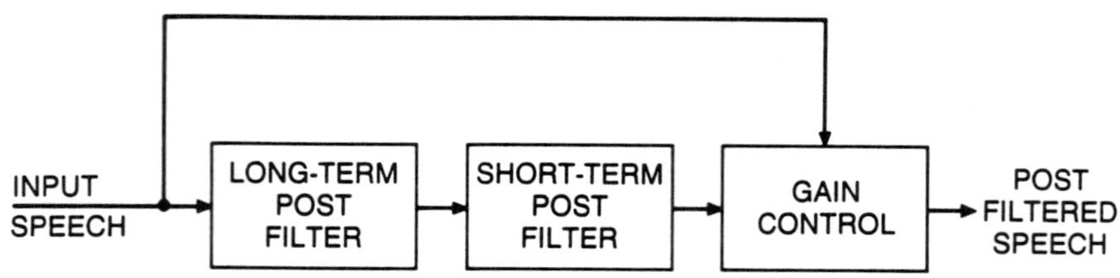

Figure 7 General arrangement for a postfilter based on long-term and short-term correlations.

1989). Moreover, postfiltering distorts the signal and therefore degrades the performance for nonspeech inputs such as modem signals.

The constraint of using fixed bit allocation and fixed frame sizes reduces the coding efficiency. For example, in segments that correspond to plosives, the long-term predictor contributes little to the signal approximation, and by disabling the predictor extra bits are available for the excitation. These bits can be used to increase the codebook size in code-excited coders or change the number of pulses or pulse spacing in the multi- and regular-pulse coders. Examples of the use of these *adaptive bit-allocation* procedures in analysis-by-synthesis schemes can be found in Kroon and Atal (1988), Wang and Gersho (1990), and Taniguchi and Unagami (1989). Although adaptive bit allocation within one or more excitation frames is conceptually sound, improvement in quality while maintaining the bit rate has not been achieved. One problem with adaptive bit allocation is the lack of an appropriate error criterion, since the least-squares error criterion used to determine the excitation is not appropriate for longer frames.

7. SOME PRACTICAL CONSIDERATIONS

For any coder used in a communication system, an important requirement besides speech quality is that of *robustness*, which is defined as the ability to cope with unexpected conditions. For example, the input to the speech coder can consist of nonspeech signals, such as voice band signals, or speech corrupted by background noise or nonlinear microphone distortions. Another form of corruption is caused by channel errors during transmission, which introduce distortions, such as pops and clicks, in the reconstructed speech. Forward error correction techniques reduce this corruption at the expense of an increase in bit rate. Simple error correction procedures only indicate the occurrence of an error but do not correct it, and recovery procedures are needed. Robustness aspects for forward-adaptive code-excited coders were examined in Cox et al. (1989), and it was concluded that these coders are robust for different types of input signals, and not significantly affected by channel errors up to 0.1%. With backward-adaptive systems one has to be concerned about propagation of channel error, but a careful design can produce good results. For example, the 16 kb/s backward-adaptive code-excited coder reported in Chen (1989) is able to cope with bit error rates of up to 0.1%, making it less sensitive to errors than ADPCM.

The algorithmic *complexity* of the analysis-by-synthesis procedures puts limitations on the choice of coder structure and bit allocations. Both multipulse and code-excited coders have been realized in real time, using one or two signal processing chips. For portable communication systems, however, cost and power consumption govern the choice of hardware and affect the selection of both speech coding algorithm and coder parameters. In general, multipulse and regular-pulse coders have a lower complexity than the code-excited coders. The complexity of code-excited coders is due to the exhaustive search needed to find both adaptive and fixed codebooks. Different approaches are possible to reduce the complexity without degradation of performance. The use of transform techniques in the search procedure of the codebook-excited coder (Trancoso and Atal, 1990) is a good example of an algorithm that produces the same results as the full-complexity version. On the other hand, changing the codebook structure by making it overlapping and sparse (Kleijn et al., 1990) and using a linear combination of a small number of basis vectors (Gerson and Jasiuk, 1990) are examples of changing the algorithm with almost no loss in performance. Another approach that involves changing both the codebook and the error criterion is based on the use of algebraic codes for the excitation (Adoul et al., 1987). For a more detailed analysis of procedures that reduce complexity, the reader is referred to Kleijn et al. (1990). One characteristic that is typical of analysis-by-synthesis coders is that the analysis is always more complex than the synthesis. Typical ratios are in the order of 80 to 20%. For some applications, such as announcements, this imbalance is a desirable feature.

The overall *delay* introduced by the combined encoding-decoding procedure is a function of different components, such as algorithmic delay, a finite processing time, and transmission time. Error correction techniques introduce an additional delay. Long coding delays, on the order of 50 ms or more, are objectionable in two-way telephone conversations, especially when they are combined with uncanceled echoes. The use of an echo canceler can reduce some of the echo problems, but at the expense of increased complexity. For non-real-time applications, such as voice mail, the coding delay is not a main concern in selecting a coder. The algorithmic delay is due to the frame-based nature of the coding algorithms. For backward-adaptive systems this delay is determined by the excitation frame size, and the overall delay is between 1 and 5 ms. For forward-adaptive systems the algorithmic delay is determined by the frame size of the short-term predictor and the corresponding interpolation scheme used and is between 50 and 100 ms.

8. PERFORMANCE AT DIFFERENT RATES

In designing a coder for a specific bit rate, one has to trade off many design parameters, such as how frequently each parameter should be updated and how accurately it should be represented. This is often a tedious process, since the impact of most parameter choices can be judged only by listening to the reconstructed speech. In addition, other factors might play a role, such as error robustness and complexity.

The coders described in this chapter are used in many applications. For example, code-excited coders are used for the U.S. federal standard 4.8 kb/s secure voice standard FS 1016 (Campbell et al., 1990) and the U.S. 8 kb/s mobile communications standard IS 54 (Gerson and Jasiuk, 1990). A backward-adaptive code-excited linear prediction (CELP) coder has been proposed as a CCITT 16 kb/s standard (Chen, 1989). The 13 kb/s coder for the European digital mobile communication network is based on a regular-pulse excitation (RPE) coder with long-term predictor (Vary et al., 1988).

Table 1 shows some features of the proposed standards. The speech quality is expressed as a mean opinion score (MOS), which is the average score obtained from listening tests in which each participant rates the speech quality produced by the coders on a scale of 1 to 5 (unsatisfactory, poor, fair, good, and excellent) (see Chapter 12). Note that these MOS figures indicate only speech quality and do not reflect other important transmission issues like robustness against transmission errors and performance for nonspeech signals.

9. CONCLUDING REMARKS AND FUTURE OUTLOOK

This chapter described members of a class of analysis-by-synthesis adaptive predictive coders. These coders use time-varying linear filters to model the short-term and long-term correlations in the speech signal. The predictor parameters are either explicitly transmitted (forward estimation) or retrieved from the reconstructed signal (backward estimation). The excitation signal is represented with a small number of bits by using a codebook representation or using a small number of pulses. For the latter approach either a small set of nonuniformly distributed pulses (multipulse) or a larger set of uniformly distributed pulses (regular pulse) is used. The resulting coders are used in a variety of applications and produce acceptable to high-quality speech in the range from 4 to 16 kb/s. Although this chapter discussed only coding of telephone-bandwidth speech, similar coding concepts can be used for the encoding of wideband speech (Jayant, 1991).

An analysis-by-synthesis approach combined with weighted mean squared error distortion has as a weak point its inflexibility to relax the waveform matching requirement. Exact time placement of events is not always required, since slight time warpings are inaudible. The mean squared error, however, places a heavy penalty on time misalignment (phase errors),

Table 1 A Comparison of Coder Parameters for Various Speech Coding Standards Used for Telephone Bandwidth Speech Sampled at 8 kHz

Parameter	Description			
Coder/Rate	FS1016 4.8 kb/s	IS54 8 kb/s	GSM 13 kb/s	CCITT 16 kb/s
Nominal bitrate	4600 b/s	7950 b/s	13000 b/s	16000 b/s
Short-term predictor	forward	forward	forward	backward
Order	10	10	8	50
Representation	lsf	rc	lar	lpc
Frame shift	30 ms	20 ms	20 ms	2.5 ms
Interpolation	yes (lsf)	yes (lpc)	yes (lar)	no
Bits	34	38	36	n/a
Long-term predictor	forward	forward	forward	none
Frame shift	7.5 ms	5 ms	5 ms	
Delay range	2.5–18.375 ms	2.5–18.375 ms	5–15 ms	
Gain (bits)	5	4[c]	2	
Delay (bits)	8/6/8/6	7	7	

Excitation	ternary shifted Gaussian codebook	trained codebook linearly dependent	regular pulse (spacing = 3)	trained codebook
Frame length	7.5 ms	5 ms	5 ms	0.625 ms
Gain (bits)	4	5.25[c]	6 + 39	2
Index + sign (bits)	10	7 + 7	2	8
Weighting filter	$\gamma_1 = 1, \gamma_2 = 0.8$	$\gamma_1 = 1, \gamma_2 = 0.8$	$\gamma_1 = 1, \gamma_2 = 0.8$	$\gamma_1 = 0.9, \gamma_2 = 0.4$
Postfilter	yes	yes	no	no
Complexity[a]	16 MIPS	12 MIPS	6 MIPS	16 MIPS
Performance[b] (MOS) 64 kbps μ-law = 4.2	3.2	4.0	3.9	4.1

Notes: FS1016 is a proposed secure voice standard for the U. S. government, IS54 is a proposed standard for the U. S. cellular phone system, GSM is the European standard for cellular phone applications, and CCITT is a proposed standard for CCITT.

[a] Floating point instructions.

[b] Scores obtained from various sources.

[c] Combined quantization of codebook gains.

resulting in an inefficient use of bits. A better understanding of the hearing process and the resulting incorporation of hearing-based error criteria should improve the coder performance.

To be able to produce high-quality speech at low bit rates, one should also address the following question. Can one, with a given configuration and a limited number of parameter values, produce the necessary outputs to cover the perceptual space of the input signals? This question becomes more complicated if one realizes that many errors in the signal space are not perceivable. The answer to this question is not known yet, and for certain bit rate and delay combinations it may be that current coders are performing close to fundamental limits.

REFERENCES

Adoul, J., Mabilleau, P., Delprat, M., and Morisette, S. (1987). Fast CELP coding based on algebraic codes. *Proc. IEEE Int. Conf. Acoustics, Speech, and Signal Processing*, pp. 1957–1960.

Atal, B. S. (1982). Predictive coding of speech at low bit rates. *IEEE Trans. Commun. COM-30*:600–614.

Atal, B. S. (1989). Speech coding and human perception. In *Working Models of Human Perception*, B. A. G. Elsendoom and H. Bouma (Eds.), Academic Press, San Diego, pp. 101–126.

Atal., B. S., and Remde, J. R. (1982). A new model of LPC excitation for producing natural-sounding speech at low bit rates. *Proc. IEEE Int. Conf. Acoustics, Speech, and Signal Processing*, April, pp. 614–617.

Atal., B. S., and Schroeder, M. R. (1979). Predictive coding of speech and subjective error criteria. *IEEE Trans. Acoustics, Speech, and Signal Processing. ASSP-27*:247–254.

Campbell, J. P., Tremain, T. E. and Welch, V. C. (1990). The proposed federal standard 1016 4800 bps voice coder: CELP. *Speech Technology*. May, pp. 58–64.

Chen, J.-H. (1989). A robust low-delay CELP speech coder at 16 k/bits. *Proc. GLOBECOM*, pp. 1237–1241.

Chen, J.-H., and Gersho, A. (1987). Real-time vector APC speech coding at 4800 bps with adaptive postfiltering. *Proc. IEEE Int. Conf. Acoustics, Speech, and Signal Processing*, pp. 2185–2188.

Cox, R. V., Kleijn, W. B., and Kroon, P. (1989). Robust CELP coders for noisy backgrounds and noisy channels. *Proc. IEEE Int. Conf. Acoustics, Speech, and Signal Processing*, pp. 739–742.

Davidson, G., and Gersho, A. (1986). Complexity reduction methods for vector excitation coding. *Proc. IEEE Int. Conf. Acoustics, Speech, and Signal Processing*, pp. 3055–3058.

Francesco, R. Di, Lamblin, C., Leguyader, A., and Massaloux, D. (1990). Variable rate speech coding with online segmentation and fast algebraic codes. *Proc. IEEE Int. Conf. Acoustics, Speech, and Signal Processing*, pp. 233–236.

Gerson, I. A. and Jasiuk, M. A. (1990). Vector sum excited linear prediction (VSELP). In *Advances in Speech Coding*, B. S. Atal, V. Cuperman, and A. Gersho (Eds.). Kluwer Academic Publishers, Boston. pp. 69–80.

Itakura, F. (1975). Line spectral representation of linear predictive coefficients of speech signals. *J. Acoust. Soc. Am. 57*(Suppl. 1):S35.

Jayant, N. S. (1991). High quality speech coding for ISDN and audio applications. In *Advances in Speech Signal Processing*, S. Furui and M. M. Sondhi (Eds.). Marcel Dekker, New York.

Jayant, N. S., and Noll, P. (1984). *Digital Coding of Waveforms*. Prentice-Hall, Englewood Cliffs, N.J.

Kleijn, W. B., Krasinski, D. J., and Ketchum, R. H. (1988). Improved speech quality and efficient vector quantization in SELP. *Proc. IEEE Int. Conf. Acoustics, Speech, and Signal Processing*, April, pp. 155–158.

Kleijn, W. B., Krasinski, D. J., and Ketchum, R. H. (1990). Fast methods for the CELP speech coding algorithm. *IEEE Trans. Acoust., Speech Signal Process. ASSP-38*:1330–1342.

Kroon, P., and Atal., B. S. (1987). Quantization procedures for 4.8 kbps CELP coders. *Proc. IEEE Int. Conf. Acoustics, Speech, and Signal Processing*, pp. 1650–1654.

Kroon, P., and Atal, B. S. (1988). Strategies for improving the performance of CELP coders at low bit rates. *Proc. IEEE Int. Conf. Acoustics, Speech, and Signal Processing*, pp. 151–154.

Kroon, P., and Atal, B. S. (1990). On improving the performance of pitch predictors in speech coding systems. In *Advances in Speech Coding*, B. S. Atal, V. Cuperman, and A. Gersho (Eds.). Kluwer Academic Publishers, Boston. pp. 321–328.

Kroon, P., and Deprettere, E. F. (1988). A class of analysis-by-synthesis predictive coders for high quality speech coding at rates between 4.8 and 16 kb/s. *IEEE J. Selected Areas Commun. SAC-6*:353–363.

Kroon, P., Deprettere, E. F., and Sluyter, R. J. (1986). Regular-pulse excitation: A novel approach to effective and efficient multipulse coding of speech. *IEEE Trans. Acoust., Speech, Signal Process. ASSP-34*:1054–1063.

Makhoul, J., and Berouti, M. (1979). Adaptive noise spectral shaping and entropy coding in predictive coding of speech. *IEEE Trans. Acoust., Speech, Signal Process. ASSP-27*:247–254.

Mano, K., and Moriya, T. (1990). 4.8 kbit/s delayed decision CELP coder using tree coding. *Proc. IEEE Int. Conf. Acoustics, Speech, and Signal Processing*, pp. 21–24.

Markel, J. D., and Gray, A. H. (1976). *Linear Prediction of Speech*. Springer-Verlag, Berlin.

Moore, B. C. J. (1982). *An Introduction to the Psychology of Hearing*, 2nd ed., Academic Press, London.

Paliwal, K., and Atal, B. S. (1990). Efficient vector quantization of LPC parameters at 24 bits/frame. *J. Acoust. Soc. Am. 87* (Suppl. 1):S39.

Rabiner, L. R., and Schafer, R. W. (1978). *Digital Processing of Speech Signals*. Prentice-Hall, Englewood Cliffs, N.J.

Ramamoorthy, V., Jayant, B. S., Cox, R. V., and Sondhi, M. M. (1988). Enhancement of ADPCM speech coding with backward-adaptive algorithms for postfiltering and noise feedback. *IEEE J. Selected Areas Commun. SAC-6*:364–382.

Schroeder, M. R., and Atal, B. S. (1985). Code-excited linear prediction (CELP): High quality speech at very low bit rates. *Proc. IEEE Int. Conf. Acoustics, Speech, and Signal Processing*, pp. 937–940.

Schroeder, M. R., Atal, B. S., and Hall, J. L. (1979). Optimizing digital speech coders by exploiting masking properties of the human ear. *J. Acoust. Soc. Am. 66*:1647–1652.

Shoham, Y. (1990). Constrained-stochastic excitation coding of speech at 4.8 kb/s. In *Advances in Speech Coding*, B. S. Atal, V. Cuperman, and A. Gersho (Eds.). Kluwer Academic Publishers, Boston. pp. 339–348.

Singhal, S. and Atal, B. S. (1989). Amplitude optimization and pitch prediction in multipulse coders. *IEEE Trans. Acoust., Speech, Signal Process. ASSP-37*:317–327.

Soong, F. K., and Juang, B.-H. (1990). Optimal quantization of LSP parameters using delayed decisions. *Proc. IEEE Int. Conf. Acoustics, Speech, and Signal Processing*, pp. 185–188.

Taniguchi, T., and Unagami, S. (1989). Multimode coding: Application to CELP. *Proc. IEEE Int. Conf. Acoustics, Speech, and Signal Processing*, pp. 156–159.

Trancoso, I. M., and Atal, B. S. (1990). Efficient search procedures for selecting the optimum innovation in stochastic coders. *IEEE Trans. Acoust., Speech, Signal Process. ASSP-38*:385–396.

Vary, P., Hofmann, R., Hellwig, K., and Sluyter, R. J. (1988). A regular-pulse excited linear predictive code. *Speech Commun.* July, pp. 209–215.

Wang, S., and Gersho, A. (1990). Phonetic segmentation for low rate speech coding. In *Advances in Speech Coding*, B. S. Atal, V. Cuperman, and A. Gersho (Eds.). Kluwer Academic Publishers, Boston. pp. 225–234

6

Low-Rate Speech Coding Based on the Sinusoidal Model

ROBERT J. McAULAY and THOMAS F. QUATIERI Lincoln Laboratory, Massachusetts Institute of Technology, Lexington, Massachusetts

1. INTRODUCTION

One approach to the problem of representation of speech signals is to use the speech production model in which speech is viewed as the result of passing a glottal excitation waveform through a time-varying linear filter that models the resonant characteristics of the vocal tract. In many applications it suffices to assume that the glottal excitation can be in one of two possible states corresponding to voiced or unvoiced speech. In attempts to design high-quality speech coders at the midband rates, generalizations of the binary excitation model have been developed. One such approach is multipulse (Atal and Remde, 1982) which uses more than one pitch pulse to model voiced speech and a possibly random set of pulses to model unvoiced speech. Code excited linear prediction (CELP) (Schroeder and Atal, 1985) is another representation which models the excitation as one of a number of random sequences or "codewords" superimposed on periodic pitch pulses. In this chapter the goal is also to generalize the model for the glottal excitation; but instead of using impulses as in multipulse or random sequences as in CELP, the excitation is assumed to be composed of sinusoidal components of arbitrary amplitudes, frequencies, and phases (McAulay and Quatieri, 1986).

A number of other approaches to analysis-synthesis that are based on sine wave models have been discussed in the literature. The phase vocoder (Flanagan and Golden, 1966) was perhaps the first attempt to represent the speech waveform by a set of narrowband functions. A set of fixed bandpass filters is used and one sine wave is assumed to pass within each filter. The frequency deviation of the sine wave from the center frequency of each band is estimated via the phase derivative of the filter output. This frequency deviation is quantized and used in the vocoder synthesis. Portnoff (1981) refined the phase vocoder by representing each sine wave component by excitation and vocal tract contributions. The sine wave frequencies in the model were constrained to be harmonically related. Another refinement of the phase vocoder was developed by Malah (1979), who also assumed a harmonic model with known pitch and made the filter bank pitch-adaptive, thus ensuring roughly one sine wave per filter. The analysis in these systems does not explicitly model and estimate the sine wave components, but rather views them as outputs of a bank of uniformly spaced bandpass filters. The synthesis waveform can be viewed as a sum of the modified outputs of this filter bank. Although speech of good quality has reportedly been synthesized using these techniques, they are not amenable to being coded at low data rates.

A different approach was taken by Hedelin (1981), who proposed a pitch-independent sine wave model for use in coding the baseband signal for speech compression. The amplitudes and phases of the underlying sine waves are explicitly estimated using Kalman filtering techniques, and each sine wave phase is defined to be the integral of the associated instantaneous frequency. As in the phase vocoder, absolute phase information is lost. Another sine wave–based speech compression system has been developed by Almeida and Silva (1984). In contrast to Hedelin's approach, their system uses a pitch estimate during voiced speech to establish a harmonic set of sine waves. The sine wave phases are computed at harmonic frequencies from the short-time Fourier transform. In a more recent implementation of this system, unvoiced speech was modeled using a set of narrowband basis functions (Marques and Almeida, 1988). Another approach to modeling unvoiced speech in the context of the sine wave model is to generate noise explicitly via the linear filtering of white noise whenever unvoiced speech components are detected in different bands. This approach, developed by Griffin and Lim (1988), is referred to as the "multiband excitation vocoder" and uses overlap-add reconstruction for synthesizing unvoiced speech in "unvoiced bands". For "voiced bands" the system uses a "bank of oscillators," which is simply another term for the sine wave analysis and synthesis scheme described in this chapter. Finally, in an attempt to improve the performance of the linear predictive coding (LPC) synthesizer, Kang and Everett (1985) used a sine wave model to represent the LPC residual as a sum of harmonically related sine waves with frequency-dependent random phase components.

In this chapter, a sinusoidal model for the speech waveform is derived which leads to a new analysis-synthesis technique that is characterized by the amplitudes, frequencies, and phases of the component sine waves. The objective is to demonstrate how a "minimal" parameter set can be derived from this representation and how this parameter set can be coded for high-quality speech at bit rates from 4.8 to 1.2 kb/s. Section 2 describes the basic sine wave analysis-synthesis framework. Section 3 gives a parametric sine wave model with a phase representation which does not use the measured sine wave phases. Section 4 briefly summarizes a sine wave–based pitch and voicing estimator which is used to replace the sine wave frequencies by a set of harmonically related sine waves. Section 5 describes the techniques used to code the sine wave amplitudes, and Section 6 gives an evaluation of the different coding schemes in the range of 4.8 to 1.2 kb/s.

2. THE BASIC SINUSOIDAL ANALYSIS-SYNTHESIS SYSTEM

In this section the sine wave representation and the corresponding analysis-synthesis system are developed. In the analysis stage, these parameter estimates are first matched and then interpolated to allow for continuous evolution of the parameters of successive frames. This sine wave analysis-synthesis system forms the basis for the remainder of the chapter.

2.1. The Model

In the speech production model illustrated in Fig. 1 (Rabiner and Schafer, 1978), the speech waveform $s(t)$ is assumed to be the output of passing a vocal cord (glottal) excitation waveform through a linear system representing the characteristics of the vocal tract. The excitation function is usually represented as a periodic pulse train during voiced speech, where the spacing between consecutive pulses corresponds to the "pitch" of the speaker and is represented as a noiselike signal during unvoiced speech. Alternatively, the binary voiced-unvoiced excitation model can be replaced by a sum of sine waves (McAulay and Quatieri, 1986; Quatieri and McAulay, 1986).

The motivation for this sine wave representation of the excitation is that the voiced excitation, when perfectly periodic, can be represented by a Fourier series decomposition in which each harmonic component of the decomposition corresponds to a single sine wave. More generally, the sine waves in the model will be aharmonic as when periodicity is not exact and

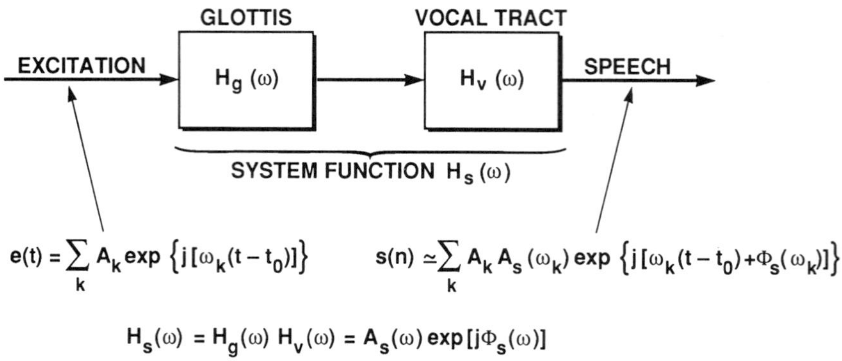

Figure 1 Speech production model for sine wave analysis-synthesis.

when the excitation is unvoiced. Passing this sine wave representation of the excitation through the time-varying vocal tract results in the sinusoidal representation for the speech waveform (McAulay and Quatieri, 1986), which, on a given analysis frame ($\approx$ 10 to 20 ms), is given by

$$s(n) = \sum_{l=1}^{L} A_l \cos(\omega_l n + \phi_l) \tag{1}$$

where A_l and ϕ_l represent the amplitude and phase of each sine wave component associated with the frequency track ω_l and L is the number of sine waves. The accuracy of this representation is subject to the caveat that the parameters are slowly varying relative to the duration of the vocal tract system response. The model in Eq. (1) will now be used to develop a speech analysis-synthesis system which will serve to test the accuracy of the sine wave representation and become the basis for sine wave–based speech coding.

2.2. Estimation of Speech Parameters

The problem in analysis-synthesis is to take a speech waveform, extract parameters that represent a quasi-stationary portion of that waveform, and use those parameters or coded versions of them to reconstruct an approximation that is "as close as possible" to the original speech. Furthermore, it is desirable to have a robust parameter extraction algorithm because the speech signal in many cases is contaminated by additive acoustic noise. The general estimation problem in which the speech signal is to be represented by multiple sine waves is a difficult one to solve analytically. The approach taken here is heuristic and is based on the observation that when the speech is perfectly periodic, the sine wave parameters correspond to the harmonic samples of the short-time Fourier transform (STFT). In this case, the model in Eq. (1) reduces to

$$s(n) = \sum_{l=1}^{L} a_l \cos(nl\omega_0 + \phi_l) \tag{2}$$

in which the sine wave frequencies are multiples of the fundamental frequency ω_0 and the corresponding amplitudes and phases are given by the harmonic samples of the STFT. If the STFT of $s(n)$ is given by

$$Y(\omega) = \sum_{n=-N/2}^{N/2} s(n) \exp(-jn\omega) \tag{3}$$

then Fourier analysis gives the amplitude estimates as $A_l = |Y(l\omega_0)|$ and the phase estimates as $\phi_l = \arg Y(l\omega_0)$. Moreover, the magnitude of the STFT (i.e., the periodogram) will have peaks at multiples of ω_0. When the speech is not perfectly voiced, the periodogram will still have a multiplicity of peaks but at frequencies that are not necessarily harmonic, and these can be used to identify an underlying sine wave structure. In this case the sine wave frequencies are simply taken to be the frequencies at which the slope of the periodogram changes from positive to negative and the amplitudes and phases are obtained by evaluating the STFT at the chosen frequencies (McAulay and Quatieri, 1986)

The above analysis implicitly assumes that the STFT is computed using a rectangular window. Since its poor side lobe structure will compromise the performance of the estimator, the Hamming window was used in all experiments. While this resulted in a very good side lobe structure, it did so at the expense of broadening the main lobes of the periodogram estimator. Therefore, in order to maintain the resolution properties that were needed to justify using the peaks of the periodogram, the window width was made at least 2½ times the average pitch period. During frames of unvoiced speech, the window is held fixed at the value obtained on the preceding voiced frame.

Once the width for a particular frame has been specified, the pitch-adaptive Hamming window $w(n)$ is computed and normalized according to

$$\sum_{n=-N/2}^{N/2} w(n) = 1 \tag{4}$$

so that the periodogram peak will yield the amplitude of an underlying sine wave. The STFT of the Hamming-windowed input speech is then taken using the fast Fourier transform (FFT). Peaks in the periodogram are obtained by finding all values which are greater than their two nearest neighbors, and peaks below a specified threshold from the maximum peak (about 80 dB) are eliminated. The locations of the selected peaks give the sine wave frequencies and the peak values give the sine wave amplitudes. Plotted in Fig. 2 is a typical periodogram for voiced speech along with the amplitudes $\{A_l\}$ and frequencies $\{\omega_l\}$ that are estimated using the above procedure. The sine wave phases $\{\phi_l\}$ are computed from the real and imaginary components of the STFT evaluated at $\{\omega_l\}$.

It should be noted that the placement of the analysis window $w(n)$ relative to the time origin is important for computing the phases. Typically, in frame-sequential processing the window $w(n)$ lies in the internal $0 \leq n \leq N$ and is symmetric about $N/2$, a placement which gives the measured phase a linear term equal to $-\omega N/2$. Since N is on the order of 100 to 400 discrete time samples, any error in the estimated frequencies results in a large random phase error and consequent hoarseness in the reconstruction. An error of one discrete Fourier transform (DFT) sample, for example, results in a $(2\pi/M)N/2$ phase error (where M is the DFT length), which could be on the order of π. To improve the robustness of the phase estimate the Hamming window is placed symmetric relative to the origin defined as the center of the current analysis frame; hence the window takes on values over the interval $-N/2 \leq n \leq N/2$.

The approximations leading to the above periodogram estimator were based on the voiced speech assumption; nowhere have the properties of unvoiced speech been taken into account. To do this in a way that results in uncorrelated amplitude samples requires use of the Karhunen–Loève expansion for noiselike signals (Van Trees, 1968). Such an analysis shows that a sinusoidal representation is valid provided the frequencies are "close enough" so that the ensemble power spectral density changes slowly over consecutive frequencies. If the window width is constrained to be at least 20 ms wide, then "on the average" there will be a set of periodogram peaks that will be approximately 100 Hz apart, and this should provide a sufficiently dense sampling to satisfy the necessary constraints. Plotted in Fig. 3 is a typical periodogram for a frame of unvoiced speech along with the amplitudes and frequencies that are estimated using this procedure.

The foregoing analysis provides a heuristic justification for the representation of the speech waveform in terms of the amplitudes, frequencies, and phases of a set of sine waves that applies to one analysis frame. As speech evolves from frame to frame, different sets of these parameters will

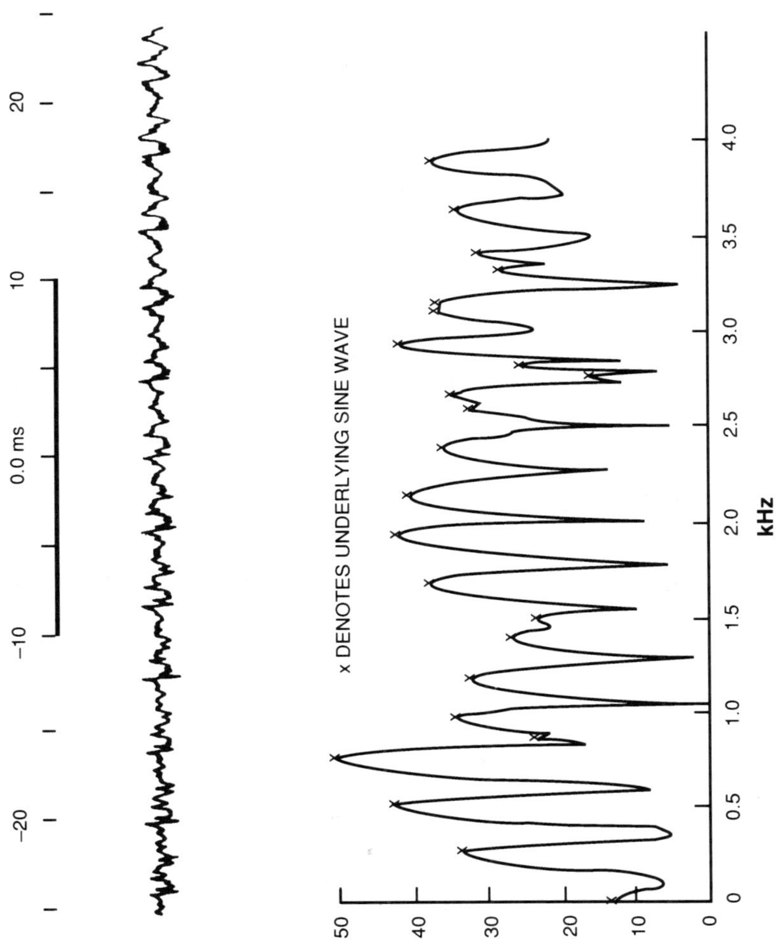

Figure 2 Typical periodogram for a frame of voiced speech and the amplitude and frequency estimates of the underlying sine waves.

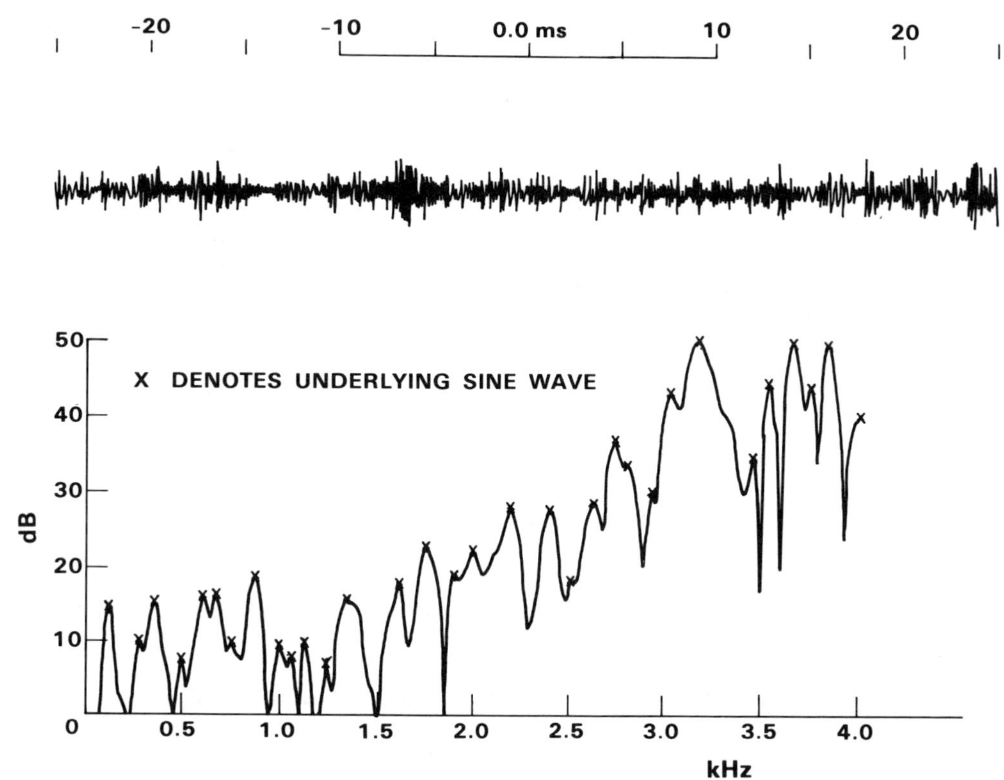

Figure 3 Typical periodogram for a frame of unvoiced speech and the amplitude and frequency estimates of the underlying sine waves.

be obtained. The next problem to address, then, is the association of amplitudes, frequencies, and phases measured on one frame with those that are obtained on a successive frame in order to define sets of sine waves that will be continuously evolving in time.

2.3. Frame-to-Frame Peak Matching

If the number of peaks were constant from frame to frame, the problem of matching the parameters estimated on one frame with those on a successive frame would simply require a frequency-ordered assignment of peaks. In practice, however, the locations of the peaks will change as the pitch changes, and there will be rapid changes in both the location and the number of peaks corresponding to rapidly varying regions of speech, such as at voiced-unvoiced transitions. In order to account for such rapid movements in the spectral peaks, the concept of "birth" and "death" of sinusoidal components is introduced. The problem of matching spectral peaks in some optimal sense while allowing for this birth-death process is generally a difficult problem. One method, which has proved to be successful for speech reconstruction, is to define sine wave tracks for frequencies that are successively "nearest neighbors." The matching procedure is made dynamic by allowing tracks to begin at any frame (a birth) and to terminate at any frame (a death), events which are determined when successive frequencies do not fall within some matching interval. The algorithm, although straightforward, is a rather tedious exercise in rule-based programming, and the interested reader can find the details in McAulay and Quatieri (1986).

An illustration of the matching algorithm showing how the birth-death procedure accounts for rapidly varying peak locations is shown in Fig. 4. The result of applying the tracker to a segment of real speech is shown in Fig. 5, which demonstrates the ability of the tracker to adapt quickly through transitory speech behavior such as voiced-unvoiced transitions and mixed voiced-unvoiced regions.

2.4. The Synthesis System

Since a set of amplitudes, frequencies, and phases is estimated for each frame, it might seem reasonable to estimate the original speech waveform on the kth frame by generating synthetic speech using the equation

$$\hat{s}(n) = \sum_{l=1}^{L} A_l^k \cos(n\omega_l^k + \theta_l^k) \tag{5}$$

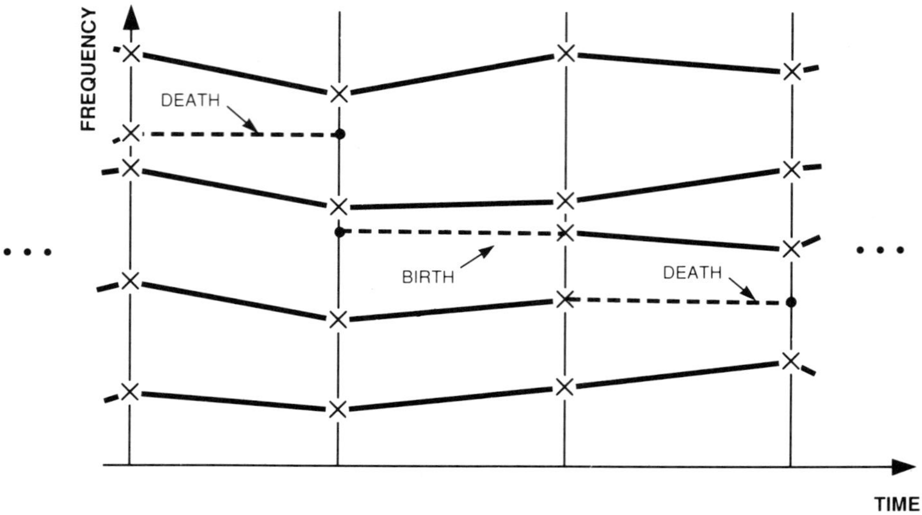

Figure 4 Illustration of frequency tracks using the birth-death frequency tracker.

where $n = 0, 1, 2, \ldots, I-1$ and I is the length of the synthesis frame. Because of the time-varying nature of the parameters, however, this straightforward approach leads to discontinuities at the frame boundaries, which seriously degrades the quality of the synthetic speech. A method will now be described for smoothly interpolating the parameters measured from one frame to those that are obtained on the next.

As a result of the frequency-matching algorithm described in the previous section, all of the parameters measured for an arbitrary frame k have been associated with a corresponding set of parameters for frame $k + 1$. Letting $\{A_l^k, \omega_l^k, \theta_l^k\}$ and $\{A_l^{k+1}, \omega_l^{k+1}, \theta_l^{k+1}\}$ denote the successive sets of parameters for the lth frequency track, an obvious solution to the amplitude interpolation problem is to take

$$\hat{A}^k(n) = A^k + (A^{k+1} - A^k)\left(\frac{n}{T}\right) \tag{6}$$

where $n = 0, 1, \ldots, T-1$ is the time sample into the kth frame. (The track subscript l has been omitted for convenience.)

Unfortunately, such a simple approach cannot be used to interpolate the frequency and phase because the measured phases θ^k and θ^{k+1} are obtained modulo 2π. Hence, phase unwrapping must be performed to ensure that the frequency tracks are "maximally smooth" across frame boundaries. The first step in solving this problem is to postulate a phase interpolation function that

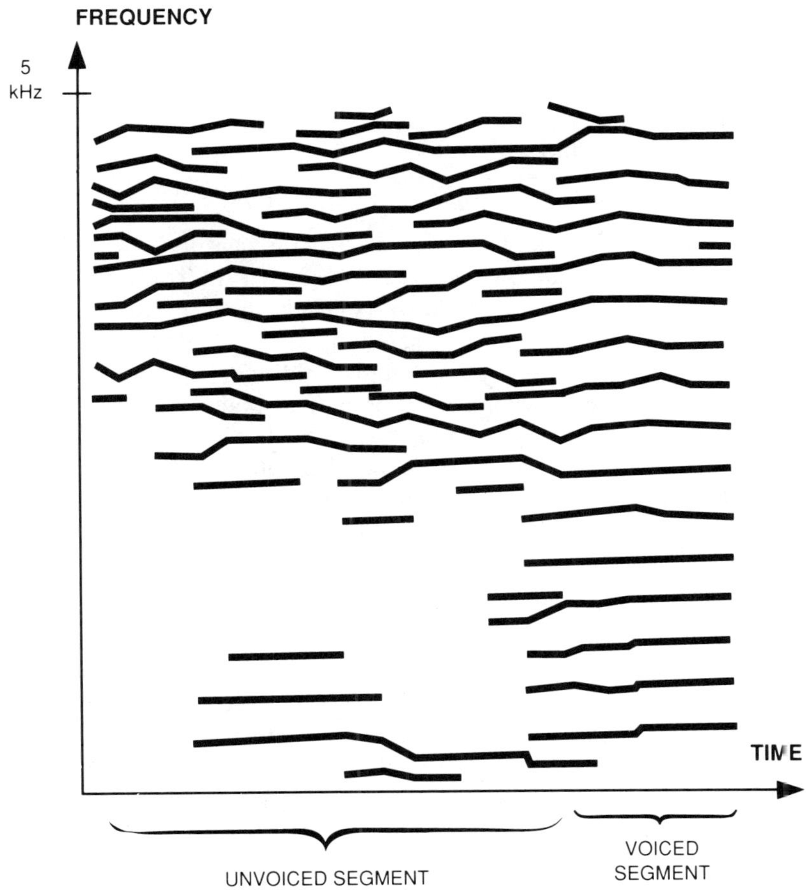

Figure 5 Typical frequency tracks for real speech.

is a cubic polynomial,* namely

$$\hat{\theta}(t) = \zeta + \gamma t + \alpha t^2 + \beta t^3 + 2\pi M \tag{7}$$

where the term $2\pi M$, m an integer, is used to account for the phase unwrapping. It is convenient to treat the phase function as though it were a function

*The idea of applying a cubic polynomial to interpolate the phase between frame boundaries was independently proposed by Almeida and Silva (1984) for use in their harmonic sine wave synthesizer.

of a continuous time variable t with $t = 0$ corresponding to the center of frame k and $t = T$ corresponding to the center of frame $k + 1$. Since the derivative of the phase is the frequency, it is necessary that the cubic phase function and its derivative equal the phase and the frequency measured at the frame k. Therefore, Eq. (7) reduces to

$$\hat{\theta}(t) = \theta^k + \omega^k t + \alpha t^2 + \beta t^3 + 2\pi M \tag{8}$$

Since only the principal value of the phase can be measured, provision must also be made for unwrapping the phase subject to the constraint that the cubic phase function and its derivative equal the measured frequency and phase at frame $k + 1$. An explicit solution can be obtained for interpolation and phase unwrapping by invoking an additional constraint requiring that the unwrapped cubic phase function be maximally smooth. The problem then reduces to finding that multiple of 2π that leads to the smoothest phase interpolation function while meeting the constraints on the frequency and phase at frame $k + 1$. In McAulay and Quatieri (1986) it is shown that these constraints are met for values of α and β that satisfy the relations

$$\begin{bmatrix} \alpha(M) \\ \beta(M) \end{bmatrix} = \begin{bmatrix} \frac{3}{T^2} & \frac{-1}{T} \\ \frac{-2}{T^3} & \frac{1}{T^2} \end{bmatrix} \begin{bmatrix} \theta^{k+1} - \theta^k - \omega^k T + 2\pi M \\ \omega^{k+1} - \omega^k \end{bmatrix} \tag{9}$$

The phase unwrapping parameter M is then chosen to make the unwrapped phase maximally smooth. An explicit rule for determining M is given in McAulay and Quatieri (1986).

As a result of this phase unwrapping procedure, each frequency track will have associated with it an instantaneous unwrapped phase which accounts for both the rapid phase changes due to the frequency of each sinusoidal component and the slowly varying phase changes due to the glottal pulse and the vocal tract transfer function. This phase model will be further refined in Section 3. Letting $\hat{\theta}_l(t)$ denote the unwrapped phase function for the lth track, the final synthetic wave form for the kth frame will be given by

$$\hat{s}(n) = \sum_{l=1}^{L} \hat{A}_l(n) \cos[\hat{\theta}_l(n)] \tag{10}$$

where $\hat{A}_l(n)$ is given by Eq. (6), $\hat{\theta}_l(n)$ is the sampled data version of Eq. (7), and L is the number of sine waves.

This completes the theoretical basis for the new sinusoidal analysis-synthesis system. Although simple in concept, the detailed analysis led to the introduction of the birth-death frequency tracker and the cubic interpolation phase unwrapping procedure chosen to ensure smooth transitions from frame to frame. The degree to which these new procedures result in the synthesis of high-quality speech will be discussed in the next section.

2.5. Experimental Results

A block diagram description of the analysis-synthesis system is given in Fig. 6. A non-real-time floating-point simulation was developed in order to determine the effectiveness of the proposed approach in modeling real speech. The speech processed in the simulation was low-pass filtered at 5 kHz, digitized at 10 kHz, and analyzed at 10-ms frame intervals. A 1024-point FFT using a pitch-adaptive Hamming window, having a width which was $2\frac{1}{2}$ times the average pitch, was found to be sufficient for accurate peak estimation. The maximum number of peaks that are used in synthesis was set to a fixed number (≈ 80) and, if excessive peaks were obtained, only the largest peaks were used. A large speech data base has been processed with this system, and it has been found that the synthetic speech was perceived to be essentially indistinguishable from the original. Visual examination of many of the

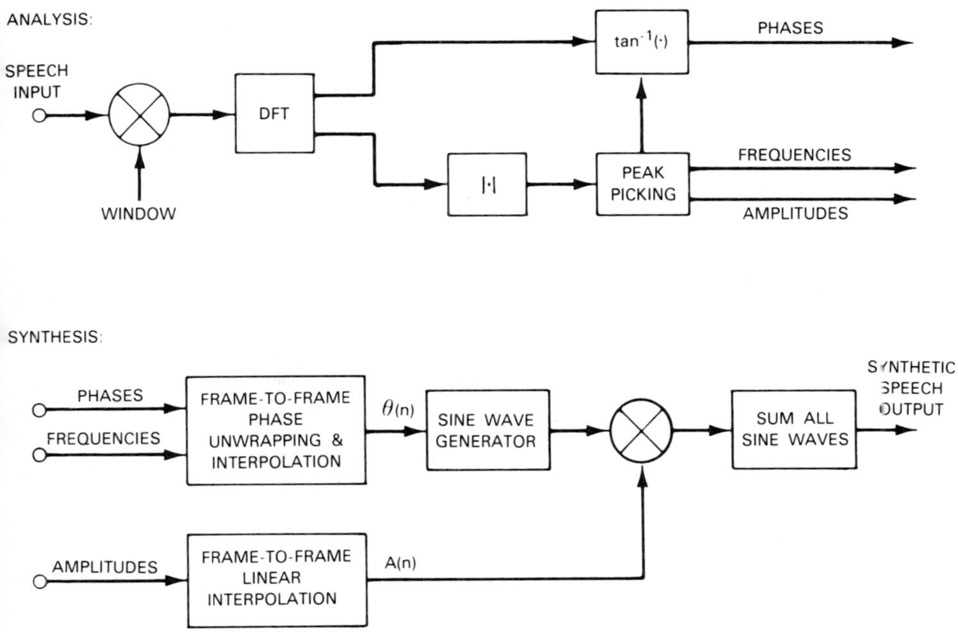

Figure 6 Block diagram of the baseline sinusoidal analysis-synthesis system.

reconstructed passages shows that the waveform structure is essentially preserved. An example of this property is shown in Fig. 7, which compares the waveforms for the original speech and the reconstructed speech during an unvoiced-voiced speech transition. This suggests that the quasi-stationarity

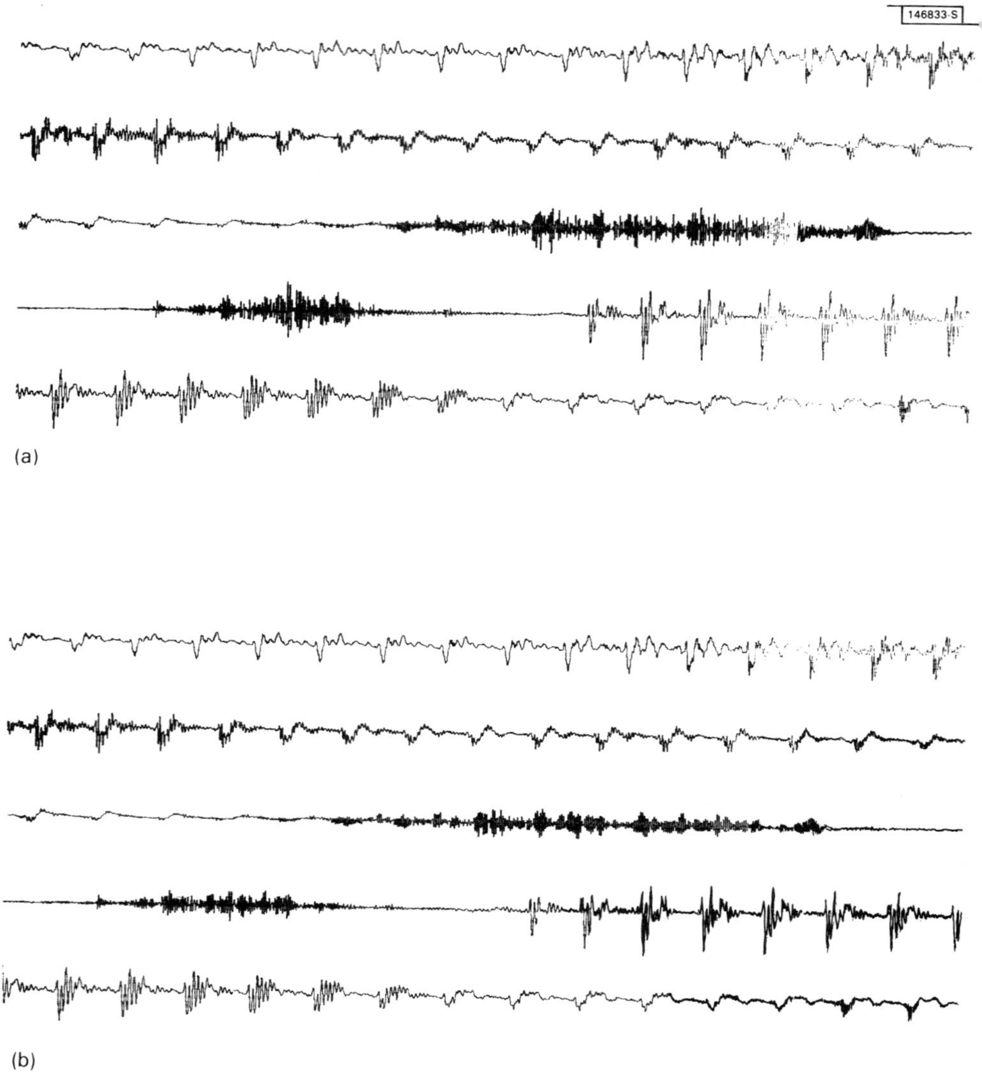

Figure 7 Sinusoidal reconstruction of speech: (a) original; (b) reconstruction.

conditions seem to be satisfactorily met and that the use of the parametric model based on the amplitudes, frequencies, and phases of a set of sine wave components appears to be justifiable for both voiced and unvoiced speech.

Although the sinusoidal model was originally designed for a single speaker, the analysis-synthesis system was found to be capable of synthesizing a broader class of signals that includes multispeaker waveforms, music, speech in a musical background, and marine biological signals such as whale sounds. Furthermore, the reconstruction does not break down in the presence of noise. The synthesized speech is essentially perceptually indistinguishable from the original noisy speech with essentially no modification of the noise characteristics. Illustrations depicting the performance of the system in the face of the above degradations are provided in McAulay and Quatieri (1986). In addition to speech coding, the subject of this chapter, the sine wave analysis-synthesis has been used in time-scale and pitch modification, peak-to-rms reduction, and two-talker separation (Quatieri and McAulay, 1986, 1989, 1990); for computer music synthesis (Smith and Serra 1987); and for the analysis of vibrato (Mahler and Beauchamp, 1989).

3. MODEL FOR THE SINE WAVE PHASES

Although the sine wave model has led to a parametric representation for speech that has been manipulated to provide new solutions to a number of speech processing problems, it is not amenable to low-rate speech coding because there are simply too many parameters to be quantized. Therefore, another more efficient parametric representation must be developed to reduce the size of the parameter set. The first step will be to develop a model for the sine wave phase by explicitly identifying the phase components due to the excitation, the glottis, and the vocal tract.

3.1. Onset Estimation

Starting from the basic vocoder model shown in Fig. 1, the sequence of excitation pitch pulses can be written in terms of sine waves as

$$e(n) = \sum_{l=1}^{L} \exp[j(n - n_0)\omega_l] \tag{11}$$

where n_0 corresponds to the time of occurrence of the pitch pulse nearest the center of the current analysis frame. The occurrence of this temporal event, called the onset time, ensures that the underlying excitation sine waves will be in phase at the time of the pitch pulse. The amplitude and phase of the excitation sine waves, not necessarily assumed to be harmonic at this point, are altered by the glottal pulse, $H_g(\omega)$, and vocal tract, $H_v(\omega)$, filters. Letting $H_s(\omega) = H_g(\omega)H_v(\omega)$ denote the composite transfer junction for these filters, called the system function, the speech signal at its output becomes

$$\hat{s}(n) = \sum_{l=1}^{L} H_s(\omega_l)\exp[j(n - n_0)\omega_l] \tag{12}$$

Since the filter is linear, the output waveform can also be represented in terms of sine waves as

$$s(n) = \sum_{l=1}^{L} A_l \exp[j(n\omega_l + \theta_l)] \tag{13}$$

The parameters in this representation can be determined using the estimation techniques described in Section 2.

In order to specify the parametric representation of the excitation waveform, the onset time must be determined, a parameter that can be estimated by choosing the value of n_0 such that $\hat{s}(n)$ is as "close as possible" to $s(n)$ according to some meaningful criterion. A reasonable criterion is to seek the minimum of the mean squared error (MSE),

$$\epsilon(n_0) = \frac{1}{N+1} \sum_{n=-N/2}^{N/2} |s(n) - \hat{s}(n; n_0)|^2 \tag{14}$$

over n_0, since this at least ensures robustness against additive white Gaussian noise (Van Trees, 1968). The MSE in Eq. (14) can be expanded as

$$\epsilon(n_0) = \frac{1}{N+1} \sum_{n=-N/2}^{N/2} \left\{ |s(n)|^2 - 2\,\mathrm{Re}[s(n)\hat{s}^*(n; n_0)] + |\hat{s}(n; n_0)|^2 \right\} \tag{15}$$

If the sinusoidal representation for $s(n)$, Eq. (13), is used in the first term of Eq. (15), then the power in the measured signal can be defined as

$$P_s = \frac{1}{N+1} \sum_{n=-N/2}^{N/2} |s(n)|^2 = \sum_{l=1}^{L} A_l^2 \tag{16}$$

Letting the system transfer function be written in terms of its magnitude, $A_s(\omega)$, and phase $\Phi_s(\omega)$, i.e.,

$$H_s(\omega) = A_s(\omega)\exp[j\Phi_s(\omega)] \tag{17}$$

and using this in Eq. (12), the second term of Eq. (15) can be written as

$$\frac{1}{N+1}\sum_{n=-N/2}^{N/2} s(n)\hat{s}^*(n;n_0) = \sum_{l=1}^{L} A_l A_s(\omega_l)\exp\{j[\theta_l + n_0\omega_l - \Phi_s(\omega_l)]\} \tag{18}$$

Finally, the third term in Eq. (15) can be written as

$$\frac{1}{N+1}\sum_{n=-N/2}^{N/2} |\hat{s}(n;n_0)|^2 = \sum_{l=1}^{L} A_s^2(\omega_l) \tag{19}$$

These relations are valid provided the sine waves are well resolved, a condition that is more or less assured by making the analysis window length, $N+1$, two and one-half times the average pitch period, which was the condition assumed in the estimation of the sine wave parameters in the first place. Combining the above manipulations leads to the following expression for the MSE:

$$\epsilon(n_0) = P_s - 2\sum_{l=1}^{L} A_l A_s(\omega_l)\cos[\theta_l + n_0\omega_l - \Phi_s(\omega_l)] + \sum_{l=1}^{L} A_s^2(\omega_l) \tag{20}$$

Equation (20) was derived under the assumption that the system amplitude $A_s(\omega)$ and phase $\Phi_s(\omega)$ were known. In order to obtain the "optimal" onset time, therefore, the amplitude and phase of the system function must next be estimated from the data. An ad hoc estimator for the magnitude of the system function is simply to apply linear interpolation between successive sine wave peaks. This results in the function

$$A_s(\omega) = A_{l-1} + \frac{A_l - A_{l-1}}{\omega_l - \omega_{l-1}}(\omega - \omega_{l-1}) \qquad \omega_{l-1} \le \omega \le \omega_l \tag{21}$$

If the system function is assumed to be minimum phase (Oppenheim and Schafer, 1975), then

$$\log H_s(\omega) = \sum_{m=-\infty}^{\infty} c_m \exp(-jm\omega) \tag{22}$$

from which it follows that

$$\log A_s(\omega) = c_0 + 2\sum_{m=1}^{\infty} c_m \cos(m\omega) \tag{23}$$

$$\Phi_s(\omega) = -2\sum_{m=1}^{\infty} c_m \sin(m\omega) \tag{24}$$

These quantities will be referred to as the cepstral amplitude and phase envelopes. The "cepstral coefficients" $\{c_m\}_{m=0}^{\infty}$ can be determined from the amplitude envelope using

$$c_m = \frac{1}{\pi}\int_0^{\pi} \log A_s(\omega)d\omega \qquad m = 0, 1, \ldots \tag{25}$$

and, hence, the system phase, $\Phi_S(\omega)$ can be determined from Eq. (24). Use of Eq. (24) is incomplete, however, since the minimum phase analysis fails to account for the sign of the input speech waveform. This is due to the fact that the sine wave amplitudes from which the system amplitude and phase are derived are the same for $-s(n)$ and $s(n)$. This ambiguity can be accounted for by generalizing the system phase $\Phi_s(\omega)$ in Eq. (24) by

$$\Phi_s(\omega) + \beta\pi \qquad \beta = 0 \text{ or } 1 \tag{26}$$

and then choosing n_0 and β to minimize the MSE simultaneously.

Using the expressions for the system amplitude and phase in Eq. (21) and Eq. (26) in Eq. (20), the MSE becomes

$$\epsilon(n_0, \beta) = P_s - 2\sum_{l=1}^{L} A_l^2 \cos[\theta_l + n_0\omega_l - \beta\pi - \Phi_s(\omega_l)] + \sum_{l=1}^{L} A_l^2 \tag{27}$$

Since only the second term depends of the phase model, it suffices to choose n_0 and β to maximize the function

$$\rho(n_0, \beta) = \sum_{l=1}^{L} A_l^2 \cos[\theta_l + n_0\omega_l - \beta\pi - \Phi_s(\omega_l)] \tag{28}$$

However, since

$$\rho(n_0, \beta = 1) = -\rho(n_0, \beta = 0) \tag{29}$$

it suffices to maximize $|\rho(n_0)|$ where now

$$\rho(n_0) = \sum_{l=1}^{L} A_l^2 \cos[\theta_l + n_0\omega_l - \Phi_s(\omega_l)] \tag{30}$$

and if $\hat{n}_o$ is the maximizing value then choose $\beta = 0$ if $\rho(\hat{n}_0) > 0$ and $\beta = 1$ otherwise.

3.2. Sine Wave Amplitude Envelope Estimation

The above results show that if the envelope of the sine wave amplitudes is known, then the MSE criterion can lead to a technique for estimating the pitch onset time under the assumption that the glottal pulse and vocal tract networks are minimum phase. This latter assumption was necessary to derive an estimate of the system phase; hence the performance of the estimator depends entirely on the system magnitude, which was taken to be linear interpolation of the measured sine wave amplitudes. The problem with such a simple envelope estimator is that the system phase will be sensitive to low-level peaks which can arise due to signal processing artifacts such as side lobe leakage. Fortunately, these problems can be avoided using the technique proposed by Paul (1981) in the development of the Spectral Envelope Estimation Vocoder (SEEVOC).

The SEEVOC algorithm depends on having an estimate of the average pitch, denoted here by $\bar{\omega}_0$. The first step is to search for the largest sine wave amplitude in the interval $[\bar{\omega}_0/2, 3\bar{\omega}_0/2]$. Having found the amplitude and frequency of that peak, labeled (A_1, ω_1), then the interval $[\omega_1 + \bar{\omega}_0/2, \omega_1 + 3\bar{\omega}_0/2]$ is searched for its largest peak, labeled (A_2, ω_2). The process is continued by searching the intervals $[\omega_{l-1} + \bar{\omega}_0/2, \omega_{l-1} + 3\bar{\omega}_0/2]$ for the largest peaks, (A_l, ω_l), until the edge of the speech bandwidth is reached. If no peak is found in a search bin, then the largest end point of the short-time Fourier transform (STFT) magnitude is used and placed at the bin center, from which the search procedure is continued. The principal advantage of this method is the fact that any low-level peaks within a pitch interval will be masked by the largest peak, presumably a peak that is close to an underlying harmonic. Moreover, the procedure is not dependent on the peaks being harmonic or on the exact value of the average pitch, since the procedure resets itself after each peak has been found. The SEEVOC envelope, the envelope on which the minimum phase analysis is based, is then obtained by applying the linear interpolation rule, Eq. (21), where now the sine wave amplitudes and frequencies are those obtained using the SEEVOC peak-picking routine.

3.3. Experimental Results

The baseline system on which the low-rate coder was developed used a 10-kHz sampling rate and a 1024-point FFT to compute the STFT. The magnitude of the STFT was computed and the underlying sine waves were identified by determining the frequencies at which a change in the slope of the periodogram occurred. The frequency of the identified sine wave was further refined using quadratic interpolation. The SEEVOC algorithm was used to estimate the envelope of the sine wave amplitudes.

A typical set of results is shown in Fig. 8. The logarithm of the STFT magnitude is shown in Fig. 8d. The SEEVOC envelope was used in Eq. (25)

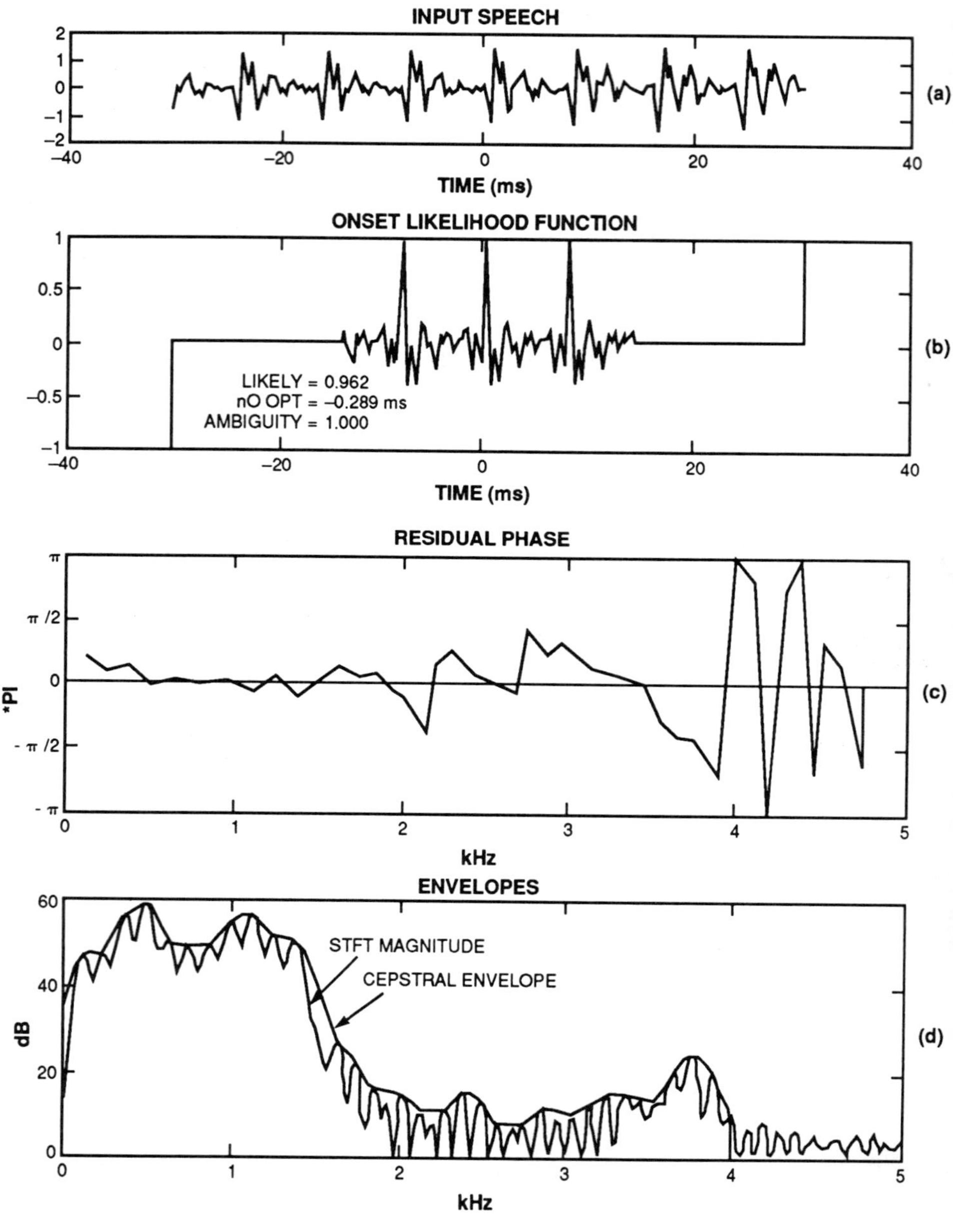

Figure 8 Typical onset time analysis for voiced speech.

to compute the cepstral envelope for 40 cepstral coefficients, and the resulting envelope is also plotted in Fig. 8d. It should be noted that the goal of the low-rate sine wave coder is to produce synthetic speech over a 4-kHz speech bandwidth. Therefore, the calculations in this section are made using a radian frequency normalized to an effective 8-kHz sampling rate, (i.e., $\omega = 2\pi f/8000$). Therefore, although the STFT provides data out to 5 kHz, the envelopes are plotted only out to 4 kHz, since they will fold over in the 4 to 5-kHz region. The cepstral coefficients were also used in Eq. (24) to compute the system phase, from which the onset discrimination function could be computed using Eq. (30). The result is shown in Fig. 8b. It is interesting to note that the peaks in the onset function occur at the points which seem to correspond to a sharp transition in the speech waveform, rather than at a peak in that waveform.

In order to develop a phase model from these results it is instructive to examine the behavior of the phase error. Letting

$$\tilde{\theta}(\omega) = -\hat{n}_0\omega + \Phi_s(\omega) + \hat{\beta}\pi \tag{31}$$

be the model of the sine wave phase at any frequency, ω, then for the sine wave at ω_l, for which the measured phase is θ_l, the phase error is

$$\epsilon(\omega_l) = \theta_l - \tilde{\theta}(\omega_l) \tag{32}$$

The phase error for the speech sample in Fig. 8a is shown in Fig. 8c. Since the residual phase errors appear to be small, it can be concluded that the phase model is quite good. The larger structured error beyond 3 kHz probably indicates the inadequacy of the minimum phase assumption. It should be noted that the best results seem to be obtained for frequencies below 3 kHz. Moreover, these results appear to hold for voiced speech in general. When the same measurements are made for unvoiced speech, an example of which is shown in Fig. 9, the phase errors appear to be random, more or less evenly distributed on the $[-\pi, \pi]$ interval. Of course, the estimate of the onset time is meaningless in this case, and this is reflected in the relatively low value of the MSE criterion.

3.4. Voicing-Dependent Phase Model

Since the foregoing results indicate that the measured phase residuals are essentially zero for voiced speech and random on $[-\pi, \pi]$ for unvoiced speech, and since experiments using the basic sine wave system have shown that random phases on $[-\pi, \pi]$ result in whispered speech, then it seems reasonable to replace the phase residuals by a voicing-dependent model in which the phase residuals would be made zero for voiced speech and random on $[-\pi, \pi]$

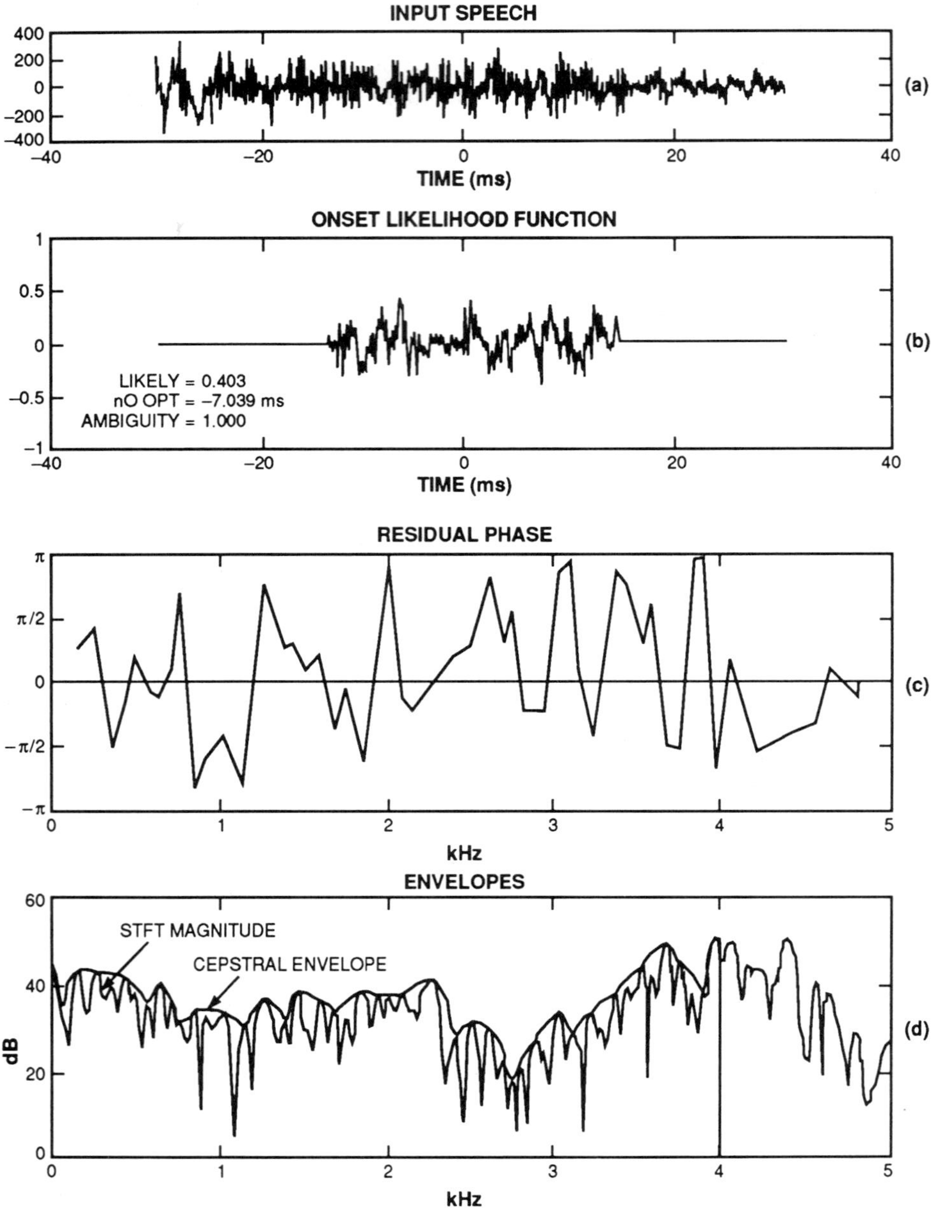

Figure 9 Typical onset time analysis for unvoiced speech.

for unvoiced speech. Minimal coding of the sine wave phases could then be achieved using only a single bit to convey the voiced-unvoiced phase information. However, as in standard low-rate "buzz-hiss" vocoders, this would render the sine wave system overly dependent on the voicing decision, causing similar artifacts to occur in the synthetic speech when this decision was made erroneously. The deleterious effects of the binary decision can be reduced significantly by using a more general mixed-excitation model of the type proposed by Makhoul et al. (1978).

In that model, a voicing transition frequency was estimated below which voiced speech was synthesized and above which unvoiced speech was synthesized. Although that work was done in the context of a conventional LPC vocoder (periodic impulse train for voiced speech, random noise for unvoiced speech), the concept is ideally suited to the sine wave model, since below the voicing transition frequency the sine wave phase residuals are made zero and above the transition they are made random. Letting

$$\omega_c(P_v) = \pi P_v \tag{33}$$

denote the voicing-dependent cutoff frequency, where P_v is the probability of voicing, then the synthetic phase residuals can be estimated by

$$\hat{\epsilon}(\omega) = \begin{cases} 0 & \text{if } \omega \le \omega_c(P_v) \\ U[-\pi, \pi] & \text{if } \omega > \omega_c(P_v) \end{cases} \tag{34}$$

where $U[-\pi, \pi]$ denotes a uniformly distributed random variable on $[-\pi, \pi]$. Using Eq. (34) in place of Eq. (32), the sine wave phase at frequency ω_l is estimated by

$$\hat{\theta}_l = -\hat{n}_0\omega_l + \Phi(\omega_l) + \hat{\beta}\pi + \hat{\epsilon}(\omega_l) \tag{35}$$

Of course, this model requires the specification of the voicing probability, which should reflect the degree to which the predictive phase model fits the measured data. Since the onset function $\rho(n_0)$, Eq. (30), reflects the accuracy of the phase model, the normalized onset function

$$\bar{\rho}(n_0) = \frac{\sum_{l=1}^{L} A_l^2 \cos[\theta_l + n_0\omega_l - \Phi_s(\omega_l)]}{\sum_{l=1}^{L} A_l^2} \tag{36}$$

might be a good basis for estimating the voicing probability. Since its value is near unity for coherent voiced speech and near 0.5 for unvoiced speech, a reasonable rule to use might be

$$P_v = \begin{cases} 1 & \tilde{\rho}(n_0) > 0.75 \\ 4[\tilde{\rho}(\hat{n}_0) - 0.5] & 0.5 \le \tilde{\rho}(\hat{n}_0) \le 0.75 \\ 0 & \tilde{\rho}(n_0) < 0.5 \end{cases} \tag{37}$$

An alternative voicing probability which measures the harmonic content of the speech signal will be defined in Section 4.

At this point the phase model is dependent on the measured onset time, which can introduce significant phase jitter, rendering the synthetic speech "rough," unless it is estimated and quantized with considerable accuracy ($\approx$ 9 bits). Since the function of the onset time is to bring the sine waves into phase at times corresponding to the sequence of pitch pulses, it is possible to achieve the same effect simply by keeping track of successive onset times generated by a succession of pitch periods that are decoded at the vocoder receiver.

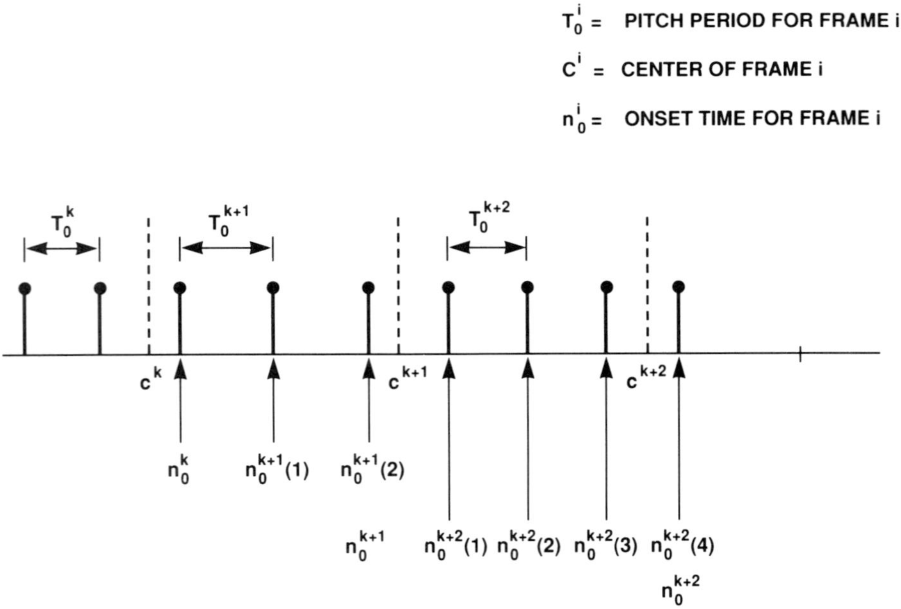

Figure 10 Determination of the synthetic onset times.

If n_0^k was the onset time on frame k and if $T_0^{k+1} = 2\pi/\omega_0^{k+1}$ was the pitch period estimated for frame $k + 1$, then a succession of onset times can be specified by

$$n_0^{k+1}(j) = n_0^k + jT_0^{k+1} \qquad j = 1,2,\ldots,J \tag{38}$$

The onset time for frame $k + 1$ is taken to be the first onset time to cross the center of frame $k + 1$.

An example of a typical sequence of onset times is shown in Fig. 10. In general, there can be more than one onset time per analysis frame, and although in theory any one of the onset times can be used, in the face of FFT quantization errors, for example, it is best to choose the onset time which is nearest the center of frame $k + 1$, since then the resulting phase errors will be minimized.

When the voicing-dependent synthetic phase model is used to replace the estimated phases in the sine wave synthesizer of Eq. (5), not only are the speaker recognition properties remarkably preserved by the synthetic speech but the speech is also of very high quality. It is notable that the synthetic speech does not have the "reverberant" quality reported in other implementations of the sine wave system, an effect which arises when the component sine waves are not forced to be phase locked.

4. THE HARMONIC MODEL FOR THE SINE WAVE FREQUENCIES

The results of the previous section have shown that it is possible to generate high-quality speech without having to code the sine wave phases. While this is a necessary first step in the development of a low-rate sine wave coder, there remains the problem of coding the amplitudes and frequencies. Almeida and Silva (1984) have developed a method for coding the sine-wave frequencies in terms of a harmonic model by using a pitch estimate during voiced speech to establish a harmonic set of sine waves. To represent unvoiced speech they used a narrowband noise representation in place of the sinusoidal model (Marques and Almeida, 1988). By contrast, the approach taken here is to use the sine wave model for voiced and unvoiced speech and, furthermore, to determine the pitch by fitting the harmonic set of sine waves to the measured set of sine waves. Moreover, the accuracy of the harmonic fit becomes an indicator of the voicing state, and this can be used to establish the probability of voicing which is used to define the frequency transition for the voicing-adaptive phase model.

4.1. Parameter Estimation for the Harmonic Sine Wave Model

As a first step in the analysis procedure, it is assumed that a frame of the input speech waveform has already been analyzed in terms of its sinusoidal components using the technique described in Section 2. The measured speech data $s(n)$ can therefore be represented as

$$s(n) = \sum_{l=1}^{L} A_l \exp[j(n\omega_l + \theta_l)] \tag{39}$$

where $\{A_l, \omega_l, \theta_l\}_{l=1}^{L}$ represent amplitudes, frequencies, and phases of the L measured sine waves. The goal is to try to represent this sinusoidal waveform by another for which all of the frequencies are harmonic. This latter waveform can be modeled as

$$\hat{s}(n;\omega_0,\theta) = \sum_{k=1}^{K(\omega_0)} \bar{A}(k\omega_0) \exp[j(nk\omega_0 + \phi_k)] \tag{40}$$

where $\omega_0 = 2\pi f_0/f_s$ is the fundamental frequency, $K(\omega_0)$ is the number of harmonics in the speech bandwidth, $\bar{A}(\omega)$ is the vocal tract envelope, and $\phi = \{\phi_1, \phi_2, \ldots, \phi_K(\omega_0)\}$ represents the phases of the harmonics. Henceforth, ω_0 will be referred to as the "pitch," although during unvoiced speech this terminology is not meaningful in the usual sense. It is desired to estimate the pitch frequency ω_0 and the phases $\{\phi_1, \phi_2, \cdots, \phi_{K(\omega_0)}\}$ such that $\hat{s}(n)$ is as "close as possible" to $s(n)$ according to some meaningful criterion.

As in the onset estimation problem described in the last section, a reasonable estimation criterion is to seek the minimum of the mean squared error,

$$\epsilon(\omega_0, \phi) = \frac{1}{N+1} \sum_{n=-N/2}^{N/2} |s(n) - \hat{s}(n;\omega_0,\phi)|^2 \tag{41}$$

over ω_0 and ϕ, since this at least ensures robustness against additive white Gaussian noise (Van Trees, 1968). The details of the analysis leading to the minimum MSE estimate of the pitch can be found in McAulay and Quatieri (1990), where it was shown that the sine wave–based pitch estimator was unambiguous and resulted in pitch-adaptive resolution, enhanced discrimination against noise, and immunity to the formant interaction problem. However, the harmonic model does not depend explicitly on the exact method used to estimate the pitch and any of the common pitch estimation techniques could be used (Hess, 1986).

4.2. Voicing Detection

In the context of the sinusoidal model, the degree to which a given frame of speech is voiced is determined by the degree to which the harmonic model fits the original sine wave data. The accuracy of the harmonic fit can be related, in turn, to the signal-to-noise ratio (SNR), defined by

$$\mathrm{SNR} = \frac{\sum_n |s(n)|^2}{\sum_n |s(n) - \hat{s}(n;\omega_0)|^2} \tag{42}$$

If the SNR is large, then the MSE is small and the harmonic fit is very good, which indicates that the input speech is most likely voiced. For small SNR, on the other hand, the MSE is large and the harmonic fit is quite poor, which indicates that the input speech is more likely to be unvoiced. Therefore, the degree of voicing is functionally dependent on the SNR. Although the determination of the exact functional form is difficult, one that has proved useful in several speech applications that use the sine wave–based pitch estimator is the following:

$$P_v = \begin{cases} 1 & \mathrm{SNR} > 10\ \mathrm{dB} \\ \frac{1}{6}(\mathrm{SNR} - 4) & 4\ \mathrm{dB} \le \mathrm{SNR} \le 10\ \mathrm{dB} \\ 0 & \mathrm{SNR} < 4\ \mathrm{dB} \end{cases} \tag{43}$$

where P_v represents the probability that speech is voiced and SNR is expressed in decibels. It is this quantity that is used to control the voicing-adaptive cutoff for the phase model derived in Section 3.4.

If some other pitch estimator is used, it will be necessary to relate its ability to detect the voicing state to the probability of voicing. For the correlation pitch estimator, for example, the pitch estimate minimizes the MSE

$$\frac{1}{N+1} \sum_{n=-N/2}^{N/2} [s(n) - s(n - \tau_0)]^2 \tag{44}$$

and, hence, its performance can be mapped directly to the expression for the SNR and in turn the voicing probability, although the constants in Eq. (43) may need to be adjusted to give the best results.

4.3. Harmonic Sine Wave Reconstruction

The simulation of the sinusoidal synthesis system was modified to force the sine wave frequencies to be harmonic in accordance with Eq. (40), and the sine wave amplitudes were replaced by the harmonic samples of the SEEVOC envelope. It was found that during unvoiced speech, meaningless pitch estimates can occur which can lead to perceptual artifacts whenever the pitch estimate is greater than about 150 Hz. This is due to the fact that, in these cases, there are too few sine waves to synthesize adequately a noise-like waveform. In some implementations of the harmonic sine wave system, a noise excitation is used in these unvoiced regions. However, this additional complexity is unnecessary because the problem can be eliminated by defaulting to a fixed low pitch ($\approx$ 100 Hz) during unvoiced speech whenever the pitch exceeds 100 Hz. The exact procedure for doing this is to define a voicing-dependent cutoff frequency, ω_c, as

$$\omega_c(P_v) = \pi P_v \tag{45}$$

which is constrained to be no smaller than 2π (1500 Hz/f_s). If the actual pitch estimate is ω_0, then the sine wave frequencies used in the reconstruction are

$$\omega_k = \begin{cases} k\omega_0 & \text{for } k\omega_0 \le \omega_c(P_v) \\ k^*\omega_0 + (k - k^*)\omega_u & \text{for } k\omega_0 > \omega_c(P_v) \end{cases} \tag{46}$$

where k^* is the largest value of k for which $k^*\omega_0 \le \omega_c(P_v)$ and where ω_u, the unvoiced pitch, corresponds to 100 Hz (i.e., $\omega_u = 2\pi(100/f_s)$). Note that if $\omega_0 < \omega_u$, then $\omega_k = k\omega_0$ for all k. The harmonic reconstruction then becomes

$$\hat{s}(n; \omega_0) = \sum_{k=1}^{K} \bar{A}(\omega_k) \exp[j(n\omega_k + \phi_k)] \tag{47}$$

Strictly speaking, this reconstruction is harmonic only during strongly voiced speech; if the speech is a voiced-unvoiced mixture then the frequencies above the cutoff, although equally spaced by ω_u, are aharmonic because they are themselves not multiples of the fundamental pitch.

If the $\{\phi_k\}$ are based on the phase of the STFT, then this model can produce synthetic speech which is of very high quality, almost perceptually equivalent to the original. Not only does this validate the performance of the harmonic model, it also shows that if the amplitudes and phases of the harmonic representation could be efficiently coded, only the pitch and voicing would be needed to code the information in the sine wave frequencies. Moreover, when the measured phases are replaced by those defined by the voicing-adaptive phase model, derived in Section 3, the synthetic speech is also of good quality, although not equivalent to that obtained using the measured harmonic phases. However, it provides an excellent basis from which to derive a low-rate speech coder.

5. SINE WAVE AMPLITUDE CODING AT LOW DATA RATES

The previous sections have shown that speech reconstruction of good quality is possible by using a harmonic representation for the sine wave frequencies and a voicing-adaptive model for the sine wave phases. Since the pitch can be coded using ≈ 7 bits and the voicing probability using ≈ 3 bits, good-quality speech at low data rates (1200 to 4800 b/s) appears to be achievable provided the sine wave amplitudes can be coded efficiently. Although the all-pole model can be adapted for coding the sine wave amplitudes, a cepstral-based approach will be pursued with the goal of achieving a more robust, less speech-dependent system.

5.1. The Cepstral Model

In Section 3 a pitch-adaptive algorithm was described for editing the sine wave amplitudes to eliminate the effects of low-level signal processing artifacts. The edited peaks were then linearly interpolated in frequency to form the SEEVOC envelope. In Section 4 this envelope was sampled at the pitch harmonics, which, when used with the STFT phases, led to high-quality synthetic speech. Therefore, it can be assumed that the SEEVOC envelope is a good starting point for the development of the low-rate amplitude coder. The first step in coding the sine wave amplitudes is to develop a parametric model for the SEEVOC envelope using the cepstral representation (Oppenheim and Schafer, 1975). Such a model can be written as

$$\log \bar{A}(\omega) = c_0 + 2\sum_{m=1}^{\infty} c_m \cos(m\omega) \tag{48}$$

$$c_m = \frac{1}{\pi}\int_0^{\pi} \log \bar{A}(\omega)\cos(m\omega)\,d\omega \qquad m = 0, 1, 2, \ldots \tag{49}$$

The cepstral representation is then truncated to M points ($M \leq 40$) and the cepstral length becomes a design parameter that will be varied depending on the coder rate. The advantage of the cepstral-based approach is that it exploits the minimum phase model, which, when smoothed, provides some additional immunity to quantization noise. More important, it provides a functional basis for warping the frequency axis prior to quantization, which can exploit perceptual properties of the ear to reduce the number of coefficients needed in the representation.

5.2. Spectral Warping

Since the ear is less sensitive to details in the spectral envelope at higher frequencies, nonlinear sampling can be used to exploit the formant broadening which occurs at the higher frequencies, thereby providing a more efficient allocation of the available bits. If $\omega' = W(\omega)$ represents the nonlinear warping of the frequency ω, and if $\omega = W^{-1}(\omega')$ represents the inverse of the warping function (an example of which is shown in Fig. 11), then a warped envelope is computed from the SEEVOC envelope using

$$\log \bar{A}_{\omega}(\omega') = \log \bar{A}[W^{-1}(\omega')] \tag{50}$$

An example of a typical SEEVOC envelope is shown in Fig. 12b, and its warped counterpart is shown in Fig. 12c. Here the SEEVOC envelope is a piecewise constant interpolation of the sine wave amplitude measurements. The cepstral envelope on the warped frequency scale is then computed using Eq. (50) in Eq. (49), in which the frequency on the warped scale, ω', is used in place of ω. An example of the warped cepstral envelope computed using 28 cepstral coefficients is shown in Fig. 12c. The cepstral envelope on the original frequency scale is then computed using

$$\log \bar{A}(\omega) = c_0 + 2\sum_{m=1}^{M} c_m \cos[mW(\omega)] \tag{51}$$

an example of which is shown in Fig. 12b. Although the standard perceptually based warping functions, such as the mel scale or the bark scale, could be used, a more general procedure was developed to allow for more flexibility in designing the coder at a multiplicity of bit rates. Basically the mapping was linear in the low-frequency region and exponential in the high-frequency region, a relationship which can be described parametrically as

$$W(\omega) = \begin{cases} \omega & 0 \le \omega \le \omega_L \\ \omega_L(1+\alpha)^{\omega-\omega_L} & \omega_L \le \omega \end{cases} \tag{52}$$

If M cepstral coefficients are used in the representation, then the warped bandwidth $W(\pi)$ is subdivided into M channels (counting a channel at DC). The spacing of the uppermost channels on the original frequency scale, β, becomes a criterion for the design of the warping function. Ideally, the linear region should be as large as possible, subject to the constraint that the channel spacing of the uppermost channels, β, is less than some reasonable amount ($\approx$ 300 Hz). Therefore, the parameters $(M, \alpha, \omega_L, \beta)$ are varied to give the best performance at a given bit rate. So far no optimal procedure has been

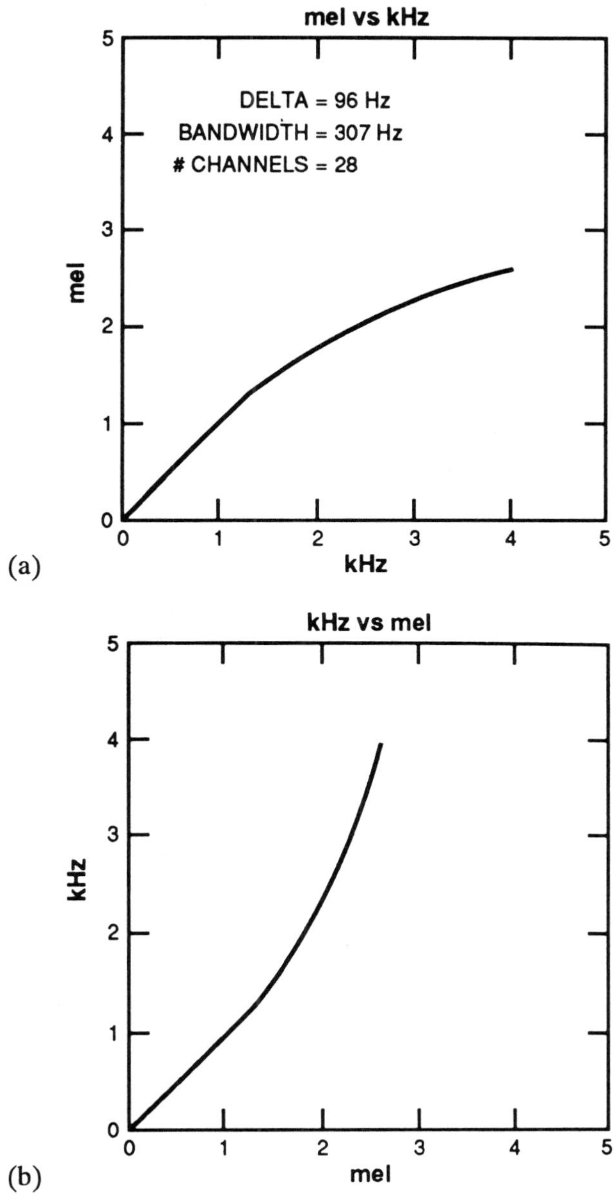

Figure 11 Typical warping function and its inverse.

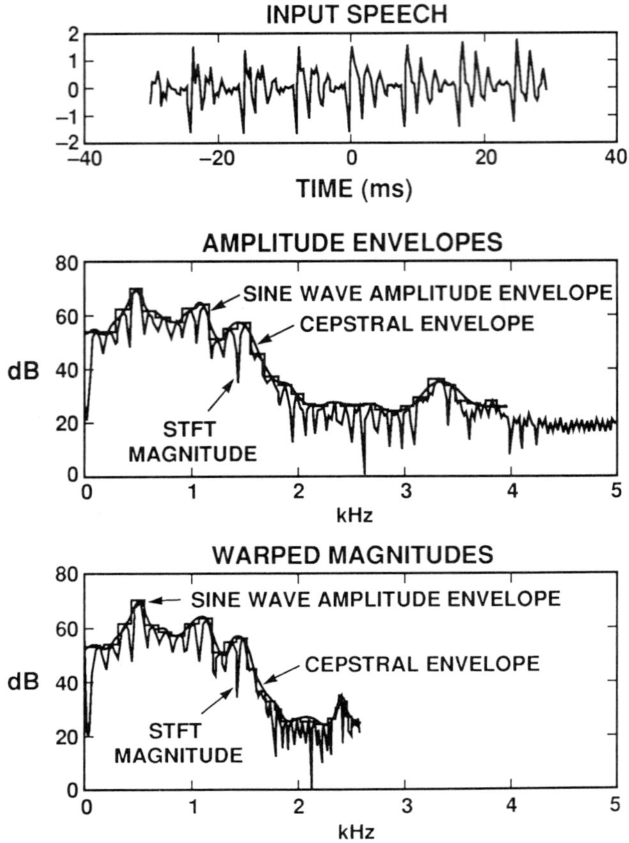

Figure 12 Typical warped cepstral amplitude envelope estimate.

found for choosing these parameters. Some typical designs which have been found to give reasonable performance are listed in Table 1.

5.3. Cepstral Transform Coding

Attempts have been made in the past to code the speech envelope using the cepstrum (Oppenheim, 1969) but these have been unsuccessful at low data rates because of the large dynamic range and uncorrelated nature of the cepstral coefficients. To avoid this problem, the truncated cepstral sequence is transformed back into the frequency domain to result in another set of

Table 1 Typical Warping Function Parameters

Data rate	M	ω_L Hz	β Hz	α
1200	28	800	300	0.05
2400	32	1000	280	0.02
4800	40	2300	250	0.01

coefficients satisfying the cosine transform pair:

$$g_k = c_0 + 2\sum_{m=1}^{M-1} c_m \cos\left(2\pi \frac{mk}{2M-1}\right) \tag{53}$$

$$c_m = g_0 + 2\sum_{k=1}^{M-1} g_k \cos\left(2\pi \frac{km}{2M-1}\right) \tag{54}$$

The advantage in using the transform coefficients $\{g_k\}_{k=0}^{M-1}$ instead of the cepstral coefficients $\{c_m\}_{m=0}^{M-1}$ is that they correspond to equally spaced samples (on the warped scale) of the low-pass filtered log magnitude and, hence, are correlated in frequency and can be quantized in units of dB in accordance with the perceptual properties of the ear. A set of these transform coefficients, which shall henceforth be referred to as channel gains, is shown in Fig. 13. Since the channel gains are the transform of the cepstral coefficients computed on the warped scale, on the original scale they are nonlinearly spaced, as shown in Fig. 13. These channel gains are then quantized using differential pulse code modulation (DPCM) techniques developed in the context of the channel vocoder (Holmes, 1980) and in the SEEVOC vocoder (Paul, 1981). The overall spectral level is set using 4–5 bits to code the first channel gain. Depending on the data rate, the available bits are partitioned among the $M-1$ remaining channel gains using step sizes that range from 1 to 6 dB. An example of a set of coded channel gains is shown in Fig. 13 for 2400 b/s operation. The reconstructed envelope is also shown in Fig. 13.

5.4. Positive Slope Overload

One problem with the DPCM coder is the fact that at low data rates its dynamic range at successive channels is limited and a positive slope overload condition can occur in which the coder fails to track the leading edge of a sharply peaked format. An example of this effect is shown in Fig. 14.

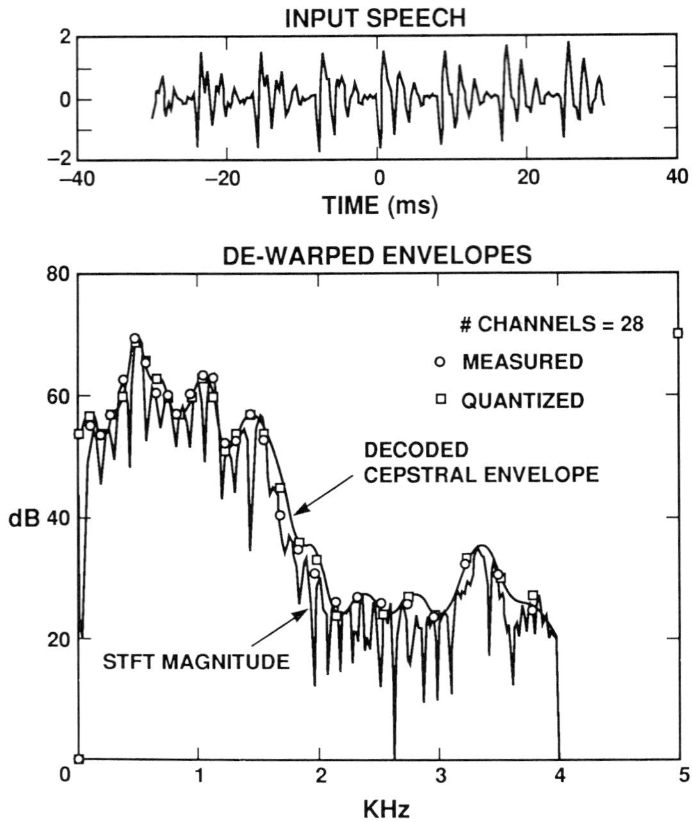

Figure 13 Typical unquantized and quantized channel gains (cepstral transform coefficients).

Once the parameters of the DPCM coder have been set (i.e., the number of channels, the number of bits per channel, and the step size per bit), it is possible to define a "reachability" condition that ensures that positive slope overload will not occur. For example, if the distance to the current amplitude from its decoded neighbor is less than the largest step size that can be taken at that channel, then positive slope overload will not occur. This condition has been used to develop a recursion for an amplitude threshold which is used to replace the DPCM quantized amplitudes by a set of amplitudes which will not experience the positive slope overload condition (provided the coder parameters were reasonably well designed in the first place) (Sabin,

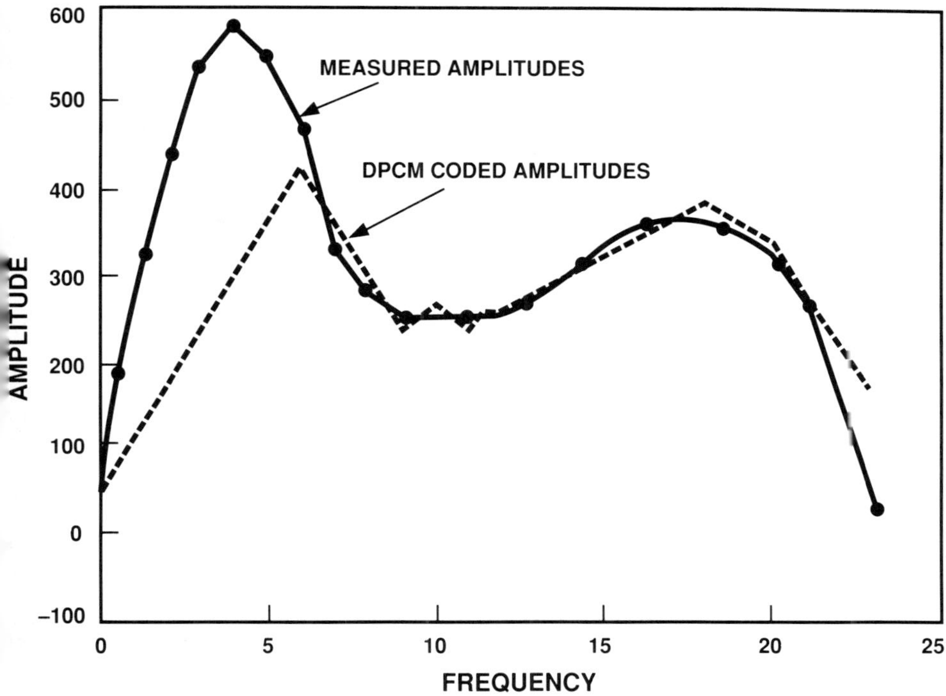

Figure 14 DPCM amplitude coding.

1991). Therefore, as the DPCM coding procedure proceeds, if an amplitude to be coded is below this threshold, it is replaced by the value at the threshold. Although this will result in some formant broadening, the formant shift will not occur, and this seems to be the more deleterious effect on synthetic speech quality. An example of the application of this algorithm to the data in Fig. 14 is shown in Fig. 15, which demonstrates the protection against the slope overload condition at the expense of the formant broadening.

5.5. Postfiltering

Although the synthetic speech produced by this system is of reasonably good quality, a slight muffling effect can be detected. This muffling increases as the amplitude envelope becomes more structured, i.e., as the number of cepstral coefficients is reduced. Such a quality loss has also been reported in code-excited LPC systems, where it has been speculated that the muffling is due to

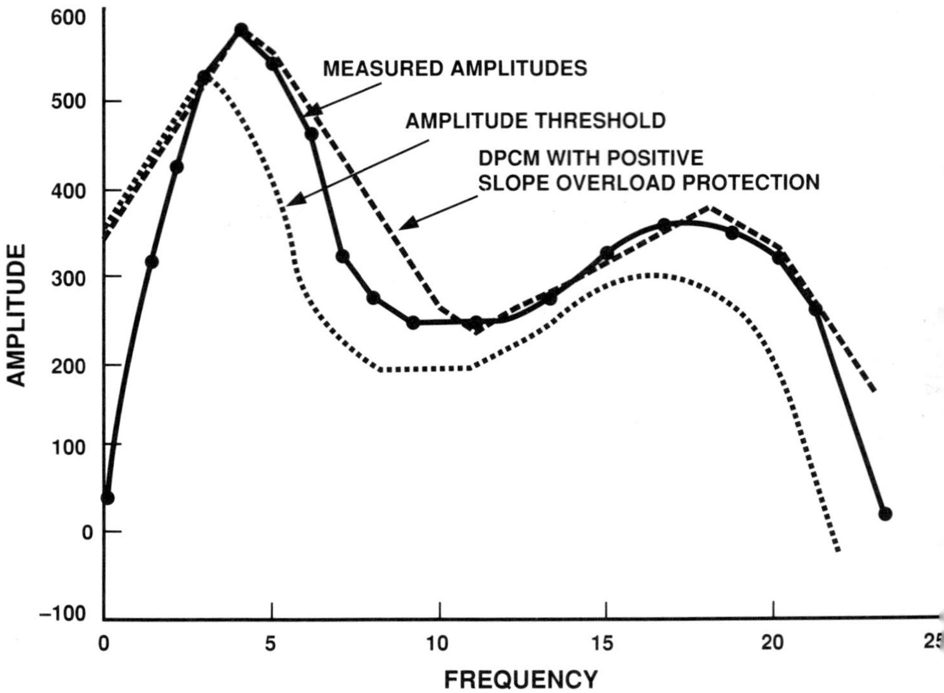

Figure 15 DPCM amplitude coding with positive slope overload protection.

coder noise showing up in the formant nulls (Chen and Gersho, 1987). Techniques have been developed for filtering out this noise by using a postfilter. In the sine wave–based system the noise-rejection postfilter is straightforward to implement, since all of the necessary data are available in the frequency domain.

The first step in the design of the postfiltering procedure is to remove the spectral tilt from the log-spectral data using the first two decoded cepstral coefficients. The spectral tilt is then given by

$$\log T(\omega) = c_0 + 2c_1 \cos \omega \tag{55}$$

and this is removed from the decoded envelope, denoted by $\log S(\omega)$, to give the residual envelope

$$\log R(\omega) = \log S(\omega) - \log T(\omega) \tag{56}$$

This is then normalized to have unity gain and compressed using a root-γ compression rule ($\gamma \cong 0.2$). If $R_{\max}$ is the maximum value of the residual envelope, then the postfilter is taken to be

$$P(\omega) = \left[\frac{R(\omega)}{R_{\max}}\right]^{\gamma} \qquad 0 \le \gamma \le 1 \tag{57}$$

At the formant peaks the normalized residual envelope in Eq. (57) will have unity gain and will not be altered by the compressor. In the formant nulls, the compressor will reduce the fractional values further so that overall a greater null depth will result.

The resulting compressed envelope then becomes the postfilter and is applied to the measured envelope to give

$$\hat{S}(\omega) = P(\omega)S(\omega) \tag{58}$$

which should cause the formants to narrow and the formant nulls to deepen, thereby reducing the effects of the coder noise. An example of the application of this procedure to real speech data is shown in Fig. 16. Unfortunately, the spectral tilt, shown in Fig. 16b, does not track the formant peaks exactly and the resulting postfilter, shown in Fig. 16c can introduce significant spectral imbalance. The effect of this imbalance can be reduced somewhat by renormalizing the postfiltered envelope so that the energy after postfiltering is equal to the energy before postfiltering. The resulting postfiltered envelope is shown in Fig. 16b, and it can be seen that the spectral nulls are indeed reduced but at the expense of some distortion in the amplitudes at the first formant peak. This problem would not have arisen had the spectral tilt provided a better fit to the formant peaks. Methods for obtaining such a tilt function are currently being investigated.

5.6. Frame-Fill Interpolation

The coding schemes described in the previous sections have proved to be quite efficient and have resulted in good-quality synthetic speech at data rates as low as 2400 b/s. A simple and straightforward approach to reducing the transmission bandwidth further is to transmit the amplitude data for every second frame, using some control information to instruct the synthesizer how to reconstruct (or "fill in") the missing information (McLarnon, 1978). A rule that has been found to be quite useful is simply to compare the amplitude data for the frame to be omitted to the amplitude data on the frames immediately preceding and succeeding. The data that "best" represents the midframe data is determined, and 1 bit can be used to signal the decision to the receiver. An additional bit can be used to allow for candidate fits that are interpolated values of the end-point data. Therefore, in the present context 2 bits are used, allowing for four frame-fill options. If $A_{k-1}(\omega)$, $A_k(\omega)$

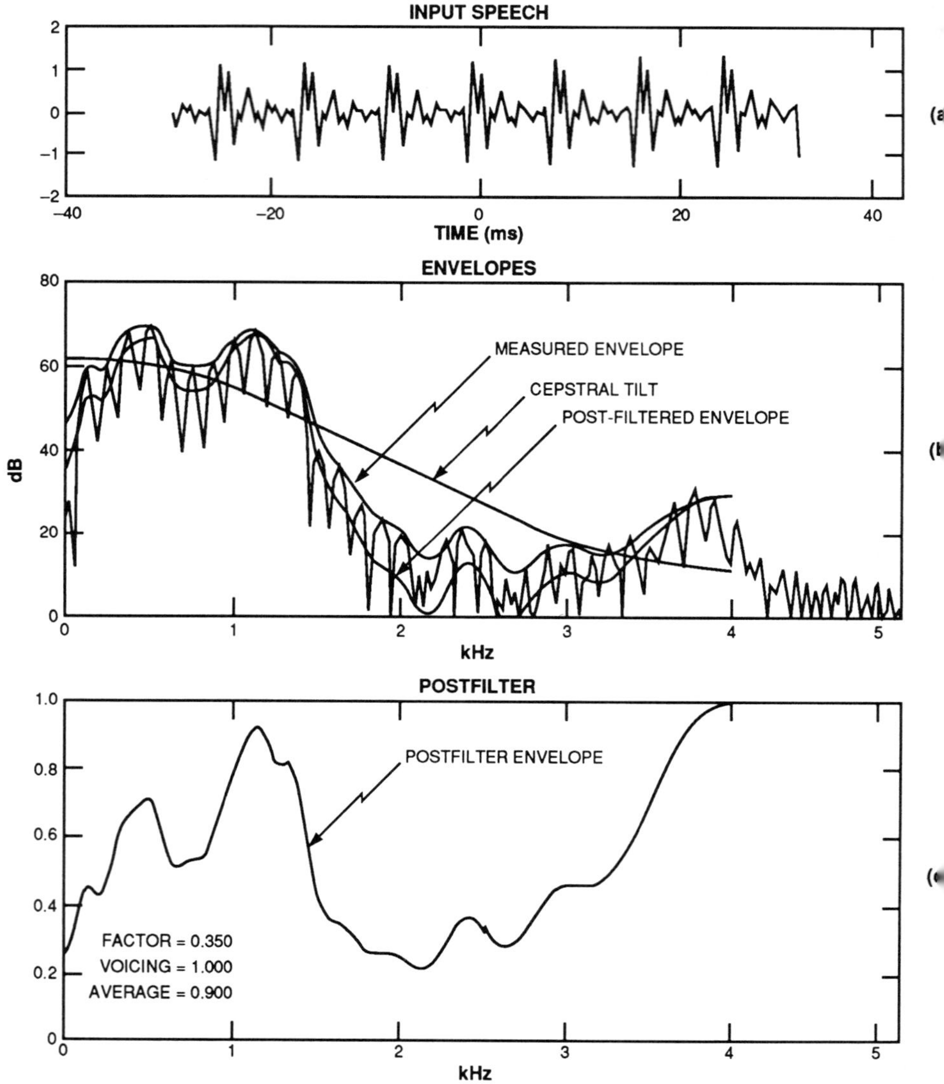

Figure 16 Frequency domain postfilter design.

$A_{k+1}(\omega)$ represent the succession of envelopes of the sine wave amplitudes, then the four frame-fill options to be used in place of $A_k(\omega)$ are

$$\log A_1(\omega) = \log A_{k-1}(\omega) \tag{59a}$$

$$\log A_2(\omega) = \log A_{k+1}(\omega) \tag{59b}$$

$$\log A_3(\omega) = 0.5 \log A_{k-1}(\omega) + 0.5 \log A_{k+1}(\omega) \tag{59c}$$

$$\log A_4(\omega) = 0.25 \log A_{k-1}(\omega) + 0.75 \log A_{k+1}(\omega) \tag{59d}$$

At the transmitter the "best" fit to the frame-k data is determined using the squared-log-error criterion:

$$d(A_i, A_k) = \frac{1}{\pi} \int_0^{\pi} [\log A_i(\omega) - \log A_k(\omega)]^2 \, d\omega \qquad i = 1, 2, 3, 4 \tag{60}$$

and the frame-fill data $\hat{A}_k(\omega)$ are determined for the condition resulting in the smallest value of $d(\,)$. In the present context, the amplitude data have been represented in terms of M cepstral coefficients; hence,

$$\log A(\omega) = c_0 + 2 \sum_{m=1}^{M-1} c_m \cos(m\omega) \tag{61}$$

from which it follows that

$$d(A_i, A_k) = (c_0^i - c_0^k)^2 + 4 \sum_{m=1}^{M-1} (c_m^i - c_m^k)^2 \tag{62}$$

Moreover, the cepstral coefficients are the transform of the channel gains,

$$c_m = g_0 + 2 \sum_{m=1}^{M-1} g_m \cos(m\omega) \tag{63}$$

hence,

$$d(A_i, A_k) = (g_0^i - g_0^k)^2 + 4 \sum_{m=1}^{M-1} (g_m^i - g_m^k)^2 \tag{64}$$

The advantage of this representation is that the channel gains are quantized and transmitted to the receiver. Therefore, if $\{\hat{g}_m^i\}_{m=0}^{M-1}$ are the quantized channel gains, then computing the metric

$$d(A_i, A_k) = (\hat{g}_0^i - g_0^k)^2 + 4 \sum_{m=1}^{M-1} (\hat{g}_m^i - g_m^k)^2 \tag{65}$$

allows the selection process to be based on the coded amplitude envelope, thereby providing more immunity against the deleterious effects of the quantization process. Moreover, since the distance computations are done on the coded channel gains, computation efficiencies can be obtained by specifying the frame-fill options in terms of the channel gains directly. Using Eqs. (62) and (64) in (60), the options are given by

$$\log g_m^1 = \log \hat{g}_m^{k-1} \quad (66a)$$
$$\log g_m^2 = \log \hat{g}_m^{k+1} \quad (66b)$$
$$\log g_m^3 = 0.5 \log \hat{g}_m^{k-1} + 0.5 \log \hat{g}_m^{k+1} \quad (66c)$$
$$\log g_m^4 = 0.25 \log \hat{g}_m^{k-1} + 0.75 \log \hat{g}_m^{k+1} \quad (66d)$$

This frame-fill method was incorporated into the coder algorithm and immediately led to a viable 1200 b/s compression system that maintains speaker naturalness and speaker identifiability. Although originally developed for the low-rate applications, the frame-fill technique has proved useful at higher rates as well, since it adds better temporal resolution to the synthetic waveform. In addition, at 2400 b/s, the frame rate can be decreased from 50 to 40 Hz, which allows the allocation of more bits to better encode the amplitude data.

6. EVALUATION RESULTS

In this chapter a sinusoidal model for the speech waveform was derived. The model led to a new analysis-synthesis technique that is characterized by the amplitudes, frequencies, and phases of the component sine waves, and a "minimal" parameter set was derived from this representation for the purpose of speech coding. In this section, evaluation of the speech coder, based on quantization of the sine wave parameter set, will be described and evaluation of different coding schemes in the range 4.8 to 1.2 kb/s will be discussed.

6.1. The Zero-Phase System

In order to evaluate the performance of this new approach to speech coding, a non-real-time floating-point computer simulation of this system was developed on a SUN-4/370 computer. The speech processed in the simulation was low-pass filtered at 5 kHz, digitized at 10 kHz, and analyzed at a 10-ms frame interval. A harmonic zero-phase version [i.e., the system phase $\Phi_s(\omega)$ in Eq. (35) was set to zero] of the system, using only pitch, voicing, and amplitude information, without coding, has achieved a diagnostic acceptability measure (DAM) of 63.0. This shows that high-quality synthetic speech can be achieved at low rates provided the sine wave amplitude information is coded efficiently.

The same harmonic zero-phase system was implemented in real time but with a 20-ms frame interval and with fixed-point arithmetic at 4800 and 2400 b/s. The real-time fixed-point implementation of these algorithms using three ADSP-2100 digital signal processing chips has been tested by Dynastat, Inc., using the DAM and the diagnostic rhyme test (DRT), and the results are given in Table 2. These results are reasonably good given the inaccuracies due to fixed-point arithmetic and algorithmic compromises that had to be made in order to achieve real-time operation (neither the warped cepstral transform coder nor the minimum phase system phase was used in this implementation). The warped cepstral transform coding technique has not yet been formally evaluated, but based on informal listening tests it is anticipated that performance substantially better than that obtained by the real-time system will likely be achieved.

6.2. Conclusions

One of the major advantages of the sine wave approach to speech coding is the way in which the parametric model lends itself to mathematical analysis. This was shown in the development of the onset estimator in Section 3. In addition, the frequency domain representation lends itself well to low-rate

Table 2 STC DAM and DRT Test Results for Zero-Phase Harmonic Real Time[a] System

Diagnostic acceptability measure (DAM)			
Rate (b/s)	6-Speaker	Male	Female
4800	62.0 (1.7)	64.0	59.8
2400	56.8 (2.0)	58.2	55.3
Diagnostic rhyme test (DRT)			
4800	92.7 (0.78)	94.4	91.0
2400	90.1 (0.77)	92.3	88.0

[a] Using three ADSP2100 fixed-point chips.

coding, as the warped cepstral transform coding technique effectively exploits some of the perceptual properties of the ear. Since this work is relatively new, it is expected that more improvements in quality will be obtained with more refined designs when methods are developed for evaluating the trade-offs between the number of channels, the warping function, and the quantization algorithms.

Additional work also needs to be done in the design of the postfilter. It was noted that as the amplitude encoding process applied more smoothing to the formant nulls, the synthetic speech became more "muffled" or "buzzy." By attenuating the formant nulls, the postfilter significantly reduce these effects. If too much suppression is applied by the postfilter, as might happen in successive tandem interconnections, the synthetic speech can begin to sound "hollow." Work is in progress to improve the design methodology of the postfilter.

Another interesting possibility is to apply the ideas behind the CELP coder to the sine wave model, in particular to the residual phase model developed in Section 3. In fact, combining the onset estimator and the warped cepstral transform coder could lead to a very efficient predictive model which might lead to residual waveforms that could be coded well using stochastic codebook methods. Therefore, it is anticipated that the sine wave–based approach will continue to lead to new approaches to speech coding which, at the least, will bring new insights into determining the essential features for successful speech coding at low data rates.

ACKNOWLEDGMENT

This work was sponsored by the Department of the Air Force.

REFERENCES

Almeida, L. B., and Silva, F. M. (1984). Variable-frequency synthesis: An improved harmonic coding scheme, *Proc. IEEE Int. Conf. Acoustics, Speech, and Signal Processing*, San Diego, p. 27.5.1.

Atal, B. S., and Remde, J. R. (1982). A new model of LPC excitation for producing natural-sounding speech at low bit rates. *Proc. IEEE Int. Conf. Acoustics, Speech, and Signal Processing*, Paris, p. 614.

Chen, J.-H., and Gersho, A. (1987). Real-time vector APC speech coding at 2800 b/s with adaptive postfiltering. *Proc. IEEE Int. Conf. Acoustics, Speech, and Signal Processing*, Dallas, p. 51.3.1.

Flanagan, J. L., and Golden, R. M. (1966). Phase vocoder. *Bell Syst. Tech. J. 45*:1493–1509.

Griffin, D., and Lim, J. S. (1988). Multiband excitation vocoder. *IEEE Trans. Acoust., Speech, Signal Process. ASSP-36*:1223–1235.

Hedelin, P. (1981). A tone-oriented voice-excited vocoder. *Proc. IEEE Int. Conf. Acoustics, Speech, and Signal Processing*, Atlanta, p. 205.

Hess, W. (1983). *Pitch Determination of Speech Signals*. Springer-Verlag, Berlin.

Holmes, J. N., The JSRU Channel Vocoder, *IEEE Proc., 127*:53–60.

Itakura, F., and Saito, S. (1970). A statistical method for estimation of speech spectral density and formant frequencies. *Electron. Commun. Jpn. 53-A*:36.

Kang, G. S., and Everett, S. S. (1985). Improvement of the excitation source in the narrow-band linear prediction vocoder. *IEEE Trans. Acoust., Speech, Signal Process. ASSP-33*:377–386.

Maher, R., and Beauchamp, J. W. (1990). An investigation of vocal vibrato for synthesis. *Appl. Acoust. (U.K.) 30*:219–245.

Makhoul, J., Viswanathan, R., Schwartz, R., and Huggins, A. W. F. (1978). A mixed-source model for speech compression and synthesis. *Proc. IEEE Int. Conf. Acoustics, Speech, and Signal Processing*, Tulsa, p. 163.

Malah, D. (1979). Time-domain algorithms for harmonic bandwidth reduction and time scaling of speech signals. *IEEE Trans. Acoust., Speech, Signal Process. ASSP-27*:121–133.

Markel, J. D., and Gray, A. H. (1980). *Linear Prediction of Speech*. Springer-Verlag, New York.

Marques, J. S., and Almeida, L. B. (1988). New basis functions for sinusoidal decomposition. *Proc. EUROCON*, Stockholm.

McAulay, R. J., and Quatieri, T. F. (1986). Speech analysis-synthesis based on a sinusoidal representation. *IEEE Trans. Acoust., Speech, Signal Process. ASSP-34*: 744–754.

McAulay, R. J., and Quatieri, T. F. (1990). Pitch estimation and voicing detection based on a sinusoidal model. *Proc. IEEE Int. Conf. Acoustics. Speech, and Signal Processing*, Albuquerque, pp. 249–252.

McAulay, R. J., Parks, T. M., Quatieri, T. F., and Sabin, M. (1989). Sine-wave amplitude coding at low data rates. IEEE Workshop on Speech Coding, Vancouver, B.C., Canada.

McLarnon, E. (1978). A method for reducing the frame rate of a channel vocoder by using frame interpolation. *Proc. IEEE Int. Conf. Acoustics, Speech, and Signal Processing*, Washington, D.C., pp. 458–461.

Oppenheim, A. V. (1969). Speech analysis-synthesis system based on homomorphic filtering. *J. Acoust. Soc. Am. 45*:459.

Oppenheim, A. V., and Schafer, R. W. (1975). *Digital Signal Processing*. Prentice-Hall, Englewood Cliffs, N.J.

Paul, D. B. (1981). The spectral envelope estimation vocoder. *IEEE Trans. Acoust., Speech, Signal Process., ASSP-29*:786–794.

Portnoff, M. (1981). Short-time Fourier analysis of sampled speech. *IEEE Trans. Acoust., Speech, Signal Process. ASSP-29*:364–373.

Quatieri, T. F., and Danisewicz, R. G. (1990). An approach to co-channel talker interference suppression using a sinusoidal model for speech. *IEEE Trans. Acoust., Speech, Signal Process ASSP-38*:56–69.

Quatieri, T. F., and McAulay, R. J. (1986). Speech transformations based on a sinusoidal representation. *IEEE Trans. Acoust., Speech, Signal Process. ASSP-34*: 1449–1464.

Quatieri, T. F., and McAulay, R. J. (1989). Phase coherence in speech reconstruction for enhancement and coding applications. *Proc. IEEE Int. Conf. Acoustics, Speech, and Signal Processing*, Glasgow, pp. 207–209.

Quatieri, T. F., and McAulay, R. J. (1991). Peak-to-rms reduction of speech based on a sinusoidal model. *IEEE Trans. Acoust., Speech, Signal Process. SP-39*:273–288.

Rabiner, L. (1977). On the use of autocorrelation analysis for pitch detection. *IEEE Trans. Acoust., Speech, Signal Process. ASSP-25*:24–33.

Rabiner, L., and Schafer, R. (1978). *Digital Processing of Speech*. Prentice-Hall, Englewood Cliffs, N.J.

Sabin, M. J. (1991). DPCM coding of spectral amplitudes without positive slope overload. *IEEE Trans. Signal Process. SP-39*, No. 3:756–758.

Schroeder, M. R., and Atal, B. S. (1985). Code-excited linear prediction (CELP): High-quality speech at very low bit rates. *Proc. IEEE Int. Conf. Acoustics, Speech, and Signal Processing*, Tampa, p. 937.

Smith, J., and Serra, X. (1987). PARSHL: An analysis/synthesis program for non-harmonic sounds based on a sinusoidal representation. *ICMC-1987: Proc. Int. Computer Music Conf.*, Computer Music Association, pp. 290–297.

Van Trees, H. (1968). *Detection, Estimation and Modulation Theory*, Part I. Wiley, New York.

7

Very Low-Bit-Rate Speech Coding

MASAAKI HONDA and YOSHINAO SHIRAKI NTT Basic Research Laboratories, Tokyo, Japan

1. INTRODUCTION

Encoding speech signals with as low a bit rate as possible while maintaining acceptable speech intelligibility is an important problem in speech coding. A low-bit-rate speech coding system comprises two major processes. One is parametric representation of speech signals by a minimal set of parameters to characterize the excitation source and a spectral shaping filter. The other is quantization of these speech parameters for conversion into discrete symbols for transmission. A linear predictive coding (LPC) vocoder is a simple parametric model of the human vocal system for characterizing speech signals in terms of the LPC coefficients (spectral parameters), gain, voiced-unvoiced decision, and a pitch value if the sound is voiced.

Many quantization techniques have been developed for representing the spectral parameters in the smallest number of bits (Viswanathan and Markel, 1975; Gray and Markel, 1976; Tohkura and Itakura, 1979). Scalar quantization techniques quantize the spectral parameters separately using uniform or nonuniform quantization in terms of appropriate LPC parameters, such as log ratios or line spectral parameters. Scalar quantization is a relatively simple procedure, but its performance in bit rate reduction is limited. If scalar quantization is applied for encoding every frame of the spectral and excitation parameters, a total of about 2.4 kilobits per second (kbps) of information is necessary to maintain acceptable speech quality.

Vector quantization (VQ) techniques quantize a set of spectral parameters jointly as a single vector (Gray, 1984; Makhoul et al., 1985; Buzo et al., 1980). Vector quantization is a pattern-matching process. Given an input vector of spectral parameters, the best match of this vector from a finite set of code vectors stored in a codebook is determined according to a minimum distortion rule. Vector quantization employs more efficiently the dependence of the components of the input vector, which cannot be captured in scalar quantization. Accordingly, it can be used to realize an 800-bps vocoder which provides the same quality as a 2.4 kbps LPC vocoder that employs scalar quantization (Wong et al., 1982).

Further bit rate reduction in the quantization of spectral parameters can be achieved by employing the interframe or time dependence of the spectral parameters. Finite-state vector quantization (FSVQ) uses an adaptive codebook dependent on the previously selected code word in quantizing an input vector (Foster et al., 1985; Dunham and Gray, 1985; Juang, 1986). This leads to a reduction in the number of code vectors used in quantizing each input vector according to the time-dependent structure of the spectral parameters. Variable frame rate VQ quantizes the spectral parameters only when the properties of the speech signal have changed sufficiently (Paul, 1983; Viswanathan et al., 1982). The spectral parameters of the nonquantized frame are regenerated through linear or some other interpolation of the parameters of the two quantized frames. These quantization techniques are useful for coding as low as 400 bps but are not sufficient for maintaining reasonable spectral distortion and speech intelligibility for data rates below about 300 bps.

Another approach using the time dependence of the spectral parameters is to perform VQ on multiple frame vectors. This method is referred to as either matrix quantization (MQ) (Wong et al., 1983; Tsao and Gray, 1985) or segment quantization (Roucos et al., 1982, 1983; Shiraki and Honda, 1988). Segment quantization is an extension of vector quantization based on the principle that rate-distortion performance is improved by using a longer block for quantization. The time dependence of consecutive frames is included implicitly in the spectral segment, unlike the situation in variable frame rate VQ (FSVQ). This approach has the potential to achieve speech coding at very low rates, below 300 bps, despite the substantial amount of computation.

In this chapter, very low bit rate speech coding techniques are presented with a focus on segment quantization. Segment quantization is classified into two types according to whether the segment length is fixed or variable. Fixed-length SQ is a straightforward extension of the ordinary VQ technique. Variable-length SQ requires additional segmentation of the spectral sequence into variable-length segments and variable-length segment matching between the input and code segments. A joint segmentation and quantization technique is considered capable of coping with this problem. In addition, a quantization procedure for finding the optimum segment boundaries and code-segment sequence which minimizes the total distortion is presented. Both fixed-length and variable-length SQ generally use a concatenated model to represent a sequence of spectral parameters. Thus, concatenating SQ is extended to a temporally overlapping SQ, which allows for temporal overlapping of code segments in representing a spectral sequence. An optimum quantization procedure of the overlapping SQ is presented. We also present design procedures for variable-length SQ for the concatenating and overlapping cases. Codebook adaptation is important in achieving a speaker-independent segment vocoder. A speaker adaptation procedure for a segment codebook based on a piecewise linear mapping method is also presented. Performance in terms of distortion and adaptation speed is compared for the speaker-independent and speaker-dependent codebook approaches.

2. VARIABLE-LENGTH SEGMENT QUANTIZATION

SQ in voice coding uses a spectral segment representing multiple frames of spectral parameters as a quantization unit. Denoting the segment by a matrix X_k which has l_k column vectors consisting of spectral parameters, the original spectral sequence can be represented as the sequence of segments $X_1, \ldots, X_k, \ldots, X_M$. Here X_1 is the first segment and X_M is the last segment in a given interval, for example, a voice interval having silence segments on both sides. Each segment length l_k can be either fixed or variable. On the other hand, the code segment Y_i $(i = 1, \ldots, N)$ is a fixed l-length segment.

If a sequence of spectral parameters is segmented into variable-length segments by some means, such as one based on detecting spectral changes, variable-length SQ is performed in a manner similar to ordinary VQ using a minimum distortion rule such as

$$\hat{X}_k = \arg\min_{Y_i \in C} d(X_k, Y_i) \tag{1}$$

Here $d(X, Y)$ represents the distance measure between the input and the code segment. Upon computation of the distance, an l-length code segment is time warped to an l_k-length segment by linear interpolation and resampled at equal-interval l vectors of the code segment. Resampling is done either uniformly in time or uniformly in distance along the trajectory in the spectral parameter space. For both cases, this procedure is represented in matrix form as

$$\Phi = YH_{l_k} \tag{2}$$

where H_{l_k} is the l by l_k time-warping matrix, whose elements are specified only by the input segment length l_k. Note that a nonlinear time warping in converting from fixed to variable segment lengths may provide better spectral matching between the original and code segments, but this needs additional information for transmitting the time-warping path.

The segment distance is usually computed as an accumulated sum of the spectral distances between the input and time-warped code vectors over the segment frames for computational simplicity and because there is no reasonable distance measure for the dynamical behavior of the speech spectrum. Several spectral distances have been proposed in both scalar and vector quantization studies. Weighted Euclidean distance of line spectral pairs (LSP's) is well suited to voice coding because it closely approximates the logarithmic spectral distortion in a simple form and because LSP parameters have good temporal continuity (Itakura, 1975; Sugamura and Itakura, 1981).

Thus, segment distance using weighted LSP distance with gain has the form (Shiraki and Honda, 1988)

$$d(X,Y) = \mathrm{tr}(X - YH_{l_k})^t W(X - YH_{l_k}) \tag{3}$$

where $x = (\theta_1, \theta_2, \ldots, \theta_p, P)^t$ and $y = (\theta_1, \theta_2, \ldots, \theta_p, P)^t$ are input and code vectors consisting of the pth order LSP parameters and speech power in dB. W is a diagonal weighting matrix, whose elements $\{w_{ii}\}_{i=1}^{y}$ are determined so as to approximate the spectral distortion in dB, and $w_{p+1,p+1}$ is determined so as to maximize the intelligibility of the coded speech.

2.1. Joint Segmentation and Quantization

Thus far, it has been assumed that segmentation of a spectral sequence into variable-length segments is performed regardless of the following quantization process. However, quantizing distortion over the entire spectral sequence depends on the segmentation. Joint segmentation and quantization determines both the sequence of code segments and their lengths or segment boundaries so as to minimize the total spectral distortion over the entire speech interval (Shiraki and Honda, 1988). This improves the quantizing distortion performance. The procedure is depicted in Fig. 1. The entire spectral sequence is generated by concatenating the code segments

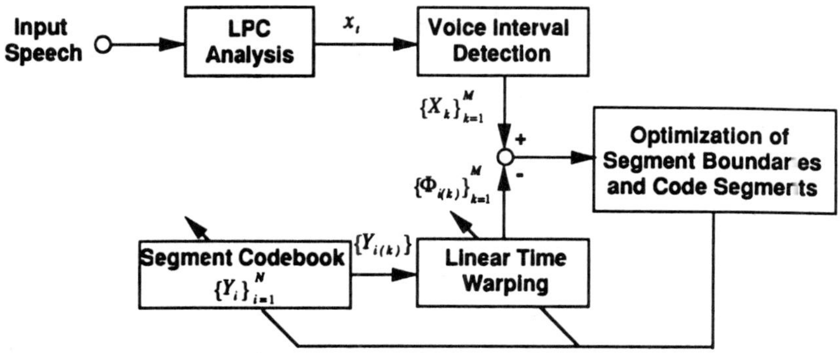

Figure 1 Block diagram of variable-length segment quantization of spectral parameters based on joint segmentation and quantization.

using fixed-to-variable length conversion and is then compared with the input spectral sequence. The optimum segment sequence for minimizing the overall distortion is given as

$$\hat{X}_k \,|_{k=1}^{M} = \arg \min_{t_k|_{k=1}^{M}} \left[\min_{Y_k|_{k=1}^{M}} \sum_{k=1}^{M} d\,(X_k, Y_i(k)) \right] \tag{4}$$

where X_k is the input segment having boundary positions T_{k-1} and t_k. $t_0 = 0$, $t_m = T$, and T is the total number of frames in a voice interval. Each segment length l_k is given by $t_k - t_{k-1}$.

This optimum solution can be efficiently obtained using the dynamic programming technique. Denoting the segment X_k by $X(t_{k-1}, t_k)$ to make clear the dependence of the input segment on the boundaries, the minimum accumulated distortion $D(t_k)$ at t_k is calculated using the recursive form

$$D(t_k) = \min_{t_{k-1} \in R_{k-1}} D(t_{k-1}) + q\,(X(t_{k-1}, t_k), Y)) \tag{5}$$

$$\beta(t_k) = \arg \min_{t_{k-1} \in R_{k-1}} (D(t_{k-1}) + q\,(X(t_{k-1}, t_k), Y))$$

$$t_k^0 - \Delta \le t_k \le t_k^0 + \Delta, k = 1, 2, \ldots, M \tag{6}$$

where

$$D(t_0) = 0$$

$$q(X, Y) = \min_{Y_i \in C} d(X, Y_i) \tag{7}$$

$$R_{k-1} = \left\{ t \mid t_{k-1}^0 - \Delta \le t \le \min\left(t_{k-1}^0 + \Delta, t_k\right) \right\}$$

Here $\beta(t_k)$ is the previous optimum segment boundary for a given boundary t_k. t_k^0 is an initial estimate of the segment boundary, and Δ is the search range for the segment boundary. $q(X,Y)$ is the minimum distortion when quantizing X using segment codebook $C = \{Y_i\}_{i=1}^{M}$. After the final accumulated distortion $D(t_M)$ has been computed, the optimum boundary positions are obtained using the backward recursion

$$t_{k-1}^* = \beta(t_k) \qquad k = M, M-1, \ldots, 2 \tag{8}$$

The search range Δ controls the performance and the computational amount of joint segmentation and quantization. As the search range becomes larger, the constraint approaches the limit of $t_{k-1} < t_k < t_{k+1}$, which provides the minimum distortion by Eq. (4). Experimental results show that nearly optimum performance can be obtained with moderate computational complexity by using initial estimates of the segment boundaries based on detecting spectral changes and by setting the search range to a value larger than 100 ms.

Using the training data of spectral parameter sequences, the variable-length SQ is designed by finding both the segment boundaries and the code segments so as to minimize

$$D = \sum_{k=1}^{M} \min_{Y_i \in C} d\,(X(t_{k-1}, t_k), Y_i) \tag{9}$$

The minimization is performed in an iterative form consisting of segment boundary updating and SQ codebook updating. The procedure is depicted in Fig. 2. The SQ codebook design algorithm is as follows.

Step 1 Initialization:
Training vector sequence $\{Y_i\}_{i=1}^{N}$, initial SQ codebook $C_0 = \{Y_i\}_{i=1}^{N}$, initial segment boundaries $\{t_k^0\}_{k=1}^{M}$, and thresholds ϵ_1 and ϵ_2 are given. Set $n = 0$ and initial distortion $D_0 = \infty$.

Step 2 Segment data updating:
Increment n. Update the training segment sequence to $\{X_k(n) = X(T_{k-1}^*, t_k^*)\}_{k=1}^{M}$, using the joint segmentation and quantization procedure with SQ codebook C_{n-1} for the training vector sequence. Then update the initial segment boundaries to $\{t_k^0(n) = t_k^*\}_{k=1}^{M}$, and compute the minimum distortion for the training segment sequence as

$$D_n(0) = \sum_{k=1}^{M} \min_{Y_i \in C_{n-1}} d\,(X_k, Y_i) \tag{10}$$

Step 3 Set $m = 0$, $C_n(0) = C_{n-1}$.

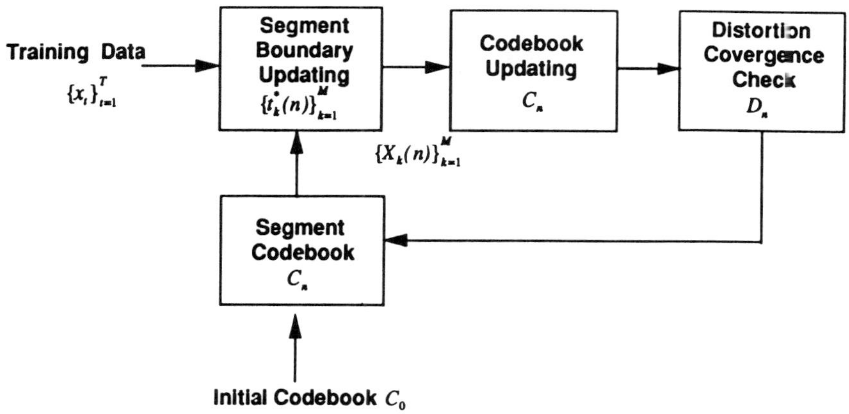

Figure 2 Training procedure of variable-length segment quantization.

Step 4 Codebook updating:

Increment m. Encode the training segment sequence using the minimum distortion rule

$$I_i = \{k \mid X_k = Y_i\} \tag{11}$$

where

$$X_k = \arg \min_{Y_i \in C_n(m-1)} d\,(X_k, Y_i)$$

For each cluster, compute the centroid using the training segment data as

$$Y_i = \left(\sum_{k \in I_i} X_k H_{l_k}^t \right) \left(\sum_{k \in I_i} H_{l_k} H_{l_k}^t \right)^+ \tag{12}$$

Here A^+ is a generalized inverse matrix of A (Rao and Mitra, 1971). Update the codebook to $C_n(m) = \{Y_i\}_{i=1}^{N}$.

Step 5 Distortion convergence check:

Compute the distortion decrease as

$$\delta = \frac{D_n(m-1) - D_n(m)}{D_n(m)}$$

where

$$D_n(m) = \sum_{Y_i \in C_n(m)} \min d\,(X_k, Y_i)$$

Go to step 4 unless $\delta \leq \epsilon_2$.
Step 6 Termination: Set $C_n = C_n(m)$, $D_n = D_n(m)$, and go to step 2 unless $(D_{n-1} - D_n)/D_n \leq \epsilon_1$.

This algorithm ensures the nonincreasing sequence of distortions as

$$D_0 \geq D_1 \geq \cdots \geq D_n \geq D_{n+1} \geq \cdots \tag{13}$$

Figure 3 shows the distortion convergence manner of the iterative codebook design procedure. The experimental results show that the design procedure converges well after three or four iterations and reduces the distortion to about 80% of the initial value after five iterations.

2.2. Temporally Overlapping Segment Quantization

Thus far, a spectral parameter sequence has been represented as a concatenation of fixed- or variable-length code segments. However, the fact that articulatory gestures between neighboring phonemes overlap suggests that an overlapping segment model could more adequately represent the spectral parameter sequence than the concatenating model (Atal, 1983). Thus, variable-length segment quantization is further extended to allow for overlap-

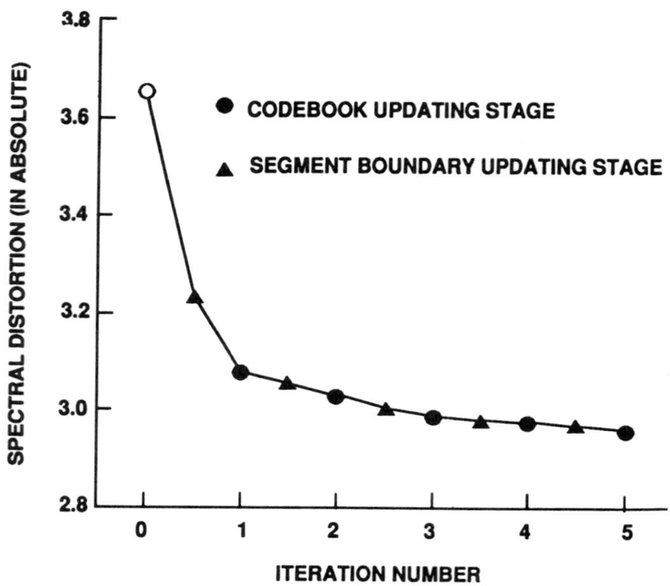

Figure 3 Distortion convergence in codebook training.

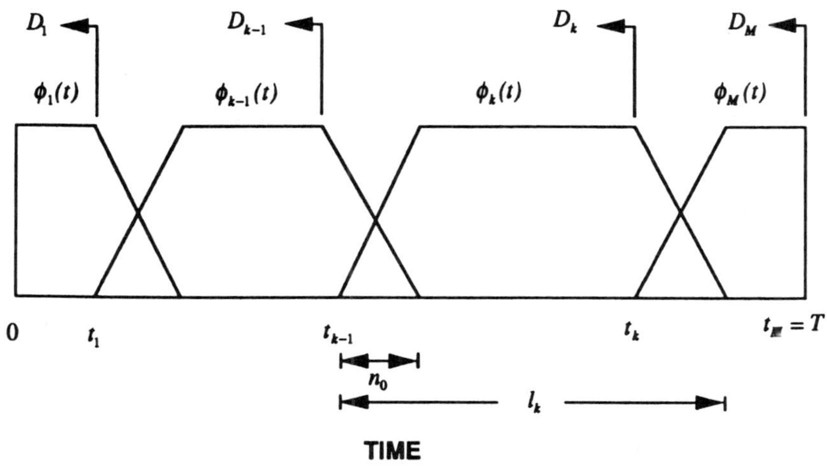

Figure 4 Overlapping segment structure.

ping of code segments in the representation of a spectral sequence (Honda, 1988, 1989).

The overlapping structure considered here is restricted to a relatively simple form by a computational consideration. The segment overlapping structure is shown in Fig. 4. Each segment overlaps only on the left and right sides with adjacent segments. Hence, single overlap is considered rather than the more general multiple overlaps. Overlap length is assumed to be fixed at N_0, which means the segment length l_k is specified by the segment positions as $t_k - t_{k-1}$, the same as in the concatenation case.

The basic configuration of overlapping SQ is similar to that shown in Fig. 1, except for the process generating the spectral sequence from the code segments. Denoting the time-warped variable-length code segment by $\Phi_k = (\phi_k(t_{k-1}+1, \ldots, \phi_k(t_k + n_0)) = Y_{i_k} H_{l_k}$ and using the assumption concerning overlapping structure, the overlapped code segment $\Phi_k^o = (\phi_k^o(t_{k-1}+1), \ldots, \phi_k^o(t_k + n_o))$ is represented as

$$\phi_k^o(t) = \begin{cases} \phi_{k-1}(t) + \phi_k(t) & t_{k-1} < t \le t_{k-1} + n_o \\ \phi_k(t) & t_{k-1} + n_o < t \le t_k \end{cases}$$

The overlapping code segment is specified by the previous code word and the previous segment positions as well as their current ones. Thus, the optimum code segment sequence in the temporally overlapping SQ is given as

$$\hat{X}_k \,|_{k=1}^{M} = \arg\min_{t_k|_{k=1}^{M}} \left(\min_{Y_k|_{k=1}^{M}} \sum_{k=1}^{M} d(X_k, \Phi_k^o(Y_{i_{k-1}}, Y_{i_k}, t_{k-2}, t_{k-1}, t_k)) \right) \tag{14}$$

The minimization problem of overlapping SQ is efficiently solved using the dynamic programing technique. The accumulated distortion to the segment position t_k is obtained in the recursive form

$$\begin{aligned} D_k\left(Y_{i_k}, t_k\right) = & \min_{\substack{Y_{i_{k-1}} \in C \\ t_{k-1} \in R_{k-1}}} D_{k-1}\left(Y_{i_{k-1}}, t_{k-1}\right) \\ & + d_k\left(X_k, \Phi_k^o\left(Y_{i_{k-1}}, Y_{i_k}, \hat{t}_{k-2}, t_{k-1}, t_k\right)\right) \qquad (k = 1, 2, \ldots, M) \end{aligned}$$

where $\hat{t}_{k-2}$ is the optimum segment boundary for a given t_k and t_{k-1}. The possible segment positions are constrained by the minimum and maximum distances between adjacent segments. Hence, segment lengths are from twice n_o to the maximum segment length $l_{\max}$:

$$R_k = \{t_k \mid t_{k-1} + n_o \le t_k \le t_{k-1} + l_{\max} - n_o\}.$$

At each stage, the accumulated distortion is a function of the codeword and the segment position at the current stage. The optimum codeword and segment position of the previous segment are stored in tables α and β:

$$\alpha_k\left(Y_{i_k}, t_k\right) = \hat{Y}_{i_{k-1}} \qquad \beta_k\left(Y_{i_k}, t_k\right) = \hat{t}_{k-1}$$

In the final stage, the minimum total distortion over the speech interval is obtained by minimizing the accumulated distortions with regard to the final segment codewords. The optimum codeword at the final segment is obtained as its argument:

$$\begin{aligned} D &= \min_{Y_{i_M} \in C} D_M\left(Y_{i_M}, T\right) \\ Y_{i_M}^* &= \arg\min_{Y_{i_M} \in C} D_M\left(Y_{i_M}, T\right) \end{aligned}$$

The optimum codeword sequence and segment positions are then obtained by backtracking the segment stages with reference to the codeword table and the position table:

$$\begin{aligned} Y_{i_{k-1}}^* &= \alpha_k\left(Y_{i_k}^*, t_k^*\right) \\ t_{k-1}^* &= \beta_k\left(Y_{i_k}^*, t_k^*\right) \qquad k = M, M-1, \ldots, 2 \end{aligned}$$

where $\hat{t}_M = T$.

Whereas the accumulated distortion is a function only of the segment position in the concatenating SQ case, it also depends on the segment code in the overlapping case. Thus, the computational complexity of overlapping SQ is N times that of concatenating SQ, where N is the number of segment codewords. The computational complexity can be reduced by eliminating the codeword and position candidates having large distortions at every stage. The experiment shows that nearly optimum distortion can be obtained by setting the number of candidates to eight.

Codebook design of overlapping SQ is performed in an iterative form, which is similar to that of the concatenating case. The design algorithm is presented below.

Step 1 Initialization:
Training vector sequence, initial codebook C_0, and thresholds ϵ_1 and ϵ_2 are given.
Set $n = 0$ and initial distortion $D_0 = \infty$.

Step 2 Segment data updating:
Perform overlapping SQ on the training data, increment n, and update the segment data X_k and the code index i_k.

Step 3 Set $m = 0$, $C_n(m) = c_{n-1}$, and $D_n(m) = D_{n-1}$.

Step 4 Codebook updating:
Cluster the training segments with the same code index:

$$I_i = \{i_k \mid i_k = i\}$$

Subtract adjacent code segments in the overlapped interval by

$$e(t) = x(t) - \phi_{k-1}(t) - \phi_{k+1}(t)$$

$$\phi_k(t) = \begin{cases} \phi(Y_{i_k}) & \text{if } i_k < i \\ \phi(Y_{i_k}) & \text{if } i_k \geq i \end{cases}$$

Compute the variable-length segment centroid by

$$Y_i = \left(\sum_{i_k \in I_i} E_k H_{l_k}^t\right)\left(\sum_{i_k \in I_i} H_{l_k} H_{l_k}^t\right)^+$$

where $E_k = (e(t_{k-1}+1), \ldots, e(t_k + n^o))$. Increment i, and repeat step 3 until $i = N$.
Increment m and update the codebook to $C_n(m) = \{Y_i\}_{i=1}^N$.

Step 5 Distortion convergence check:
If $(D_n(m-1) - D_n(m))/D_n(m) \leq \epsilon_2$, go to step 6. Otherwise, perform overlapping SQ with fixed segment boundaries, update the code index, and go to step 4.

Step 6 Termination:

Set $C_n = C_n(m), D_n = D_n(m)$, and go to step 2 unless $(D_{n-1} - D_n) \leq \epsilon_1$.

2.3. Evaluation of Segment Vocoder

2.3.1. Rate-Distortion Performance

Distortion performance of the segment quantization techniques presented in the previous subsections was examined using ordinary vector quantization as a reference. Speech data were taken from several sentences uttered at normal speed by a single male speaker. The 12th-order LSP parameters and speech power were calculated using ordinary linear prediction analysis for a frame of 10 ms with a 30-ms-long Hamming window and an 8-kHz sampling rate. The original sequence was divided into a training sequence of 118,500 frames and a test sequence of 14,840 frames, excluding silence frames. In variable-length SQ, initial estimates of the segment boundaries were obtained using a procedure based on detecting spectral change. The spectral change was obtained at every frame as the average slope of the regression lines determined by the least mean square error fitting the LPC cepstrum block (Sagayama and Itakura, 1979). Hence, the segment rate varied according to the spectral change of the input speech. The average segment rate for this speech data base is 14.3 segments per second, which is approximately equal to the phoneme rate. There are totals of 16,976 segments for the training sequence and 2141 for the test sequence. The code segment length is fixed at 10.

Figure 5 shows the rate-distortion performance in quantization of spectral parameters for the open data by SQ using joint segmentation and quantization (joint SQ), temporally overlapping SQ, SQ with separate segmentation (separate SQ), and VQ, as a function of spectral information per second. In variable-length SQ, the segment length is quantized with 4 bits using uniform scalar quantization in the range of 10 to 160 ms, and the information is included in the spectral information.

Spectral distortion of joint SQ with a 10-bit codebook, where the spectral information rate is 200 bps, is comparable with that of VQs at 700 bps at a frame rate of 10 ms. Joint SQ provides better distortion performance than separate SQ. A spectral distortion by overlapping SQ with an 8-bit codebook is nearly the same as that of 10-bit separate SQ. Further improvement is achieved by temporally overlapping SQ. The spectral distortion by overlapping SQ with a 7-bit codebook is equal to the distortion by the joint SQ with an 8-bit codebook, which leads to a 50% reduction in codebook size.

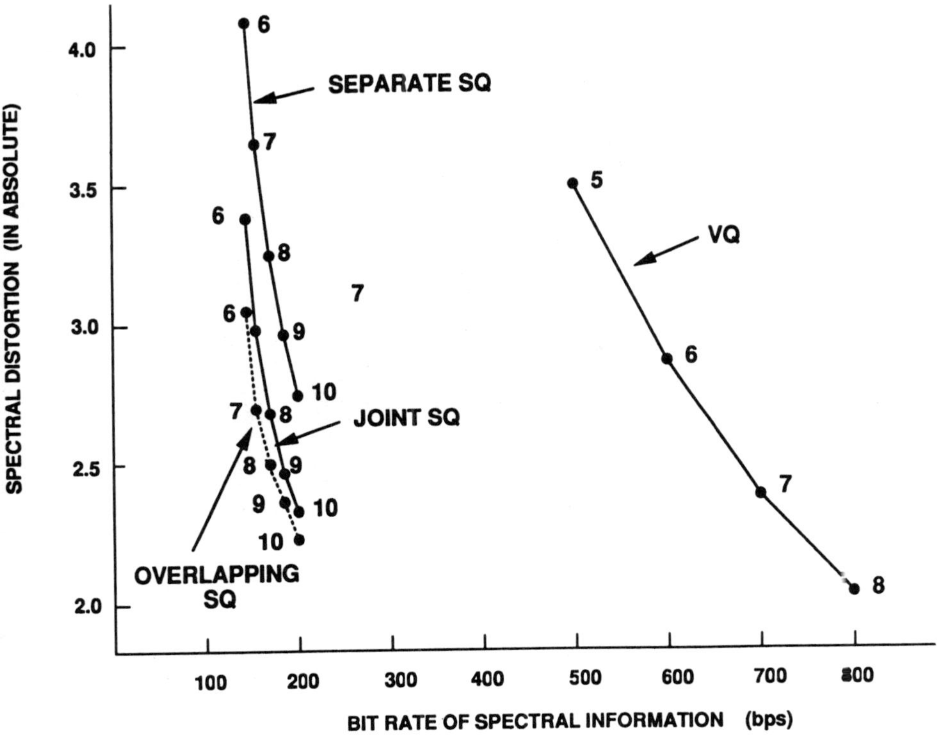

Figure 5 Rate-distortion performance of variable-length segment quantization.

2.3.2. Speech Intelligibility

The performance of the segment vocoder should be tested in terms of speech quality. Intelligibility of the coded speech is a minimal requirement which a very low bit rate coder must maintain. Speech samples taken from 100 isolated Japanese CV syllables were used for a speech intelligibility test. Speech was encoded using a 10-bit codebook on data for 20,000 training segments taken from sentence speech in which the test speech data were not included. A voice-unvoiced decision was assigned once for every segment automatically. When a segment vocoder is designed with assignments of 10 bits for the segment codebook, 2.7 bits for segment length (using entropy coding) and 4 bits for pitch (using differential coding), the total bit rate is 150 bps for the average segment rate of 9 segments/s for the test speech data. The result is shown in Fig. 6. The articulation score of speech coded by the seg-

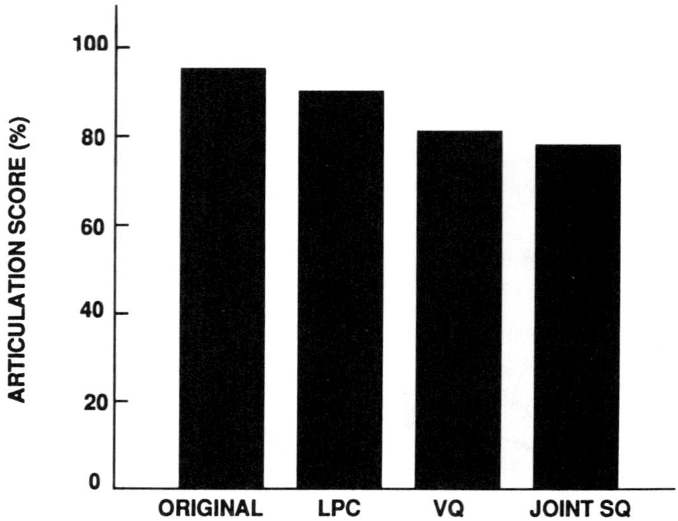

Figure 6 Speech intelligibility of segment vocoder.

ment vocoder is 78%. This is slightly better than that for the VQ vocoder using a 7-bit codebook. Although the speech intelligibility of the segment vocoder degrades from the maximum performance attainable by an unquantized LPC vocoder (articulation score of 90%), it yields intelligibility sufficient for correct recognition of sentences.

3. SPEAKER ADAPTATION IN SEGMENT QUANTIZATION

A speaker-dependent codebook does not work for speakers other than the training speaker. Using a speaker-independent codebook is a simple way to achieve a speaker-independent segment vocoder. However, this approach requires a larger codebook—four to eight times the size of a speaker-dependent codebook. This leads to increased computation in the quantization procedure and to a larger training data set to maintain performance for the open data set. Codebook adaptation is another useful approach for a speaker-independent segment vocoder. There are two adaptation techniques: forward adaptation and backward adaptation. Backward adaptation updates the codebook using only past transmitted codewords. FSVQ is a kind of backward adaptation. Forward adaptation updates the codebook using incoming data. A simple forward adaptation is to select from among multiple codebooks the codebook which performs best on incoming data (Paul, 1983; Viswanathan et al., 1982; Wong et al., 1983; Tsao and Gray, 1985; Roucos et al., 1982, 1983; Shiraki and Honda, 1988; Itakura, 1975; Sugamura and Itakura, 1981; Rao and Mitra, 1971; Atal, 1983; Honda, 1988, 1989; Sagayama and Itakura, 1979; Rosenberg and Soong, 1986). Another approach is to modify the code segments or code vectors to give a better fit to the data (Roucos and Wilgus, 1984)

In the following sections, a forward adaptation of a segment codebook is presented for speaker adaptation. The adaptation performance is discussed in terms of adaptation speech and the distortion for an open set of speakers.

3.1. Speaker Adaptation Using Piecewise Linear Mapping

Generally speaking, codebook adaptation is done to determine the mapping function from the original code segments to the adapted ones. The form of the mapping function for codebook adaptation must differ with the local spectral characteristics in the spectral segment space. Thus, multiple mapping functions, each representing a local spectral correspondence between the original code segment and the input segments of a new speaker, perform better than a single mapping function. Such multiple mapping is realized by a piecewise linear mapping (PLM) function (Shiraki and Honda, 1990) The PLM function is represented as a linear combination of difference vectors between centroid vectors of the code vectors and input spectral vectors in a subcluster of the speaker's spectral space. The continuity of the mapping function is considered when the linear combination is determined.

The procedure is depicted in Fig. 7. Each code vector is modified according to

$$c'_m = c_m + \delta c_m = c_m + \sum_{i=1}^{L} w_{im} \Delta_i \tag{15}$$

where c'_m is the adapted code segment, c_m is the original code segment, Δ_i is the PLM function or vector, and w_{im} is a weighting coefficient of the linear combination shown to be inversely proportional to the distance between each code vector $c_n m$ and the local spectral centroid u_i:

$$w_{im} = \frac{d_{im}}{\sum_{i=1}^{L} d_{im}} \tag{16}$$

Here $d_{im} = 1/d(c_m, u_i)$ and $d(x,y)$ is the Euclidean distance between x and y. Since the modification vector δc_m is calculated from Eqs. (15) and (16), the local continuity of the spectral distribution of the original code vectors is

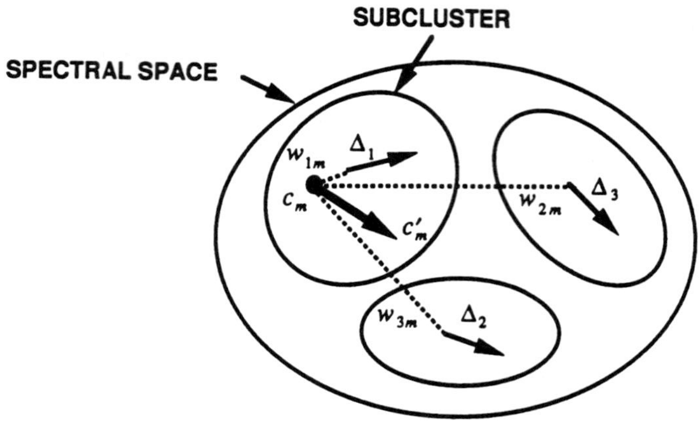

Figure 7 Speaker adaptation procedure based on piecewise linear mapping (PLM) method.

maintained. Two algorithms for determining PLM functions are presented below.

3.1.1. Simplified Algorithm

The simplified algorithm uses spectral code vectors generated by the LBG algorithm (Linde et al., 1989), as representative vectors of the spectral vector space of an unknown speaker.

Step 1 Initialization:
Number of clusters L, p-dimension training vectors x_n ($n = 1, 2, \ldots, N$), and initial code words c_m ($m = 1, 2, \ldots, M$) are given.

Step 2 Subpattern generation:
Training vectors x_n are divided into clusters using the LBG algorithm. Then subpatterns v_i ($i = 1, 2, \ldots, L$) are calculated as the centroid of each cluster.

Step 3 Centroid shifting:
The centroid of the codeword is shifted to the centroid of the subpatterns:

$$c'_m = c_m + \Delta_0 \tag{17}$$

where

$$\Delta_0 = \frac{1}{L}\sum_{i=1}^{L} v_i - \frac{1}{M}\sum_{m=1}^{M} c_m$$

Step 4 Codebook clustering:
Centroid vectors u_i are calculated by clustering codewords c'_m using each subpattern v_i as a seed vector:

$$u_i = \frac{1}{|I_i|} \sum_{m \in I_i} c'_m \tag{18}$$

where

$$i(m) = \arg \min_{1 \le i \le L} d(c'_m, v_i) \tag{19}$$

$$I_i = \{m \mid i(m) = i\}$$

and $|I_i|$ is the number of elements in an integral set I_i.

Step 5 PLM calculation:
For each cluster, PLM is calculated as the deviation vector Δ_i between u_i and v_i:

$$\Delta_i = v_i - u_i \tag{20}$$

Step 6 For each code vector c'_m, weighting factors w_{im} are calculated from Eq. (16).

Step 7 Codebook updating:
Each code vector c'_m is updated according to Eq. (15).

3.1.2. Minimum Distortion Algorithm

In the simplified algorithm, quantized distortion for training vectors using the adaptive codebook is not guaranteed to be less than or equal to that using the original codebook. The minimum distortion algorithm determines the PLM functions so as to minimize the quantized distortion for training data given by

$$D = \sum_{n=1}^{N} \min_m \|x_n - (c_m + \delta c_m)\|^2 \tag{21}$$

where $\delta c_m = \sum_{i=1}^{L} w_{im} \Delta_i$. The minimization is carried out by iterating k-means clustering and the least-squares technique.

Step 1 Initialization:
Given p-dimension training vectors x_n, initial code words $c_m^{(0)}$, number of clusters L, and thresholds ϵ, set initial distortion $D(-1) = \infty$ and $k = 0$.

Step 2 Quantize x_n by $c_M^{(k)}$ and compute code index i_n and quantized distortion $D(k)$:

$$i_n = \arg \min_{1 \le m \le M} d(x_n, c_m^{(k)}) \tag{22}$$

$$D(k) = \sum_{n=1}^{N} d(x_n, c_{i_n}^{(k)}) \tag{23}$$

Step 3 Centroid calculation:
Cluster code vector $c_m^{(k)}$ using the LBG algorithm and calculate the centroid u_i of each cluster.

Step 4 For each codeword $c_m^{(k)}$, calculate the weighting factors w_{im} from Eq. (16).

Step 5 PLM calculation:
By differentiating the quantized distortion D with respect to each Δ_i, calculate $\boldsymbol{\Delta}$ from

$$\boldsymbol{\Delta} = EW^t(WW^t)^{-1} \tag{24}$$

where $\boldsymbol{\Delta} = (\Delta_1, \Delta_2, \ldots, \Delta_L)$, $E = (e_1, e_2, \ldots, e_N)$, e_n represents the quantized error vector $e_n = x_n - c_{i_n}^{(k)}$ and W is a matrix whose elements are given by

$$W = \begin{pmatrix} w_{1i_1} & \cdots & w_{1i_N} \\ w_{2i_1} & \cdots & w_{2i_N} \\ \vdots & & \vdots \\ w_{Li_1} & \cdots & w_{Li_N} \end{pmatrix} \tag{25}$$

Step 6 Codebook updating:
Update each code vector according to the equation

$$c_m^{(k+1)} = c_M^{(k)} + \sum_{i=1}^{L} w_{im} \Delta_i \tag{26}$$

Step 7 Convergence check:
If $(D(k-1) - D(k))/D(k-1) \leq \epsilon$, stop; otherwise increment k and go to step 2.

3.2. Adaptation Performance

Adaptation performance of the PLM method is compared with speaker-independent and speaker-dependent codebook approaches. Speech data were taken from sentences spoken by 19 male speakers and divided into 45-min-long data from 15 speakers for the speaker-independent codebook training and 15-min-long data from 4 speakers for the speaker-dependent codebook training. Other 15-min-long data from the test speakers were used to examine the distortion performance for the test data. Both speaker-independent and -dependent segment codebooks were designed using the joint SQ training procedure described in Section 2.1. A speaker-adaptive codebook was obtained by updating the speaker-independent codebook using the PLM adaptation procedure for the training data set from the test speakers.

The distortion performance of the 8-bit adaptive codebook as a function of the amount of training data is compared with that of the joint SQ training procedure in Fig. 8. When training data for the test data are longer than 3 min, an 8-bit adaptive codebook applying the simplified method has distortion performance comparable to that of a 9.5-bit speaker-independent codebook and a 7-bit speaker-dependent codebook. For the same training data, a minimum distortion-type-adaptive codebook has distortion performance comparable to that of an 11-bit speaker-independent codebook and

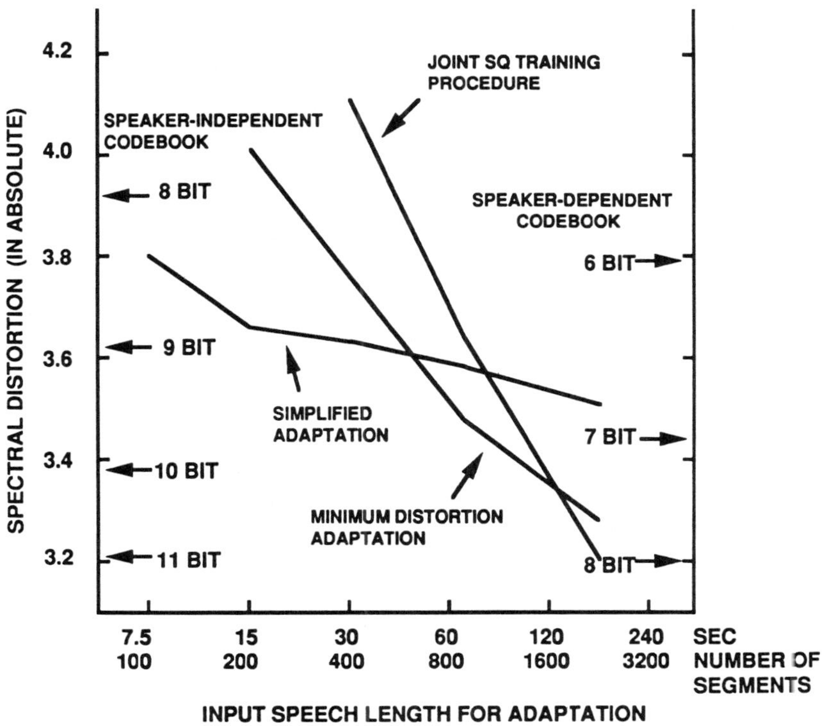

Figure 8 Distortion performance of speaker adaptation method as a function of training data size.

an 8-bit speaker-dependent codebook. This indicates that the speaker-independent codebook approach requires a codebook 8 times larger than the speaker-dependent or adaptive codebook approach to maintain the same level of distortion. In addition, the attainable distortion performance of the minimum-distortion methods is nearly the same as for the joint SQ training procedure. If only limited training data are available, namely the training data under 128 s, the adaptive codebook provides substantially better distortion performance than the speaker-dependent codebook obtained by the joint SQ training procedure. If the training data are extremely limited, the simplified method is superior to the minimum distortion method.

4. CONCLUSION

We have presented a very low bit rate speech coding technique with a focus on segment quantization of spectral parameters. A variable-length segment quantization technique has been considered for concatenating and overlapping segment representation of a spectral sequence. For both cases, the optimum segment quantization procedure based on joint segmentation and quantization and the codebook training procedure have been presented. The rate-distortion performance of variable-length segment quantization and the speech intelligibility of a segment vocoder at 243 bps have been investigated experimentally. It has been shown that the segment vocoder is able to produce speech with sufficient speech intelligibility to allow understanding of the meaning of a spoken sentence and is thus closely comparable to a 1000-bps VQ coder.

We have also presented a speaker adaptation procedure for segment quantization based on piecewise linear mapping of code segments. Experimental results show that the segment codebook adaptation can substantially reduce the distortion for an unknown speaker without increasing codebook size and can achieve the same performance as the LBG codebook design procedure if the training data size is sufficiently large. Further study is needed to improve speech intelligibility at very low bit rates and to lower the computational complexity with little degradation of the performance.

REFERENCES

Atal, B. S. (1983). Efficient coding of LPC parameters by temporal decomposition. *Proc. IEEE Int. Conf. Acoustics, Speech, and Signal Processing*, pp. 81–84.

Buzo, A., Gray, A. H., Gray, R. M., and Markel, J. D. (1980). Speech coding based upon vector quantization. *IEEE Trans. Acoust., Speech, Signal Process. ASSP-28*: 562–574.

Dunham, M. O., and Gray, R. M. (1985). An algorithm for the design of labeled-transition finite-state vector quantizers. *IEEE Trans. Commun. COM-23*:83–89.

Foster, J., Gray, R. M., and Dunham, M. O. (1985). Finite-state vector quantization for waveform coding. *IEEE Inform. Theory IT-31*:348–359.

Gray, R. M. (1984). Vector qunatization. *IEEE ASSP Mag. 1*(2):4–29.

Gray, A. H., and Markel, J. (1976). Quantization and bit allocation in speech processing. *IEEE Trans. Acoust., Speech, Signal Process. ASSP-24*:459–473.

Honda, M. (1988). Temporally overlapping segment quantization for speech coding. *Acoust. Soc. Am. 84*(Suppl. 1):G7.

Honda, M. (1989). Speech analysis based on temporally overlapping spectral segment model. Trans. Acoust. Soc. JIEICE Technical Report SP88-140, pp. 45–52.

Itakura, F. (1975). Line spectral representation of linear predictor coefficients of speech signals. *J. Acoust. Soc. Am. 57*(Suppl. 1):S35.

Juang, B. H. (1986). Design and performance of trellis vector quantizers for speech signals. *Proc. IEEE Int. Conf. Acoustics, Speech, and Signal Processing*, pp. 437–440.

Linde, Y., Buzo, A, and Gray, R. M. (1980). An algorithm for vector quantizer design. *IEEE Trans. Commun. COM-28*:84–95.

Makhoul, J., Roucos, S., and Gish, H. (1985). Vector quantization in speech coding. *Proc. IEEE 73*:1551–1588.

Paul, D. B. (1983). An 800 bps adaptive vector quantization vocoder using a perceptual distance measure. *Proc. IEEE Int. Conf. Acoustics, Speech, and Signal Processing*, pp. 73–76.

Rao, C. R., and Mitra, S. K. (1971). *Generalized Inverse of Matrices and Its Applications*. Wiley, New York.

Rosenberg, A. E., and Soong, F. K. (1986). Evaluation of a vector quantization talker recognition system in text-independent and text-dependent modes. *Proc. IEEE Int. Conf. Acoustics, Speech, and Signal Processing*, pp. 17.4.1–17.4.4.

Roucos, S., and Wilgus, W. M. (1984). Speaker normalization algorithm for very-low-rate speech coding. *Proc. IEEE Int. Conf. Acoustics, Speech, and Signal Processing*, pp. 1.1.1–1.1.4.

Roucos, S., Schwarts, R., and Makhoul, J. (1982). Segment quantization for very low rate speech coding. *Proc. IEEE Int. Conf. Acoustics, Speech, and Signal Processing*, pp. 1565–1568.

Roucos, S., Schwarts, R., and Makhoul, J. (1983). A segment vocoder at 150 B/S. *Proc. IEEE Conf. Acoustics, Speech, and Signal Processing*, pp. 61–64.

Sagayama, S., and Itakura, F. (1979). On individuality in a dynamic measure of speech. *Trans. Acoust. Soc. Jpn.* 3-2-7, June.

Shiraki, Y., and Honda, M. (1988). LPC speech coding based on variable-length segment quantization. *IEEE Trans. Acoust., Speech, Signal Process. ASSP-36*: 1437–1444.

Shiraki, Y., and Honda, M. (1990). Speaker adaptation algorithms based on piecewise moving adaptive segment quantization method. *Proc. IEEE Int. Conf. Trans. Acoustics, Speech, and Signal Processing,* pp. 657–660.
Sugamura, N., and Itakura, F. (1981). Line spectrum representation of linear predictor coefficients of speech signal and its statistical properties. *Trans. Inst. Electron. Commun. Eng. Jpn. J64-A*:323–340.
Tohkura, Y., and Itakura, F. (1979). Spectral sensitivity analysis of PARCOR parameters for speech data compression. *IEEE Trans. Acoust., Speech, Signal Process. ASSP-27*:273–280.
Tsao, C., and Gray, R. M. (1985). Matrix quantizer design for LPC speech using the generalized Lloyd algorithm. *IEEE Trans. Acoust., Speech, Signal Process. ASSP-33*:537–545.
Wong, D. Y., Juang, B. H., and Gray, A. H. (1982). An 800 bit/s vector quantization LPC vocoder. *IEEE Trans. Acoust., Speech, Signal Process. ASSP-30*:770–780.
Wong, D. Y., Juang, B. H., and Cheng, D. Y. (1983). Very-low data rate speech compression with LPC vocoder and matrix quantization. *Proc. Int. Conf. Acoustics, Speech, and Signal Processing,* pp. 65–68.
Viswanathan, R., and Markel, J. (1975). Quantization properties of transmission parameters in linear predictive coding. *IEEE Trans. Acoust., Speech, Signal Process. ASSP-23*:309–320.
Viswanathan, V., Makhoul, J., Schwarz, R., and Huggins, A. W. F. (1982). Variable-frame rate transmission: A review of methodology and application to narrowband LPC speech coding. *IEEE Trans. Commun. COM-30*:674–686.

8

Speech Coding Based on Physiological Models of Speech Production

JUERGEN SCHROETER and M. MOHAN SONDHI AT&T Bell Laboratories, Murray Hill, New Jersey

1. INTRODUCTION

In a physiological model, speech is characterized not by the properties of the speech signal but by the properties of the vocal apparatus that produced the signal. By properties of the vocal apparatus we mean the positions of the articulators, the coupling of the nasal cavity, the pressure in the lungs, the tension of the vocal cords, etc. The models which we will consider here discard the simplified source-filter description of speech production, which is at the basis of all coding schemes in use today. Instead, they include the nonlinear interaction between the excitation source (vocal cord vibration and/or turbulence) and the time-varying input impedance of the vocal/nasal tract.

It has long been conjectured that such models will ultimately provide the most economical description of a speech signal and will also produce the most natural-sounding speech. There are several reasons for belief in this conjecture:

1. In such a model, speech is produced from the same slowly varying parameters as those controlled by a human being. Hence it is reasonable to assume that these parameters are better candidates for coding

than parameters describing the properties of the speech signals. It is easy to construct examples where complicated and fast variations in the spectrum of the speech signal can result from simple, easily specified changes in the shape of the vocal tract (for example, a uniform tube with a constriction which moves slowly from one end to the other). It seems, therefore, more natural to code the changes in the shape of the vocal tract than to code the corresponding spectral changes in the resulting speech.

2. Since in a physiological model the vocal cord motion is affected by the loading due to the vocal/nasal tract, changes in the glottal waveform and in the pitch occur automatically and in a natural manner. In the conventional models these changes have to be specified in detail to maintain naturalness.
3. A physiological model facilitates interpolation of parameters. For economical coding one would like to specify parameters infrequently and obtain their values at intermediate times by interpolation. Parameters specifying the positions of articulators vary slowly and therefore allow accurate interpolation. Also, the interpolation yields reasonable values for these parameters. This is not true of parameters in general. (For instance, interpolation of a linear predictive coding (LPC) vector between two stable LPC vectors may result in a unstable filter.)

In spite of the reasonableness of the conjecture, so far no one has been able to prove its assertions. However, during the past several years, coding on the basis of such models has received considerable attention. One of the reasons for this renewed interest is that all known coding algorithms (for some of the most recent algorithms refer to, for example, Chapters 3 and 5) give unacceptably poor speech quality at low bit rates--for example, under 4000 bits/s. A coding scheme which mimics human speech production may have advantages at these low bit rates. Another reason for the renewed interest is the fact that digital signal processing is now possible at a scale unimaginable only a decade ago. As will become evident in the following sections, the algorithms for coding on the basis of physiological models are so complex that it would have been practically impossible to implement them just a few years ago. In order to become useful practical algorithms, their complexity would eventually have to be reduced. However, in the present learning phase at least, a high degree of computational complexity appears to be unavoidable.

In the following sections of this chapter we will describe such physiological models and the way in which they may be used for synthesis and coding of speech signals. In Section 2 we will describe the building blocks of the models and present some ideas that have been proposed for implementing them. In Section 3 we will tackle the more difficult problem of deriving the parameters to control such a synthesizer so as to mimic a given speech signal. Finally, in Section 4 we will outline some future directions of research.

2. COMPONENTS OF THE SPEECH PRODUCTION MODEL

As far as acoustical properties are concerned, there are basically three aspects of the speech production mechanism that need to be modeled. First, the geometry of the vocal and nasal tracts needs to be parametrized. Second, a model must be selected to describe wave propagation in the tract. Third, the sound sources (vocal cord vibration and turbulent air flow) and their interactions with the tract must be modeled. Let us briefly consider each of these three components.

2.1. Modeling the Geometry of the Tract

For nonnasal speech sounds we need consider only the vocal tract, i.e., the variable-area tube from the glottis to the lips. For nasal sounds this tube is coupled to the nasal tract through the opening of the velum. The sizes of these two branches are roughly of the same order of magnitude, so the problems of approximating their acoustical properties are similar. The nasal tract is essentially constant, so it can be described by a fixed filter (with the exception of the variable coupling to the oral tract). The vocal tract, however, is continuously varying and its shape must be specified every few milliseconds.

2.1.1. The Vocal Tract

The geometry of the vocal tract is conveniently described in terms of the positions of the articulators—the tongue, lips, glottis, etc. Models based on such a description are called articulatory models. Such models have been around for about 25 years. The earliest ones that we know of are those proposed by Coker and Fujimura (1966) and by Henke (1967). Two of the most celebrated and widely used models are those of Mermelstein (1973) and Coker (1976). Both are offshoots and refinements of the model described in Coker and Fujimura (1966). Other models have been proposed by, for example, Liljencrants (1971),* Perkell (1974), Shirai and Honda (1976), Harshman et al.,

*Also see Schroeder (1967).

(1971), Kiritani et al. (1977), and Maeda (1979, 1990). All these models are primarily based on x-ray measurements, most of which were made during the 1950s and 1960s.

In our own work (e.g., Schroeter et al., 1987) we have mostly used a version of Mermelstein's model, which was derived from an implementation by Bocchieri (1983). Unlike Mermelstein's original model, we keep the jaw distance s_j constant at 11.3 cm and allow the tongue to have a variable radius. Also, with the exception of the velum opening, we *linearly transform* each parameter so as to have a nominal range from 0 to 2. Under extreme conditions, the parameters are allowed to vary from −0.5 to 2.5. The variable parameters of this model are identified in Fig. 1 and listed in detail in Table 1. Table 1 gives the unscaled extreme parameter values corresponding to mapped values of −0.5 and 2.5.

From the articulatory description, data representing the cross-sectional area as a function of distance along the tract from glottis to lips are computed by a procedure described by Maeda (1972). This area function is an adequate

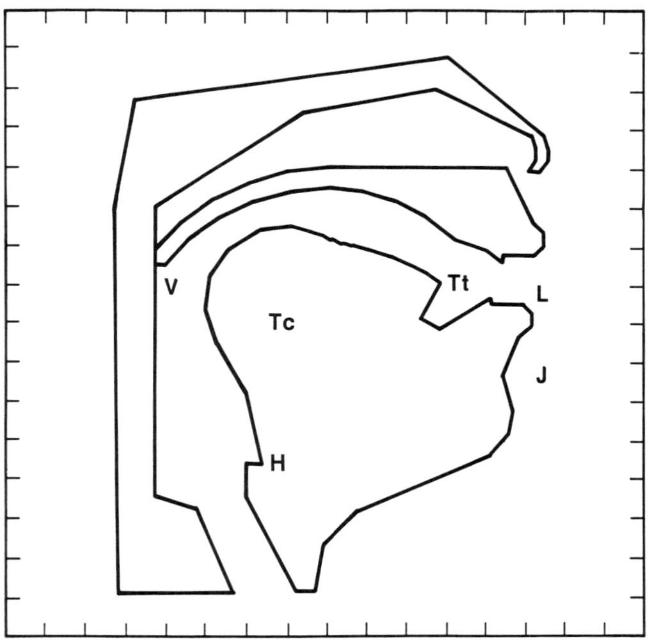

Figure 1 Parameters of the Mermelstein articulatory model.

Table 1 Identification of Parameters and Unscaled Limits Corresponding to Scaled Values of −0.5 and 2.5 (except vel) for the Mermelstein Articulatory Model

No.	Name	Meaning	Lower limit	Upper limit
1	rtp	tongue radius [cm]	0.5	3.5
2	tejp	jaw angle [deg]	10.9	28.0
3	xcp	tongue center x [cm]	4.0	10.0
4	stp	tongue blade length [cm]	2.0	4.4
5	xclp	lip protrusion x [cm]	−0.36	1.89
6	xhp	hyoid position x [cm]	5.20	7.60
7	ycp	tongue center y [cm]	1.48	7.48
8	telp	tongue elevation [deg]	68.2	−96.8
9	cylp	lip height y [cm]	−0.93	1.32
10	yhp	hyoid position y [cm]	7.43	9.83
11	vel	velum opening [cm^2]	0	2

description of the shape of the tract because, as discussed in Section 2.2, the wave motion in the tract is invariably assumed to be planar.

2.1.2. The Nasal Tract

The nasal tract is essentially fixed during speech production. As in the case of the vocal tract, the wave motion in the nasal tract is assumed to be planar. Studies of the geometric and acoustic properties of the nasal tract were made, for instance, by Bjuggren and Fant (1964). (Also see summary in Fant, 1970, chapter 2.4.) For large (e.g., 2-cm^2) velar openings, modeling the nasal tract by just a variable-area tube gives too high a frequency for the first pole–zero pair of the nasal transfer function (Fujimura and Lindqvist, 1971). For this case, more realistic values result if the sinus cavities are included (Lindqvist and Sundberg, (1972). For medium velar openings of about 0.6 cm^2, however, Feng et al. (1985) reported that they were able to match the lowest nasal pole frequency *without* modeling the sinuses, even for the difficult-to-nasalize low vowels (e.g., /a/). A nasal tract including a model of the sinus maxillaries was devised by Maeda (1981, 1982a, 1982b) and can be modeled efficiently by a Helmholtz resonator coupled to the nasal tract (Sondhi and Schroeter, 1987).

2.2. Wave Propagation in the Vocal/Nasal Tract

For the analysis of the acoustical properties of the vocal tract it is generally assumed that (1) the linear wave equation applies; (2) the tract can be "straightened out" and approximated as a variable-area tube; and (3) the motion in this modified tube is planar, i.e., is identical at all points in a plane perpendicular to the axis of the tube. Each of these assumptions needs justification.

The assumption of linearity is valid as long as the Mach number of the particle velocity is small compared to unity. This is the case except when the tract is extremely constricted, e.g., just before and just after a plosive. If, at a future date, feasible methods can be found to solve the nonlinear equation, it might be possible to use them for these specific cases.

It is difficult to evaluate the effect of the curvature. Sondhi (1986) published some computations for the case of a tube of *constant* cross section bent approximately to the curvature of the vocal tract. The computations show that the formants change only by a few percent compared to those of a straight tube with the same cross section. This suggests, though by no means proves, that the curvature may be neglected.

The assumption of plane waves also is justifiable at frequencies at which the half-wavelength is larger that the cross dimensions of the tract. For speech signals with bandwidth corresponding to the telephone channel (i.e., below 3500 Hz) the assumption is reasonable.

Some attempts have been made (e.g., Ranade and Sondhi, 1981; Hegerl, 1987; Liljencrants, 1990) to compute the general two- and three-dimensional wave propagation. It appears unlikely, however, that nonplanar waves will be incorporated into a coding scheme in the foreseeable future. This is because the computational complexity increases by more than an order of magnitude. Also, it is difficult to acquire adequate data to specify the complicated shape of the time-varying vocal tract, although some work has been done in this respect (see, e.g., Baer et al., 1990, 1991; Stone, 1990).

Once the above simplifying assumptions have been made, the acoustical properties of the vocal/nasal tract can be computed. First, the geometry of the vocal and nasal tracts is specified in terms of the parameters of an articulatory model (as discussed in Sections 2.1.1 and 2.1.2). From this specification the cross-sectional area function is derived. Under the assumption of planar motion, the acoustical properties of the variable-area tube can be computed in a variety of ways, e.g., by computing its formant frequencies (Coker, 1976; Lin, 1990), by representing the tube as a wave-digital filter (Kelly and Lochbaum, 1962; Maeda, 1982c; Strube, 1982; Liljencrants, 1985; Meyer et al., 1989) or as a lumped-parameter network (Portnoff, 1973; Flanagan et al., 1975), or by a chain matrix relating the pressure and volume velocity at one end of the tube to those at the other end (Allen and Strong, 1985; Sondhi and Schroeter, 1987). Of these, the chain matrix approach is particularly convenient and versatile. We will therefore briefly consider that approach.

The chain matrix (also known as the *ABCD* matrix) is a frequency-domain representation. For any given portion of the vocal or nasal tract, we have

$$\begin{pmatrix} P_{\text{out}} \\ U_{\text{out}} \end{pmatrix} = \begin{pmatrix} A & B \\ C & D \end{pmatrix} \times \begin{pmatrix} P_{\text{in}} \\ U_{\text{in}} \end{pmatrix} = K \times \begin{pmatrix} P_{\text{in}} \\ U_{\text{in}} \end{pmatrix}$$

where, by convention, we may call the glottal side the input end and the side toward the lips or nostrils the output end. In Eq. (1) all the quantities are functions of the frequency ω.

The chain matrix K can represent any portion of the tract, of any length or variable cross-sectional area. However, the computation of the elements of K is particularly simple if the tube has a constant cross section. Therefore the cross-sectional area function is computed at some finite number of points N (usually between 10 and 20) and the tube approximated as a concatenation of piecewise constant segments. The chain matrix for any portion of the tract is then obtained by multiplying the 2×2 matrices for the sequence of elementary homogeneous segments comprising it. The elements of K can be chosen to include losses as well as the effect of yielding walls. One particularly simple model for the losses and wall vibration is that of Sondhi (1974). With losses assumed to be of the form proposed in that reference, the elements of K for a homogeneous tube of length Δl and area *area* are given by

$$A = \cosh\left(\frac{\sigma\Delta l}{c}\right), \qquad B = -\frac{\rho c}{area}\gamma \sinh\left(\frac{\sigma\Delta l}{c}\right)$$
$$C = -\frac{area}{\rho c}\frac{\sinh(\sigma\Delta l/c)}{\gamma}, \qquad D = \cosh\left(\frac{\sigma\Delta l}{c}\right) \tag{1a}$$

Here, the complex variables σ and γ are defined as

$$\gamma = \sqrt{\frac{\alpha + j\omega}{\beta + j\omega}} \tag{1b}$$

and

$$\sigma = \gamma(\beta + j\omega) \tag{1c}$$

where

$$\alpha = \sqrt{j\omega c_1} \tag{1d}$$

and

$$\beta = \frac{j\omega\omega_0^2}{(j\omega + a)j\omega + b} + \alpha \tag{1e}$$

The last equation differs from Eq. (29) in Sondhi (1974) because we have also taken wall compliance into account. The parameters ω_0^2, a, and b are defined by the equations

$$\omega_0^2 = \frac{\rho c^2}{area\, L'_w} \tag{1f}$$

$$a = \frac{R'_w}{L'_w} \tag{1g}$$

$$b = \frac{1}{L'_w C'_w} \tag{1h}$$

where L'_w, C'_w, and R'_w are the mass, compliance, and resistance of the wall per unit length, respectively. Note that as in Sondhi (1974) we assume the wall parameters to vary with *area* in such a way that ω_0^2, a, and b are independent of *area*. Also ω_0 is interpreted as the lowest angular resonance frequency of the tract when it is closed at both ends. Data for all variables of the vocal and nasal tract used are given in Table II of Sondhi and Schroeter (1987). In the current version of the model, the value of the nasal cavity loss factor c_1 is one-half and that of the nasal resonance frequency ω_0 is double the value in the above reference.

Before leaving the subject of modeling the vocal tract, let us mention that all the (linear) properties of the tract can be derived in terms of the components of the chain matrix. The two most important properties that we shall need later are the transfer function and the input impedance of the tract when it is terminated with some load impedance. Let the load impedance be Z_{out}; that is, $(P_{\text{out}}/U_{\text{out}}) = Z_{\text{out}}$. Then it follows from Eq. (1) that the transfer function is given by

$$H = \frac{P_{\text{out}}}{U_{\text{in}}} = \frac{Z_{\text{out}}}{A - CZ_{\text{out}}} \tag{2}$$

Similarly, the input impedance is

$$Z_{\text{in}} = \frac{P_{\text{in}}}{U_{\text{in}}} = \frac{DZ_{\text{out}} - B}{A - CZ_{\text{out}}} \tag{3}$$

In our current version of the synthesizer that runs at a sampling rate of 8 kHz, we low-pass H (Eq. 2) by the filter $(1 + z^{-1})/(1 + 0.95z^{-1})$. The transfer function and the input impedance will play an important role in Section 2.4, where we describe the synthesis algorithms.

2.3. Modeling the Excitation

Except for certain click sounds used in some African languages, the excitation of the vocal/nasal tract is mainly provided by two mechanisms. For voiced sounds the excitation is due to the quasi-periodic interruption of the air flow from the lungs by the vibrating vocal cords (glottis). For fricative and aspirated sounds the excitation is provided by turbulent motion of air flowing through a narrow constriction. In some sounds these two forms of excitation occur simultaneously.

2.3.1. Model for Voiced Excitation

The model we use for voiced excitation in our synthesizer is basically the well-known two-mass model of Ishizaka and Flanagan (1972). An outline of this model is shown in Fig. 2. The glottal volume velocity $u_g(t)$ satisfies the differential equation

$$R_{\text{tot}} u_g + L_{\text{tot}} \frac{du_g}{dt} = \Delta p \tag{4}$$

Here Δp is the pressure drop across the glottis, that is, the difference between subglottal pressure p_s and supraglottal pressure p_1 (see Fig. 2). We discretize

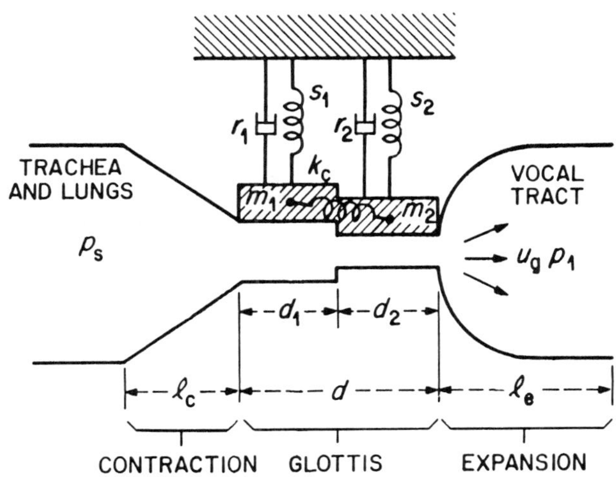

Figure 2 The two-mass model of the vocal cords. (After Ishizaka and Flanagan, 1972.)

Eq. (4) using backward differences. For sampling instant n this yields

$$u_g(n) = \frac{\Delta p(n)t_s + L_{\text{tot}}(n)u_g(n-1)}{t_s R_{\text{tot}}(n) + L_{\text{tot}}(n)} \tag{5}$$

Here t_s is the sampling interval, and R_{tot} and L_{tot} are the total quasi-stationary resistance and inductance representing contraction, glottis, and expansion, as depicted in Fig. 2. These variables are given by

$$L_{\text{tot}} = \rho\left(\frac{d_1}{A_{g1}} + \frac{d_2}{A_{g2}}\right) \tag{6}$$

and

$$R_{\text{tot}} = \frac{\rho}{2}\left[\frac{0.37}{A_{g1}^2} + \frac{1 - 2\frac{A_{g2}}{area_1}\left(1 - \frac{A_{g2}}{area_1}\right)}{A_{g2}^2}\right]|u_g| + 12\mu l_g^2\left(\frac{d_1}{A_{g1}^3} + \frac{d_2}{A_{g2}^3}\right) \tag{7}$$

where $area_1$ is the area of the first section of the vocal tract. In eqs (6, 7) the quantities A_{g1}, A_{g2}, $area_1$, and u_g are all time varying. To simplify the expressions, their dependence on n is not shown explicitly. The glottal areas A_{g1} and A_{g2} are related to the lateral displacements x_1 and x_2 by

$$A_{g1}(n) = A_{g01} + 2l_g x_1(n) \tag{8a}$$

and

$$A_{g2}(n) = A_{g02} + 2l_g x_2(n) \tag{8b}$$

where A_{g01} and A_{g02} are commonly set to the same glottal rest area A_{g0}. The displacements x_1 and x_2 in turn are obtained as the solutions of a coupled pair of second-order nonlinear oscillators, driven by the transglottal pressure and the glottal volume velocity. For details concerning the oscillator equations and the discretization, see Sondhi and Schroeter (1987). The current version of this model runs at a sampling rate of 16 kHz. The resulting glottal flow is low-pass–filtered with a 41-sample linear-phase finite impulse response (FIR) filter and then down-sampled to 8 kHz.

The somewhat unreasonably large ripples on the glottal flow u_g, which are generated by source-tract interaction, are reduced when additional losses are incorporated. Previously, the losses due to area variations inside the glottal duct ("vertical phasing") were disregarded, and only the losses due to area changes at the inlet and outlets of the glottis, as well as the viscous losses, were taken into account (Ishizaka, 1985, 1989). The losses due to a phase difference between the two masses can be modeled by another flow-dependent resistor

$$R_{12} = \frac{\rho}{2}\eta_{12}\left(\frac{1}{A_{g1}} - \frac{1}{A_{g2}}\right)^2 |U_g| \tag{9a}$$

where

$$\eta_{12} = \begin{cases} 0.4 & \text{if} A_{g1} \geq A_{g2} \\ 1.0 & \text{if} A_{g1} < A_{g2} \end{cases} \tag{9b}$$

R_{12} is added to R_{tot} of Eq. (7). Also, note that $R_{12} = 0$ for $A_{g1} = A_{g2}$.

In the two-mass model outlined above, the lung pressure p_s is assumed constant, which is equivalent to the assumption that the acoustic impedance of the subglottal structure (trachea and lungs) is zero. This is not strictly true. Data on the subglottal impedance is available in the literature (e.g., Fant et al., 1972; Ishizaka et al., 1976; Hudde and Slatky, 1989). Its incorporation into the model would improve the model in cases of extended periods of maximum opening and in cases of glottal "leakage," that is, incomplete closure during the glottal cycle. Both these conditions are believed to cause the percept of "breathiness" (see Klatt and Klatt, 1990).

The incorporation of a glottal "chink," that is, a permanently open portion of the glottal duct, into the model was studied by Ishizaka (1989). He found that a chink has a dramatic effect on the resulting pitch F_0, which decreases with increasing chink opening A_c. He also found that the waveform envelopes of synthesized /h/-to-vowel transitions were more realistic—the amplitude increases more gradually—than without a glottal chink.

Recently, we have started to experiment with area-driven versions of the glottal model (Gupta and Schroeter, 1990). In such a modified model, the acoustic glottal flow is still computed according to Eq. (5), but the self-oscillating mechanical part of the original model is replaced by a parametric model of the glottal area function. Such a model is reasonable because the acoustic load interacts with the flow much more than with the mechanical motion of the cords (Guérin, 1985). One of the advantages of using parametrized glottal area functions is that acoustic features such as spectral tilt can be controlled independently of other features such as pitch. These features are highly interdependent in the self-oscillating model, which makes it very difficult to match the model to certain speakers. Another advantage is that the instants of glottal closure can be positioned precisely. This is not possible with the self-oscillating model. On the other hand, area-driven glottal models may need more parameters than self-oscillating models and, most likely, also a higher parameter update rate. These issues are currently being investigated.

2.3.2. *Model for Turbulent Excitation*

In natural speech unvoiced sounds are produced either by frication (at a constriction of the tract) or by aspiration (at the glottis). Both these cases are treated in our synthesizer; that is, noise of the correct amplitude is automatically injected at the appropriate place within the tract.

Let us consider frication first. Initially, we followed the suggestions of Fant (1970, pp. 272–275), Flanagan (1972, p. 251), and Flanagan et al. (1975) and introduced a noise pressure source whenever the squared Reynolds number Re^2 at the constriction exceeded a critical value $\mathrm{Re}^2_{\mathrm{crit}}$. The amplitude of the noise source was set to $\mathrm{Re}^2 - \mathrm{Re}^2_{\mathrm{crit}}$. This configuration is depicted in Fig. 3a. In this figure, the impedance Z_1 is the backward input impedance seen toward the glottis at the constriction. Assuming a wide-open glottis and zero subglottal impedance, we obtain

$$Z_1 = \frac{B_{\mathrm{fric}}}{D_{\mathrm{fric}}} \tag{10}$$

where B_{fric} and D_{fric} are elements of the chain matrix representing the tract from glottis to the constriction. Also, the impedance Z_2 in Fig. 3a is given by

$$Z_2 = \frac{D_L Z_L - B_L}{A_L - C_L Z_L} \tag{11}$$

where $A_L, B_L, \ldots$ are the elements of the $ABCD$ matrix K_L that represents the portion of the tract between the constriction and the lips, and Z_L is the radiation impedance at the lips. The squared Reynolds number Re^2 is given

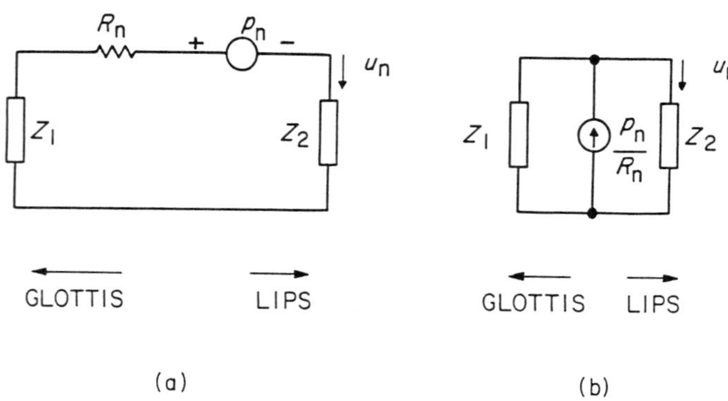

Figure 3 Implementing constriction noise sources in the vocal tract model. Impedances Z_1 and Z_2 represent the residual vocal tract from the reference plane to the glottis and to the lips, respectively. (a) Series pressure source model; (b) parallel volume velocity source model.

by

$$\mathrm{Re}^2(n) = \frac{4\rho^2}{\pi\mu^2}\frac{\bar{u}_C^2(n)}{area_C} \tag{12}$$

where $\bar{u}_C$ is a low pass–filtered version of the volume velocity u_C at the constriction, and $area_C$ is the constriction area. The source resistance of the noise pressure source is

$$R_n(n) = \frac{\rho|\bar{u}_C(n)|}{2\,area_C^2} \tag{13}$$

In our current version of the synthesizer, the flow at the constriction u_C is assumed to be identical to the glottal flow u_g (velum assumed closed).

It turned out that the exact placement of the series noise pressure source was very critical. For different fricatives, we needed to inject noise at different places downstream from the glottis. More robust performance was obtained by introducing the noise source at a "reference plane" one section downstream of the constriction and by converting the series pressure source into a shunt volume velocity source, as shown in Fig. 3b. Shadle (1983) has shown that if the noise source is represented by a volume velocity source, then its position is not very critical. Following the procedure outlined by Liljencrants (1985, p. 5–4 eq. 5.109), we computed the volume velocity of the noise source as $u_n = p_n/R_n$, where p_n is the amplitude of the series pressure source of Fig. 3a, and R_n is the resistance given in Eq. (13). Thus,

$$u_n(n) = G \times \mathrm{random}(n)\frac{\mathrm{Re}^2 - \mathrm{Re}_{\mathrm{crit}}^2}{R_n(n)} \tag{14}$$

where random(n) is a uniformly distributed random number between -0.5 and 0.5. The critical squared Reynolds number $\mathrm{Re}_{\mathrm{crit}}^2$ is 2070^2; a good choice for the gain factor G is 0.5×10^{-6}.

The transfer function between noise flow U_n and radiated noise pressure at the lips can be derived by inspection of Fig. 3b and by using, for example, Eq. (2). In the frequency domain (capitalized variable names) we get

$$H_N = \frac{P_L}{U_n} = \frac{Z_L}{A_L - C_L Z_L}\frac{Z_1}{Z_1 + Z_2} \tag{15}$$

For aspiration, we basically follow the same approach. Here, the best way we found to compute the squared Reynolds number is

$$\mathrm{Re}^2(n) = \left[\frac{l_g\rho}{\mu A_{g2}}u_g(n)\right]^2 \tag{16}$$

where μ is the viscosity of air ($\mu = 1.86 \times 10^{-4}$ dyn s/cm^2), ρ is the density of air ($\rho = 1.14 \times 10^{-3}$g/cm^3), and l_g is the width of the glottal duct ($l_g = 1.4$ cm). Then, an aspiration noise flow is computed in analogy to Eq. (14), in this case using the constants $G = 0.5 \times 10^{-7}$ and $\text{Re}_{\text{crit}}^2 = 2700^2$, and using R_{tot} of Eq. (7) instead of R_{n} of Eq. (13) as the noise source impedance.

2.4. The Synthesizer

The details of the algorithm for producing a speech signal in terms of the components described above can be found in Sondhi and Schroeter (1987). Here we will content ourselves with just a description of the sequence of steps required.

From the specification of the articulatory parameters, the chain matrix from the glottis to the lips is computed as described in Section 2.2. From the A, B, C, D components of this matrix, the transfer function and the input impedance are obtained from Eqs. (2) and (3). The inverse fast Fourier transform (FFT) of the transfer function H and the input reflectance $R_{\text{in}} = (Z_{\text{in}} - Z_0)/(Z_{\text{in}} + Z_0)$ with Z_{in} of Eq. (3) give the corresponding time functions $h(n)$ and $r_{\text{in}}(n)$. These functions are computed every 20 ms, and the intermediate values are obtained by linear interpolation.

The current pressure $p_1(n)$ at the input to the vocal tract is then computed by convolving r_{in} with a linear combination of past values of the glottal volume velocity u_g and p_1. For details, see Eq. (18) in Sondhi and Schroeter (1987). With p_1 known, the pressure $\Delta p = P_s - p_1$ on the right-hand side of Eq. (4) is known. The same pressure also drives the coupled oscillator equations governing the motion of the vocal cords. Thus a new value is computed for the displacements of the two masses and the glottal volume velocity. These give a new value for p_1 and so on. The glottal volume velocity obtained in this way is convolved with the transfer function $h(n)$ to produce voiced speech.

In order to introduce frication, the chain matrix of the tract is actually derived as the product of two matrices—from the glottis to the narrowest constriction and from the constriction to the lips, as discussed in Section 2.3.2. This enables us to compute the volume velocity at the constriction and thus introduce a noise source on the basis of the Reynolds number (see Section 2.3.2).

Finally, to produce nasal sounds, the chain matrix for the nasal tract is also computed (see Section 2.2). This gives us the input impedance of the nasal tract at the velum and the transfer function of the nasal tract from the velum to the nostrils. The input impedance of the nasal tract can be represented as a special chain matrix (Sondhi and Schroeter, 1987), which is inserted in the chain of matrices from the glottis to the lips. The volume velocity at the velum when convolved with the nasal transfer function, gives the nasal output, which is added to the output from the lips to give the total speech signal.

3. CODING BASED ON SPEECH PRODUCTION MODELS

In the previous section we described details of the speech production models we use. In this section we will address the remaining difficult problem of how to derive the parameters for a mimic, that is, a complete analysis and (re-)synthesis of speech. Obtaining these control parameters has proved to be a nontrivial task. Let us briefly review other work on the problem before concentrating on our own approach.

Flanagan et al. (1980) were the first to demonstrate the feasibility of deriving the control parameters in a complete analysis-synthesis procedure—at least for all-voiced, nonnasal speech and a highly simplified articulatory model. What they demonstrated was a fully automatic analysis procedure based on optimization (i.e., analysis by synthesis) utilizing a cepstrally smoothed short-time log-amplitude spectrum for matching original and synthetic speech. For this purpose, the Hooke-and-Jeeves optimization procedure was used (see, e.g., Kuester and Mize, 1973). Some control parameters could not successfully be estimated using the spectral match alone. For example, the tension parameter q of the Ishizaka–Flanagan glottal model had to be determined by extracting and matching the location of the cepstral peak corresponding to the current pitch frequency. One of the drawbacks of this early approach was that every trial synthesis started from a quiescent synthesizer; that is, the steady state had to be reached before the features used for optimization could be extracted. Other drawbacks include the rather high sampling rate of 20 kHz, which was needed to ensure stable operation of the glottal model, and the fact that the mouth opening area was obtained by photography and planimetry rather than estimated from the speech signal.

In Japan, very interesting work on the parameter estimation problem has been reported by Shirai and his colleagues at Waseda University. However, their interest in this problem has been exclusively for the purpose of speech recognition. Shirai and Honda (1976) introduced an articulatory model based on a linear factor analysis of x-ray data. Early attempts at control parameter estimation were reported in Shirai and Honda (1978). Further work on the application of this problem to speech recognition was reported in Shirai and Kobayashi (1986) and Shirai (1988). The authors solve the speaker adaptation problem by adapting the nominal tract length of their model (much like our approach described in Section 3.2.2 below). They deal with coarticulation by estimating step functions representing neural motor commands as features for recognition. More recently, Shirai has studied neutral networks for mapping acoustic cepstral features to articulatory parameters (Shirai, 1988; Shirai and Kobayashi, 1990).

Strube and colleagues at the University of Göttingen, Germany, realized the drawbacks of earlier frame-by-frame optimization attempts—rough parameter trajectories due to noise and bias on the measured spectral features—and therefore decided to pursue an iterative Kalman-filter approach to the nonlinear estimation problem (Meyer et al., 1989). Problems with this approach (Wilhelms, 1987) include the generally rather poor spectral match of the original and the resynthesized speech; failure of convergence for fricatives and sounds with mixed excitation; and the need for extensive side information, for example, the need to set externally the times and amount of velum opening in order to match vowel-nasal-vowel transitions. More recently, the group recorded a large database of tongue-related articulatory data measured by electromagnetic means, as well as lip opening data measured by video camera, velum opening data measured by hot-wire anemometer, and natural speech sampled at 10 kHz (Panagos, 1989; Krueger, 1989). The data base was used to explore both directions of the articulatory-to-acoustic mapping. The acoustic features used were log-area ratios computed from a bark-warped eighth-order LPC (Strube, 1980), The mappings were obtained by quadratic regression or, alternatively, a single neural net, followed by linear and nonlinear variants of Kalman filtering. The Kalman filtering employed linear or nonlinear multidimensional predictors; the latter were based on Volterra series expansions.

Panagos and Krueger judged their results to be very preliminary. The main pitfalls encountered were the un(re-)solved ambiguities in both mappings, the large bias in the acoustic features introduced by glottal variability, the unsolved issue of speaker adaptation (judged to be required even between different recordings of the *same* speaker!), and the need for very large data sets for training the nonlinear predictors/nonlinear Kalman filters.

For the remainder of this chapter we will concentrate on our own research while referencing relevant studies reported in the literature. Our work is obviously related to the work reviewed above, although it has a somewhat different focus. Our main target application is speech coding; therefore, the intelligibility and the naturalness of the resynthesized speech are our most important goals. Also, input speech may come from almost any speaker. The possibility that an algorithm works for only a small number of speakers (or speech sounds) and fails drastically for the majority of others needs to be addressed. For these reasons, we have to take into account issues like robustness and completeness of coverage of the wide temporal-spectral range of natural speech. In the following, we will describe some of the steps taken toward these goals.

3.1. Estimation of Model Parameters

The problem we address in this section is as follows: Given a segment of speech, we wish to estimate the sequence of vocal tract and glottal configurations that produced the given speech.

In the past, attempts have been made to derive the vocal tract shape directly from the LPC vector (or other similar representations; see, e.g., Wakita, 1973; Atal and Hanauer, 1971). These attempts have not been successful because such a reconstruction strongly depends on the assumptions made about the excitation source and the boundary condition at the lips. None of these is known. For a lossless tract it can be shown that even if this information were known *exactly*, the vocal tract shape could not be determined because the transformation from the vocal tract transfer function to vocal tract shape is inherently nonunique (Sondhi, 1979). Although this nonuniqueness has not been proved theoretically for a lossy vocal tract, we have observed it quite often.

In our approach the model parameters are derived using an analysis-by-synthesis optimization procedure and also employing the Hooke-and-Jeeves optimization method mentioned earlier. Our procedure is outlined in Fig. 4.

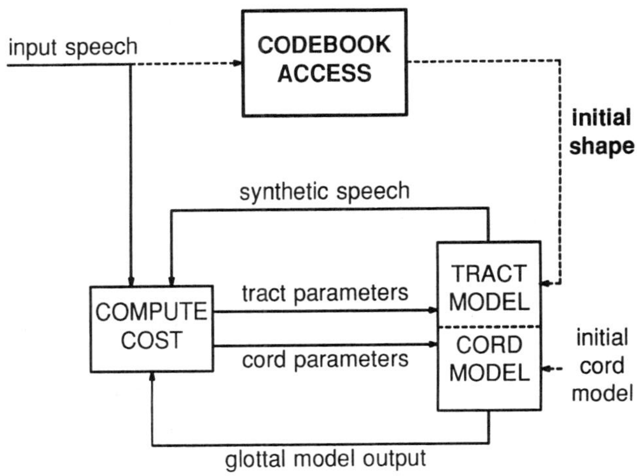

Figure 4 Optimization scheme used to obtain tract shapes and cord (glottal) parameters.

As shown there, the parameters are iteratively adjusted so as to minimize a cost function. The choice of cost function is crucial to the success of the method. It is through selection of this function that the nonuniqueness mentioned above is alleviated. Our current cost function will be discussed in the next section.

From the input speech we compute a short-term spectral representation, a voicing parameter, pitch, and energy, much as in a vocoder application. Unlike the vocoder case, however, the short-term spectrum is used in our approach to access a codebook of vocal tract shapes. The purpose of the codebook is to provide a good starting shape for the optimization, thus reducing the probability of getting trapped in a local minimum of the objective cost function. The starting values for the glottal parameters are obtained by appropriate interpretation of the signal energy, the voicing parameter, and pitch in terms of the respective physiological parameters of a glottal model; for example, for the two-mass model these are P_S (lung pressure), A_{g0} (glottal rest area), and q (see Section 2.3.1).

The ideal distance measures used in defining the cost function should be perceptually meaningful; that is, they should be sensitive to distortions perceived by the ear and should be insensitive to details of the synthetic waveform which are irrelevant to the human ear. It turns out, however, that the specifics of the task also affect our choice of distance measure(s). For example, the ideal distance measure for optimizing the vocal tract should be sensitive to changes in the tract but insensitive to changes in glottal parameters. The opposite is true for optimization of the glottal parameters. Also, the best distance measure for accessing the codebook is not necessarily optimal for estimating the vocal tract shape.

3.1.1. *The Cost Function to be Minimized*

Let us briefly review the current status of the frame-by-frame optimization procedure originally described in Schroeter et al. (1987). In the mimic described there, we use a fixed frame size of 20 ms and a fixed frame shift of 10 ms. In any frame, the original speech and synthesized speech are windowed using an asymmetric Hamming window with the maximum weight placed at 0.9 of the current frame length. Autocorrelations normalized by the residual energy, tenth-order LPC vectors, and autocorrelations of the LPC vectors are computed for both the original and synthesized speech. These are used to compute the cost function used to evaluate the match between the original and synthetic signals. The cost function has four components as follows.

The first component is the symmetrized likelihood ratio D_{sp} given by

$$D_{sp}(\mathbf{a}_R, \mathbf{a}_T) = \tfrac{1}{2}\left[d_{sp}(\mathbf{a}_R, \mathbf{a}_T) + d_{sp}(\mathbf{a}_T, \mathbf{a}_R)\right] \tag{17}$$

where

$$d_{sp}(\mathbf{a}_R, \mathbf{a}_T) = \frac{\mathbf{a}_R' V_T \mathbf{a}_R}{\mathbf{a}_T' V_T \mathbf{a}_T} - 1 \tag{18}$$

is the likelihood-ratio distance between the two LPC vectors $\mathbf{a}_R$ (the LPC vector for the original speech) and $\mathbf{a}_T$ (the LPC vector for the synthetic speech), and V_T is the signal autocorrelation matrix that gave rise to LPC vector $\mathbf{a}_T$.

The second component compares the energies of synthesized and original speech. It is computed as

$$D_{\text{energy}} = \left| 10\log_{10}\left(\frac{R(0)_{\text{ori}}}{R(0)_{\text{syn}}}\right)\right| \tag{19}$$

where $R(0)_{\text{ori}}$ is the zeroth-order autocorrelation of the original speech and $R(0)_{\text{syn}}$ is the zeroth-order autocorrelation of the synthesized speech.

The third cost component compares the synthetic ($\dot{u}_g$) and the estimated ($\hat{\dot{u}}_g$) time derivative of the glottal excitation. The latter is obtained by inverse filtering the original speech, that is, by convolving it with the impulse response of the filter $\mathrm{j}\omega/H$, where H is given by Eq. (2) and j is the imaginary unit. The distance between $\dot{u}_g$ and $\hat{\dot{u}}_g$ is computed using the normalized correlation between them:

$$\rho(k) = \frac{\langle \dot{u}_g(i)\hat{\dot{u}}_g(i-k)\rangle - \langle \dot{u}_g(i)\rangle\langle \hat{\dot{u}}_g(i-k)\rangle}{\left(\langle \dot{u}_g^2(i)\rangle - \langle \dot{u}_g(i)\rangle^2\right)^{1/2}\left(\langle \hat{\dot{u}}_g^2(i)\rangle - \langle \hat{\dot{u}}_g(i)\rangle^2\right)^{1/2}} \tag{20}$$

which is insensitive to amplitude differences. In Eq. (20), k is a positive window shift on the inverse filtered speech and $\langle\ \rangle$ denotes the time average of the enclosed expression for i ranging over the window length. The maximum $\rho_{\max} = \max_k \rho(k)$ is then used in the distance measure:

$$D_{u_g} = 1 - \rho_{\max} + D_{sp}(\mathbf{b}_T, \mathbf{b}_R) \tag{21}$$

The symmetrized likelihood ratio distance D_{sp} (see Eq. 17) is also computed using the LPC vectors $\mathbf{b}_T$ and $\mathbf{b}_R$, which are obtained from the time derivative of the synthesized glottal flow and the inverse-filtered speech signal, respectively. The D_{sp} component in D_{u_g} helps ensure the correct spectral tilt on the excitation, thus alleviating ambiguities due to source-tract compensation. Note that D_{u_g} is computed only for voiced speech; otherwise it is set to zero.

The fourth component of the cost function is the parameter distance d_{geo} defined as

$$d_{\text{geo}}(S_n, S_{n-1}) = \left\{ \frac{1}{N_{\text{par}}} \sum_{i=1}^{N_{\text{par}}} [S_n(i) - S_{n-1}(i)]^2 \right\}^{1/2} \qquad (22)$$

where n and $n-1$ denote the current and the previous frame, respectively, $S_n(i)$ is the ith articulatory parameter for frame n, and N_{par} is the number of parameters (11). The inclusion of d_{geo} in the cost function alleviates the problem of acoustic ambiguity and of defining the shape during short silent periods—for example, during the production of plosives.

These four components are weighted as follows:

$$C = w_{\text{acoust}} \left[0.6 D_{\text{sp}}(\mathbf{a}_T, \mathbf{a}_R) + 0.4 D_{u_g}\right] + w_{\text{geom}} d_{\text{geo}} + 0.4 D_{\text{energy}} \qquad (23)$$

The weight w_{acoust} is set to a maximum of 1.0 for the highest-energy, voiced portions and is set to a minimum of 0.5 in unvoiced or silent portions of the input speech. The weight w_{geom} is set to $1 - w_{\text{acoust}}$. Since we do not demand synchrony of original and synthesized pitch epochs for voiced speech, we allow the optimization window to slide by about $\pm$ half a pitch period over the original speech s in order to ensure an equivalent position in relation to the corresponding window applied on the synthesized speech $\hat{s}$. The optimum position is found by an exhaustive search maximizing the normalized correlation between the two speech signals.

Tract and glottal parameters can be optimized jointly. It is also possible to deviate from this joint optimization and optimize both models separately. In the latter case, for each frame we first optimize the tract shape, using the initial glottal parameters, because we generally expect the initial estimates for the glottal parameters to be more accurate than those for the tract areas. In a second stage of optimization, we use the best tract shape of stage 1 and optimize the three parameters of the two-mass glottal model. The signals compared in this stage are the glottal flow derivative $\dot{u}_g$ from the two-mass model and the inverse-filtered input speech $\hat{\dot{u}}_g$. This scheme of sequential optimizations can be iterated. It was found that usually four iterations are sufficient. Comparing joint and iterated separate optimizations, we found that the joint optimization takes about half the computations of the sequential procedure but achieves an optimal cost about 80 to 100% larger.

In Schroeter et al. (1988), we showed that better spectral matches can be obtained by optimizing tract areas directly, thus eliminating any articulatory model. Spatial smoothness can be ensured by adding a "smoothness" component to the composite cost (Eq. 23). An effective smoothness distance is

$$d_{\mathrm{sm}}(\mathbf{A}) = \frac{1}{\mathbf{A}'\mathbf{A}} \sum_{i=1}^{N_{\mathrm{sec}}-1} (A_{i+1} - A_i)^2$$

where **A** is the vector of N_{sec} cross-sectional areas of the tract at equispaced points along the tract.

3.1.2. Multiframe Estimation of Parameters

Several parameters need somewhat special treatment because they cannot be determined reliably (or uniquely) from a single frame of speech. One such parameter is the nominal length of the vocal tract. It is well known that the first three of four formants of the speech spectrum can be matched by adjusting the shape of a vocal tract of arbitrary length. Furthermore, the shape required to match the formants stays quite reasonable even if the assumed tract length is quite different from (e.g., 20% off) the correct value.

The easiest way to change the length of the vocal tract is to change the parameter Δl, the section length. Thus the problem is to find an estimate of that parameter. Other parameters which cannot be identified reliably from a single frame are lung pressure and the scale factor that relates the given speech signal to the physical pressure in the acoustic wave to which it corresponds.

The way in which we estimate these "global" scale factors is to conduct an initial one-parameter (or at most two-parameter) optimization on these scale factors over fairly long sections of speech (about 10 or 15 frames). These scale factors are then kept constant for the segments over which they have been estimated. The details of the optimization procedure may be found in (Schroeter et al., 1988). The frequency with which these global parameters ought to be updated is under study.

3.2. Initialization of the Tract Optimization

The cost function described above is not a simple quadratic function of the parameters to be optimized. It has large numbers of local extrema; therefore the success of the optimization procedure depends strongly on the initial values chosen for the parameters. Unless the initial parameter vector is close to the optimum, convergence will be to a local (possibly very bad) minimum. Our proposal for dealing with this problem for the tract optimization is discussed in the following section.

3.2.1. Articulatory Codebook and Distance Measures for Codebook Access

We use a codebook to supply the starting shape of the vocal tract. Articulatory codebooks consist of geometric parameter vectors describing vocal tract shapes that are linked to corresponding acoustic parameter vectors representing vocal tract filters. (Some ideas similar to the use of such codebooks appear also in Shirai and Kobayashi, 1986.) To start the optimization of the vocal tract, we use acoustic parameters obtained from natural speech as a key to retrieve corresponding tract shapes. For good performance, two conditions have to be fulfilled: (1) the acoustic space spanned by the codebook has to cover the space of natural speech spectra, and (2) the distance measure used to access the codebook must ignore differences due to glottal excitation (e.g., spectral tilt).

Articulatory codebooks are generated by clustering a training set of *representative* tract configurations. In order to generate a codebook of representative vocal tract shapes, consideration must be given to the method of sampling the articulatory space, the parameter representation and distance measure used, and the method for clustering the shape vectors. In Larar et al. (1988) we described our initial approach to codebook generation. There, we generated a training set of tract shapes by interpolating articulatory parameters between 20 so-called root shapes. These root shapes were created interactively with a graphical version of the Mermelstein articulatory model (see Section 2.1.1) acting as front end to our synthesizer. Then we clustered such a training set in two steps: first, we used the geometric distance of Eq. (22) to obtain initial clusters; second, we clustered each initial cluster into several subclusters, this time using the acoustic distance defined in Eq. (17). The resulting codebooks, however, did not contain phonemes (e.g., /r/, /w/). Also, the clustering method used could not handle the millions of training vectors which we felt were necessary to provide adequate sampling. Therefore, we later developed a different method which we will explain in the following. Details are given in Schroeter et al. (1990).

In light of condition (1) above, let us describe how we achieved a large coverage. Unlike the "root-shape interpolation" technique employed in the generation of our earlier codebooks, we now use uniformly distributed random numbers to compose articulatory training vectors. We use the Mermelstein and the Coker articulatory models to convert articulatory vectors to tract-area vectors. Random sampling of a region somewhat larger than that spanned by the nominal ranges of the parameters ensures that our training vectors cover as much formant space as allowed by the particular articulatory model.

We partition the three-dimensional space of formant frequencies into a grid of frequency bins, each of which has a width of 5% of its center frequency in each of the three dimensions. For each tract-area vector in the training data, we compute the first three formants and the index of the corresponding bin. The training vector is placed in that bin unless a geometrically similar vector is already in that bin, that is, if the geometric distance between the two is smaller than a predefined threshold. This measure guarantees that we retain all tract shapes that result in close formant frequencies but are geometrically different. The geometric distance measure used for pruning is

$$d_{\text{geo}}(\mathbf{A}, \mathbf{B}) = (A_V - B_V)^2 + 1/N \sum_{i=1}^{N} [\log(A_i) - \log(B_i)]^2 \qquad (25)$$

where **A** and **B** are the augmented tract area vectors $(A_1, \ldots, A_N, A_V)$ and $(B_1, \ldots, B_N, B_V)$ to be compared, $N = 20$ is the number of tract sections, and A_N and B_V are the velum opening areas. For pruning, geometric thresholds between 0.01 and 2.0 were used. Using one specific threshold, the training vectors retained in all the formant bins composed a single codebook. Different codebooks—different in the geometric threshold, the articulatory model, the limits on the articulatory parameters, or the number of random training vectors used for generation—were then combined to form larger codebooks, the largest of which at the present time has more than 200,000 entries. Figure 5 shows a comparison of the formant coverage of a midsized, nonnasal codebook with formant data for sonorants and vowels published in the literature. The symbols such as LL refer to different sources. For details see Schroeter et al. (1990). From the figure, it is clear that the new codebooks encompass the required formant-frequency space.

In regard to condition (2) above, let us now focus on the important issue of what acoustic distance measure to use for codebook access. Originally, we chose the codebook entry such that the symmetrized likelihood ratio (Eq. 17) between it and the given speech frame was a minimum. This distance, however, is strongly dependent on the spectral tilt and bandwidth broadening due to the glottal excitation signal. Shirai and his colleagues (Shirai and Honda, 1978; Shirai and Kobayashi, 1986) deal with this problem by employing an adaptive first-order preemphasis filter. Instead, we minimize a weighted cepstrum distance defined as

$$d(\mathbf{c}, \hat{\mathbf{c}}) = \sum_{k=1}^{k_{\max}} w_k^2 (c_k - \hat{c}_k)^2 \qquad (26)$$

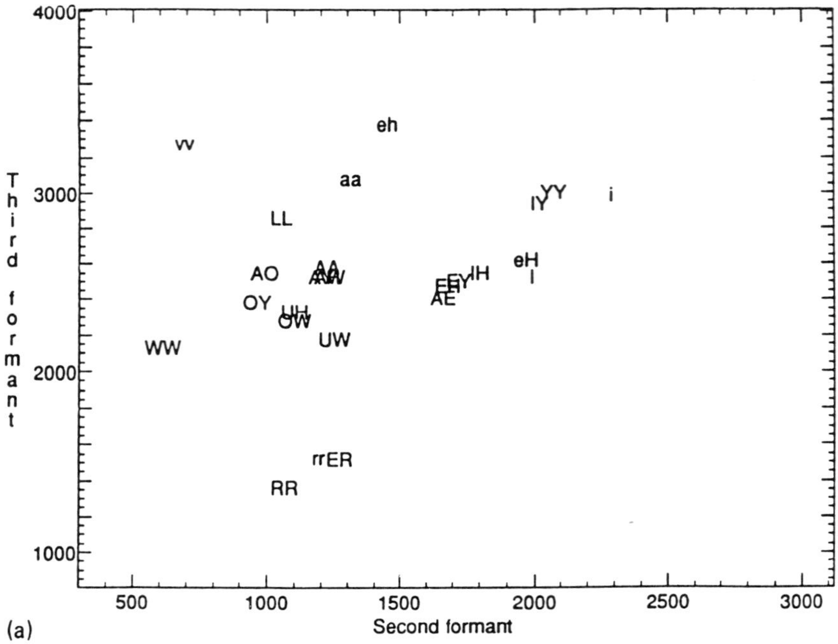

(a)

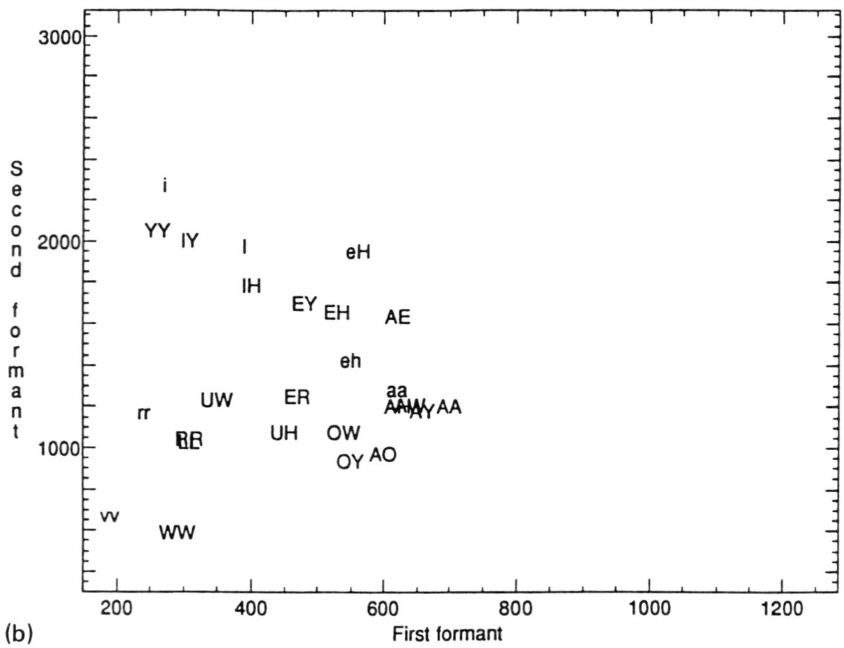

(b)

Figure 5 Formant space of natural speech (a,b) versus formant space of one of our nonnasal codebooks (c,d), F_1/F_2 (b,d), and F_2/F_3 (a,c).

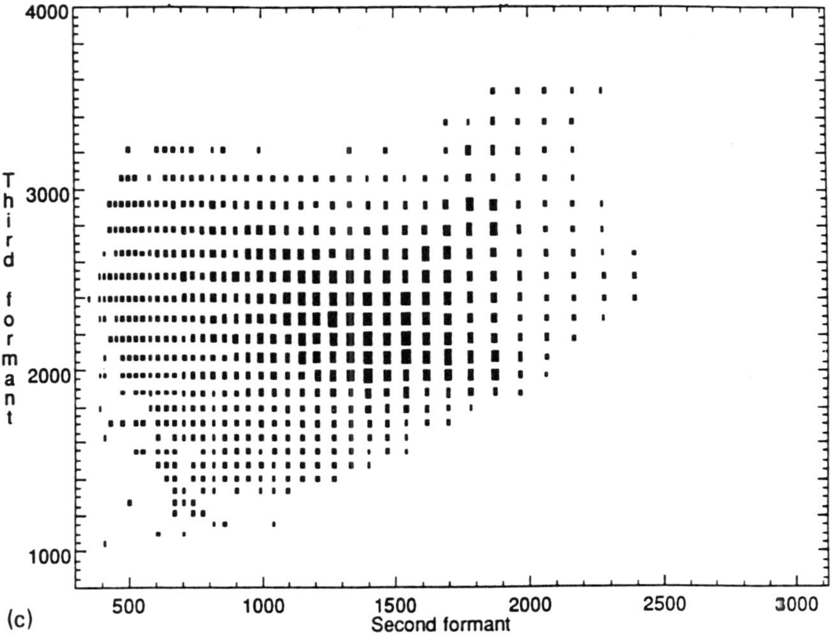
4000
3000
2000
1000
Third formant
500
1000
1500
2000
2500
3000
Second formant

(c)

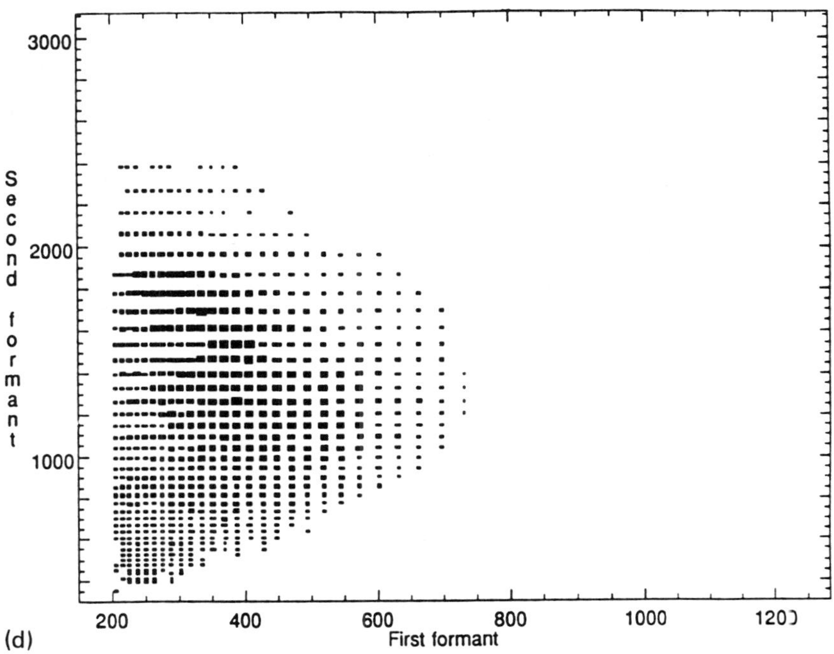
3000
2000
1000
Second formant
200
400
600
800
1000
1200
First formant

(d)

Here $\mathbf{c} = (c_1, \ldots, c_{k_{\max}})$ is a vector of cepstral coefficients derived from the speech frame and $\hat{\mathbf{c}} = (\hat{c}_1, \ldots, \hat{c}_{k_{\max}})$ is the vector of cepstral coefficients derived from the codebook. The cepstral vectors can be derived from the LPC vectors or from the FFT of the speech signals. $\mathbf{w} = (w_1, \ldots, w_{k_{\max}})$ is the weighting vector or lifter. This distance measure degenerates into a spectral slope distance if we choose a linear lifter $w_i = k$ (Tohkura, 1986).

To find the best lifter $\mathbf{w}$, we conducted a study in which we evaluated 11 candidate lifters (i.e., weight vectors) on speech *synthesized* from known tract shapes with various glottal excitations (Meyer et al., 1991).

A useful bandpass lifter is the raised sine lifter described by Juang et al. (1987):

$$w_k = 1 + 0.5k_{\max} \sin\left(\frac{k_\pi}{k_{\max}}\right) \quad \text{for } 0 < k \le k_{\max} \tag{27}$$

In Meyer et al. (1991) we derived the following shape of an optimal FFT lifter ($k_{\max} = 30$):

$$w_k = \begin{cases} \left(\dfrac{k}{20}\right)^{0.4} & 0 < k \le 20 \\ 0.5 + 0.5\cos\left(\dfrac{\pi(k-21)}{20}\right) & 20 < k \le k_{\max} \end{cases} \tag{28}$$

We found that FFT-based cepstral distance measures outperform LPC-based cepstral distance measures. In addition, we introduced cepstral low-pass filtering (log spectral weighting) to improve the balance between errors incurred for different formants. Kobayashi et al. (1985), as well as Shirai and Kobayashi (1986), also report improved results by applying a cepstral low-pass filter. We found the filtered and liftered cepstral distance superior to the likelihood-ratio distance and at least as good as distance measures based on a critical-band filter bank.

Most recently, we performed extensive tests using *natural* speech. For natural speech, of course, the "correct" tract shape is difficult to acquire. Instead of measuring the speaker's articulator positions (e.g., as in Panagos, 1989), we decided to identify the correct shape through use of the formant just-noticeable-difference (JND) box measure of Ghitza and Goldstein (1986). For a tract shape to be labeled as "correct," we require it to produce formant frequencies and bandwidths which are not farther away from the original than one JND. Ghitza and Goldstein's results for continuous speech are summarized by the following equation:

$$\mathrm{JND}(F_i)[\%] = \begin{cases} 14.1F_i^{-0.41} & 0\ \mathrm{kHz} < F_i \leq 1.5\ \mathrm{kHz} \\ 6.3F_i^{1}.58 & 1.5\ \mathrm{kHz} < F_i \leq 4.0\ \mathrm{kHz} \end{cases} \tag{29}$$

Here F_i in kHz is the frequency of the ith formant.

To evaluate the distance measures in terms of JNDs, we used the following procedure: Let M be the number of voiced frames in an utterance and let n_i be the number of these frames for which the frequency of the ith formant is in error by more than a JND. Then the quantity

$$\bar{O} = \frac{1}{3} \sum_{i=1}^{3} \frac{n_i}{M} \cdot 100 \tag{30}$$

gives the average percentage of frames for which the retrieved shape is deemed to be incorrect.

We compared 96 different cepstral distance measures and the symmetrized likelihood-ratio distance used in Schroeter et al. (1987) on a speech data base of eight utterances obtained from male speakers. In this study, we investigated four aspects of the access distance measure used for codebook access given one frame of speech: length of a Hamming window applied to the speech signal in the time domain, weighting of the log spectrum obtained through FFT, lifter type, and lifter length. The results can be summarized a follows: First, we found that a fixed-length Hamming window of 20 ms performed better than a variable-length window spanning two pitch periods. Second, bandpass weighting of the log spectrum which rolls off sharply below 250 Hz and emphasizes the frequency range 250 to 1500 Hz while tapering off gradually above 1500 Hz leads to better results than using either no weighting or just the low-pass part of the optimal weighting. Third, we found that a lifter of 5 ms outperformed shorter lifters. Fourth, we found a slight advantage of the Meyer lifter over the Juang lifter, except in cases where codebook and speaker match extremely well. In total, our best cepstral measure scored 11.1% errors (Eq. 30) for our old small codebook of 1882 tract shapes (Larar et al., 1988) compared to 28.5% using the likelihood-ratio distance. Our best new codebook reduced this error further to 4.5%. Details are given in Schroeter et al. (1990).

3.2.2. Dynamic Codebook Access

For any given acoustic feature vector derived from a single speech frame, there are usually several candidate tract shapes with almost identical acoustic distances to the given input speech frame. This is due to the nonuniqueness of the acoustic-to-articulatory transformation or to model inadequacies, or both. Therefore, the resulting sequence of tract shapes retrieved from an articulatory codebook using acoustic match as the only criterion often varies erratically, and the speech resynthesized from these tract shapes is of poor quality. It is intuitively reasonable to expect that parameter trajectories could be made smoother if *sequences* of tract shapes were retrieved from the codebook so as to optimally match *sequences* of speech frames. Such an optimization, of course, requires that the decision on the sequence of tract shapes be delayed until the entire sequence of speech frames is available. For this reason we call this delayed-decision codebook access.

Various methods may be used to ensure smooth parameter trajectories. One way is to use Kalman filtering (e.g., Hoege and Lang, 1981; Meyer et al., 1989). Krueger (1989) in his mapping experiments done with Strube found that their iterative Kalman filter introduced a delay of three to four frames (about 80 ms). In our work we apply dynamic programming (DP) to minimize the sum of *acoustical* and *geometrical* cost components, accumulated over several speech frames. For the examples given later, we used a DP length of 200 ms with an overlap between successive DPs of 100 ms (i.e., a delay of 100 ms). In several examples this gave results identical to those obtained by using the whole speech utterance. Let us now elaborate on how to access an articulatory codebook with the constraint of smooth parameter trajectories.

The problem we consider is as follows: We are given a sequence of vectors $\mathbf{s}_t$, $t = 1, 2, \ldots, T$, which provide some convenient representation of the spectra of T successive frames of a speech signal. We are also given a codebook of M pairs of entries—the prototypical spectra $\hat{\mathbf{s}}_j$ and corresponding vocal tract shapes $\mathbf{A}_j$, $j = 1, 2, \ldots, M$. The index $j \in [1, M]$ may be used to specify the pair $(\hat{\mathbf{s}}_j, \mathbf{A}_j)$. We are required to find a sequence of integers $j(t)$, $t = 1, 2, \ldots, T$, representing a sequence of selections from the codebook, which is optimal for the sequence $\mathbf{s}_t$ according to some given criterion. The optimality criterion is specified as the minimization of a "cost" associated with every sequence. The problem may be conveniently described with reference to the $M \times T$ matrix of nodes shown in Fig. 6. The nodes in the tth column represent the M possible selections from the codebook at time t. As illustrated, a selected sequence may be represented by a path connecting a node in column $t - 1$ to a node in column t, for $t = 2, 3, \ldots, T$.

The naive approach would be to compute the cost for each possible path and select the one for which the cost is a minimum. However, this requires a prohibitive amount of computation. The number of possible paths is M^T, and we are interested in situations in which M may be several thousand and T may be 10 or more. Fortunately, we can specify useful cost criteria for

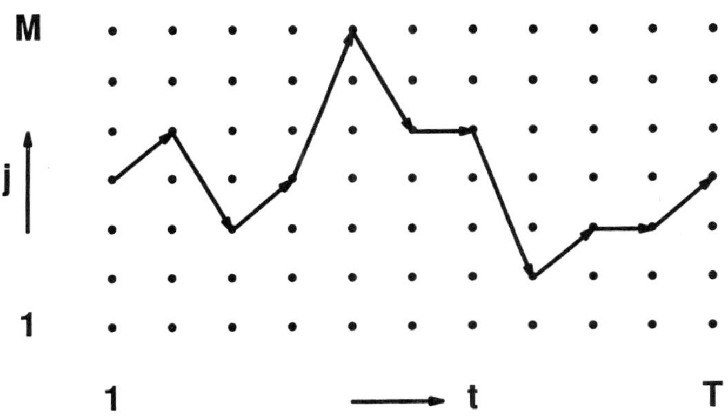

Figure 6 Example of a path representing a sequence of selected vocal tract shapes.

which the search can be accomplished with a computational burden which is proportional to M^2T.

One property we would like the optimal sequence to have is that the spectra of the synthetic and the original speech should match closely. Thus, for the sequence of codebook entries $j(t)$ we would like to minimize the total spectral mismatch between $\hat{\mathbf{s}}_{j(t)}$ and $\mathbf{s}_t$. If $d(\hat{\mathbf{s}}, \mathbf{s})$ is a measure of spectral distance, we might therefore define the cost of the sequence as

$$C_1 = \sum_{t=1}^{T} d(\hat{\mathbf{s}}_{j(t)}, \mathbf{s}_t) \tag{31}$$

With this definition of cost, the computation of the optimal path requires only on the order of MT computations. At each time t, the entry $j(t)$ is selected which minimizes $d(\hat{\mathbf{s}}_{j(t)}, \mathbf{s}_t)$. We will call such a cost function *static* because it can be minimized independently for each frame.

However, minimizing C_1 does not deal with the problem of erratic changes in vocal tract shape mentioned above. For this we need to include a *dynamic* cost component which penalizes a large change during a transition from one frame to the next. At present we use the change in the vocal tract shape as a measure of this dynamic component. Thus if $d_{\text{geo}}(\mathbf{A}_j, \mathbf{A}_k)$ is some measure of the distance between two given shapes $\mathbf{A}_j$ and $\mathbf{A}_k$, then $\sum_2^T d_{\text{geo}}(\mathbf{A}_{j(t-1)}, \mathbf{A}_{j(t)})$ is a measure of the total frame-to-frame variation in shape. In order to penalize this we may redefine the cost to be

$$C_2 = \sum_{1}^{T} d(\hat{\mathbf{s}}_{j(t)}, \mathbf{s}_t) + \sum_{2}^{T} d_{\text{geo}}(\mathbf{A}_{j(t-1)}, \mathbf{A}_{j(t)}) \tag{32}$$

Rewriting this in the form

$$C_2 = d(\hat{\mathbf{s}}_{j(1)}, \mathbf{s}_1) + \sum_{2}^{T-1} [d_{\text{geo}}(\mathbf{A}_{j(t-1)}, \mathbf{A}_{j(t)}) + d(\hat{\mathbf{s}}_{j(t)}, \mathbf{s}_t)] \tag{33}$$

the DP algorithm can be written down almost by inspection. Suppose that one has determined the optimal path to every node in the column at $(t-1)$; i.e., for every node $(j, t-1), j = 1, 2, \ldots, M$. Let $C_2(j, t-1), j = 1, 2, \ldots, M$, be the total accumulated costs for these paths, respectively. Then it is seen from the expression for C_2 that the optimal path that ends at node (k, t) is the one coming from the node $(j^*, t-1)$, where j^* is given by

$$j^* = \underset{j \in [1,M]}{\arg\min} [C_2(j, t-1) + d_{\text{geo}}(A_j, A_k)] \tag{34}$$

The total accumulated cost for this path is

$$C_2(k, t) = C_2(j^*, t-1) + d_{\text{geo}}(A_{j^*}, A_k) + d(\hat{s}_k, s_t) \tag{35}$$

Repeating for each value of k, we see that the optimal paths up to $t-1$ can be extended to t with M^2 comparisons.

For $t = 1$ the optimal paths are trivially known. Therefore, iterating the above procedure for $t = 2, 3, \ldots, M$ gives the optimal path ending at each of the M nodes in the Tth column. At T, the path ending at the node for which $C_2(k, T)$ is a minimum is selected. Note that at each value of t and k, the previous node from which the optimal path arrives at (k, t) is known. Therefore the complete path is recovered by backtracking.

3.3. Sample Results

Figure 7 shows spectrograms of one original utterance, "*Why were you away a year, Roy*?" (top), the utterance resynthesized from optimized articulatory parameters using Mermelstein's articulatory model (middle), and the utterance resynthesized from optimized tract areas (bottom). The most obvious differences between the results from articulatory and area optimization can be found in the /r/ around 2.00 s, where the articulatory optimization failed to match the formants F_2 and F_3, while area optimization had no trouble. The mean squared spectral distortions based on preemphasized signals were 3.21 dB and 2.23 dB for the articulatory and area optimization, respectively. These results are in agreement with informal listening tests which revealed that the area optimization produced speech significantly closer to the original.

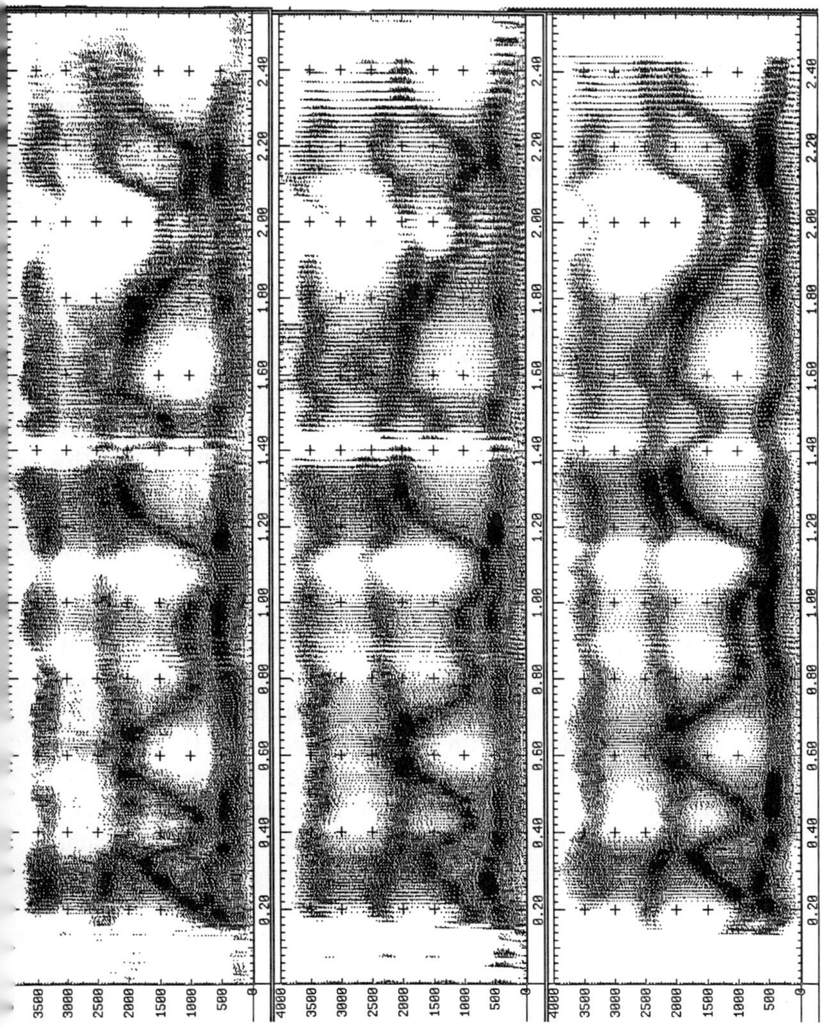

Figure 7 Spectrograms of the sentence *Why were you away a year, Roy?* spoken by a male speaker. (Top), Original, (middle) result of optimizing parameters of the Mermelstein articulatory model, and (bottom) result of optimizing tract areas directly.

4. FUTURE

Although significant progress has been made toward using models for speech production in speech coding, we are still far from having implemented a working coder. To reach this goal, more work needs to be done to improve the tract and glottal models and to improve the parameter estimation methods. It is obvious that these aspects are closely interwoven.

Let us consider parameter estimation first. The estimation problem is difficult mainly for two reasons. First, more than one combination of parameters can yield the same low value for the cost function. This is essentially the nonuniqueness problem. A good example of this is the estimation of tract length. For a single frame of speech, a fairly good match to the spectrum can be realized with a vocal tract of any length in a wide range. Second, the sensitivity of our cost functions to change in one parameter may be very much larger than to change in another. (For example, the spectrum changes dramatically if the area or location of a narrow constriction is changed slightly, but changes much less for a similar change at a point where the area is larger.)

Our work (Schroeter et al., 1988) has shown that the nonuniqueness can be alleviated by searching for smooth parameter trajectories over periods of about 100 ms. We also have learned that in a frame-by-frame optimization considerable smoothing is needed to overcome the inherent noise of the estimation process. Based on one short frame of, for example, 20 ms, we cannot expect to be able to estimate slowly varying speech production parameters which have a low-pass characteristic with a cutoff of under 30 Hz. Therefore, we expect to improve parameter estimation by synthesizing larger portions of speech at a time (about 100 to 200 ms) and accumulating the spectral distance over multiple overlapping short analysis windows. Here the frame-by-frame parameter variation is computed by suitable interpolation between properly chosen segmentation points. This method has been successful in articulatory text-to-speech synthesis (Parthasarathy and Coker, 1991). Note that so-called segment vocoders (e.g., Roucos et al., 1983; Soong, 1989) use a similar approach, but interpolate signal parameters instead of speech production parameters. For example, Soong interpolates linearly line-spectral frequencies over 50 ms with transparent quality. Roucos et al. interpolate log-area ratios over time periods of up to 100 ms. For our work, we need to explore whether we can interpolate tract areas sufficiently well, of whether we have to use articulatory parameters as in Parthasarathy and Coker (1991).

What are some of the remaining deficiencies of the models described in this chapter? In our analysis-synthesis system, we use a concatenation of articulatory models, (acoustic) tract models, and glottal and fricative (excitation) models. First of all, it is important to note that each model has its own inaccuracies and limitations which affect the operation of the other models and the resulting speech quality. We would like the optimization to be constrained by the models to ensure "speechlike" synthesis, that is, avoid artifacts such as the reverberation and noise commonly seen in low-bit-rate coders. At the same time we would like the models to avoid unnecessary constraints. Optimization cannot overcome limitations of overconstrained models. For example, transitions into and out of /r/ are hard to produce with the articulatory model we have used (see Fig. 7), while direct optimization of tract areas eliminates this problem. If the use of articulatory parameters turns out to be unavoidable because of their more favorable interpolation properties (see above), we will have to improve the articulatory model. The model used to compute the acoustic properties of the vocal tract from tract and nasal areas (Sondhi and Schroeter, 1987) needs to be improved by incorporating variable losses (Parthasarathy and Coker, 1991) and, perhaps, by incorporating more than one fricative noise source. So far, we have been unable to obtain satisfactory fricatives in our analysis-synthesis experiments. Finally, we found that the original version of the self-oscillating glottal model of Flanagan and Ishizaka cannot provide the large range of spectral tilts needed to match all speakers and speaking modes. We are currently experimenting with parametric models for the glottal opening.

After robust parameter estimation methods have been found and necessary model improvements have been made, we will need to find efficient strategies for coding glottal and tract parameters. For this, we will need to train glottal and tract codebooks from natural speech.

REFERENCES

Allen, D. R., and Strong, W. J. (1985). A model for the synthesis of natural sounding vowels. *J. Acoust. Soc. Am. 78*(1):58–69.

Atal, B. S., and Hanauer, S. L. (1971). Speech analysis and synthesis by linear prediction of the speech wave. *J. Acoust. Soc. Am. 50*:637–655.

Baer, T., Grecco, L. C., and Nye, P. W. (1991). MRI imaging of the vocal tract. Second Seminar of Speech Production, Leeds, Great Britain, May 13–15, 1990; *J. Phonet.*, in press.

Baer, T., Gore, J. C., Grecco, L. C., and Nye, P. W. (1991). Analysis of vocal tract shape and dimensions using magnetic resonance imaging: Vowels. Submitted to *J. Acoust. Soc. Am.*, also see *J. Acoust. Soc. Am.* 84(Suppl. 1):S125, 1988.

Bjuggren, G., and Fant, G. (1964). The nasal cavity structures. Speech Transmission Laboratory—Quarterly Progress and Status Report (STL-QPSR) 4/1964, Royal Institute of Technology, Stockholm, 5–7.

Bocchieri, E. L. (1983). An articulatory speech synthesizer. Ph.D. thesis, University of Florida.

Coker, C. H. (1976). A model of articulatory dynamics and control. *Proc. IEEE 64*(4): 452–460.

Coker, C. H., and Fujimura, O. (1966). A model for specification of vocal tract area function. *J. Acoust. Soc. Am. 40*:1271(A).

Fant, G. (1970). *Acoustic Theory of speech production*, 2nd printing. Mouton, The Hague.

Fant, G., Ishizaka, K., Lindqvist, J., and Sundberg. J. (1972). Subglottal formants. Speech Transmission Laboratory—Quarterly Progress and Status Report (STL-QPSR) 1/1972, Royal Institute of Technology, Stockholm, pp. 1–12.

Feng, G., Abry, C., Guérin, B. (1985). "How to cope with nasal vowels?" Séminaire Franco-Suédois, Grenoble, France, April, 22–24; printed in *Recueil d'articles parus en 1984–1985*, Institut de la Communication Parlée, Grenoble Cedex, France (in English).

Flanagan, J. L. (1972). *Speech Analysis Synthesis and Perception*, 2nd ed. Springer-Verlag, New York.

Flanagan, J. L., Ishizaka, K., and Shipley, K. L. (1975). Synthesis of speech from a dynamic model of the vocal cords and vocal tract. *Bell Syst. Tech. J. 45*(3):485–506.

Flanagan, J. L., Ishizaka, K., and Shipley, K. L. (1980). Signal models for low bit-rate coding of speech. *J. Acoust. Soc. Am. 68*(3):780–791.

Fujimura, O., and Lindqvist, J. (1971). Sweep-tone measurements of vocal-tract characteristics. *J. Acoust. Soc. Am. 49*(2, Part 2):541–558.

Ghitza, O., and Goldstein, J. L. (1986). Scalar LPC quantization based on formant JND's. *IEEE Trans. Acoust., Speech, Signal Process. ASSP-34*:697–708.

Guérin, B. (1985). Effects of the source-tract interaction using vocal fold models. In *Vocal Fold Physiology*, J. R. Titze and R. C. Scherer, eds. Denver Center for the Performing Arts, Denver, pp. 482–499.

Gupta, S., and Schroeter, J. (1990). Pitch-synchronous analysis/synthesis using models of speech production. *J. Acoust. Soc. Am. 87*(Suppl. 1):S39.

Harshman, R., Ladefoged, P., and Goldstein, L. (1977). Factor analysis of tongue shapes. *J. Acoust. Sac. Am. 62*(3):693–707.

Hegerl, H. C. (1987). Numerische Simulation von Stroemungen im Bereich der Glottis. *Fortschritte der Akustik—DAGA'87*, pp. 625–628 (in German).

Henke, W. L. (1967). Preliminaries to speech synthesis based upon an articulatory model. *Proc. IEEE Conf. Speech Commun. Process.*, pp. 170–182.

Hoege, H., and Lang, M. (1981). Estimation of vocal tract parameter from the speech signal. *Proc. European Conf. on Circuit Theory and Design*, The Hague, May 1981, 25–28, R. Boite and P. Dewilde, eds. North-Holland, Amsterdam.

Hudde, H., and Slatky, H. (1989). The acoustical input impedance of excised human lungs—measurements and model matching. *J. Acoust. Soc. Am. 86*(2):475–492.

Ishizaka, K. (1985). Air resistance and intra-glottal pressure in a model of the larynx. In *Vocal Fold Physiology*, J. R. Titze and R. C. Scherer, eds. Denver Center for the Performing Arts, Denver. pp. 414–424.

Ishizaka, K. (1989). Improved two-mass model of the vocal folds with the glottal chink. AT&T Bell Laboratories Technical Memorandum.

Ishizaka, K., and Flanagan, J. L. (1972). Synthesis of voiced sounds from a two-mass model of the vocal cords. *Bell Syst. Tech. J. 51*(6):1233–1268.

Ishizaka, K., Matsudaira, M., and Kaneko, T. (1976). Input acoustic-impedance measurement of the subglottal system. *J. Acoust. Soc. Am. 60*(1):190–197.

Juang, B. H., Rabiner, L. R., and Wilpon, J. G. (1987). On the use of bandpass liftering in speech recognition. *IEEE Trans. Acoust., Speech. Signal Process., ASSP-35*: 947–954.

Kelly, J. L., and Lochbaum, C. C. (1962). Speech synthesis. *Proc. Fourth Int. Congr. Acoust.*, Paper G-42, pp. 1–4; reprinted in *Speech Synthesis*, J. L. Flanagan and L. R. Rabiner, eds. Dowden, Hutchinson & Ross, Stroudsburg, Pa., pp. 127–130, 1973.

Kiritani, S., Sekimoto, S., and Imagawa, H. (1977). Parameter description of tongue point movements in vowel production. *Ann. Bull. RIHLP 11*:31–37.

Klatt, D. H., and Klatt, L. C. (1990). Analysis, synthesis, and perception of voice quality variations among female and male talkers. *J. Acoust. Soc. Am. 87*(2):820–857.

Kobayashi, T., Yazawa, J., and Shirai, K. (1985). Evaluation of Spectral Distance Measure for the Estimation of Articulatory Motion by the Model Matching Method, *Trans. Inst. Electron. Commun. Eng. Jpn. J68-A*(2):210–217 (in Japanese).

Krueger, E. (1989). Optimierung von akustisch-artikulatorischen Schaetzungen mit Hilfe eines Systemmodells. Doctoral thesis, University of Goettingen, Germany (in German).

Kuester, J. O., and Mize. J. H. (1973). *Optimization Techniques with Fortran*, McGraw-Hill, New York.

Larar, J. N., Schroeter, J., and Sondhi, M. M. (1988). Vector quantization of the articulatory space. *IEEE Trans. Acoust., Speech. Signal Process. ASSP-36*:1812–1818.

Liljencrants, J. (1971). Fourier series description of the tongue profile. Speech Transmission Laboratory—Quarterly Progress and Status Report (STL-QPSR) 4/1971, Royal Institute of Technology, Stockholm, pp. 9–18.

Liljencrants, J. (1985). Speech synthesis with a reflection-type line analog. Dissertation, Royal Institute of Technology, Stockholm.

Liljencrants, J. (1990). Airflow simulations. Second Seminar of Speech Production, Leeds, Great Britain, May 13–15, 1990; *J. Phonet.*, in press.

Lin, Q. (1990). Speech production theory and articulatory speech synthesis. Dissertation, Royal Institute of Technology, Stockholm.

Lindqvist. J., and Sundberg, J. (1972). Acoustic properties of the nasal tract. Speech Transmission Laboratory—Quarterly Progress and Status Report (STL-QPSR) 1/1972, Royal Institute of Technology, Stockholm, pp. 13–17.

Maeda, S. (1972). Conversion of midsaggital dimensions to vocal tract area function. *J. Acoust. Soc. Am. 51*(A):89–90.

Maeda, S. (1979). Une modèle articulatoire de la langue avec des composantes linéaires. 10èmes Journées d'étude sur la Parole, Grenoble, May 30–June 1 (in French).

Maeda, S. (1981). Better acoustic model of the vocal tract. *J. Acoust. Soc. Am.* 70 (Suppl. 1):S42.

Maeda. S. (1982a). Acoustic cues for vowel nasalization: A simulation study. *J. Acoust. Soc. Am.* 72(Suppl. 1):S102.

Maeda, S. (1982b). The role of the sinus cavities in the production of nasal vowels. *Proc. IEEE Int. Conf. Acoustics, Speech, and Signal Processing*, pp. 911–914.

Maeda, S. (1982c). A digital simulation method of the vocal-tract system. *Speech Commun. 1*:199–229.

Maeda, S. (1990). Compensatory articulation during speech: Evidence from the analysis and synthesis of vocal-tract shapes using an articulatory model. In *Speech Production and Speech Modeling* (NATO Advanced Study Institute Series), W. J. Hardcastle and A. Marchal (eds.). Kluwer Academic Publishers, Boston, pp. 131–149.

Mermelstein, P. (1973). Articulatory model for the study of speech production. *J. Acoust. Soc. Am. 53*(40):1070–1082.

Meyer, P., Wilhelms, R., and Strube, H. W. (1989). A quasiarticulatory speech synthesizer for German language running in real time. *J. Acoust. Soc. Am. 86*(2): 523–539.

Meyer, P., Schroeter, J., and Sondhi, M. M. (1991). Design and evaluation of optimal cepstral lifters for accessing articulatory codebooks. *IEEE Trans. Acoust., Speech, Signal Process.*, in press.

Panagos, G. (1989). Messung artikulatorischer Parameter und Schaetzung des artikulatorisch-akustischen Zusammenhangs. Doctoral thesis, University of Goettingen, Germany (in German).

Parthasarathy, S., and Coker, C. H. (1991). On automatic estimation of articulatory parameters in a text-to-speech system. *Comput. Speech Language*, in review.

Perkell, J. S. (1974). A physiologically-oriented model of tongue activity in speech production. Ph.D. thesis, MIT, Cambridge, Mass.

Portnoff, M. R. (1973). A quasi-one-dimensional digital simulation for the time-varying vocal-tract. S.B/S.M. thesis, MIT, Cambridge, Mass.

Ranade, A., and Sondhi, M. M. (1981). Two-dimensional wave motion in a variable-area duct: Time-domain computations using the finite element method. AT&T Bell Laboratories Technical Memorandum.

Roucos, S., Schwartz, R. M., and Makhoul, J. (1983). A segment vocoder at 150 B/s. *Proc. IEEE Int. Conf. Acoustics, Speech, and Signal Processing*, pp. 61–64.

Schroeder, M. R. (1967). Determination of the geometry of the human vocal tract by acoustic measurements. *J. Acoust. Soc. Am. 41*(4, Part 2):1002–1010.

Schroeter, J., Larar, J. N., and Sondhi, M. M. (1987). Speech parameter estimation using a vocal tract/cord model. *Proc. IEEE Int. Conf. Acoustics, Speech, and Signal Processing*, pp. 308–311.

Schroeter, J., Larar, J. N., and Parthasarathy, S. (1988). Vocal-tract areas versus articulatory parameters in speech production modeling. *J. Acoust. Soc. Am. 84*(Suppl. 1):S127.

Schroeter, J., Meyer, P., and Parthasarathy, S. (1990). Evaluation of improved articulatory codebooks and codebook access distance measures. *Proc. IEEE Int. Conf. Acoustics, speech, and signal processing*, pp. 393–396.

Shadle, C. (1983). Turbulence noise in the vocal tract. *Proc. 11th Int. Congr. Acoustics*, Paris, vol. 4, pp. 171–174.

Shirai, K. (1988). Feature extraction of articulatory motion based on the speech production model. *J. Acoust. Soc. Am.* Suppl. 1 *84*:S146.

Shirai, K., and Honda, M. (1976). An articulatory model and the estimation of articulatory parameters by nonlinear regression method. *Trans. Inst. Electron. Commun. Eng. Jpn. E59*(8):17–18 (in Japanese).

Shirai, K., and Honda, M. (1978). Estimation of articulatory parameters from speech waves. *Trans. Inst. Electron. Commun. Eng. Jpn. 61A*(5):409–416 (in Japanese).

Shirai, K., and Kobayashi, T. (1986). Estimating articulatory motion from speech wave. *Speech Commun. 5*:159–170.

Shirai, K., and T. Kobayashi (1990). Estimation of articulatory motion using neural networks. Second Seminar of Speech Production, Leeds, Great Britain, May 13–15, 1990; *J. Phonet.*; in press.

Sondhi, M. M. (1974). Model for wave propagation in a lossy vocal tract. *J. Acoust. Soc. Am. 55*(5):1070–1075.

Sondhi, M. M. (1979). Estimation of vocal-tract areas: The need for acoustical measurements. *IEEE Trans. Acoust. Speech Signal Process. ASSP-27*:268–273.

Sondhi, M. M. (1986). Resonances of a bent vocal tract. *J. Acoust. Soc. Am. 79*(4): 1113–1116.

Sondhi, M. M. and Schroeter, J. (1987). A hybrid time-frequency domain articulatory speech synthesizer. *IEEE Trans. Acoust., Speech, Signal Process. ASSP-35*:955–967.

Soong, F. (1989). A phonetically labeled acoustic segment (PLAS) approach to speech analysis-synthesis. *Proc. IEEE Int. Conf. Acoustics, Speech, and Signal Processing*, pp. 584–587.

Stone, M. (1990). A three-dimensional model of tongue movement based on ultrasound and X-ray microbeam data. *J. Acoust. Soc. Am. 87*(5):2207–2217.

Strube, H. W. (1980). Linear prediction on a warped frequency scale, *J. Acoust. Soc. Am. 68*(4):1071–1076.

Strube J. W. (1982). Time-varying digital filter and vocal tract models. *Proc. IEEE Int. Conf. Acoustics, Speech, and Signal Processing*, pp. 923–926.
Tohkura, Y. (1986). A weighted cepstral distance measure for speech recognition. *Proc. IEEE Int. Conf. Acoustics, Speech, and Signal Processing*, pp. 761–764.
Wakita, H. (1973). Direct estimation of the vocal tract shape by inverse filtering of the acoustic speech waveforms. *IEEE Trans. Audio Electroacoust. AU-21*:417–427.
Wilhelms, R. (1987). Schaetzung von artikulatorischen Bewegungen eines stilisierten Artikulatormodells aus dem Sprachsignal. Doctoral thesis, University of Goettingen, Germany (in German).

9

Signal Processing for Cochlear Implants

TOHRU IFUKUBE Hokkaido University, Hokkaido, Japan

1. INTRODUCTION

Many speech processing strategies for cochlear implants have been proposed as aids for the profoundly deaf whose auditory hair cells are defective (Parkins, 1983, 1985; Miller et al., 1984; ASHA, 1985, Feigenbaum, 1987). In most cases, a single electrode or an electrode array is implanted in the scala tympani of the cochlear so that it can stimulate the peripheral auditory nerves by electric current through the lymph inside the cochlea.

In general, the speech waveforms are directly converted to current waveforms, and all models incorporate the function of dynamic range compression for the single-channel implants. Most of the multichannel implants use a filter bank approach to extract the spectral information. Recently, however, digital signal processing methods using a diagnostic and programming system (DSP) have been developed to extract precisely the pitch and the formants, and new algorithms have been proposed to compensate for the wide current spread inside the cochlea.

The perceptual system may be described only in very broad terms, such as the implantee's age, period of deafness, and previous linguistic and musical experience and the duration and type of training with electrical stimulation.

Consequently, speech communication performance with a particular processing system cannot be simply compared with that of other systems (Miller et al., 1984). This chapter does not describe speech communication performance, but it focuses on some of the speech processing strategies designed or proposed in the past decade with reference to recently published related papers.

2. ELECTRODE CONFIGURATIONS AND PERCEPTUAL CHARACTERISTICS ELICITED BY ELECTRICAL STIMULATION TO AUDITORY NERVES

2.1. Electrode Configurations

The signal processing method depends on the electrode configuration. The many electrode configurations are classified into six categories (Eddington, 1980) as schematically represented in Fig. 1.

The single extracochlear monopolar electrode (Fig. 1a) is usually positioned on the promontory or the round window membrane of the cochlea (Vernon, 1974). The single intracochlear monopolar electrode (Fig. 1b) is inserted into the cochlea through the round window or through a hole drilled into the cochlear bony wall (House and Urban, 1973). Configurations using a single stimulating electrode with a remote ground lead tend to excite large groups of fibers and may be incapable of generating the complex spatial patterns of activity thought to be important in coding stimuli at the level of the auditory nerve. It is necessary to stimulate the auditory nerve by converting the complex spatial patterns into the time-sequential patterns in the case of the single-channel electrode.

The system of multiple intracochlear monopolar electrodes (Fig. 1c) consists of several electrodes inserted into the scala tympani through the round window (Hochmair-Desoyer and Hochmair, 1980) or through a window drilled into a more apical region of the cochlea. Because the multichannel stimulation is expected to excite a distinct group of auditory peripheral nerves, complex tones can be discriminated according to place pitch in addition to frequency pitch. However, the use of a greater number of stimulating channels does not usually convey significantly more information because of the overlap by the wide current spread inside the cochlea (Ifukube and White, 1987b). It is thus necessary to design a speech processor which functions to compensate for the current spread.

A variation on this configuration, which is thought to reduce the current spread along the cochlea, is to set ground electrodes close to the active electrodes. In the multiple intracochlear compartmental electrodes (Fig. 1d),

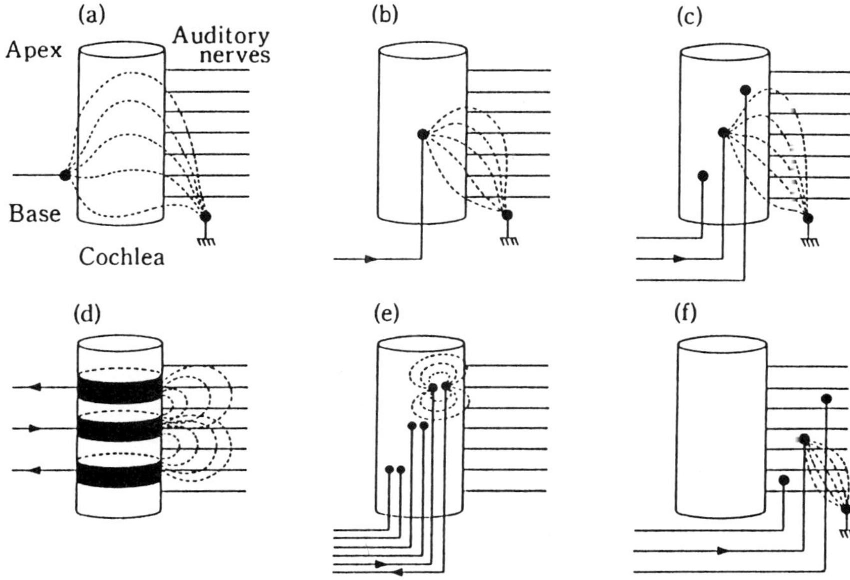

Figure 1 Schematic illustration of six categories of electrode configurations used for auditory prostheses. The auditory nerve is represented by five horizontal lines and the cochlea by a trapezoid. Electrodes are represented by filled circles and insulated electrode leads by line segments connected to the electrodes. (a) Single extracochlear monopolar electrode; (b) single intracochlear monopolar electrode; (c) multiple intracochlear monopolar electrodes; (d) multiple intracochlear compartmental electrodes; (e) multiple intracochlear bipolar electrodes; (f) multiple modiolar monopolar electrodes. (From Eddington, 1980.)

windows are drilled through the bony wall of the cochlea at several locations and an electrode is inserted through each window (Banfai et al., 1985). Another attempt to reduce the current spread in the cochlea involves the use of multiple intracochlear bipolar electrodes (Fig. 1e). Pairs of electrodes embedded in a Silastic carrier are inserted through the round window into the scala tympani, and the stimulus current is passed between the closely spaced pair electrodes (Merzenich, 1985). In the Merbone method, the intracochlear electrode array consists of 22 platinum rings spaced at 0.75-mm intervals along a flexible Silastic holder. The electrode array expands approximately 24 mm into the cochlea, within a region that corresponds to 600 to 8000 Hz in the normal cochlea. Electrical pulses are applied between pairs of electrodes separated by 1.5 mm, giving 20 pairs altogether. A very different approach is the use of multiple modiolar monopolar electrodes (Fig. 1f) inserted directly into the auditory nerve (Simmons et. al., 1984).

2.2. Perceptual Characteristics Elicited by Electrical Stimulation to Auditory Nerves

Every signal processor for implants is designed based on the perceptual characteristics of electrical stimulation to the auditory nerves. Virtually all of the reported threshold contours for sinusoidal electrical stimuli assume the same general shape, as illustrated schematically in Fig. 2 (Pfingst, 1984). Typically, the lower break frequency (BF_1) is at about 50 to 150 Hz. Above BF_1, the threshold increases as a function of the frequency at rates between 5 and 15 dB per octave until the second break frequency (BF_2) is reached, above which the slopes are typically above 4 ± 1 dB per octave. Although values for BF_2 vary considerably, most commonly they range from 250 to 500 Hz.

The upper limit of the dynamic range for electrical stimulation is defined as the maximum comfortable loudness. The dynamic contour is idealized in Fig. 2. This contour typically has a slope near zero at frequencies below BF_a and a slope of about 3 dB per octave above BF_a. The value of BF_a is typically

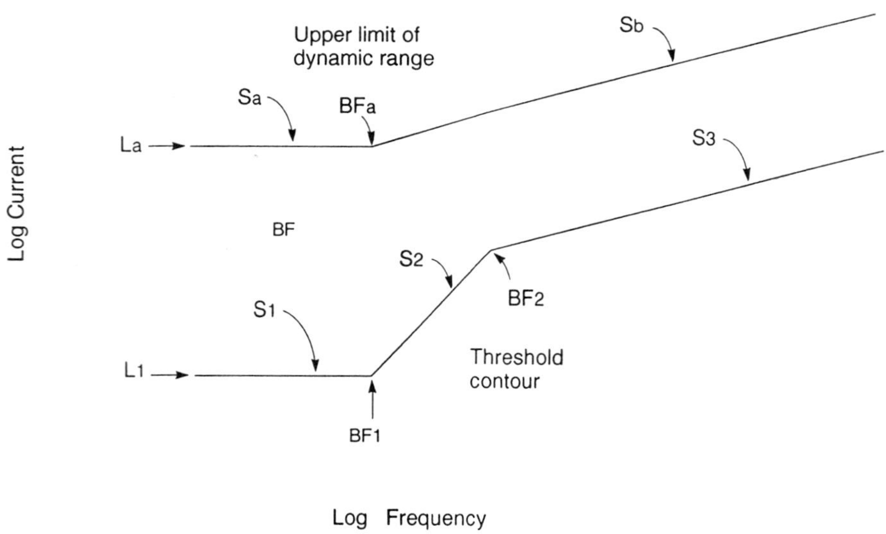

Figure 2 Schematic representation of threshold contour, upper limit of dynamic range, and damage limit for sinusoidal electrical stimulation of cochlea. BF, Break frequency; S, slope; L, level. (From Pfingst, 1984.)

near or somewhat higher than that of BF_1, ranging from about 100 to 250 Hz. Typical dynamic range values for long-duration (> 200 ms) stimuli to widely spaced electrodes are 20 to 40 dB below BF_1 and 10 to 25 dB above BF_a.

The perceived pitch presents almost the same pattern of threshold characteristics. The pitch versus rate data for three electrodes in the eighth nerve for a postlingually otoxically deafened subject are shown in Fig. 3 (White, 1982).

The growth of loudness as a function of initial stimulus intensity has been measured using loudness estimation procedures in humans. Most investigators have found that log loudness grows linearly as a function of log current, at least at higher intensities of stimulation (Fig. 4, S_H). However, several investigators have also shown functions with a lower slope at low intensities (Fig. 4, S_L).

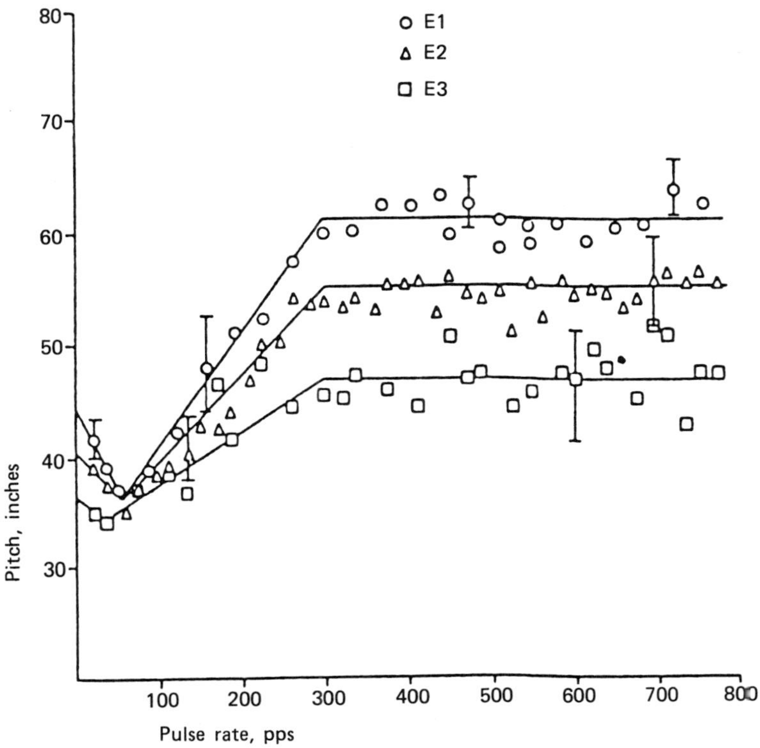

Figure 3 Data on perceived pitch versus electrode stimulation rate for three electrodes in the eighth nerve. (From Simmons et al., 1984.)

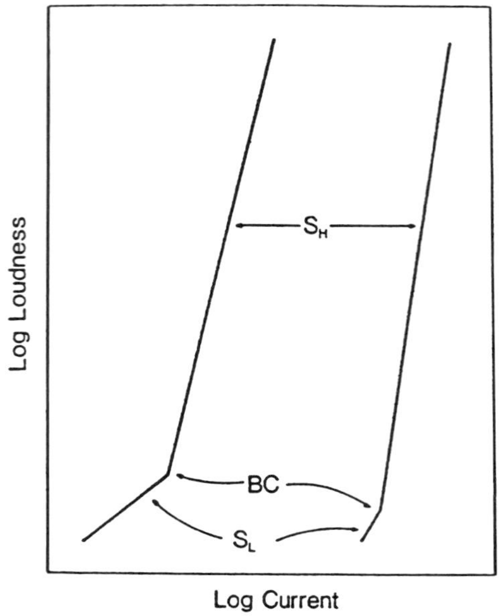

Figure 4 Schematic representation of loudness growth functions for low-frequency (left curve) and high-frequency (right curve) sinusoids. BC, Break current; S_H, high-intensity slope; S_L, low-intensity slope. (From Pfingst, 1984.)

Auditory sensation elicited by a single-channel stimulation to the promontory of the cochlea was found to have very high time resolution (Ifukube et al., 1990). In the experiment, two time-sequential patterns with a double-pulse series (Fig. 5) were presented to profoundly deaf subjects. The experimental results showed that the subjects could easily discriminate the pattern having a time difference of 1000 μs from the pattern having a time difference of 1200 μs. The discrimination with respect to the time resolution (Δt) of the double pulse was found to be about 150 μs. This high time resolution makes it possible to perceive timbre in addition to pitch and intensity.

The earlier work performed with several different versions of implants showed that stimulation using a somewhat modified analog speech signal results in superior speech understanding compared with that achieved with pulsatile stimulation schemes. From these experiments it may be concluded that the waveform of the electrical stimulus does, to some extent, determine the percept.

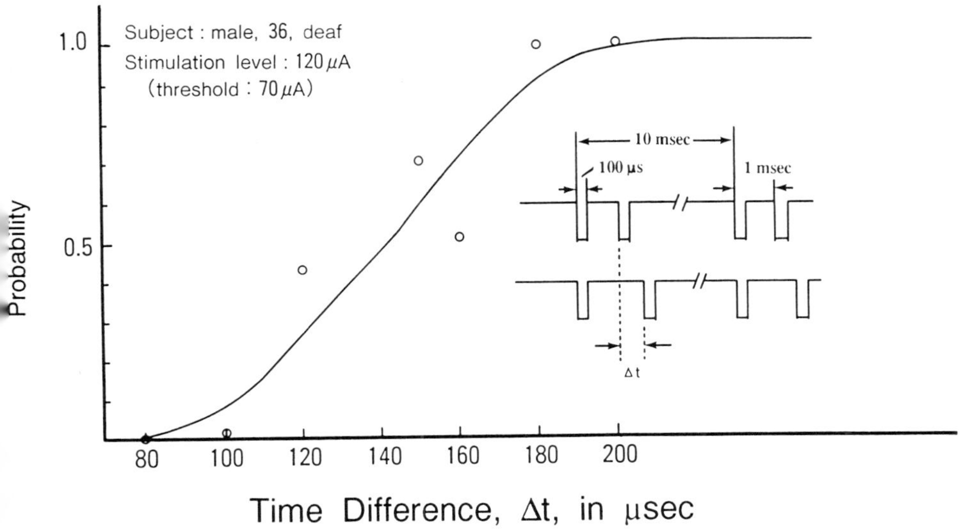

Figure 5 Probability at which the subject could discriminate two stimulation patterns applied to the promontory of the cochlea as a function of time difference Δt. (From Ifukube et al., 1990.)

Two trends appear in the pitch scaling data of Fig. 6 (Eddington, 1980) in the multichannel electrode. First, for any one repetition rate, pitch increases from the most apical electrode (1) to the most basal electrode (6). Thus, the relationship between the pitch of the sounds elicited and the position of the stimulated electrode is consistent with the presumed tonotopic organization of the cochlea. The second trend is that for any one electrode, pitch increases with the stimulus repetition rate. This result is consistent with those of experiments using other electrode configurations, which confirms that both the pitch and the difference limen of the pitch increase when the repetition rate is increased between 80 and 500 Hz. Above 400 to 500 Hz, the difference limen for pitch is so large that increases in the stimulus repetition rate have little effect on the perceived pitch. The pitches of percepts elicited by different electrodes can overlap when the stimulus repetition rates are different.

Historically, the main features of interest in speech processing have been formant positions, pitch, and energy. Indeed, the most successful multichannel systems use an analog processor to extract formant frequencies and then pulse the nearest electrode in the cochlea at a rate corresponding to the pitch.

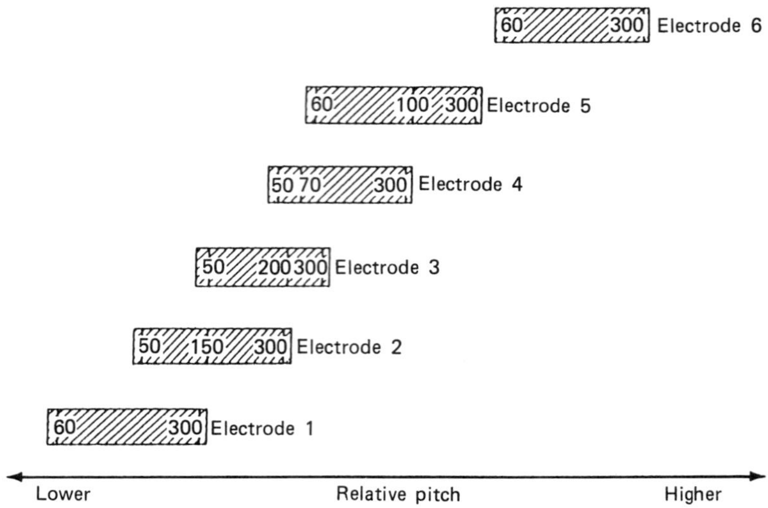

Figure 6 The pitch scaling data with stimulus repetition rate as the parameter. Intracochlear monopolar electrodes were used. (From Eddington, 1980.)

Channel interaction, however, occurs when two or more electrodes simultaneously produce current inside the cochlea because of the wide current spread. The current pattern becomes extremely complex outside the cochlea because of the spiral shape of the cochlea. The use of a greater number of stimulating channels does not usually convey significantly more information because of the overlap by the wide current spread inside the cochlea (Ifukube and White, 1987b). It is thus necessary to design a speech processor which incorporates a function that can compensate for the current spread.

A great deal of psychophysical data is now available to aid in the design of speech processing schemes. Some signal processing methods are described in the next section.

3. SOME METHODS OF SIGNAL PROCESSING FOR SINGLE-CHANNEL COCHLEAR IMPLANTS

The House single-channel cochlear implant (CI) was developed by House and Urban in the early 1970s and has been successfully used by over several hundred profoundly deaf patients. In the House Ear Institute method, a signal can be traced from the microphone, through the processor, to the implanted coil as shown in Fig. 7 (Edgerton et al., 1983). The frequency response of the input stage of the processor is relatively flat over the range 200 to 4000 Hz at moderate input levels.

From the output of the filter section, the signal is routed to a modulator control and then to a 16-kHz amplitude modulator. The modulator control serves to vary the sensitivity of the device. That is, it allows the patient to increase or decrease the signal level entering the modulator. Increased sensitivity is usually obtained at the level of increased background noise.

After leaving the modulator control, the signal enters the modulator. At this stage the 16-kHz signal is amplitude modulated using a voltage divider network with a field effect transistor (FET). The modulator is highly nonlinear, as shown in Fig. 8. In this figure the voltage amplitude at the external coil is plotted as a function of sound pressure level (SPL) at the microphone. The shape of this curve can be varied by the patient using the volume and modulation controls and by the audiologist using controls inside the stimulator.

Again in Fig. 8, three distinct regions in the function can be identified. The first region (A) occurs when the input levels are less than about 45-dB SPL, usually below the threshold. In the region between the 45- and 65-dB SPL inputs (B), the unit output increases dramatically with small changes in the input level. The third region (C) occurs for input signals greater than

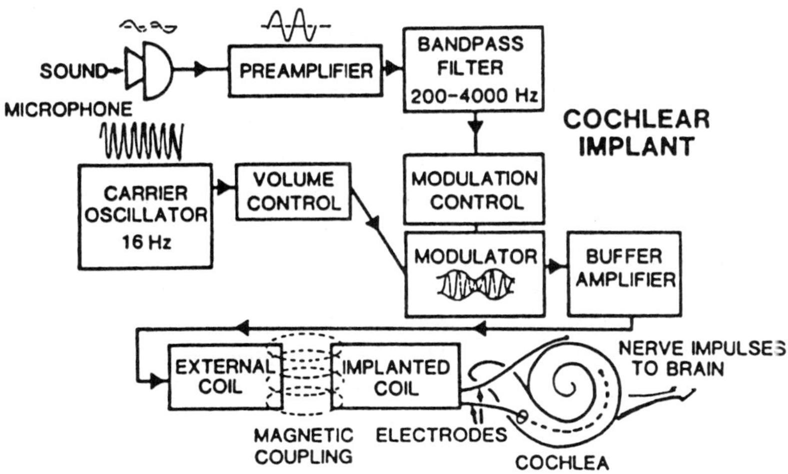

Figure 7 Block diagram of the House–Urban single-channel cochlear implant system. (From Edgerton et al., 1983.)

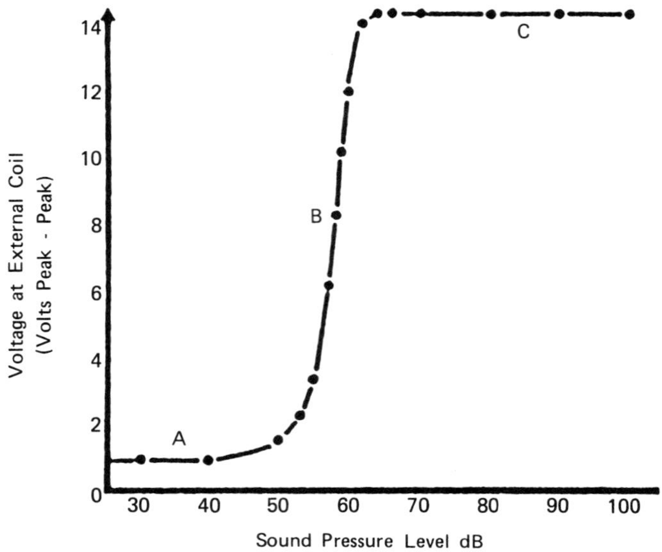

Figure 8 Input-output function for the House–Urban cochlear implant stimulator. Input is a 1-kHz sine wave at sound pressure levels from 30 to 100 dB. Output is in peak-to-peak volts at the external coil. (From Edgerton et al., 1983.)

65-dB SPL. In this region the peak amplitude does not increase beyond the clipping level, but there is a slight increase in the duty cycle of the 16-kHz burst.

The modulator is also a buffer amplifier, which drives the external coil with an amplified version of the modulator output. The patient's volume control functionally varies the gain of this buffer amplifier. The buffer drives the coil differentially. The effects of signal processing by the House–3M cochlear implant on consonant perception have been described by members of the House Ear Institute (Edgerton and Brimacombe, 1984).

The necessary processing of the signal is quite straightforward in the Vienna method (Hochmair and Hochmair-Desoyer, 1983). It essentially consists of amplitude compression and frequency equalization. This is in contrast to various types of feature extractions performed for pulsatile stimulation. Hence, they have obtained a simple system that easily fits the power and volume constraints of a wearable stimulator and that results in more or less natural-sounding percepts, which makes rehabilitation easier.

They believe that, besides the dependence of percepts on wave shape, there is another reason why pulsatile stimulation signals are not well suited for speech coding (Hochmair and Hochmair-Desoyer, 1983). The coding process involves a considerable amount of feature extraction, the consequences of which are not very well defined. Furthermore, coding that involves speech feature extraction frequently destroys the natural appearance of environmental sounds. Finally, because of this lack of naturalness, a longer period of rehabilitation is needed.

The simple sound processing performed in the wearable speech processor is detailed in Fig. 9. The amplified audio signal is compressed by a syllabic compressor, that is, a gain-controlled amplifier (GCA) with a very short attack time (0.5 ms) and a release time of approximately 20 ms. The compression ratio is about 6. Adjustments according to the patient's dynamic range can be made. This compression ratio derives from a comparison of the Stevens function's exponents for electrically and acoustically induced hearing (approximately 3.5 and 0.6, respectively).

The exact shape of the amplitude compression function over the whole dynamic range is important in everyday life, with its varying amplitude levels and its constantly changing amount of background noise. In contrast, in similar devices suggested for hearing aids, compression sets in only at a fairly high output level.

The residual frequency response is corrected by an adjustable equalization circuit following the amplitude compressor. It is adjusted to the subject's characteristics by a fitting procedure that locates three poles and one zero

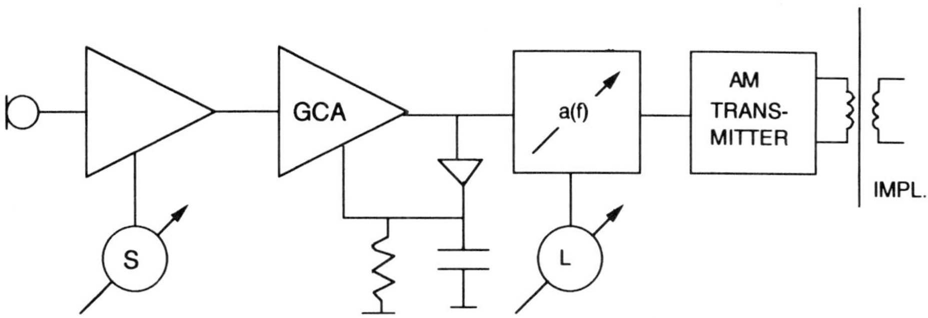

Figure 9 Single-channel sound processor including an AM transmitter for transcutaneous transmission. The gain-controlled amplifier (GCA) performs the dynamic range compression. Preamplifier and modulation amplifier gains are variable to control sensitivity and loudness, respectively. The frequency response of the equalization network, $a(f)$, is detailed in Fig. 10. (From Hochmair and Hochmair-Desoyer, 1983.)

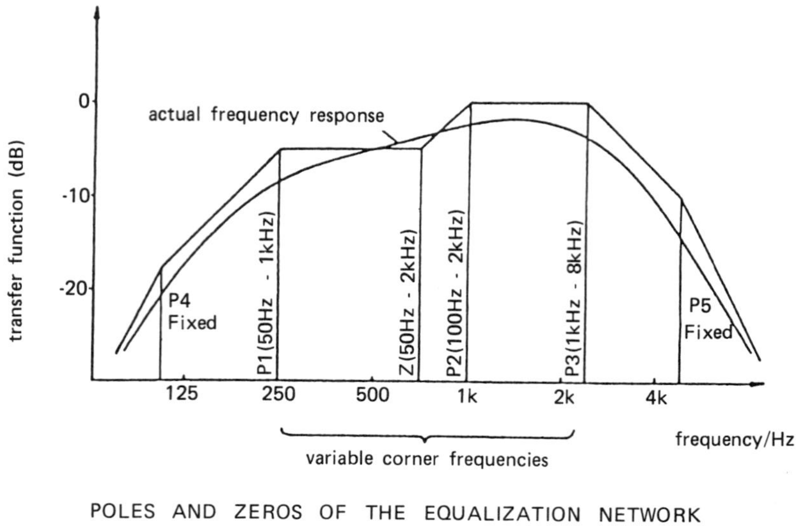

Figure 10 Example of the transfer function of the equalization network. Poles $P1 \ldots P3$ and zero (Z) are varied for the best equalization when the sound processor is fitted to the patient. (From Hochmair and Hochmair-Desoyer, 1983.)

in such a way that the 40% equal-loudness contour (most comfortable loudness) corresponds to a constant microphone output voltage for frequencies from 100 Hz to 4 kHz (Fig. 10). This fitting of the frequency response is very important for speech understanding.

A research group at Hokkaido University in Japan has designed a new model of an extracochlear single-channel prosthesis (Ifukube et al. 1990). The configuration of their auditory prosthesis is shown in Fig. 11. They have also designed a stimulatory electrode covered with polyvinyl alcohol gel and a small-size telemetry method. This is a less invasive alternative to insertion in the scala tympani. It is possible that their cochlear prosthesis may be applied even to small children. They have further proposed a new concept for a speech coding method for transmitting second-formant information besides pitch by the prosthesis.

In the speech processor, a digital signal processor is used to extract the pitch frequency and the second-formant frequency in real time. These frequencies are converted into time-sequential signals. In the speech coding method, an additional stimulation pulse is inserted among the pitch pulses and the time delay from the pitch to the additional pulse is set in proportion to the second-formant frequency.

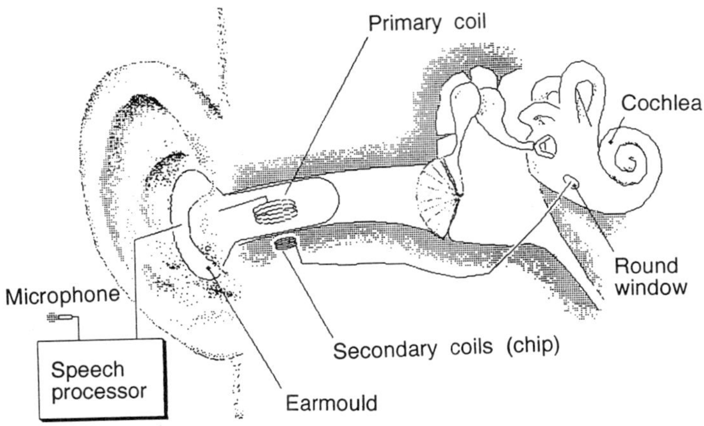

Figure 11 Configuration of single-channel extracochlear prosthesis designed for profoundly deaf infants. (From Ifukube et al., 1990.)

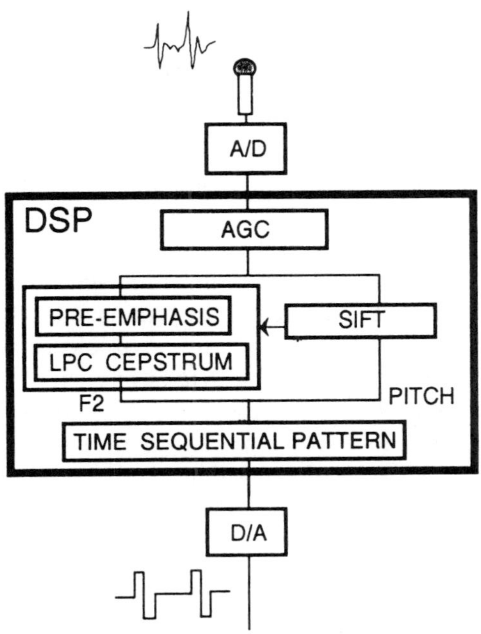

Figure 12 Block diagram of the speech processor. This processor can simultaneously transmit the pitch and the second formant frequencies by converting the speech spectrum into time-sequential patterns. (From Ifukube et al., 1990.)

Figure 12 presents a block diagram of the speech processor. The input voice picked up by a microphone is sampled with a 10-kHz sampling frequency and the pitch period is then detected by obtaining the autocorrelation of the residual signals extracted by the linear predictive coding (LPC) algorithm. Next, the second-formant frequency is estimated from LPC cepstrum. These two signals are combined to obtain a time-sequential signal with double pulses.

A digital signal processor (μPD77230, 3.3 MHz clock, NEC Corporation), and an analog-to-digital (A/D) converter (AD7772, 10 kHz, 12 bit, AMD Corporation), and a digital-to-analog (D/A) converter are used in this processor. The five Japanese vowels strongly depend on the second-formant frequency, so that time differences between the pitch and additional pulse change depend on them. From the experimental results for vowel discrimination, the average identification rate using this prosthesis was found to be 60%.

4. SOME SPEECH CODING METHODS FOR MULTICHANNEL COCHLEAR IMPLANTS

4.1. Direct Conversion of Spectrum Pattern to Electrode Array

With most multichannel cochlear implants, the sound processor receives the incoming sound signal and separates it into several frequency ranges. The frequencies in these ranges are compressed and then amplified by different amounts.

In the Eddington method (Eddington, 1983), the speech waveform is converted to current waveforms for intracochlear four-channel stimulation using the scheme outlined in Fig. 13. Each of the four channels consists of a bandpass filter, a voltage-to-current converter, and an electrode. The output of each of the four bandpass filters is converted to a current waveform and delivered to an electrode. The relative gains of the four channels (1, 2.6, 8.4, and 19.6 for channels 1 to 4) correspond to the relative current thresholds of their electrodes as measured using sinusoidal stimuli. A multielectrode cochlear implant has been shown to increase a subject's ability to discriminate between selected synthetic speech sounds. It represents a clear improvement over single-channel schemes, as shown in Fig. 14.

Hochmair has designed almost the same signal processor as in the Eddington method. In his strategy the complete signal, with a bandwidth of 100 to approximately 5000 Hz, is then transmitted to one electrode channel of the implant in their method (Burian et al., 1986). The four AM receiver/demodulators are arranged in two disks and employ carrier fre-

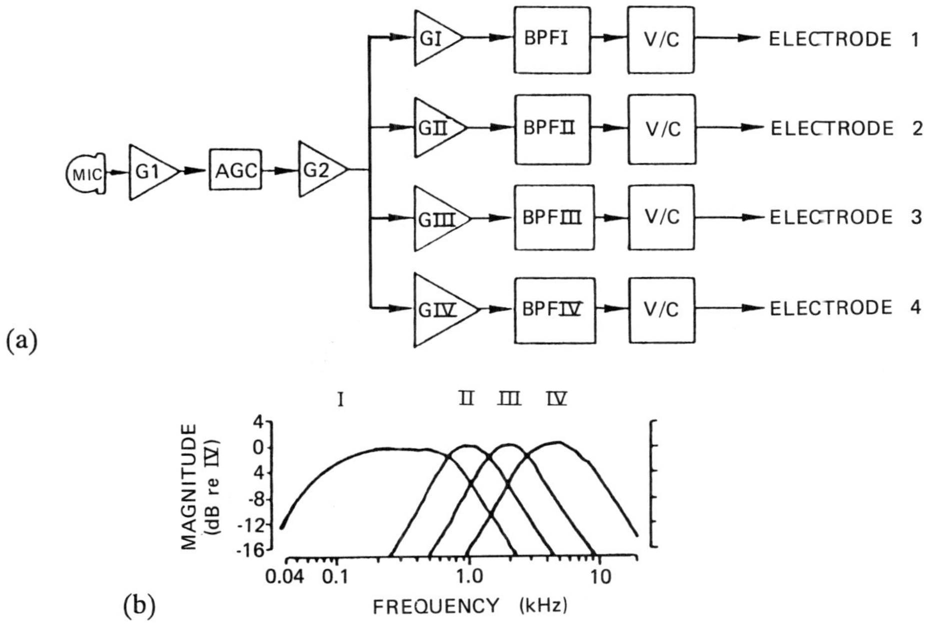

Figure 13 (a) Four-channel processing scheme used to transform speech stimuli to electrical stimuli. The acoustic waveform is transmitted to an electrical analog by the microphone (MIC). This waveform passes through two gain stages (G1 and G2) and an automatic gain control stage (AGC) before it is presented as input to each of the four processing channels. Each processing channel consists of a gain stage (GI), bandpass filter stage (BPF I), and voltage-to-current source converter (V/C I). (b) Magnitude of the frequency response for each of the four bandpass filters of the processing system. (From Eddington, 1983.)

quencies for the transcutaneous signal transmission of 12 and 31 MHz, respectively. With this implant, the sound processor activates one of the four channels, which produces the best understanding of speech (Hochmair and Hochmair-Desoyer, 1981).

A block diagram of the speech processor (CA) used in the UCSF/Storz prosthesis is shown in Fig. 15 (Wilson et al., 1988). In the initial stage of the processor, the input signal is attenuated at 6 dB/octave for frequencies below 1.2 kHz. This gentle attenuation acts to flatten the speech spectrum so that the otherwise intense components at low frequencies will not dominate the percepts elicited by speech inputs. The filtered signal is then compressed to map it into the narrow dynamic range of electrically evoked hearing.

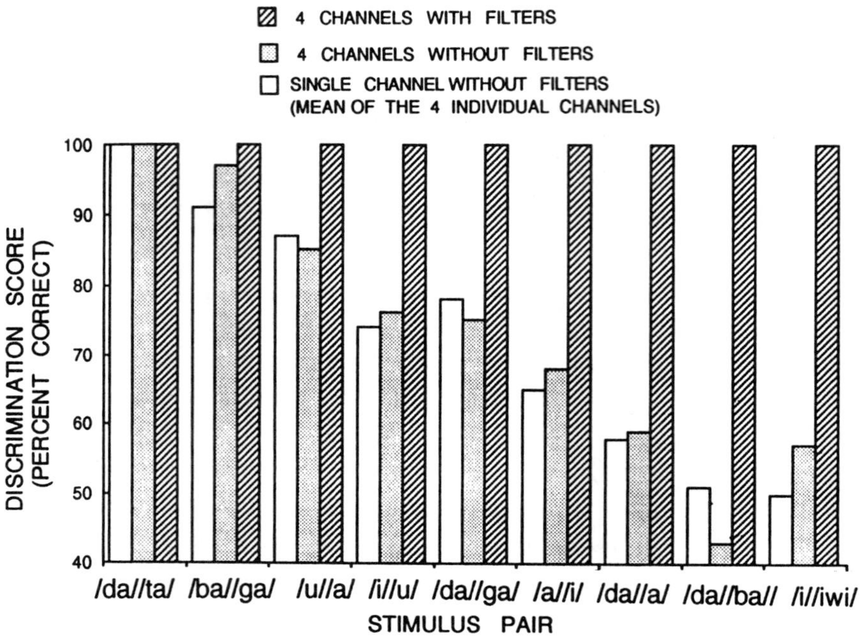

Figure 14 Histogram of the discrimination scores for a stimulus pair of synthetic speech sounds. Represented here is a four-channel system showing a clear improvement over single-channel schemes. (From Eddington, 1980.)

The frequency ranges of the bandpass and high-pass filters following the compressor encompass the first and higher formants of speech. In particular, the frequency ranges are selected so that the first formant is represented in channel 1 and the third and higher formants are represented in channel 4. The second formant is divided between channels 2 and 3 in order to emphasize discrimination of this essential component of speech. The high-pass filter in the chain for channel 1 provides a first-order equalization of loudness for frequency components below 300 Hz.

Specifically, the speech processor is partially compacted for the large differences in the thresholds for low-frequency and high-frequency stimuli. The adjustable clippers in the chains for channels 3 and 4 limit the peak intensities of waveforms in those channels. These clippers are sometimes necessary because the dynamic ranges for high-frequency stimuli are generally less than the dynamic ranges for low-frequency stimuli. Appropriate adjustment of the clippers in such cases ensures that the maximum levels for comfortably loud

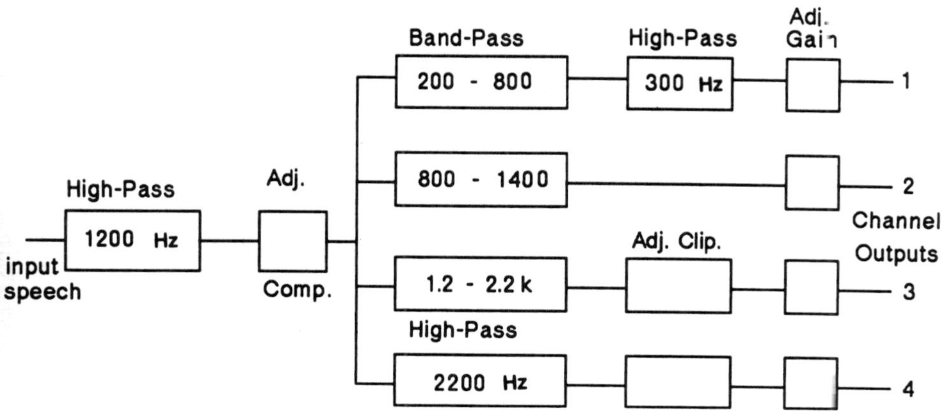

Figure 15 Block diagram of a four-channel, compressed analog (CA) processor used in the UCSF/Storz prosthesis. (From Wilson et al., 1988.)

percepts are not exceeded in the upper channels. Finally, adjustable gain controls are provided for each channel so that speech features signaled in that channel's band can be made clearly audible. For example, in the fitting of the prosthesis, the gain of channel 4 is increased until "s" sounds are heard.

Wilson et al. (1988) have designed another type of speech processor (interleaved pulse, IP) as illustrated in Fig. 16. The design of the IP processor is quite different from that of CA processors. In the IP processor an automatic gain control (AGC) continuously adjusts the level of speech input so that a steady level is presented to subsequent stages of the processor. Typical attack and release times for the AGC are 8 and 20 ms, providing slow AGC action. The level-adjusted signal is then high-pass filtered as in the CA processor to reduce the amplitudes of speech components below 1.2 kHz.

The output of the high-pass filter is fed to a bank of bandpass filters, whose center frequencies span the combined range of the first three formants of speech along a logarithmic scale. The root-mean-square (rms) energy in each band is sensed by a full-wave rectifier and low-pass filter. Next, a post-processor is programmed to scan the RMS outputs on a periodic basis. The output of a filter band channel is coded for stimulation of its assigned electrodes only if the RMS energy delivered to the selected channel is calculated with a logarithmic mapping law of the form

$$\text{Pulse amplitude} = A \log(\text{RMS level}) + k \qquad (1)$$

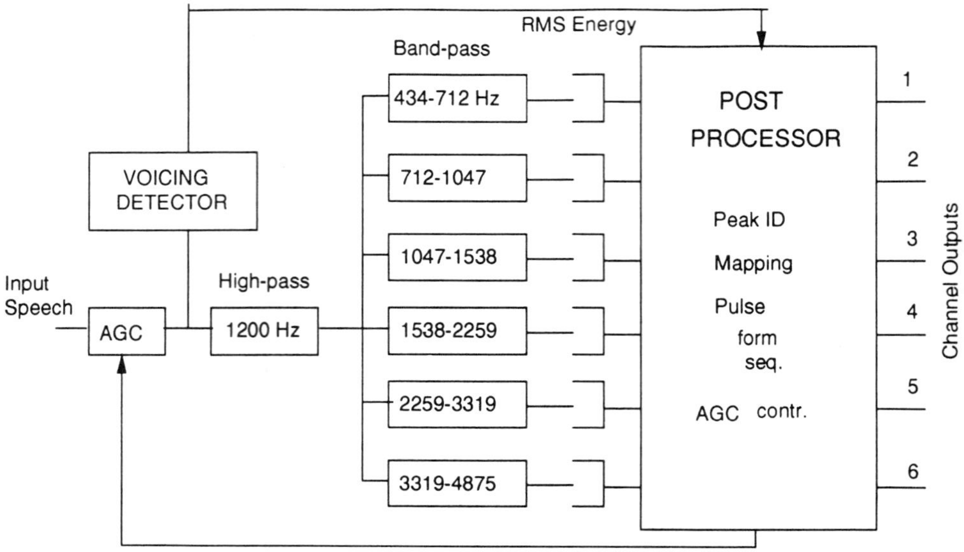

Figure 16 Block diagram of a six-channel, interleaved pulse (IP) processor used in the UCSF/Storz prosthesis for comparison with the CA processor. (From Wilson et al., 1988.)

where the parameters A and k are determined for each channel on the basis of threshold and most comfortable loudness level measurements. Finally, the voicing detector distinguishes voiced and unvoiced intervals of speech and estimates the fundamental frequency of the waveform within the voiced intervals. The output of the voicing detector can be used by the postprocessor to control the timing of stimulation cycles.

From the experimental results, Wilson et al. concluded that IP processors were far superior to other processors for at least two patients with psychophysical signs of poor nerve survival, whereas processors other than the IP processors may be superior for patients with good nerve survival.

In the French 12-channel cochlear implant (Weber et al., 1984; Chouard et al., 1984), the acoustic signal is cut between 40 and 100 dB in order to enhance speech understanding. It is then divided into 12 channels by filters as shown in Fig. 17. The different energies of each filter are selected one after the other with a multiplexer in 3-ms sampling. Then a semilogarithmic conversion with a compression rate of 60 dB and the same 3-ms time sampling is done to adapt the electrical signal to the ear characteristics. Each of these 12 energy levels is moderated as a function of each electrode-nerve inter-

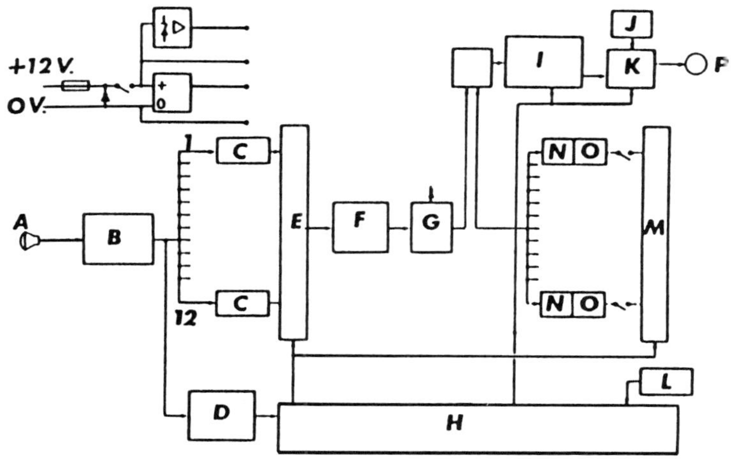

Figure 17 Block diagram of a French 12-channel cochlear implant. A, Microphone; B, preamplifier; C, filter; D, detection of minimum level; E, filter scanning; F, compressor; G, volume (front panel potentiometer): H, logic transmission sequence; I, conversion voltage, duration; J, clock; K, power; L, normal stimulation; M, balance scanning; N, intensity unit; O, intensity ten; P, antenna; I, high filter; 12, low filter. (From Weber et al., 1984.)

face. These energy levels are subsequently converted to impulses of variable duration.

In this way successive chains of impulses are produced, each chain having a start signal and an impulse of variable duration corresponding to each channel. The first impulses correspond to the high frequencies, the last impulses to the low frequencies. Even if the 12 electrodes are stimulated at 100 dB, the cycle duration does not exceed 3 ms. Its frequency is constant, that is, equal to 300 Hz. The elementary cycle functions only when a signal must be sent, which greatly diminishes consumption. All of these chains are modulated by a carrying wave of a 3.2-MHz energy vector, permitting bypassing of the cutaneous wall by means of an antenna.

The auditory prosthesis can be subdivided essentially into three structural parts in the Banfai extracochlear 16-channel electrode system, as shown in Fig. 18 (Banfai et al., 1986). In this extracochlear system, the frequencies are transformed into particular cochlear surface locations as shown in Fig. 19 (Hortmann et al., 1984).

The first part includes the microphone with the preamplifier and a 16-channel filter bank. The sound events received by the condenser directional

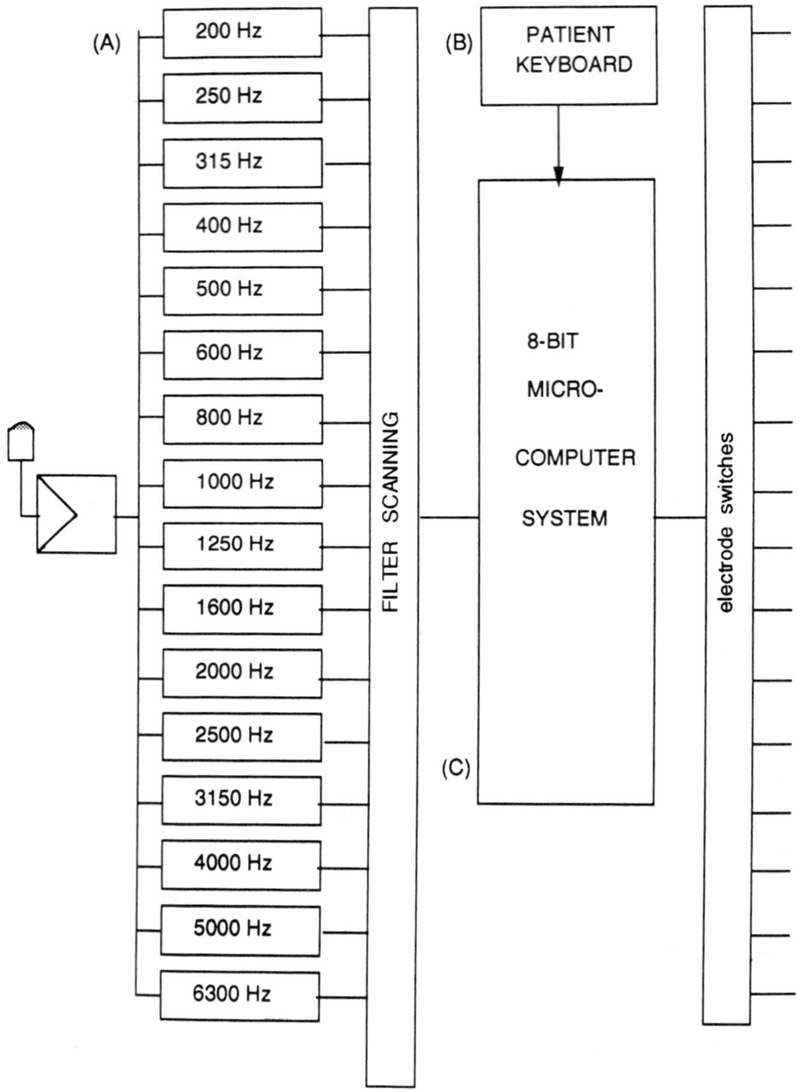

Figure 18 Block diagram of Banfai extracochlear system consisting of three structural parts. (A) Hardware; (B) adjustment of sensitivity; (C) 8-bit microprocessor. (From Banfai et al., 1986.)

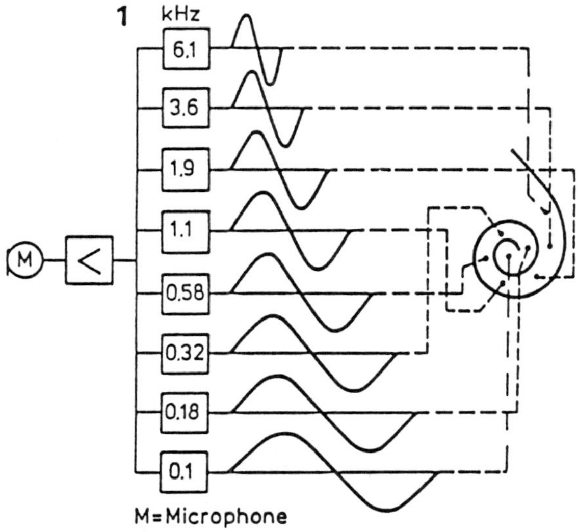

Figure 19 Principle of frequency-into-place transformation in Banfai system. (From Hortmann et al., 1984.)

microphone are amplified and then broken down into their frequency constituents. The mutual conductance of the filters is 18 dB per octave. The sensitivity can be adjusted. In the "wide" switch position, the microprocessor recognizes signals which are louder than 55-dB SPL. In the "near" switch position, sound events must reach a level of 65 dB. The patient can switch this to adjust for his or her optimal auditory impression.

The core of the auditory prosthesis is an 8-bit microprocessor. This queries the filter bank for the frequency distribution of the signals. The microprocessor recognizes when the volume of a frequency range changes from a value under 55 dB to a value above this. The filter outputs are interrogated in the series from channel 1 (200 Hz) up to channel 16 (6300 Hz). This promotes the possibility of switching on the deep channels and is advantageous for prelingual children because the basic frequencies of speech are converted more emphatically into electrical impulses.

In another version, the filters are interrogated in the direction opposite to that in the above version, that is, first channel 16, then channel 15, and finally channel 1. In this way, the upper frequencies and thus the formant ranges are preferred. The hearing sensation with this version is described as higher and clearer, whereas the first version sounds duller. The second version is especially suitable for postlingual patients. The same instrument can be used advantageously by changing software for different groups of patients. All the programs and data necessary to operate the auditory prosthesis are stored in a 2-kbyte EPROM. The third part of the circuitry is the stimulus output stage. With the assistance of a D/A converter and two demultiplexers, the microprocessor is able to stimulate optional electrode pairs with bipolar stimulus impulses.

4.2. Feature-Extracting Strategies

The Nucleus 22-electrode cochlear implant is one of the most successful prosthetic devices. The fundamental frequency combined with second-formant information ($F_0 - F_2$) in this device is significantly better than the speech processor extracting the fundamental frequency (F_0) alone, based on results of both closed- and open-set tests (Clark et al., 1984). The $F_0F_1F_2$ speech coding strategy has been shown to result in improved understanding of speech compared with the F_0F_1 strategy. Significant improvement in closed-set speech recognition in the presence of background noise is an additional advantage of the $F_0F_1F_2$ strategy (Franz et al., 1987). The $F_0F_1F_2$ speech coding strategy of the Nucleus 22-channel prosthesis is described below.

As shown in Fig. 20 (Blamey et al., 1987), a hand-held microphone is used in the wearable speech processor and 6-dB per octave preemphasis above 1000 Hz is switched into the circuit. The automatic gain control (AGC) performs two functions. First, it conditions the signal so that it falls within the dynamic range of the parameter extraction circuitry. Second, it adjusts the maximum signal output of the preamplifier to a level that roughly corresponds to the full scale of the analog-to-digital converter (ADC).

The AGC is arranged so that the maximum output corresponds to a comfortable level. It is then necessary only to ensure that stimulation is close to the threshold level during quiet periods. To achieve this, a sensitivity switch is provided to allow the patient to control the maximum gain that the AGC would reach during breaks, avoiding unnecessary stimulation due to background noise. The maximum gain is variable in five steps: 32, 56, 66, 74, and 80 dB. The attack time of the AGC is about 1 ms and the decay time between 25 and 50 ms. An amplitude compression ratio in excess of 25 : 1 is used.

First, the signal is full-wave rectified to eliminate, as much as possible, the periodicity of the F_1 component. Rectification is followed by low-pass filtering to produce a smooth speech signal envelope. The dc component, or absolute value of this signal, is then removed and the undulations are retained. These envelope undulations are then passed through a zero-crossing

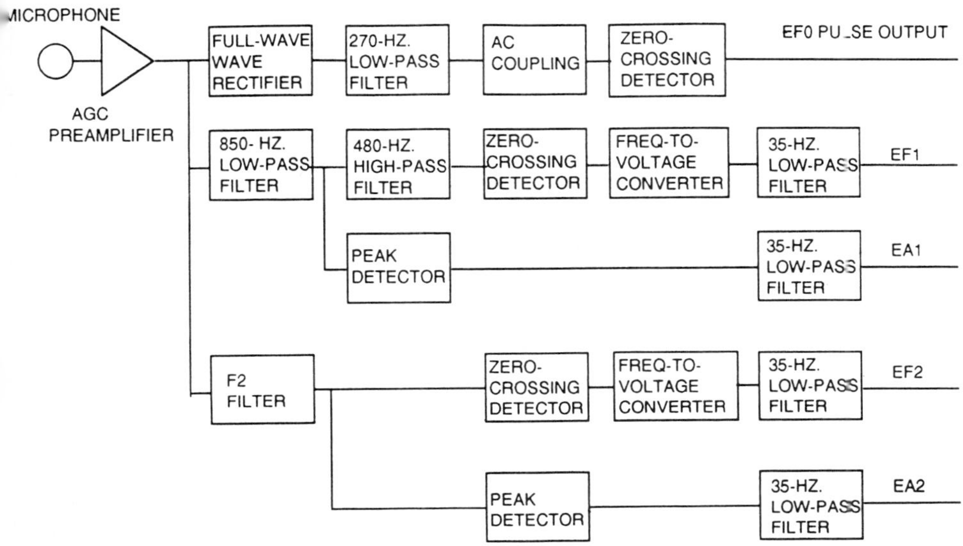

Figure 20 Block diagram of the wearable speech processor of the Nucleus 22-channel cochlear implant. (From Blamey et al., 1987.)

detector that has a hysteresis level determined from their peak level. This eliminates the false zero crossings produced by noise and intrusion of F_1 components. The output of the zero-crossing detector is used as an estimate of F_0. During unvoiced periods, the pitch extractor continues to examine the zero crossings of the signal envelope, but these excitations of the cochlea also occur at random intervals, producing a noiselike percept.

The first-formant frequency is estimated by low-pass filtering with an 850 Hz, two-pole filter and then high-pass filtering by a single pole at 480 Hz to remove the direct F_0 component. The filtered signal is passed through a zero-crossing detector, a frequency-to-voltage converter, and a 35-Hz low-pass filter to convert the zero-crossing rate linearly to a dc level representing EF_1.

The second-formant frequency is estimated using a four-pole high-pass filter with poles placed at 850 and 1000 Hz so that, for an average of a large number of speakers, F_2 predominates over the other formants. This filter is more effective than a simple high-pass filter because in some cases both F_1 and F_2 are located in the high-pass filter. After being shaped, the signal is passed through a zero-crossing detector and the resulting pulse train is fed to a frequency-to-voltage converter. The analog output of this converter is low-pass filtered to 35 Hz, so that the smoothed EF_2 is able to vary with a response time of approximately 12 ms. As in the case of F_0 estimation, the same circuitry is used for voiced and unvoiced sounds. Unvoiced sounds result in a high zero-crossing rate and produce stimulation on one of the more basal electrodes.

Amplitude estimates are obtained by taking the outputs of the formant filters, propagating them through peak detectors, and then finally smoothing the result with a 35-Hz low-pass filter. The use of the same type of filter as for the formant frequency estimation ensures that amplitude and frequency information are accurately synchronized.

Each electrode pair is used to represent a frequency range of EF_1 and EF_2 in the speech coding scheme. The five most apical electrode pairs represent frequencies up to 1000 Hz. Their upper frequency boundaries are equally spaced on a logarithmic scale from 300 to 1000 Hz. The remaining 15 electrode pairs have upper-frequency boundaries equally spaced on a logarithmic scale from 1000 to 4000 Hz. There is effectively an upward translation of frequency so that an EF_1 of 300 Hz produces stimulation in a region of the cochlea that would normally correspond to about 600 Hz, and an EF_2 of 4000 Hz would correspond to 8000 Hz in the cochlea.

The EF_0 is coded by the electrical pulse rate presented to the patients. Whenever a zero crossing is detected by the F_0 estimation circuit, two electrical pulses are produced. The first pulse is applied to an electrode pair chosen according to EF_2 and the second pulse is applied to an electrode pair chosen on the basis of EF_1. The pulse representing EF_1 starts 800 μs after the pulse representing EF_2.

Amplitudes are converted to current levels using a power law relationship in such a way that a 30-dB range of amplitude corresponds to electrical stimuli between the threshold and the maximum comfortable level. If EA_1 is greater than EA_2, then pulses of equal loudness, corresponding to EA_1, are applied to the two electrode pairs. If EA_2 is greater than EA_1, then the current levels for the two pulses vary independently.

In addition, the diagnostic and programming system (DSP) as depicted in Fig. 21 (Crosby et al., 1984) is used for conducting psychophysical tests on implant patients and for configuring the patient's wearable speech processor on the basis of these test results. Vowel and consonant confusions obtained for three good patients using a speech processor that coded EF_0, EF_1, EF_2, EA_1, and EA_2 confirmed that these parameters were used for phonemic recognition (Blamey et al., 1987).

The filter bank approach is, however, sensitive to variations in speech excitation, such as changes in the fundamental frequency, between utterances. These variations are inevitable and unavoidable in natural speech and, as a

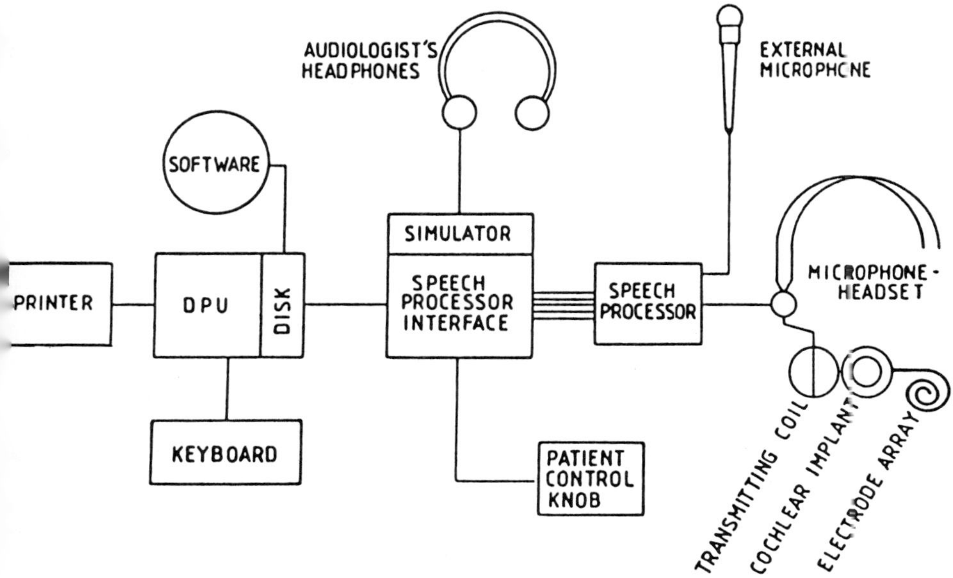

Figure 21 Block diagram of diagnostic and programming system for Nucleus system. (From Crosby et al., 1984.)

result, become the main factor that makes this method of spectral measurement unreliable. This anomaly must be compensated for if filters are used to track the fundamental frequency.

LPC-based speech modeling techniques preserve sufficient information to recreate the original speech signal, and the speech spectral envelope is simply one transformation of the LPC parameters. Furthermore, LPC analysis effectively separates the excitation and vocal tract transfer function, which thus gives reliable and accurate spectral and pitch information.

The hardware for the Canadian cochlear implant system (Morris and Barszczewski, 1989) under development (Fig. 22) consists of a programmable digital speech processor (DSP:TMS32010), a stimulator chip, and an electrode array. Computers play an essential role in system creation as well as system operation because the stimulator is a programmable device. The speech processor extracts speech features, formats 24-bit instructions, and transmits them to the implanted stimulator, which, in turn, responds by sending pulses of varying amplitudes, durations, and polarity to the electrode array.

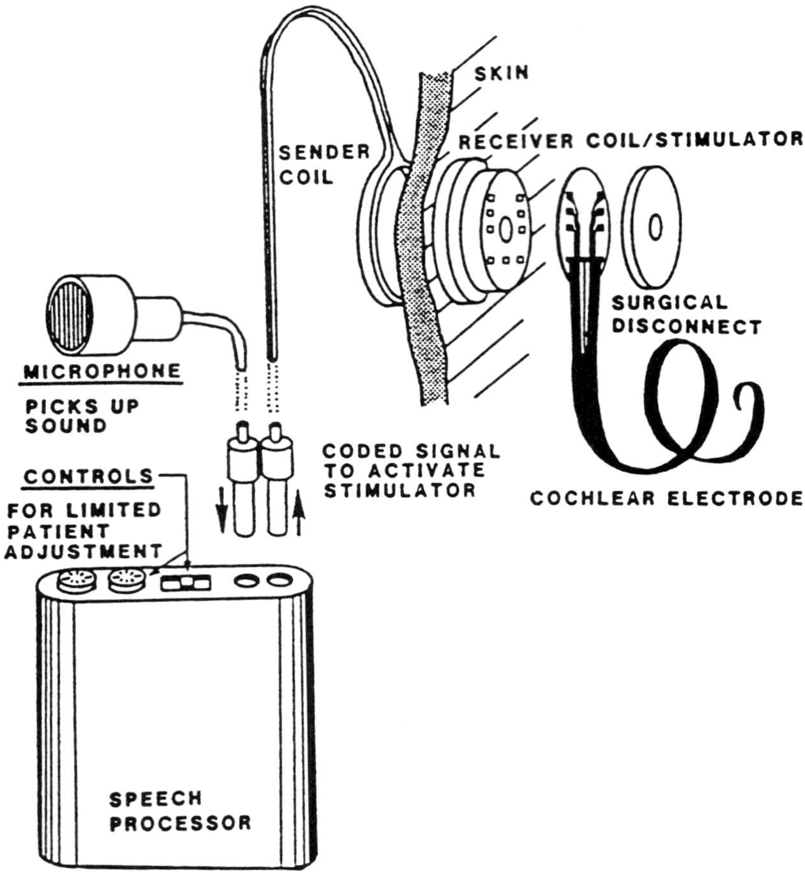

Figure 22 System hardware components of Canadian cochlear implant system. (From Morris and Barszczewski, 1989.)

Digital signal processing techniques are used to perform feature extraction from speech. These features include pitch and spectral envelope, the latter including formants. Together, this information is used to decide how the electrodes fed by the stimulator should be excited. The positions of the first three formants are used to choose the electrodes for excitation, and the relative amplitudes of the formants are used to choose the excitation strength for the selected electrodes. The selected electrodes are pulsed at a rate proportional to the pitch. If the short-time speech segment is unvoiced, however, the electrodes will be excited randomly with a total excitation strength proportional to the speech segment's energy.

The pitch extraction algorithm currently being employed is the simplified inverse filter tracking (SIFT) algorithm. A brief description of the program's operations for pitch extraction is shown in Fig. 23a. Incorporated in this algorithm are pitch extraction decision rules, which are used to determine whether the samples collected are for voiced or for unvoiced speech. These heuristic rules are zero-crossing counting, autocorrelation peak height, and the computation of input energy as their basis for decision. The 82 output points shown in Fig. 23a are output to the D/A converter.

In formant extraction, the spectral envelope is first calculated according to the algorithm shown in Fig. 23b. Using the obtained spectrum, a peak-picking algorithm is all that is needed to extract the first three formants for the voiced speech. The system is suitable both for laboratory-based research and as a take-home portable device. In addition to its small size and low power consumption, its software is easily modified in response to the patient's experience and thus can be rapidly adapted to new techniques of speech feature mapping and/or stimulation.

4.3. Attempts to Reduce Current Spread

Even with the best design, a significant current spread is still present. The function of lateral inhibition is expected to be useful for the speech processing of the multichannel cochlea implant because it may sharpen the wide current spread and because it can also compensate for the small dynamic range. A speech processor including lateral inhibition is proposed for a monopolar 8-channel cochlear implant designed by a combined research group from Stanford University and Hokkaido University (Ifukube and White, 1987a).

The input voice is picked up by microphone with preamplification, and the higher-frequency components of the speech signal are emphasized by a high-pass filter (6 dB/octave). The output of the preemphasizer is filtered by 16-channel bandpass filter for which the bandwidth and center frequencies are chosen in relation to the auditory critical bands as shown in Fig. 24. The output of each filter is transferred to the envelope detector, which is composed of a rectifier and a low-pass filter.

Vowel spectrum patterns, however, change remarkably with the input voice level. In particular, two of the formants are fused at the higher voice level of about 70 dB (S.L.). This fusion phenomenon is remarkable in cochlear implants, since the wide current spread does not allow the formants to separate.

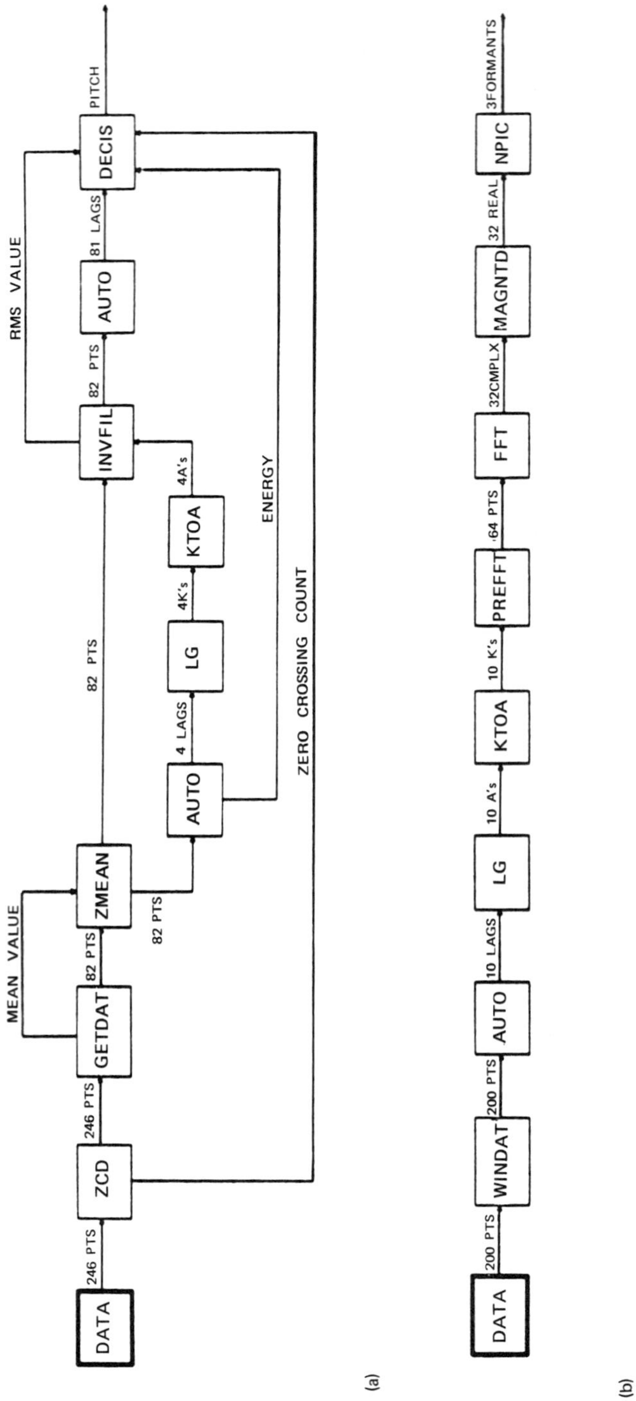

Figure 23 System software block diagram. (From Morris and Barszczewski, 1989.)

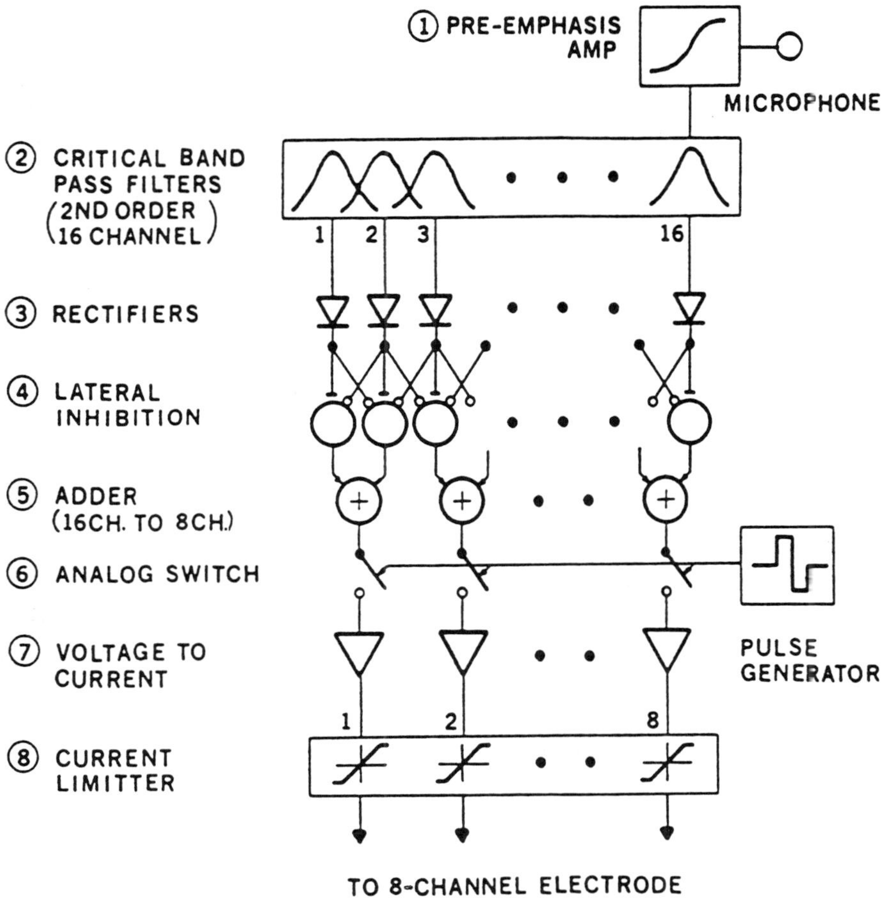

Figure 24 Block diagram of speech processor for eight-channel cochlear implant with lateral inhibition. (From Ifukube and White, 1987a.)

In lateral inhibition, a conversion method is adopted which is modeled after the nervous system, which, in turn, has a similar function to a spatial bandpass filter. Output level $Y(i)$ of channel i is obtained in accordance with Eq. (2):

$$Y(i) = \sum_{j=1}^{16} H(j)X(i+j) \qquad Y(i) > Y_0 \tag{2}$$

Here $X(i)$ is the input level of channel i and $H(i)$ is the function shown in the middle of Fig. 25. Coefficients a, b, m, and n represent the weight factor of excitation, that of inhibition, the number of channels in the excitatory area, and that in the inhibition area, respectively. These coefficients were chosen to maximize the spectral difference among the five spectral patterns of Japanese

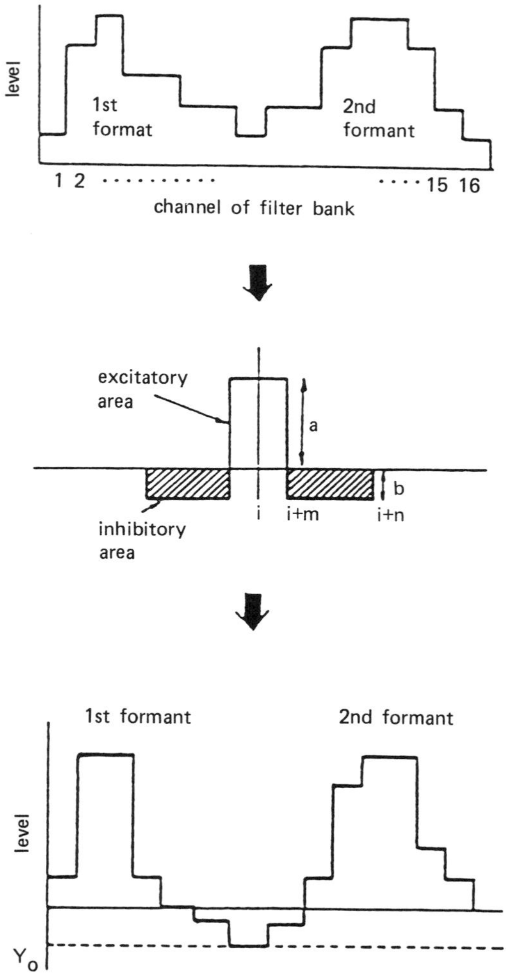

Figure 25 Schematic representation of the function of lateral inhibition. (top) The original spectral pattern of a vowel; (bottom) the output pattern of the lateral inhibition. (From Ifukube and White, 1987a.)

vowels. The maximum value was obtained at $m = 1, n = 4, a = 1$, and $b = 0.33$ in the case of the higher voice levels.

In the processor, these 16 components are reduced to 8 components by adding two adjacent channels in order to match the electrode channel. As shown in Fig. 24, each output of the adder is converted to a series of biphasic pulses having 200-μs duration by an analog switch. Each output voltage is converted to a current level which is proportional to the voltage using a voltage-to-current converter. Each output current is limited to avoid excess current using a current limiter so that the maximum levels become $\pm 140\ \mu$A.

How lateral inhibition can compensate for the wide current spread and for the small dynamic range was investigated by testing subjects using synthetic vowels. From the experimental results, the lateral inhibition was found to be useful for distinguishing vowels such as /i/ and /e/, which have two separated formants, the first and the second formants. It became clear that lateral inhibition functions as an AGC which could control the total current.

Another way to reduce the current spread for any given electrode configuration is by deconvolution (Townshend and White, 1987). The first step in this technique is to determine the function which maps the current levels at the electrodes to the excitation levels at each of the possible neural sites. This function can then be inverted to give electrode current patterns, each of which produces excitation at only one neural site.

The current pattern in the n_e electrodes of the implant can be represented by a vector $\boldsymbol{i}$. When these currents are presented to the subject, a current density field is established in the subject's cochlea. The corresponding current density j is represented by S_i, which has n_s rows as shown in Eq. (3). Each J represents the current density at a region of neural receptors that are affected by a stimulus.

$$\begin{pmatrix} j_1 \\ j_2 \\ \vdots \\ j_{n_s} \end{pmatrix} = \begin{pmatrix} S_{11} & S_{21} & \cdots & S_{n_e 1} \\ S_{12} & S_{22} & \cdots & S_{n_e 2} \\ \vdots & \vdots & \ddots & \vdots \\ S_{1n_s} & S_{2n_s} & \cdots & S_{n_e n_s} \end{pmatrix} \begin{pmatrix} i_1 \\ i_2 \\ \vdots \\ i_{n_e} \end{pmatrix} \tag{3}$$

Brent determined the matrix S to be the spread matrix. For the 8-electrode Stanford array, the spread matrix has eight columns, one for each electrode. The number of neural sites, which is also the number of rows in the spread matrix, depends on the subject and on the precision of the data collected from the subject. Calculation of the current spread matrix is based on the subject's psychophysical thresholds. The data used in this method were collected from two subjects implanted with 8-channel electrodes in the scala tympani. Deconvolved patterns were generated by inverting the spread patterns calculated using all previous threshold data.

The S matrix allows us to determine the level required for any pattern of stimulus to produce a neural threshold at some site. Assuming that the psychophysical threshold occurs approximately when at least one neural site is above the threshold level, then, for a given i, the threshold will occur if and only if $j = S_i$ has any components greater than one.

For each trial of the threshold paradigm used, a set of current vectors i was obtained. For some of these, the subject reported hearing something. This implies that the value of j at some site must have been greater than one. At levels at which the subject did not hear anything, all elements of j must have been less than one. Based on these observations, an error function can be defined that indicates how well a given S accounts for a set of n_i threshold observations. Once the error function is specified, the problem is reduced to finding the elements of S such that the error is minimized. Once the spread matrix is known, the next step is to determine the patterns of current i^* so that only one site at a time is stimulated. This procedure for obtaining the S matrix and current f^* using the error function is described in detail by Townshend and White (1987).

Table 1 is an example of the matrices obtained by applying their algorithm to the threshold data collected from one of the subjects. In this matrix, row values show the predicted normalized current densities at each of the sites when the site of interest has exactly unity current density. If this matrix was

Table 1 Results of Multiplying the Derived Current Profiles by the Spread Matrix Using Data from an Implantee[a]

Electrode current pattern	Site number					
	0	1	2	3	4	5
0	1.00	−0.20	−0.08	−0.09	−0.09	−0.08
1	−0.14	1.00	−0.28	−0.13	−0.14	−0.13
2	−0.17	−0.17	1.00	−0.21	−0.26	−0.17
3	−0.28	−0.28	−0.29	1.00	−0.30	−0.28
4	−0.28	−0.28	−0.28	−0.34	1.00	−0.28
5	−0.22	−0.19	−0.17	−0.18	−0.17	1.00

[a] Row values represent the current densities at each site that result from applying the current pattern. (From Townshend and White, 1987.)

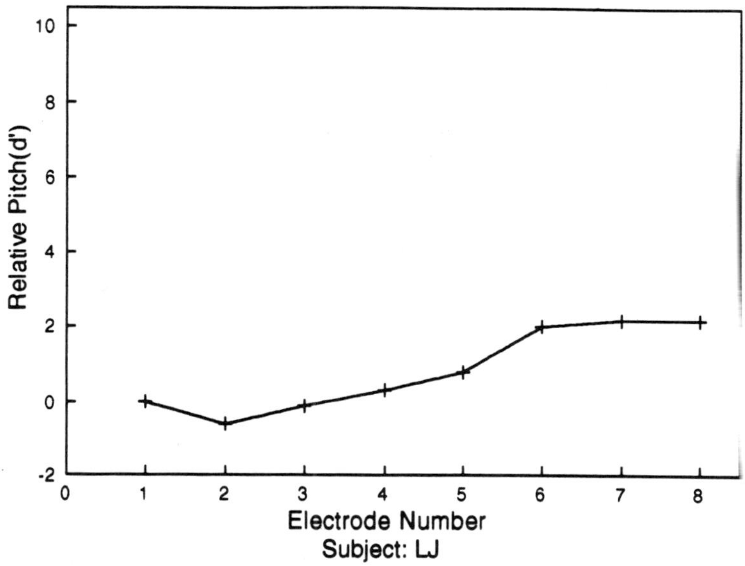

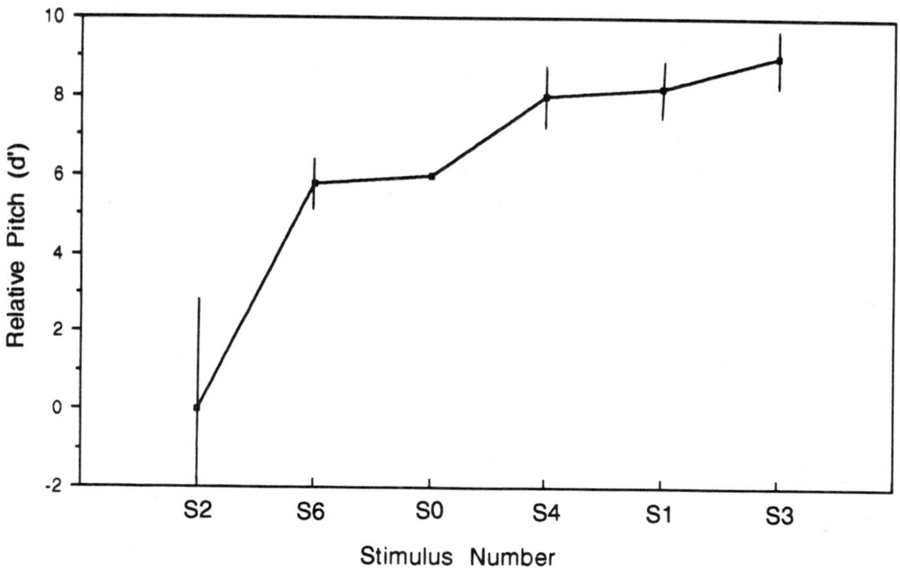

Figure 26 Pitch ranking results. (Top) Pitch ranking for the monopolar stimuli; (bottom) that using the deconvolved stimuli. (From Townshend et al., 1987.)

an identity matrix, it would indicate that the inverse patterns could fully deconvolve the current spread. For this subject, all off-diagonal elements are negative, indicating that stimulation on any site will result in inhibition of stimulation of all other sites. Thus, no matter how many sites are stimulated, no additional sites can reach the threshold unintentionally. The patterns have, in theory, perfect selectivity.

The corresponding pitch-ranking results obtained from the other subjects are shown in Fig. 26. For the monopolar stimuli, the perceived pitch spans a range of delta = 3.0. Using the deconvolved stimuli, the range increased to 6.4 units (Townshend et al., 1987).

The use of sharpened stimuli can easily be added to most speech encoding strategies. Instead of stimulating a single electrode or bipolar pair with the output of a multichannel speech processor, a stimulation pattern can be activated. Research along this path can be used for both vocoder and formant extraction strategies.

4. CONCLUSION

Several strategies of signal processing for cochlear implants were reviewed by referring to reports presented in the past decade. Some successful results were reported using both feature extraction and analog presentation methods. It is apparent that single-channel coding can be maximized to produce some degree of speech discrimination in certain selected patients and that multichannel coding can produce better results than single-channel stimulation. However, speech perception results for a particular processing system cannot be simply compared with those for other systems because of the variety of implantees. It is important not to adopt a single philosophical strategy of speech processing for auditory prostheses because different processing strategies can produce widely different outcomes for individual implant patients.

A number of problems still remain to be solved before signal processing can truly replace the functions of the cochlea. However, new devices such as digital signal processors and microprocessors will be very useful tools for constructing a speech processor which, by changing software, can easily adapt to the patient's characteristics. Algorithms such as LPC and deconvolution will also be used in the near future to extract features more precisely and to reduce the current spread inside the cochlea.

Research to ascertain optimal signal processing will contribute to analyzing the mechanism of the cochlea and the auditory nervous system. These findings will be useful in designing a better cochlear prosthesis capable of functioning more like the cochlea.

REFERENCES

ASHA. (1985). Cochlear implant: Five companies respond to ASHA survey. ASHA, May, pp. 27–34.

Banfai, P., Kubik, S., Karczag, A., and Lüers, P. (1985). Our technique of electrode implantation: Experience with forty-six cases of electrode implantation. *Am. J. Otol.* *6*:295–304.

Banfai, P., Karczag, A., Kubik, S., Lüers, P., and Sürth, W. (1986). Extracochlear sixteen-channel electrode system. *Otolaryngol. Clin. North Am.* *19*:371–408.

Blamey, P. J., Dowell, R. C., and Clark, G. M. (1987). Acoustic parameters measured by a formant-estimating speech processor for a multiple-channel cochlear implant. *J. Acoust. Soc. Am.* *82*:38–47.

Burian, K., Hochmair-Desoyer, I. J., and Eisenwort, B. (1986). The Vienna cochlear implant program. *Otolaryngol. Clin. North Am.* *19*:313–328.

Chouard, C. H., Meyer, B., Chabolle, F., Alcaras, N., and Gegu, D. (1984). Sound signal processing. *Acta Otolaryngol. (Stockh.) Suppl.* 411:85–104.

Clark, G. M., Tong, Y. C., and Dowell, R. C. (1984). Comparison of two cochlear implant speech-processing strategies. *Ann. Otol. Rhinol. Laryngol.* *93*:127–131.

Crosby, P. A., Seligman, P. M., Patrick, J. F., Kuzma, J. A., Money, D. K., Rider, J., and Dowel, R. (1984). The Nucleus multi-channel implantable hearing prosthesis. *Acta Otolaryngol. (Stockh.) Suppl.* *411*:111–114.

Eddington, D. K. (1980). Speech discrimination in deaf subjects with cochlear implants. *J. Acoust. Soc. Am.* *68*:885–891.

Eddington, D. K. (1983). Speech recognition in deaf subjects with multichannel intracochlear electrodes. *Ann. NY Acad. Sci.*, 241–258.

Edgerton, B. J., and Brimacombe, J. A. (1984). Effects of signal processing by the House–3M cochlear implant on consonant perception. *Acta Otolaryngol. (Stockh.) Suppl.* *411*:115–123.

Edgerton, B. J., Doyle, K. J., Brimacombe, J. A., Danley, M. J., and Fretz, R. J. (1983). The effect of signal processing by the House–Urban single-channel stimulator on auditory perception abilities of patients with cochlear implants. *Ann. NY Acad. Sci.* 311–322.

Feigenbaum, E. (1987). Cochlear implant device for the profoundly hearing impaired. *IEEE BME-6*:10–21.

Franz, B. K.-H. G., Dowell, R. C., Clark, G. M., Seligman, P. M., and Patrick J. F. (1987). Recent development with the Nucleus 22-electrode cochlear implant: A new two formant speech coding strategy and its performance in background noise. *Am. J. Otol.* *8*:516–518.

Hochmair, E. S., and Hochmair, Desoyer, I. J. (1981). An implanted auditory eight channel stimulator for the deaf. *Med. Biol. Eng. Comput.* *19*:141–148.

Hochmair, E. S., and Hochmair-Desoyer, I. J. (1983). Percepts elicited by different speech-coding strategies. *Ann. NY Acad. Sci.* 268–279.

Hochmair-Desoyer, I. J., and Hochmair, E. S. (1980). An eight-channel scala tympani electrode for auditory prosthesis. *IEEE Trans. BME-27*:44–50.

Hortmann, G., Banfai, P., and Sürth, W. (1984). Sound signal processing. *Acta Otolaryngol. (Stockh.) Suppl. 411*:131–134.

House, W. F., and Urban, P. E. (1973). Long term results of electrode implantation and electronic stimulation of the cochlea in man. *Ann. Otol. Rhinol. Laryngol. 82*: 504–517.

Ifukube, T., and White, R. L. (1987a). A speech processor with lateral inhibition for an eight channel cochlear implant and its evaluation. *IEEE Trans. BME-34*: 876–882.

Ifukube, T., and White, R. L. (1987b). Current distributions produced inside and outside the cochlea from a scala tympani electrode. *IEEE Trans. BME-34*:883–890.

Ifukube, T., Hirata, Y., and Matsushima, J. (1990). A new model of auditory prosthesis using a digital signal processor. *J. Microcomput. Appl. 13*:219–227.

Merzenich, M. M. (1985). UCSF cochlear implant device. In *Cochlear Implants*, R. A. Schindler and M. M. Merzenich (Eds.). Raven, New York, pp. 121–129.

Miller, J. B., Tong, Y. C., and Clark, G. M. (1984). Speech processing for cochlear implant prosthesis. *J. Speech Hearing Res. 27*:280–296.

Morris, L. R., and Barszczewski, P. (1989). Algorithms, hardware, and software for a digital signal processor microcomputer-based speech processor in a multielectrode cochlear implant system. *IEEE Trans. BME-36*:573–584.

Parkins, C. W. (1983). Cochlear implant: A sensory prosthesis frontier. *Eng. Med. Biol. Mag.*, June, pp. 18–27.

Parkins, C. W. (1985). The bionic ear: Principle and current status of cochlear prosthesis. *Neurosurgery 16*:853–865.

Pfingst, B. E. (1984). Operating ranges and intensity psychophysics for cochlear implants—implication for speech processing strategies. *Arch Otolaryngol. 110*: 140–144.

Simmons, F. B., Herdon, M. H., Atlas, L. E., White, R. L., and Dent, R. J. (1984). Multielectrode modiolar stimulation: Some selected psychophysical and speech results. *Adv. Audio. 2*:163–169. Karger, Basel.

Townshend, B., and White, R. L. (1987). Reduction of electrical interaction in auditory prosthesis. *IEEE Trans. BME-34*:891–897.

Townshend, B., Cotter, N., Compernolle, D., and White, R. L. (1987). Pitch perception by cochlear implant subjects. *J. Acoust. Soc. Am. 82*:106–115.

Vernon, J. (1974). Round window stimulation in man. In *Proceedings of the First International Conference on Electrical Stimulation of the Acoustic Nerve as a Treatment for Profound Sensorineural Deafness in Man*, M. M. Merzenich, R. A. Schindler, and F. A. Sooy (Eds.). San Francisco, pp. 63–78.

Weber, J. L., Chouard, C. H., and Alcaras, N. (1984). Description of the French 12 channel cochlear implant. *Acta Otolaryngol. (Stockh.) Suppl. 411*:140–143.

White, R. L. (1982). Review of current status of cochlear prostheses. *IEEE Trans. BME-29*:233–237.

Wilson, B. S., Finley, C. C., Farmer, J. C., Lawson, D. T., Weber, B. A., Wolford, R. D., Kenan, P. D., White, M. W., Merzenich, M. M., and Schindler,

R. A. (1988). Comparative studies of speech processing strategies for cochlear implants. *Laryngoscope 98*:1069–1077.

Part II

Enhancement and Quality Assessment

10

Speech Enhancement in the 1980s: Noise Suppression with Pattern Matching

STEVEN F. BOLL ITT Aerospace/Communications Division, San Diego, California

1. MOTIVATION

A detailed study was conducted in 1987 to assess the status of noise stripping for improved intelligibility (Makhoul and McAulay, 1989). This study considered the major two application areas requiring noise suppression: (1) two-way communications by voice, both coded and uncoded, and (2) transcription of previously recorded material. The panel concluded that there was currently no approach which improves intelligibility as measured by a closed set intelligibility test such as the diagnostic rhyme test (DRT).

The need for noise suppression in two-way conversation was also emphasized in reports prepared by Kang (Kang and Fransen, 1987) and in an earlier study by Teacher and Coulter (1979). Although the speech coding technology using linear predictive coding (LPC) has improved substantially since 1979, the DRT scores, as shown in Table 1, dramatically emphasize the need for speech enhancement in military platforms. Currently, a DRT intelligibility score greater than 85 is considered necessary for effective two-way communications.

Why has no one found a way to eliminate noise from speech in order to improve intelligibility? One of the most interesting and simple answers

Table 1 DRT Scores

Platform	Uncoded DRT score	Coded (LPC) DRT score
Saipan ship	96	78
Jeep	94	78
P-3C aircraft	96	80
M-60A tank	87	68
RH-53 helicopter	89	58
LVTP-7 Amtrak	72	46

is attributed to Bishnu Atal, who noted, "We call it noise because we know nothing about it!" Attempting to discover more about the noise and, more importantly, how it can be modeled with the speech has been the ongoing goal of research for the past 15 years.

As noted by a National Academy of Sciences panel, "Because broadband continuous noise overlaps the speech signal in both time and frequency, it is the most difficult type of noise to remove. Unfortunately, it is also the most ubiquitous type of noise and has received the greatest attention in the development of noise reductions techniques" (Makhoul and McAulay, 1989).

This chapter reviews the early attempts, highlights their strengths and weaknesses, and shows the evolution of these methods toward the more and more powerful approaches currently under test today. At the heart of all these methods is the apparently impossible objective of trying to "get under the noise." Have we made progress? Are we closer to the objective of improving intelligibility? The conclusion of this chapter is guarded optimism if one is willing to pay the price of more computation and more algorithm training.

2. SCOPE

The scope of this chapter has been limited to the following assumptions:

Microphone selection: Recordings are assumed to be made with a single microphone without microphone modification. This requires our investigation to ignore two microphone approaches (Widrow et al., 1976). These have been shown to offer a potential for improving the intelligibility under specially designed laboratory conditions (Boll and Pulsipher, 1980; Chabries and Christiansen, 1982; Feder et al., 1989).

It is also assumed that the microphone is not under the listener's control and that one cannot alter the transducer to account explicitly for noise.

Adding more recording ports has been successful in designing improved noise-canceling microphones (Brouns, 1981; Viswanathan and Henry, 1989).

Noise Characterization: Only additive noise uncorrelated with the speech will be considered. The noise is assumed to be "continuous" and wideband. Impulsive or tonal noise can be successfully removed with straightforward gating and notch filters (Cupples and Foelker, 1987).

This chapter will not discuss the problem of automatic estimation of the background noise statistics. It will be assumed that an accurate estimate of the noise statistics can be obtained by processing nonspeech data which has been validated to ensure that it statistically represents the same type of noise that was added to the speech.

Nonspeech Distortion: This chapter will not consider other distortions such as speaker stress and fatigue, recording channel distortion, Lombard effect, and multipath echo (Rajasekaran et al., 1986; Hansen, 1989). Compensation for these effects remains an unsolved problem and is certainly worthy of being the subject of a separate survey.

Cochannel Distortions: Finally, this chapter will not address the effects due to an interfering talker (sometimes called the cochannel problem) (Parsons, 1976; Hanson and Wong, 1984; Naylor and Boll, 1987). Reducing cochannel interference faces the difficult challenge of suppressing nonstationary, wide- and narrowband signals, which can be correlated with the speech to be separated (same prosody, same phonetics, etc.).

3. SINGLE-FILTER METHODS

This section summarizes the major approaches developed in the period from roughly 1975 to 1985. An excellent survey through 1981 is given in Lim (1983).

3.1. Short-Time Fourier Transform Approaches

These methods are generally characterized by the algorithm elements shown in Fig. 1:

The original work using a fast Fourier transform (FFT) approach was developed by Weiss et al. (1974). They developed three algorithms for suppressing tones, clicks, and broadband additive noise. A real-time, Very High Speed Integrated Circuit (VSHIC) processor implementation of their approach called the Speech Enhancement Unit is currently being developed by the U.S. Air Force (Cupples and Foelker, 1987). This pioneering approach was followed by a number of methods classified under the general category of "spectral subtraction." They were motivated in part by the attempt to replace

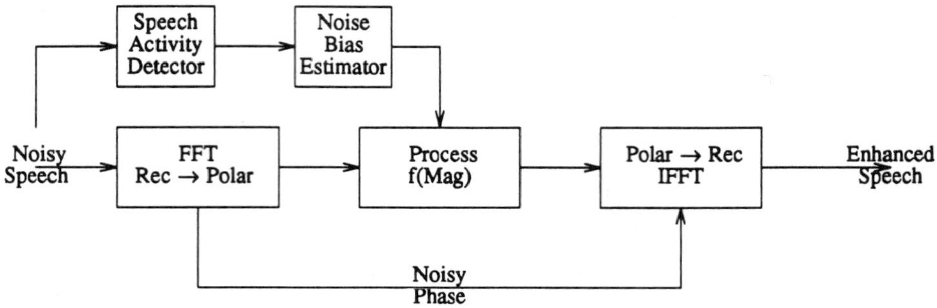

Figure 1 Short-time Fourier transform.

the spectral estimates used in the Wiener filter with instantaneous functions of FFT spectra (Boll, 1979; Schwartz et al., 1979). The basic idea was to subtract a scaled version of the noise floor from a compression-like function of the power spectrum and combine the difference with the noisy phase to form the enhanced speech. The use of the noisy phase was justified in a number of studies (Makhoul and McAulay, 1989; Ephraim and Malah, 1984). The spectral subtraction method (Boll, 1979) improved by 9 points the DRT intelligibility score for LPC-10 coded speech recorded in the RH-53 helicopter. No improvement in intelligibility was measured for uncoded speech. This was the first of many examples in which the noise was reduced by 10 to 15 dB without increasing the DRT intelligibility scores.

3.2. MAP Parameter Estimation and All-Pole Modeling of Degraded Speech

A number of approaches to noise suppression were developed on the basis of two powerful techniques for parameter estimation (Van Trees, 1968).

1. Maximum likelihood estimation, ML: Maximize prob(observation | parameters). That is, estimate the parameter values which most likely resulted in producing the observation.
2. Maximum a posteriori estimation, MAP: Maximize prob(parameters | observation). That is, estimate the parameter values which maximize the conditional probability of the parameters given the observation.

The MAP approach was used to estimate an all-pole spectrum which was then used to form a Wiener filter (Lim and Oppenheim, 1978, 1979). A block diagram depicting this approach is shown in Fig. 2.

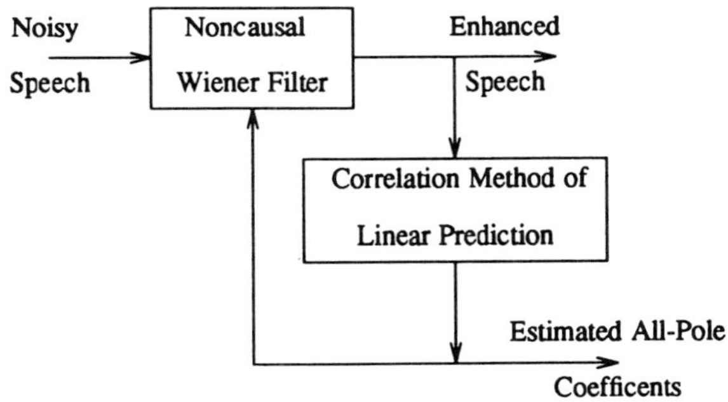

Figure 2 All-pole MAP estimation.

This is an iterative approach which

1. Estimates an LPC all-pole spectrum from the current estimate of the enhanced speech (starting with the noisy speech),
2. Uses this spectrum to form a Wiener filter,
3. Uses the Wiener filter to suppress the noise, and then
4. Repeats the LPC estimation–Wiener filtering process until convergence.

The all-pole modeling approach of Lim and Oppenheim was later enhanced by incorporating inter- and intraframe spectral constraints (Hansen and Clements, 1987, 1988).

3.3. Minimum Mean Square Estimation (MMSE)

The other major technique for parameter estimation is to use the expected value of the conditional probability density of the clean parameters given the noisy observations: E[parameters | observation] These parameter values minimize the mean square error between the estimate and the actual value (Van Trees, 1968).

3.3.1. *MMSE Using Two States*

This approach (McAulay and Malpass, 1980) was a forerunner to the pattern-matching approaches developed in the late 1980s. Two states were assumed: state H_1 denoting speech present and state H_0 denoting speech absent. The MMSE estimation equation is given by:

$$\hat{S} = E(S \mid Y) = E(S \mid Y, H_1)P(H_1 \mid Y) + E(S \mid Y, H_0)P(H_0 \mid Y)$$

where S is the clean spectral envelope and Y is the measured noisy envelope. Since the expected value of the spectral envelope during nonspeech is zero, this estimator reduced to a single-filter approach:

$$\hat{S} = E(S \mid Y, H_1)P(H_1 \mid Y)$$

A real-time system was constructed with this approach and used as a preprocessor for speech coding and recognition. No formal intelligibility tests were reported.

3.3.2. *MMSE Using Gaussian Models*

This approach estimates the spectral magnitude by evaluating

$$\hat{S}_k = E[S_k \mid Y_k]$$

where $\hat{S}_k$ is the MMSE estimator of the kth discrete Fourier transform (DFT) spectral magnitude S_k, and Y_k is the noisy spectral component. It is assumed that each spectral component can be modeled as a statistically independent, Gaussian random variable (Ephraim and Malah, 1983, 1984). Their approach produced speech which was claimed to produced *colorless* residual noise, in contrast to earlier single-filter methods. No formal intelligibility testing was conducted.

3.3.3. *MMSE Trained from Speech*

In parallel with the development of the Gaussian model MMSE approach, Porter and Boll (1984) developed a least-squares estimate for functions of each spectral component (magnitude, power, and log magnitude), without assuming a form for the distribution. The spectral estimate $\hat{S}$ was calculated as the expected value:

$$\hat{S}_k = E[f(S)]_{s|y} = \int f(s)p(s \mid y)\,ds$$

Simplifying this expression gave

$$\hat{S}_k = \frac{E[f(s)p_n(y-s)]_s}{E[p_n(y-s)]_s}$$

The least-squares estimate is calculated as the ratio of the expected value of two random variables with respect to the distribution of the clean speech spectra s. This approach had the advantage of not assuming any specific distribution for the speech. However, it had to be trained on clean speech (this apparent restriction is now common with the later methods).

For each spectral component, a table was constructed mapping the noisy spectral magnitude into an estimate of the clean spectral magnitude. The effectiveness (as well as the limitation) of this approach can be highlighted by examining a typical scatter plot showing the noisy versus clean spectral magnitudes superimposed with the MMSE estimator in Fig. 3. It shows that the estimator properly passes through the conditional expected value of the clean magnitude given the noise magnitude. It also highlights the inherent shortcoming of this approach: all speech was used to form the estimator. There are no distinctions among difference acoustic categories (one estimator fits all!).

This noise suppression technique was not tested using the DRT. It was used as a preprocessor for a speaker-dependent, connected-digit recognizer and found to restore noisy performance to within 3% of the noise-free case using MMSE estimates from noisy template data and MMSE estimates from noisy unknown data. This was one example of many which followed, showing that noise suppression preprocessing was able to increase the intelligibility for machine recognition in contrast to not improving intelligibility for human recognition.

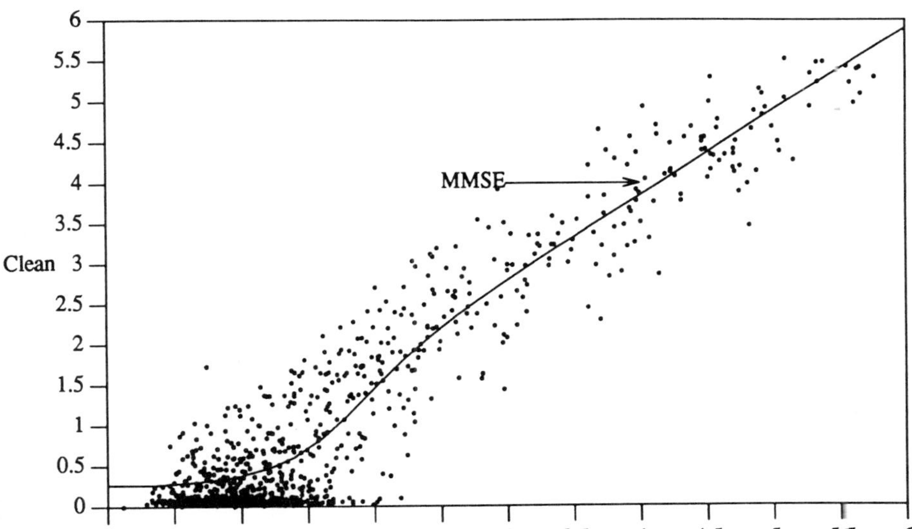

Figure 3 Example of MMSE estimator.

3.4. Problems and Solutions

A number of problems were shared by all of the single-filter approaches. These include the following.

Residual noise: In all of these methods there was usually a level of residual, unnatural, non-Gaussian, background noise accompanying the speech. This annoying noise has been called the "musical noise" or the "doodle-lee-doos."* Eliminating this noise by increasing the suppression level invariably reduced or eliminated the presence of low-energy speech segments.

Destructive processing: Care had to be taken when implementing these schemes to prevent more harm than good. In the absence of noise, the processor should reduce to an identity system. More important, the processing should not reduce the intelligibility below that measured in the noise. This negative effect was always possible with spectral subtraction implementations if the selected noise floor was larger than the actual value. This resulted in excessive attenuation of low-energy frequency bins with a corresponding reduction in intelligibility. (This destructive effect is also a problem for speech recognizers. When they are trained with noisy speech and then operated in a noise-free condition they inevitably produce suboptimal performance.)

Speech characteristics: These single-filter methods ignored the fact that speech is nonstationary. As noted in the National Academy of Sciences report, "Judicious use should be made of additional known short-term properties of speech signals, noise, and human auditory perception" (Makhoul and McAulay, 1989).

At a minimum, a set of estimators are required which are dependent on the type of speech waveform under analysis. Further, there is a need to incorporate the attribute that accurate speech models can be trained which predict the temporal and spectral characteristics of speech. This predictive power should be able to look both ahead and behind.

4. PATTERN-MATCHING APPROACHES

This section describes a number of recent attempts to introduce more complexity into the enhancement process by matching the noise suppressor to the speech signal with pattern-matching techniques.

---

*This descriptive phrase is attributed to T. G. Stockham, who coined it in the early 1970s while directing a project aimed at trying to remove background noise from Caruso recordings.

4.1. MMSE Using Additional Parameters

A study was conducted by Porter (1989) to improve the MMSE estimator by conditioning the estimation on more information than just the noisy speech, the average speech, and the average noise power. He noted that spectral error was highly correlated with relative frame energy, defined as the difference between the current frame energy and the smoothed maximum energy of nearby frames. This energy measure was small for vocalics and large for intervocalic events such as nasals, fricatives, and unvoiced segments in general. He designed a spectral estimator based on the conditional mean value of the spectrum given the noisy magnitude, the speech and noise power, *and* the relative frame energy. This "event parameter" was estimated using the conditional mean of the clean value given the noisy value. This approach significantly reduced the mean square error of the spectral estimator, but for improving performance in a word spotting experiment using noisy speech it proved inconclusive. However, it did establish that conditioning the estimator with event-sensitive information could reduce filter bank parameter estimation error. The issue became one of selecting an approach for modeling event information and then, equally important, developing a reliable method for selecting the event using the *noisy* speech.

4.2. MAP Estimation with Hidden Markov Modeling

This approach models the clean speech and noise processes as ergodic hidden Markov models (HMMs) (Ephraim, 1988, 1989; Ephraim et al., 1989). The HMM is initially trained on clean speech. Enhancement of the noisy speech is accomplished by a two-step process. First, the most likely state and its observation sequence from the HMM are estimated using Viterbi decoding. From each state and mixture within the state, a set of LPC coefficients (autoregressive, AR process) are selected and used to form a Wiener filter constructed from the all-pole filters. A new estimate of the clean speech is calculated by filtering the noisy speech using the all-pole filters associated with the most likely sequence of states. This clean estimate is then used to find another mixture, in turn supplying another Wiener filter, etc. This iterative process continues until a preset convergence criterion is reached. Thus the HMM selects the most likely all-pole filter needed by the Wiener filter. Figure 4 shows a block diagram of the approach.

This approach is similar to that proposed by Lim and Oppenheim (1978) with two important improvements:

1. The enhancement is conditioned and enhanced by the effect of the adjacent frames (Markovian model).

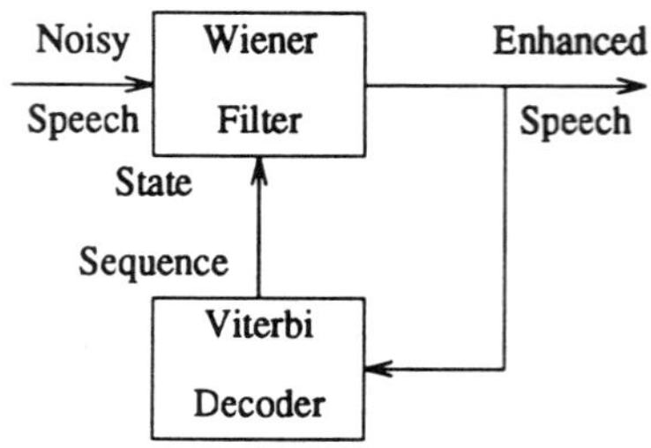

Figure 4 HMM application.

2. The estimate of the AR process is constrained to be from a predesigned set of AR models *trained* from the clean speech.

The Wiener filter is implemented in the frequency domain using the noisy phase, and the reestimation algorithms are initialized using the noisy speech.

This approach made a significant step in overcoming many of the limitations of earlier approaches. However, it still contained two drawbacks: (1) the MAP estimation forced it to select only a single filter, and (2) this filter was selected based on an approximate version of the clean speech, namely the Wiener-filtered noisy speech. The next approach removes both of these limitations.

4.3. Minimum Mean Square Estimation with Hidden Markov Modeling

This approach uses the same Gaussian AR HMMs to model the clean speech and a low-order Gaussian AR to model the noise (Ephraim, 1989). The estimator now consists of a filterbank of Wiener filters, one for each individual mixture component within each state. Each Wiener filter is similarly constructed from the all-pole filter corresponding to the AR process for that state and mixture. The output of each filter is now weighted by the probability of the state and mixture component occurrence at each time instant.

This approach differs from the MAP approach in the following important ways:

1. The probabilities and AR parameters are selected from the *noisy* speech, whereas in the MAP approach the weighting factors and parameters are calculated from the probabilities of the state and mixtures given an estimate of the *clean* speech.
2. No attempt is made to select a particular estimator. The MMSE includes all estimators for all states and all mixtures. This is done at the

expense of a considerable increase in computing over the single-filter cases.

A block diagram of this method is shown in Fig. 5, where $y_t = s_t + n_t, x_t$ is the state at time t, and h_t is the mixture at time t.

The expected values are implemented as Wiener filters in the frequency domain using all-pole filters from the AR process and the noisy phase. The filter output is weighted by $p(x_t = i, h_t = j \mid y_t)$, the probability of state i and mixture j at time t. Typical values are 8 states and 64 mixtures. Ephraim also develops MMSE estimators for functionals of the spectrum including magnitude, power, and log magnitude.

4.4. Probabilistic Vector Mapping

Researchers at BBN (Gish et al., 1990) have developed a template-matching approach for noise suppression which calculates an estimate of the clean cepstrum using a combination of MAP and MMSE. They note that since the conditional probability of the clean parameters given both the noisy parameters and an energy event parameter is non-Gaussian, they must divide the clean speech parameter domain using vector quantization and create a different Gaussian model for each VQ symbol. Their method assumes that both the clean speech and noisy speech with a 10-dB signal-to-noise ratio (SNR) are available. The clean speech is vector quantized into 256 regions. For each VQ region, the noisy data are grouped and used to estimate the mean and diagonal covariance of the conditional probability of the noisy speech given the clean speech and a new parameter, a measure of instantaneous SNR (similar to the Porter event parameter). They assume that the clean and noisy conditional probability density functions are Gaussian. This allows them to postulate a linear model between the clean and noisy cepstral values as

$$s = A(k, \gamma)y + b(k, \gamma) + \epsilon$$

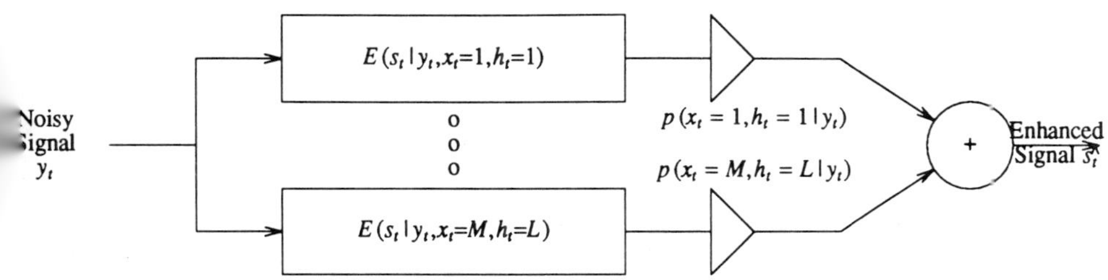

Figure 5 HMM MMSE.

where s is the clean cepstral vector, y is the noisy cepstral vector, k is the selected VQ symbol, γ is the instantaneous SNR, A and b are the linear model terms, and ϵ is a zero-mean Gaussian vector. The linear mapping relating the noisy parameters to the clean parameters is estimated using MMSE. Thus the estimate of the clean cepstrum is found in a two-step process: (1) they find the VQ symbol $\hat{k}$ as the maximum of $p(k \mid y, \gamma)$, and (2) they obtain the clean vector by selecting the linear model corresponding to $\hat{k}$. The linear model parameters $A(k, \gamma)$ and $b(k, \gamma)$ are estimated by collecting pairs of clean and noise vectors and using a minimum mean square error criterion.

This approach differs from the prior Ephraim approach in that

1. It uses VQ instead of HMM.
2. It uses a MAP approach and finds the most likely VQ estimate rather than calculating all MMSE estimates and weighting them by their probability of occurring.
3. It uses a regression equation to filter the parameter set rather than a DFT-based Wiener filter (they are concerned with recognition, not synthesis).

This approach provided a substantial improvement in word-spotting recognition accuracy. It was not used in an analysis-synthesis framework.

4.5. MMSE with Vector Quantization on Filter Bank Data

Researchers at SRI (Erell and Weintraub, 1990) have extended the MMSE filter bank estimation procedures [Port84] to get around the "one estimator fits all" problem. They model the speech distribution as consisting of clusters and assume that, within each cluster, the energies at different frequency channels are significantly more uncorrelated than speech as a whole. This allows them to generate MMSE estimators for each channel in each cluster. The estimator for channel k is given by

$$\hat{S}_k = \sum_n \hat{S}_k \mid nP(n \mid Y_n)$$

where $\hat{S}_k \mid n$ is the MMSE estimator in the nth cluster and $P(n \mid Y_n)$ is the probability of the nth cluster.

Their approach is similar to the MMSE approach taken by Ephraim, except that they use only $n = 30$ clusters rather than 512 mixtures. They report substantial improvement when the approach is used as a preprocessor with a speaker-independent, large-vocabulary speech recognizer.

4.6. Noise Suppression Using Sine Wave Analysis-Synthesis

Researchers at Lincoln Laboratory (Quatieri and McAulay, 1990) have applied their sine wave modeling approach, [Quat86] to the problem of noise suppression. Noisy speech is matched against a set of templates representing sine wave models. The noisy signal is replaced the output of a harmonic, zero-phase, sine wave analysis-synthesis system. They use MMSE to estimate their parameters. Specifically, the estimate of the clean signal $\hat{s}$ is given by

$$\hat{s} = E[s \mid y] = \sum_{i=1}^{M} E[s \mid y, H_i] p[H_i \mid y]$$

where H_i represent the ith hypothesis (template) and $p[H_i \mid y]$ is the probability of the hypothesis given the observation. Assuming that each hypothesis is equally likely and using Bayes' rule, $\hat{s}$ can be shown to be

$$\hat{s} = \sum_{i=1}^{M} E[s \mid y, H_i] w_i$$

where

$$w_i = \frac{p[y \mid H_i]}{\sum_{j=1}^{M} p[y \mid H_j]}$$

represents the importance of each template. It assumes that if the noise is Gaussian, then $p[y \mid H_i]$ is Gaussian, and

$$w_i = \frac{\exp[-\|y - H_i\|^2/2\sigma]}{\sum_{j=1}^{M} \exp[-\|y - H_j\|^2/2\sigma]}$$

To improve performance, they investigated using noisy templates and estimating the template weights using only the nearest-neighbor templates rather than all M templates.

4.7. Observations

The pattern-matching approaches have a number of similarities as well as differences. All methods assume that the conditional distributions are non-Gaussian and must therefore be divided using VQ or HMM modeling. Likewise, each approach accounts for the fact that noisy speech must be used to select the VQ region by some means, such as using noisy templates (LL), noise-generated distributions (BBN), or noise-conditioned autocorrelations in the Itakura–Saito metric (AT&T). With the exception of the Lincoln

Laboratory approach (which uses template substitution), all methods estimate the clean parameters using an MMSE estimate of the noisy parameters. The MMSE estimators are implemented in the frequency domain using discrete Fourier transforms of the autoregressive processes, in the parameter domain using a matrix mapping, or in the filter bank domain using analytic representations based on Gaussian statistics.

The approaches primarily differ in the following areas:

- *VQ versus HMM*: the use of HMM offers the advantage of incorporating information from the adjacent frames under the Markovian assumption at the cost of increase storage, training, and computation.
- *MAP versus MMSE*: The use of the MAP estimate forces the selection of a single filter, whereas the use of MMSE allows the contribution of all filters weighted by the probabilities of their state occurring.
- *Parameterization*: The use of a specific parameter set, e.g., cepstral, filter bank, and LPC, was dependent on the application: machine recognition versus human recognition.

5. FUTURE AREAS

It now appears that the solution to improved noise suppression demands effective speech recognition, and thus the techniques found effective for machine recognition can be applied to this problem. Successful efforts in the area of recognition using noisy speech include the following.

- *Robust parameterizations and better metrics*: Alternative robust front ends have been designed based on perceptually based criteria. Ghitza has investigated time synchrony models (Ghitza, 1987), O'Shaughnessy (1988) has developed techniques using a formant metric, which he claims are more robust because decisions are based on regions having a high SNR. A number of improved metrics and matching techniques have been developed for machine recognition of noisy speech. Examples of these are given in Mansour and Juang (1989), Holmes (1986), Soong and Sondhi (1988), and Varga et al. (1988).
- *Neural networks*: Efforts have been made to train artificial neural networks for noise suppression (Tamura and Waibel, 1988; Tamura, 1989). Hidden layers are trained using noisy speech as input and clean speech as output. Early results show that these nets are insensitive to noise floor shifts and perform similarly to single-filter methods. Expanding the node structure to support the 10^3 events modeled by

HMMs may provide improvement due to the absence of modeling assumptions.

6. SUMMARY AND CONCLUSIONS

This survey has attempted to summarize the significant efforts in noise suppression during the past 15 years. It started with single-filter approaches motivated by replacing the clean spectra in the Wiener filter with spectral subtraction estimates and all-pole estimates. These approaches lead to investigations in which direct MMSE estimators were calculated with different assumptions about the speech distributions. From these have come multiple event estimators using MAP first and then returning to MMSE.

Continued improvements will draw from the successes achieved in speech recognition and high quality, low rate coding. In these areas, better performance has come at the expense of more complex models supported by greater and greater computing requirements. Just like with everything else in digital speech processing, ... "nothing comes easy, and nothing comes cheap!"

ACKNOWLEDGMENT

The author would like to thank J. Porter for his helpful insight and support.

REFERENCES

Boll, S. F. (1979). Suppression of acoustic noise in speech using spectral subtraction. *IEEE Trans. Acoust., Speech, Signal Process. ASSP-27*:113–120.

Boll, S. F., and Pulsipher, D. C. (1980). Suppression of acoustic noise in speech using two microphone adaptive noise cancellation. *IEEE Trans. Acoust., Speech, Signal Process. ASSP*-28:752–753.

Brouns, A. (1981) Second-order gradient noise canceling microphone. *Proc. Int. Conf. Acoustics, Speech, and Signal Processing.* Atlanta, vol. 81CH1610-5, pp. 786–789.

Chabries, D. M., and Christiansen, R. W. (1982). Application of the LMS adaptive filter to improve speech communication in the presence of noise. *ICASSP 82*, Paris, May, vol. 1, p. 148.

Cupples, E. J., and Foelker, J. L. (1987). Air Force Speech Enhancement Program, Military Speech Tech., Rome Air Development Center, Washington D.C.

Ephraim, Y. (1988). On the application of hidden Markov models for enhancing noisy speech. *ICASSP 88*, New York, April, vol. 1, p. 533.

Ephraim, Y. (1990). Minimum mean square error approach for speech enhancement. *ICASSP 90*, Albuquerque, April, Vol. 2, p. 829.

Ephraim, Y., and Malah, D. (1983). Speech enhancement using optimal non-linear spectral amplitude estimation, *ICASSP 83*, Boston, April, vol. 1, p. 1118.

Ephraim, Y., and Malah, D. (1984). Speech enhancement using a minimum mean-square error short-time spectral amplitude estimator, *IEEE Trans. Acoust., Speech Signal Process. ASSP-32*:1109–1122.

Ephraim, Y., Malah, D., and Juang, B. H. (1989). On the applications of hidden Markov models for enhancing noisy speech. *IEEE Trans. Acoust., Speech, Signal Process. ASSP-37*:1846–1856.

Erell, A., and Weintraub, M. (1990). Estimation using log spectral distance criteria for noisy robust speech recognition. *ICASSP-90* Albuquerque, April, vol. 2, p. 853.

Feder, M., Oppenheim, A. V., and Weinstein, E. (1989). Maximum likelihood noise cancellation using the EM algorithm. *IEEE Trans. Acoust., Speech. Signal Process. ASSP-37*:204–216.

Ghitza. O. (1987). Robustness against noise: The role of timing-synchrony measurement. *ICASSP 87* Dallas, April, vol. 1, p. 2372.

Gish, H., Chow, Y., and Rohlicek, J. R. (1990). Probabilistic vector mapping of noisy speech parameters for HMM word spotting. *ICASSP*, Albuquerque, April, vol. 1, p. 117.

Hansen. J. H. (1989). Stress compensation and noise reduction algorithms for robust speech recognition. *ICASSP 89*, Glasgow, May. vol. 1, p. 266.

Hansen. J., and Clements. M. (1987). Iterative speech enhancement with spectral constraints. *ICASSP 87*, Dallas, April, vol.1, p. 189.

Hansen. J. H., and Clements, M. A. (1988). Constrained iterative speech enhancement with application to automatic speech recognition. *ICASSP 88*, New York, April, vol. 1, p. 561.

Hanson, B. A., and Wong, D. Y. (1984). The harmonic magnitude suppression (HMS) technique for intelligibility enhancement in the presence of interfering speech. *ICASSP 84*, San Diego, March, vol. 1, p. 18A.5.

Holmes, J. N. (1986). noise compensation for speech recognition using probabilistic models. *ICASSP 86*, Tokyo, April, vol. 1, p. 741.

Kang, G., and Fransen, L. (1987). Navy programs to improve LPC-processed speech in noise. Military Speech Tech., Naval Research Laboratory, Washington D.C.

Lim, J. S. (1983). *Speech Enhancement*. Prentice-Hall, Englewood Cliffs, N.J.

Lim, J. S. and Oppenheim, A. V. (1978). All-pole modeling of degraded speech. *IEEE Trans. Acoust., Speech, Signal Process. ASSP-26*:197–209.

Lim, J. S. and Oppenheim, A. V. (1979). Enhancement and bandwidth compression of noisy speech, *Proc. IEEE 67*:1586–1604.

Makhoul, J. and McAulay, R. (1989). *Removal of Noise from Noise-Degraded Speech Signals*, National Academy Press, Washington, D. C.

Mansour, D. and Juang, B. H. (1989). The short-time modified coherence representation and noisy speech recognition. *IEEE Trans. Acoust., Speech, Signal Process. ASSP-37*:795–804.

McAulay, R. J. and Malpass, M. L. (1980). Speech enhancement using a soft-decision noise suppression filter. *IEEE Trans. Acoust., Speech, Signal Process. ASSP-28*: 137–145.

Naylor, J. A., and Boll, S. F. (1987). Techniques for suppression of an interfering talker in co-channel speech. *ICASSP 87*, Dallas, April, vol. 1, p. 205.

Parsons, T. (1976). Separation of Speech From Interfering Speech By Means of Harmonic Selection. *J. Acoust. Soc. Am. 60*:

Porter, J. (1989). Event sensitive minimum mean square error estimators. Multi-Channel Gisting Final Report, Rome Air Development Center.

Porter, J. E., and Boll, S. F. (1984). Optimal estimators for spectral restoration of noisy speech. *ICASSP 84*, San Diego, March, vol. 1, p. 18A.2.

Quatieri, T. F., and McAulay, R. J. (1986). Speech transformations based on a sinusoidal representation. *IEEE Trans. Acoust., Speech, Signal Process. ASSP-34*, p. 1499.

Quatieri, T. F., and McAulay, R. J. (1990). Noise reduction using a soft-decision sine-wave vector quantizer. *ICASSP*, Albuquerque, April, vol. 2, p. 821.

Rajasekaran, P. K., Doddington, G. R., and Picone, J. W. (1986). Recognition of speech under stress and in noise. *ICASSP 86*, Tokyo, April, vol. 1, p. 733.

Schwartz, R., Berouti, M., and Makhoul, J. (1979). Enhancement of Speech Corrupted by Acoustic Noise, Washington, p. 208, ICASSP 79.

Shaughnessy, D. O. (1988). Speech enhancement using vector quantization and a formant distance measure. *ICASSP*, New York, April, vol. 1, p. 549.

Soong, F., and Sondi, M. M. (1988). A frequency-weighted Itakura spectral distortion measure and its application to speech recognition in noise. *IEEE Trans. Acoust., Speech, Signal Process. ASSP-36*:41–48.

Tamura, S. (1989). An analysis of a noise reduction neural network. *ICASSP 89*, May, pp. 2001–2004.

Tamura, S., and Waibel, A. (1988). Noise reduction using connectionist models. *ICASSP 88*, New York, April, vol. 1, p. 553.

Teacher, C., and Coulter, D. (1979). Performance of LPC vocoders in noisy environment. *ICASSP 79*, Washington, April, p. 216.

Van Trees, H. L. (1968). *Detection, Estimation and Modulation Theory*. Wiley, New York.

Varga, A., Moore, R., Bridle, J., Ponting, K., and Russell, M. (1988). Noise compensation algorithms for use with hidden Markov model based speech recognition. *ICASSP 88*, New York, April, vol. 1, p. 481.

Viswanathan, V. R., and Henry, C. M. (1989). Noise-immune multisensor speech input: Formal subjective testing in operational conditions. *ICASSP 89*, Glasgow, May, vol. 1, p. 373.

Weiss, M. R., Aschkenasy, E., and Parsens. T. W. (1974). Study and development of the INTEL technique for improving speech intelligibility. Report NSC-FR/4023, Nicolet Scientific Corp., December.

Widrow, B., McCool, J. M., Larimore, M. G., and Johnson, C. R., Jr. (1976). Stationary and nonstationary learning characteristics of the LMS adaptive filter. *Proc. IEEE 64*:1162–1172.

11

Adaptive Echo Cancellation for Speech Signals

M. MOHAN SONDHI and WALTER KELLERMANN AT&T
Bell Laboratories, Murray Hill, New Jersey

1. INTRODUCTION

With rare exceptions, conversations take place in the presence of echoes. We hear echoes of our speech waves as they are reflected from the floor, walls, and other neighboring objects. If a reflected wave arrives a very short time after the direct sound, it is perceived not as an echo but as a spectral distortion, or reverberation. Most people prefer some amount of reverberation to a completely anechoic environment, and the desirable amount of reverberation depends on the application. (For example, much more reverberation is desirable in a concert hall than in an office.) The situation is very different, however, when the reflection arrives a few tens of milliseconds after the direct sound. In such a case, it is heard as a distinct echo. Such echoes are invariably annoying, and under extreme conditions they can completely disrupt a conversation. It is such distinct echoes that we will be concerned with in this chapter.

Echoes have long been the concern of architects and designers of concert halls. However, since the advent of telephony, they have also been the concern of communication engineers, because echoes can be generated *electrically*, due to impedance mismatches at points along the transmission medium. Such echoes are called line echoes.

If the telephone connection is between two handsets, then the only echoes encountered are line echoes. These echoes are not a problem in local telephone calls because the sources of echo are insignificant, and the echoes, if any, occur after very short delays. In a long-distance connection, however, distinct echoes may be heard. A significant source of line echoes in such circuits is a device called a hybrid, which we shall discuss briefly in Section 2.

Echoes at hybrids have been a potential source of degradation in the telephone network for many decades, and many solutions have been devised to overcome them. Of particular interest to us in the present chapter are devices known as adaptive echo cancelers. Interest in such devices arose during the 1960s, in anticipation of telephone communications via satellites (Sondhi, 1967; Thies and Zmood, 1967). As satellite communication gained an ever-increasing share of telephone traffic during the 1970s, considerable development of echo cancelers took place (Campanella et al., 1972; Demytko and Mackechnie, 1973; Kato et al., 1973; Horna, 1977). Their widespread use began around 1980, with the arrival of a very large scale integration (VLSI) implementation (Duttweiler and Chen, 1980).

When the telephone connection is between one or more hands-free telephones or between two conference rooms, a major source of echoes is the acoustic coupling between the loudspeaker and the microphone at each end. Such echoes have been called *acoustic echoes*, and interest in adaptive cancellation of such echoes has attracted much attention during the past decade. This aspect of the problem of echo cancellation will be the major new topic which we shall discuss in this chapter. A caveat to the reader is in order at this point. Although we will be dealing with *acoustically* generated echoes, we will only consider cancellation of these echoes in the electrical portion of the circuit. We will not discuss the related, but much more difficult, problem of canceling echoes acoustically.

In the next two sections we will briefly discuss the problem of line echoes and adaptive cancellation of such echoes. We refer the reader to review articles (Sondhi and Berkley, 1980; Messerschmidt, 1984) for a more detailed account. Besides introducing the reader to the echo problem, this preliminary discussion will also lay the groundwork for the more modern problem of canceling acoustic echoes. The latter problem is the one in which significant advances have been made in the recent past, so we will concentrate on that. However, the algorithms used for this problem have many features in common with those used in line echo cancelers. In Section 4 we will discuss acoustic echoes and the changes and improvements in the echo canceler required to deal with them. Section 5 will summarize our conclusions. Let us turn, then, to the line echo problem.

2. LINE ECHOES

As mentioned in the last section, the main source of line echoes is the device known as a hybrid. Figure 1 illustrates, in a highly simplified manner, the function and placement of hybrids in a typical long-distance telephone connection.

Every telephone in a given geographic area is connected to a central office by a two-wire line, called the customer's loop, which serves for communication in either direction. A local call is set up by simply connecting the two customers' loops at the central office. When the distance between the two telephones exceeds about 35 miles, amplification becomes necessary. Therefore a separate path is needed for each direction of transmission. The device that connects the four-wire part of the circuit to the two-wire portion at each end is known as a hybrid (or a hybrid transformer). With reference to Fig. 1, the purpose of the hybrids is to allow signals from A to go along path L1 to B, and to go from B along path L2 to A. However, they must prevent signals in path L1 from returning along path L2 back to A. Similarly, the signal in path L2 is to be prevented from returning along path L1 back to B.

We do not wish to go into the detailed workings of a hybrid. Further information can be found in Sondhi and Berkley (1980) and other references cited there. Suffice it to say here that a hybrid is a bridge network which can achieve the objectives stated above, provided the impedance of the customer's loop can be exactly balanced by an impedance located at the

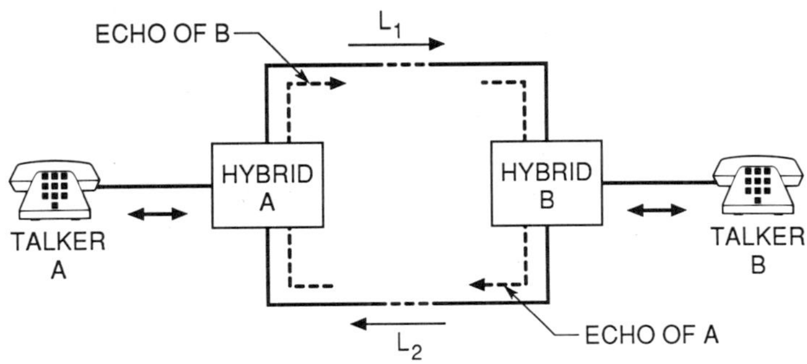

Figure 1 A long-distance connection.

hybrid. Unfortunately, this is not possible in practice because there are far fewer four-wire circuits than there are two-wire circuits. Therefore a hybrid may be connected to any of the customer loops served by the central office. By their very nature, customer loops have a wide variety of characteristics—various lengths, type of wire, type of telephone, number of extension phones, etc. It appears, therefore, that the echo at the hybrid cannot be completely eliminated. A compromise impedance is used to balance the bridge, and the average attenuation (in the United States) from input to output of the hybrid is 11 dB with a standard deviation of 3 dB. This amount of attenuation is not adequate for satisfactory communication on circuits with long delays (Sondhi and Berkley, 1980).

2.1. The Echo Suppressor

The problem of such echoes has been around ever since the introduction of long-distance communication. On terrestrial circuits, the device most widely used to control line echoes is the echo suppressor (Sondhi and Berkley, 1980). Again, we will not describe echo suppressors in detail, but merely mention that they are voice-operated switches whose object is to remove the echo of the talker's speech and yet allow the listener to interrupt as in normal conversation. The principle of the echo suppressor can be explained by referring to Fig. 2, which shows the end B of the telephone circuit of Fig. 1 with an echo suppressor included. Suppose A has been talking for a while. Based on the level of signals in paths L1 and L2, a decision is made as to whether the signal in L2 is an interruption by B trying to break into the conversation or an echo of A's speech. If the decision is the latter, then the circuit L2 is opened (or a large loss is switched in). A similar switch at the other end prevents B's echo from returning to B.

If the decision mechanism were to behave flawlessly, the echo suppressor would be an ideal form of echo control. The decision, however, cannot be perfect. The two signals that have to be distinguished are both speech signals, with more or less the same statistical properties. Essentially the only distinguishing property is the level. Therefore, sometimes a high level of echo is returned, and sometimes when the speech level is low (or during initial and final portions of speech bursts) the interrupter's speech is mutilated. However, with considerable ingenuity, echo suppressors have been designed to keep such malfunctions at an acceptable level.

2.2. The Line Echo Canceler

Echo suppressors have served well for over 60 years on circuits with round-trip delays of 100 ms or less. With the advent of commercial communications

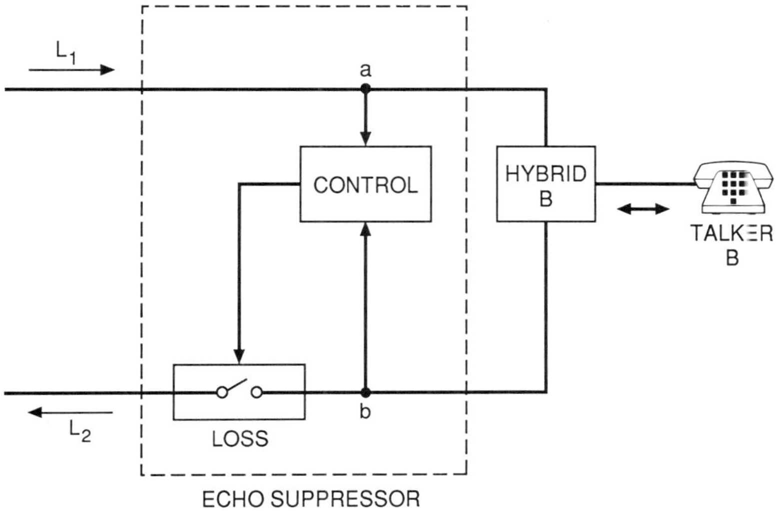

Figure 2 Illustrating an echo suppressor.

satellites in 1965, however, the situation changed significantly. A synchronous satellite (i.e., one which is stationary with respect to the earth) must have an orbit which is about 23,000 miles above the earth's surface. A telephone connection via such a satellite will have a round-trip delay of 500 to 600 ms (Sondhi and Berkley, 1980). With such long delays, echo suppressors fail to function satisfactorily. The long delay induces a change in the pattern of conversation in such a way as to increase the number of errors significantly. New methods of echo control were proposed for circuits with such long delays. Of these, the most versatile, and the one in widespread use, is the adaptive echo canceler (Sondhi, 1967). The unique feature that makes it so attractive is that, unlike the other forms of echo control, the echo canceler does not tamper with the path carrying the echo. Therefore, it never mutilates the speech of the interrupting party. The basic idea of the echo canceler is illustrated in Fig. 3. Again we show only one canceler located at the end B of the telephone circuit of Fig. 1. A similar canceler is symmetrically located at the other end.

As illustrated in Fig. 3, instead of interrupting path L2, a synthetic echo is generated from A's speech and subtracted from the signal going out on path L2. The synthetic echo is generated by passing the signal of path L1 through a filter whose impulse response (or transfer function) matches that of the echo path from point "a" to point "b" via hybrid B.

As mentioned in Section 2., the echo path is highly variable, so the filter in Fig. 3 cannot be a fixed filter. It must be estimated for the particular local

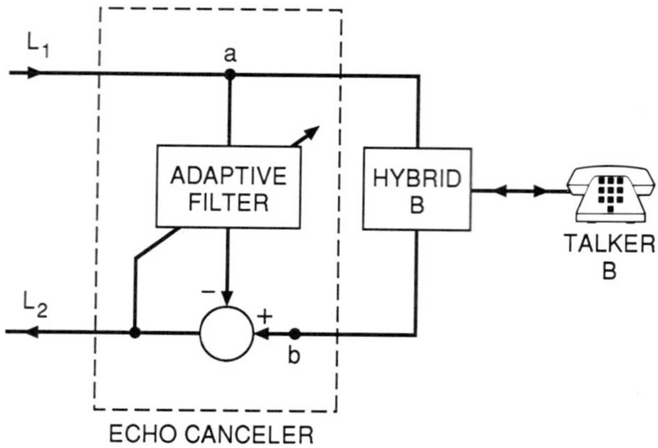

Figure 3 Illustrating an echo canceler.

loop to which the hybrid gets connected. One simple way to derive the filter is to measure the impulse response of the echo path and then approximate it with some filter structure, e.g., a tapped delay line. However, the echo path is, in general, not stationary. Therefore, such measurements would have to be made repeatedly during a conversation. Clearly, this is highly undesirable. To eliminate the need for such measurements the filter is made adaptive. An algorithm is implemented which uses the residual error to adapt the filter to the characteristics of the local loop and to track slow variations in these characteristics. In the next few sections we will discuss some basic adaptation algorithms in some detail.

3. ADAPTIVE CANCELLATION

To implement a filter which approximates the echo path, the first step is to choose a representation of the filter in terms of a finite number of parameters. Assuming the echo path to be linear, this can be achieved by finding an expansion of the impulse response of the echo path in terms of a set of basis functions. The problem then reduces to the estimation of the expansion coefficients. If $w_n(t)$, $n = 0, 1, 2, \ldots, N-1$ is the (truncated) set of basis functions, then the expansion can be implemented by the set of N filters illustrated in Fig. 4a. The output of the filter bank, $y(t)$, is related to the input $x(t)$ by the relation

$$y(t) = x(t) * \sum_{n=0}^{N-1} g_n W_n(t)$$

$$= \sum_{n=0}^{N-1} g_n x_n(t)$$
$$= \mathbf{g}'\mathbf{x}. \tag{1}$$

Here $*$ indicates convolution, $x_n(t)$ is the output of the nth filter component, and g_n is the nth expansion coefficient. In the last line of Eq. (1) we have introduced matrix notation which will be useful later. The boldface quantities $\mathbf{g}$ and $\mathbf{x}$ are column vectors with dimension $N \times 1$, and $'$ denotes matrix transpose. Also, for simplicity of notation, we will suppress the dependence of quantities on the time t, except where it helps avoid confusion.

In the special case when $w_n(t) = \delta(t - n\Delta)$, the filter becomes an n-tap transversal filter (tapped delay line) with a delay Δ between taps, as illustrated in Fig. 4b. This is the most commonly used filter structure, although other structures—e.g., when the $w_n(t)$'s are Laguerre functions—or truncated (or damped) sinusoids have been tried (Sondhi, 1967).

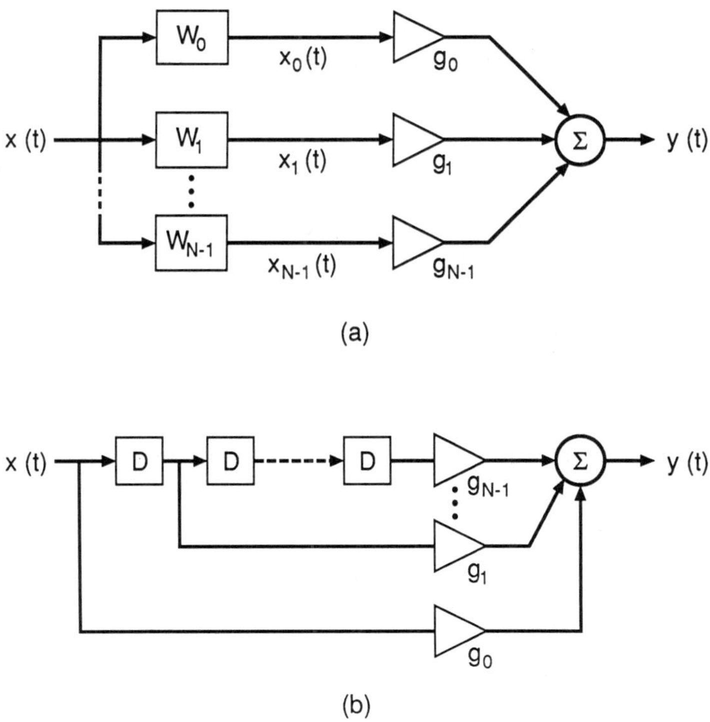

Figure 4 Illustrating a filter expansion.

The general properties of the adaptation algorithms, which we shall discuss presently, arise mainly from the fact that the output depends linearly on the parameters g_n. Therefore, our discussion will apply for any choice of functions $w_n(t)$ (although, of course, the rate of convergence will depend strongly on that choice). The general features will, in fact, be valid even if the $x_n(t)$'s are nonlinearly filtered versions of $x(t)$. This fact allows one to handle a class of nonlinear echo paths by the same methods. A proposal to do this was made (Thomas, 1971) but, to our knowledge, has never been used in echo cancellation for speech signals.

3.1. The Stochastic Gradient Algorithm

By far the most popular algorithm for adapting the filter structure of Eq. (1) to the echo path is the stochastic gradient algorithm. It is now popularly known as the least mean square (LMS) algorithm and was first introduced around 1960 for adaptive switching (Widrow and Hoff, 1960). It was first used for echo cancelers in the mid-1960s (Sondhi, 1967) and has, since then, found application in many other areas such as noise control and adaptive antenna design. The basic idea of the stochastic gradient algorithm is quite simple. Let $x(t)$ and $z(t)$ be the signals in branches L1, L2, respectively, of Fig. 3. Let us assume that

$$z(t) = y(t) + n(t), \tag{2}$$

where $y(t)$ is an echo of the input signal $x(t)$, and $n(t)$ is an added noise component which may include talker B's speech. We will assume that $y(t)$ has the representation given in Eq. (1) for some (unknown) coefficient vector $\mathbf{g}$. If this is not strictly true, then $n(t)$ will include the residual modeling error as well.

Suppose an estimate of the echo $\hat{y}(t) = \mathbf{h}'\mathbf{x}$, is formed with a trial coefficient vector $\mathbf{h}$. We wish to implement an algorithm to improve $\mathbf{h}$, i.e., bring it closer to the vector $\mathbf{g}$. Since $\mathbf{g}$ is unknown, we must evaluate the goodness of $\mathbf{h}$ indirectly. One measure of the performance of $\mathbf{h}$ is the error $e(t) = z(t) - \hat{y}(t)$. Since the objective is to make the vector $\mathbf{h}$ approximate the vector $\mathbf{g}$, one might search for that vector $\mathbf{h}$ which minimizes the expected value of the squared error $e^2(t)$. A natural way is to move $\mathbf{h}$ in the direction opposite to the gradient of this expected error. Thus one might try the algorithm

$$\begin{aligned} \frac{d\mathbf{h}}{dt} &= -\mu \nabla_h E\left[(z(t) - \hat{y}(t))^2\right] \\ &= -\mu \nabla_h E\left[e^2(t)\right], \end{aligned} \tag{3}$$

where μ is a parameter that controls the rate of change, E denotes mathematical expectation, and ∇_h is the gradient with respect to $\mathbf{h}$. Equation (3) is just one form of gradient search for the location of the minimum of a function of several variables. What the stochastic gradient algorithm does is to replace the *expected value* of the squared error by the *instantaneous value*. As we shall see, even such a crude estimate of the gradient is adequate, under certain reasonable conditions, to make **h** approach **g**.

The stochastic gradient version of Eq. (3) is

$$\begin{aligned}\frac{d\mathbf{h}}{dt} &= -\mu\nabla_h\left[e^2(t)\right]\\ &= -2\mu e(t)\nabla_h\left[e(t)\right]\\ &= 2\mu e\mathbf{x}. \end{aligned} \tag{4}$$

Figure 5 illustrates the block diagram of a circuit to implement the adaptation according to Eq. (4). The circuit shows an analog implementation. Most current implementations are digital and are obtained by sampling all the functions at the appropriate (Nyquist) rate and replacing the derivative by a first difference. We will, however, stay with the analog representation in this section. This is partly for historical reasons [the earliest echo canceler

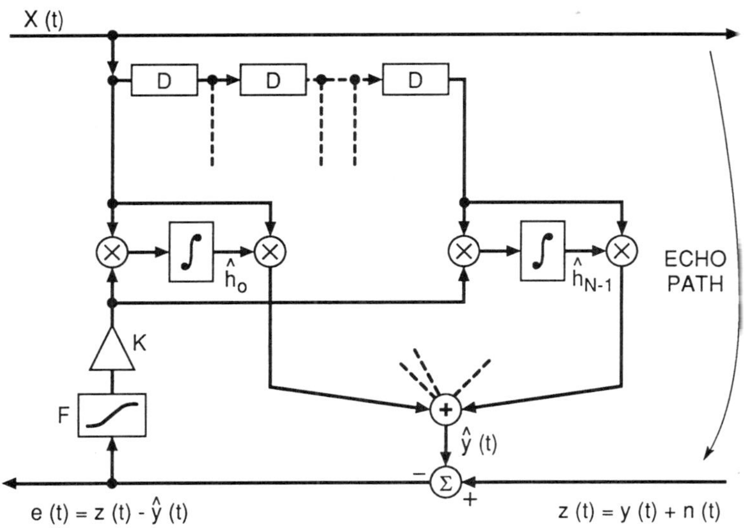

Figure 5 A stochastic gradient echo canceler.

was implemented as an analog device (Sondhi, 1967)] but also because the basic properties are easiest to describe in the continuous version.

The circuit of Fig. 5 includes a function F which equals the identity function when implementing Eq. (4). The introduction of F allows one to handle a more general criterion than the squared error. For instance if the expectation of the *magnitude* of the error is to be minimized, then $F(.) = \text{sign}(.)$ must be chosen. This choice of F has been used in some implementations of the algorithm.

3.1.1. *Convergence in the Ideal Case*

Suppose first that the echo path is perfectly stationary, the model represents the echo path exactly, and there is no noise or interrupting speech. Under these ideal conditions we have $z(t) = \mathbf{g}'\mathbf{x}(t)$, and the error $e(t)$ is given by

$$e(t) = (\mathbf{g} - \mathbf{h})'\mathbf{x} \\ = \mathbf{r}'\mathbf{x}, \tag{5}$$

where $\mathbf{r}$ is the misalignment vector. Since $\mathbf{g}$ is assumed constant, the time derivatives of $\mathbf{h}$ and $-\mathbf{r}$ are identical. Therefore Eq. (4) may be rewritten as

$$\frac{d\mathbf{r}}{dt} = -2\mu e\mathbf{x}. \tag{6}$$

Premultiplying both sides of Eq. (6) by $2\mathbf{r}'$, and noting that $2\mathbf{r}'(d\mathbf{r}/dt) = d(\mathbf{r}'\mathbf{r})/dt$, we get

$$\frac{d}{dt}\mathbf{r}'\mathbf{r} = -4\mu e^2(t), \tag{7}$$

which shows that the length of the misalignment vector $\mathbf{r}$ is nonincreasing. It is strictly decreasing as long as there is an uncanceled echo. Another important piece of information that can be gathered from Eq. (7) is that $e^2(t)$ eventually goes to zero. This is seen by integrating both sides of Eq. (7) with respect to t from 0 to T, which yields

$$\mathbf{r}'\mathbf{r}\,|_{t=0} - \mathbf{r}'\mathbf{r}\,|_{t=T} = \int_0^T e^2(t)\,dt. \tag{8}$$

Since the left-hand side is a finite number ($\leq$ the initial value of $\mathbf{r}'\mathbf{r}$) it follows that $e^2(T)$ must, in the limit, go to zero.

We cannot, however, be satisfied with the error going to zero. What we require is that the misalignment vector $\mathbf{r}$ should go to zero. Unfortunately, this is not provable even in the ideal situation considered in this section without imposing conditions on the input signal $x(t)$. The reason is that $e(t) = 0$ does not imply that $\mathbf{r} = 0$, but only that $\mathbf{r}$ is orthogonal to $\mathbf{x}$.

Sufficient conditions for the convergence of $\mathbf{r}$ to zero are derived in Sondhi and Mitra (1976) and we will not discuss them here. However, intuitively speaking, the conditions assure that the time-varying vector $\mathbf{x}(t)$ does not stay confined to a subspace of dimension less than N for too long; i.e., $\mathbf{x}$ should evolve in time in such a way as to cover the entire n-dimensional space. Even when the above "mixing" condition is satisfied, it is a difficult matter to get accurate estimates of the rate of convergence of $\mathbf{r}$. Suppose, for instance, that $x(t)$ is a member of a stationary ergodic process. One would expect that in this case the expected convergence rate could easily be computed. This is not the case. If, for instance, the expectation of both sides of Eq. (7) is taken, it does not help because the right-hand side depends on $\mathbf{r}$ itself. However, if μ is very small, one can assume that on the right-hand side $\mathbf{r}$ and $\mathbf{x}$ are independent. (For small μ, $\mathbf{r}$ changes slowly, and one may assume that the expectation on the $\mathbf{x}$ ensemble, E_x, can be taken with $\mathbf{r}$ assumed quasi-constant.) Under this assumption we see that

$$\begin{aligned} E_x\left(e^2(t)\right) &= E_x(\mathbf{r}'\mathbf{x}\mathbf{x}'\mathbf{r}) \\ &= \mathbf{r}'\mathbf{R}\mathbf{r}, \end{aligned} \tag{9}$$

where $\mathbf{R}$ is the correlation matrix of $\mathbf{x}$. Then the expected value of Eq. (7) gives the exponentially decaying upper and lower bounds

$$\exp(-4\mu\lambda_{\max}t) \leq \mathbf{r}'\mathbf{r} \leq \exp(-4\lambda_{\min}t). \tag{10}$$

Fortunately, for convergence rates of interest in general, these bounds are useful. Nevertheless, it is important to remember that the bounds are *not* valid for large μ. For instance, if $\lambda_{\min} > 0$, the upper bound shown above implies that $\mathbf{r}$ can be made to go to zero as fast as desired by merely increasing μ. This just is not the case. In fact, the following simple argument shows that the convergence rate must start *decreasing* when μ is *increased* beyond a certain value. Note from Eq. (6) that $\mathbf{r}$ changes in a direction such as to make it more orthogonal to $\mathbf{x}$. If μ is so large that $\mathbf{r}$ can change must faster than $\mathbf{x}$, it is intuitively clear that $\mathbf{r}$ rapidly becomes perpendicular to $\mathbf{x}$ and from here on does not change in length appreciably. (If $\mathbf{x}$ were a strictly constant vector, $\mathbf{r}$ would not change at all once it became perpendicular to $\mathbf{x}$).

The argument of the last paragraph shows that the convergence rate goes to zero as $\mu \to \infty$, and it obviously goes to zero as $\mu \to 0$. Thus there is some optimum value of μ which gives the most rapid convergence. There is no known way to derive this optimum even for a simple (e.g., stationary ergodic) input signal $x(t)$, let alone a speech signal. A good setting can only be found experimentally. However, some theoretically derived bounds, and a more rigorous derivation of the intuitive arguments presented above, may be found in Sondhi and Mitra (1976).

Although the convergence rates are difficult to estimate, it is clear from Eq. (10) that the convergence rate can fluctuate quite a lot if the spread of eigenvalues of the correlation matrix **R** is large. To reduce these fluctuations one would ideally want to "whiten" the speech signal (i.e., make all the eigenvalues equal and constant). Now speech is a nonstationary signal whose spectral properties change much more rapidly than does the echo path. However, one source of variability of the eigenvalues can be eliminated rather easily. This is the variability due to the change in signal level. Since the eigenvalues are proportional to the variance (or power) of the input signal, this objective can be accomplished by dividing the right-hand side of Eq. (4) by a local estimate of power. One simple way is to modify Eq. (4) to

$$\frac{d\mathbf{h}}{dt} = -2\mu \frac{e}{\mathbf{x}'\mathbf{x}} \mathbf{x}. \tag{11}$$

Except for experimental apparatus, all line echo cancelers in use today implement a time-discrete version of Eq. (11); that is,

$$\mathbf{h}_{n+1} = \mathbf{h}_n - K \frac{e_n}{\mathbf{x}_n'\mathbf{x}_n} \mathbf{x}_n, \tag{12}$$

where K is a new constant, and the subscript n indicates the value of a quantity at time t equal to n times the sampling interval. Because of the division by the input power, this algorithm is called the *normalized* LMS (NLMS) algorithm.

3.1.2. *Convergence in the Nonideal Case*

The convergence process, in practice, is even more complicated than described in the previous section. Detailed discussion of the nonideal case is beyond the scope of this chapter. However, in Sondhi and Mitra (1976) it is shown that under essentially the same restrictions on $x(t)$, theoretical bounds can be derived in the nonideal case as well. If the only perturbation is an additive noise, the vector **r** converges to lie within a sphere around the origin whose radius is proportional to the root-mean-square (rms) value of the noise. If the echo path is not constant, the radius of the sphere is also proportional to the rate of change of the impulse response.

The most severe situation arises during intervals of "double talking" (i.e., intervals during which the speech from both speakers, A and B, is present simultaneously at the echo canceler). If the echo canceler has converged to a small misalignment, the interfering speech signal from B can be much louder than the uncanceled echo and can completely misalign the canceler in a very short time. About the only effective way of dealing with this problem is to use a system similar to the echo suppressor (see Section 2.1) to detect the occurrence of double talking. However, instead of breaking the return path, just the adaptation loop is temporarily disabled during these intervals.

3.2. Other Algorithms

Before turning to a discussion of acoustic echo cancelers, let us briefly discuss three algorithms which attempt to improve on the simple stochastic gradient algorithm discussed so far. They are (1) a canceler based on two echo path models, (2) the least-squares (LS) algorithm, and (3) the recursive least-squares (RLS) algorithm. None of these is in current use for line (or acoustic) echo cancellation, although prototypes have been implemented. Algorithm (1) has not found wide application mainly because of the difficulty in designing a decision algorithm (see below) needed for its implementation. As for the LS and RLS algorithms, the main reason is the added computational complexity. In view of the rapid advancement of digital technology, these and other more complex algorithms will, no doubt, be used in the future.

3.2.1. Two Echo Path Models

In the discussion of the stochastic gradient algorithm we defined the error signal $e(t) = \mathbf{h}'\mathbf{x} - z$, which was used to control the adaptation of the canceling filter $\mathbf{h}$. In the usual implementation of the echo canceler, the same error signal is the one sent to the remote speaker. However, this is not imperative. The returned signal could be derived differently, say as $\tilde{e} = \tilde{\mathbf{h}}\mathbf{x} - z$, where the filter $\tilde{\mathbf{h}}$ is derived from but not identical to $\mathbf{h}$. Ochiai et al. (1977) argued that this observation can be used to advantage. Since the gradient used in the LMS algorithm is only a very crude estimate of the gradient of the mean squared error, every adaptive step does not necessarily improve $\mathbf{h}$. Their suggestion is to monitor both e and $\tilde{e}$ and to copy the coefficients of $\mathbf{h}$ into $\tilde{\mathbf{h}}$ whenever $\mathbf{h}$ performs consistently better than $\tilde{\mathbf{h}}$ over some specified time interval. The decision on when to transfer coefficients may be based on observation of e, $\tilde{e}$, $\mathbf{h}$, $\tilde{\mathbf{h}}$, $\mathbf{x}$, and z. It appears that although the method seemed promising in laboratory experiments, a robust decision mechanism was not found. Nevertheless, it is an intriguing idea which might become useful if such an algorithm can be found.

3.2.2. The Least-Squares Algorithm

In Section 3.1 we noted that the impulse response $\mathbf{h}$ can be estimated by minimizing the expectation of the squared error e^2. The stochastic gradient algorithm sidesteps the problem of estimating the *expected* value of e^2 by taking an incremental step in the direction which reduces its *instantaneous* value. Instead of this, the LS algorithm minimizes a deterministic approximation to the expected value. Specifically, considering the time-discrete formulation, the LS algorithm computes the vector **h** which minimizes the arithmetic mean of e_m^2 for some range of sampling instants m. This is equivalent to minimizing

$$E = \sum_{m=T-M+1}^{T} e_m^2. \tag{13}$$

Upon substituting for e in terms of **h** and **x**, the problem reduces to minimizing

$$E = \mathbf{h}'\mathbf{X}'\mathbf{X}\mathbf{h} - 2\mathbf{b}'\mathbf{h} + \sum_{T-M+1}^{T} z_m^2. \tag{14}$$

Here **X** is an $M \times N$ matrix whose ith row $(1 \leq i \leq M)$ is $\mathbf{x}'_{T-M+i}$, and the vector $\mathbf{b} = \sum_{T-M+1}^{T} z_m \mathbf{x}_m$. Setting the gradient of E in Eq. (14) to zero shows that **h** is the solution of

$$\mathbf{X}'\mathbf{X}\mathbf{h} = \mathbf{b}, \tag{15}$$

and can be computed by a single matrix inversion. Since the matrix to be inverted is of size $N \times N$, its inversion would ordinarily require $O(N^3)$ computations. (Recall that N is the length of the vector **h**.) However, if the adaptive structure is a transversal filter, then the matrix **X** is Toeplitz (i.e., all entries on any diagonal are identical). Taking advantage of this property, the inversion can be performed in $O(N^2)$ operations (Marple, 1981).

The solution depends on both T and M. The matrix $\mathbf{X}'\mathbf{X}$ is invertible if and only if **X** has independent columns. If this is not the case, then the solution is not unique, and a pseudoinverse of $\mathbf{X}'\mathbf{X}$ may be used to select the minimum norm solution. If the input and the echo path were stationary, the dependence on T would be eliminated, and a single matrix inversion would be required. However, because of the time variation of **h**, the impulse response must be updated as often as the available hardware allows.

3.2.3. *The Recursive Least-Squares Algorithm*

The least-squares algorithm is a *block* processing algorithm. An optimum estimate of **h** is derived from a block of data (of length M in the above description). This optimum estimate is assumed to be valid until the next block of data is processed to give a new estimate of **h**, and so on. There is an alternative algorithm in which an optimal estimate of **h** is obtained recursively at *every* time instant. The algorithm is a deterministic version of the Kalman filter.

At every instant, the estimated **h** minimizes a weighted sum of the squared errors at all past instants of time. In order to be able to track slowly varying impulse responses, the weighting is chosen such that errors in the remote past do not affect the current estimate. Recursive algorithms can be derived for several weighting functions which achieve this objective. One convenient error measure is

$$E(T) = \sum_{-\infty}^{T} \lambda^{T-n} e_n^2 \tag{16}$$

with λ chosen in the range $0 < \lambda < 1$. The value chosen for λ determines the effective duration of the past input that is used to derive the current estimate of **h**.

In terms of x and **h** Eq. (16) can be rewritten as

$$E(T) = \mathbf{h}'\mathbf{\Phi}_T\mathbf{h} - 2\mathbf{b}_T'\mathbf{h} + \sum_{-\infty}^{T} \lambda^{T-n} z_n^2. \tag{17}$$

Here the matrix $\mathbf{\Phi}_T = \sum_{-\infty}^{T} \lambda^{T-n}\mathbf{x}_n\mathbf{x}_n'$ and the vector $\mathbf{b}_T = \sum_{-\infty}^{T} \lambda^{T-n} z_n\mathbf{x}_n$. Thus at time T the optimal impulse response vector $\mathbf{h}_T$ is the solution of

$$\mathbf{\Phi}_T\mathbf{h}_T = \mathbf{b}_T. \tag{18}$$

From the definitions of $\mathbf{\Phi}_T$ and $\mathbf{b}_T$ it is straightforward to show that they satisfy the recursions

$$\mathbf{\Phi}_T = \lambda\mathbf{\Phi}_{T-1} + \mathbf{x}_T\mathbf{x}_T' \tag{19}$$

and

$$\mathbf{b}_T = \mathbf{b}_{T-1} + s_T\mathbf{x}_T. \tag{20}$$

Because of the recursion in Eq. (19), $\mathbf{\Phi}_T^{-1}$ can be omitted by updating $\mathbf{\Phi}_{T-1}^{-1}$ through use of the matrix inversion lemma (Haykin, 1986, p. 385). The optimal estimate $\mathbf{h}_T$ is thus obtained recursively from $\mathbf{h}_{T-1}$. The update algorithm is rather cumbersome, although simple in principle. We refer the reader to Haykin (1986, chapter 8) for the details.

Recursion based on Eqs. (19 and 20) requires $O(N^2)$ operations per iteration. Although much less than the $O(N^3)$ computations which would be required without the recursions, the computational load is still much more than the $2N$ multiplications per iteration required by the LMS algorithm. The advantage gained by the extra computations is the highly improved convergence.

As in the case of the LS algorithm, the computational requirements can be reduced dramatically if the adaptive structure is a transversal filter. Algorithms that accomplish this, known as fast RLS or fast transversal filter algorithms, have been developed by Ljung et al. (1978), Falconer and Ljung (1978), Cioffi and Kailath (1984), and others. They achieve the good convergence properties of the RLS algorithm at a computational requirement which grows linearly with N.

The main limitation of these fast algorithms is that they tend to be numerically unstable unless multiple precision arithmetic is used. In practical implementations the algorithms have to be reset periodically. Nevertheless, as mentioned in Section 4.1.1 below, the fast transversal filter algorithm has recently been successfully implemented for subband acoustic cancellation.

4. ACOUSTIC ECHO CANCELLATION

The problem of canceling acoustical echoes in hands-free telephony and teleconferencing differs from the cancellation of line echoes mainly because of the different nature of the echo paths. Instead of the mismatch of the hybrid, a loudspeaker-room-microphone system needs to be modeled in these applications (Fig. 6).

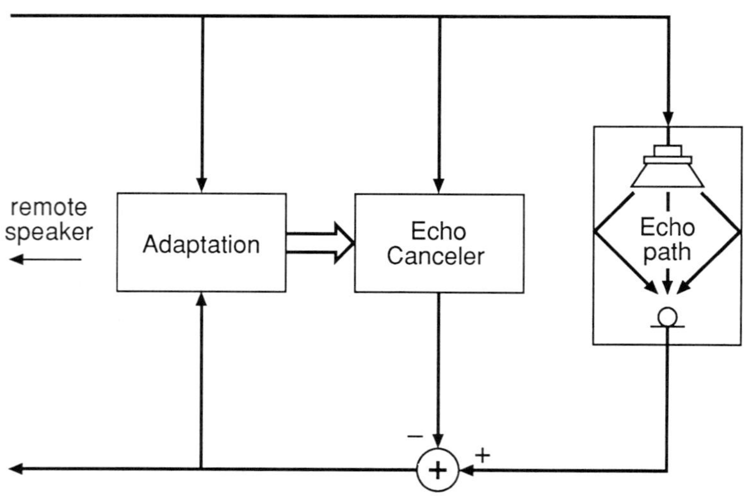

Figure 6 Cancellation of acoustic echoes.

Comparing typical impulse responses of these two kinds of echo paths, it becomes obvious that acoustic echo cancellation is a far more challenging task than line echo cancellation. The duration of the impulse response of the acoustic echo path is usually several times longer (100 to 400 ms) and it may change rapidly (e.g., due to an opening door or a moving person) at any time. To achieve even a modest improvement (say a misalignment 20 dB below the uncanceled room response) requires a transversal filter with over 1000 taps at 8 kHz for a typical office (Schultheiss et al., 1984). Some commercial products foresee requiring as many as 4000 taps at a 16-kHz sampling rate. Perhaps the number of parameters could be reduced significantly by modeling the acoustic echo path as a recursive, infinite impluse response (IIR) filter. However, no one has demonstrated this so far. Therefore, adaptive transversal finite impluse response (FIR) filters still are the ubiquitous choice.

For line echo cancellation, it is sometimes possible to switch to a slower-converging tracking mode after an initial rapid convergence (with, for example, a training signal). This is because for certain circuits it is possible to assume that the echo path varies only very slowly. As mentioned above, this is not a good assumption for acoustic echo cancellation, so one cannot switch to a slowly converging algorithm even in the tracking mode.

We see, therefore, that acoustic echo cancelers require more computing power than line cancelers for two reasons: first, just because of the length of the impulse response to be compensated, and second, because faster-converging algorithms are desirable. Current products (and prototypes) only partly meet these requirements, mainly because they use LMS-type adaptation algorithms. Such algorithms are known to perform badly for long impulse responses and with speech as the input signal (Haykin, 1986).

As mentioned in Section 3.2, LS and RLS algorithms can increase convergence speed. They have been successfully used to implement line echo cancelers. However, they are still infeasible for acoustic echo cancellation because of the long impulse responses involved. Other methods have therefore been explored to improve the convergence speed of LMS-type algorithms. One direction aims at "whitening" the speech signal for the adaptation. One way to do this is to use the LPC inverse filter, and this leads to faster convergence with little increase in complexity (Schultheiss, 1987). A simpler version of this idea has been proposed, which is essentially a continuously updated first-order linear predictor (Yasukawa et al., 1989). Another direction foresees nonuniform step size factors for the different taps of the LMS-adapted FIR filter. A general method for doing this, not specifically tailored for the acoustic echo cancellation problem, was suggested in Harris et al. (1986). Another approach exploits the structure of the impulse response of the acoustic echo path and assigns different step sizes to different sections of the echo path impulse response (Koizumi and Makino, 1989). The idea is that, ideally, the impulse response samples with large values are adapted with large step size while those with small values get a small step size. This should result in a faster overall convergence. Obviously, the efficiency of this method depends highly on the a priori knowledge concerning the current echo path impulse response and the ability to adjust the step size accordingly.

While all these approaches attempt to improve convergence speed without adding too much computational complexity, the subband concept, which is discussed in more detail in the following, reduces computational complexity while at the same time providing more favorable circumstances for fast convergence.

4.1. Subband Approach

Subband structures were first proposed in 1984 by Furukawa (1984) for teleconferencing purposes and by Kellermann (1984, 1985) as a general framework for acoustic echo cancellation. The fundamental structure is depicted in Fig. 7. With reference to that figure, it is seen that the input and output signals of the echo path are passed through identical analysis filter banks A producing vectors of N subband signals which are sampled at a re-

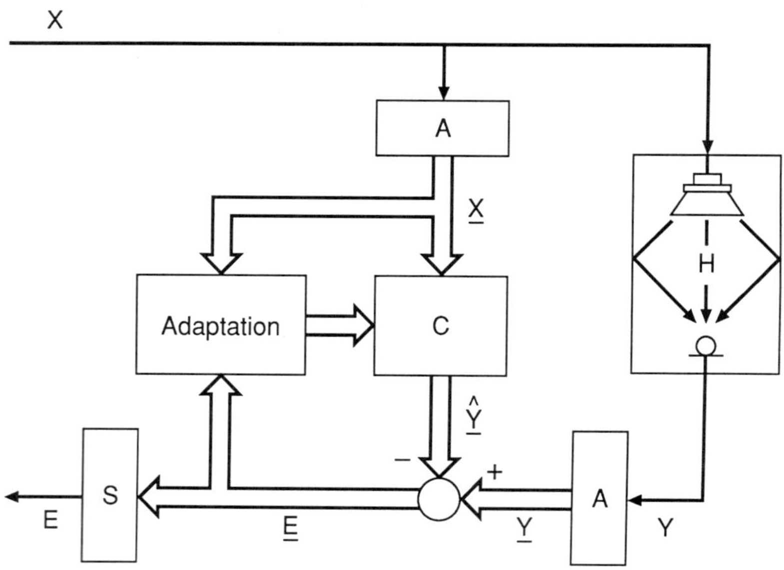

Figure 7 Block diagram of a subband echo canceler.

duced rate. The cancellation structure C forms a vector of subband signals $\hat{\mathbf{Y}}$ to approximate the corresponding subband echo signals **Y**. The resulting subband errors **E** are passed through a synthesis system S to give a full band signal E, which is then transmitted back to the remote speaker. The adaptation block utilizes the vector of subband error signals and input signals to adjust the input-output characteristics of the cancellation unit so as to drive the vector of error signals toward zero.

An alternative configuration of the subband canceler is depicted in Fig. 8. The difference between this and the system of Fig. 7 is that the adaptation unit works with the full band input and error signals. The advantages and disadvantages of these two configurations will be discussed in Sections 4.1.1 and 4.1.2.

Intuitively, the promise of the subband structure is twofold: On the one hand, the computational complexity is reduced because of the downsampling; and on the other hand, the convergence of the adaptation algorithms is expected to speed up for LMS-type algorithms because of the decomposition of the input speech signal. To illustrate the advantage in computational complexity consider the case in which the canceler has N parallel adaptive filters and all subband signals are downsampled by the same factor R. Assume that

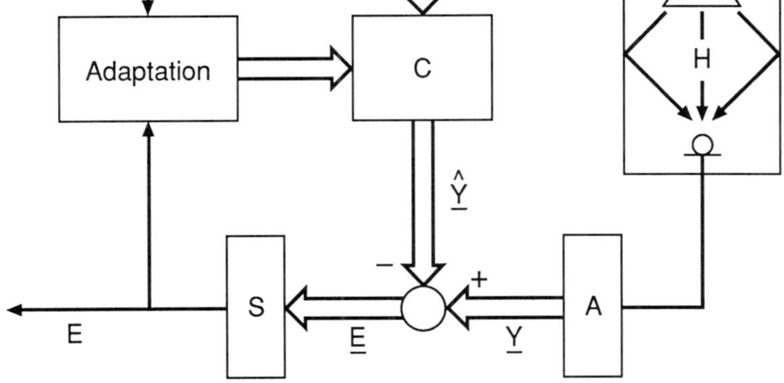

Figure 8 Synthesis-dependent subband canceler.

the impulse response of each of the subbands has the same duration as the full band impulse response. Then the number of coefficients needed for its representation is fewer by a factor R compared to the full band representation because of the downsampling. Further, filtering and adaptation of the subband cancelers are performed at the reduced sampling rate. Therefore, the computational complexity (measured as computations per second) for one subband canceler is $1/R^2$ of that of the full band canceler. Taking into account all N subbands, a gain factor of approximately R^2/N can be expected, assuming that the computational load for the analysis and synthesis systems is comparably small. The gain in computational complexity can be exploited in several ways: the overall system bandwidth or the duration of the impulse response to be modeled can be increased, more complex adaptation algorithms can be employed, or, most obviously, hardware can be saved to reduce cost. The price paid for these advantages can also be inferred from a comparison of Figs. 6 and 7: in the subband canceler the microphone signal is delayed before it is sent to the far end by the group delay of the cascade of analysis and synthesis filters. A suggestion for dealing with this problem of delay will be described in Section 4.1.2.

Assuming that all systems involved are linear and considering a fixed but arbitrary instant in time, a matrix notation in the z-domain can be used to describe the subband structure conveniently (Kellermann, 1988). For analysis and synthesis we allow the general class of systems that can be represented as filter banks with a time-invariant filtering part, and downsampling or upsampling, respectively (see, e.g., Vetterli, 1987). Although extension of the description to nonuniform downsampling factors is possible (Kellermann, 1989a), we use here the same downsampling factor R for all subbands for the sake of simplicity. Then the general cancellation condition in the absence of local signals,

$$E(z) \equiv 0, \tag{21}$$

implies that (Kellermann, 1988)

$$S(z)C(z^R)A(z)X(z) = S(z)A(z)H(z)X(z), \tag{22}$$

where the matrices $A(z)$, $S(z)$, $H(z)$, and $C(z^R)$ correspond to analysis, synthesis, echo path, and cancellation unit, respectively, and $X(z)$ corresponds to the input signal. For signal-independent solutions we must have

$$S(z)C(z^R)A(z) = S(z)A(z)H(z). \tag{23}$$

Solutions satisfying Eq. (23) will be called *synthesis-dependent* because, in general, the fact that the full band error is zero does not imply that the vector of subband errors is also zero. The latter condition is met by the subset of solutions of Eq. (23) which satisfy the stricter condition

$$C(z^R)A(z) = A(z)H(z). \tag{24}$$

Solutions of Eq. (24) will be called *synthesis-independent.*

We turn now to a discussion of cancelers based on these two types of solutions.

4.1.1. Synthesis-Dependent Solutions to the Cancellation Condition

Cancelers based on such solutions have the general configuration of Fig. 8. The main advantage of this configuration, as we shall see presently, is that it allows very efficient implementations when the impulse responses are long. A major shortcoming of this configuration is that the error information is available to the adaptation algorithm only after having been delayed by the synthesis filter bank. Thus the adaptation feedback loop (cancellation unit–synthesis–adaptation–cancellation unit) has an inherent delay in it. This has a deleterious effect on convergence, especially when the echo path response is changing rapidly. Another disadvantage of this configuration is apparent when we observe that the cancellation unit is a set of N parallel adaptive filters each with a single input and a single output, and the adaptation algorithm must adapt all these filters on the basis of a common error signal. The components of the error signal outside the frequency range of each filter thus act as a noise on the adaptation process for that filter. As we shall see in the next section, these two disadvantages are overcome in the case of the synthesis-independent configuration of Fig. 7.

In principle there is a variety of synthesis-dependent solutions of Eq. (23) (Kellermann, 1989b). However, because of a lack of suitable adaptation algorithms, only a few solutions have been considered so far. These are all based on fast convolution which basically utilizes a fast Fourier transform (FFT) as the analysis system and an inverse FFT for synthesis. Each of the subband cancelers consists of a single, generally complex-valued, multiplier. In order to achieve low computational complexity, the downsampling rate R can be chosen as large as $\lfloor (N + 1)/2 \rfloor$ (Kellermann, 1988). The application of this structure to acoustic echo cancellation was first proposed by Sommen et al. (1987). An adaptive algorithm suitable for this structure is the *frequency domain adaptive filter* (FDAF), which had been proposed much earlier in other applications (Ferrara, 1980; Mansour and Gray, 1982; Clark et al., 1983). The FDAF algorithm is commonly viewed as discrete Fourier transform (DFT) domain implementation of the LMS algorithm and provides a dramatic decrease in computational requirement. For an impulse response of length L samples, the number of real multiplications required for one output sample is only on the order of $2\log_2(L)$, compared to $2L$ for the LMS (Clark et al., 1981). For $L = 1000$ this is a hundredfold improvement. However, the delays which are introduced into the signal path and into the adaptation loop, due to the block processing of the FFT, are both as long as the echo path impulse response which is modeled. This makes this concept inapplicable when usual signal delay constraints for telephone systems have to be met. Also, as mentioned above, such a large delay in the adaptation loop makes it difficult to track rapid variations in the impulse response.

The *frequency bin adaptive filter* (FBAF) was proposed by Asharif et al. (1987) as a method of reducing this delay. The idea is to take the FFT on a shorter block, but make the canceler in each subband an FIR filter rather than a single multiplier. As shown in Kellermann (1988), however, the optimal filter coefficients of the subband filters are in general not unique. Thus the proposed adaptation algorithm exhibits impaired convergence because the error surface has more than one minimum. As for the delay in the adaptation loop, this method reduces it considerably compared to the FBAF method but, of course, cannot completely eliminate it. Also, the algorithm is not as computationally efficient as the FDAF because of the reduced downsampling factor R necessitated by the shorter transform length.

A similar approach called *double convolution* has been proposed by Xu and Grenier (1989). It avoids the nonuniqueness of the approach of Asharif et al. (1987) but, of course, shares the other drawbacks of the FBAF mentioned above.

4.1.2. *Synthesis-Independent Solutions*

Although it has to meet a stricter requirement, the class of synthesis-independent solutions to the cancellation problem (i.e., the class with the configuration of Fig. 7) has attracted far more attention than the synthesis-dependent class. This may partly be due to the fact that the basic idea of the subband concept seems to be more directly applicable to this class. Also, as is clear from Fig. 7, the problem of delay in the adaptation loop is avoided in implementing this class of solutions. Much work has been carried out to develop cancelers using only two or four bands (Furukawa, 1984; Itoh et al. 1987; Yasukawa et al., 1987, 1989). However, other authors have tried to explore the concept to a greater extent (Gilloire, 1987; Kellermann, 1988, 1989a, 1989b; Chen et al., 1988; Gilloire and Vetterli, 1988; Hätty, 1989).

Let us consider some of the implications of the solution to the cancellation condition Eq. (24). As before, assuming that the cancellation unit consists of a set of N parallel (adaptive) filters, Eq. (24) has analytically well-defined solutions for given analysis-synthesis systems as derived in Kellermann (1988, 1989a). There it is shown that solutions exist if and only if the analysis produces aliasing-free subband signals X_ν. Obviously, this requires bandpass filters with infinite stopband attenuation. Also, the downsampling factor R must be chosen such that the passband and transition region of the modulated versions of the analysis bandpass do not overlap in the frequency domain after downsampling. If this requirement is met, each subband canceler has to model an ideally bandpass-filtered and downsampled version of the full band echo path impulse response. The passband of the ideal bandpass corresponds to the transition and passband region of the corresponding analysis bandpass. This solution may be called a *frequency subband solution*, as each subband covers its own portion of the frequency range and does not interfere with others. However, this solution can obviously not be realized in the strict sense. First, bandpass filters with infinite stopband attenuation can only be approximated, and second, filtering a finite impulse response with an ideal bandpass leads to an impulse response which extends from $-\infty$ to $+\infty$ in the time domain and, therefore, is not realizable either. Thus, two approximation problems are linked to the realization of the frequency-subband concept, and it is important to consider the extent to which the approximation errors impair its usefulness. This is briefly outlined in the sequel, where the minimum misalignment that can be obtained for a given frequency subband configuration is considered (for further details see Kellermann, 1989a or 1989b).

Approximation of the Frequency Subband Solution Let us consider the approximation of Eq. (24) that is possible with ideal settings of the cancelers. In the next section we will consider the adaptation aspects. We will use a quadratic distance measure to estimate system misalignment. This may be interpreted as the transmission loss factor of the corresponding system with white noise as the input signal. The misalignment of the whole structure has two components: the misalignment due to the residual aliasing within the subband signals and the misalignment of the subband cancelers due to the truncation of their impulse responses. Numerical investigation shows that the overall system misalignment is determined by the larger of these two contributions until the overall system misalignment reaches a certain threshold (e.g., about -25 to -30 dB for $N = 16$ and $R = 12$) (Kellermann, 1989b). For smaller misalignment, the truncation errors of the subband cancelers have little impact and the misalignment of the whole system is mainly determined by the aliasing due to the finite stopband attenuation of the analysis filter. The reason for this becomes obvious from the following consideration of the spectral distribution of the truncation error. If only a small number of coefficients is used to approximate the response of the ideal subband canceler, the truncation will cause more or less uniformly distributed deviations from the ideal frequency response of the subband canceler. Beyond a certain number of coefficients the truncation error in the inner region of the subband will be small and the misalignment of the subband canceler will be determined mainly by the amount of error concentrated at the band edges, where the infinitely steep slope has to be approximated. This contribution to the misalignment of the subband canceler, however, has little influence on the misalignment of the overall system. This is because the misalignment due to the subband canceler is heavily weighted down at the band edges by the analysis and synthesis bandpass filters. Thus, once a certain concentration of the truncation error toward the band edges has been achieved, there is little to be gained by increasing the length of the subband canceler. This observation is important for efficient design of the canceler. For instance, the delay needed to model "noncasual" coefficients of the subband canceler impulse response can be kept small.

Once the truncation error has been reduced to a low value, it is the stopband attenuation that controls the residual misalignment. The stopband attenuation must be chosen sufficiently high to keep the misalignment due to aliasing small. A conservative rule of thumb for $N \leq 64$ is

$$\begin{aligned}&\text{overall misalignment [dB]}\\&\qquad = \text{stopband amplitude [dB]} + ((1 + 3 \times \log_2(N)) + 5)\ \text{dB}\end{aligned}$$

The effect of the aliasing within the subband signals also explains the problems with the otherwise attractive choice of critical sampling ($R = N$). The simple two- and four-band systems implemented so far employ analysis filters which introduce only the stopband aliasing described above. However, they thereby create gaps in the overall frequency response of the analysis-synthesis system. For larger numbers of subbands the number of gaps would increase and would not be tolerable from the point of view of speech quality. Filter bank designs methods that provide perfect reconstruction using critical sampling do not exhibit these gaps but they cause at least transition regions of the analysis filters to be aliased in the subband signals (see, e.g., quadrature mirror filter (QMF) banks). This causes severe misalignment of the overall subband structure. [This was observed—although not explained—by Gilloire (1987) with a canceler implemented with QMF banks.] For these reasons it is necessary to choose $R < N$ to decrease the aliasing of the transition regions, even though this sacrifices some computational efficiency.

Adaptation of the Frequency Subband Structure As pointed out earlier, the canceler in each subband is essentially independent of the others. Therefore, the frequency subband structure can utilize any of the adaptation algorithms developed for full band cancelers. We will discuss a few of them, emphasizing the differences compared to their application in full band systems.

To date, the normalized LMS algorithm is the most widespread adaptation algorithm for both full band and subband systems. For this algorithm, it has been observed that the convergence is faster in the subband canceler than with the full band implementation. As discussed in Section 3.1, the convergence speed of the LMS algorithm directly depends on the eigenvalue spread of the autocorrelation matrix of the input signal. Therefore, it is often concluded that the eigenvalue spread of the subband signals must be smaller than that of the full band signal. However, examining the actual eigenvalues shows that their spread is, in fact, larger for the subband signals than for the full band signal. This is because the subband signals are not really "whiter." The slopes of the frequency responses of the analysis filters cause notches in the subband spectra at the band edges, thus creating some very small eigenvalues. Indeed, the convergence behavior of the subband canceler itself is in general not better for a subband signal than for the full band signal. The improvement which is observed for the subband structure as a whole is the result of the same masking property that determines the effect of the truncation errors of the subband cancelers: as mentioned above, for the overall system misalignment the band edges of the subband canceler spectra get little weight due to the characteristics of the analysis and synthesis filters. These parts of the subband canceler spectra are the same ones which are affected by the small eigenvalues of the subband signal and, therefore, correspond to the slower-converging modes. But as these modes have little influence on the overall system misalignment, the latter converges faster than a full-band system because the eigenvalue spread in the inner region of the subband spectra is indeed smaller than that of the full band signal. The increased convergence speed for "nonwhite" input signals is, therefore, the result of an only indirect "whitening" effect.

Skipping all the incremental improvements that are possible to speed up the normalized LMS algorithm (see the previous section on synthesis-dependent approaches) in the subband structure, let us mention briefly the most sophisticated adaptation algorithm used so far for acoustic echo cancellation. This is the RLS implementation as proposed by Hätty (1989). In this implementation the number of subbands is $N = 16$, the decimation rate is $R = 13$, and the subband filters have about 100 taps each. The adaptation algorithm is a modified version of the fast transversal filter algorithm of Cioffi and Kailath (1984) mentioned in Section 3.2. As discussed in that section, such algorithms need to be periodically reinitialized to guarantee numerical stability. To avoid sudden large increases of misalignment, the subband filters are reset one at a time. The computational complexity is only on the order of a full band implementation of the LMS algorithm, whereas the initial convergence speed is comparable to a full band RLS algorithm.

4.1.3. Extensions of the Frequency Subband Structure

There are at least two areas in which subband cancelers could be improved. First, the cancelers discussed above realize the cancellation condition of Eq. (24) with a bank of adaptive filters, in which aliasing is kept small by choosing a decimation rate R less than the number of filters N. The choice $R = N$ is not recommended for $N > 4$, because of the spectral gaps needed to avoid aliasing. In an attempt to realize this maximum decimation, Gilloire and Vetterli (1988) propose a structure in which adaptive cross-filters are used to cancel the influence of aliasing in each subband. However, it appears doubtful that this structure would have satisfactory convergence properties, except, perhaps, in some special cases. The main subband filter and the corresponding cross-filters have to use the same subband error signal to steer their respective adaptation by the LMS algorithm. As such, the adaptation algorithms for the different filters have no indication of the contribution of each to the total error. Therefore, it appears to be difficult to achieve stable and fast initial convergence. Also, the fact that additional cross-filters have to be adapted reduces the computational efficiency, and the gain over a slightly oversampled system (e.g., $R = 3N/4$) might be marginal.

Besides computational complexity, there is another very undesirable aspect of subband cancelers. That is the delay introduced in the path from the near end talker to the remote listener (the delay in the analysis and synthesis filters shown in the bottom portion of Fig. 7). This delay may be 10 to 20 ms. In many applications this much delay may not be tolerable.

An interesting variant of the subband canceler, which eliminates this delay, has been proposed by Chen et al. (1988). Their idea is depicted in Fig. 9. The analysis and synthesis filters are removed from the return path, and an extra synthesis filter bank is included in the path that generates the canceling signal. Notice that the delay in the return path has been eliminated; however, because of the delay in the adapted path, it is clear that the initial portion of the impulse response cannot be canceled. Therefore the additional full-band adaptive filter marked AF in Fig. 9, is necessary. With that filter included, the system is in principle able to cancel the echoes without introducing delay. Note, however, that the adaptation algorithm is now forced to work with the full band error signal. As mentioned in the discussion of Fig. 8 in Section 4.1.1, this structure has certain disadvantages. In particular, an in-

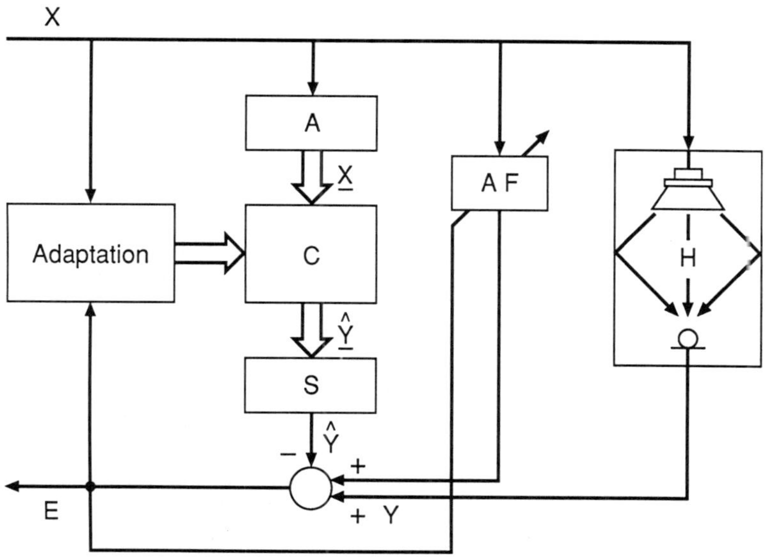

Figure 9 A proposal to eliminate delay in the return path of a subband canceler.

herent delay is introduced in the feedback loop. This might be a drawback in applications in which the impulse response is changing rapidly.

5. CONCLUDING REMARKS

When adaptive echo cancelers were first proposed about 25 years ago, many people expressed doubts about their economic feasibility and about the feasibility of cancelers with more than 50 to 100 adapted parameters. Advances in digital technology have proved these doubts to be unfounded. Since the appearance of the first VLSI implementation of cancelers in 1980, line echo cancelers have been ubiquitous on the telephone network. Cancelers based on similar principles have also found widespread use in data communication (although we have not dealt with that application in this chapter). The most modern application of voice echo cancelers is to the cancellation of acoustic echoes, e.g., for conference telephony. In the preceding section we have described several approaches to this problem. Hardware implementations based on these proposals have just begun to appear within the past year or so. Present technology allows such implementations for canceling impulse responses typical of small offices. Given the pace of development of digital technology, the next decade may well see general use of acoustic echo cancelers for offices and larger conference rooms.

REFERENCES

Asharif, M. R., Amano, F., Unagami, S., and Murano, K. (1987). Acoustic echo canceler based on frequency bin adaptive filtering (FBAF). *Proc. IEEE GlobeCom Conf.*, pp. 49.2.1–49.2.5.

Campanella, S. J., Suyderhoud, H. G., and Onufry, M. (1972). Analysis of an adaptive impulse response echo canceller. *Comsat Tech. Rev. 2*:1–36.

Chen, J., Bes, H., Vandewalle, J., and Janssens, P. (1988). A new structure for sub-band acoustic echo canceler. *Proc. IEEE Int. Conf. Acoustics, Speech, and Signal Processing*, pp. 2574–2577.

Cioffi, J. M., and Kailath, T. (1984). Fast, recursive-least-squares transversal filters for adaptive filtering. *IEEE Trans. Acoust., Speech, Signal Process. ASSP-32*:304–337.

Clark, G. A., Mitra, S. K., and Parker, S. R. (1981). Block implementation of adaptive digital filters. *IEEE Trans. Acoust., Speech, Signal Process. ASSP-29*:744–752.

Clark, G. A., Parker, S. R., and Mitra, S. K. (1983). A unified approach to time- and frequency-domain realization for FIR adaptive digital filters. *IEEE Trans. Acoust., Speech, Signal Process. ASSP-31*:1073–1083.

Demytko, N., and Mackechnie, L. K. (1973). A high speed digital adaptive echo canceller. *Austral. Telecomm. Rev. 7*:20–28.

Duttweiler, D. I., and Chen, Y. S. (1980). A single chip VLSI echo canceller. *Bell Syst. Tech. J. 59*:149–160.

Falconer, D. D., and Ljung, L. (1978). Application of fast Kalman estimation to adaptive equalization. *IEEE Trans. Commun. Com-26*:1439–1446.

Ferrara, E. R. (1980). Fast implementation of LMS adaptive filters. *IEEE Trans. Acoust., Speech, Signal Process. ASSP-28*:474–475.

Furukawa, I. (1984). A design of canceller of broad band acoustic echo. *Int. Teleconference Symposium*, Tokyo, pp. 1/8–8/8.

Gilloire, A. (1987). Experiments with sub-band acoustic echo cancellers for teleconferencing. *Proc. IEEE Int. Conf. Acoustics, Speech, and Signal Processing*, pp. 49.2-1–49.2-4.

Gilloire, A., and Vetterli, M. (1988). Adaptive filtering in sub-bands. *Proc. IEEE Int. Conf. Acoustics, Speech, and Signal Processing*, pp. 1572–1575.

Harris, R. W., Chabries, D. M., and Bishop, F. A. (1986). A variable step (VS) adaptive filter algorithm. *IEEE Trans. Acoust., Speech, Signal Process. ASSP-34*: 309–316.

Hätty, B. (1989). Block recursive least squares adaptive filters using multirate systems for cancellation of acoustical echoes. *IEEE Conf. Record of the ASSP Workshop on Application of Digital Signal Processing to Audio and Acoustics*, New Paltz, N.Y.

Haykin, S. (1986). *Adaptive Filter Theory*. Prentice-Hall, Englewood Cliffs, N.J.

Horna, O. A. (1977). Echo canceller with adaptive transversal filter utilizing pseudologarithmic coding. *Comsat Tech. Rev. 7*:393–428.

Itoh, Y., Maruyama, Y., Furuya, N., and Araseki, T. (1987). An acoustic echo canceller for teleconference. *Proc. Int. Conf. on Communications*, pp. 46.6-1–46.6-5.

Kato, Y., Chiba, S., Ishiguro, T., Sato, Y., Tajima, M., Ogihara, T., Campanella S. J., Suyderhoud, H. G., and Onufry, M. (1973). A digital adaptive echo canceller. *NEC Res. Dev. 31*:32–41.

Kellermann, W. (1984). Kompensation akustischer Echos in Frequenzteil-bändern. *Aachener Kolloquium 1984*, Aachen, FRG, pp. 322–325 (in German).

Kellermann, W. (1985). Kompensation akustischer Echos in Frequenzteil-bändern. *Frequenz 39* (7/8):209–215 (in German).

Kellermann, W. (1988). Analysis and design of multirate systems for cancellation of acoustical echoes. *Proc. IEEE Int. Conf. Acoustics, Speech, and Signal Processing*, pp. 2570–2573.

Kellermann, W. (1989a). Zur Nachbildung physikalischer Systeme durch parallelisierte digitale Ersatzsysteme im Hinblick auf die Kompensation akustischer Echos. Ph.d. thesis, Technische Hochschule Darmstadt, Darmstadt, FRG (in German).

Kellermann, W. (1989b). Some aspects of the frequency-subband approach to the cancellation of acoustical echoes. *IEEE Conf. Record ASSP Workshop on Application of Digital Signal Processing to Audio and Acoustics*, New Paltz, N.Y.

Koizumi, N., and Makino, S. (1989). Acoustic echo canceller algorithm based on room acoustic characteristics. *IEEE Conf. Record ASSP Workshop on Application of Digital Signal Processing to Audio and Acoustics*, New Paltz, N.Y.

Ljung, L., Morf, M., and Falconer, D. D. (1978). Fast calculation of gain matrices for recursive estimation schemes. *Int. J. Control 27*:1–19.

Mansour, D., and Gray, A. H., Jr. (1982). Unconstrained frequency domain adaptive filter. *IEEE Trans. Acoust., Speech, Signal Process. ASSP-30*:726–734.

Marple, S. (1981). Efficient least-squares FIR system identification. *IEEE Trans. Acoust., Speech, Signal Process. ASSP-29*:62–73.

Messerschmitt, D. G. (1984). Echo cancellation in speech and data transmission. *IEEE Trans. Special Topics Commun. COM-32*:283–297.

Ochiai, K., Araseki, T., and Ogihara, T. (1977). Echo canceller with two echo path models. *IEEE Trans. Commun. COM-25*:589–595.

Schultheiss, U. (1987). Über die Adaption eines Kompensators für akustischer Echos. Ph.D. thesis, Technische Hochschule Darmstadt, Darmstadt, FRG (in German).

Schultheiss, U., Becker, T., and Hänsler, E. (1984). Probleme bei der Kompensation akustischer Echos. *Frequenz 38*:143–148 (in German).

Sommen, P., Gerwen, P. J., Kotmans, H. J., and Janssen, A. J. E. M. (1987). Convergence analysis of a frequency domain adaptive filter with exponential power averaging and generalized window function. *IEEE Trans. Circuits Syst. 34*: 788–798.

Sondhi, M. M. (1967). An adaptive echo canceller. *Bell Syst. Tech. J. 46* (3):497–511.
Sondhi, M. M., and Berkley, D. A. (1980). Silencing echoes on the telephone network. *Proc. IEEE 68*:948–963.
Sondhi, M. M. and Mitra, D. (1976). New results on the performance of a well known class of adaptive filters. *Proc. IEEE 64*:1583–1597.
Thies, A. W., and Zmood, R. B. (1967). New ways of echo suppression. *Austral. Telecomm. Rev. 1*:14–19.
Thomas, E. J. (1971). An adaptive echo canceller in a nonideal environment (nonlinear or time variant). *Bell Syst. Tech. J. 50*:2779–2795.
Vetterli, M. (1987). A theory of multirate filter banks. *IEEE Trans. Acoust., Speech, Signal Process. ASSP-35*:356–372.
Widrow, B., and Hoff, M. E., Jr. (1960). Adaptive switching circuits. *IRE Wescon Conf. Rec.*, Part 4, pp. 96–104.
Xu, M., and Grenier, Y. (1989). Time-frequency domain adaptive filters. *Proc. IEEE Int. Conf. Acoustics, Speech, and Signal Processing*, pp. 1154–1157.
Yasukawa, H., Shimada, S., and Furukawa, I. (1987). Acoustic echo canceller with high speech quality. *Proc. IEEE Int. Conf. Acoustics, Speech, and Signal Processing*, pp. 49.8-1–49.8-4.
Yasukawa, H., Furukawa, I., and Ishiyama, Y. (1989). Acoustic echo control for high quality audio teleconferencing. *Proc. IEEE Int. Conf. Acoustics, Speech, and Signal Processing*, pp. 2041–2044.

12

Quality Assessment of Coded Speech

NOBUHIKO KITAWAKI NTT Telecommunication Networks Laboratories, Tokyo, Japan

1. INTRODUCTION

This chapter describes the speech quality assessment of waveform coding and analysis-synthesis. Waveform coding strives for facsimile reproduction of the signal waveform, while speech analysis-synthesis aims at reproducing a signal frequency spectrum, taking the human speech generation mechanism into account. Hybrid coding, which combines some of the advantages of both, is also presented. Since these new speech codings have progressed greatly, their introduction in such technologies as commercial telephone networks is under consideration. This introduction may take various forms: integrated service digital networks (ISDNs), public switched telephone networks (PSTNs), private networks, special service networks such as mobile radio networks, and voice storage services.

Reasonable speech quality assessment methodologies are required for the diagnostic assessment and fair comparison of coding algorithms, performance evaluation of a system equipped with coding, and optimum designs for telecommunications networks that apply speech coding.

The relationship between the speech transmission rate and speech quality is portrayed in Fig. 1 (Flanagan et al., 1979). The criteria from the viewpoint of technical qualifications are broadcast (or commentary) quality at bit rates exceeding 64 kb/s, toll quality at bit rates from 64 to 12 kb/s, communications quality at bit rates from 12 to 4 kb/s, and synthetic quality at bit rates below 4 kb/s. Broadcast-quality codings have wider frequency bands than the conventional commercial telephone bandwidth. Toll quality corresponds to commercial telephone speech quality, which is achieved mostly by speech waveform coding. Communications quality speech is highly intelligible but has noticeable quality reduction, some detectable distortion, and perhaps reduced speaker recognition. Communications-quality coding is a hybrid with the merits of both speech waveform and analysis-synthesis. Synthetic-quality speech is substantially less natural than toll quality but is essentially intelligible. It is achieved by speech analysis-synthesis or by the speech synthesis-by-rule method.

Speech quality, in terms of various quality factors (e.g., intelligibility and naturalness), must be satisfactory for good communications (e.g., Richards,

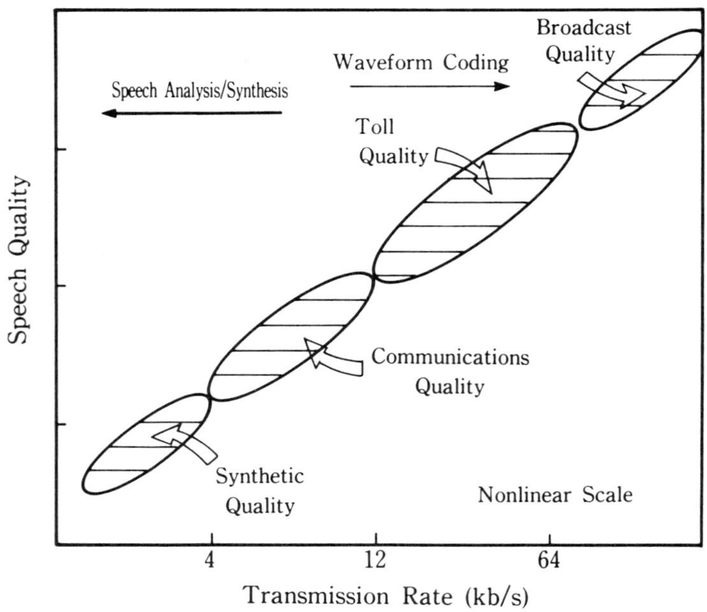

Figure 1 Quality of speech coding transmission rates. (From Kitawaki and Nagabuchi, © 1988 IEEE.)

1973). Considering the level of current coding techniques, however, naturalness of the speech is particularly essential in the higher quality ranges and intelligibility of the speech is fundamental in the lower quality ranges.

2. CONCEPT OF SPEECH QUALITY ASSESSMENT

2.1. Speech Communication Model

Figure 2 shows a speech communication model for speech quality assessment. This model features production, transmission, and listening components. Sending quality (*S*) in speech production implies the degree to which a talker's intended meaning is expressed. Listening quality (*L*), in the receiving component, indicates the degree to which listener understands. Transmission quality (*T*) implies the degree of transmission degradation. Whole speech quality (*Q*) includes *S*, *L*, and *T*.

Most speech quality studies have involved quantitative investigation of how physical transmission impairment factors affect subjective quality. These factors can be controlled by the experimenter, after normalization of *S* and *L*, to eliminate talker and listener factors including the acoustic environment such as noise, reverberation, and competing speech. To normalize sending and receiving quality, test circuit quality can be relatively expressed by reference circuit quality. However, speech synthesis by rule is itself a speech source, and speech recognition is itself an intelligent activity. Therefore, since

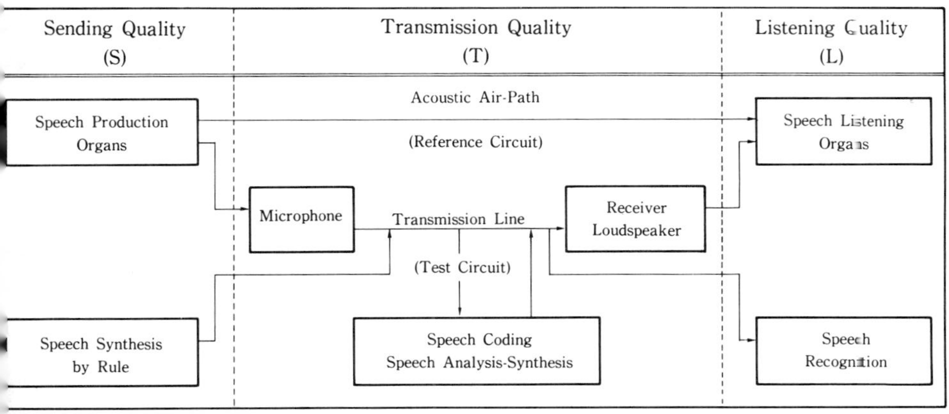

Figure 2 Speech communication model. (From Kitawaki and Nagabuchi, © 1988 IEEE.)

S cannot be normalized to assess synthesized speech by rule and L cannot be normalized to assess speech recognition, their quantitative assessments are more difficult.

2.2. Naturalness Assessment Methods

This section describes subjective quality assessment methods for coded speech as related to naturalness. Natural communication is especially required for toll-quality waveform codings. Relative to such communication, the opinion rating and paired comparison types of unidimensional measure recommended by the CCITT (International Telegraph and Telephone Consultative Committee) are discussed (CCITT, 1988c). This is followed by a description of the diagnostic acceptability measure (DAM), a kind of multidimensional measure and a relatively new way to measure quality (Voiers, 1977).

2.2.1. Opinion Rating

Opinion rating, which is one of the successive categories methods in psychometric measurements, is employed to assess directly the naturalness of speech quality. In the case of telecommunications systems, five grades of speech quality are often distinguished: excellent (E), good (G), fair (F), poor (P), and unsatisfactory (U). Their weighted mean values are calculated using 5 to 1, respectively, for each grade. This value is called the mean opinion score (MOS).

One advantage of the MOS method is that different impairment factors can be assessed simultaneously. Also, because customer opinions can be directly assessed and trained subjects are not required, an opinion test is easy to perform. Consequently, the opinion method has been used frequently to evaluate coded speech quality. On the negative side, however, this method has some problems, since several items affect opinion rating results: selection and order of presentation of the testing conditions, presentation of the exercise conditions, subject selection (especially nationality), instructions to the subjects, and other factors (Goodman and Nash, 1982). Therefore, when using the direct method, attention to detail is essential.

The equivalent noise method is frequently used to normalize the sending and receiving quality diagrammed in Fig. 2. In this method, reference signals, which have several different signal-to-noise ratios (SNRs), are included in the opinion testing conditions. The quality of the test circuit is expressed on the basis of the SNR ratio (dB) of the reference signal, whose quality is equivalent to that of the test circuit expressed by a 5-point scale.

2.2.2. *Paired Comparison*

In assessing a high-quality test circuit using the MOS method, the assessment value choices are sometimes concentrated in the E or G grade so that fine differences are indistinguishable. Therefore. a paired comparison test in psychometric measurement methods is more appropriate for precise subjective tests in high quality ranges. In this test, a signal pair, consisting of test signals A and B among the designated stimuli, is presented to the listeners, who are then asked to choose the better signal. Every combination of the stimuli is tested. The preference score is defined as the percentage of the designated test signal chosen as the preferred one. Isopreference contours specifying a pair of two testing parameter values—for example, loudness and circuit noise—yield equal preference scores for the two testing parameter values of the designated signal (Munson and Karlin, 1962), and a psychological scale value derived from the preference scores is also available to express speech quality (Guilford, 1954).

In the equivalent noise paired comparison test, a pair consists of a reference signal with a varied SNR and a test signal. The quality of the test circuit is defined as the SNR (dB) of the reference signal that corresponds to the intersection of the regression curve of the preference scores at the 50% level.

The advantage of the paired comparison test is that judgment is precise. However, this has the disadvantage that it is difficult to judge which signal of a signal pair is preferable when each has a different kind of impairment factor. The method is also time consuming because of the large amount of pairs involved.

2.2.3. *Diagnostic Acceptability Measure*

The other approach to assessing coded speech quality is the DAM incorporating a multidimensional measure, which evaluates a speech signal on 16 separate scales, each with a range of 0 to 100 points. These scales are classified according to signal, background, and total quality categories. The composite acceptability scale summarizes the DAM scale information in a single unidimensional scale and is simply a weighted average scale.

2.3. Intelligibility Assessment Methods

This section describes subjective quality assessment methods for coded speech as related to speech intelligibility.

2.3.1. *Articulation*

Articulation is the ratio of nonsense syllables sent from the sending end to those correctly received (Fletcher and Steinberg, 1929). The rate of correctly heard monosyllables is syllabic articulation (%). Sound articulation (%) is defined as the ratio of phonemes sent to those correctly received.

The equivalent loss method presented in Fig. 3 is frequently used to normalize sending and receiving quality. In this method, articulation measurements are made on reference and test circuits for various junction attenuation values. When the ARAEN (which complies with pertinent CCITT recommendations) is used as the reference system, equivalent articulation loss (called AEN) is defined as the difference in attenuation values that corresponds to 80% sound articulation between the reference and test circuits (CCITT, 1987). Here, the ARAEN design simulates the characteristics of a 1-m acoustic air path between a talker's lips and listener's ear with the latter facing the talker. An 80% sound articulation yields 100% sentence intelligibility for 50% of the people tested.

(a)

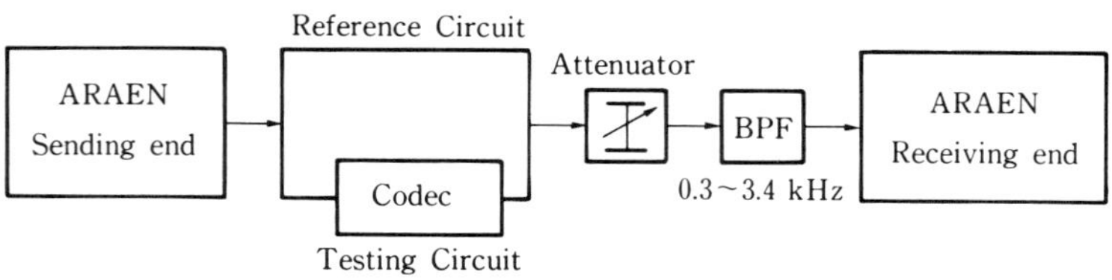

(b)

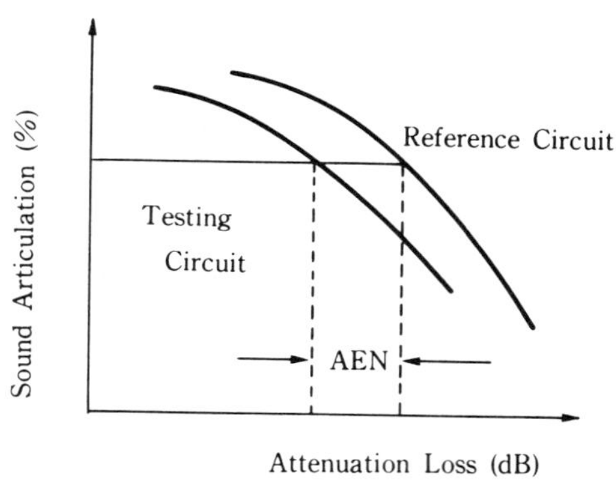

Figure 3 Measurement circuit and definition for articulation equivalent loss (AEN). (From Kitawaki et al., © 1984 IEEE.)

2.3.2. *Rhyme Tests*

The rhyme test has been proposed to counter the difficulty in generating articulation lists, especially in English (Fairbanks, 1958). This test is a kind of articulation test using a set of rhyming words that are composed of monosyllables, /C-V-C/, where C and V stand for consonant and vowel sounds. In a word list containing 50 monosyllabic words, the trailing vowel-consonant is specified and the leading consonant is varied. The subjects can distinguish the leading consonants. In this way, each word tests the intelligibility of a single consonantal phoneme.

To make it easy to perform, the basic rhyme test was modified into the modified rhyme test (MRT) (House et al., 1965). For each stimulus word among 50 meaningful words in a rhyming list, a closed set of 6 alternatives is constructed. That is for each test 50 times, one of the 6 words is chosen as the stimulus. The subjects are asked to choose one from among the set of alternatives. As a final note, the distinctive features (Jakobson et al., 1967) of the leading phoneme can be distinguished in the diagnostic rhyme test (DRT) (Voiers, 1977, 1982).

2.3.3. *Word Intelligibility*

Word intelligibility is the ratio of sent word sounds to correctly received ones. For an assessment of synthesis quality by rule, the syllabic articulation measurement method may be insufficient because the overall naturalness of concatenated sounds cannot be assessed, and the meaningful word intelligibility measurement may be more desirable. Since the contents of the word lists are important in this measurement, they are constructed taking into consideration phonetic balance (PB), word length, stress position, word importance, and so on. For example, a small set of words is selected from a Japanese dictionary containing 40,000 words so as to coincide with both distributions of the feature parameters and is then formed into a word list (Watanabe et al., 1988).

3. SUBJECTIVE QUALITY ASSESSMENT OF CODED SPEECH

Presented in this section is the speech quality measurement of waveform coding in terms of speech naturalness in communications, which is most important in the toll-quality range of speech coding. To normalize sending and receiving quality in Fig. 2, the equivalent noise method, used for the opinion test shown in Fig. 4A or for the paired comparison test, is generally used

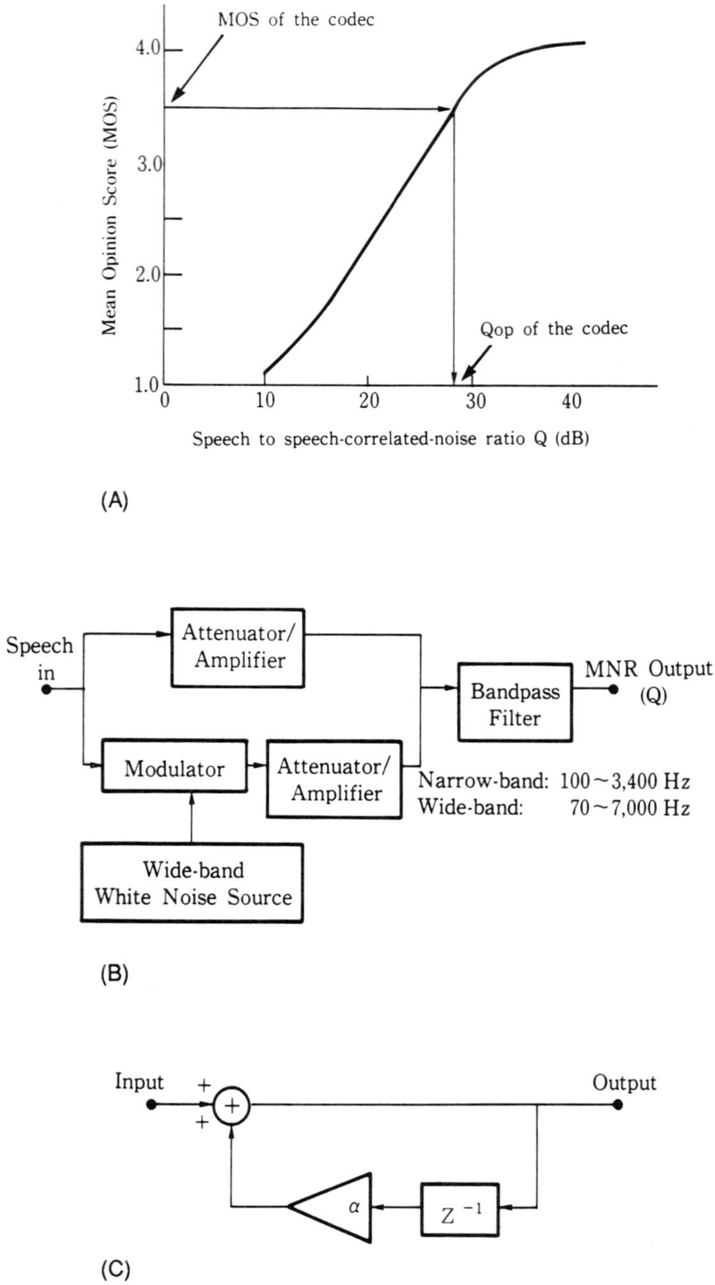

Figure 4 Arrangement of the MNRU and opinion equivalent Q determination. (From Kitawaki et al., © 1984 IEEE.)

to assess speech waveform coding. CCITT recommends that a specific configuration for a reference signal generation device be adapted for use in the equivalent noise method (CCITT, 1988c). This device is called the modulated noise reference unit (MNRU). Standard opinion and standard paired comparison test methods are also recommended using the MNRU.

3.1. Reference Conditions for Subjective Measurements

Figure 4B shows an MNRU arrangement that outputs a speech signal and speech amplitude–correlated noise with a flat frequency spectrum to assess waveform coding. The signal to speech-correlated noise ratio is called Q (dB). The quality of speech waveform coding is thus expressed as "opinion equivalent Q" or "preference equivalent Q" and is equivalent to the signal to speech correlated noise ratio in terms of the subjective quality of the coding.

Considerable progress has been made in using highly efficient speech waveform coding techniques other than the conventional pulse code modulation (PCM). Typical examples are 32 kb/s ADPCM (adaptive differential PCM) and 64 kb/s sub band CODEC (coder-decoder), which have been standardized in CCITT recommendations (CCITT, 1988b). Since the former has a conventional telephone bandwidth (3.4 kHz) and the latter has a broad bandwidth (7 kHz), they are respectively classified as narrowband and wideband CODECs. Corresponding to these CODEC classifications, narrowband and wideband MNRUs are defined in the CCITT recommendation to allow more stable and precise assessment of speech quality. The narrowband MNRU has a frequency bandwidth of 100 Hz to 3.4 kHz, and the wideband MNRU has one of 70 Hz to 7 kHz.

The wideband MNRU, which has been provisionally recommended by CCITT, outputs a speech signal and speech amplitude–correlated noise with a flat frequency spectrum. In general, the speech signal power of a higher-frequency component is lower than that of a lower-frequency component. Thus their SNRs are not equal to that of the MNR signal having a flat frequency spectrum. The SNR of the higher-frequency component is inferior to that of the lower-frequency component. Therefore, in subjective equivalent noise measurement, higher-frequency components having lower SNRs are more strongly perceived than lower-frequency ones, and noise perception accordingly varies across the frequency range. To avoid this and to provide a more stable and precise assessment, a noise-shaping filter has been proposed for use in the wideband MNRU and is to be further studied by CCITT. This noise-shaping filter is inserted between the modulator and the attenuator in the MNRU noise path. An example of the noise-shaping filter is shown in Fig. 4C.

3.2. Quality Measures for Network Planning

For measurements using MNRUs, decibel representation of the ratio is called Q_N for the narrow-band version and Q_W for the wideband version. These Q values represent the speech quality of waveform coding and constitute suitable quality criteria for national and international network planning. Figure 5 illustrates an example of coded speech quality in terms of opinion equivalent Q. These speech waveform codings are APC-AB (adaptive predictive coding with adaptive bit allocation) (Honda et al., 1982), ATC (adaptive transform coding) (Zelinski and Noll, 1977), and ADPCM (Cummiskey et al., 1973), and the speech analysis-synthesis methods are LSP (line spectrum pair) (Sugamura and Itakura, 1981) and PARCOR (partial autocorrelation) (Kitawaki et al., 1978). As is clear from Fig. 5, speech quality degrades as bit rates decrease. The ranking order of speech quality for the same bit rate is APC-AB, ATC, ADPCM, and PCM in waveform codings and LSP and PARCOR in speech analysis-synthesis.

Speech quality for CODEC tandem connection should be studied by introducing CODECs into commercial telephone networks. CODEC tandem connection methods are asynchronous coding in analog transmission channels and synchronous coding in digital transmission channels. If a transcoding algorithm is optimally designed in synchronous coding, then quantizing distortion caused by tandem connections does not accumulate. On the other

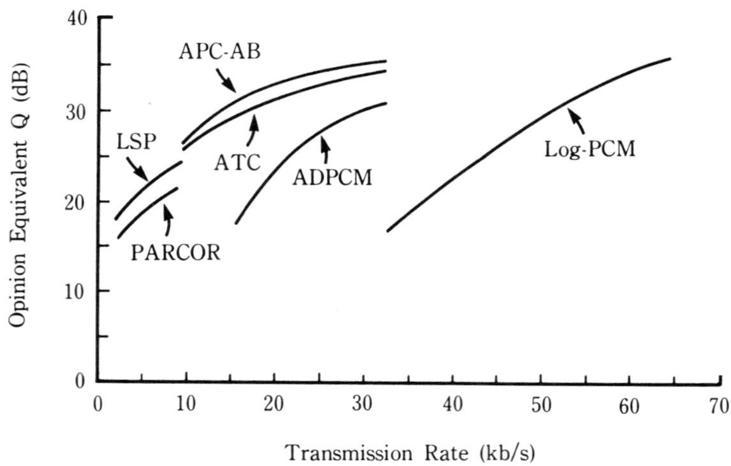

Figure 5 Examples of subjective evaluation in opinion equivalent Q (Q_{op}) for various kinds of codings. (From Kitawaki et al., © 1984 IEEE.)

hand, speech quality degradation due to asynchronous coding is unavoidable. Opinion equivalent Q in asynchronous coding degrades according to the logarithmic law for the number of connections. Also, the additive law has been demonstrated between the coding distortion expressed by Q values and the analog noise that occurs in mixed analog-digital networks (Watanabe et al., 1987).

Another quality criterion for easier network planning is the quantizing distortion unit (QDU) recommended by CCITT (1988a). Assigning appropriate weights to digital processes facilitates network planning of telephone speech transmission. One method is to consider that one unit of impairment is assigned to an average 8-bit A- or μ-law CODEC pair to offset quantizing distortion. A planning rule that has been provisionally agreed to allows 14 units of impairment for each overall international connection, with up to 5 units for each national extension and 4 units for international circuits.

Table 1 gives planning values for quantizing distortions where the QDU rule allows 14 tandem unintegrated 8-bit processes. Until further information is available it is recommended provisionally that three units of impairment (3 QDUs) should be assigned to a 7-bit PCM CODEC pair and 3.5 QDUs should be assigned to the 32 kb/s ADPCM (including PCM-ADPCM-PCM conversion for interfacing transmission circuit), according to CCITT recommendations on speech transmission quality. These preliminary conclusions are based on a limited amount of information, and the weights may have to be revised as new information becomes available. Also, the linear additive law in the region of larger QDU numbers should be further studied.

Table 1 Planning Values for Quantizing Distortion

Distortion process	Quantizing distortion units
8-bit PCM CODEC pair	1
7-bit PCM CODEC pair	3
Transmultiplexer pair based on 8-bit PCM	1
32 kb/s ADPCM (combination of PCM-ADPCM-PCM)	3.5
Digital loss pad (PCM)	0.7
A/μ or μ/A converter	0.5
A/μ/A law tandem conversion	0.5
μ/A/μ law tandem conversion	0.25
PCM to ADPCM to PCM tandem conversion	2.5
8-7-8 bit transcoding	3

The quality characteristics of speech codings resulting from transmission bit error are particularly important for their application to mobile radio communication. The speech quality resulting from transmission bit error for sophisticated speech codings with adaptation techniques was studied (Kitawaki et al., 1984). Speech quality degrades at a bit error rate of over 5×10^{-4} for sophisticated codings and at a lower bit error rate for 64 kb/s PCM. This is because PCM speech degradation is perceived as a sharp clicking noise and sophisticated codings are perceived as dull clicking sounds.

4. OBJECTIVE QUALITY ASSESSMENT OF CODED SPEECH

Speech quality has previously been assessed by subjective methods using listener judgments. However, speech codings and communications networks designed for subjective quality assessment alone are inefficient because subjective assessments are very expensive and time consuming. They also require huge apparatuses, including master reference systems such as ARAEN that are pertinent to the CCITT recommendations, sending and receiving booths, and a number of test subjects. Therefore, it is quite common in telecommunications to estimate speech quality by objective methods using mathematical models, the results of which will correspond well with subjective evaluations (Quackenbush et al., 1988). This section discusses some objective methodologies for speech waveform codings from the principal viewpoint of speech naturalness. On the other hand, objective articulation measurement methods have been studied for telephone transmission circuit (Collard, 1930; Pocock, 1939) and room acoustics (Houtgast and Steeneken, 1973, 1984). However, applications of these methods for coded speech quality estimation remain to be further investigated.

4.1. Objective Quality Measurement System

Figure 6 shows a typical approach to the objective measurement of coded speech quality, which is composed of an objective quality criterion and a testing signal (Itoh et al., 1984). When the testing signal is applied to the codings, the criterion is calculated and then the subjective quality is estimated from the obtained criterion. Here, distortion between the input and output signals of the coding is one of objective quality criteria which approximate well the subjective measurements. Several kinds of objective quality measures of

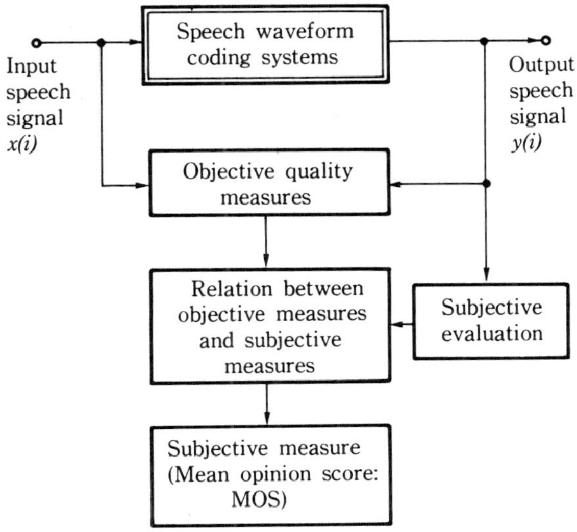

Figure 6 Subjective value estimation method from objective quality measures.

the distortion are described and are classified according to the time and frequency domains shown in Fig. 7. Moreover, frequency domain measures are classified in two categories, depending on whether the speech spectrum contains whole spectrum characteristics or the spectrum envelope only. Thus, the objective measures for coding quality assessment are waveform distortion, spectral distortion, and spectral envelope distortion.

Another approach to objective measurement is based on an expert pattern recognition system (Quincy, 1987) which processes speech, extracts parameter values, classifies the speech into one of five classes in opinion rating, and provides the reasoning necessary for classification. Its performance objective is to minimize the average error probability when objective classifications are compared to subjective panel classifications on training data.

4.2. Waveform Distortion Measures

Objective measures in the time domain are defined as distortion between the input and output speech waveforms. Signal-to-noise ratio measures are appropriate only for speech waveform codings.

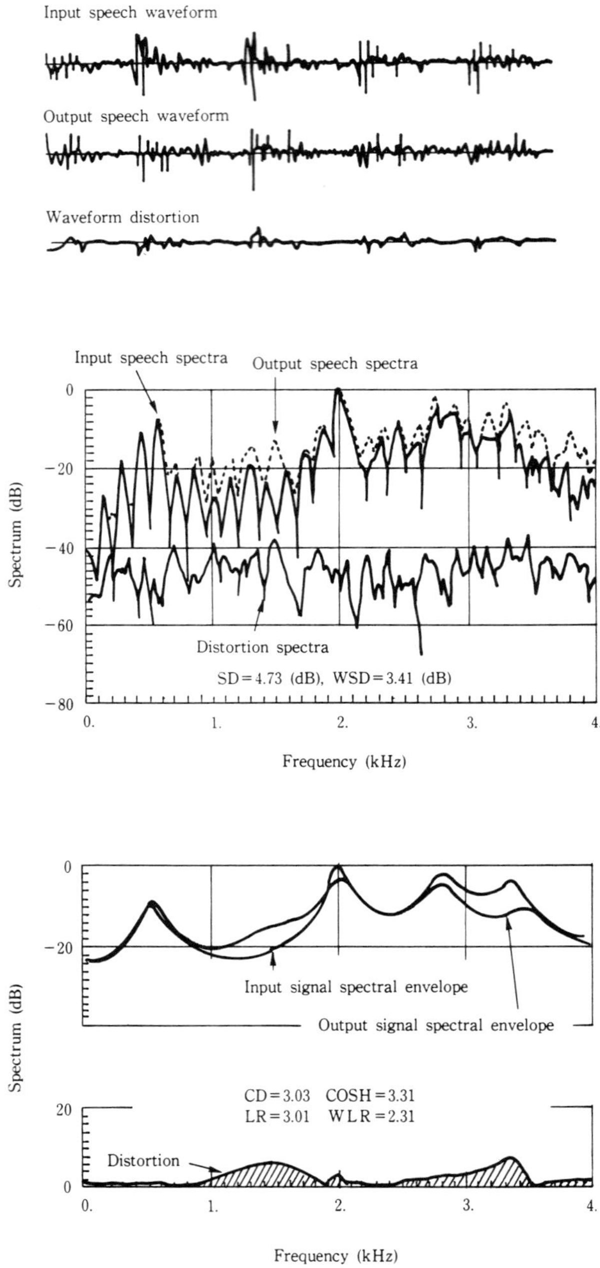

Figure 7 An example of objective quality measures for speech waveform codings in time and frequency domains.

4.2.1. *Signal-to-Noise Ratio Measures*

The conventional SNR is defined in Eq. (1), which has been used as one of common objective quality measures:

$$\mathrm{SNR} = 10\log_{10}\frac{\sum_{j=1}^{M} x(j)^2}{\sum_{j=1}^{M}[x(j)-y(j)]^2} \tag{1}$$

Here, $x(j)$ is an undistorted input speech signal, $y(j)$ is the distorted output speech signal of the waveform coding, and M is the number of samples in the signal. In a conventional PCM system, a sinusoidal signal or band-limited Gaussian noise, instead of a speech signal, is applied and the signal-to-noise ratio is called SNR_q. The most important point here is that various kinds of codings can be uniformly evaluated by a unique objective measure. However, the SNR_q is not sufficiently large to measure various kinds of codings with the new adaptation techniques. This is because a sinusoidal signal with a single frequency spectrum or band-limited Gaussian noise with a flat frequency spectrum cannot measure the quantizing noise of the new sophisticated codings with nonuniform frequency characteristics.

4.2.2. *Segmental Signal-to-Noise Ratio Measures*

The SNR measure characterizes the ratio of long-term average speech power to long-term average quantizing noise power. The long-term calculation method weights a larger-power sentence section more heavily than a smaller-power one. Therefore, a smaller-power speech section cannot reflect the SNR measure as well as a larger-power section, in spite of its importance, such as the consonant and transient periods that occur during speech.

This situation is improved by taking the same measurement over short segments of the speech waveform and then summing the measurements for all segments. This is called segmental SNR ($\mathrm{SNR}_{\mathrm{seg}}$) and is defined as

$$\mathrm{SNR}_{\mathrm{seg}} = \frac{1}{N}\sum_{f=1}^{N}\mathrm{SNR}_f \tag{2}$$

Here, N is the number of segments (called frames) and SNR_f is the signal-to-noise ratio of frame f. The frames are extracted by windowing speech having a length of 16 to 32 ms. Thus, $\mathrm{SNR}_{\mathrm{seg}}$ is an average value of the SNR calculated for short intervals using real speech as the test signal. The objective measure weights are equal to both the larger-power and smaller-power speech sections. The segmental SNR exhibits better common correspondence to the subjective values than does classical SNR for various kinds of speech codings.

4.3. Spectral Distortion Measures

Spectral distortion measures in the frequency domain are defined as distortion between the input and output whole speech spectra.

4.3.1. *Spectral Distance Measures*

The spectral distance (SD) measure is defined as an objective measure containing the entire speech spectrum characteristics:

$$\mathrm{SD} = \left[\frac{1}{w} \int_0^w \left\{ S_x(w) - S_y(w) \right\}^2 dw \right]^{1/2} \tag{3}$$

Here, $S_x(w)$ and $S_y(w)$ are input and output speech spectra, respectively, and w is the signal frequency band. For example, the speech spectrum containing its entire characteristics is computed by fast Fourier transform (FFT). Thus, SD contains strict characteristics for spectral differences.

The coherence function (CF) measure is an another approach to calculating spectral distortion by Fourier transform analysis (CCITT, 1982). It utilizes the cross-correlation of the entire spectrum between the input and output speech signals of the codings and its coherence function as in Eq. (4);

$$\hat{\gamma}^2(f) = \frac{\left| \sum_{n=1}^{N} X_n(f) Y_n^*(f) \right|^2}{\sum_{n=1}^{N} |X_n(f)|^2 \sum_{n=1}^{N} |Y_n(f)|^2} \tag{4}$$

Here, $X_n(f)$ and $Y_n(f)$ are again input and output speech spectra, and Y_n^* is the complex conjugate of Y_n. Finally, the coherence function measure is defined as the ratio of the coherent output power spectrum to the noncoherent output power spectrum in Eq. (5):

$$\mathrm{SDR}(f)_{\mathrm{dB}} = 10 \log_{10} \frac{\hat{\gamma}^2(f)}{1 - \hat{\gamma}^2(f)} \tag{5}$$

4.3.2. *Weighted Spectral Distortion Measure*

The weighted spectral distortion (WSD) measure is a frequency-weighted spectral distance. The WSD in Eq. (6) is based on the articulation index to the spectral distance calculated by Eq. (3):

$$\mathrm{WSD} = \frac{1}{B} \sum_{b=1}^{B} W_b \mathrm{SD}_b \tag{6}$$

Here, SD_b is spectral distortion in subband b, B is the number of subbands, and W_b is the weighting coefficient according to the articulation index. The speech band is often divided into 16 subbands in proportion to W_b.

4.4. Spectral Envelope Distortion Measures

The speech spectral envelope contains macroscopic spectral information, which constitutes the most important speech signal characteristic. One of the calculation methods for the speech spectral envelope is based on linear predictive coding (LPC) analysis. The objective measures by LPC analysis are classified into several categories by the difference in weighting function to spectral distortion between the input and output signals to the coding system. The optimum weighting function should be determined by comparative study of the subjective and objective assessment values.

4.4.1. Likelihood Ratio Measures

The likelihood ratio (LR) is a measure of the spectral similarity between the all-pole models of the reference and coded speech (Itakura, 1975). This measure assumes that short speech intervals can be represented by the pth all-pole model given in Eq. (7):

$$\begin{aligned} S_v(\theta) &= \frac{\sigma^2}{2\pi} \frac{1}{|1 + \alpha_1^{(n)} z + \alpha_2^{(n)} z^2 + \cdots + \alpha_p^{(n)} z^p|^2} \\ &= \frac{\sigma^2}{2\pi} \frac{1}{A_0 + 2A_1 \cos\theta + 2A_2 \cos 2\theta + \cdots + 2A_p \cos(p\theta)} \\ &\qquad (z = e^{-j\theta}, \theta = 2\pi fT) \end{aligned} \tag{7}$$

Here, $A_y(i)$ is the maximum likelihood coefficient of the LPC analysis shown in Eq. (8):

$$A_y(i) = \sum_{j=0}^{p=-|j|} \alpha_i \alpha_{j-|i|} \tag{8}$$

where α_j is the linear predictive coefficient.

Thus, LR is presented in terms of the autocorrelation method for linear prediction analysis shown in Eq. (9):

$$\mathrm{LR} = \left\{ 2\log\left[\frac{1}{R_x}\sum_{i=-p}^{p} A_y(i) r_x(i)\right]\right\}^{1/2} \tag{9}$$

Here, R_x, $r_x(i)$, and p are the normalized residual for the input signal, autocorrelation coefficient of the input signal, and prediction order, respectively.

The peak neighborhood of the spectral envelope is considered to be important to speech perception. The weighted likelihood ratio (WLR) measure is defined in Eq. (10) taking perceptual knowledge into account (Sugiyama and Shikano, 1981):

$$\mathrm{WLR} \approx \left\{ 2 \sum_{i=1}^{k} [r_x(i) - r_y(i)][C_x(i) - C_y(i)] \right\}^{1/2} \tag{10}$$

Here, $r_x(i)$ and $r_y(i)$ are LPC autocorrelation coefficients, $C_x(i)$ and $C_y(i)$ are LPC cepstrum coefficients of the input and output coding signals, respectively, and k is coefficient order.

4.4.2. *Cepstrum Distance Measure*

The cepstrum distance (CD) measure, which is computed by cepstrum analysis, is equal to the logarithmic spectral envelope distance of Parseval's theorem concerning the Fourier coefficient (Atal, 1974). The CD based on the LPC technique is given in Eq. (11):

$$\mathrm{CD} \cong \left\{ 2 \sum_{i=1}^{K} [C_x(i) - C_y(i)]^2 \right\}^{1/2} \tag{11}$$

4.4.3. *COSH Distance Measure*

The COSH measure is defined as the logarithmic spectral envelope distance weighted by cosh function as shown in Eq. (12) (Gray and Markel, 1976):

$$\mathrm{COSH} = \left[2 \frac{\sum_{i=-p}^{p} A_y(i) r_x(i) \sum_{i=-p}^{p} A_x(i) r_y(i)}{R_x R_y} - 2 \right]^{1/2} \tag{12}$$

4.5. Comparison of Objective Quality Measures

Comparisons of some kinds of objective measures for coded speech quality assessments have been reported (Kitawaki et al., 1982). The quantizing noise of PCM is in proportion to the input speech signal. However, those of the others, which utilize the new adaptation techniques, are not. Therefore, waveform distortion measures are poorer estimators for adaptive codings. Frequency domain measures have been experimentally demonstrated to reflect subjective quality better than do time domain measures.

It has also been demonstrated that in frequency domain measures, spectral envelope distortion measures have better correspondence to subjective assessment values than do spectral distortion measures that represent a whole speech spectrum. The CD measure is appropriate for MOS estimation, and the LR measure is appropriate for articulation estimation. However, these results should be confirmed by more experiments for various kinds of coded speech.

4.6. Artificial Voice for Objective Quality Assessment

A test signal is indispensable for the objective quality assessment of speech codings and is classified into real speech and artificial signals. Either a sinusoidal signal or frequency band-limited noise, which are artificial signals, has been used as the test signal in conventional PCM quality assessment. However, the performance of sophisticated codings using human speech production and listening mechanisms cannot be assessed by such conventional simple test signals because these codings do not have the flat-spectrum quantizing distortion of PCM but have speech-correlated complicated quantizing distortion.

Real speech samples are often used to calculate the coding distortion. However, there are difficulties in choosing real speech samples, because human voice varies greatly according to the individual and the language used. To make universal choice of the test signals easy, an artificial voice signal which reflects the characteristics of the human voice and is generated in the same manner by an algorithm is preferable. In the first stage, the average characteristic is important because it facilitates measurement of average performance, which is the most important index for network planning and coding-algorithm research. An artificial voice for evaluating the coded speech quality of the sophisticated codings must simulate the short-term average characteristics of the human voice as well as the long-term ones. Therefore, the following real speech characteristics should be reproduced by an artificial voice: long-term and short-term average spectrum, instantaneous amplitude distribution, level distribution and spectrum of segmental speech power, and voiced-unvoiced structure of the speech waveform.

As mentioned above, the artificial voice should be generated by an algorithm designed to achieve standardization. Speech synthesis based on a human speech production model is an effective method for controlling many kinds of speech features. Figure 8 illustrates the basic configuration of the artificial voice generation process (Kitawaki et al., 1989). The artificial voice is generated by applying two excitation signals, a glottal signal and random noise with a uniform distribution, to a spectral shaping filter. The glottal excitation signal is a periodic pulse train whose fundamental period determines the pitch frequency of the artificial voice. The sounds generated by the glottal excitation signal and the random noise correspond to voiced and unvoiced

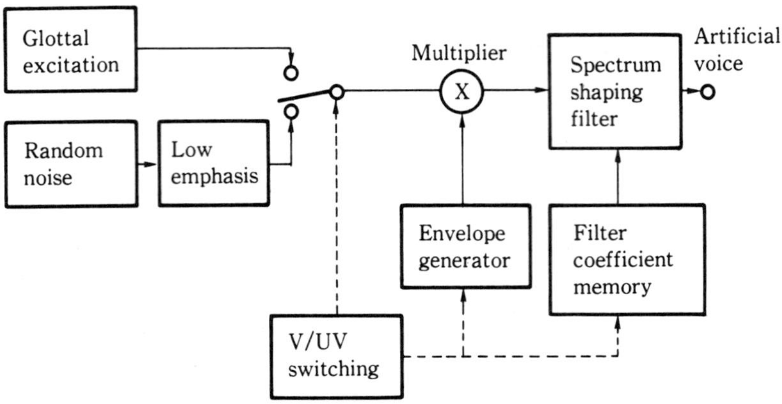

Figure 8 Artificial voice generation process. (From Kitawaki et al., © 1989 IEEE.)

sounds. The spectral shaping filter is a PARCOR speech synthesis filter having a frequency response that simulates the transmission characteristics of the vocal tract. Variations in the short-term spectral characteristics of real speech signals are reproduced by this filter, the coefficients of which can vary within a short time.

The required characteristics for the artificial voice were extracted from real speech samples composed of 20 kinds of languages, taking language dependence into account. The short-term spectral variations, which are especially required for objective quality assessment of sophisticated speech codings, were classified into 16 patterns by means of a vector quantization algorithm (Linde et al., 1980). The feature parameters for controlling the artificial voice generation circuit were determined so as to reproduce the extracted characteristics of real speech.

5. SPEAKER DEPENDENCE OF CODED SPEECH QUALITY

In low-bit-rate speech coding other than conventional PCM, the quality varies greatly from the speaker compared with PCM. This dependence on the speaker is often observed in waveform coding at rates of less than 32 kb/s and even more so in speech analysis-synthesis (Nagabuchi, 1988).

5.1. Variation in Subjective Speech Quality

Figure 9 shows variations in subjective quality, in terms of opinion equivalent Q, obtained using 15 speakers and three waveform codings (ADPCM, APC-AB, ATC) as compared to log-PCM. The results confirm that the variations in opinion equivalent Q over speakers in low-bit-rate coding other than PCM are larger than in PCM coding. In the quality evaluation of low-bit-rate codings other than PCM, the speaker is thus clearly an important factor.

5.2. Criterion for Speaker Selection

In adaptive speech codings using LPC techniques, the coded speech quality is considered to correspond to the prediction gain, namely, the flatness of the speech spectrum. Therefore, the power of prediction error for speech weighted by speech power, G, in Eq. (13), is an effective measure for

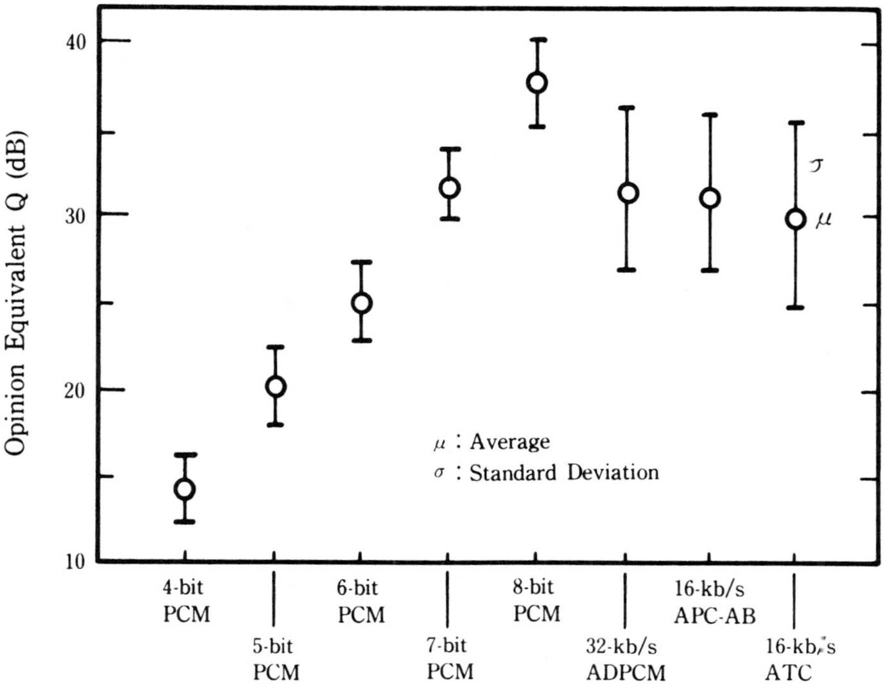

Figure 9 Deviation of opinion equivalent Q for 15 speakers. (From Kitawaki et al., © 1988 IEEE.)

estimating variations in coded speech quality across a range of different speakers:

$$G(\gamma, \mathrm{IP}) = -10\log \frac{\sum_{i=1}^{N} V_{0i}^{\gamma} U_i(\mathrm{IP})}{\sum_{i=1}^{N} V_{0i}^{\gamma}} \tag{13}$$

where $U_i(\mathrm{IP}) = \dfrac{V_i(\mathrm{IP})}{V_{0i}} = \prod_{j=1}^{\mathrm{IP}}(1 - k_{ij}^2)$

V_{0i} = power of speech signal at ith segment
$V_i(\mathrm{IP})$ = power of prediction error signal of speech at ith segment
IP = prediction order
K_{ij} = jth order PARCOR parameter at ith segment
N = total number of segments
γ = weighting coefficient

G depends on two parameters: prediction order and weighting coefficient. Their optimum values are 4 and 0.5, respectively, in terms of maximizing the correlation between objective measure G and the subjective quality obtained for 45 speakers. Thus, G with optimal parameter values (G_{opt}) can be used as an effective index to select speakers in the quality assessment of low-bit-rate speech coding.

6. CODED SPEECH QUALITY FOR ATM NETWORKS

The asynchronous transfer mode (ATM) technology is expected to enable future broadband integrated services digital networks (B-ISDNs) to handle integrated traffic ranging from narrowband voice and data services to broadband video services. In ATM networks, speech is coded and processed via "cell" packetization in constant short lengths. This section describes a type of speech coding quality for ATM networks, taking cell processing into account, that should improve service quality.

6.1. Quality Factors in ATM Networks

An ATM network consists of cell processing nodes and digital transmission links. ATM factors affecting speech quality are speech coding distortion, speech detection error, cell loss distortion, and system-dependent delays. To compress the speech transmission bit-rate, speech coding techniques are used. In the ATM approach, a variable-bit-rate coding technique which makes use of the bursty nature of speech is available. This is because transmission of the speech information depends only on the cell information and does not require synchronous signaling. Variable-rate coding can achieve optimum coding while adjusting to the transmission channel rate.

Speech segments are detected and retained while inactive segments are ignored to achieve low-bit-rate speech transmission. Consequently, however, speech detection errors lead to clipping distortion at the beginning and end of utterances, and silent period compression causes differences in background noise between the transmitted and reconstructed periods at the receiving end when the silent periods are reconstructed.

Short, bursty interruptions due to the discarding of cells during busy hours or the misdelivery of cells and system-dependent delays due to cell assembly or queuing are typical in ATM networks.

6.2. Speech Codings for ATM Networks

When new speech codings are applied to ATM networks, special technical features should be developed. One is a variable-rate coding capability for encoding burst speech signal sources or controlling the speech coding bit rate in response to user requests. This feature gives speech coding techniques a rate-independent capability in ATM networks.

Another technical feature is missing-cell recovery. Cell discarding, which can occur at congested network nodes, is a network impairment peculiar to ATM networks and can seriously degrade speech quality. In low-bit-rate codings, the degradation caused by cell loss is greater than that for PCM, since the decoding process for a low-bit-rate coder indexed to the current time is performed using previously reconstructed output signals in the decoder.

To apply these speech coding techniques, missing-cell recovery techniques that minimize the quality degradation caused by cell discarding should be developed.

6.3. Missing-Cell Recovery Methods

The quality of reconstructed output speech that has experienced cell loss greatly depends on the contents of the missing cells. When one of two consecutive cells is lost, the signal discarded after cell disassembly is distributed according to the arrangement of the code words in the cell.

A substitution technique used in missing-cell recovery methods is one in which the signal in the discarded cell is replaced by another signal, such as all zeros or the previous waveform. An interpolation technique is one in which the missing samples can be reconstructed when the missing code words appear alternately in the disassembled code word. The substitution and interpolation techniques cannot be applied to low-bit-rate codings because a large amount of distortion is caused by the mistracking operation.

The other method is an embedded coding technique (Goodman, 1980). A code word is divided into two segments which contain the most significant bits (MSBs) and the least significant bits (LSBs) of the code word, and these segments are assembled into separate cells. If necessary, each segment is controlled to drop only the least significant cells. Since the embedded coding accommodates bit dropping with a graceful and smooth degradation of speech quality, it can readily provide variable-rate coding.

6.4. Speech Quality of Embedded Codings

Figure 10 exemplifies the speech quality obtained using embedded coding techniques for 64 kb/s PCM and 32 kb/s ADPCM, as recommended by CCITT. In PCM, opinion equivalent Q (Q_{op}) for the LSB dropping method is higher than for the all-bit dropping method for higher missing-cell rates. In ADPCM, embedded coding with LSB dropping is superior to ADPCM with all-bit dropping. In embedded ADPCM, the quality degrades gradually as the missing-cell rate increases. Thus, at higher missing-cell rates, the quality

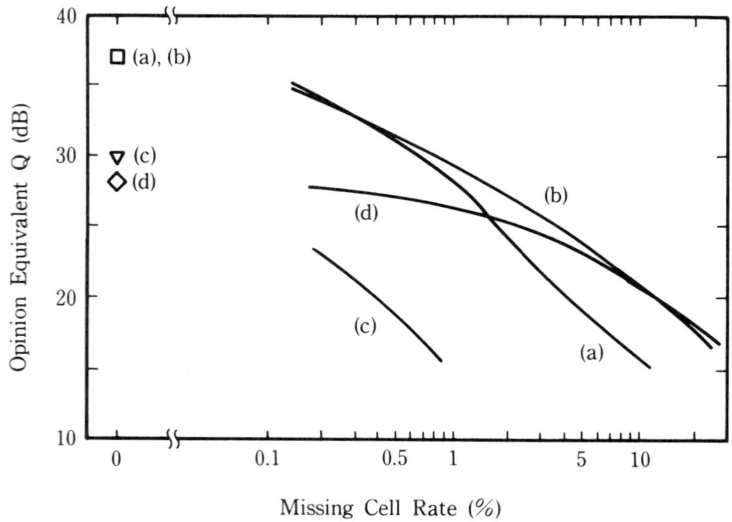

Figure 10 Decrease in opinion equivalent Q with increasing cell loss. (From Kitawaki et al., © 1984 IEEE.)

exceeds that of PCM with all-bit dropping. The original coding distortion of embedded ADPCM is larger than that of ADPCM and, of course, PCM, but embedded ADPCM is resistant to the distortion resulting from increases in cell loss. Thus, the embedded coding with LSB dropping applied to ADPCM is one of the most promising techniques for speech coding in ATM networks.

7. QUALITY OF SPEECH ANALYSIS-SYNTHESIS

7.1. Subjective Quality Assessment

For speech analysis-synthesis, the same subjective quality assessment methodologies are available as for speech waveform codings. Figure 5 showed an example of coded speech quality in terms of opinion equivalent Q for typical speech waveform codings and speech analysis-synthesis. For the speech analysis-synthesis in the synthetic quality region (see Fig. 1), however, the naturalness is substantially less but the speech is intelligible. Intelligibility of the speech is more essential, considering the level of current coding techniques. The articulation equivalent loss (AEN) for such codings is given in Fig. 11. The AEN for the waveform codings approaches 0 dB with a higher bit rate. The ranking order of speech quality at the same bit rates is the same as in Fig. 5 in terms of Q_{op}.

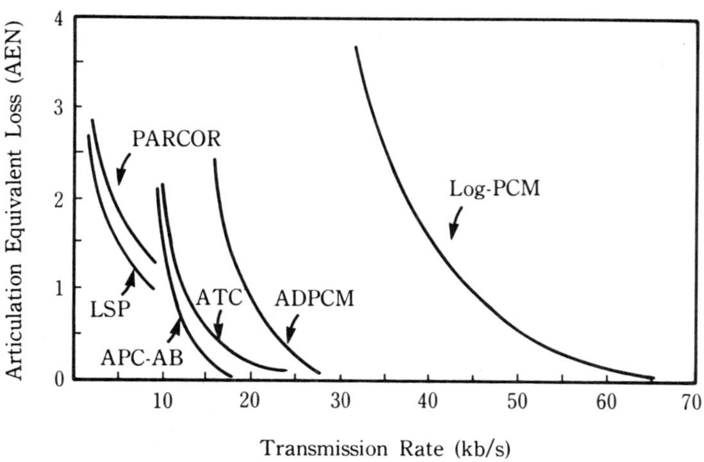

Figure 11 Examples of subjective evaluation in terms of articulation equivalent loss (AEN) for various kinds of CODECs. (From Kitawaki et al., © 1984 IEEE.)

7.2. Optimum System Parameter Design

Coded speech quality is preferably assessed by objective methods whose results correspond well with the subjectively determined results. Objective methods for speech analysis-synthesis have not yet been developed, however, because of the difficulties concerning appropriate objective measures when the speech analysis-synthesis quality is affected by exciting the source and spectral parameters of the speech synthesis circuit.

In approach the spectral parameters in speech analysis-synthesis are encoded by an objective method using the spectral envelope distance to minimize the spectral distortion (Kitawaki et al., 1978). After using the objective method, a subjective method may be necessary for perceptual confirmation because of the lack of total optimization, including the excitation source parameters of the speech synthesis.

Spectral envelope distortion in speech analysis-synthesis consists of the distortion caused by updating the spectral parameters, the period of which is called the frame period, and quantizing the spectral parameters within a frame. Encoding is optimized at each of the following steps.

1. Determine the constant updating rate and the number of quantizing bits for a set of spectral parameters which minimizes the spectral envelope distortion with a given total number of bits.
2. Assign the number of quantizing bits determined in step 1 to each of the spectral parameters nonuniformly to minimize quantizing distortion.
3. Readjust the updating rate of each spectral parameter. Since the time-variant characteristics of the spectral parameters are different from one another, each spectral parameter can be updated at a different rate to decrease spectral distortion.

7.3. Required Quality for Speech Analysis-Synthesis System

For synthesized speech, since the intelligibility of specific sounds is somewhat different from that in natural speech, measuring only the average intelligibility of the sounds is insufficient. Thus, the intelligibility of synthesized speech is evaluated by individual sound articulation as well as by the average value (Kitawaki et al., 1981). Required average sound articulation is generally 80% from the viewpoint of sentence intelligibility. The word intelligibility decreases greatly for individual sound articulation scores below about 15 to 20%. Thus, an individual sound articulation of 15 to 20% can be a criterion for the intelligibility requirement of speech analysis-synthesis systems, along with a word intelligibility of 80%. This attempt should be confirmed by further experiments.

8. CONCLUSION

This chapter described the quality assessment of coded speech from the viewpoint of practical use in telecommunications. The quality assessment methodologies of speech waveform codings have made considerable progress and are being studied for standardization by CCITT to facilitate comparison of assessment values. Objective quality measurement methodologies are now under investigation to avoid expensive and time-consuming subjective quality assessment. The quality assessment of speech synthesis has scarcely been reviewed, but study has begun.

REFERENCES

Atal, B. S. (1974). Effectiveness of linear prediction characteristics of the speech wave for automatic speaker identification and verification. *J. Acoust. Soc. Am. 55*:

CCITT, COM XII-60 (1982). Evaluation of non-linear distortion via the coherence function. 1981–1984 (BNR).

CCITT. (1987). Handbook on Telephonometry.

CCITT Recommendations. (1988a). General characteristics of international telephone connections and circuits. Recommendations G.101–G.181, Blue Book, vol. III, Fascicle III.1, IXth Plenary Assembly.

CCITT Recommendations. (1988b). Digital networks, transmission systems, and multiplexing equipment. Recommendations G.700–G.956, Blue Book, vol. III, Fascicle III.3, IXth Plenary Assembly.

CCITT Series P Recommendations. (1988c). Telephone transmission quality. Blue Book, vol. V, IXth Plenary Assembly.

Collard, J. (1930). Calculation of the articulation of a telephone circuit from the circuit constants. *Electron. Commun. 8*:141.

Cummiskey, P., Jayant, N. S., and Flanagan, J. L. (1973). Adaptive quantization in differential PCM coding of speech. *Bell Syst. Tech. J. 52*:1105–1118.

Fairbanks, G. (1958). Test of phonemic differentiation: The rhyme test. *J. Acoust. Soc. Am. 30*(7):596–600.

Flanagan, J. L., Schroeder, M. R., Atal, B. S., Crochiere, R. E., Jayant, N. S., and Tribolet, J. M. (1979). Speech coding. *IEEE Trans. Commun. COM-27*(4):610–737.

Fletcher, H., and Steinberg, J. C. (1929). Articulation testing methods. *Bell Syst. Tech. J. 8*:806–854.

Goodman, D. J. (1980). Embedded DPCM for variable bit rate transmission. *IEEE Trans. Commun. COM-28*(7):1040–1046.

Goodman, D. J., and Nash, R. D. (1982). Subjective quality of the same speech transmission conditions in seven different countries. *IEEE Trans. Commun. COM-30*(4):642–654.

Gray, A. H., and Markel, D. (1976). Distance measure for speech processing. *IEEE Trans. Acoust., Speech, Signal Process. ASSP-24*:

Guilford, J. P. (1954). *Psychometric Methods*. McGraw-Hill, New York.

Honda, M., Kitawaki, N., and Itakura, F. (1982). Adaptive Bit Allocation Scheme in Predictive Coding of Speech. *Proc. 1982 IEEE Int. Conf. Acoustics, Speech, and Signal Processing*, S14.3.

House, A. S., Williams, C. E., Hecker, M. H. L., and Kryter, K. D. (1965). Articulation testing methods: Consonantal differentiation with a closed-response set. *J. Acoust. Soc. Am. 37*(1):158–166.

Houtgast, T. and Steeneken, H. J. M. (1973). The modulation transfer function in room acoustics as a predictor of speech intelligibility. *Acoustica 28*:

Houtgast, T. and Steeneken, H. J. M. (1984). A multi-language evaluation of the RASTI-method for estimating speech intelligibility in auditoria. *Acoustica 54*:

Itakura, F. (1975). Minimum prediction residual principle applied to speech recognition. *IEEE Trans. Acoust., Speech, Signal Process. ASSP-28*:

Itoh, K., Kitawaki, N., and Kakehi, K. (1984). Objective quality measures for speech waveform coding systems. *Rev. Electr. Commun. Lab. NTT, Japan 32*(2):220–228.

Jakobson, R., Fant, C. G. M., and Halle, M. (1967). *Preliminaries to Speech Analysis: The Distinctive Features and Their Correlates*. MIT Press, Cambridge, Mass.

Kitawaki, N., Itoh, K., and Itakura, F. (1978). PARCOR speech analysis-synthesis system. *Rev. Electr. Commun. Lab., NTT, Japan 26*(11–12):1439–1455.

Kitawaki, N., Itoh, K., and Kakehi, K. (1981). Speech quality measurement methods for synthesized speech. *Rev. Electr. Commun. Lab., NTT, Japan 29*(9–10):895–906.

Kitawaki, N., Itoh, K., Honda, M., and Kakehi, K. (1982). Comparison of objective speech quality measures for voiceband codecs. *Proc. IEEE ICASSP '82*, S9.5.

Kitawaki, N., Honda, M., and Itoh, K. (1984). Speech quality assessment methods for speech coding systems. *IEEE Commun. Mag. 22*(10):26–33.

Kitawaki, N., Nagabuchi, H., and Itoh, K. (1988). Objective quality evaluation for low-bit-rate speech coding systems. *IEEE JSAC 6*(2):

Kitawaki, N., and Nagabuchi, H. (1988). Quality assessment of speech coding and speech synthesis systems. *IEEE Commun. Mag. 26*(10):36–44.

Kitawaki, N., Itoh, K., Irii, H., and Nagabuchi, H., (1989). Artificial voice signal for objective quality evaluation of speech coding systems. *Proc. IEEE ICC '89* S12.2.

Kitawaki, N., Nagabuchi, H., Taka, M., and Takahashi, K. (1990). Speech coding technology for ATM networks. *IEEE Commun. Mag. 28*(1):

Linde, Y., Buzo, A., and Gray, R. (1980). An algorithm for vector quantizer design. *IEEE Trans. Commun. COM-28*(1):

Munson, W. A., and Karlin, J. E. (1962). Isopreference method for evaluating speech-transmission circuits. *J. Acoust. Soc. Am. 34*(6):762–774.

Nagabuchi, H. (1988). Talker-dependency of quality in low-bit-rate speech coding. *Trans IEICE Jpn. J71-A*:601–607 (in Japanese).

Pocock, L. C. (1939). The calculation of articulation for effective rating of telephone circuits. *Electron. Commun. 18*:120.

Quackenbush, S. R., Barnwell, T. P., III, and Clements, M. A. (1988). *Objective Measures of Speech Quality*. Prentice Hall, Englewood Cliffs, N.J.

Quincy, E. A. (1987). Prolog-based expert pattern recognition system shell for technology independent, user-oriented classification of voice transmission quality. *Proc. IEEE ICC '87*.

Richards, D. L. (1973). *Telecommunication by Speech*. Butterworths, London.

Sugamura, N., and Itakura, F. (1981). Speech data compression by LSP speech analysis-synthesis technique. *Trans. IECE Jpn. J64-A*:8 (in Japanese).

Sugiyama, M., and Shikano, K. (1981). LPC peak weighted spectral matching measures. *Trans. IECE Jpn. J64-A*:5 (in Japanese).

Voiers, W. D. (1977). Diagnostic acceptability measure for speech communication systems. *Proc. 1977 IEEE ICASSP*, pp. 204–207.

Voiers, W. D. (1982), Measurement of intrinsic deficiency in transmitted speech: The diagnostic discrimination test. *Proc. 1982 IEEE ICASSP*, pp. 1004–1007.

Watanabe, T., Nagabuchi, H., and Kitawaki, N. (1987). Law of addition for subjective quality of voiceband CODECs. *Rev. Electr. Commun. Lab., NTT, Japan 35*(4): 435–441.

Watanabe, T., Nagabuchi, H., and Kitawaki, N. (1988). A word-selecting method for intelligibility assessment of synthesized speech by rule. *Trans. IEICE Jpn. 71-A*(3): 616–623 (in Japanese).

Zelinski, R., and Noll, P. (1977). Adaptive transform coding of speech. *IEEE Trans. Acoust., Speech, Signal Process. ASSP-25*:299–309.

13

Quality Assessment of Text-to-Speech Synthesis by Rule

LOUIS C. W. POLS Institute of Phonetic Sciences, University of Amsterdam, Amsterdam, The Netherlands

1. INTRODUCTION

The present speech quality of text-to-speech synthesis-by-rule systems is certainly good enough for specific applications but is far from being anywhere near the quality of natural speech. Some deficiencies of present-day systems at different levels of processing, from text input to spoken output are listed below; this list is taken from Pols (1988a), but similar remarks can be found in Klatt (1987).

- Written texts are prepared mainly for reading purposes; it is not yet known what the requirements would be for optimal text structures meant to be spoken.
- Semantic and pragmatic knowledge is not used because it cannot be deduced automatically from text.
- Complex grammatical structures cannot be identified.
- Word stress and sentence accent are not well indicated.
- Variable speaking rates and appropriate pauses are not realized.
- The system generally speaks with only one (male) voice.

- Various voice characteristics, speaking styles, and emotional states cannot be identified or realized.
- Appropriate intonation contours are frequently not achieved.
- Other prosodic aspects need more attention.
- Synthetic speech is generally overarticulated, whereas unstressed syllables are not treated as such.
- Segmental characteristics are suboptimal.
- Systems should have a more natural speech quality, allowing for extended listening, also in noise, over the telephone, and by the hearing impaired and children.
- At least in Europe, a universal approach for many different languages would be desirable.

For these and probably many other reasons, more research is required to improve the quality of text-to-speech systems. This is as much a linguistic problem as an acoustic-phonetic and a technical problem. Various disciplines should work together to achieve optimal results.

The present chapter is concerned with the *methodology* of performance assessment as well as with the actual performance *results*. A very interesting type of assessment is that of diagnostic evaluation: one can learn from the errors, and one can subsequently try to improve the system under consideration. There are other useful types of evaluations that are mainly of a descriptive, comparative, and global nature, resulting in statements such as: the naturalness of system/algorithm A is judged to be better than that of system/algorithm B.

Since, after all, it is the human user who has to listen to the synthesized speech, it is the judgment of the listener that counts most. This implies that, contrary to speech recognition assessment, subjective assessment by listeners is required. Nevertheless, efforts are also under way to develop objective evaluation methods, which ideally should give similar results. Here one can note an analogy with the quality assessment of coded speech and speech communication channels, where methods such as articulation index (AI), speech transmission index (STI), and signal-to-noise ratio (SNR) sometimes provide good alternatives to subjective evaluation by speakers and listeners.

At present there is growing interest in the topic of speech technology evaluation. We refer to the proceedings of two workshops in this field, the first one in 1982 in Gaithersburg, Maryland (Pallett, 1982), and the second in Noordwijkerhout, The Netherlands in 1989. Esprit project 2589—multilingual speech input/output assessment, methodology, and standardization (SAM for short)—has an active working group on speech output assessment (Fourcin et al., 1989). Overviews can be found in O'Malley and Caisse (1987), Pascal (1987), Pisoni et al. (1985), van Bezooijen and Pols (1989a), and Calliope (1989, chapter XVI). In various national research programs attention is also given to the topic of speech output evaluation (e.g., Pols, 1988a).

In subsequent sections I will concentrate on various aspects of quality assessment of *rule-synthesized* speech. The limitation of rule-synthesized speech will be clear, considering the presence of chapter 12 in this book on quality assessment of coded speech. I will make a distinction between *subjective* assessment (via listener judgments) and *objective* assessment (via physical means). Another distinctive dimension is that of *diagnostic* versus *global* evaluation. Furthermore, I will differentiate between *segmental* and *suprasegmental* evaluation. Here, the segmental level is mainly limited to phoneme intelligibility, whereas the suprasegmental level concentrates on (word and) sentence aspects, including prosody. The higher the quality of the acoustic-phonetic part of a text-to-speech synthesizer, the more attention can be devoted to the *linguistic* components, such as text preprocessing, grapheme-to-phoneme conversion, and stress assignment, which also require some form of evaluation. A final distinction could be that between *laboratory* tests and *field* tests or application-specific tests, implying a fully integrated system, dialogs with inexperienced users, noisy environment, and the like.

2. SHORT OVERVIEW OF ACOUSTIC AND LINGUISTIC ASPECTS OF RULE SYNTHESIS

Although chapter 24 explores text-to-speech synthesis systems, I will start by describing briefly the three basic front-end types of synthesizers, mainly with respect to the specific system characteristics that might have implications for the test material and test method used (Sections 2.1, 2.2, and 2.3). Consider, for instance, the segmental evaluation of a diphone-based system. The commonly used set of CVC test words is insufficient for testing the full diphone inventory, which includes, apart from CV- and VC-type units, also VV- and CC-type units (V stands for vowel, C for consonant).

Next I will summarize the various steps that have to be taken to get from unrestricted text to natural-sounding speech (Section 2.4). Each step, such as text preprocessing, morphological parsing, or stress assignment, in principle requires a separate evaluation. See Klatt (1987) for an outstanding review of all aspects of text-to-speech conversion.

2.1. Articulatory-Based Synthesis

Articulatory-based synthesis is mainly of research interest. As far as I know, no commercial products have been developed along this line. However, the concept is very attractive because the control parameters are specified in terms of the voice and articulation mechanism itself, implying that natural boundaries and interactions are assured. The best-known laboratory systems along this line are those developed by Coker (1973) and Scully (Alwood and Scully, 1982; Scully, 1987). These systems would, in principle, allow variation in source characteristics and systematic evaluation of source-filter interaction.

2.2. Diphone-Based Synthesis

I use the term "diphone" here for all the various prestored (and parameterized) units used as building blocks in synthesis, whether these are diphones, demisyllables, syllables, dyads, transemes, or other elements one has chosen, or whether one has used other names. The general characteristics of such systems are:

1. Rather high basic quality, because the difficult-to-control phoneme transitions are already included in the units themselves. (However, to get beyond that basic quality is very difficult.)
2. Uniqueness to one voice. For every new speaker a new diphone inventory has to be prepared. (Semiautomatic procedures have been developed for this purpose.)
3. Difficulty in introducing unstressed syllables, rate changes, or changes in articulation quality.

For quite a few different languages one or more diphone vocabularies and synthesis systems have been developed (for an overview see Pols, 1989). Since the diphone vocabulary itself is of crucial importance for the overall quality of these systems, a systematic evaluation is worthwhile. In order to include all diphones, the choice of CVVC- and VCCV-type nonsense words seems to be required.

2.3. Allophone-Based Synthesis

Potentially, allophone-based synthesis is a very powerful method: the better the rules, the better the speech quality is. Every phoneme, with its allophonic variations, is described in some parametric form, generally by using formant parameters. Every transition is rule-governed, as are changes related to stress, accent, speaking rate, and so on. In principle, it should also be possible to control articulation quality, voice, and emotional aspects by applying appropriate rules. The problem here lies on the analysis side and is related to our limited knowledge of the whole process of speech production and perception.

The best-known system following this approach is MITalk (Allen et al., 1987), together with its various, partly commercialized, versions (e.g., Bernstein and Pisoni, 1980). The MITalk system has been very well evaluated at various levels, mainly by Pisoni and his colleagues. The knowledge put in the MITalk system by D. Klatt cannot easily be transferred to other languages, although various attempts are being made, for instance, for German (Rigoll, 1987). Several other multilingual systems exist, such as Infovox (Carlson et al., 1982) and Multivox (Olaszy, 1989). (See also Javkin et al., 1989, and van Rijnsoever, 1988.)

2.4. From Text to Speech

Ordinary text input will certainly contain elements that require *text preprocessing*. This implies at least abbreviations and all kinds of numerals. A performance evaluation of such algorithms can be done on paper.

Once every element is fully written out, the next step is *grapheme-to-phoneme conversion*. This can be done via lexicon search (always incomplete) and/or by rules (never perfect), with or without morphological decomposition. The syllable boundary and word stress assignment are other aspects of this conversion step. For certain applications this whole component can be bypassed by simply applying a pseudophonetic input. Evaluating all aspects of this conversion step is not an easy task. At the word level, a phonematized dictionary might be the criterion to test against, but this will not suffice for between-word effects and grammatical ambiguities. An evaluation on paper alone may not tell the whole story, since acoustic realization may mask as well as enhance certain errors.

The *prosodic parser* should provide information about phrasing (syntactic boundaries) and accentuation (sentence accent). Since syntactic and semantic analysis is still in its infancy, these syntactico-prosodic parsers are generally simply based on recognizing what the function words (closed set) and the content words are and on punctuation marks.

The *intonation grammar* should provide the actual pitch contour on the basis of intonation markers derived earlier.

Phonetic context, sentence accent, and syntactic boundaries all influence the *duration module*. The limited set of rules originally developed by Klatt (1979) for American English is still the best that is available at this point.

Finally, the *units* should be chosen for the actual acoustic realization. This is especially relevant for advanced diphone-based systems in which stressed as well as unstressed diphones are available and certain phoneme combinations are stored separately as triphones or even quadrophones.

It is rather difficult to evaluate individual components, such as the prosodic parser or the duration module, in an independent way, since the acoustic realization also always implies all other aspects. Subjects in listening experiments generally have great difficulty basing their judgments on one signal aspect only, while ignoring all other aspects.

3. WAYS TO EVALUATE THE QUALITY OF TEXT-TO-SPEECH SYSTEMS

3.1. Text Preprocessing, Grapheme-to-Phoneme Conversion

There is very little experience in evaluating the *text preprocessing* part of a text-to-speech synthesizer. For instance, McAllister et al. (1986) give some specifications of the preprocessor module of the CSTR text-to-speech system, though without quantifying the performance of that module. It is actually not always clear what the final result should be. In proofreading, one might prefer to have some form of pronunciation for every character in the text, but in normal reading one would probably prefer full pronunciation of abbreviations such as Dr., but not of others such as MIT, whereas, for instance, the parentheses around (1986) would normally not be pronounced at all.

Grapheme-to-phoneme conversion has received more attention, but again performance figures are scarce. Emerard and Lienard (1980) give some preliminary results for a number of grapheme-to-phoneme conversion approaches for French. Marty and Hart (1985) compiled a list of words and sentences presenting the main conversion difficulties in French. It might be possible to evaluate the performance of grapheme-to-phoneme conversion by comparing its output against a large phonematized dictionary of that specific language, supposing that the algorithm used did not already use the same dictionary. Goudie-Marshall (1984) describes such a dictionary for American English computed at TI. For the 11,000 most frequently used words, one or more (maximum 12) pronunciations are stored. Van den Broecke et al. (1987) describe a similar phonematized corpus of the 12,116 most frequently used words in Dutch taken from a corpus of 2.3 million word forms. However, they gave no alternative pronunciations.

Most phonematized dictionaries presently available give one ideal phonetic transcription only, sometimes with stress markers included, whereas one would like to see a number of realistic alternative pronunciations. An even more serious problem is the fact that these dictionaries can be used only for words in isolation, since they do not give alternatives for between-word interaction.

A practical problem related to that of grapheme-to-phoneme conversion is that of the *pronunciation of names* (family names, given names, place names, business listings), with the additional difficulty of the foreign origin of many of these names, which is sometimes respected in the pronunciation but frequently not. Furthermore, there is substantial individual and regional variation.

Spiegel (1985) presents the results of an evaluation of text-to-speech synthesizers for the pronunciation of surnames; at that time synthesizers pronounced only about 75% of the 2000 most common American surnames correctly. Meanwhile, progress has been made in improving that performance (Spiegel et al., 1989c). A similar development is under way for Swedish (Carlson et al., 1989).

Larreur and Sorin (1977) compared the intelligibility of 100 natural and diphone-synthesized French surnames over the telephone line, both at the word and at the segmental level. The intelligibility of synthesized surnames appeared to improve by 10% when they were embedded in carrier sentences of the type "Le № de téléphone de Mme X a été modifié." Due to the specific nature of the tested words, this improvement cannot be attributed to lexical predictions but probably reflects the short-term adaptation of the listener to synthetic speech.

An interesting multilingual test in grapheme-to-phoneme conversion, although limited to the isolated word level, has been performed in Esprit project 291/860, "Linguistic analysis of European languages." In the near future a full report about this test and many other aspects of this project will become available in book form. For seven different European languages, four test corpora each were defined, consisting of:

1. Newspaper texts of at least 100,000 words each
2. 150 names of national towns with the highest number of inhabitants
3. 31 city names of the European capitals in the equivalent spelling for each language
4. 150 most frequent Christian names for each language

The results of the conversions are summarized in Table 1. One should not just compare figures over languages, however, because the approaches differed widely. For some languages they were just state of the art, whereas for other languages the latest developments were involved. For Greek, Italian, and Spanish, almost pure sets of letter-to-sound rules were used, whereas for

Table 1 Grapheme-to-Phoneme Conversion Accuracy for Seven Different European Languages[a]

Language	Newspaper	National towns	Capitals	First names
Dutch	98.9	85.3	96.7	89.4
English	90.3	46.0	58.0	58.0
French	96.9–94.5	77.3	74.2	84.7
German	93.0–90.0	81.0	61.0	80.0
Greek	98.7	97.3	93.5	97.3
Italian	85.2	85.3	80.6	86.9
Spanish	98.9	95.3	96.6	98.0

[a]The newspaper score is weighted for the frequency of word occurrence. The higher, first score for German involves the use of an exception list. The higher score for French is a rerun with exclusion of abbreviations, acronyms, digits, etc.

the other languages some form of preanalysis (such as decomposition of compounds, subdivision of words into affixes and root) was performed on each word before such rules were applied. For English, the conversion was mainly based on a not-yet-complete morph dictionary, which caused great problems for the names.

3.2. Segmental Evaluation

As long as the phoneme intelligibility for most rule synthesizers is as low as it is at present, there is ample reason for continuing to perform segmental intelligibility tests in order to improve the performance at that level. The phoneme intelligibility can be considered the basis for the final speech quality of any rule synthesizer.

Long before speech synthesizers were developed, there was already a need for well-defined speech material for evaluating the intelligibility of communication channels and coded speech or for testing the performance of hearing-impaired listeners. This resulted in articulation testing methods (Fletcher and Steinberg, 1929; Egan, 1948), the rhyme test (Fairbanks, 1958), the modified rhyme test (House et al., 1965), the diagnostic rhyme test (Voiers, 1977a,b, 1982), the use of phonetically balanced (PB) CVC nonsense words, and various speech audiometric word lists (Bosman, 1989). From the time when such tests were also used in evaluating rule synthesizers, several more word types have been suggested: VCV (Carlson and Granström, 1989),

CVVC and VCCV (Pols et al., 1987), CV syllables (Nusbaum et al., 1984a), and the rhyme and disyllable test designed to evaluate the Chinese tonal language (Zhou, 1986). For an overview of various word types used in segmental tests, see Table 2.

Various characteristics of these tests deserve attention (Pols, 1989):

- Nonsense words vs. meaningful words
- Isolated words vs. words in carrier phrases
- Open vs. closed response set
- Frequency of occurrence of phonemes
- Free vs. fixed word forms such as CVC
- Language (in)dependence

On the one hand, one can imagine the situation with close-to-natural speech material, untrained listeners, and more interest in overall performance than in diagnostics, such as in speech audiometry. This leads to the use of meaningful words with few response alternatives. There can be a need, however, especially in diagnostic comparative evaluating of rule synthesizers, for diagnostic word material. This leads to the use of nonsense words with an equal frequency of occurrence of all phonemes, an open response condition, and, frequently, some training of the subjects.

In the Esprit project SAM, the speech output assessment working group has decided to agree on "a test which is language-specific but comparable across languages in order to evaluate in a diagnostic way the intelligibility of processed speech at the phoneme level." This has resulted in a segmental test comprising two parts: a small subset fixed per language, and a variable

Table 2 Various Word Types Used in Segmental Intelligibility Tests

Syllable/word type	Meaningful vs. nonsense	Reference as an example
CV and VC syllables	Nonsense	Nusbaum et al., 1984a
VCV	Nonsense	Carlson and Granström, 1989
PB words (CVC)	Nonsense	Schwab et al., 1985
CVC	Meaningful	MRT and DRT
CVC	Both	Bosman, 1989
CVVC and VCCV	Nonsense	Pols et al., 1987
C^nVC and CVC^n	Both	van Bezooijen, 1988
Polysyllabic words	Meaningful	Hirsh et al., 1954
Rhyme and disyllable	Both	Zhou, 1986

application-specific set. The fixed subset consists of a limited set of VCV, VC, and CV words where *all* single consonants per language are tested in initial, medial, and final positions, in combination with a close front, a close back, and an open vowel, such as /a, u, i/. For languages such as Italian, in which closed VC syllables are virtually nonexistent, it may be better to limit the word types to VCV and CV.

The second part of the segmental test can be more flexibly tuned to system characteristics and test requirements. Only the general structure is defined, not the actual content. In its most extended form the second part may be a full set with complex structures such as CVC, CVVC, VCCV, or even C^nVC, or CVC^n, in which not just all phonemes but also all phoneme combinations are represented. Despite the fact that the majority of these words will be nonsense words, phoneme combinations should follow the phonotactic constraints in each specific language.

Table 3 shows some results of using CVVC and VCCV words to evaluate the segmental intelligibility of an Italian, a French, and a Dutch diphone-based system together with several reference conditions. One should be careful in interpreting these results, since conditions over languages are very different, for instance, with respect to language complexity, subjects, number of words, and interstimulus intervals. Per language, however, one can

Table 3 Phoneme and Word Score for CVVC- and VCCV-Type Words for Various Italian, French, and Dutch Synthesis Systems[a]

Phoneme/ word	Italian			French		Dutch		
	PCM	LPC-15	Rule	LPC-10	Rule	LPC-10	Allophone	Diphone
V	89.1	90.2	89.5	65.1	62.1	91.9	70.8	74.3
C	86.9	79.4	68.3	63.2	52.8	76.6	35.0	59.8
C	94.0	91.2	78.1	76.9	63.6	81.5	31.5	60.7
V	79.7	79.9	87.8	65.6	62.1	96.3	80.7	92.8
VCCV	57.8	52.5	45.0	22.9	17.6	58.3	10.1	31.5
C	94.3	88.4	74.4	78.8	60.6	74.7	33.3	67.0
V	90.4	90.8	86.5	63.4	45.1	87.5	65.2	90.8
V	84.1	85.3	84.9	61.5	42.2	77.7	64.3	62.5
C	88.2	87.5	76.4	81.8	64.4	84.5	47.6	80.4
CVVC	63.2	60.2	44.4	32.2	13.6	47.0	6.8	3.3

[a] For more details, refer to the original publications by van Son et al. (1988), Pols et al. (1987), and van Bezooijen and Pols (1987), respectively.

compare over conditions. Apart from the overall correct scores, additional information can be derived from the confusions. Some systems also use triphones and even quadrophones, mainly for difficult transitions, such as VhV, and consonant clusters. So far, there is very little experience in evaluating the intelligibility and naturalness of consonant clusters (van Bezooijen, 1988; Spiegel et al., 1989b).

Once the test method and the test material are well defined, it is possible to use computer-controlled procedures to generate the word lists in appropriately randomized form, to control the presentation of the stimuli, to score the responses (preferably by using keystroke input) as well as to process the responses (semi-)automatically, and to present the data in graphical and statistical forms. Several programs have been developed to align the given responses with the presented stimuli, at both the phoneme and the word level. They generally make use of some form of dynamic time warping and sometimes also of phoneme-distance matrices to allow for judging whether a certain error most probably is a substitution or an insertion or deletion (Picone et al., 1986; Lefèvre et al., 1988; Maret, 1987).

I highly promote the use of *open responses* in diagnostic tests. Logan et al. (1989) present convincing data to support this view, as indicated in Table 4. All systems tested with an open response set show about four times higher error scores than in the closed response condition (six alternatives). Although it generally does not change the position of the system on a good-to-bad scale, it gives a much better idea of the remaining deficiencies of the segmental quality, for instance, by studying the confusion matrices. Such a matrix shows *all* occurring confusions, whereas with a closed response set the confusions are limited to the two (DRT) or six (MRT) alternatives presented. Logan et al. (1989) present only percentage error scores per consonant for the five most frequently occurring errors, but the patterns there already differ for the open- vs. closed-response conditions. Actually, their open-response condition was not even completely open because the stimulus words were all meaningful and therefore, most probably, the responses given by the listeners will have been meaningful as well.

In the monosyllabic test suggested by Spiegel et al. (1989b) this criticism is taken into account by combining nonsense words and meaningful words in one test. In Pols and Olive (1983) another example is given of the potential danger of using fixed responses: the most frequent error of all (from b to v) in that specific test is barely covered in the six-alternative MRT and would therefore have remained undetected if MRT had been used instead of an open-response test with nonsense CVC words.

Further arguments against closed-response tests such as the DRT have been put forward by Greenspan et al. (1989). They point at possible ceiling

Table 4 Percentage Error Rates for Initial and Final Consonants Using the Modified Rhyme Test for Several Synthesis Systems as Well as a Natural Voice[a]

	Six Alternatives			Open response set
System	Initial	Final	Overall	Overall
Natural speech	0.50	0.56	0.53	2.78
DECtalk 1.8, Paul	1.56	4.94	3.25	12.92
DECtalk 1.8, Betty	3.39	7.89	5.72	17.50
MITalk-79	4.61	9.39	7.00	24.56
Prose 2000 3.0	7.11	4.33	5.72	19.42
Amiga 500	13.89	10.61	12.25	42.89
Infovox SA 101	10.00	15.00	12.50	37.14
TSI prototype 1	10.78	24.72	17.75	?
Votrax Type'n'Talk	32.56	22.33	27.44	56.89
Echo	35.56	35.56	35.56	73.97

[a]The overall error rate is the average over initial and final consonants. For a subset of the systems, the overall error rate using an open-response format is also given in the last column. These data are taken from Logan et al. (1989).

effects for the DRT and even more importantly, to the fact that the DRT examines only single-feature confusions, whereas frequently multiple-feature confusions occur when running open-response consonant identification tests.

For completeness, Tables 5 and 6 show an overview of various segmental intelligibility tests of synthesizers using DRT or MRT. After all, Pisoni et al. (1985) qualify the modified rhyme test as the "de facto standard measure of the segmental intelligibility of synthetic speech." I disagree with them, not just for the reasons already mentioned but also because of the fact that the MRT, just as the DRT, is "standardized" only for American English, although (comparable?) versions have been developed for a few other languages (Peckels and Rossi, 1973; Sotscheck, 1982; Steeneken, 1982).

In contrast to natural speech, synthetic speech is very sensitive to distortion such as *masking noise*. Pratt (1986) shows 30% and more reduction in DRT percentage correct scores at a signal-to-noise ratio of 0 dB, both for good systems such as DECtalk and for simpler systems. For natural speech this drop in DRT score is only about 15%.

Table 5 Two Phoneme Intelligibility Studies for Speech Synthesizers Using the Two-Alternative Diagnostic Rhyme Test (DRT)

Systems and conditions	Language	Reference
DECtalk, three voices, two runs + six other synthesizes + human; clear as well as noise masked	British English	Pratt, 1986
Three CNET diphone systems	French	Graillot et al., 1983

Another type of distortion is the *telephone channel.* Although there are potentially many useful applications of the use of synthetic speech over the telephone, it is surprising to realize that very few tests so far have been done for this type of condition. Spiegel and his colleagues are among the very few working in this area (Spiegel et al., 1989a,b,c). They also developed an interesting test corpus for segmental evaluation, consisting of monosyllabic CVC-type words in which the initial or the final consonant can also be a cluster, representing onsets and plain or affixed offsets. One half of the corpus consists of words and one half of nonwords. Listeners are encouraged to write down exactly what they hear (open response). Averaged over two male speakers, the percentage correct word score for natural speech was 88.3%, which is, of course, lower than for the laboratory-condition, closed-set, single-consonant MRT score for natural speech of about 99%. The tested Bellcore demisyllable synthesizer yielded 72.4%, which is nice but still some 15% lower than for natural speech. Therefore the authors, rightly, conclude that the basic hurdle of consonant intelligibility in practical environments has not yet been overcome by current synthesis systems.

Table 6 Some Phoneme Intelligibility Studies for Speech Synthesizers, Using the Six-Alternative Modified Rhyme Test (MRT)

Systems and Conditions	Language	Reference
Haskins lab. system + natural	American English	Nye and Gaitenby, 1973
Ten systems + natural speech (closed as well as open, respectively)	American English	Logan et al., 1989
CNET-IPG diphone system	German	Schnabel, 1986
Two × Votrax and Infovox (several test series)	German	Ruehl and Kulas, 1985

3.3. Evaluation at the Word and Sentence Level

The quality of synthesized speech in terms of aspects like word frequency, predictability, syllable structure, and grammatical structure is judged at this level of evaluation (word and sentence intelligibility). Another important aspect at the sentence level, namely the prosody (word stress, word duration, sentence accent, pauses, intonation), will be discussed separately in the next section.

A straightforward extension of the mainly monosyllabic words mentioned in the preceding section would be the use of multisyllabic words. However, apart from the spondaic words sometimes used in speech audiometry, I know of no examples of that. Instead, short, phonetically balanced sentences have been developed, for instance, for English (Egan, 1948), French (Combescure, 1981), and Dutch (Plomp and Mimpen, 1979). These are fixed, limited sets of well-designed, meaningful sentences for which the percentage of correctly identified (key) words is scored. Because of strong learning effects, the sentences can be used only once per listener.

The 100 English, so-called Harvard psychoacoustic sentences, have been extended by Nye and Gaitenby (1974) with 100 semantically anomalous, so-called Haskins, sentences. These last have a fixed grammatical structure of the form "The ADJ NOUN1 VERB the NOUN2" (e.g."The late voice knew the table"). They have been used repeatedly by Pisoni and colleagues to measure word identification in sentence context for various synthesis systems (Pisoni and Hunnicutt, 1980; Greene et al., 1984; Manous et al., 1984). (See Table 7.) For the (natural) Harvard sentences there is some danger of ceiling effects. For the much less predictable Haskins sentences the difference in quality between natural and synthetic speech becomes immediately clear again.

In the Esprit SAM project, suggestions have been made to extend the idea of semantically anomalous sentences in various ways:

1. By using five different structures, instead of one fixed grammatical structure
2. By making the test multilingual by choosing identical, or at least similar, structures over languages
3. By fixing the form of the sentences but not the sentences themselves, thus creating a virtually unlimited reservoir of different sentences

For more details, see van Erp and Grice (1989), Benoit et al. (1989), and Grice (1989). The general structure and example sentences for English, Dutch, and French are given in Table 8. Each time, under (a), the functional sentence

Table 7 Word Recognition in Two Types of Sentence Context[a]

System	Harvard	Haskins	Reference
MITalk-79	93.3	78.7	Pisoni and
Natural Speech	99.2	97.7	Hunnicutt, 1980
DECtalk			
Paul	95.3	86.8	Greene et al., 1984
Betty	90.5	75.1	
Speech Plus	83.7	64.5	Manous et al., 1984
Prose-2000 prototype			

[a]Harvard sentences (syntactically correct and meaningful) and Haskins sentences (syntactically correct but semantically anomalous) for several synthesis systems as well as for natural speech.

elements, such as subject, verb, and direct object, are specified; these are constant over languages. Under (b) the syntactic categories, such as determiner, adjective, noun, and verb, are given. For these syntactic categories minor local order differences between languages can occur. For languages in which determiners, adjectives, and nouns are marked for gender, local congruency is not violated (e.g., "*La* rue vert*e*" and not "Le rue vert"). Results of tests in which these sentences were used to evaluate various synthesis systems for English and French can be found in Hazan and Grice (1989), Benoit (1989a), and van Son and Pols (1989) (See Table 9). The approach is interesting but still requires further study because of its apparent sensitivity to such factors as repeated presentation, training, and word frequency. These are not necessarily drawbacks because they can also be manipulated in a systematic way. Benoit (1989b) tries to quantify the amount of context redundancy in these sentences. It will be clear from the results in Table 9 that, for this type of sentence also, natural speech has a substantially better word intelligibility than any tested rule synthesizer.

Manous et al. (1985) used sentence material in a slightly different way, namely by applying the "sentence verification task." Subjects had to judge as fast as possible whether presented three- or six-word sentences were true statements (e.g., "Mud is dirty"; "Birds fly south for the winter") or false (e.g., "Rockets move slowly"; "Beer is a popular contact sport"). After the true/false reaction, they also had to transcribe the sentence. Results (in terms of verification response latency, verification accuracy, and sentence transcription accuracy) showed that this task, like most reaction time tasks, is a very

Table 8 The Five Different Grammatical Structures Used, Each with Its Functional Sentence Elements, Syntactic Categories, and a Semantically Anomalous Example in Three Different Languages[a]

I. Declarative with adverbial				
(a)	Subject	Verb	Adverbial	
(b)	Det-noun	intrans. verb	Prep-det-adj-noun	
English:	The table	walked	through the blue truth	
Dutch:	De stoel	loopt	door een lief huis	
French:	La robe	entre	vers la science rouge	
II. Declarative (Haskins structure)				
(a)	Subject	Verb	Direct object	
(b)	Det-adj-noun	trans. verb	det-noun	
English:	The strong way	drank	the day	
Dutch:	Een warm bot	drinkt	de dag	
French:	Le verre vrai	ouvre	le coin	
III. Imperative				
(a)	Adverbial-verb	Direct object		
(b)	*Adverb*-trans. verb	Det-noun-conj-det-noun		
English:	*Never* draw	the house and the fact		
Dutch:	Eet *vaak*	het boek en de pen		
French:	Tourne *peu*	la date ou la main		
IV. Wh question				
(a)	Q-adverbial	Verb-subject	Direct object	
(b)	Q-adv	(*Aux*)-det-noun-*trans. verb*	Det-adj-noun	
English:	How	*does* the day *love*	the bright word?	
Dutch:	Hoe	*eet* het woord	een snel glas?	
French:	Quand	le texte *pose-t-il*	la fille crue?	
V. Declarative with relative clause				
(a)	Subject	Verb	Complex direct object	
(b)	Det-noun	trans. verb	Det-noun	Rel pron-intrans. verb
English:	The plane	closed	the fish	that lived
Dutch:	De vloer	sloot	de vis	die liep
French:	La chose	lance	le train	qui pense

[a]*Source*: Grice (1989).

sensitive procedure for studying comprehension processes. In Pisoni et al. (1987) the predictability of the last word in such three- or six-word sentences was varied. Subjects were always faster for natural speech and always faster for highly predictable sentences. Very few transcription errors were made, and response accuracy was consistently high across all conditions. Still, the

Table 9 Percentage of Correct Words in Semantically Anomalous Sentences for a Number of Different Rule Synthesizers

Type of synthesizer	Language	Natural	Synthetic	Reference
JSRU synthesis by rule	British English	98.8	83.4	Hazan and Grice, 1989
OROS diphone synthesis	French	?	80.7	van Son and Pols, 1989
PSOLA-KDG				
flat prosody	French	98.3	88.4	Benoit, 1989a
prosody model			91.2	
Formants				
flat prosody			71.6	
prosody model			72.5	

results show that even high-quality synthetic speech is not perceived in the same way as natural speech.

Pisoni and Koen (1981) investigated the effect of *noise masking*, and Bacri and Graillot (1986) studied the effect of noise on sentence intelligibility, not just for two French rule synthesizers but also for LPC-resynthesized speech, as well as for natural speech. Manous et al. (1984) explored the effect of signal duration in a word gating paradigm.

Noise, or more specifically the signal-to-noise ratio, can also be used in an indirect way to evaluate and classify various synthesis systems. Bacri and Graillot (1986) studied the ability of such a method based on identifying a set of phonetically balanced sentences (Combescure, 1981) in noise (equivalent signal-to-noise-ratio method). The four tested systems were natural speech, LPC-16 resynthesized speech, and two LPC diphone-based synthesis systems, one with a male and the other with a female voice. The results suggest that the method is sensitive enough to make a fine classification of various synthesis systems, without being diagnostic, of course. The main advantage of this method is that is furnishes a "comprehension facility" score expressed in "equivalent SNR," a measure already used for evaluating speech coders. This method was recently proposed by the French PTT Administration (1987) to the Commission of the International CCITT.

3.4. Prosodic Evaluation

To the best of my knowledge, no standard procedures have yet been developed for evaluating the prosodic characteristics of rule-synthesized speech.

The higher the basic phoneme, word, and sentence intelligibility becomes, the more important prosody will become in the future for further improvement of the synthesized speech in terms of acceptability and naturalness. So far, mainly paired comparison procedures have been used: preference judgments for one sentence over another, mainly by introducing small specific changes at a time. (See for instance Collier et al., 1989, and Terken and Collier, 1989.)

Martin (1978) investigated the effect of different F_0 *contours* on the judged quality of synthesized sentences using the CNET diphone synthesis system. Seven subjects judged the naturalness of many different contours on a five-point scale.

Monaghan and Ladd (1989), as well as van Bezooijen and Pols (1989b), asked subjects to judge adequacy or appropriateness while intonation contours or *sentence accent* assignments were manipulated.

Quazza et al. (1989) report some initial results concerning the ability of a prosodic preprocessor to correctly position *pauses* as prosodic markers on the basis of a syntactic analysis of sentences. The prosodic parser of Larreur et al. (1989) for French is said to be able to specify pauses, segment duration, and pitch patterns but no performance figures have been reported yet.

A preliminary evaluation of the *duration rules* developed by Klatt (1979) was performed by Carlson and Granström (1979). They asked nine phonetically trained subjects to order various versions of nine short sentences according to "naturalness of the durational structure" and then to assign a number between 0 and 100 to each version. The rule-synthesized version obtained a score very close to that for the reference condition (synthesis by rule using the measured durations from natural speech). An intelligibility experiment was also done with 18 inexperienced subjects (for six versions only). On average, 12 of the 122 possible words (excluding articles) in these nine sentences were misidentified for the rule-synthesized version. A maximum of 24 misidentified words was found when only a very simple duration rule was applied. There was a clear correlation between the word identification rate and the naturalness score.

In some preliminary experiments the *speaking rate*, as one of the characteristics of rule synthesis, was systematically varied (Luce et al., 1983; Carlson et al., 1976; O'Shaughnessy, 1986). Simpson and Marichonda-Frost (1984) used semantic differential scales to measure the effects of speech rate and pitch on the acceptability of synthetic speech generated by the VOTRAX ML-1 synthesizer.

One of the suggestions made by U. Jekosch (personal communication, 1989) is to ask subjects to *imitate* synthetic speech and then to study those productions as an indication of the appropriateness of prosody.

3.5. Overall Quality

Especially in evaluating the quality of speech coders, various overall measures have been introduced, such as the diagnostic acceptability measure (DAM) (Voiers, 1977a,b) and the mean opinion score (MOS) (Goodman and Nash, 1982). In evaluating the overall performance of rule synthesizers, we might also end up using such scores, although their diagnostic value and their precision are very limited.

There is some interest in direct comparison of systems in order to evaluate whether a change in the synthesis procedure has an effect on the listener's attitude to a specific system. One possibility is direct scaling of intelligibility, naturalness, and acceptability with the magnitude estimation technique. Pavlovic et al. (1989a,b,c) systematically evaluated this technique of direct numerical estimation of the sensory magnitudes on different systems and different distortions. They found little effect of number and range of test conditions or of practice. The magnitude estimates of intelligibility showed a high correlation with actual intelligibility scores (number of words correct). They also found good correlation between magnitude estimates of the "quality" of several synthesis systems with different prosodic rules and pair-comparison quality preference results. However, van Bezooijen (1988) found a poorer correlation between measured consonant cluster intelligibility and intelligibility judged on a 10-point scale.

3.6. Linguistic and Psychological Evaluation

The reason for introducing this extra level is that one does not just want to study the acoustic aspects of the generated words and sentences, but also the linguistic and psychological aspects: Can the sentences be reproduced? How large is the memory load? Do differences exist in the perception of natural and synthetic speech? Can one listen to synthetic speech for extended periods of time? Can one reproduce the gist of a story? What about the surface properties? Can one comprehend the prose? Do children have more difficulty with synthetic speech? Pisoni and his co-workers at Indiana University have the most experience with this type of testing, although it is still in its infancy.

Nusbaum and Pisoni (1985) address the question of *why* natural and synthetic speech are perceived differently. Several studies related to *processing time* for words and nonwords give a partial answer to that question.

In a lexical decision task (classifying a stimulus as fast as possible as either a "word" or a "nonword"), the mean response times for correct responses were significantly faster for natural words (903 ms) and natural nonwords (1046 ms) than for synthetic words (1056 ms) and synthetic nonwords (1179 ms) (Slowiaczek and Pisoni, 1981). Thus, the phonological encoding of synthetic speech appears to require more cognitive effort. Also the naming latency for synthetic words and nonwords is much larger than for natural ones (Pisoni, 1982b), again suggesting more processing time for synthetic speech.

Short-term memory (STM) experiments (recalling digits and recalling words in serial order; Luce et al., 1983) demonstrate that synthetic speech requires more STM capacity than natural speech. This additional load on STM was also clear from a decreased priming effect. Also, word monitoring performance was better for subjects listening to natural speech than for those listening to synthetic speech (Ralston et al., 1989).

Nusbaum and Pisoni (1985) suggested "that the problems in perception of synthetic speech are largely tied to the processes that encode and recognize the acoustic-phonetic structure of words." Synthetic speech is also less redundant and therefore a larger part of the stimulus is required before a word can be identified in a word gating task (Manous et al., 1984). Results of Slowiaczek and Nusbaum (1983) have shown that "the contribution of suprasegmental structure to the intelligibility of synthetic speech was quite small compared to the effects of degrading acoustic-phonetic structure." This might be true for intelligibility, but according to the present author's opinion this is certainly not true for acceptability or naturalness. Nusbaum et al. (1984a) concluded from a consonant identification experiment with CV syllables produced by three text-to-speech systems and a natural male voice at SNR = +28, 0, −5, and −10 dB that synthetic speech can hardly be interpreted as "noisy" speech but that it really has a deviating acoustic-phonetic structure.

In order to formulate some idea of *listening comprehension*, Pisoni and Hunnicutt (1980) selected 15 narrative passages and appropriate sets of multiple-choice test questions taken from several standardized adult reading comprehension tests. One group of subjects listened to the synthetic version of these passages, whereas a second group listened to natural speech and a third group read the passages. All three groups answered the same set of test questions immediately after each passage. Reading gave highest performance with an overall advantage of 7% over the synthetic speech group. However, this difference appeared only in the first half of the test. Natural speech showed no better performance than synthetic speech in this specific task (see also Luce, 1981).

Several authors also studied the effect of *training or practice* of the listeners on various intelligibility tasks (Pisoni and Hunnicutt, 1980; Schwab et al., 1985; Greenspan et al., 1985a,b). Green et al. (1984, 1986) studied the perception of synthetic speech by nonnative speakers of English.

3.7. Field Tests

For a few rule-synthesis systems, controlled field tests have been executed. One example is the study conducted by Hampshire et al. (1982). Texttalk (Carlson et al., 1982) was used to generate a spoken version of a weekly newspaper consisting entirely of "small ads." This was presented to 75 visually handicapped subjects, who rated the quality and gave many useful comments. More recently, a similar full-size experiment with a complete spoken daily newspaper was performed with very favorable results (Hjelmquist, 1989). Pilot experiments for Dutch were less promising, probably because they lacked the interactive, real-time mode (van Bezooijen, 1989).

Emerard et al. (1978) studied the intelligibility and acceptability of 45 synthesized directory sentences (CNET diphone synthesis) under actual user conditions. The spoken sentences had the structure "Les informations que vous avez demandées sont les suivantes: votre correspondant *DUPONT, Jean*; je répète *D, U, P, O, N, T*. profession: *charcutier*; domicilié: *7, rue des Genêts rouges*, a *BEAULIEU*, a pour indicatif le *32*, suivi de *47, 29, 33*. The experiment was run over the French telephone network and 101 listeners participated. The names were recognized completely correctly 56.4% of the time.

Under the same heading of "field tests" one could also include the research necessary to study the integration of speech input, speech storage and speech output facilities as in dialog systems. One should furthermore keep in mind the ergonomic, or human factors, aspects (Cooper, 1987).

3.8. Objective Speech Quality Evaluation

Thus far I have talked exclusively about the possibilities for evaluating the speech quality of rule synthesizers by using listener judgments (subjective methods). However, in telecommunications it is quite common to use *objective* methods as well, such as a signal-to-noise measurement, the articulation index (Kryter, 1962), or the speech transmission index (Steeneken and Houtgast, 1980; Steeneken, 1987). In my view it is certainly worth considering the use of some objective method for evaluating rule synthesizers as well. With all the present knowledge in automatic speech recognition and in the area of speech segmentation and labeling, it might well be possible to develop methods for analyzing synthetic speech and for comparing its segmental and suprasegmental characteristics against natural utterances or against certain predefined standards. Some preliminary results have been achieved for F_0 distributions and temporal characteristics (Barry et al., 1989).

4. CONCLUSIONS

As with many other technologies, good progress has been made in speech synthesis technology. In science fiction this technology is incorporated as being perfect already. Many system developers know their system's voice so well that they accept its speech quality as being sufficient. However, when it comes to inexperienced users and actual applications, the performance and the public acceptance are disappointing. That is why systematic, comparative, diagnostic evaluation is still much needed, both in the design state and in the application state.

Most of what has been said in the preceding sections has to do with the design state. Laboratory tests can establish benchmarks of performance for comparative and diagnostic purposes. However, the better the systems become, the less need there will be for routine standardized intelligibility tests. Specific applications will then impose specific demands and constraints. Appropriate tests will have to be developed to study total system performance under these conditions. The human factor issue, including full dialog situations with speech input and speech output, will then require much more attention.

REFERENCES

Allen, J., Hunnicutt, S., and Klatt, D. H. (1987). *From Text to Speech: The MITalk System*. Cambridge University Press, Cambridge, UK.

Alwood, E., and Scully, C. (1982). A composite model of speech production. *Proc. IEEE-ICASSP82*, pp. 932–935.

Bacri, N., and Graillot, P. (1986). Intelligibilité de phrases en parole naturelle, codées et de synthèse. Note technique CNET, NT/LAA/TSS/314.

Barry, W., Grice, M., Hazan, V., and Fourcin, A. (1989). Excitation distributions for synthesised speech. *Proc. Eurospeech '89*, Paris, vol. 1, pp. 353–356.

Benoit, C. (1989a). Intelligibility test for the assessment of French synthesisers using semantically unpredictable sentences. *Noordwijkerhout Proceedings*, pp. 1.7.1–1.7.4.

Benoit, C. (1989b). Towards the perceptual quantification of context redundancy in sentences. *Noordwijkerhout Proceedings*, pp. 5.13.1–5.13.4.

Benoit, C., Erp, A. van, Grice, M., Hazan, V., and Jekosch, U. (1989). Multilingual synthesiser assessment using semantically unpredictable sentences. *Proc. Eurospeech '89*, Paris, vol. 2, pp. 633–636.

Bernstein, J. (1982). Evaluating synthetic speech. *Proc. Workshop on Standardization for Speech I/O Technology*, National Bureau of Standards, pp. 87–91.

Bernstein, J., and Pisoni, D. B. (1980). Unlimited text-to-speech system: Description and evaluation of a microprocessor based device. *Proc. IEEE-ICASSP80*, pp. 576–579.

Bezooijen, R. van (1988). Evaluation of the quality of consonant clusters in two synthesis systems for Dutch. *Proc. Speech '88*, 7th FASE Symposium, Edinburgh, Book 2, pp. 445–452.

Bezooijen, R. van (1989). Evaluation of the suitability of Dutch text-to-speech conversion for application in a digital daily newspaper. *Noordwijkerhout Proceedings*, pp. 6.3.1–6.3.4.

Bezooijen, R. van, and Pols, L. C. W. (1987). Evaluation of two synthesis-by-rule systems for Dutch. *Proc. European Conf. on Speech Technology*, Edinburgh, vol. 1, pp. 183–186.

Bezooijen, R. van, and Pols, L. C. W. (1989a). Evaluation of text-to-speech conversion for Dutch: From segment to text. *Noordwijkerhout Proceedings*, pp. 3.4.1–3.4.4.

Bezooijen, R. van, and Pols, L. C. W. (1989b). Evaluation of a sentence accentuation algorithm for a Dutch text-to-speech system. *Proc. Eurospeech '89*, Paris, vol. 1, pp. 228–231.

Bosman, A. (1989). Speech perception by the hearing impaired. Ph.D thesis, Utrecht University.

Broecke, M. P. R. van den, Aerts, A., Reizevoort, J., Veenhof, T., Lammens, J., and Elstrodt, M. (1987). Type- and token-frequencies of word classes, phonemes, and phoneme pairs in Dutch. *PRIPU*, vol. 11, no. 1, pp. 1–7 + appendices.

Calliope (1989). *La parole et son traitement automatique*. Collection scientifique et technique des Télécommunications. Masson, Paris.

Carlson, R., and Granström, B. (1979). Some notes on the perception of temporal patterns in speech. *Proc. 9th Int. Congress of Phonetic Sciences*, Copenhagen, vol. II, pp. 260–267.

Carlson, R., and Granström, B. (1989). Evaluation and development of the KTH text-to-speech system on the segmental level. *Noordwijkerhout Proceedings*, pp. 1.3.1–1.3.4.

Carlson, R., Granström, B., and Larson, K. ((1976). Evaluation of a text-to-speech system as a reading machine for the blind. STL-QPRS 2-3, pp. 9–13.

Carlson, R., Granström, B., and Hunnicutt, S. (1982). A multilingual text-to-speech module. *Proc. IEEE-ICASSP82*, Paris, vol. 3, pp. 1604–1607.

Carlson, R., Granström, B., and Lindström, A. (1989). Predicting name pronunciation for a reverse directory service. *Proc. Eurospeech '89*, Paris, vol. 1, pp. 113–116.

Coker, C. H. (1973). Synthesis by rule from articulatory parameters. In *Speech Synthesis*, J. L. Flanagan and L. R. Rabiner (Eds.), Dowden, Hutchinson & Ross, Stroudsburg, Pa., pp. 396–399.

Collier, R., Zitter, A. de, and Terken, J. (1989). On the perceptual salience of melodical variations and its consequences for intonation synthesis. *Proc. Eurospeech '89*, Paris, vol. 2, pp. 108–111.

Combescure, P. (1981). 20 listes de dix phrases phonetiquement equilibrees. *Rev. Acoust. 56*:34–38.

Cooper, M. (1987). Human factor aspects of voice input/output. *Speech Technol. 3*(4):82–85.

Egan, J. P. (1948). Articulation testing methods. *Laryngoscope 58*:955–991.

Emerard, F., and Lienard, J. S. (1980). Quelques aspects de la traduction phonétique et prosodique du texte écrit. *Rev. Acoust. 53*:97–101.

Emerard, F., Graillot, P., Dalila, E., and Roumiquière, J. P. (1978). Résultats d'un test sur l'intelligibilité des lignes d'annuaire en parole de synthèse. Note Technique CNET, NT/DAS/ETA/53.

Erp, A. van, and Grice, M. (1989). Multi-lingual syntactic structures for semantically unpredictable sentences and guidelines for compiling lexicons for Dutch, English, French, German, Italian, and Swedish. In Esprit Project 1541 (SAM) Extension Phase Final Report, pp. 43–60.

Fairbanks, G. (1958). Test of phonemic differentiation: The rhyme test. *J. Acoust. Soc. Am. 30*:596–600.

Fletcher, H., and Steinberg, J. C. (1929). Articulation testing methods. *Bell Syst. Tech. J. 8*:806–854.

Fourcin, A. J., Harland, G., Barry, W., and Hazan, V. (Eds.) (1989). *Speech Input and Output Assessment. Multilingual Methods and Standards*. Ellis Horwood, Chichester.

French PTT. (1987). An "objective" evaluation of difficulty in understanding voice synthesis devices. CCITT Com. XII, Study Group XII, Contribution No. FR4-E.

Goodman, D. J., and Nash, R. D. (1982). Subjective quality of the same speech transmission conditions in seven different countries. *IEEE Trans. Commun. 30*: 642–654.

Goudie-Marshall, K. M. (1984). Defining and developing intelligibility in a text-to-speech system. *J. Acoust. Soc. Am. 76*(Suppl. 1):S2 (A).

Graillot, P., Emerard, F., and Le Bras, J. (1983). Tests de rime appliqués à des systèmes de synthèse par diphones. Note Technique CNET, NT/LAA/TSS/180.

Greene, B. G. (1983). Perception of synthetic speech by children. Research on Speech Perception, Progress Report 9, Indiana University, pp. 335–347.

Greene, B. G. (1986). Perception of synthetic speech by nonnative speakers of English. *Proc. Human Factors Soc.*, pp. 1340–1343.

Greene, B. G., Manous, L. M., and Pisoni, D. B. (1984). Perceptual evaluation of DECtalk: A final report on Version 1.8. Research on Speech Perception, Progress Report 10, Indiana University, pp. 77–127.

Greene, B. G., Logan, J. S., and Pisoni, D. B. (1986). Perception of synthetic speech produced automatically by rule: Intelligibility of eight text-to-speech systems. *Behav. Res. Methods Instrum. Comput. 18*(2):100–107.

Greenspan, S., Nusbaum, H. C., and Pisoni, D. B. (1985a). Perception of synthetic speech generated by rule: Effects of training and attentional limitations. Research on Speech Perception, Progress Report 11, Indiana University, pp. 263–287.

Greenspan, S. L., Nusbaum, H. C., and Pisoni, D. B. (1985b). Generalization of training with synthetic words and sentences. Research on Speech Perception, Progress Report 11, Indiana University, pp. 361–376.

Greenspan, S. L., Bennett, R. W., and Syrdal, A. K. (1989). A study of two standard speech intelligibility measures. *J. Acoust. Soc. Am. 85*(Suppl. 1):S43 (A).

Grice, M. (1989). Syntactic structures and lexicon requirements for semantically unpredictable sentences in a number of languages. *Noordwijkerhout Proceedings*, pp. 1.5.1–1.5.4.

Hampshire, B., Ruden, J., Carlson, R., and Granström, B. (1982). Evaluation of centrally produced and distributed synthetic speech. STL-QPSR 2-3, pp. 18–23.

Hazan, V., and Grice, M. (1989). The assessment of synthetic speech intelligibility using semantically unpredictable sentences. *Noordwijkerhout Proceedings*, pp. 1.6.1–1.6.4.

Hirsh, I. J., Reynolds, E. G. and Joseph, M. (1954). Intelligibility of different speech materials. *J. Acoust. Soc. Am. 26*:530–538.

Hjelmquist, E. (1989). Spoken newspaper for the blind. *Proc. ESCA Tutorial Day on Speech Input/Output Assessment and Speech Databases*, Noordwijkerhout, The Netherlands, pp. 13–19.

House, A. S., Williams, C. E., Hecker, M. H. L., and Kryter, K. D. (1965). Articulation testing methods: Consonantal differentiation with a closed response set. *J. Acoust. Soc. Am. 37*:158–166.

Javkin, H., et al. (1989). A multi-lingual text-to-speech system. *Proc. IEEE-ICASSP-89*, Glasgow, vol. 1, pp. 242–245.

Klatt, D. H. (1979). Synthesis by rule of segmental durations in English sentences. *Proc. 9th Int. Congress of Phonetic Sciences*, Copenhagen, vol. II, pp. 290–297.

Klatt, D. H. (1987). Review of text-to-speech conversion for English. *J. Acoust. Soc. Am. 82*:737–793.

Kryter, K. D. (1962). Methods for the calculation and use of the articulation index. *J. Acoust. Soc. Am. 34*:1689–1697.

Laddaga, R., Sanders, W. R., and Suppes, P. (1981). Testing intelligibility of computer-generated speech with elementary-school children. In *University-Level Computer-Assisted Instruction at Stanford: 1968–1980*, P. Suppes (Ed.). Inst. for Math. Studies in the Social Sciences, Stanford University, pp. 377–398.

Larreur, D., and Sorin, C. (1977). Résultats de quelques tests sur l'intelligibilité des noms propres synthétisés par diphones. Note Technique CNET, NT/DAS/ETA/44.

Larreur, D., Emerard, F., and Marty, F. (1989). Linguistic and prosodic processing for a text-to-speech synthesis system. *Proc. Eurospeech '89*, Paris, vol. 1, pp. 510–513.

Lefèvre, J. P., Auberge, V., and Maret, D. (1988). Alignement automatique optimal de deux chaines phonetiques. *Proc. GALF Meeting.*

Logan, J. S., Greene, B. G., and Pisoni, D. B. (1989). Segmental intelligibility of synthetic speech produced by rule. *J. Acoust. Soc. Am. 86*(2):566–581.

Lucas, J. P., Roumiguière, J. P., and Emerard, F. (1988). Test de 4 algorithmes de traitement de signal pour la synthèse de parole à partir du texte. CNET Doc. NT/LAA/TSS/375, Tome I.

Luce, P. A. (1981). Comprehension of fluent synthetic speech produced by rule. Research on Speech Perception, Progress Report 7, Indiana University, pp. 229–241.

Luce, P. A., Feustel, T. C., and Pisoni, D. B. (1983). Capacity demands in short-term memory for synthetic and natural speech. *Hum. Factors 25*:17–32.

Manous, L. M., Greene, B. G., and Pisoni, D. B. (1984). Evaluation of Prose—the Speech Plus text-to-speech system. I. Phoneme intelligibility and word recognition in meaningful sentences. Speech Research Lab. Technical Note 84-04, Indiana University.

Manous, L. M., Pisoni, D. B., Dedina, M. J., and Nusbaum, H. C. (1985). Comprehension of natural and synthetic speech using a sentence verification task. Research on Speech Perception, Progress Report 11, Indiana University, pp. 33–57.

Maret, D. (1987). Comparisons de chaines de charactères, accès lexicaux tolérants et applications. Ph.D. thesis, Grenoble.

Martin, P. (1978). Perception des séquences de contours prosodiques de phrases synthétisées. 9èmes J.E.P., GALF, Lannion, pp. 21–29.

Marty, F., and Hart, R. S. (1985). Computer programs to transcribe French text into speech: Problems and suggested solutions. University of Illinois.

McAllister, J. M., Laver, J., and McAllister, M. J. (1986). The preprocessor module of the CSTR text to speech system. *Proc. Inst. of Acoustics*, vol. 8, part 7, pp. 371–376.

Monaghan, A I. C., and Ladd, D. R. (1989). Evaluating intonation in the CSTR text-to-speech system. *Noordwijkerhout Proceedings*, pp. 3.6.1–3.6.4.

Nusbaum, H. C., and Pisoni, D. B. (1985). Constraints on the perception of synthetic speech generated by rule. *Behav. Res. Methods Instrum. Comput. 17*:235–242.

Nusbaum, H. C., Schwab, E. C., and Pisoni, D. B. (1983). Perceptual evaluation of synthetic speech: Some constraints on the use of voice response systems. Research on Speech Perception, Progress Report 9, Indiana University, pp. 283–294. Also published in *Proc. Voice Data Entry Systems Appl. Conf.*, Chicago, pp. 1–14.

Nusbaum, H. C., Dedina, M. J., and Pisoni, D. B. (1984a). Perceptual confusions of consonants in natural and synthetic CV syllables. Research on Speech Perception, Progress Report 10, Indiana University, pp. 409–422.

Nusbaum, H. C., Schwab, E. C., and Pisoni, D. B. (1984b). Subjective evaluation of synthetic speech: Measuring preference, naturalness and acceptability. Research on Speech Perception, Progress report 10, Indiana University, pp. 391–407.

Nye, P. W., and Gaitenby, J. H. (1973). Consonant intelligibility in synthetic speech and in natural speech control (modified rhyme test results). Haskins Labs. SR-33, pp. 77–91.

Nye, P. W., and Gaitenby, J. H. (1974). The intelligibility of synthetic monosyllabic words in short syntactically normal sentences. Haskins Labs. SR-37/38, pp. 169–190.

Olaszy, G. (1989). Multivox—a flexible text-to-speech system for Hungarian, Finnish, German, Esperanto, Italian and other languages for IBM-PC. *Proc. Eurospeech '89*, Paris, vol. 2, pp. 525–528.

O'Malley, M. H., and Caisse, M. (1987). How to evaluate text-to-speech systems. *Speech Technol. 3*(4):66–75.

O'Shaughnessy, D. (1986). The effects of speaking rate on formant transitions in French synthesis-by-rule. *Proc. IEEE-ICASSP86*, pp. 2027–2030.

Pallett, D. H. (Ed.) (1982). *Proc. of Workshop on Standardization for Speech I/O Technology*, National Bureau of Standards, Gaithersburg, Md.

Pascal, D. (1987). Méthodologies d'évaluation subjective de la qualité des systèmes de communication. In Dossier: La qualité des services de communication, Bulletin No. 28 de l'IDATE.

Pavlovic, C. V., Sorin, C., Roumiguière, J. P., and Lucas, J. P. (1989a). A comparative analysis of the magnitude estimation and the pair comparisons techniques for use in assessing quality of text-to-speech synthesis. *Noordwijkerhout Proceedings*, 3.1.1–3.1.3.

Pavlovic, C. V., Rossi, M., and Espesser, R. (1989b). Subjective assessment of acceptability, intelligibility and naturalness of text-to-speech synthesis. *Noordwijkerhout Proceedings*, pp. 3.2.1–3.2.4.

Pavlovic, C. V., Rossi, M., and Espesser, R. (1989c). Direct scaling of the performance of text-to-speech synthesis systems. *Proc. Eurospeech '89*, Paris, vol. 2, pp. 644–647.

Peckels, J. P., and Rossi, M. (1973). Le test de diagnostic par paires minimales. Adaptation au francais du "Diagnostic Rhyme Test" de W. D. Voiers. *Rev. Acoust. 27*: 245–262.

Picone, J., Goudie-Marshall, K. M., Doddington, G. R., and Fisher, W. (1986). Automatic text alignment for speech systems evaluation. *IEEE Trans. Acoust., Speech, Signal Process. ASSP-34*(4):780–784.

Pisoni, D. B. (1982a). Perceptual evaluation of voice response systems: Intelligibility, recognition & understanding. *Proc. of Workshop on Standardization for Speech I/O Technology*, National Bureau of Standards, Gaithersburg, Md.

Pisoni, D. B. (1982b). Perception of Speech: The human listener as a cognitive interface. *Speech Technol. 1*(2):10–23.

Pisoni, D. B. and Hunnicutt, S. (1980). Perceptual evaluation of MITalk: The MIT unrestricted text-to-speech system. *Proc. IEEE-ICASSP80*, pp. 572–575.

Pisoni, D. B. and Koen, E. (1981). Some comparison of intelligibility of synthetic and natural speech at different speech-to-noise ratios. Research on Speech Perception, Progress Report 7, Indiana University, pp. 243–253.

Pisoni, D. B., Nusbaum, H. C., and Greene, B. G. (1985). Perception of synthetic speech generated by rule. *Proc. IEEE 73*(11):1665–1676.

Pisoni, D. B., Manous, L. M., and Dedina, M. J. (1987). Comprehension of natural and synthetic speech: effects of predictability on the verification of sentences controlled for intelligibility. *Comput. Speech Language 2*:303–320.

Pisoni, D. B., Greene, B. G., and Logan, J. S. (1989). An overview of ten years of research on the perception of synthetic speech. *Noordwijkerhout Proceedings*, pp. 1.1.1–1.1.4.

Plomp, R., and Mimpen, A. M. (1979). Improving the reliability of testing the speech reception threshold for sentences. *Audiology 8*:43–52.

Pols, L. C. W. (1988a). A joint Dutch research program for developing a high-quality text-to-speech synthesis system. *J. Acoust. Soc. Am. 84*(Suppl. 1):S190 (A). Full version published in Preprints booklet of that Session GGG; also *Proc. Institute of Phonetic Sciences*, Amsterdam, IFA 12, pp. 9–17.

Pols, L. C. W. (1988b). Improving synthetic speech quality by systematic evaluation. *Proc. 2nd Symp. on Advanced Man-Machine Interface through Spoken Language*, Makaha, Hawaii, pp. 17.1–17.9; also *Proc. Inst. of Phonetic Sciences*, Amsterdam, IFA 12, pp. 19–27, and *Proc. Tutorial Day Speech Input/Output Assessment and Speech Databases*, Noordwijkerhout, The Netherlands, 1989, pp. 3–11.

Pols, L. C. W. (1989). Assessment of text-to-speech synthesis systems. Chapter III in Fourcin et al. (1989), pp. 53–81, 251–266.

Pols, L. C. W., and Olive, J. P. (1983). Intelligibility of consonants in CVC utterances produced by dyadic rule synthesis. *Speech Commun. 2*:3–13.

Pols, L. C. W., Lefèvre, J.-P., Boxelaar, G., and Son, N. van (1987). Word intelligibility of a rule synthesis system for French. *Proc. European Conf. on Speech Technology*, Edinburgh, vol. 1, pp. 179–182.

Pratt, R. L. (1984). The assessment of speech intelligibility at RSRE. *Proc. Inst. of Acoustics*, vol. 6, part 4, pp. 439–443.

Pratt, R. L. (1986). On the intelligibility of synthetic speech. *Proc. Inst. of Acoustics*, vol. 8, part 7, pp. 183–192.

Pratt, R. L. (1987). Quantifying the performance of text-to-speech synthesizers. *Speech Technol. 3*(4):54–64.

Proc. of ESCA Workshop on Speech Input/Output Assessment and Speech Databases, Noordwijkerhout, The Netherlands, September, 20–23, 1989.

Quazza, S., Varesa, G., and Vivalda, E. (1989). Syntactic pre-processing for high quality text-to-speech. *Proc. Eurospeech '89*, Paris, vol. 1, pp. 506–509.

Ralston, J. V., Mullennix, J. W., Greene, B. G., and Lively, S. E. (1989). On-line measures of comprehension of natural and synthetic speech. *J. Acoust. Soc. Am. 85*:S125 (A).

Rigoll, G. (1987). The DECtalk system for German: A study of the modification of a text-to-speech converter for a foreign language. *Proc. IEEE-ICASSP87*, pp. 1450–1453.

Rijnsoever, P. A. van (1988). A multilingual text-to-speech system. IPO Annual Progress Report 23, pp. 34–40.

Robert, J.-M., Choiniere, A., and Descout, R. (1989). Subjective evaluation of naturalness and acceptability of three text-to-speech systems in French. *Proc. Eurospeech '89*, Paris, vol. 2, pp. 640–643.
Ruehl, H.-W., and Kulas, W. (1985). Eine vergleichende Messung der Lautverständlichkeit für Sprachsynthetisatoren. *Fortschritte der Akustik DAGA '85*. Stuttgart.
Sagisaka, V., and Sato, H. (1986). Word identification method for Japanese text-to-speech conversion system. *Proc. IEEE-ICASSP86*, pp. 2411–2414.
Schnabel, B. (1986). Evaluation de la qualité de l'allemand synthétisé par diphones. 15èmes J.E.P., GALF, Aix-en-Provence, pp. 19–20.
Schwab, E. C., Nusbaum, H. C., and Pisoni, D. B. (1985). Some effects of training on the perception of synthetic speech. *Hum. Factors 27*:395–408.
Scully, C. (1987). Linguistic units and units of speech production. *Speech Commun. 6*:77–142.
Simpson, C. A., and Marichonda-Frost, K. (1984). Synthesized speech rate and pitch effects on intelligibility on warning signals for pilots. *Hum. Factors 26*:509–517.
Slowiaczek, L. M., and Nusbaum, H. C. (1983). Effects of speech rate and pitch contour on the perception of synthetic speech. Research on Speech Perception, Progress Report 9, Indiana University, pp. 109–133. Also *Hum. Factors 27*: 1985: 701–712.
Slowiaczek, L. M., and Pisoni, D. B. (1981). Effect of practice on speeded classification of natural and synthetic speech. Research on Speech Perception, Progress Report 7, Indiana University, pp. 255–262.
Son, N. van, and Pols, L. C. W. (1989). Intelligibility of words in isolation and words in semantically unpredictable sentences. A study of two diphone-based speech synthesis systems in French. Esprit-SPIN deliverable.
Son, N. van, Pols, L. C. W., Sandri, S., and Salza, P. L. (1988). First quality evaluation of a diphone-based speech synthesis system for Italian. *Proc. SPEECH '88, 7th FASE Symp.*, Edinburgh, Book 2, pp. 429–436.
Sorin, C. (1982/83). Evaluation de la contribution de F_0 à l'intelligibilité. *Rech. Acoust. CNET 7*:141–155.
Sotscheck, J. (1982). Ein Reimtest für Verstandlichkeits-messungen mit deutscher Sprache als ein verbessertes Verfahren zur Bestimmung der Sprachgüteübertragung. *Fernmeldeingenieur 36*:1–84.
Spiegel, M. (1985). Automatic pronunciation of names. Bell Core Technical Report TM-ARN-000453.
Spiegel, M., Altom, M. J., Macchi, M., and Wallace, K. (1989a). Using a monosyllabic test corpus to evaluate the intelligibility of synthesized and natural speech. *Proc. AVIOS*.
Spiegel, M., Altom, M. J., Macchi, M., and Wallace, K. (1989b). A monosyllabic test corpus to evaluate the intelligibility of synthesized and natural speech. *Noordwijkerhout Proceedings*, pp. 1.2.1–1.2.4.
Spiegel, M. F., Macchi, M. J., and Gollhardt, K. D. (1989c). Synthesis of names by a demisyllable-based speech synthesizer (Spokesman). *Proc. Eurospeech '89*, Paris, vol. 1, pp. 117–120.

Steeneken, H. J. M. (1982). Ontwikkeling en toetsing van een nederlandstalige diagnostische rijmtest voor het testen van spraakkommunikatiekanalen. TNO Inst. for perception, Report IZF 1982-13.
Steeneken, H. J. M. (1987). Diagnostic information from subjective and objective intelligibility tests. *Proc. IEEE-ICASSP87*, pp. 5–8.
Steeneken, H. J. M., and Houtgast, T. (1980). A physical method for measuring speech-transmission quality. *J. Acoust. Soc. Am. 67*:318–326.
Stella, M. (1983). Fabrication semi-automatique de dictionnaires de diphones. *Rech. Acoust. CNET 7*:51–63.
Terken, J. M. B., and Collier, R. (1989). Automatic synthesis of natural-sounding intonation for text-to-speech conversion in Dutch. *Proc. Eurospeech '89*, Paris, vol. 1, pp. 357–359.
Voiers, W. D. (1977a). Diagnostic evaluation of speech intelligibility. In *Benchmark Papers in Acoustics*, vol. 11, *Speech Intelligibility and Speaker Recognition*, M. Hawley (Ed.). Dowden, Hutchinson & Ross, Stroudsburg, pp. 374–387.
Voiers, W. D. (1977b). Diagnostic acceptability measure for speech communication systems. *Proc. IEEE-ICASSP77*, pp. 204–207.
Voiers, W. D. (1982). Measurement of intrinsic deficiency in transmitted speech: The Diagnostic Discrimination Test. *Proc. IEEE-ICASSP82*, pp. 1004–1007.
Voiers, W. D. (1983). Evaluating processed speech using the Diagnostic Rhyme Test. *Speech Technol. 1*(4):30–39.
Zhou, K. C. (1986). Preliminary evaluation of a Chinese text-to-speech system. *Proc. SST-86*, pp. 162–167.

Part III

Recognition and Verification

14

Spectrum Distance Measures for Speech Recognition

KIYOHIRO SHIKANO NTT Human Interface Laboratories, Tokyo, Japan

FUMITADA ITAKURA Nagoya University, Aichi, Japan

1. INTRODUCTION

Several kinds of spectral matching measures have been studied. In particular, spectral matching measures based on linear predictive coding (LPC), which seem to be superior to matching measures based on the filter bank or on the fast Fourier transform (FFT), have been shown to be very effective in speech recognition, speech analysis, speech synthesis, and speech coding.

The *LPC cepstrum distance* (CEP) was introduced by Atal (1974). Itakura (1975) introduced an efficient measure called the *Itakura distance* or the *likelihood ratio measure* and showed its usefulness for spoken word recognition. Moreover, he clarifies its meaning as a spectral matching measure in the frequency domain. Gray and Markel (1976) summarized these LPC-based measures, together with the Cosh measure, from the standpoint of matching in the frequency domain and successfully applied these measures to vector quantization.

The LPC cepstrum distance, the Itakura distance, and the Cosh measure show the same characteristics in matching between similar spectra (Gray and Markel, 1976). According to our evaluation of these measures by spoken word recognition (Shikano and Sugiyama, 1982) and vowel verification

(Shikano and Kohda, 1980), the Itakura distance is somewhat inferior to the LPC cepstrum distance and the Cosh measure because of its asymmetrical spectral matching in the frequency domain. The LPC cepstrum distance and the Cosh measure demonstrate quite similar performances for word recognition and vowel recognition. These measures, however, lack two characteristics of frequency domain spectral matching. One is a local peak-weighted matching like formant matching. The other is a frequency axis weighting such as provided by Mel scaling.

Sugiyama and Shikano introduced a group of new measures which possess these two characteristics (Sugiyama and Shikano, 1981; Sugiyama, 1985). These measures are called peak-weighted measures. Furthermore, the *weighted likelihood ratio* (WLR) measure was shown to yield better performance than other peak-weighted and conventional measures in its application to vowel verification, word recognition, and noisy speech recognition for Japanese utterances (Shikano and Sugiyama, 1982; Sugiyama and Shikano, 1981, 1983).

Nocerino, Soong, and others compares various LPC distance measures using a telephone digit and alphabet data base uttered in English by four male speakers (Nocerino et al., 1985). In this experiment, the Itakura distance showed better performance than CEP and WLR. The digit and alphabet data base used in their studies is so restricted in phonetic units as well as in frequency bandwidth that it is difficult to generalize these results. These reports, however, suggest that the performance of LPC distance measures indeed depends on each phonetic category.

In the first part of this chapter, the distance measures CEP and WLR are evaluated with attention to phonetic unit recognition rates. For these experiments, the Harvard sentence (Egan, 1944) data base uttered by three male speakers is used. WLR shows better performance than CEP in the vowel recognition experiments. CEP shows better performance than WLR in the consonant recognition experiments. These results suggest that combining CEP and WLR might provide better overall results.

In order to attain good performance by a combination of CEP and WLR, explicit parameters which discriminate between vowels and consonants are needed. Here, several simple parameters for this purpose are proposed:

1. Power value: Large power indicates vowel parts, and small power indicates consonant parts.
2. Flatness measure: Vowels have spectra with a clear formant structure (i.e., the spectra are not flat), and consonants are likely to have flat spectra.

3. Power dip: A power dip usually indicates the presence of consonant parts.
4. Zero crossings: Consonants are likely to have large values.

In this chapter, the power value is adopted as the simplest discrimination parameter between vowels and consonants. The combined measure using CEP and WLR, which is called CCW, shows better recognition rates than CEP and WLR separately.

A bilinear transform by a recursive formulation is introduced to realize frequency axis warping (Oppenheim and Johnson, 1972). Two types of bilinear transforms are tested: the bilinear transform on digitized speech and the bilinear transform on LPC cepstrum coefficients. CEP, WLR, and CCW are adopted as distance measures between spectra. There are two specific experimental results:

1. Vowel recognition rates are improved by frequency warping of the bilinear transform.
2. Consonant recognition rates are improved slightly by frequency warping, although the improvement is marginal compared to that of the vowels.

These results suggest that recognition should be performed differently for vowels and consonants, if not for finer subcategories.

Next, spectral matching measures are augmented with difference spectra (Furui, 1984) and power information (Aikawa and Shikano, 1985; Rabiner et al., 1984). Spectral distance represents the dynamic features from the viewpoint of perception. Spectral matching measures augmented with difference spectra and power information attain considerable improvement for phonetic unit recognition.

To confirm these results, the same recognition experiments are carried out on phonetic units using additional speech data from the Harvard sentence data base uttered by three female speakers.

The second part of this chapter focuses on the robustness approach to distance measures. The noise-robust distance in noisy environments must be investigated to realize a better speech recognition system. Several distance measures have been reported which deal with noisy input through the capability of global peak weighting in order to realize noise-tolerant speech recognition. They are:

1. Weighted likelihood ratio measure (WLR) (Sugiyama and Shikano, 1983)
2. Weighted cepstrum distance (WCEP) (Tohkura, 1986)

3. Root-power sum distance (RPS) (Hanson and Wakita, 1987)
4. Smoothed group delay spectrum distance (SGDS) (Itakura and Umezaki, 1987)
5. Weighted group delay distance (WGD) (Mitsui and Matsumoto, 1988)

In this chapter, the robustness approach to distance measures is studied with attention given to low-bit representation of robust phoneme features based on SGDS.

For speech recognition, extraction of the relevant features of the speech signal, commonly called analysis, is of vital importance for the recognition process. Important speech information removed at this stage cannot easily be recovered later on.

Analysis and the subsequent similarity calculation have been intensively studied over the past years. One common standard is the LPC cepstrum, which performs very well in well-behaved environments. On the other hand, under adverse conditions, such as variable frequency characteristics (different microphones, changing transmission lines, etc.) or additive noise, LPC-based measures are strongly affected.

To overcome this problem, weighting of the cepstral coefficients has been proposed, with the objective of emphasizing the spectral peaks and separating the influence of neighboring poles. One of these proposed analysis methods is the smoothed group delay spectrum (Itakura and Umezaki, 1987). The Fourier coefficients $G(n)$ of the group delay spectrum were shown to be identical to the cepstral coefficients weighted by a weighting function $W(n) = n$. To avoid overemphasis on the cepstral components having a large index n and problems with truncation effects due to a finite number of coefficients, a generalized weighting function $W(n)$ having the form of a Gaussian window has been proposed:

$$W(n) = n^s \exp\left(-\frac{n^2}{2\tau^2}\right)$$

Multiplication with this Gaussian window in the frequency domain is equivalent to smoothing with a Gaussian window in the logarithmic spectral domain.

The necessity for smoothing the group delay spectrum can easily be seen from Fig. 1. The peaks at the formant frequencies of the group delay spectrum shown in Fig. 1d are very sharp, which may be appropriate to the estimation of formant frequencies. For speech recognition, on the other hand, broadening of the peaks is indispensable, since just a slight deviation in formant frequencies between the two patterns would result in a large

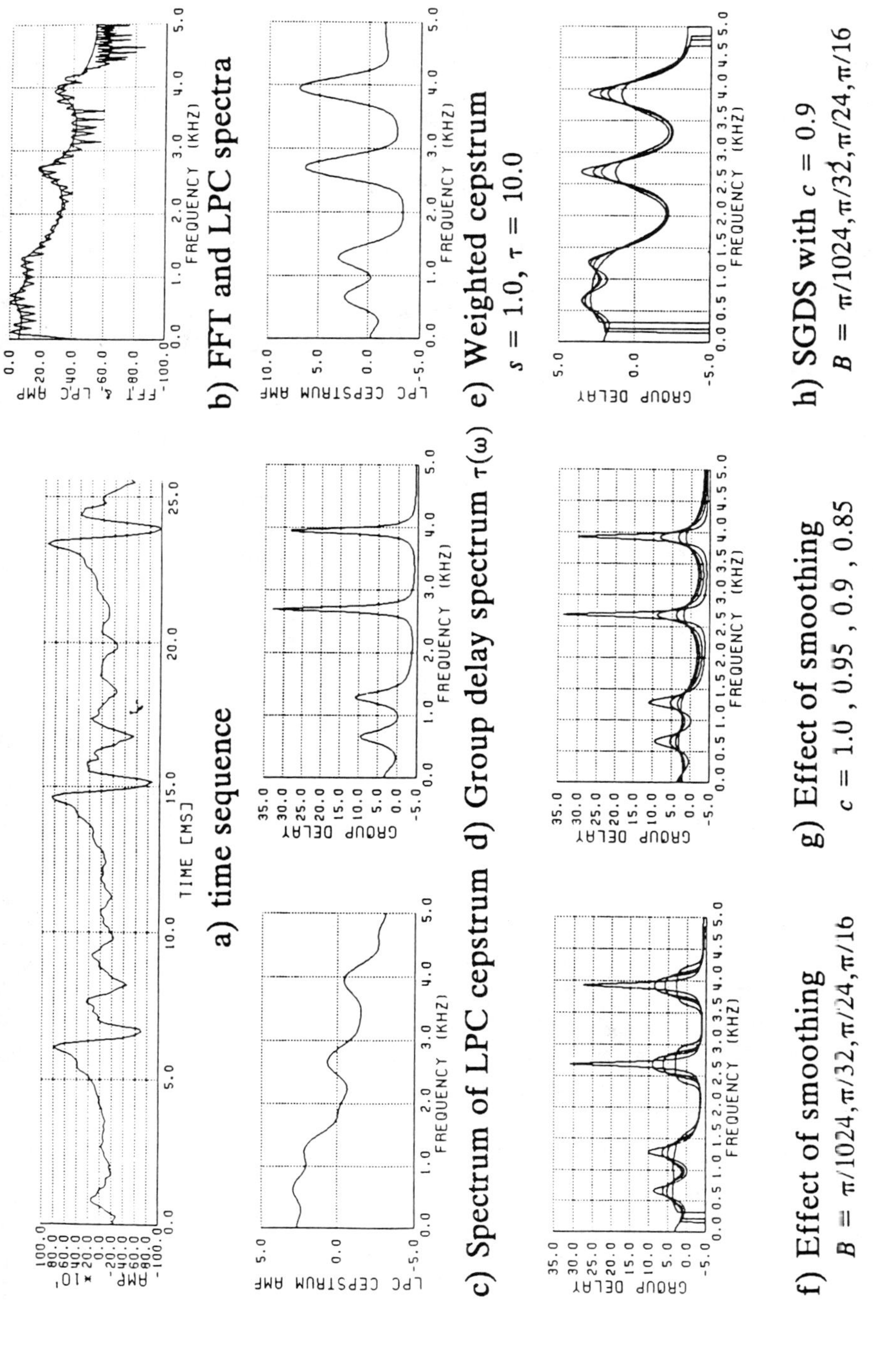

Figure 1 Second vowel /a/ in MIKASA.

distance value. We can see that this smoothing is achieved by using the Gaussian weighting $W(n)$ according to the previous equation (Fig. 1e). Optimal recognition results for this smoothed group delay spectrum were obtained for $\tau = 10.0$ (corresponding to a window bandwidth of 300 Hz) and $s = 1.0$. This distance measure performs very well even under noisy conditions.

In Section 2.5 we show that a similar (not identical) smoothed group delay function can be directly calculated from the LPC coefficients. We introduce the basic algorithm. In Section 4 we also give details about the data bases and the pattern-matching algorithm (standard DTW) and report recognition results, comparing the new method with the above-mentioned smoothed group delay weighted LPC cepstrum and the standard LPC cepstrum (CEP). For ease of explanation we will henceforth call the smoothed group delay spectrum calculated by the Gaussian weighted LPC cepstrum simply the weighted cepstrum (WCEP) to distinguish it from the directly calculated smoothed group delay spectrum (SGDS).

In Section 4 we report the results of modifications of the basic algorithm. A rough quantization into 1 to 3 bits per parameter and the use of the perceptually based Bark scale are investigated. Finally, we discuss our findings.

2. LPC SPECTRAL MATCHING MEASURES

In speech recognition, LPC analysis is one of the most powerful spectral analysis techniques. In this section, LPC analysis, the LPC spectral matching measures, the bilinear transform, and the difference spectrum are reviewed as a basis for the subsequent sections.

2.1. LPC Analysis

LPC analysis with incomplete preemphasis is performed on speech data as shown in the following procedure. This LPC analysis does not minimize the prediction error in the total frequency domain, but the accuracy in the low-frequency domain is improved substantially.

Procedure

1. AD conversion (16 kHz)
2. Hamming windowing (24-ms window for every 6-ms step)
3. Autocorrelation analysis (13-order)
4. Incomplete preemphasis ($1 - 0.5z^{-1}$)
5. LPC analysis (12-order)

6. Bilinear transform on LPC cepstrum coefficients (optional stage)

2.2. LPC Spectral Matching Measure

Conventional LPC distance measures have been applied to speech recognition and speech coding (Itakura, 1975; Atal, 1974; Gray, 1976; Lowerre, 1976). They include the *LPC cepstrum distance* (CEP), the *maximum likelihood ratio measure* (Itakura distance), and the *Cosh measure*. Gray and Markel (1976) compared these measures from the viewpoint of spectral matching and distortion. They concluded that the conventional LPC distance measures have almost the same characteristics in their function as a distortion measure. Shikano and Kohda (1980) carried out a comparative study of these measures from the viewpoint of vowel verification. The results showed that the Itakura distance is slightly inferior to the Cosh measure and the LPC cepstrum distance because of the asymmetry of the Itakura distance. These results coincide with Gray's conclusion.

In this chapter, CEP is taken as a representative of conventional distance measures. CEP is easily calculated from the LPC cepstrum coefficients as follows.

$$\begin{aligned} \text{CEP} &= \int_{-\pi}^{\pi} [\log(f) - \log(g)]^2 \frac{d\lambda}{2\pi} \\ &= \sum_{i=-\infty}^{\infty} (C_i - C_i^*)^2 \end{aligned}$$

where f is the LPC spectrum of a reference and g is the LPC spectrum of an input. LPC cepstrum coefficients are denoted by C_i and C_i^*, where C_i corresponds to the reference spectrum f and C_I^* corresponds to the input spectrum g. Since C_0 and C_0^* are only the power values and $C_i = C_{-i}$, we obtain the following equation by setting $C_0 = C_0^*$ to minimize CEP.

$$\text{CEP} = 2\sum_{i=1}^{\infty}(C_i - C_i^*)^2$$

Since LPC spectrum envelopes are smooth enough to truncate the high-order LPC cepstrum coefficients, we can redefine CEP as

$$\text{CEP} = 2\sum_{i=1}^{16}(C_i - C_i^*)^2$$

CEP can be easily and efficiently computed through LPC analysis. Its disadvantage is that the distance is not weighted at local peaks such as formants.

Sugiyama and Shikano (1981; Sugiyama, 1985) proposed the LPC peak-weighted measures, which are capable of local peak weighting and frequency weighting. In these measures, the *weighted likelihood ratio* has been shown to achieve excellent performance for vowel recognition and word recognition (Shikano and Sugiyama, 1982; Sugiyama and Shikano, 1983; Sugiyama, 1985). WLR is derived from the likelihood ratio measure sometimes called the Itakura–Saito distortion measure (Itakura, 1975; Gray and Markel, 1976) and is defined as

$$\begin{aligned} \mathrm{WLR} &= \int_{-\pi}^{\pi} \left\{ \frac{[\log(f/g) + g/f - 1]f}{u} + \frac{[\log(g/f) + f/g - 1]g}{u^*} \right\} \frac{d\lambda}{2\pi} \\ &= \int_{-\pi}^{\pi} \left(\frac{f}{u} - \frac{g}{u^*} \right) [\log(f) - \log(g)] \frac{d\lambda}{2\pi} + \left(\frac{u}{u^*} + \frac{u^*}{u} - 2 \right) \\ &= \sum_{i=-\infty}^{\infty} (r_i - r_i^*)(C_i - C_i^*) + \left(\frac{u}{u^*} + \frac{u^*}{u} - 2 \right) \end{aligned}$$

where LPC autocorrelation coefficients and the power of f are denoted by r_i and u, and r_i^* and u_i^* correspond to the input spectrum g.

Therefore, by setting $u = u^*$ to minimize the measure value, we obtain

$$\mathrm{WLR} = \sum_{i=-\infty}^{\infty} (r_i - r_i^*)(C_i - C_i^*)$$

Since LPC spectrum envelopes are smooth enough to truncate the high-order LPC coefficients for r_i and C_i, and since r_{-i} and C_{-i} take the same value as r_i and C_i, respectively, we can redefine WLR as

$$\mathrm{WLR} = \sum_{i=1}^{16} (r_i - r_i^*)(C_i - C_i^*)$$

A combined measure using CEP and WLR, which is expected to incorporate the merits of both, is proposed as a linear combination of them. We abbreviate this combined measure as CCW. CCW uses a power value as the linear combination weighting factor and is defined by the following equation:

$$\mathrm{CCW} = \alpha \mathrm{WLR} + \beta(1 - \alpha)\mathrm{CEP}$$

where α is a linear combination weighting factor which is determined by a power value, and β is a scaling adjusting parameter between WLR and CEP.

2.3. Bilinear Transform

The bilinear transform is a method for transforming the linear frequency axis into a warped one by use of the following all-pass filter:

$$Z_{\text{new}}^{-1} = \frac{z^{-1} - a}{1 - az^{-1}} \qquad (-1 < a < 1)$$

$$\omega_{\text{new}} = \omega + 2\tan^{-1}\left(\frac{a \sin\omega}{1 - a\cos\omega}\right)$$

where a is a parameter of frequency axis warping, ω is the sampling frequency expressed by the normalized angular frequency, and ω_{new} is the converted sampling frequency. This transformation can be done in a recursive way according to Oppenheim and Johnson (1972). A positive a value converts the frequency axis into the low-frequency weighted one by lengthening the low-frequency axis and shortening the high-frequency axis.

There are at least three possible ways to use the bilinear transform in LPC analysis:

1. The bilinear transform directly on digitized speech: There is no problem with stability of LPC analysis, but it requires a large amount of computation.
2. The bilinear transform on autocorrelation coefficients of LPC autocorrelation coefficients: Stability of LPC analysis following the bilinear transform is sometimes violated.
3. The bilinear transform of LPC cepstrum coefficients: Since the transformed cepstrum coefficients do not agree with the LPC model, we cannot calculate the corresponding LPC autocorrelation coefficients from the cepstrum coefficients.

We will examine the performance of some of these methods in this chapter.

2.4. Difference Spectrum

The movement of spectra such as formant change is considered to be important not only in perception but also in phonetic unit recognition. Furui (1984) confirmed the effectiveness of CEP with difference spectra in word recognition experiments. He derived the cepstrum coefficients of difference spectra through a linear interpolation of the linear equations of LPC cepstrum coefficients in the adjacent frames. In the following, the difference spectra are defined by the difference of two LPC spectra not only to reduce the amount of computation but also to reduce the smoothing effect of the linear interpolation. Hence, the difference spectrum f_i^d of the ith frame is defined as

$$\log(f_i^d) = \log(f_{i+n}) - \log(f_{i-n})$$
$$f_i^d = \frac{f_{i+n}}{f_{i-n}}$$

where f_{i+n} is the LPC spectrum of the $(i + n)$th frame. The frame difference value n will be determined by phonetic unit recognition experiments in Section 3.4.

The difference spectrum of the ith frame is represented by cepstrum coefficients C_j^{di}, $0 \leq j \leq 16$, as

$$C_j^{di} = C_j^{i+n} - C_j^{i-n}$$

where C_j^{i+n} is the jth LPC cepstrum coefficient of the $(i + n)$th frame.

2.5. Smoothed Group Delay Spectrum (SGDS)

In the LPC analysis each frame of speech can be represented by a pth-order all-pole filter with the transfer function

$$H(z) = \frac{1}{1 + \sum_{k=1}^{p} a_k z^{-k}}$$

The LPC coefficients a_k are calculated with the autocorrelation method.

With

$$H(\omega) = H(z)\,\big|_{z=e^{j\omega}} \qquad \text{and} \qquad A(\omega) = A(z)\,\big|_{z=e^{j\omega}} = \sum_{k=0}^{p} a_k e^{-j\omega k}$$

The phase $\Theta(\omega)$ of the transfer function $H(z)$ can be written as

$$\Theta(\omega) = -\arctan \frac{\mathrm{Im}\,\{A(\omega)\}}{\mathrm{Re}\,\{A(\omega)\}}$$

For the calculation of $\Theta(\omega)$, we used a discrete time Fourier transform (DTFT) with the Goertzel algorithm (Oppenheim and Schafer, 1975).

The group delay spectrum is defined as the derivative of the phase

$$\tau(\omega_i) = -\left.\frac{\partial\Theta(\omega)}{\partial\omega}\right|_{\omega=\omega_i}$$

Instead of this derivative we used the finite difference

$$\hat{\tau}(\omega_i) = -\frac{1}{\omega_i - \omega_{i-1}} \int_{\omega_{i-1}}^{\omega_i} \tau(\omega)\,d\omega = -\frac{\Theta(\omega_i) - \Theta(\omega_{i-1})}{\omega_i - \omega_{i-1}}, \qquad 1 \leq i \leq L$$

where L is the number of channels of SGDS and $\omega_i > \omega_{i-1}$ with $\omega_0 \geq 0, \omega_L \leq \pi$.

The averaging operation in the above equation is equivalent to multiplication of the cepstral coefficients with a sine window. The logarithmic spectrum of the LPC cepstrum and the group delay spectrum can be expressed as

$$\log|H(\omega)| = \sum_{k=0}^{\infty} C_k \cos(k\omega), \qquad \tau(\omega) = \sum_{k=0}^{\infty} kC_k \cos(k\omega)$$

where C_k are the cepstral coefficients. The phase $\Theta(\omega)$ is therefore

$$\Theta(\omega) = -\int_0^{\omega} \tau(x)\,dx = -\int_0^{\omega} \sum_{k=0}^{\infty} kC_k \cos(kx)\,dx$$

$$= -\sum_{k=0}^{\infty} kC_k \frac{\sin(k\omega)}{k} = -\sum_{k=0}^{\infty} C_k \sin(k\omega)$$

If we define $\hat{\tau}(\omega, B)$ as the difference equation smoothed group delay function, where the bandwidth is $2B$, we get

$$\hat{\tau}(\omega,B) = \frac{\Theta(\omega+B) - \Theta(\omega-B)}{2B}$$

$$= \frac{\sum_{k=0}^{\infty} C_k \sin[k(\omega+B)] - \sum_{k=0}^{\infty} C_k \sin[k(\omega-B)]}{2B}$$

$$= \frac{\sum_{k=0}^{\infty} C_k \cos(k\omega)\sin(kB)}{B} = \sum_{k=0}^{\infty} \frac{\sin(kB)}{B} C_k \cos(k\omega)$$

If we compare the equations for $\tau(\omega)$ and $\hat{\tau}(\omega, B)$, we see that we have in fact weighted the cepstral coefficients C_k with a sine function $\sin(kB)/B$. For 16 sampling points, B would be $\pi/32$. Use of a sine window for smoothing was also advocated by Juang (1986), although the window length is different from that in our approach (here the window length is not limited), The effect of different values of B can be seen in Fig. 1f.

An additional smoothing effect is achieved by multiplying the coefficients a_k with an exponentially decaying window prior to the DTFT:

$$\hat{a}_k = c^k a_k, \qquad c < 1.0$$

which moves the poles z_k of $h(z)$ away from the unit circle. The poles of $\hat{H}(z)$ are then located at $\hat{z}_i = z_i c$. The effect of this smoothing for different values of c on the group delay spectra is shown in Fig. 1h, where both types of smoothing were used with $c = 0.9$ and different values for B.

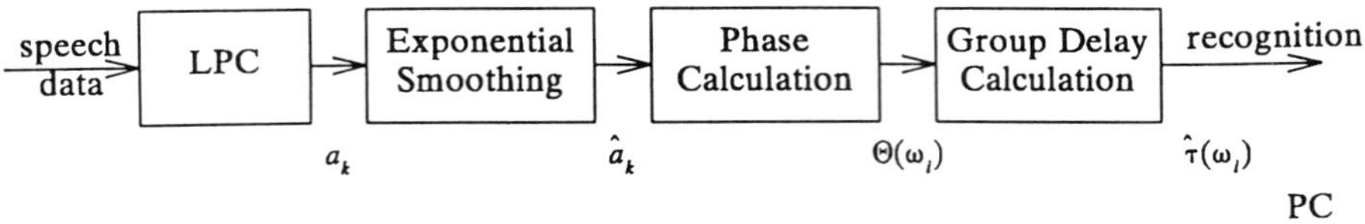

Figure 2 Block diagram of the basic algorithm. LPC parameters a_k, smoothed LPC parameters $\hat{a}_k$, phase $\Theta(\omega_i)$, and smoothed group delay $\hat{\tau}(\omega_i)$.

The complete basic algorithm is depicted as a block diagram in Fig. 2. In the recognition phase, we define the distance metric of the SGDS between two speech frames as

$$d = \sum_{i=0}^{L} [\hat{\tau}_R(\omega_i) - \hat{\tau}_T(\omega_i)]^2$$

where $\hat{\tau}_R(\omega_i)$ and $\hat{\tau}_T(\omega_i)$ are the values of the ith channel of the reference pattern and the test pattern, respectively.

3. EVALUATION OF DISTANCE MEASURES FOR PHONETIC UNIT RECOGNITION

In this section the speech data base used to perform the phonetic unit recognition experiments and the procedure for carrying out phonetic unit recognition are described first. Second, the WLR measure and the LPC cepstrum are compared. Third, the bilinear transformed LPC cepstrum measure is evaluated. Finally, difference spectra and power information are introduced to improve distance measures.

3.1. Speech Data Base

Harvard sentence speech data uttered by three male speakers were used as the data base. Each speaker uttered 10 different Harvard sentences. The Harvard sentence speech data (16-kHz sampling) and phonetic unit labels (ptlola files) were stored in disk files. Ptlola files contained phonetic labels hand-labeled by expert spectrogram readers. Ptlola files included about 60 phonetic units and several modifiers as listed in Table 1. All modifiers were ignored in the phonetic unit recognition experiments.

Linear matching between spectrum sequences was adopted for calculating the distance. The linear matching allowed only linear warping with

Table 1 Phonetic Units and Modifiers

(Phonetic Units)	
Vowels	
front vowels: /iy/, /ih/, /eh/, /ae/,	back vowels: /aa/, /ao/, /uh/, /uw/
central vowels: /ix/, /ax/, /ah/,	diphthongs: /ey/, /ay/, /oy/, /aw/, /ow/, /o/
front rounded vowels: /ux/, /oi/, /oe/,	r-colored vowels: /er/
Stops	
voiceless stops: /p/, /t/, /k/,	voiced stops: /b/, /d/, /g/
Fricatives	
voiceless fricatives: /f/, /th/, /s/, /sh/,	voiced fricatives: /v/, /dh/, /z/, /zh/
Affricates: /ch/, /jh/	
Semivowels	
semivowels: /y/, /w/,	liquids: /l/, /r/
Nasals: /m/, /n/, /ng/	
Other events:	
transition: /epi/ between a fricative and a vocalic sound.	
aspirates: /hh/, /hv/	
closure, et al.: /cl/, /*cl/, /dx/, /nx/, /rx/, /q/, /voi/ ; /cl/-voiceless closures, /*cl/-voiced closures	
(Modifiers): -b,-h,-r,-l,-n,-p,-w,-q; All modifiers are ignored in the experiments.	

fixed start and end points. Nonlinear alignment by dynamic time warping was tested, but linear warping showed almost the same performance because multitemplates of phonetic units were likely to include the warped templates. The total matching distance divided by the average number of frames per input phonetic unit (k) and reference phonetic unit (l), $(k+l)/2$, was employed in phonetic unit recognition experiments. Speaker-dependent recognition of phonetic units was carried out using 10 Harvard sentences uttered by each speaker. All phonetic units specified by the label files excluding the input phonetic unit to be recognized were used as phonetic unit templates. The nearest-neighbor decision rule by multitemplates was adopted.

3.2. WLR Measure and LPC Cepstrum Distance

The peak weighting measure WLR has been applied successfully to vowel recognition, word recognition, and noisy speech recognition for Japanese utterances. In this subsection, the WLR performance for English utterances is investigated through comparison with the LPC cepstrum distance (CEP). Since WLR is capable of peak weighting and frequency weighting, optimization of these weighting schemes is also tried.

The process of LPC analysis for WLR calculation is listed below. In the case of CEP calculation, step 5 may be omitted because the values of CEP are relatively independent of the spectral overall tilt. In filtering for frequency weighting in step 5, $A = 0.0$ is used for low-frequency weighting, $A = +0.2$ is used for weak low-frequency weighting, and $A = +0.6$ is used for weak high-frequency weighting.

LPC analysis process for WLR

1. AD data, 16-kHz sampling
2. Hamming windowing, 6-ms step, 24-ms window width
3. Filtering $(1 - 0.5z^{-1})$
4. Autocorrelation calculation and power normalization, 13-order
5. Filtering for frequency weighting $(1 - Az^{-1})$
6. LPC analysis, 12-order LPC

Phonetic unit recognition experiments are carried out using the Harvard sentence data base uttered by three male speakers following the procedure described in the previous subsection. The recognition results for broad phonetic unit categories are summarized in Table 2. These results show that the effectiveness of peak weighting and frequency weighting depends on phonetic

Table 2 Phonetic Unit Recognition Results in WLR, CEP and CCW

	Number of correctly recognized phonetic units				
	WLR				
	$A = 0.0$	$A = +0.2$	$A = +0.6$	CEP	CCW
Vowels (total 269)	83	84	77	61	89
Consonants (total 267)	105	107	108	114	114
Nasals (total 57)	15	17	14	18	19
Semivowels and liquids (total 69)	22	24	24	25	31
Closures (total 154)	65	68	79	90	91
Total (816)	290	300	302	308	344
%	35.5	36.8	37.0	37.7	42.2

categories. The results for fine phonetic categories were also studied. In brief, we found that

1. Peak weighting by WLR is effective for all vowels and voiceless stops.
2. CEP is effective for voiced stops, voiceless fricatives, and closures.

A combined measure using CEP and WLR, CCW, which has the merits of both, is proposed as a linear combination of WLR and CEP, using a normalized power value. The normalized power value is one of the simplest parameters for discriminating vowel-like phonetic units from consonant-like phonetic units. CCW is defined by

$$\mathrm{CCW} = \alpha \mathrm{WLR} + \beta(1-\alpha)\mathrm{CEP}$$

where α is the linear combination weighting factor, which is determined by a normalized power value, and β is a scaling adjusting parameter between WLR and CEP. The parameters α and β are determined through phonetic unit recognition experiments. Phonetic unit recognition results in CCW are compared with those in CEP and WLR as shown in Table 2. These results show that CCW has merits of both CEP and WLR in phonetic unit recognition. CCW is superior to CEP and WLR for all kinds of broad phonetic unit categories.

3.3. Bilinear Transform

The bilinear transform (Oppenheim and Johnson, 1972) is an efficient way to convert the linear frequency axis into a Mel scale–like one. There are three possible ways to apply the bilinear transform:

1. Digitized speech
2. LPC autocorrelation coefficients
3. LPC cepstrum coefficients

Since warped autocorrelation coefficients do not always satisfy the positive definite condition, we apply the bilinear transform to digitized speech and LPC cepstrum coefficients.

A typical analysis of the bilinear transform application to digitized speech proceeds as follows:

1. AD data, 16-kHz sampling
2. Hamming windowing, 6-ms step, 24-ms window width
3. Preemphasis filtering $(1 - 0.98z^{-1})$
4. Frequency warping by the bilinear transform on digitized speech
5. Autocorrelation calculation and power normalization, 12-order

6. LPC analysis, 12-order LPC

Phonetic unit recognition results for frequency axis warping parameter a = 0.0, 0.4, 0.6, and 0.8 are shown in Table 3. The phonetic recognition results using the bilinear transform of digitized speech are summarized as follows.

1. CEP warped by $a = 0.6$ gives the best results. CCW and WLR cannot attain better results than the CEP warped by $a = 0.6$.
2. For vowel recognition, WLR using $a = 0.4$ and CEP using $a = 0.6$ are promising. In particular, CEP with $a = 0.6$ shows considerable improvement. These results show that frequency warping or frequency weighting is very useful for vowel recognition.
3. For consonant recognition, frequency warping for WLR and CEP showed almost no improvement on the average. These results suggest that high-frequency information is sometimes very significant for consonant recognition.

LPC cepstrum coefficients represent the smoothed LPC spectrum envelope. The warped LPC cepstrum coefficients also represent the smooth

Table 3 Phonetic Unit Recognition Results in WLR, CEP, and CCW Through the Bilinear Transform on Digitized Speech

	Number of correctly recognized phonetic units										
	WLR				CEP				CCW		
Warping parameter	0.0	0.4	0.6	0.8	0.0	0.4	0.6	0.8	0.0	0.4	0.6
Vowels (total 269)	84	95	81	71	61	84	99	73	89	94	94
Consonants (total 267)	107	96	105	91	114	116	115	81	114	108	110
Nasals (total 57)	17	21	22	19	18	21	22	18	19	21	20
Semivowels and liquids (total 69)	24	23	19	19	25	31	32	22	31	26	26
Closures (total 154)	68	87	84	74	90	85	94	79	91	92	94
Total (816)	300	322	311	274	308	337	362	273	344	341	344
%	36.8	39.5	38.1	33.6	37.7	41.3	44.4	33.5	42.2	41.8	42.2

frequency-warped spectrum, but they do not satisfy the LPC model. Since these warped cepstrum coefficients do not satisfy the LPC model, we cannot obtain the LPC autocorrelation coefficients using the recursive equation between the LPC cepstrum coefficients and the LPC autocorrelation coefficients. The cepstrum distance can be calculated from cepstrum coefficients, however, because it does not require the autocorrelation coefficients. A typical analysis of the bilinear transform of cepstrum coefficients proceeds as follows:

1. AD data, 16-kHz sampling
2. Hamming windowing, 6-ms step, 24-ms window width
3. Pre-emphasis filtering $(1 - 0.98z^{-1})$
4. Autocorrelation calculation and power normalization, 12-order
5. LPC analysis, 12-order LPC
6. Calculation of LPC cepstrum coefficients up to the 16th to 64th order
7. Frequency warping by the bilinear transform of LPC cepstrum coefficients

Phonetic unit recognition results for $a = 0.0, 0.2, 0.4, 0.6$, and 0.8 are shown in Table 4. The effects of the bilinear transform on LPC cepstrum coefficients are summarized as follows.

1. For vowel, semivowel, and liquid recognition, strong frequency warping by $a = 0.6$ and 0.8 is effective. These results are similar to those for the bilinear transform on digitized speech.
2. For consonant recognition, frequency warping has no adverse effect. It seems that the weak frequency warping by $a=0.2$ is somewhat effective.

3.4. Difference Spectra and Power Information

Spectrum movement is useful for recognition of some phonetic units. In this subsection, the spectral matching measures CEP, WLR, and CCW are augmented with the distance terms of difference spectra and difference power as well as power and duration information:

$$\begin{aligned}\text{New measure} &= \text{temporal spectral distance}\\ &\quad + q(\text{difference spectrum distance}) + s(\text{difference power distance})\\ &\quad + w(\text{power distance}) + d(\text{duration distance})\end{aligned}$$

In this subsection, the weighting parameters q, s, w, and d and the frame number difference for calculating difference spectra are determined based on phonetic unit recognition experiments.

Table 4 Phonetic Unit Recognition Results in CEP Through the Bilinear Transform on LPC Cepstrum Coefficients

	Number of correctly recognized phonetic units				
	CEP (cepstrum distance)				
Warping parameter	0.0	0.2	0.4	0.6	0.8
Vowels (total 269)	61	65	77	89	98
Consonants (total 267)	114	123	123	124	117
Nasals (total 57)	18	17	16	19	19
Semivowels and liquids (total 69)	25	28	33	32	35
Closures (total 154)	90	92	92	95	88
Total (816)	308	325	341	359	357
%	37.7	39.7	41.8	44.9	43.8

First, we determined the temporal difference n and the weighting parameter q for CEP augmented with the cepstrum distance of difference spectra by means of phonetic unit recognition experiments. The temporal difference n varied from 2 frames (12 ms) to 16 frames (96 ms). The weighting parameter q varied from 0.2 to 1.0. The number of correctly recognized phonetic units for various sets of the temporal difference length n and the weighting parameter q is shown in Table 5. The experimental results in Table 5 show that the temporal difference n at 8 frames (48 ms) to 14 frames (84 ms) is promising and that the weighting parameter q at 0.3 and 0.5 is promising. We set the temporal difference n to 8 frames (48 ms). The weighting parameter q was fixed at 0.5 for CEP and at 0.2 for WLR and CCW according to the same phonetic unit recognition experiments. Phonetic unit recognition results in CEP, WLR, and CCW are summarized in Table 6. These results show that significant improvement is attained for all measures of temporal spectra and for all phonetic unit broad categories.

The weighted parameter s for the difference power term, the parameter w for the power term, and the weighting parameter d for the duration term

Table 5 Number of Correctly Recognized Phonetic Units for Various Parameter Sets in CEP Augmented with Cepstrum Distance of Difference Spectra

Frame length of difference (n)	Number of correctly recognized phonetic units[a]			
	CEP (cepstrum distance)			
	Weighting parameter q for difference spectrum term			
	0.2	0.3	0.5	1.0
0	308	308	308	308
2 (12 ms)	—	309	309	304
4 (24 ms)	—	333	333	331
6 (36 ms)	—	332	338	332
8 (48 ms)	340	343	347	343
10 (60 ms)	332	342	347	333
12 (72 ms)	346	343	337	333
14 (84 ms)	343	343	337	328
16 (96 ms)	—	336	330	303

[a]Total number of phonetic units is 816.

were also determined through phonetic unit recognition experiments. The phonetic unit recognition results are also summarized in Table 6.

3.5. Summary of Phonetic Unit Recognition Results

To confirm the results of the previous subsections obtained by three male speakers' data, more phonetic unit recognition experiments were carried out using another subset of the Harvard sentence data base that was uttered by three female speakers. These results, summarized in Table 7, roughly agree with those obtained with the data base of the three male speakers.

Results obtained with three male speakers and three female speakers for LPC spectral matching measures of CEP, WLR, and CCW are summarized in Table 8.

In the beginning of this section, CEP and WLR were viewed to clarify the performance difference for phonetic unit categories. The experimental results showed that WLR is better than CEP in vowel recognition and that CEP is better than WLR in consonant recognition. In conclusion, based on these results, a linear combination of CEP and WLR (CCW) was introduced.

Table 6 Phonetic Unit Recognition Results in WLR, CEP, and CCW Augmented with Spectrum Difference and Power Information

	Number of correctly recognized phonetic units								
Temporal spectra	WLR	WLR	WLR	CEP	CEP	CEP	CCW	CCW	CCW
Difference spectra	—	CEP	CEP	—	CEP	CEP	—	CEP	CEP
Weighting (q)	—	0.2	0.2	—	0.5	0.5	—	0.2	0.2
Difference power (s)	—	—	0.02	—	—	0.03	—	—	0.02
Power (w)	—	—	0.005	—	—	0.04	—	—	0.005
Duration (d)	—	—	0.2	—	—	0.3	—	—	0.2
Vowels (total 269)	84	92	98	61	84	95	92	95	107
Consonants (total 267)	107	122	125	114	122	129	114	126	130
Nasals (total 57)	17	26	29	18	18	22	19	23	27
Semivowels and liquids (total 69)	24	30	33	25	32	33	31	32	35
Closures (total 154)	68	77	90	90	91	92	91	90	92
Total (816)	300	347	375	308	347	371	344	366	391
%	36.8	42.5	46.0	37.7	42.5	45.5	42.2	44.9	47.9

CCW yielded better recognition rates than CEP and WLR for both vowels and consonants.

To improve the performance of these measures, the bilinear transform was introduced to realize frequency axis warping. CEP with the bilinear transform showed substantial improvement in the phonetic unit recognition experiments. Finally, difference spectra and power information were used to improve the recognition results. The use of difference spectra and power information attained measurable improvement.

The results of these phonetic unit recognition experiments showed the following.

1. WLR attains better recognition rates than CEP in vowel recognition, and CEP attains better recognition rates than WLR in consonant recognition. This means that local peak weighting and low–frequency weighting are essential for vowel recognition and that high–frequency information is useful for consonant recognition.

Table 7 Summary of Phonetic Unit Recognition Performance for Three Female Speakers

Measure	Bilinear trans-form	Diff. spec. and diff. power	Power and dura-tion	Average recognition rate (%)	Vowel recognition rate (%)	Consonant recognition rate (%)
CEP	—	—	—	43.6	34.9	50.4
WLR	—	—	—	42.7	36.8	46.4
CCW	—	—	—	44.7	37.2	51.1
CEP	—	used	—	49.3	38.0	54.7
WLR	—	used	—	48.7	41.5	53.3
CCW	—	used	—	47.8	41.5	53.6
CEP	—	used	used	51.2	42.6	57.2
WLR	—	used	used	49.3	41.1	55.8
CCW	—	used	used	48.4	42.2	56.2
CEP	used	used	used	50.4	43.4	55.4

Table 8 Summary of Phonetic Unit Recognition Performance for Three Male Speakers and Three Female Speakers

Measure	Bilinear trans-form	Diff. spec. and diff. power	Power and dura-tion	Average recognition rate (%)	Vowel recognition rate (%)	Consonant recognition rate (%)
CEP	—	—	—	40.8	28.8	46.6
WLR	—	—	—	39.7	34.0	43.3
CCW	—	—	—	43.4	35.2	46.9
CEP	—	used	—	47.1	36.1	51.5
WLR	—	used	—	46.8	38.1	51.6
CCW	—	used	—	47.6	39.4	52.1
CEP	—	used	used	48.5	39.0	52.8
WLR	—	used	used	47.6	38.8	51.3
CCW	—	used	used	48.5	39.7	54.0
CEP	used	used	used	49.2	42.5	52.6

2. CCW attains the best results among the temporal spectra matching measures.
3. Difference spectra information substantially improves the recognition rate of CEP, WLR, and CCW.
4. The bilinear transform improves vowel recognition rates by CEP substantially.
5. The bilinear-transformed CEP with all added information achieves a phonetic recognition rate of 49.2%, which is 8.4% better than that of the CEP.

4. SMOOTHED GROUP DELAY ANALYSIS AND ITS APPLICATION TO ISOLATED WORD RECOGNITION

The effectiveness of the smoothed group delay distance measure was demonstrated in previous work (Itakura and Umezaki, 1987; Umezaki et al., 1989). The coefficients of the smoothed group delay spectrum were then calculated by multiplying the LPC cepstrum coefficients with a smoothing weight function; i.e., the representation was in the time domain.

In this section we calculate the SGDS coefficients using a discrete time Fourier transform (DTFT) of the linear prediction coefficients; i.e., the representation is in the frequency domain. We report isolated word recognition experiments with a low-bit quantization of these SGDS coefficients. We show that recognition accuracy can be maintained using only 26 bits per frame as compared to the conventional calculation with floating-point accuracy. Using a Bark-scale representation, the error rate can be reduced even further.

4.1. Data Bases and Matching Algorithms

Data base 1 consists of 550 Japanese city names recorded twice by five Japanese male speakers. The second utterances were recorded 1 week after the first utterances.

Sixty-eight easily confusable pairs (Singer et al., 1989) were chosen for the recognition experiments. In the preliminary experiments we found that nearly all errors occurred as a confusion of these pairs. Therefore, we performed DTW matching only between these pairs to speed up our experiments.

Data base 2 consists of 15 digits (Singer et al., 1989) recorded once by 50 Japanese male speakers.

The pattern matching algorithm is a standard end point–constrained DTW algorithm. Data base 1 was used for intraspeaker recognition (speaker dependent) with reference patterns and test patterns from the same speaker. We also investigated interspeaker recognition with reference and test patterns from different speakers.

Data base 2 was used only for speaker-independent recognition. Ten speakers were designated as reference speakers and their utterances stored as reference templates. The other 40 speakers were test speakers. A variant of the popular K-means algorithm was used: An "unknown" test pattern was DTW-matched with all 10×15 reference templates and the calculated distances were stored. For each of the labels the three best results were averaged ($K = 3$). The label with the best averaged result was chosen as the final result.

To investigate the robustness of the distance measure under the noisy condition, we added multiplicative signal-dependent white noise as follows:

$$\hat{S}_n = S_n(1 + AR_n) \qquad \text{with } A = \sqrt{3}\,10^{-\,\mathrm{SNR(dB)}/20}$$

S_n is the "clean" speech signal, $\hat{S}_n$ is the noisy speech signal, A is the relative noise amplitude, SNR(dB) is the desired signal-to-noise ratio (SNR), and R_n is a uniform distributed random number between -1 and 1. We investigated only two cases, namely SNR $= \infty$ dB and SNR $=$ 20 dB.

4.2. Comparison of Experimental Results

First, we attempted to find an optimal value for the smoothing factor c. The number of SGDS channels was fixed at 16. In accordance with Umezaki et al. (1989), an increase in the number of channels did not improve recognition accuracy. Figure 3 shows the error rate for data base 1 for changing values of c. Both intraspeaker and interspeaker experiments were also performed with added segmental noise (see Section 4.1.). As a consequence, we choose $c = 0.9$ for all subsequent experiments (if not explicitly mentioned otherwise). This value was also confirmed by Umezaki et al. (1989).

We also found that not using the higher channels improved the recognition rate considerably (see Fig. 4). The best results were obtained using only channels 1 to 11. At a 10-kHz sampling rate and an evaluation of $\tau(\omega_i)$ at 16 equally spaced frequency points, the bandwidth of channel 1 corresponds to 5 kHz/16 $\approx$ 300 Hz. Not using channels 12 to 16 is thus equivalent to a low-pass filter having a cutoff frequency at 3.3 kHz. The relevant information is apparently confined to the region up to around 3.3 kHz. This result is interesting in view of the fact that telephonic speech uses the frequency range between 300 Hz and 3.4 kHz.

We then compared our results with the standard LPC cepstrum (CEP) and weighted LPC cepstrum ($s = 1.0, \tau = 10.0$ from the Gaussian window weighting equation in Section 1), all with 16 parameters (channels) per frame.

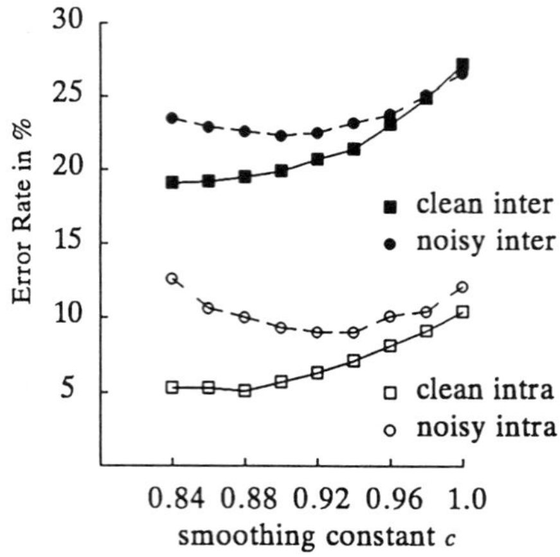

Figure 3 Dependence of error rate on smoothing constant c for interspeaker and intraspeaker experiments.

Figure 5 presents the recognition results for data base 1 (intraspeaker only) and data base 2. Without noise, the three methods yield comparable results. With noise added, the score for CEP drops considerably. Especially data base 1, with its many confusable consonants, is strongly affected by noise. Similar results have been reported by Umezaki et al. (1989). On the other hand, we could show that both WCEP and SGDS are quite robust against the influence of white noise.

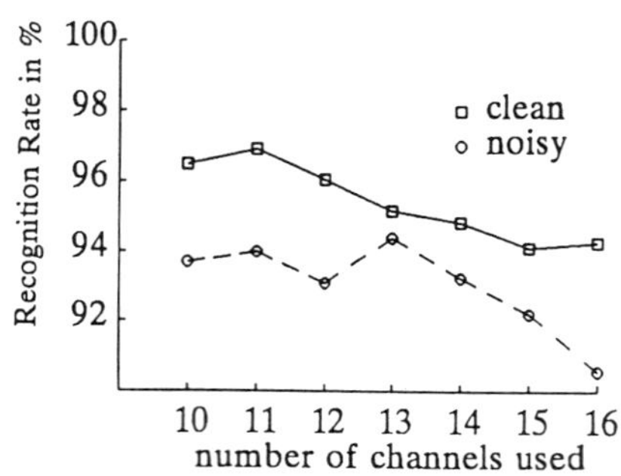

Figure 4 Effect of not using SGDS parameters of higher frequencies.

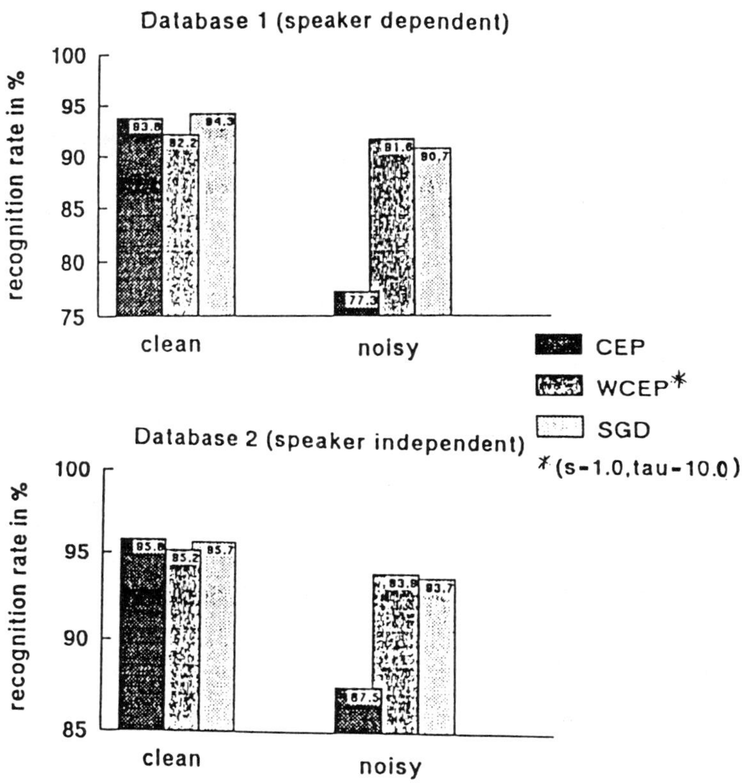

Figure 5 Comparison of cepstrum, weighted cepstrum, and SGD.

4.3. Low-Bit Quantization

Under the assumption that the essential features for speech recognition are frequencies of the formants, we can further reduce the information contents by low-bit quantization of the SGDS parameters. Three problems have to be addressed:

- How many bits per channel are necessary?
- Where should the quantization thresholds be set for optimal recognition results?
- Which value should be assigned to each quantization level?

We tried to solve these problems heuristically, since minimum distortion is not necessarily equivalent to the maximum recognition rate. Quantization also works as an additional smoothing effect, which can even raise the recognition score. Optimal quantization thresholds for quantization into 1, 2, and 3 bits are shown in Table 9. Note that the LPC order $p = 10$. The theoretical minimum for the group delay $\tau(\omega_i)$ is shown to be $-p/2$ (here $-p/2 = -5$). It is obvious that the threshold values also depend on the smoothing factor c (here $c = 0.9$).

Figure 6 demonstrates how the SGDS channels $\tau(\omega_i)$ were quantized according to Table 5. Figure 7 shows how the formant structure is preserved even by quantization with 1 bit.

The recognition results for data bases 1 and 2 are shown in Fig. 8. Obviously, quantization with 1 bit is too rough, whereas 2 bits give comparable results to calculation with floating-point precision.

Only using channels 1 to 11, as explained in Section 2.3, also gives better results than using all 16 channels for the quantized SGDS. This signifies that one frame of speech can be represented by 22 bits. A hypothetical large vocabulary recognition system with 5000 words, having an average word length of 0.5 s and 125 frames/s, would thus necessitate $5000 \times 0.5 \times 125 \times 22/8$ bytes $\approx$ 1 Mbyte of memory space. SGDS is thus well suited for the implementation of an inexpensive large vocabulary recognition system.

4.4. Bark Scale and Mel Scale

Many researchers suggest a perceptually based approach, using filter banks which model human hearing. In this chapter, we simply use bark scale, which

Table 9 Thresholds for Quantization

Bits	Thresholds
1	0
2	0, 2.5, 5
3	−2.5, 0, 1, 2, 3, 4, 5

Figure 6 Quantization of $\tau(\omega_l)$. The continuous line represents the SGD spectrum of the vowel /a/ in KITAMI, using only the smoothing of Eq. (9), $\tau(\omega)$ was calculated according to Eq. (4). The crosses represent SGDS parameters according to Eq. (5). The numbers at the right margin stand for the output of each quantization range, the numbers on the bottom for the actual quantizer output.

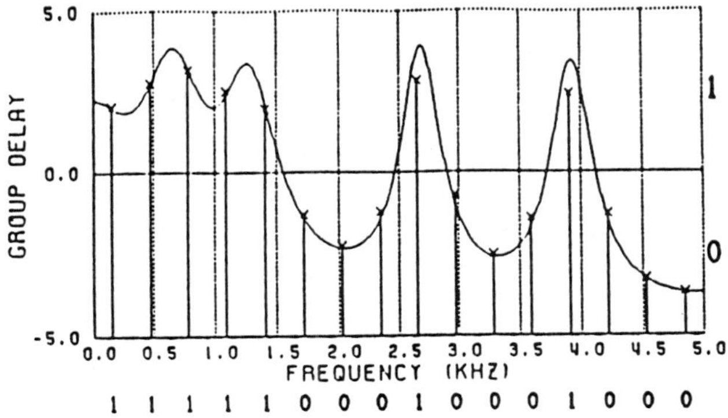

Quantization with 1 bit

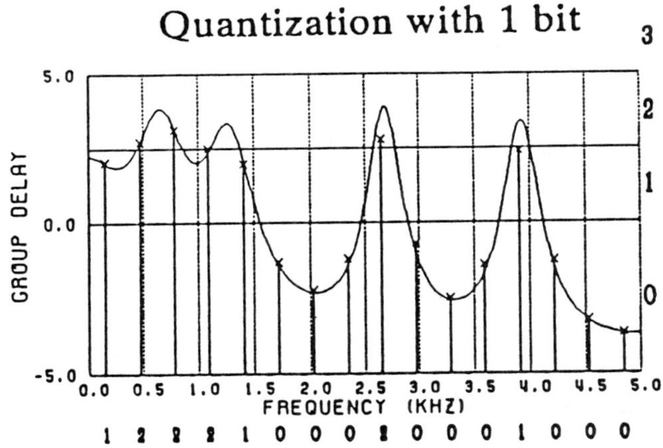

Quantization with 2 bit

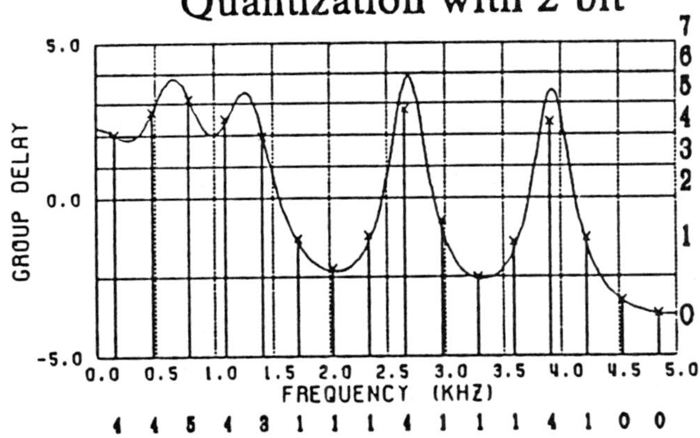

Quantization with 3 bit

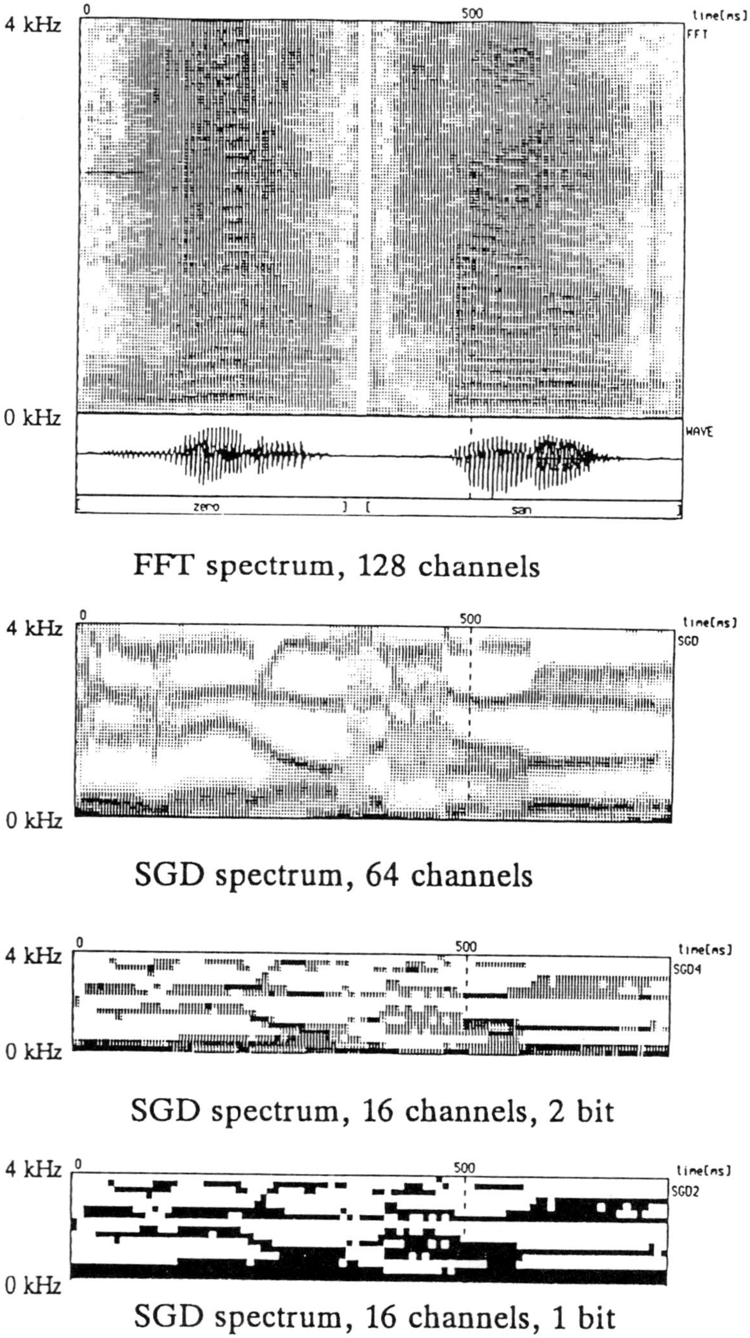

Figure 7 Time frequency pattern of FFT and SGD spectra.

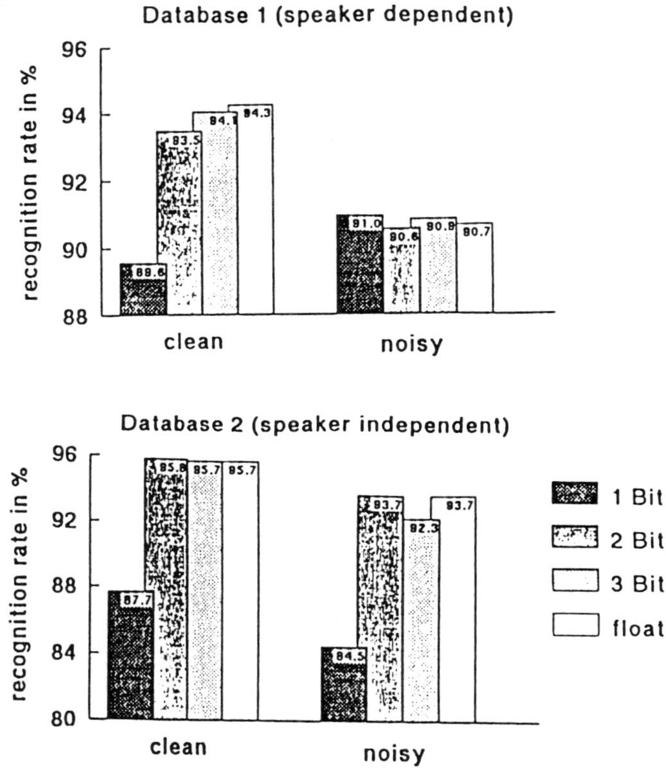

Figure 8 Recognition results for quantized SGDS.

models the critical band width in the cochlea. Bark scale conversion can be applied either in the time (or quefrency) domain or in the frequency domain. We choose the frequency domain; that is, the frequencies f_i are now chosen equally spaced on bark scale. Bark scale to frequency scale conversion is performed by

$$f_i = 600 \sinh\left(\frac{m_i}{6}\right), \qquad m_i = 6 \operatorname{arcsinh}\left(\frac{f_i}{6}\right)$$

with f_i in Hz and m_i in bark.

Figure 9 illustrates the positive effect of "bark scale sampling." The first and second formants, which would have been merged into one peak if sampled linearly on the frequency scale, are clearly distinct if bark-scale sampling is applied.

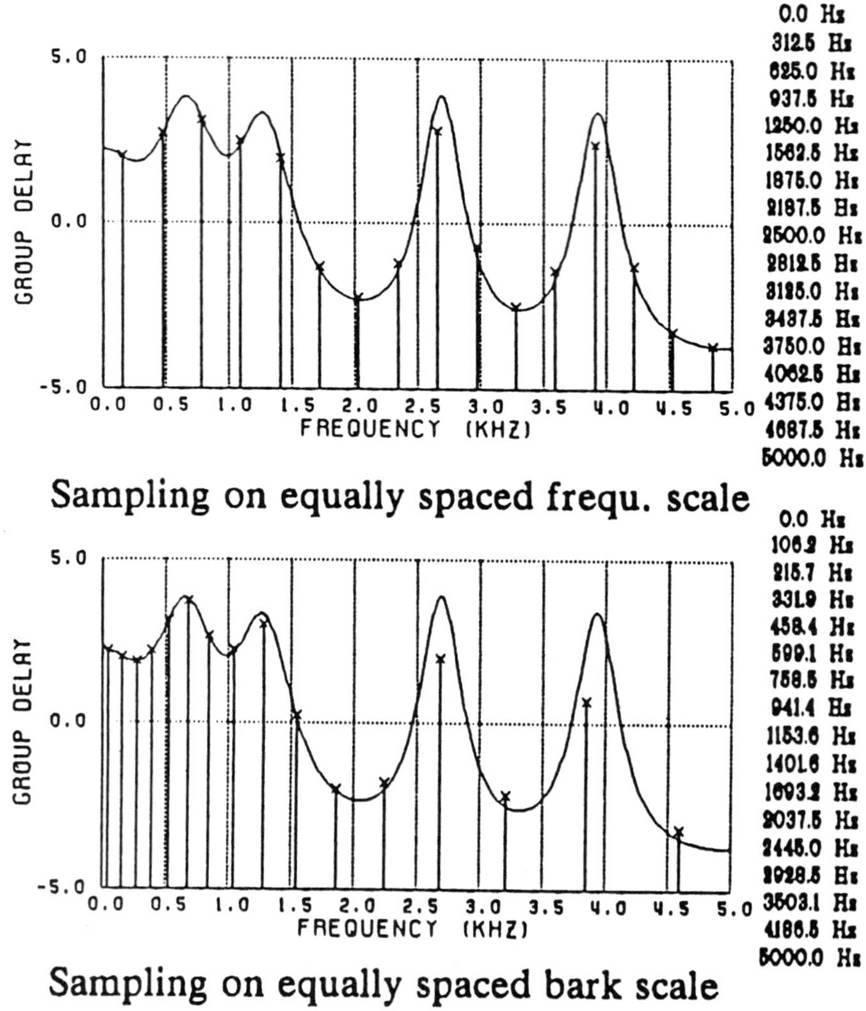

Figure 9 Linear frequency scale versus bark scale.

We also found that not using the 1st, 15th and 16th channels gave equal or better results. Since the data have not been high-pass filtered, deleting this first parameter is equivalent to a high pass with a 100-Hz cutoff frequency. (For linear frequency spacing this equivalent high pass would have a cutoff

Table 10 Recognition Rate (%) on Data Base 1 (Speaker Dependent)

	SGD				LPC-CEP
SNR	Float	1 bit	2 bit	3 bit	Float
∞ dB					
Bark	97.2	91.9	95.6	96.9	96.0
Linear	96.9	90.3	94.8	95.7	93.8
20 dB					
Bark	95.9	90.3	93.7	95.4	86.2
Linear	94.0	90.0	91.5	90.9	77.3

frequency of about 300 Hz, also eliminating information concerning the first formant.) Not using channels 15 and 16 is equivalent to a low pass with cutoff frequency of about 3500 Hz, a result consistent with that in Section 4.2.

Table 10 compares the results for data base 1. For linear-frequency sampling (linear) only channels 1 to 11 were used, whereas channels 2 to 14 were used for bark-scale sampling. Quantization thresholds were chosen according to Table 9. The Mel cepstrum was calculated by warping the LPC cepstrum coefficients according to an all-pass filter having warping constant $a = 0.45$ (Kobayashi et al., 1987).

4.5. Summary

Our results confirm that (after carefully tuning various parameters like quantization thresholds) recognition rates are better for bark scale than for the linear frequency scale. We can thus conclude that frequency domain techniques (bark scaling, frequency weighting, and low-bit quantization) can easily be used with the SGDS. Contrary to the cepstral representation, the SGDS can also be easily visualized as a time-frequency pattern and therefore easily interpreted.

The low-bit representation enables high-speed computation in the pattern matching stage of the recognizer. Furthermore, considerable (one order of magnitude) saving of memory space can be achieved, allowing inexpensive implementation of a large vocabulary speech recognizer.

5. CONCLUSION

Spectrum distance measures for speech recognition were described focusing on phonetic unit recognition and robust speech recognition.

In phonetic unit recognition, the LPC cepstrum distance (CEP) and the weighted likelihood ratio measure (WLR) were mainly discussed, and the bilinear transform was introduced. Moreover, the difference spectrum, power, and duration information were introduced to augment distance measures. These refinements resulted in 9% phonetic unit recognition rate improvement, that is from 40 to 49% on the average for three male and three female speakers.

In robust speech recognition, the smoothed group delay spectrum distance was proposed to extract the relevant features of the speech signal. SGDS made possible the representation of speech in very low bits.

ACKNOWLEDGMENT

We are deeply indebted to Mr. Harald Singer and Dr. Taizo Umezaki for the smoothed delay spectrum research.

REFERENCES

Aikawa, K., and Shikano, K. (1985). Spoken word recognition using vector quantization in power-spectrum vector space. *Inst. Electr. Commun. Eng. Jpn. Trans. D J68-D*(3): (in Japanese).

Aikawa, K., Sugiyama, M., and Shikano, K. (1984). Spoken word recognition based on top-down phoneme recognition. *Inst. Electr. Commun. Eng. Jpn. Trans. A J67-A*(6):693–700(in Japanese).

Atal, B. S. (1974). Effectiveness of linear prediction characteristics of the speech wave for automatic speaker identification and verification. *J. Acoust. Soc. Am. 55*:1304–1312.

Egan, J. (1944). Articulation testing method II. OSRD Report, U.S. Department of Commerce, November, P1322848.

Furui, S. (1984). Spoken word recognition using dynamic features of speech spectrum. *Trans. Comm. Speech Res. Acoust. Soc. Jpn. S84-65*:509–516 (in Japanese).

Gray, A. H., Jr., and Markel, J. D. (1976). Distance measure for speech processing. *IEEE Trans. Acoust. Speech Signal Process. ASSP-24*:380–391.

Hanson, B. A., and Wakita, H. (1987). Spectral slope distance measures with linear prediction analysis for word recognition in noise. *IEEE Trans. Acoust. Speech Signal Process. ASSP-35*:968–973.

Itakura, F. (1975). Minimum prediction residual principle applied to speech recognition. *IEEE Trans. Acoust. Speech Signal Process. ASSP-23*:67–72.

Itakura, J., and Umezaki, T. (1987). Distance measure for speech recognition based on the smoothed group delay spectrum. *Proc. ICASSP*, April, pp. 1257–1260.

Juang, B. H. (1986). On the use of bandpass liftering in speech recognition. *Proc. ICASSP*, April, pp. 765–768.

Kobayashi, T., Kondo, S., and Imai, S. (1987). Evaluation of generalized cepstral distance measures for isolated word recognition, *IECE*, June, vol. SP87-18 (in Japanese).

Lee, K. (1985). Network representation of templates in word recognition. Paper presented at a meeting of the Acoustical Society of America, April.

Lowerre, B. T. (1976). The HARPY speech recognition system. Ph.D. thesis, Carnegie-Mellon University.

Mitsui, H., and Matsumoto, H. (1988). A study on a modified WLR distance metric for noise robustness. *Trans. Comm. Speech Res. Inst. Electr. Commun. Eng. Jpn. SP88-49*: (in Japanese).

Nocerino, N. S. (1985). A comparative study of distance measures for speech recognition. Master's thesis, Massachusetts Institute of Technology

Nocerino, N., Soong, F., Rabiner, L., and Klatt, D. (1985). Comparative study of several distortion measures for speech recognition. *Proc. IEEE Int. Conf. Acoust. Speech Signal Processing*, March.

Oppenheim, A. V., and Johnson, D. H. (1972). Discrete representation of signals. *Proc. IEEE 60*:681–691.

Oppenheim, A. V., and Schafer, R. W. (1975). *Digital Signal Processing*, Prentice-Hall, Englewood Cliffs, N.J.

Rabiner, L. R., Pan, C. K., and Soong. F. K. (1984a). On the performance of isolated speech recognition using vector quantization and temporal energy contours. *AT&T Bell Lab. Tech. J. 63*(7):1245–1260.

Rudnicky, A. I., Cole, R. A., Anderson, S., Weaver, J., and Weide, R. (1984). Notes on spectrogram labeling. Internal Memo, Speech Group, Computer Science Department, Carnegie-Mellon University, September.

Shikano, K. (1985). Text-Independent speaker recognition experiments using codebooks in vector quantization. Paper presented at a meeting of the Acoustical Society of America, April.

Shikano, K., and Kohda, M. (1980). On the LPC distance measures for vowel recognition in continuous utterances. *Inst. Electr. Commun. Eng. Jpn. Trans. E J63-D*(5): 394–401 (in Japanese).

Shikano, K., and Sugiyama, M. (1982). Evaluation of LPC spectral matching measures for spoken word recognition. *Inst. Electr. Commun. Eng. Jpn. Trans. D J65-D*(5):535–541 (in Japanese).

Singer, H., Umezaki, T., and Itakura, F. (1989). Low-bit quantization of the smoothed group delay spectrum for speech recognition. *Proc. ICASSP*, 55S15a.3.
Soong, F. K., and M. M. Sondhi (1987). A frequency-weighted Itakura distortion measure and its application to speech recognition in noise. *IEEE Trans. Acoust. Speech Signal Process. ASSP-36*(1):41–48.
Sugamura, N., Shikano, K., and Furui, S. (1983). Isolated word recognition using phoneme-like templates. Paper presented at a meeting of the IEEE International Conference on Acoustics, Speech, and Signal Processing, April.
Sugiyama, M. (1985). Spectral matching measure in speech recognition. Ph.D. thesis, Tohoku University.
Sugiyama, M., and Shikano, K. (1982). Frequency weighted LPC spectral matching measures. *Inst. Electr. Commun. Eng. Jpn. Trans. A J64-A*(5):409–416 (in Japanese)
Sugiyama, M., and Shikano, K. (1983). WLR measure applied to word recognition. *Inst. Electr. Commun. Eng. Jpn. Trans. D J66-D*(3):385–391 (in Japanese).
Tohkura, Y. (1986). A weighted cepstrum distance measure for speech recognition. *Proc. ICASSP*, pp. 761–764.
Umezaki, T., and Itakura, F. (1989). Speech analysis by group delay spectrum of all-pole filters and its application to the spectrum distance measure for speech recognition. *IECE Trans. J72-D-II*(8) (in Japanese).
Umezaki, T., Singer. H., and Itakura, F. (1989). Using low bit quantization of the smoothed group delay spectrum (SGDS) for speech recognition. Research Report No. PASL 1-1-1 (in Japanese).

15

Auditory Nerve Representation as a Basis for Speech Processing

ODED GHITZA AT&T Bell Laboratories, Murray Hill, New Jersey

1. INTRODUCTION

"Speech", according to Webster's Dictionary, is the "communication or expression of thoughts in spoken words." The speech signal is the means by which this information is carried from one destination to another. It is generated by the speech production organs of the human speaker with the purpose of being processed by the auditory system of the human listener. Philosophically, one might wonder if natural selection played a role in determining the rules of speech production, to match the needs of the auditory system. After all, the auditory system served the human being in a variety of tasks long before speech communication was developed. If indeed speech evolved that way, one might assume that acoustic properties of the speech signal are shaped in a unique way to ensure an appropriate auditory representation of the speech to higher parts of the nervous system.

In building speech processing systems, we make the axiomatic assumption that the speech waveform contains the entire information produced by the human speech production mechanism. In a given speech processing system, only part of this information is used, depending on the purpose for which the system was designed. Traditionally, the task-related information is extracted either from the speech waveform or from its Fourier representation, using statistical inference methods. But in view of the unique nature of speech and its possible link to auditory limitations, we can explore alternative ways which utilize knowledge of auditory mechanisms to extract task-related information from speech.

The success of utilizing this approach depends on the amount of knowledge we have about the auditory system. This knowledge is achieved by combining data that have been collected in both psychological and physiological studies of the auditory system. Little is known, at present, about the functional operations of the auditory nervous system. However, studies of the population response of single auditory nerve fibers in the cat to speech-like signals provide a rich source of information concerning the principles by which such sounds are encoded in the auditory nerve (Sachs and Young, 1979; Young and Sachs, 1979; Delgutte and Kiang, 1984a, 1984b, 1984c, 1984d; Carney and Geisler, 1986). As the sole afferent path from the cochlea to the central nervous system, the auditory nerve, by necessity, contains all the information available to the more central regions of the auditory pathway. Thus, from the perspective of auditory speech representation it is sufficient to retain only those properties of the speech signal that are encoded in the firing patterns of the auditory nerve.

Several studies have been published that examine how the responses of a stimulated auditory nerve may be processed to provide relevant representation of the speech signal. Each study utilizes a computational model to simulate either the direct firing activity or another related representation of the auditory nerve. However, the manner in which this information is processed differs among the studies, reflecting differences in the structural properties of the central processor hypothesized by each study. These structural properties can be described by using a two-component characterization: the place-nonplace component, which indicates whether the central processor utilizes explicit knowledge about the fibers' tonotopic place of origin in the cochlear partition, and the rate-temporal component, which indicates whether the central processor uses instantaneous firing rate measurements alone or higher-order statistics (e.g., the interspike interval statistics). Using this two-component criterion, the following categories are traditionally used: (1) rate-place category, where the central processor possesses explicit knowledge of place and uses only instantaneous rate information (Delgutte, 1986; Lyon, 1986; Sachs et al., 1988), (2) temporal-place category, where place information is used with detailed temporal information of local neural responses (Goldstein and Srulovicz, 1977; Seneff, 1988; Shamma, 1988; Deng et al., 1988), and (3) temporal-nonplace category, where place information is omitted altogether and the only sources of information are the temporal properties of the global neural response (Allen, 1985; Ghitza, 1988). For an excellent overview of current auditory models, the reader is referred to Greenberg (1988).

This chapter focuses on a model that belongs to the last category. Specifically, it is assumed that (1) neural circuits beyond the auditory nerve have a place-independent structure and (2) these circuits operate on detailed timing information conveyed in the auditory nerve fibers, irrespective of their tonotopic place of origin in the cochlear partition. Detailed physiological modeling of the auditory periphery (up to the auditory nerve level) is used to simulate the firing activity of the auditory nerve. Then a plausible transformation is applied on the simulated firing patterns to account for the higher stages in the auditory pathway. In Section 3 of this chapter, a representation of this timing information is described as an ensemble interval histogram (EIH). Conceptually, the EIH is a measure of the spatial (tonotopic) extent of coherent neural activity across the simulated auditory nerve. Mathematically, it is the short-term probability density function of the reciprocal of the intervals between successive firings, measured over the entire simulated auditory nerve.

In Section 4 two key questions are addressed with respect to this form of speech representation: (1) does the EIH preserve all of the phonetically relevant information contained in the speech signal? and (2) is the EIH representation more resistant to noise than the Fourier spectrum? In order to answer these questions, the Fourier power spectrum and the EIH were each used as the input to a speech recognition system. The comparison demonstrates that in the noise-free case recognition scores with the EIH and with the Fourier power spectrum are comparable. It also demonstrates that with increasing levels of background white noise, the recognition scores based on the EIH are superior to those using the Fourier power spectrum.

In Section 5 the EIH model is viewed as a general processing method, without reference to its auditory origins. We examine how sensitive the EIH is to changes in the main components of the model, namely the *filter bank*, which governs the properties of the simulated firing patterns, and the *analysis method*, which transforms these firings into an EIH. The study shows that the robustness of the EIH in the presence of noise is due mainly to the properties of the analysis method (the timing-synchrony analysis) rather than to the unique shape of the cochlear filters. The study also shows that the characteristics of the filter bank determine the fine structure of the EIH representation. To achieve an EIH with properties that correspond to human psychophysics, the correct cochlear filters must be used.

Before introducing the EIH model, let us briefly describe the structure of the mammalian auditory system. This will put into perspective the state of our knowledge and what part of it is being utilized in our current models.

2. A BRIEF DESCRIPTION OF THE AUDITORY PATHWAY

The auditory system is symmetrically organized around a midline between the left and the right sides. The most peripheral part at each side consists of the *external and middle ears*, the *cochlea*, and the *auditory nerve*. At the cochlea, the speech waveform is decomposed into spectral bands via the mechanical properties of the basilar membrane. One can view the basilar membrane as a unique realization of a continuum of filters. This continuum is sampled by the inner hair cells (IHCs), which transform the mechanical motion of the basilar membrane into a receptor potential in the cell body. Neurotransmitter release by the IHCs results in firings in the auditory nerve fibers that innervate the IHCs. Each IHC is innervated by some 10 fibers, and altogether the auditory nerve contains some 40,000 fibers.

The auditory nervous system, which receives its inputs from both left and right auditory nerves, consists of several neural nuclei that can be grouped into three major parts—the auditory brain stem, the auditory midbrain, and the auditory cortex. Each nucleus can be divided into different zones, characterized by their morphological structure, neurophysiological response, and input-output mappings. The main information flow is along an ascending pathway that begins at the external ears and ends in the auditory cortex. In parallel, there is an information flow in a descending pathway that begins at the auditory cortex and can be traced all the way down to the middle ears.

Figure 1 shows the major nuclei of the mammalian auditory system. The auditory nerve at each side projects to its corresponding *cochlear nucleus*, which is the first nucleus in the auditory brain stem. The projections of various divisions in each cochlear nucleus form two parallel branches, one on the same side of the cochlear nucleus and the other across the midline. These projections, on both sides, proceed either directly, or indirectly via other brain stem nuclei (the *trapezoid body* and the *superior olivary complex*), to the auditory midbrain nuclei (the *lateral leminiscus* and the *inferior colliculus*). The inferior colliculus provides the major ascending path to the *medial geniculate body*, which in turn provides inputs to the *primary auditory cortex*.

For an introductory survey of the auditory system the following references are recommended: Pickles (1982) and Kiang and Peake (1988).

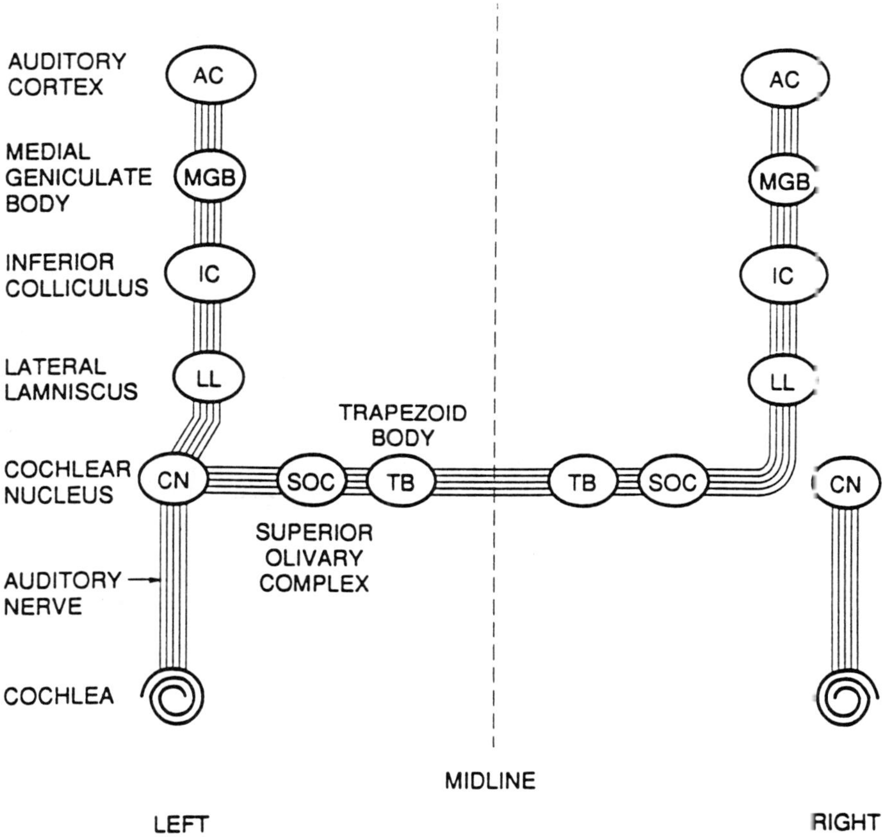

Figure 1 Major nuclei of the mammalian auditory system.

3. THE ENSEMBLE INTERVAL HISTOGRAM REPRESENTATION

3.1. Description of the Model

The model is illustrated schematically in Fig. 2. Its first stage represents the auditory periphery up through the level of the auditory nerve. The mechanical motion of the basilar membrane, at any given place, is simulated as the output signal of a bandpass filter whose frequency response reflects the mechanical tuning characteristics at that place (e.g., Allen, 1985). The mapping of place along the basilar membrane onto frequency is approximately logarithmic, and the distribution of the inner hair cells along the cochlear partition is relatively uniform. The location of an IHC along the basilar membrane is

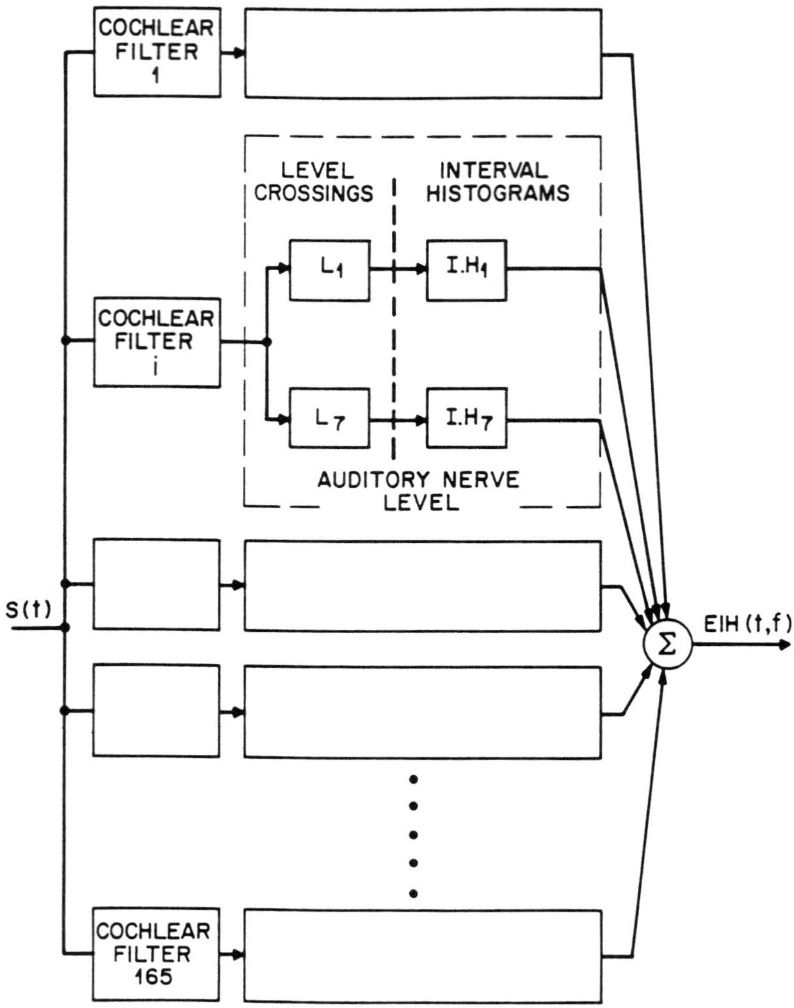

Figure 2 The ensemble interval histogram (EIH) computational model. The cochlear component consists of 165 filters (channels), whose center frequencies are equally spaced on a log-frequency scale, between 150 and 7000 Hz. The level crossings are measured at positive threshold crossings. The positive threshold levels are pseudorandomly distributed, on a log scale, over the dynamic range of the signal. The multidimensional point process derived from the level-crossing detectors simulates the auditory nerve firing patterns. The interval histogram is created by distributing the inverse of the detected intervals across 128 bins equally spaced on a linear frequency scale, between 0 and 4000 Hz. Only the most recent 20 intervals were used in the computation (an interval is defined as the time between two adjacent positive-going level crossings). The ensemble histogram is the sum of the corresponding histogram bins over all of the simulated fibers in the array.

specified by the corresponding "characteristic" frequency (CF). The motion of the basilar membrane is reflected in the AC component of the IHC receptor potential. The transformation from mechanical motion to receptor potential involves a number of nonlinearities, the most relevant to the present discussion being the half-wave rectification imposed as a consequence of the unidirectional depolarization of the IHC. Each IHC is innervated by approximately 10 auditory nerve fibers, whose spontaneous activity ranges between 0 and 100 discharges per second. The spontaneous rate is highly correlated with the fiber diameter and the size of the synaptic region between the fiber and inner hair cell. The spontaneous discharge rate is also correlated with the sensitivity of response. For any given CF region, fibers with a high spontaneous rate tend to be between 5 and 20 dB more sensitive than units with very low rates of background activity (Liberman, 1978, 1982; Geisler et al., 1985). Occasionally, low-spontaneous units may be as much as 40 to 60 dB less sensitive than high-spontaneous units of comparable CF (Liberman, 1978).

In the present model, the mechanical motion of the basilar membrane is sampled by 165 IHC channels, equally spaced along a log-frequency axis between 150 and 7000 Hz. The corresponding cochlear filters have been stimulated in detail, using actual neural tuning curves for cats collected by M. C. Liberman (unpublished data). The amplitude responses of 20 (one every eight) filters are shown in Fig. 3. Their phase characteristic is minimum phase and their relative gain, measured at their center frequencies, reflects the cat's middle ear transfer function. The ensemble of nerve fibers innervating a single IHC is simulated with an array of level-crossing detectors at the output of each cochlear filter (i.e., each level-crossing detector is equivalent to a fiber of specific threshold). A neural firing is simulated as the positive-going level crossing. The detectors are pseudorandomly distributed across a range of positive levels (Fig. 4). The value assigned to the level j of every filter is a random Gaussian variable, with a mean L_j and a standard deviation $\sigma_j = 0.2L_j$. The mean values $L_1, \ldots, L_j, \ldots, L_7$ are uniformly distributed on a log scale over the amplitude range characteristic of speech sounds. The random values of the jth level across the cochlear channels reflect the fact that diameters and synapse-connection sizes of fibers innervating the same side of different IHCs along the cochlear partition incorporate a certain amount of intrinsic variability (which is characteristic of most physiological systems). The model omits certain properties of the auditory nerve function. By modeling the nerve fiber firing activity as a point process produced by a level-crossing detector, the probabilistic nature of the neural firing mechanism is essentially neglected. Thus, the level crossings should be interpreted as the *joint* firing activity of a collection of fibers originating in different IHCs located in a range along the basilar membrane small enough to ensure

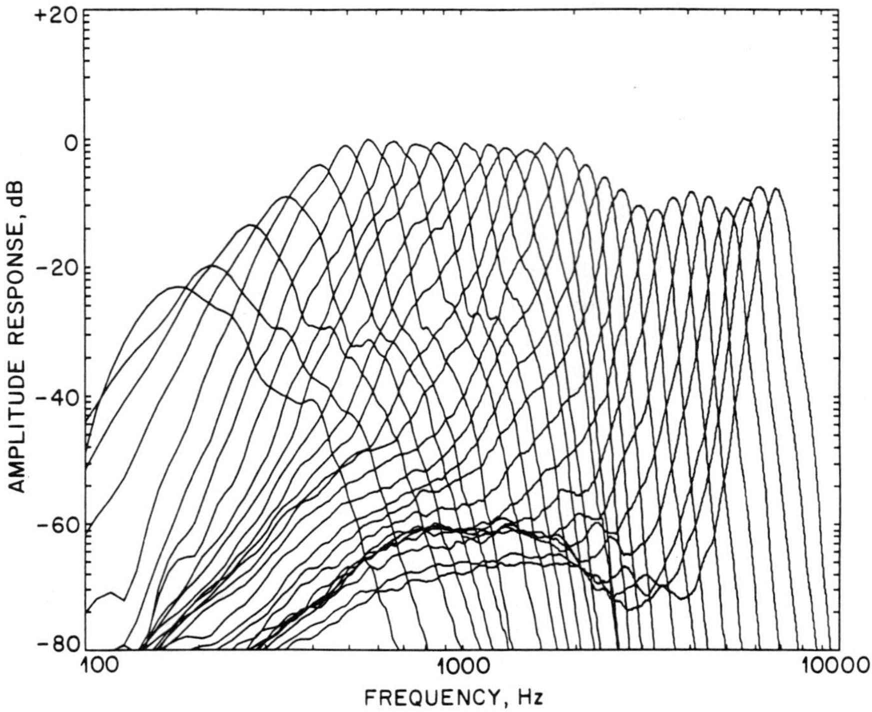

Figure 3 The amplitude responses of 20 (one every eight) simulated cochlear filters, plotted in a logarithmic frequency/decibel scale. The filters' amplitude responses are the actual neural tuning curves for cats collected by M. C. Liberman (unpublished data). Note the high degree of overlap among filters.

similar cochlear tuning characteristics. The adaptation of nerve fiber discharge (Smith and Zwislocki, 1975) is also omitted from the model. This omission may not affect the basic representation since the amount of adaptation is reduced considerably in the presence of noise, even at noise levels where human performance is still high (Sachs et al., 1983). If it is assumed that the principles by which the central auditory system operates remain constant over a wide range of acoustic conditions, then one may conclude that the information conveyed by rate adaptation does not play a significant role in the processing of speech.

The output of the level-crossing detectors represent the discharge activity of the auditory nerve fibers. Figure 5 shows simulated auditory nerve activity for the first 60 ms inside the vowel [o] in the word "gob." The abscissa

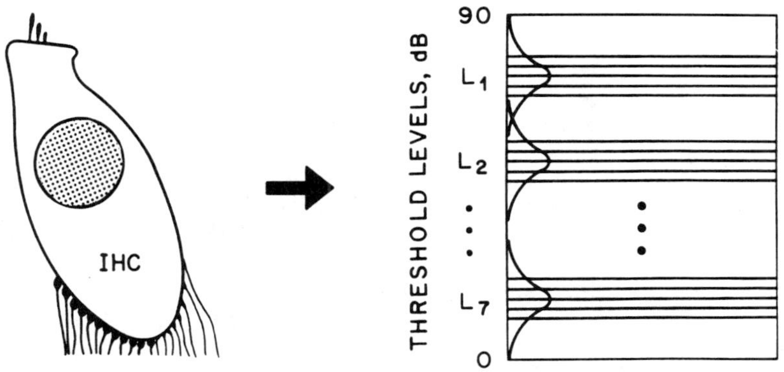

Figure 4 Assignment of threshold levels to the simulated auditory nerve fibers. The positive threshold levels are pseudorandomly distributed. The values assigned to the level j of the ith filter is a random Gaussian variable with a mean L_j and a standard deviation $\sigma_j = 0.2L_j$. The mean values $L_1, \ldots, L_j, \ldots, L_7$ are uniformly distributed on a log scale over the dynamic range of the speech signal.

represents time and the ordinate represents the characteristic frequency of the IHC channels. Note the logarithmic scale of the characteristic frequency, which represents the place-to-frequency mapping on the basilar membrane. In the figure, a level-crossing occurrence is marked as a single dot, and the output activity of each level-crossing detector is plotted as a separate trace. Each IHC channel contributes seven parallel traces, with the lower trace representing the lower-threshold level-crossing detector. If the magnitude of the filter's output signal is low, only one level will be crossed, as is the case for the very top channels of Fig. 5. However, for large signal magnitudes, several levels will be activated, creating a "darker" area of activity.

The level-crossing patterns represent the auditory nerve activity This activity serves, in turn, as the input to a second, more central stage of neural processing, as discussed below. This higher-level stage in the model conforms to a number of processing principles drawn from analysis of firing patterns of auditory nerve fibers. Instead of using all of the available information in the discharge patterns (via linear operations), the model attempts to extract, through certain nonlinear operations, only specific information of relevance to speech encoding. The model also emphasizes simplicity of operation, on the assumption that neural processing in the central auditory pathway is based on a large number of simple operators functioning in parallel. Obviously, these assumptions are intended as a heuristic since our knowledge of higher-level auditory processing is quite limited at the present time.

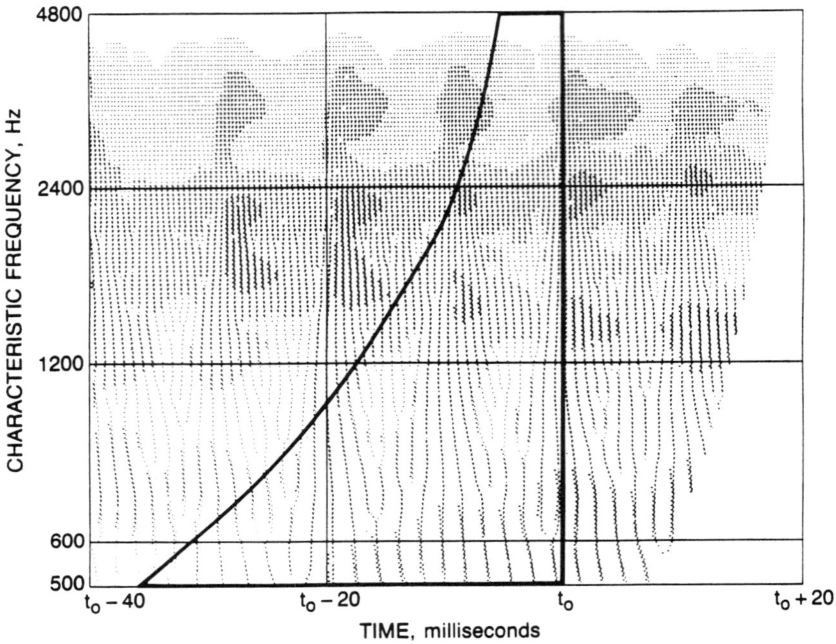

Figure 5 Simulated auditory-nerve activity for the first 60 ms inside the vowel [o] in the word "gob." The abscissa represents time and the ordinate represents the characteristic frequency of the IHC channels. Note the logarithmic scale of the characteristic frequency, which represents the place-to-frequency mapping at the basilar membrane. In the figure, a level-crossing occurrence is marked as a single dot, and the output activity of each level-crossing detector is plotted as a separate trace. Each IHC channel contributes seven parallel traces, with the lower trace representing the lower-threshold level-crossing detector. If the magnitude of the filter's output signal is low, only one level will be crossed, as is the case for the very top channels. However, for large signal magnitudes, several levels will be activated, creating a "darker" area of activity. The figure also illustrates how the length of the analysis window in each channel is related to its center frequency (CF). Since the histograms are constructed using the most recent 20 intervals, the equivalent window length in the analysis is different for each channel and is roughly proportional to 1/CF.

Measurements of the firing responses of cats' auditory nerve fibers to speechlike stimuli (Sachs and Young, 1979; Young and Sachs, 1979; Delgutte and Kiang, 1984a, 1984b, 1984c, 1984d) suggest that average firing rate is an insufficient carrier of speech information and that the use of temporal information from the firing patterns is necessary. It is also evident from these measurements that as the sound pressure level increases, more fibers fire in synchrony with certain temporal properties of the stimulus waveform. For this reason, the spatial (tonotopic) extent of such coherent neural activity can be used to provide estimates of the frequency and relative intensity of the underlying frequency components (Ghitza, 1985). In this sense, the model is based on the temporal, nonplace approach advocated by Carlson et al. (1975), since it operates solely on neural timing information, irrespective of its tonotopic origin in the cochlear partition. It should be noted, however, that tonotopic information is present *implicitly*, since the information conveyed by each fiber is by itself place dependent due to the tuning characteristics of the basilar membrane.

For purposes of computational implementation, the amount of coherent neural activity across the simulated fiber array is measured by determining the similarity in the short-term interval probability density functions of individual levels. An estimate of the interval probability density function of a given level is obtained by computing a histogram of the inverse intervals from the point process data produced by the level-crossing detector. This is accomplished by plotting the reciprocal of the intervals (i.e., converting the intervals into units of frequency) in a histogram consisting of 128 linearly spaced bins ranging from 0 to 4000 Hz. The 20 most recent intervals are used in the histogram's computation to ensure a reliable estimate of the interval statistics. To measure the extent of coherent neural activity across the fiber array, the similarity between the interval probability density functions is determined. This is achieved by collecting the individual histograms into one ensemble interval histogram, by summing the corresponding histogram bins for all fibers. The resulting representation is the EIH. The tonotopic extent of the coherent neural activity generated by an underlying frequency component is encoded as the magnitude of the corresponding bin in the EIH.

3.2. Characteristics of the Model

Several properties of the EIH are of interest with respect to speech coding. First, the window length of the interval histogram differs across channels as a consequence of the analysis being restricted to the 20 most recent intervals. Because the cochlear filters are quite sharply tuned, the spectrum of their outputs is largest near the CF of the channel. Therefore, the window length for any given channel is roughly proportional to 1/CF. For each point in time,

t_0, the EIH is a composite of histograms from the entire fiber array. If each channel responds to frequencies near the center of its filter, the corresponding window lengths will vary with CF, as illustrated in Fig. 5. A consequence of this behavior is that the frequency resolution of low-CF channels is superior to that of the high-frequency channels. Conversely, temporal resolution is superior in the high-CF channels.

As a consequence of the multilevel crossing detectors, the EIH representation preserves information about the signal's overall energy. Consider the case where the input signal is $s(t) = A \sin(2\pi f_0 t)$ and the CF of a given channel is equal to f_0 (Fig. 6a). For a given intensity A, the cochlear filter output will activate only some low level-crossing detectors. For a given detector, the time interval between two successive positive-going level crossings is $1/f_0$. Since the histogram is scaled in units of frequency, this interval contributes a count to the f_0 bin. For the input signal in the illustrated example, all of the intervals are the same, resulting in a histogram where the magnitude of each bin, save one (f_0), is zero. As the signal amplitude A increases, more levels are activated. As a result, this cochlear filter contributes additional counts to the f_0 bin of the EIH. Since the crossing levels are equally distributed on a log-amplitude scale, the magnitude of any EIH bin is related, in some fashion, to decibel units. However, this relation is not a straightforward one since there are several sources contributing counts to the f_0 bin in a nonlinear manner. Figure 6b shows an input signal $s(t) = A \sin(2\pi f_0 t)$ driving five adjacent cochlear filters with an amplitude response $|H_i(f)|$ and a phase response $\phi_i(f)$, $i = 1, 2, \ldots, 5$. Due to the shape of the filters, more than one cochlear channel will contribute to the f_0 bin. In fact, all the cochlear filters which produce $s_i(t) = A|H_i(f_0)| \sin(2\pi f_0 t + \phi_i(f_0))$ will contribute to the f_0 bin of the EIH, provided that $A|H_i(f_0)|$ exceeds any of the level-crossing thresholds. In Fig. 6b only cochlear filters 2, 3, and 4 are contributing nonzero histograms to the EIH. The number of counts is different for each filter, depending on the magnitude of $A|H_i(f_0)|$.

Another feature of the EIH representation is its inherent dominance property (which might be related to the psychophysical notion of masking). Because of the shape of the cochlear filters, a high-amplitude spectral component f_1 is capable of dominating the output of a cochlear filter whose CF is appreciably greater than f_1, even in the presence of a second less intense components, f_2, equal in frequency to the filter CF. When such "synchrony capture" occurs, the output of this channel contributes counts to the EIH at bin f_1 rather than bin f_2. A consequence of this synchrony dominance is that when the stimulus is a vowel, a high-amplitude component f_1 in the first formant region will dominate the synchrony patterns (and hence the interval histograms) of a large number of channels. The precise number of counts

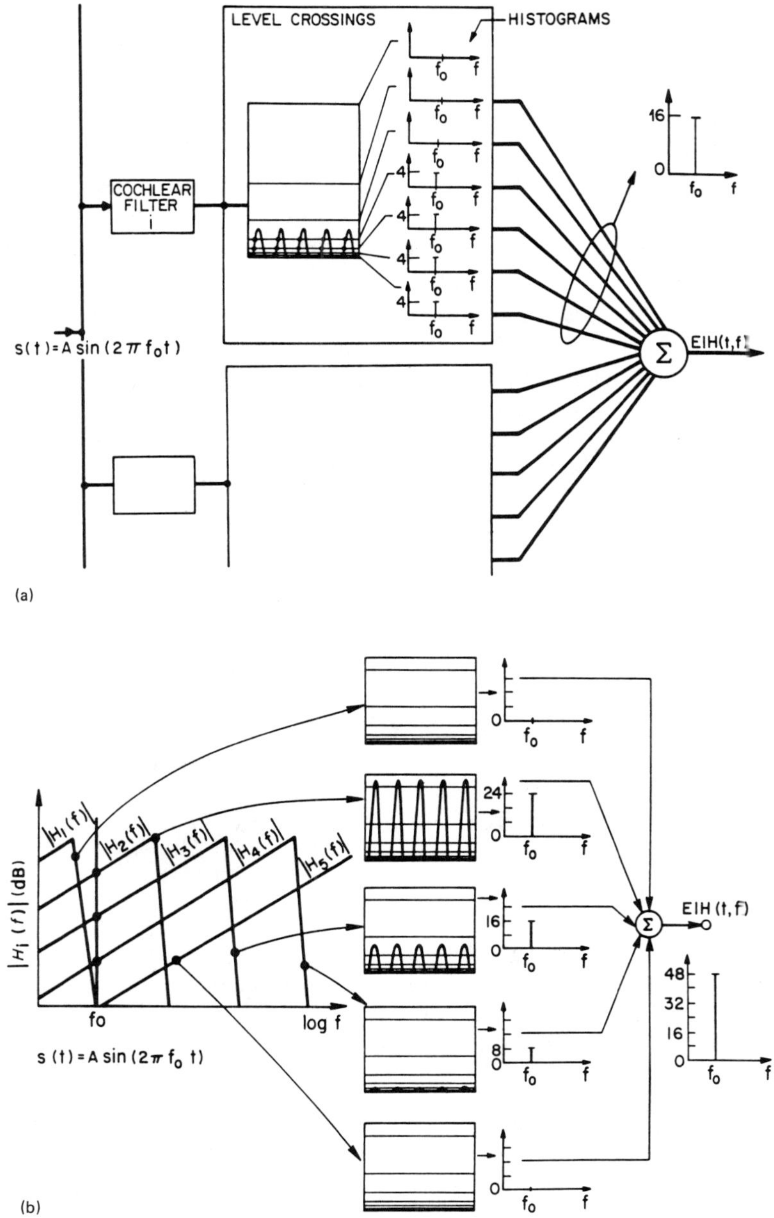

Figure 6 (a) The contribution of the ith channel to the ensemble interval histogram for an input signal $s(t) = A \sin(2\pi f_0 t)$. Only the lowest four level-crossing detectors are contributing nonzero histograms to the ensemble. (b) The contribution of five successive channels $|H_i(f)|$, $i = 1, 2, \ldots, 5$, to the EIH, for an input signal $s(t) = A \sin(2\pi f_0 t)$. Channel i contributes to the f_0 bin of the EIH provided that $A|H_i(f_0)|$ exceeds any of the level-crossing thresholds.

contributed by any given channel to the f_1 bin of the EIH depends on the degree to which its output is dominated by that frequency.

The EIH representation does not significantly deteriorate in the presence of high levels of background noise. The ability to maintain the EIH over a wide range of signal-to-noise ratios is due to a number of nonlinear properties of the model. Perhaps the most important of these is the measurement of coherent neural activity across a broad range of simulated fibers.

So far we have examined the EIH properties in view of its conceptual meaning. That is, the EIH is a measure of the spatial (tonotopic) extent of coherent neural activity across the simulated auditory nerve. However, in view of the way it is constructed, we can provide a mathematical definition for the EIH. At time t_0, the EIH is the short-term probability density function of the reciprocal of the intervals, measured over the time-frequency domain illustrated in Fig. 5.

3.2. Analysis Conditions

A set of speech signals was used to demonstrate the properties of the EIH. These signals were low-pass filtered to 3600 Hz and sampled at an 8-kHz rate. In order to increase the measurement accuracy in the high-CF cochlear channels, the original digital samples were upsampled by a factor of 5, yielding a sampling rate of 40 kHz. When an upward-going level crossing had been found, a linear interpolation was used to compute the precise time of the crossing. EIH analysis was performed every 3.2 ms.

3.4. Examples of the EIH

In Fig. 7, the EIH representation is compared with the Fourier magnitude spectrum for different speech frames. The Fourier magnitude spectra were computed by a 128 points discrete Fourier transform, using 20-ms Hamming-window weighted frames. The frequency resolution of the analysis was 31.25 Hz.

Figure 7 illustrates two examples of steady-state portions of vowels spoken by a male speaker. The signal's spectral fine structure becomes fuzzy in the high-frequency region of the EIH. This reflects the fact that the EIH encodes with high resolution information concerning the fine structure of the low-frequency range but is unable to encode the details of spectral fine structure at high frequencies.

Figure 8 illustrates the manner in which rapid temporal changes are represented in the EIH. After an initial silent period (such as a pause or

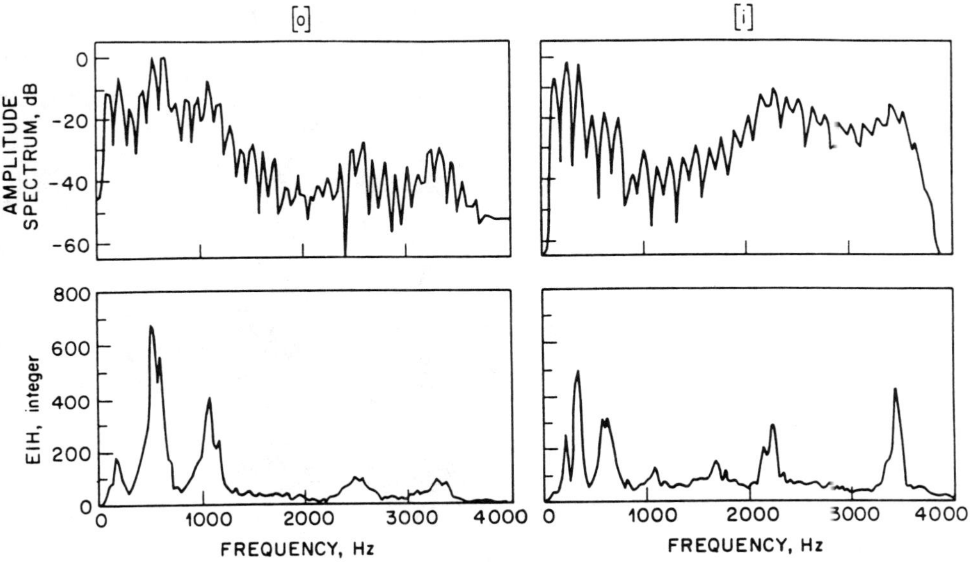

Figure 7 Spectral representation of the vowels [o] and [i] produced by a male speaker. The upper plot is the high-resolution Fourier magnitude spectrum computed by a 128 points discrete Fourier transform, using a 20-ms Hamming window (frequency resolution = 31.25 Hz). The plot uses a linear frequency-decibel scale. The lower plot is the EIH representation of the same signal. The EIH plots are on a linear frequency-EIH scale.

a stop closure), the mid- to high-frequency channels respond quickly (due to the high temporal resolution of these channels). A little later the low-frequency harmonic structure of the signal appears, as the low-CF channels respond. Figure 8 also illustrates the form in which gross amplitude changes are preserved from frame to frame in the EIH representation.

Figure 9 demonstrates the robustness of the EIH representation at a global SNR of 0 dB (global SNR is defined in Section 4.2). The right column illustrates the EIH representation and the left column displays the Fourier spectrum. The curve fitting for the Fourier representation is a 12th-order cepstral fit. The smoothed curves for the EIH are a 19th-order cepstral fit (the smoothing considerations for the EIH are discussed in Section 4.1). Because the EIH representation is less sensitive to noise, there is better agreement between the curves fit to the signal representations in quiet and in noise than is the case for the curves fit to the Fourier spectra.

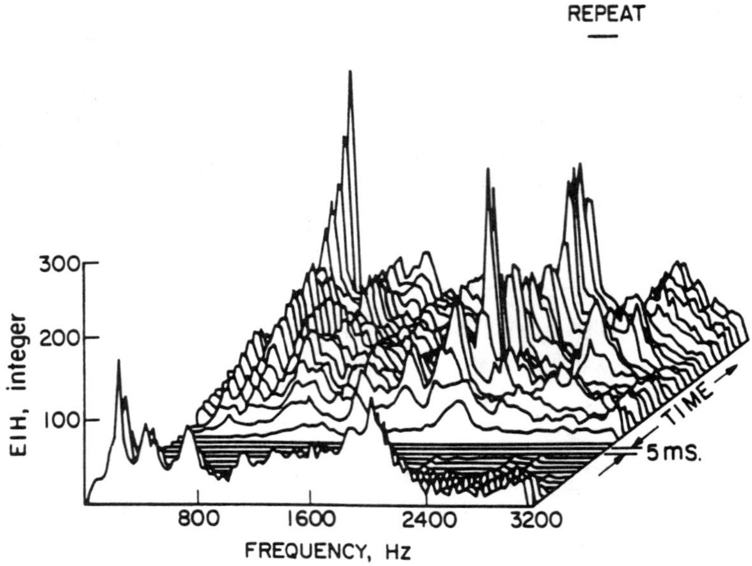

Figure 8 EIH display of successive frames during the [p] release in the word "repeat." The EIH measurement occurs at 5-ms intervals.

4. THE EIH REPRESENTATION AS A FRONT END TO A SPEECH RECOGNIZER

In Section 3, the EIH was discussed in a schematic manner by comparing it with the Fourier spectrum in quiet (Figs. 7 and 8) and in noise (Fig. 9). In this section it is shown through a quantitative analysis that the EIH can encode phonetically relevant information in quiet and outperforms the Fourier representation in the presence of background noise. The analysis compares the performance of a recognition system using either the Fourier spectrum or the EIH as the front-end component.

4.1. Experimental Procedure

The EIH, like the conventional Fourier spectrum, contains information about the formant pattern as well as about the spectral fine structure. However, only part of this information is generally used in speech recognition systems. The recognition tests to be described here were performed using a dynamic time warping (DTW) recognizer (e.g., Wilpon and Rabiner, 1985) that utilizes only spectral envelope variations.

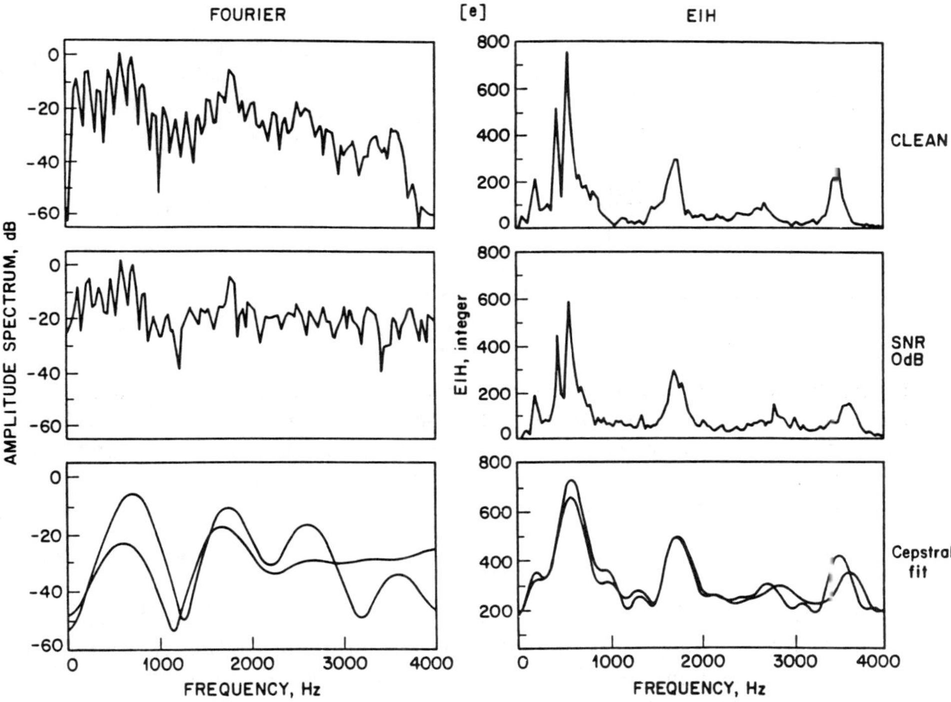

Figure 9 Illustration of the robustness of the EIH representation in the presence of background white noise. The global SNR is 0 dB. The left-hand column displays the Fourier magnitude spectra of a vowel [e] produced by a male speaker. The right-hand column shows the EIH for the same signal. The curve fitting is a 12th-order cepstral fit for the Fourier spectrum and a 19th-order cepstral fit for the EIH.

Figure 10 illustrates the system that was used in the recognition tests. The inputs to the existing alignment procedure are the first P autocorrelation coefficients, representing the spectral envelope. For the Fourier-based front end these coefficients can be computed directly from the input signal. In the present implementation, P was set to 9 to achieve optimal performance for the Fourier power spectrum (Wilpon and Rabiner, 1985). A source-filter model, which treats the EIH as the output log spectrum of a linear system, was used to distinguish the fine structure from the envelope in the EIH. Although the assumptions underlying this technique may not be entirely appropriate, the source-filter model provides a simple mechanism for adapting the EIH to the DTW recognizer. It was also assumed that the underlying system could

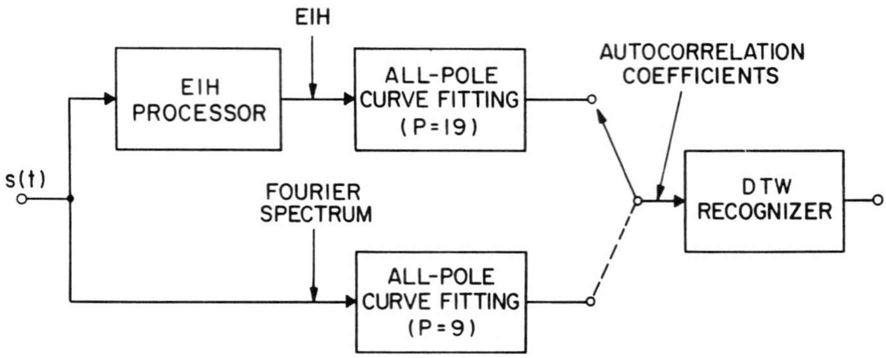

Figure 10 Applying the EIH model as a front end to a standard DTW speech recognizer. With the Fourier-based front end, the spectral envelope is represented by the first 9 autocorrelation coefficients. The EIH envelope is represented by the first 19 autocorrelation coefficients. A larger order for the all-pole polynomial fit is required because of the greater dynamic range of the EIH.

be modeled by an all-pole filter. If this assumption is valid, the EIH "spectrum" envelope can be determined using conventional frequency domain, linear prediction approximation methods. Since the EIH is measured in logarithmic units, the antilog of the EIH is first computed in order to plot it on a linear scale. The inverse discrete Fourier transform (DFT) of this representation is then computed in order to obtain its autocorrelation function. The degree of smoothing depends on the number of autocorrelation coefficients that are included. For the EIH, an appropriate envelope fit is achieved by applying a 19th-order all-pole polynomial fit (Fig. 9). This order of fit was required because of the larger dynamic range of the EIH (compared to that of the Fourier representation).

4.2. Data Base and Analysis Conditions

The data base for the recognition experiments was a 39-word alpha-digit vocabulary. This data base includes the letters of the alphabet, the digits "zero" through "nine," and the control words "stop," "error," and "repeat." Two male speakers repeated the vocabulary list a total of 13 times. Five of the repetitions were marked as the training set and the other eight repetitions were used as the test set. Recordings were made in a soundproof booth over a standard telephone line. The signals were bandpass filtered between 100 and 3200 Hz and sampled at a rate of 6.67 kHz. Each of the words was digitally edited in order to mark the end points. In order to increase the measurement accuracy in the high-CF cochlear channels, the digital waveform was upsampled by a factor of 6, again yielding a sampling rate of 40 kHz. When an upward-going level crossing had been found, linear interpolation was used to compute the precise time of the crossing. The EIH was computed at 5-ms intervals.

The "noisy" versions of the data base were created by adding white noise to the original ("clean") signals. The overall signal-to-noise ratio was defined using *global* measurements. First, the total energy, E_T, of the original (noise-free) word was computed. Then the average energy per digital sample, E, was determined by dividing E_T by the number of sample points in the signal. E was used to set the variance of a white noise generator to a level dependent on the desired global SNR. This definition of global SNR overestimates the actual signal-to-noise ratio during the consonantal segments because the magnitude of the noise is largely dependent on the amplitude of the vocalic portion of each word.

4.3. Recognition System

A speaker-dependent recognition test was performed. The recognizer used two reference templates for each word, created from the uncorrupted training set. No clustering technique was used in creating the templates. Instead, the distance between every pair of repetitions belonging to the training set was computed and the most mismatched pair was chosen as the template set.

Given the templates for each word, the recognition was performed using standard dynamic time warping techniques. A weighted cepstral distance measure was used, where the cepstral coefficients were derived from the LPC parameters that represent either the envelope of the Fourier power spectrum or the EIH envelope. The weighting coefficients were determined by the following expression (Juang et al., 1987):

$$w_k = 1 + 0.5L \sin\left(\pi \frac{k}{L}\right), \qquad k = 1, 2, \ldots, L \tag{1}$$

where L is the number of cepstral coefficients. The word corresponding to the template that produced the smallest distance from the test word was chosen as the recognized word.

4.4. Experimental Results

Figure 11 summarizes the average recognition scores for the two male speakers (eight repetitions each) in the test set. Under noise-free conditions, the

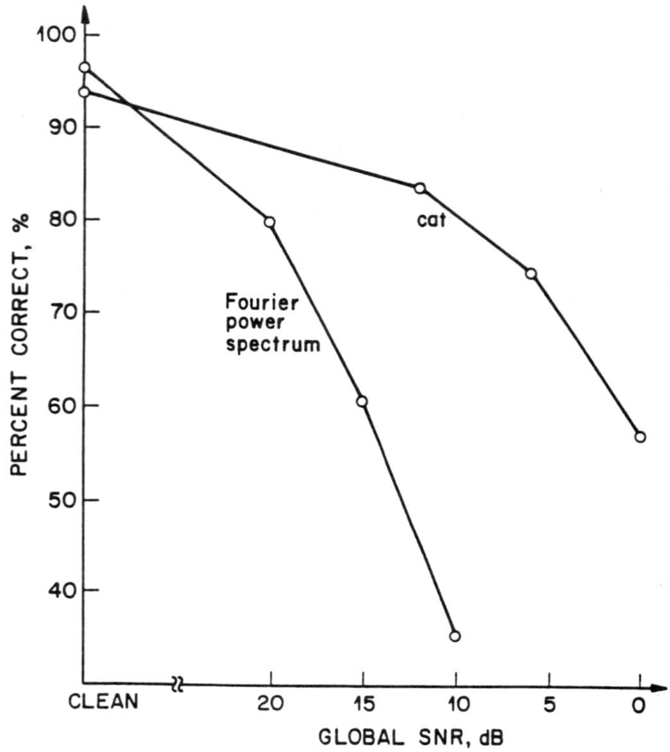

Figure 11 Recognition scores as a function of signal-to-noise ratio for the EIH-based and Fourier-based front ends to the DTW system. The data base consisted of a 39-word alpha-digit vocabulary spoken by two male speakers.

Fourier-based front end is slightly superior. However, with increasing levels of background noise, the recognition scores based on the Fourier spectrum decline more markedly than those using the EIH front end. At an SNR of 6 dB, the recognition scores with the EIH front end are equivalent to those obtained with the Fourier-based front end at 18 dB SNR, a gain of 12 dB.*

*Note the existence of other front ends which are robust against noise (Ephraim et al., 1987; Mansour and Juang, 1988a, 1988b). However, for the purpose of validating the ability of the EIH in representing speech, we chose the Fourier power spectrum as the reference representation because of its well-understood behavior as a signal representation.

5. STRUCTURAL PROPERTIES OF THE EIH

In previous sections we introduced the EIH and demonstrated, on the basis of speech recognition experiments, that it is capable of preserving relevant phonetic information in quiet and outperforming the Fourier power spectrum in the presence of additive white noise. In this section, the EIH model is viewed as a general processing method without regard to its auditory origins. We examine how sensitive the EIH is to changes in the two main components of the model, namely the filter bank and the analysis method that transforms the simulated firings into an EIH. We term this analysis the *timing-synchrony analysis*.

5.1. Role of the Timing-Synchrony Analysis

Following the results of the recognition experiments in Section 4 (Fig. 11), a question arises: What are the differences between a Fourier power spectrum measurement and the EIH measurement that cause the EIH to be more robust against noise?

Figures 12a and b show a block diagram description of a Fourier power spectrum measurement and a EIH measurement, respectively. An M-point Fourier power analysis on a Hamming-windowed signal can be viewed as a two-step spectral estimation procedure. At first, the signal is filtered by M identically shaped filters, equally spaced in a frequency band determined by the sampling frequency. The frequency response of the prototype filter is the Fourier transform of the Hamming window, and its bandwidth is determined by the duration of the Hamming window. We term these filters the "Hamming" filters. The power spectrum estimate is achieved by computing the short-term power at the output of each filter.

Similarly, the EIH model is also a two-stage procedure, where the first stage is a cochlear filter bank and the second stage performs a timing-synchrony analysis on the output signals of the filter bank. The EIH measurement thus differs from the Fourier power spectrum measurement both in the filter bank part and in the analysis part. The filter banks are different in two ways. First, the cochlear filters are equally spaced on a logarithmic scale, as opposed to the Hamming filters, which are equally spaced on a linear scale. Second, the shape of the cochlear filters is very different. Roughly, their frequency responses have an incline of +20 dB per octave on the low-frequency side of the center frequency and a very sharp roll-off on the high-frequency end (e.g., −120 dB per octave). As for the differences in the analysis part, in contrast to the power analysis in the Fourier power spectrum,

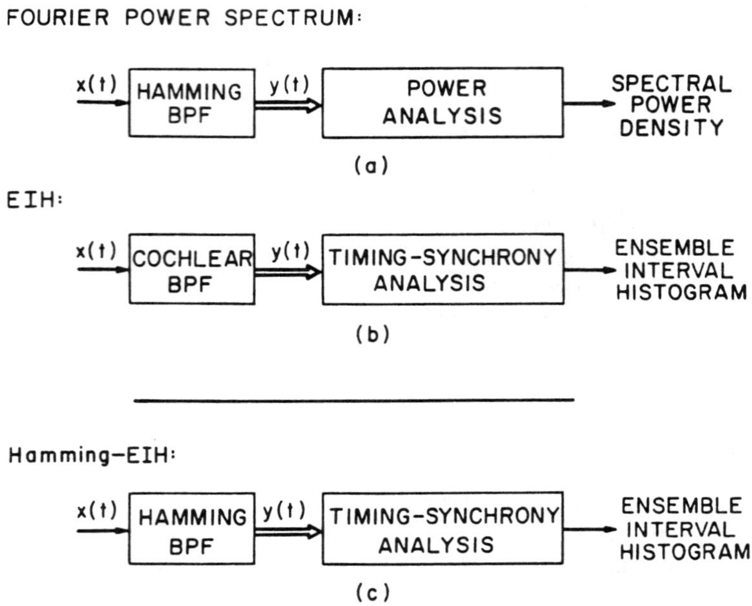

Figure 12 Block diagram of (a) a Fourier power spectrum measurement, (b) an EIH measurement, and (c) a Hamming-EIH measurement. The Fourier power spectrum and the Hamming-EIH differ in their analysis part.

the timing-synchrony analysis is based on measuring coherence across a multidimensional point process produced by passing the filter bank output signals through an array of multilevel detectors. Mathematically, the power computation is replaced by computing the short-term probability density function of the inverse intervals over a time-frequency region according to Fig. 5.

In view of the differences between the Fourier power spectrum and the EIH, it is possible to relate the robustness of the EIH against noise to the unique shape of the cochlear filters, to the properties of the timing-synchrony analysis, or to both. In order to examine the relative contributions of the cochlear filters and the timing-synchrony analysis separately, the EIH system in Fig. 12c has been constructed. We term this system the "Hamming-EIH." Here, the timing-synchrony analyzer operates on the output signals produced by the Hamming filters. Thus, the Fourier power spectrum measurement and the Hamming-EIH measurement differ only in the analysis part.

To obtain a quantitative comparison of the two measurement methods, the Hamming-EIH was examined as a front end to a DTW recognizer under the same conditions as in Section 4. The filter bank for the Hamming-EIH is shown in Fig. 13 in a logarithmic frequency/decibel scale (same as Fig. 3). It consists of 100 Hamming filters, equally spaced in linear frequency scale, 40 Hz apart. The −3 dB points bandwidth of each filter is 430 Hz.

The recognition scores for the Hamming-EIH front end are presented in Fig. 14, combined with the previous results for the Fourier-based and the cochlear EIH front ends. The Hamming-EIH and the EIH with the cochlear filters perform in a comparable way across all SNR conditions. Hence, the following conclusions can be made: (1) timing-synchrony analysis maintains all the speech information which is relevant for speech recognition tasks, (2) timing-synchrony measurement is significantly more robust against noise than power measurement, and (3) robustness is due mainly to properties of

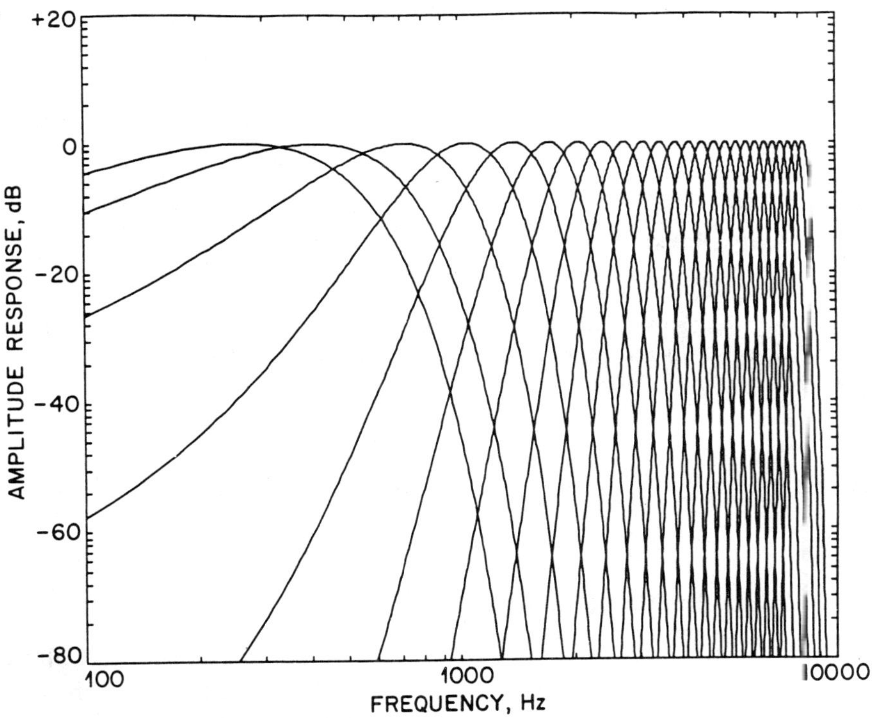

Figure 13 The filter bank for the Hamming-EIH, plotted in a logarithmic frequency/decibel scale (same as in Fig. 3). It consists of 100 Hamming filters, equally distributed in linear frequency scale, 40 Hz apart. The −3 dB points bandwidth of each filter is 430 Hz.

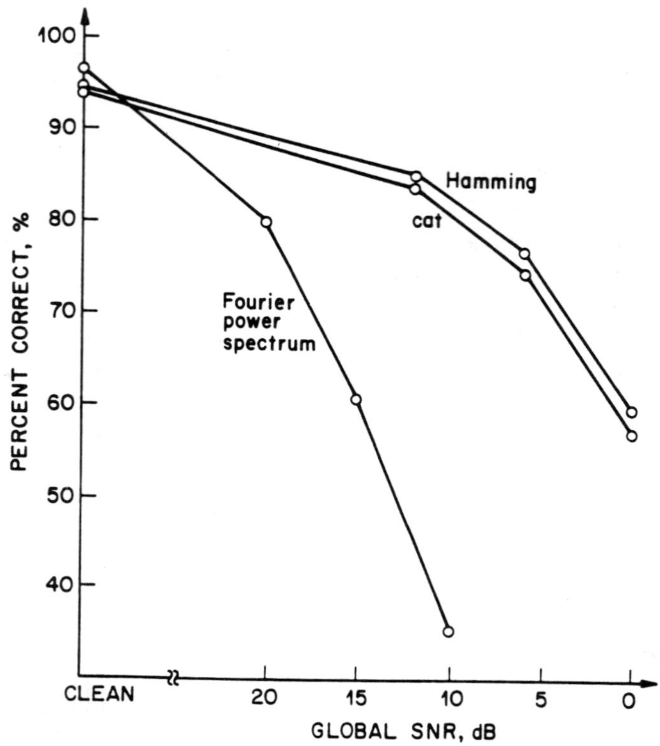

Figure 14 Recognition scores for the Hamming-EIH front end, combined with the previous results for the Fourier-based and the cochlear EIH front ends (from Fig. 11).

the timing-synchrony analyzer and not to the unique shape of the cochlear filters.

5.2. Role of the Filter Characteristics

So far we have discussed the behavior of the EIH as a function of the analysis method while keeping the filter bank the same. The Hamming filter bank was used, combined with two analysis methods: traditional power analysis and physiology-based timing-synchrony analysis. In this section we shall study the complementary situation in which the filter characteristics are changed while the analysis method remains the same. The analysis method of interest is the timing-synchrony analysis, because of its relations to physiology.

Figure 15 (replotted from Fig. 14) shows recognition scores obtained by two systems that use the same timing-synchrony analysis but differ in their filter bank. The filter banks under consideration are the cat's cochlear filter bank (Section 4.3) and the Hamming filter bank (Section 5.1). As far as recognition is concerned, both systems perform in a similar way. Since the recognizer in these experiments was presented with spectral envelope information alone, we can conclude that relevant phonetic information is equally preserved by both filter banks.

Although the phonetic information is equally preserved, substantial differences are observed in the fine-structure properties of the EIHs produced by the two filter banks. Figure 16a shows "EIHgrams" produced by the cat's filter bank (middle) and the Hamming filter bank (upper) for the word "zed"

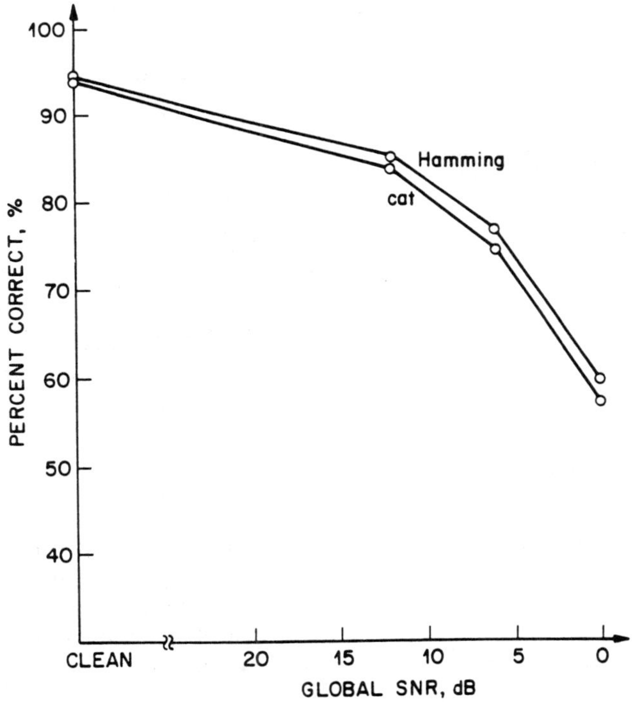

Figure 15 Recognition scores obtained by two systems that use the same timing-synchrony analysis but differ in their filter bank. The filter banks under consideration are the cat's cochlear filter bank and the Hamming filter bank (from Fig. 14).

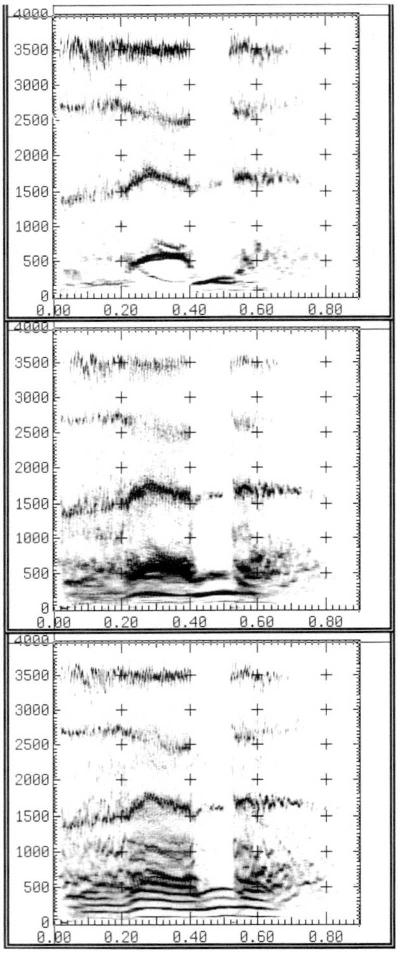

(a)

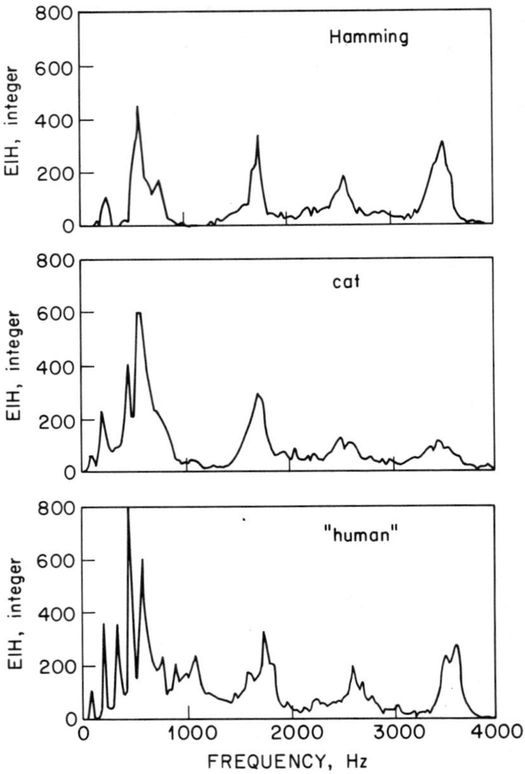

(b)

Figure 16 (a) EIHgrams produced by the Hamming filter bank (upper), the cat's filter bank (middle), and the "human" filter bank (lower) for the word "zed" spoken by a male speaker. The analysis method is timing-synchrony analysis, with level distributions and short-term interval statistics described in Section 3. (b) Slices of the EIHgrams of part (a), taken from inside the vowel region.

spoken by a male speaker. In both cases, the analysis method is timing-synchrony analysis, with level distributions and short-term interval statistics as in Section 3. Figure 16b shows slices of the EIHgrams taken from inside the vowel region. Comparing the figures, both systems produce similar EIHs in the higher frequency range (beyond, say, 1000 Hz). In the lower frequency range, however, EIHs with the cat's cochlear filters have finer frequency resolution. While EIHs with the Hamming filters possess one mode in that frequency range, which can be associated with the first formant, EIHs with the cat's filters show more detailed representation. This difference between the EIHgrams is due to the different bandwidth characteristics of the filter banks. Figure 17 shows the bandwidth as a function of center frequency for different families of filters. The bandwidth is in terms of the equivalent rectangular bandwidth (ERB) used by Moore and Glasberg (1983). The

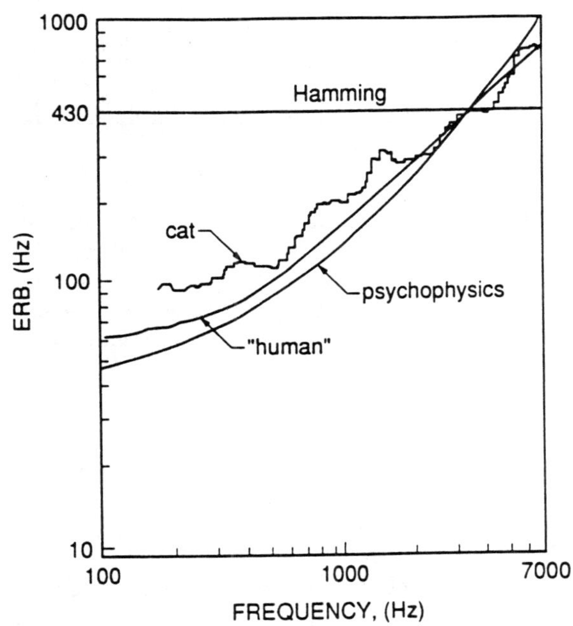

Figure 17 Bandwidth as a function of center frequency for different families of filters. The bandwidth is in terms of the equivalent rectangular bandwidth (ERB) used by Moore and Glasberg (1983). The psychophysical curve is according to a formula suggested by Moore and Glasberg (1983), based on psychophysical experiments. The ERBs for the cat's cochlear filters and the "human" filters are calculated from the amplitude responses of Figs. 3 and 18, respectively.

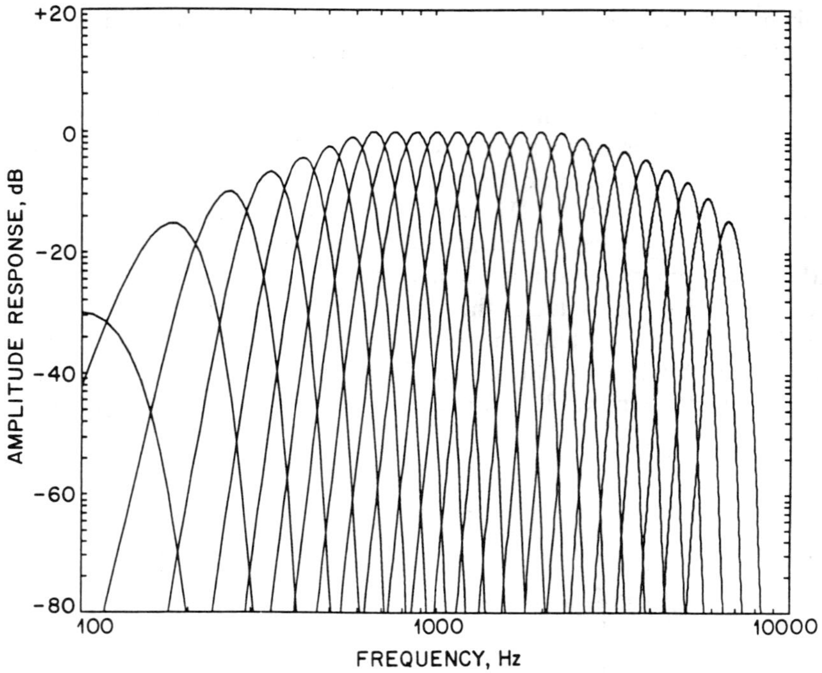

Figure 18 Amplitude response of the "human" filters. They were designed as Gaussian-shaped filters and their center frequency is distributed along the logarithmic frequency axis, similarly to the cat's filters. Their ERB bandwidth characteristics are close to the psychophysical ERB (Fig. 17), their phase characteristic is minimum phase, and their relative gain, measured at their center frequencies, roughly follows the cat's middle ear transfer function.

bandwidth of each filter in the Hamming filter bank is 430 Hz. The bandwidth of the cat's cochlear filters changes with frequency and is less than 200 Hz for filters with center frequency in the first-formant frequency range. Since the bandwidth of a first formant is in the range of 100 Hz, every Hamming filter has a bandwidth wider than the formant bandwidth. Consequently, the whole first formant is passing through the filter, causing the interval statistics between the level-crossing events to be related to the center frequency of the formant. This results in a clear representation of the first formant in the Hamming EIHgram of Fig. 16. On the other hand, the bandwidth of a cat's cochlear filters in the first-formant frequency range is similar in value to the bandwidth of the first formant, resulting in better frequency resolution in this frequency range.

The lower part of Fig. 16a shows an EIHgram that was produced by yet another filter bank, termed the "human" filter bank. The reason for referring to these artificially designed filters as "human" is that their bandwidth characteristics are close to the psychophysical ERBs, as shown in Fig. 17. Figure 18 shows the amplitude response of the "human" filters. Their distribution along the log-frequency axis is similar to that of the cat filter bank, their phase characteristic is minimum phase, and their relative gain, measured at their center frequencies, roughly follows the cat's middle ear transfer function. In the first-formant frequency range, the "human" filters are much narrower than the bandwidth of the first formant. Thus, individual harmonics in that frequency range are resolved. There is no clear display of the first formant. Instead, the EIH representation reflects a pitch–first formant interaction with properties that are yet to be explored. In the frequency range beyond 1000 Hz, the bandwidths of the "human" filters are similar to those of the cat, resulting in EIHs with similar properties.

To evaluate the "human" EIH in a quantitative manner, an attempt was made to measure its performance in preserving relevant phonetic information by repeating the recognition experiment described in Section 4.3. This attempt failed to produce meaningful results because of our inability to decompose the pitch–first formant interaction using available smoothing techniques. This may suggest a need for alternative decomposition methods based upon post auditory nerve processing principles.

Our present difficulty in extracting the first formant from the "human" EIH profiles, as opposed to the success in doing so with the cat's cochlear filters, leads to a fundamental problem, yet to be resolved, that relates back to the auditory origins of the EIH model: do auditory nerve representations of speech produced by a cat and by a human behave in a similar way? At present, we have considerable data about the way speech is represented at the cat's auditory nerve (Sachs and Young, 1979; Young and Sachs, 1979; Delgutte and Kiang, 1984a, 1984b, 1984c, 1984d; Carney and Geisler, 1986), but the more relevant information, namely how speech is represented in the human's auditory nerve, is obviously not available. However, properties of simulated auditory nerve firing patterns, as reflected in the EIHgrams of Fig. 16, may point out substantial differences between the cochleas of human and cat, especially in the frequency range below 1000 Hz. If we hypothesize that development of neural processing structures beyond the auditory nerve is influenced by properties of the auditory nerve firing patterns, and if our purpose is to mimic the actual representation of speech in the human auditory pathway, the available data about the way speech is represented at the cat's auditory nerve may not assist us in modeling speech related post auditory nerve functions. Meaningful models of these functions may need a correct description of the human cochlear filters.

However, it is not possible to mimic the actual representation of speech in the human auditory pathway without knowledge of the correct description of the human cochlear filters.

6. SUMMARY

In view of the unique nature of speech and the possible link between its acoustical properties and human auditory limitations, it is reasonable to explore auditory processing principles as a guide for speech representation. By doing so, it is hoped that the performance of future speech processing systems can be significantly improved. Speech recognition systems, for example, may achieved better discrimination of confusable sounds. In speech coding systems, where the total number of bits available is limited, auditory representation can be used to determine the bit allocation that is required to produce reconstructed speech with the highest possible quality.

This chapter has presented a computational model that uses detailed physiological modeling of the auditory periphery to simulate the firing activity of the auditory nerve. A heuristic analysis method was used to account for the higher stages in the auditory pathway. In particular, the analysis stage measures the extent of coherent activity in the simulated firing activity in terms of the short-term interspike-interval statistics, taken over the entire simulated auditory nerve. The resulting EIH shows properties that are different from properties of other spectral representations (e.g., the Fourier power spectrum). Quantitatively, it properly encodes the relevant phonetic information in quiet and outperforms the Fourier power spectrum representation in the presence of high levels of background white noise. The robust performance in noise is due mainly to properties of the timing-synchrony analysis and not to the unique shape of the cochlear filter bank. However, the characteristics of the filter bank play an important role in determining the fine-structure properties of the EIH. For the banks of filters that were examined in this chapter, the low frequency range (below 1000 Hz) is affected the most. If the filters in this frequency range are sufficiently wide compared to the bandwidth of the first formant, the first formant is uniquely presented but the fine-structure information is lost. On the other hand, if the bandwidth is reduced to follow the psychophysical bandwidth characteristics, the EIH displays a pitch–first formant interaction with detailed fine-structure information. Hence, for different speech processing tasks, desirable filter characteristics can be chosen to provide an adequate representation. From the perspective of the actual representation of speech in the human auditory pathway, however, the correct description of the mechanical tuning characteristic of the human cochlea is needed.

REFERENCES

Allen, J. B. (1985). Cochlear modeling. *IEEE ASSP Mag.*, vol. 2, No. 1, 3–29.

Carlson, R., Fant, G., and Granstrom, B. (1975). Two formant models, pitch and vowel perception. In *Auditory Analysis and Perception of Speech*, G. Fant and M. Tatham (Eds.). Academic Press, London, pp. 55–82.

Carney, L. H., and Geisler, C. D. (1986). A temporal analysis of auditory-nerve fiber responses to spoken stop consonant-vowel syllables. *J. Acoust. Soc. Am. 79*:1896–1914.

Delgutte, B. (1986). Analysis of french stop consonants using a model of the peripheral auditory system. In *Invariance and Variability in Speech Processes*, J. S. Perkell and D. H. Klatt (Eds.). Lawrence Erlbaum, Hillsdale, New Jersey, pp. 163–177.

Delgutte, B., and Kiang, N. Y. S. (1984a). Speech coding in the auditory nerve: I. Vowel-like sounds. *J. Acoust. Soc. Am. 75*:866–878.

Delgutte, B., and Kiang, N. Y. S. (1984b). Speech coding in the auditory nerve: III. Voiceless fricative consonants. *J. Acoust. Soc. Am. 75*:887–896.

Delgutte, B., and Kiang, N. Y. S. (1984c). Speech coding in the auditory nerve: IV. Sounds with consonant-like dynamic characteristics. *J. Acoust. Soc. Am. 75*:897–907.

Delgutte, B., and Kiang, N. Y. S. (1984d). Speech coding in the auditory nerve: V. Vowels in background noise. *J. Acoust. Soc. Am. 75*:908–918.

Deng, L., Geisler, C. D., and Greenberg, S. (1988). A composite model of the auditory periphery for the processing of speech. *J. Phonet. 16*:93–108.

Ephraim, Y., Wilpon, J. G., and Rabiner, L. R. (1987). A linear predictive front-end processor for speech recognition in noisy environments. *Int. Conf. Acoustics, Speech and Signal Processing—ICASSP '87*, vol. 3, pp. 1324–1327.

Geisler, C. D., Deng, L., and Greenberg, S. (1985). Thresholds for primary auditory fibers using statistically defined criteria. *J. Acoust. Soc. Am. 77*:1102–1109.

Ghitza, O. (1985). A measure of in-synchrony regions in the auditory nerve firing patterns as a basis for speech vocoding. *Int. Conf. Acoustics, Speech and Signal Processing—ICASSP '85*, vol. 2, pp. 505–508.

Ghitza, O. (1988). Temporal non-place information in the auditory-nerve firing patterns as a front-end for speech recognition in a noisy environment. *J. Phonet. 16*: 109–124.

Goldstein, J. L., and Srulovicz, P. (1977). Auditory nerve spike intervals as an adequate basis for aural spectrum analysis. In *Psychophysics and Physiology of Hearing*, E. F. Evans and J. P. Wilson (Eds.). Academic Press, London, pp. 337–346.

Greenberg, S. (1988). Representation of speech in the auditory periphery, *J. Phonet. 16*(1).

Juang, B. H., Rabiner, L. R., and Wilpon, J. G. (1987). The use of bandpass liftering in speech recognition. *IEEE Trans. Acoust. Speech and Signal Process. ASSP-35*: 947–954.

Kiang, N. Y. S., and Peake, W. T. (1988). Physics and physiology of hearing. In *Stevens Handbook of Experimental Psychology*, 2nd ed., vol. 1. Wiley, New York, pp. 227–326.

Liberman, M. C. (1978). Auditory-nerve response from cats raised in a low-noise chamber. *J. Acoust. Soc. Am. 63*:442–455.

Liberman, M. C. (1982). Single-neuron labeling in the cat auditory nerve. *Science 216*:1239–1241.

Lyon, R. F. (1986). Experiments with a computational model of the cochlea. *Int. Conf. Acoustics, Speech and Signal Processing—ICASSP '86*, vol. 3, pp. 1975–1978.

Mansour, D., and Juang, B. H. (1988a). A family of distortion measures based upon projection operation for robust speech recognition. *Int. Conf. Acoustics, Speech and Signal Processing—ICASSP '88*, vol. 1, pp. 36–39.

Mansour, D., and Juang, B. H. (1988b). The short-time modified coherence representation and its application for noisy speech recognition. *Int. Conf. Acoustics, Speech and Signal Processing—ICASSP '88*, vol. 1, pp. 525–528.

Moore, B. C. J., and Glasberg, B. R. (1983). Suggested formula for calculating auditory-filter bandwidth and excitation patterns. *J. Acoust. Soc. Am. 74*:750–753.

Pickles, J. O. (1982). *An Introduction to the Physiology of Hearing*. Academic Press, London.

Sachs, M. B., and Young, E. D. (1979). Encoding of steady-state vowels in the auditory nerve: Representation in terms of discharge rate. *J. Acoust. Soc. Am. 66*: 470–479.

Sachs, M. B., Voigt, H. F., and Young, E. D. (1983). Auditory nerve representation of vowels in background noise. *J. Neurophysiol. 50*:27–45.

Sachs, M. B., Blackburn, C. C., and Young, E. D. (1988). Rate-place and temporal-place representations of vowels in the auditory nerve and anteroventral cochlear nucleus. *J. Phonet. 16*:37–54.

Seneff, S. (1988). A joint synchrony/mean-rate model of auditory speech processing. *J. Phonet. 16*:55–76.

Shamma, S. (1988). The acoustic features of speech sounds in a model of auditory processing: Vowels and voiceless fricatives. *J. Phonet. 16*:77–92.

Smith, J. C., and Zwislocki, J. J. (1975). Short-term adaptation and incremental responses of single auditory-nerve fibers. *Biol. Cybernet. 17*:169–182.

Wilpon, J. G., and Rabiner, L. R. (1985). A modified K-means clustering algorithm for use in isolated word recognition. *IEEE Trans. Acoust. Speech Signal Process. ASSP-33*:587–594.

Young, E. D., and Sachs, M. B. (1979). Representation of steady-state vowels in the temporal aspects of the discharge patterns of populations of auditory-nerve fibers. *J. Acoust. Soc. Am. 66*:1381–1403.

16

Dynamic Programming–Based Speech Recognition Algorithms

HIROAKI SAKOE Kyushu University, Fukuoka, Japan

1. INTRODUCTION

Dynamic programming (DP; Bellman, 1957) is used for two principal problems in speech recognition: time warping problems in both isolated and continuous speech recognition and optimum word sequence search problems in continuous speech (connected word) recognition. Since dynamic programming was first applied to speech recognition around 1970 (Vintsyuk, 1968; Slutsker, 1968; Velichko and Zagoruyko, 1970; Sakoe and Chiba, 1971), numerous efforts have been made with different intentions. These include recognition performance (accuracy) optimization, efficiency enhancement, and extensive application to continuous speech or connected word recognition.

Sakoe and Chiba (1978) and Myers et al. (1980) investigated the performance trade-off in the DP-based time warping algorithm. The main problems they studied were the optimum whole word distance (dissimilarity) evaluation function and smoothing or slope constraint techniques for the warping function.

For improving computation speed, the global path constraint or search range limiting techniques were first used (Sakoe and Chiba, 1978; and Rabiner et al., 1978).

Based on these investigations, several dedicated large scale integrated circuits (LSI) were developed (Banetre et al., 1982; West et al., 1983; Kawakami et al., 1985). Some of them are currently actually used in practical recognizers.

A more efficient technique for improving speed is beam search pruning. Lowerre (1976) demonstrated its usefulness in the HARPY speech recognition system. Later, Bridle et al. (1982), Ney et al. (1987), and Sakoe et al. (1988) investigated detailed techniques in their DP-based systems, confirming that the required computation time is reduced by more than one order of magnitude. Vector quantization of the reference patterns is another successful speed enhancement technique. Sugamura et al. (1983) showed that only a 0.3% local distance calculation was sufficient for retaining the same performance as the standard DP. The storage requirement for the reference patterns was also reduced to 2% of that required by the standard DP. Sakoe et al. (1988) presented an efficient integration of beam search and vector quantization.

Regarding the algorithm generalization for continuous speech or connected words, while Vintsyuk's pioneering work (Vintsyuk, 1971a,b) was not known in the West until Bridle et al. (1982) first explicitly quoted it, Sakoe (1979, 1982) and Myers and Rabiner (1981) proposed different recognition algorithms which search for the optimum reference pattern sequences by DP. In fact, Sakoe's two-level algorithm was actually used in commercialized speech recognizers. Efforts have continued to seek a more efficient recognition algorithm.

In the 1980s, Bridle et al. (1982), Sakoe and Watari (1981), and Sakoe et al. (1988) formulated a frame synchronous algorithm in which DP calculation is done framewise, i.e., synchronized with frame input. Ney (1984) studied the same algorithm. Their methods, called one-pass DP, one-stage DP, or frame synchronous DP, depending on the case, are essentially equivalent and historically are duplicates of Vintsyuk's method. One of the most significant improvements made by these recent researchers is the incorporation of syntax control in the frame synchronous algorithm. A typical frame synchronous DP algorithm will be described in Section 2.

One of the most recent topics is merging the DP algorithm into the neural network model. Combining the high time alignment capability of DP with the flexible learning function of the neural network is expected to lead to an advanced recognition model applicable to speaker-independent recognition problems. In Section 3 two relevant trials made by Sakoe et al. (1989) and Iso and Watanabe (Iso, 1989; Iso and Watanabe, 1990) will be presented.

When DP is used in the framework of the hidden Markov model (HMM) it is referred to as the Viterbi algorithm. Many analogies can be noted between the DP algorithms mentioned above and the Viterbi-type HMM; Juang (1984) and Levinson (1985) discussed the relationship between DP and HMM.

2. FRAME SYNCHRONOUS DP

Frame synchronous DP is one of the most efficient continuous speech (connected word) recognition algorithms in the sense that, in addition to the small computation requirement, calculation for time warping and optimum word sequence search is carried out synchronously with the speech pattern input. The algorithm can be applied to general (large vocabulary, for example) continuous speech recognition problems by using the subword reference pattern. The principle of frame synchronous DP is extensively applicable to Viterbi-type HMM systems without the need for significant modification. In this section, a frame synchronous DP algorithm with syntax control by finite state automaton (FSA) is described according to Sakoe et al. (1988).

Let

$$\mathbf{A} = \mathbf{a}_1\mathbf{a}_2 \cdots \mathbf{a}_i \cdots \mathbf{a}_I \tag{1}$$

be a speech pattern. Consider the word set

$$\Sigma = \{n \mid n = 1, 2, \ldots, N\} \tag{2}$$

as the vocabulary. For each word n, a reference pattern

$$\mathbf{B}^n = \mathbf{b}_1^n\mathbf{b}_2^n \cdots \mathbf{b}_j^n \cdots \mathbf{b}_{J^n}^n \tag{3}$$

is prepared. We consider a syntax control mechanism based on FSA:

$$\text{FSA} = \langle \Pi, \Sigma, \Delta, p_o, f \rangle \tag{4}$$

where Π = a set of states $\{p\}$
Σ = a set of words $\{n\}$
Δ = a set of state transition rules $\{(p,q,n)\}$
where $\{(p,q,n)\}$ means a transition from a state p to a state q by a word n
p_0 = an initial state (denoted by a numeral 0 in the following)
f = a final state

Our recognition problem is to find a word sequence that is accepted by the FSA and that optimally approximates the input speech. Time warping is embedded in the distance estimation for each word. The problem is equivalently depicted by the optimum path search problem in Fig. 1. In this figure, each cylinder stands for each state of the FSA. Time axis i is allocated in the longitudinal direction of each cylinder. For each cylinder pair between which the state transition (p,q,n) is defined in the FSA, a plane (i,j) is spanned, where j means a time in the reference pattern $\mathbf{B}_n$. Our problem is then to search for the best path on which the summation of local distances is minimum among the paths which, starting at state $p = 0$ and time $i = 0$, travel over (i,j) planes, terminating at $p = f$ and $i = I$.

[Work area]

In order to solve this problem we prepare the following tables for cumulative distances and back pointers.

$g(p,n;i,j)$: Minimum cumulative distance on a path from $i = 0, f = 0$ up to (i,j) on the plane from a state (cylinder) p with word n. This plane is considered only for the state-word pair (p,n) for which

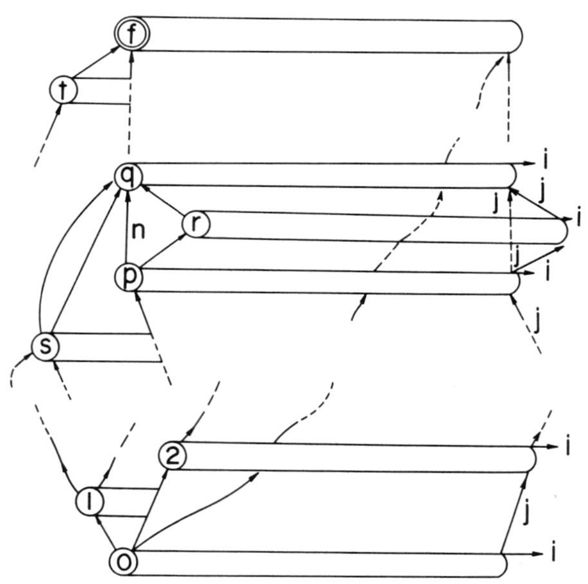

Figure 1 Search space in connected word recognition.

(p,q,n) is defined in Δ for some q. We call such a state-word pair a significant plane or significant state-word pair (p,n).

$l(p,n;i,j)$: The starting time i of the above path on the significant plane (p,n). [To realize the planes (i,j) spanning between states p and q with a word n, tables $g(p,q,n;i,j)$ and $l(p,q,n;i,j)$ should be used, instead of $g(p,n;i,j)$ and $l(p,n;i,j)$ respectively. However, the values appearing in these tables are independent of terminating state q. Therefore, index q is omitted to simplify the tables.]

$T(q;i)$: The minimum cumulative distance on the above path up to state q and time i.

$P(q;i)$: The directly preceding state p on the path arriving at state q at time i.

$L(q;I)$: The time when the path departed from the directly preceding state.

$N(q;i)$: The word n which caused the transition to state q, terminating at time i. (These four tables are allocated for each cylinder q with longitudinal coordinate i in Fig. 1).

[Initialization]

Prior to running the frame synchronous algorithm, the following initialization is performed.

(i) $T(0;0) = 0$,
(ii) $g(p,n;0,j)$ = sufficiently large value.

The main part of the algorithm is the following frame synchronous DP calculation.

[Initialization in a frame]

(i) $T(q;i)$ = sufficiently large value.

[Frame synchronous process]

(ii) Do (iii) to (v) for all significant pairs (p,n).

(Intraword operation)

(iii) $g(p,n;i-1,-1)$ = sufficiently large value
$g(p,n;i-1,0) = T(p;i-1)$
$l(p,n;i-1,0) = i-1$

(iv) Calculate DP equation (5a) and accompanied substitution (5b) for $j = 1$ to $j = J^n$:

$$g(p,n;i,n) = d(n;i,j) + \min \begin{bmatrix} g(p,n;i-1,j) \\ g(p,n;i-1,j-1) \\ g(p,n;i-1,j-2) \end{bmatrix} \quad (5a)$$

where $d(n;i,j)$ is the local distance between an input vector $\mathbf{a}_i$ and a reference vector $\mathbf{b}_j^n$.

$$\begin{aligned} &[1] \quad l(p,n;i,j) = l(p,n;i-1,j) \\ &[2] \quad l(p,n;1,j) = l(p,n;i-1,j-1) \\ &[3] \quad l(p,n;i,j) = l(p,n;i-1,j-2) \end{aligned} \quad (5b)$$

where $[x]$ means that each substitution is activated when the xth value is minimum in the preceding minimization.

(Interword operation)

(v) For $(p,q,n) \in \Delta$ execute

$$\min[T(q;i), g(p,n;i,J^n)] \longrightarrow T(q;i) \quad (6a)$$

[1] No operation

[2] $N(q;i) = n$

$P(q;i) = p$

$L(q;i) = l(p,n;i,J^n)$

The frame synchronous operations (i) to (v) are repeated each time a new input vector $\mathbf{a}_i$ is given. After the final vector ($i = I$) is processed, the following backtracking is run.

[Decision process]

(i) $i = I, q = f$.

(ii) While $i > 0$, repeat (iii) and (v).

(iii) Recognition result $n = N(q;i)$.

(iv) Inverse state transition $p = P(q;i)$,
Inverse time transition $l = L(q;i)$.

(v) $i = l$, $q = p$.

The inverted sequence of n in step (iii) gives the recognition result word sequence.

As can be seen, the calculation of Eqs. (5) is on the innermost loop. The repetition time of this loop is proportional to the number of significant state-word pairs (p,n). This means that the computation amount required for the algorithm rapidly increases as the vocabulary size and the FSA complexity increase. For this problem, Bridle et al. (1982) noted the usefulness of the beam search technique. Later, Ney et al. (1987) and Sakoe et al. (1988) discussed the practical implementation using lists and a dynamically controlled work area. Ney's implementation is effective especially when a pipeline processor is used, while Sakoe's implementation is suitable for a general-purpose processor. These implementations reduce not only the required computation but also the memory size for tables g and l.

When the vector quantization of the reference pattern is incorporated into the frame synchronous DP, the local distances are obtained from the distances between the input vector and the code vectors. For the beam search version algorithm, it is evident that calculating the distances to the code vectors corresponding to active (escaped from pruning) grid points $(n;j)$ is sufficient, at each frame cycle i. Furthermore, a simple consideration leads us to the idea that such a set of code vectors must include some overlaps. Removing these overlaps ensures the reduction of the distance calculation. This idea was confirmed by Sakoe et al. (1988).

In this section, recent efforts to enhance the efficiency of the connected word recognition algorithm were described. As a result of these improvements, moderate size recognition tasks which needed special DP hardware several years ago can now be executed in real time by recent general-purpose personal computers. It should also be noted that some large-scale recognition problems which cannot be solved without using a bottom-up operation (segmentation, phoneme recognition, or word spotting) can now be solved by top-down strategies.

3. DP ALGORITHMS IN THE NEURAL NETWORK

After a long inactive period since the 1960s, the neural network has recently witnessed a revival in a wide range of technical areas. Currently, it is attracting the interest of many speech recognition researchers. The reasons for this are that the neural network has a highly flexible learning capability based on back-propagation (Rumelhart and McClelland, 1986), can realize sufficiently complex mapping (between input pattern and category symbols, for example), and provides high speed and reliability through a parallel processing architecture.

A number of activities have already been reported using the neural network applied to speech recognition problems. These include phoneme recognition, phoneme recognition error correction, and speech parameter mapping for speaker adaptation. In this section two activities closely related to DP are described.

3.1. Dynamic Programming Neural Network—Incorporation of DP-Based Time Warping in the Neural Network*

3.1.1. Network Architecture

Although DP is effective for problems related to the time axis and word sequence, it is not especially capable of assimilating a wide variety of speaker-dependent spectrum pattern variations. However, using a neural network which has been trained with sufficient speaker-dependent spectrum variety can be an effective solution for speaker-independent recognition problems other than time axis problems. Therefore, a neural network with DP-based warping capability is expected to be a high-performance speaker-independent recognizer. The dynamic programming neural network (DNN) is one realization of such a DP–neural network integration.

Different from a general neural network, the DNN has a special structure suitable for handling the time sequence pattern as shown in Fig. 2. The input layer and the hidden layer are organized in the frame structure with time axis j. Each input frame contains K input units (j,k), while each hidden frame contains L hidden units (j,l). Several (two, for example) time-delayed input frames (j) and $(j-1)$ are connected to each hidden frame (j). This set of input frames is called an input block (j). This local organization of the network is based on the same concept as that of the time-delayed neural network (TDNN) of Waibel et al. (1989) used for phoneme recognition. The architecture allows the network to detect local features including the dynamic features by the input block to the hidden unit connections. These local features are then summarized by the output unit. It should be noted that the number of connections is significantly reduced by the block structure as compared to the general neural network.

The input layer receives the input signal

$$\mathbf{X} = \mathbf{x}_1\mathbf{x}_2 \cdots \mathbf{x}_j \cdots \mathbf{x}_J \tag{7}$$

where

$$\mathbf{x}_j = [x_{j1}, x_{j2}, \ldots, x_{jk}, \ldots, x_{jK}]^t$$

is an input frame vector. An input unit (j,k), upon receiving x_{jk}, transfers it to the hidden units.

*See Sakoe et al. (1989).

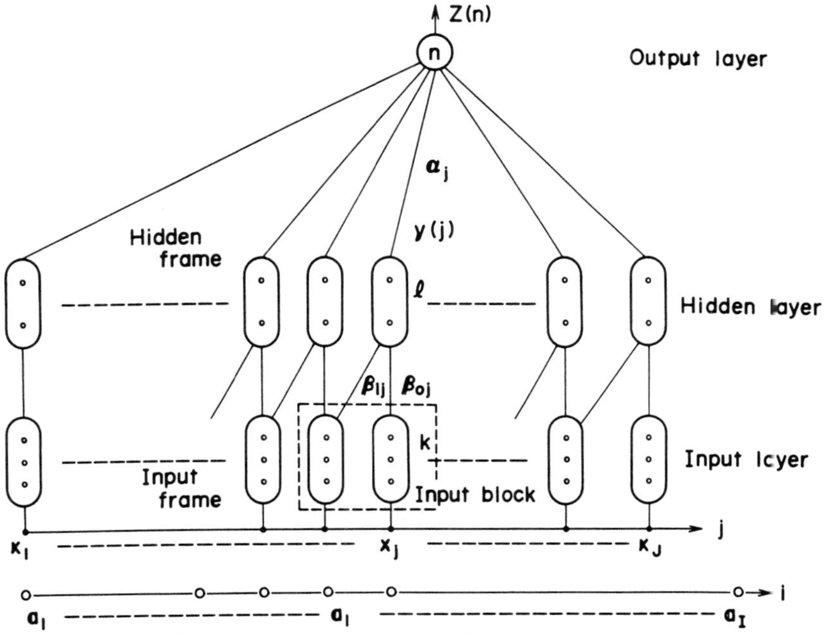

Figure 2 Architecture of dynamic programming neural network.

The hidden unit (j, l), upon receiving signals from the input units in the accompanying input block (j), calculates a weighted summation according to the weighting coefficients. The weighting coefficient for the connection from the input unit (j, k) is given by $\beta_{lk}(j, 0)$, while the weighting coefficient for the connection from the input unit $(j-1, k)$ is given by $\beta_{lk}(j, 1)$. The weighted summation is then

$$\text{net}_j(l) = \sum_k \beta_{lk}(j,0)x_{jk} + \sum_k \beta_{lk}(j,l)x_{j-1,k} \tag{8}$$

The output of this hidden unit is given by

$$y_j(l) = f(\text{net}_j(l)) \tag{9}$$

where $f(\cdot)$ is a sigmoid function. Outputs $y_j(l)$ from a hidden frame (j) form a hidden frame output vector

$$\mathbf{y}_j = [y_j(1), \ldots, y_j(l), \ldots, y_j(L)]^t \tag{10}$$

The above function, for deriving vector $\mathbf{y}_j$ from the input vectors $\mathbf{x}_j$ and $\mathbf{x}_{j-1}$, is formally denoted by

$$\mathbf{y}_j = \mathbf{f}(\boldsymbol{\beta}(j,0)\mathbf{x}_j + \boldsymbol{\beta}(j,1)\mathbf{x}_{j-1}) \tag{11}$$

where $\boldsymbol{\beta}(j,0) = [\beta_{lk}(j,0)]$ and $\boldsymbol{\beta}(j,1) = [\beta_{lk}(j,1)]$ are coefficient matrices.

For each connection to an output unit from the hidden unit (j,l), a weighting coefficient $\alpha(j,l)$ is given. Consider the coefficient vector

$$\boldsymbol{\alpha}_j = [\alpha_j(1) \cdots \alpha_j(l) \cdots \alpha_j(L)] \tag{12}$$

The weighted summation at the output unit is then given by

$$\begin{aligned} \text{Net} &= \sum_j \boldsymbol{\alpha}_j \cdot \mathbf{y}_j \\ &= \sum_j \boldsymbol{\alpha}_j \cdot \mathbf{f}(\boldsymbol{\beta}(j,0) \cdot \mathbf{x}_j + \boldsymbol{\beta}(j,1) \cdot \mathbf{x}_{j-1}) \end{aligned} \tag{13}$$

This gives the output value or the output unit, or equivalently the network output:

$$Z = \text{Net} \tag{14}$$

(In ordinary neural networks, a sigmoid function is employed at the output unit. The function is omitted here since it is monotonic. Note that the sigmoid function is employed in the learning phase described later.)

3.1.2. Time Warping

The summation form in Eq. (13) implies the applicability of the DP-based time warping function into the network. Let

$$\mathbf{A} = \mathbf{a}_1, \mathbf{a}_2, \ldots, \mathbf{a}_i, \ldots, \mathbf{a}_I \tag{15}$$

be an input pattern to the neural network defined above.

Figure 3 shows a time warping scheme. Consider the warping function

$$i = i(j) \tag{16}$$

between input pattern time i and the neural net input block (j). When input vector $\mathbf{a}_{i(j)}$ is allotted to the input frame (j), the input vector $\mathbf{a}_{i(j)-1}$ is then allotted to the input frame $(j-1)$ by the function, or

$$\text{if } \mathbf{x} = \mathbf{a}_{i(j)}, \quad \text{then } \mathbf{x}_{j-1} = \mathbf{a}_{i(j)-1} \tag{17}$$

According to Eqs. (13) and (14), the neural network output is

$$Z = \sum_j \boldsymbol{\alpha} \cdot \mathbf{f}(\boldsymbol{\beta}(j,0) \cdot \mathbf{a}_i + \boldsymbol{\beta}(j,1) \cdot \mathbf{a}_{i-1}) \tag{18}$$

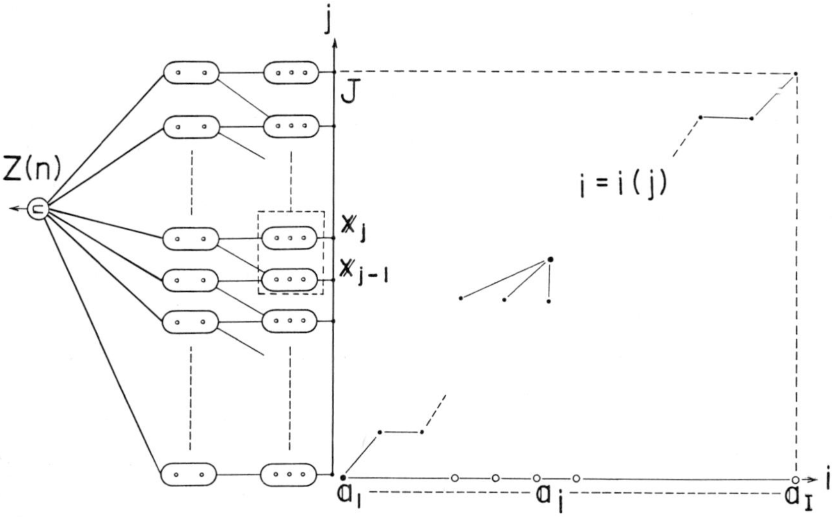

Figure 3 Time warp for DNN.

Let

$$r(i,j) = \alpha \cdot \mathbf{f}(\beta(j,0) \cdot \mathbf{a}_i + \beta(j,1) \cdot \mathbf{a}_{i-1}) \tag{19}$$

The time warping problem is now formulated as

$$Z = \max_{i=i(j)} \left[\sum_j r(k,j) \right] \tag{20}$$

This maximization problem is solved by the following very familiar DP algorithm:

[Initial condition]

$$g(1,1) = r(1,1) \tag{21}$$

[DP equation]

$$g(i,j) = r(i,j) + \max \begin{bmatrix} g(i,j-1) \\ g(i-1,j-1) \\ g(i-2,j-1) \end{bmatrix} \tag{22}$$

[Network output or similarity]

$$Z = g(I,J) \tag{23}$$

This model is called the DP neural network or DNN for short. The model is prepared for each word category. When appropriately trained, the model holds the abstracted feature of the word as connection coefficients $\boldsymbol{\alpha}$ and $\boldsymbol{\beta}$. when recognition is conducted, time alignment is achieved by DP, yielding the similarity Z of input pattern A evaluated by the DNN.

3.1.3. Learning Strategy

The back-propagation learning method is a well-known network training method. The method is directly applied to the DNN. It should be noted, however, that the DNN has a sufficient time alignment capability. It is, therefore, unnecessary to learn all the time distortion patterns. This, along with the reduced number of weighting coefficients, makes the learning procedure quantitatively easier than when a general network is trained by a massive volume of training speech data required for covering the time structure distortion. The training algorithm described below uses DP time alignment to generate most effective training data at each stage of training.

Consider two word categories, n and $\bar{n}$. As a training sample set, $\{\mathbf{A}_m\}$ is prepared:

$$\mathbf{A}_m = \mathbf{a}_m^1, \mathbf{a}_m^2, \ldots, \mathbf{a}_m^i, \ldots, \mathbf{a}_m^I \tag{24}$$

where $\mathbf{A}_m \in n$, or $\mathbf{A}_m \in \bar{n}$. Determine a scale reference pattern $\mathbf{A} = \mathbf{a}_1, \mathbf{a}_2, \ldots, \mathbf{a}_u, \ldots, \mathbf{a}_I$ for category n. ($\mathbf{A}$ may be a centroid of the subset $\{\mathbf{A}_m \in n\}$.) The time structure of $\mathbf{A}$ is considered to be a standard time base of word n. Time align all $\mathbf{A}_m \in n$ to the time base of the scale reference pattern $\mathbf{A}$ using DP, generating a time-aligned data base $\{\mathbf{AA}_m\}$.

After these preparations, the DNN for category n is trained by the following "adaptive time warp learning procedure" (Figs. 4 and 5):

1. Instead of training data $\mathbf{A}_m \in n$, time-aligned data $\mathbf{AA}_m \in n$ is fed to the DNN and back propagation with the target $f(Z) = 1$ is conducted.
2. For inhibitory training, $\mathbf{A}_m \in \bar{n}$ is warped by DP, yielding an $\mathbf{AA}_m$ most inseparable by the current DNN. ("Most inseparable" means that network output Z is maximum.) Using these data, back propagation with the target $f(Z) = 0$ is conducted.

Giving the most inseparable training pattern in (2) is considered desirable for training efficiency. Figure 5 explains this idea. Each ellipse shows

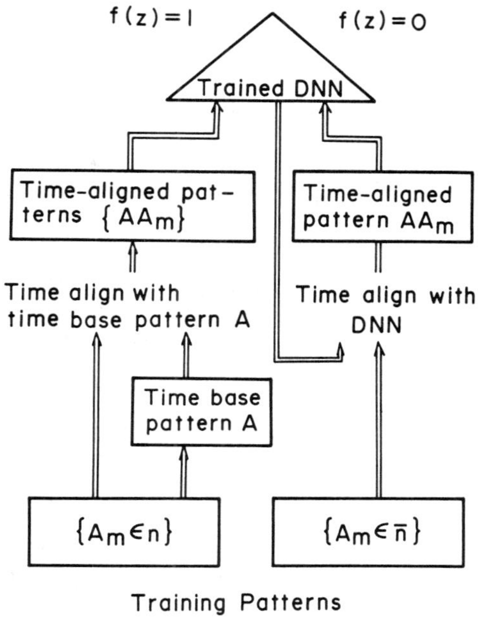

Figure 4 Training procedure for DNN.

a set of time-warped patterns from training data $\mathbf{A}_m$. As long as the DNN executes the time warping in the recognition phase, all the time-warped patterns in these ellipses belonging to $\mathbf{A}_m \in \bar{n}$ must be separated from the set $\{\mathbf{AA}_m \in n\}$. It is evident that the training data inside the ellipses contribute less to training than do those on the edge. Furthermore, among these time-warped patterns on the edge, the data most inseparable by the current DNN (or current discriminating hyperplane) play the most effective role for improving the DNN. On the other hand, in (i), fixed training data $\{\mathbf{AA}_m \in n\}$ are used to retain the training stability.

As a result, the discriminating hyperplane, attained through the above learning procedure, separates the time-aligned training patterns $\{\mathbf{AA}_m \in n\}$ into the positive side $[f(Z) = 1]$ and all time-warped training patterns generated from $\{\mathbf{A}_m \in \bar{n}\}$ into the negative side $[f(Z) = 0]$.

3.1.4. Preliminary Experiment

In order to examine the effectiveness of the DNN, a speaker-independent isolated Japanese digit word recognition experiment was conducted using

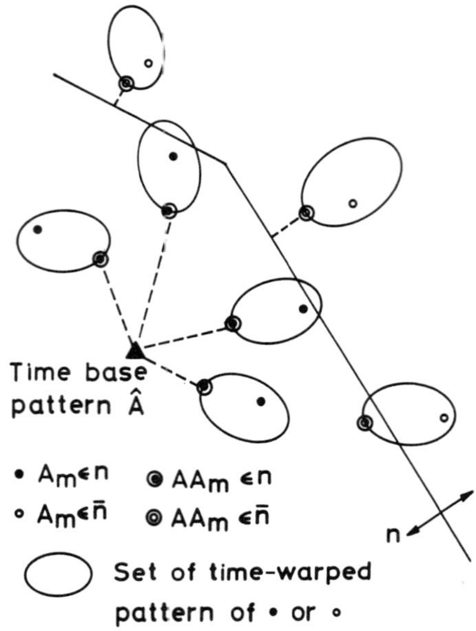

Figure 5 Adaptive warp training principle.

speech data uttered by 107 speakers (62 male and 45 female speakers). Each speaker uttered each word once through a telephone set. The speech wave was digitized at an 8.0-kHz sampling rate, from which 10-dimensional mel-scale cepstrum coefficients were calculated at every 16-ms frame period. The changing ratio parameter for the amplitude was also calculated. A total of 570 samples uttered by 57 speakers was used for DNN training, and a total of 500 samples uttered by 50 speakers was used for the recognition test. The DNN with the same number of frames as the number of frames of time scale patterns was prepared for each word. For each input frame, 10 (when only mel-cepstrum coefficients were used) or 11 (when the amplitude changing ratio was used in addition to the above 10 coefficients) input units were prepared. The number of units in a hidden frame was four.

Figure 6 presents the learning curve of the DNN when 10 mel-scale cepstrum coefficients were used. Here "iteration" means the number of cycles in which the total 570 training data were given to the 10 DNNs. An overlearning phenomenon is observed. The best recognition rate for the test data was 98.9%, which occurred around 50 iterations.

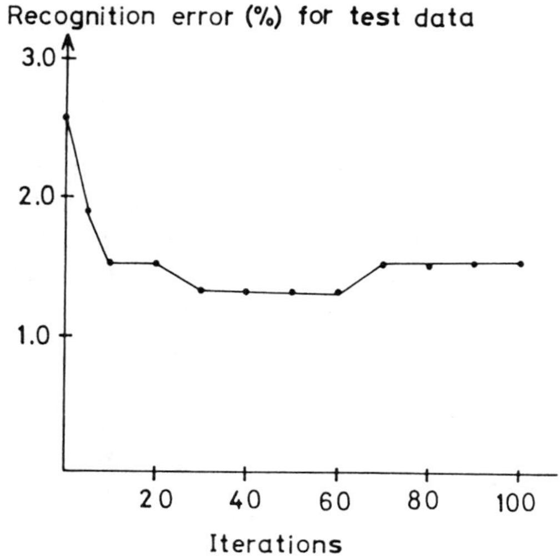

Figure 6 DNN learning curve.

Table 1 compares the results of the experiments with those of conventional multitemplate DP. (Two templates per word selected by clustering algorithm were used.) For both parameter conditions, the DNN showed comparable or slightly superior performance to multitemplate DP. For more detailed experimental results, see Sakoe et al. (1989). These results demonstrate the potential effectiveness of integrating DP with the neural network.

Table 1 DNN Evaluation Results

Method	Error (%) Without amplitude	Error (%) With amplitude
DNN	1.2	0.7
Conventional DP[a]	1.6	1.1

[a]Two reference patterns were used for each word.

3.2. Neural Prediction Model Embedded in DP Time Warping*

3.2.1. Neural Prediction Model

It is will known that a neural network with sufficient structure complexity can learn and simulate a complex input signal–to–output signal mapping as long as a specific relationship exist between the signals (Funahashi, 1989). Iso (1989) and Iso and Watanabe (1990) considered that the speech feature in the input pattern is predictable from the preceding speech features in the pattern, under the hypotheses of word and time position in that word. They also proposed a novel technique for using the above mapping function of the neural network for this prediction.

Figure 7 depicts their prediction model. This model is prepared at each time n of a word c. Upon receiving the information of word hypothesis c and time hypothesis n, and the preceding features $\mathbf{a}_{t-1}$ and $\mathbf{a}_{t-2}$, the model predicts the current feature $\mathbf{a}_t$ of the present input pattern by

$$\mathbf{h}_t(c,n) = \mathbf{f}(\mathbf{U}(c,n)\cdot\mathbf{a}_{t-1} + \mathbf{V}(c,n)\cdot\mathbf{a}_{t-2}) \quad (25a)$$

$$\hat{\mathbf{a}}_t(c,n) = \mathbf{W}(c,n)\cdot\mathbf{h}_t(c,n) \quad (25b)$$

Here, $\mathbf{U}(c,n)$, $\mathbf{V}(c,n)$, and $\mathbf{W}(c,n)$ are coefficient matrices, $\mathbf{h}_t(c,n)$ is the hidden layer output vector, and $\mathbf{f}(\cdot)$ is a function which generates a vector by applying the sigmoid function to each component of the argument vector. (Here, the threshold parameter of the unit is omitted to simplify the notation.) This model is called the neural prediction unit (NPU).

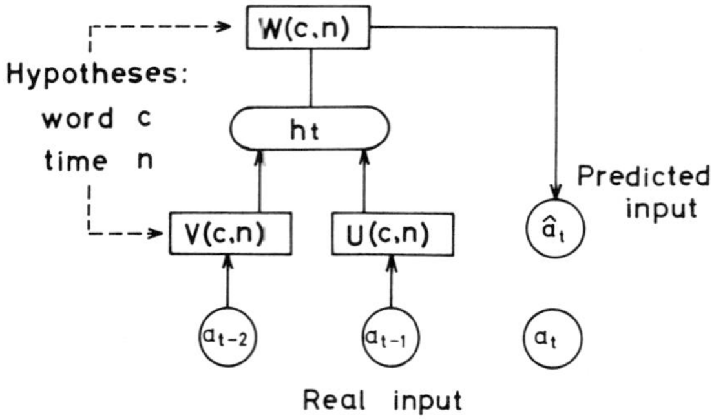

Figure 7 Neural prediction unit (NPU).

*See Iso (1989) and Iso and Watanabe (1990).

3.2.2. Whole-Word Prediction Model and DP

An NPU sequence forms the neural prediction model of word c, or NPM(c). When it is applied to a whole input word, it gives a whole-word prediction sequence.

The distance measure

$$d(c,t,n) = \|\hat{\mathbf{a}}_t(c,n) - \mathbf{a}_t\|^2 \tag{26}$$

evaluates the local prediction error of NPU at an input pattern time t and a hypothesized time n of the NPM(c). Considering a warping function

$$n = n(t) \tag{27}$$

where

$$n(1) = 1, \quad n(T) = N_c, \quad \text{and} \quad 0 \le n(t) - n(t-1) \le 1 \qquad (1 < t \le T)$$

a whole-word prediction error is then defined by

$$D(c) = \min_{n(t)} \sum_t d(c,t,n) \tag{28}$$

This whole-word prediction error stands for the word distance of DP time warping, and is calculated on the t-n plane in Fig. 8 by the DP equation

$$g(c,t,n) = d(c,t,n) + \min \begin{bmatrix} g(c,t-1,n) \\ g(c,t-1,n-1) \end{bmatrix} \tag{29}$$

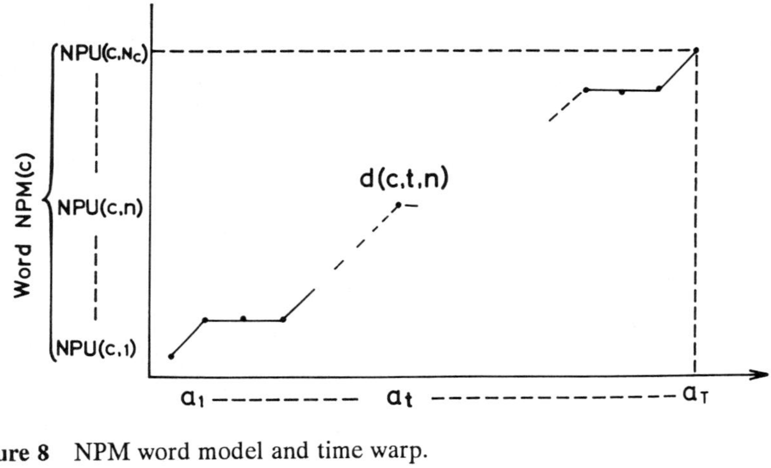

Figure 8 NPM word model and time warp.

The value $D(c) = g(c,T,N_c)$ gives the whole-word prediction error caused by the optimally time-aligned word model NPM(c).

3.2.3. *Learning Procedure*

The NPM is trained based on DP time alignment and back-propagation coefficient adjustment. The procedure is as follows.

1. Initialize the NPM of word c.
2. Repeat (3) to (4) until the stop condition is attained.
3. Perform DP between the NPM and a training pattern of category c obtaining the optimum time alignment $n(t)$.
4. Conduct back-propagation training for each NPU($c,n(t)$) with $\mathbf{a}_t$ as the desired output.

In order to retain the learning stability, the adjustment window is effectively used, with an initially narrow and then a gradually widened warp range as the learning progresses.

3.2.4. *Some Experimental Results*

Iso and Watanabe studied the performance of the NPM under several model specifications through speaker-independent digit word recognition experiments. The speech data used in their experiments were the same as those used in Section 3.1.

Nine hidden units gave the best performance. As for N_c, the number of NPUs in an NPM, the best condition was to use NPUs equal to half of the average frame number of training patterns of the category. At this optimum condition an average recognition rate as high as 99.8% was achieved. Table 2 presents a comparison with several typical conventional methods. For more extensive experiments, see Iso (1989) and Iso and Watanabe (1990). These

Table 2 NPM Evaluation Results

Method	Recognition rate (%)
Conventional DP[a]	98.9
DNN	99.3
NPM	99.8

[a] Two reference patterns were used for each word.

experimental results demonstrate that the NPM is a very promising speech recognition method.

3.2.5. *Related Activities*

Levin (1990) independently presented a similar model to that of Iso and Watanabe and applied it to multispeaker connected digit recognition. Tebeleskis and Waibel (1990) generalized these models and formulated a subword unit-based model.

4. CONCLUDING REMARKS

The progress made in using DP for speech recognition in the past 10 years has been summarized. The first topic focused on was efficiency improvement of the DP algorithm for connected word recognition with low-level (finite state automaton) syntax control. Frame synchronization, beam search, and vector quantization have markedly improved the algorithm's efficiency. Through these improvements, the speech recognizer, which necessitated dedicated DP hardware several years ago, is now readily designed using the general-purpose microprocessor. In addition, large-scale experiments which required a bottom-up (or approximation) strategy to bypass the search problem complexity can now be conducted by a strict top-down approach.

Syntax control by finite state automaton is considered to be an established method. Also, the use of a higher-level grammar such as context-free grammar or dependency grammar is now being studied (Nakagawa, 1987; Ney, 1987; Ozeki, 1987; Matsunaga and Kohda, 1988).

The second topic described was DP algorithm incorporation into neural networks. Two typical models were presented. These models are expected to solve the problems of speaker-independent recognition through its time alignment capability and high learning capability for parameter fluctuations. Studies using DP in the framework of the neural network (or using the neural network in the framework of DP) are relatively new. Possibilities are also unfolding in which some novel and more effective models will be formulated for application to various speech recognition problems.

REFERENCES

Banetre, J., Frison, P., and Quinton, P. (1982). A systoric array for connected word recognition. *Proc. ICASSP '82*, pp. 1243–1236.

Bellman, R. (1957). *Dynamic Programming*. Princeton University Press, Princeton, N.J.

Bridle, J., Brown, M. D., and Chamberlain, R. M. (1982). An algorithm for connected word recognition. *Proc. ICASSP '82*, pp. 899–902.

Funahashi, K. (1989). On the approximate realization of continuous mapping by neural networks. *Neural Networks 2*:183.

Iso, K. (1989). Speech recognition using neural network prediction model. IEICE Technical Report, 89, pp. 81–87 (in Japanese).

Iso, K., and Watanabe, T. (1990). Speaker-independent word recognition using a neural prediction model. *Proc. ICASSP '90*, pp. 441–444.

Juang, B. H. (1984). On the hidden Markov model and dynamic time warping for speech recognition—a unified view. *AT&T Bell Lab. Tech. J. 63*(7):1213.

Kawakami, Y., Ishizuka, H., Watari, M., Sakoe, H., Hoshi, T., and Iwata, T. (1985). A microprocessor for speech recognition. *IEEE J. Selected Areas Commun. SAC-3*(2):369.

Levin, E. (1990). Word recognition using hidden control neural architecture. *Proc. ICASSP '90*, pp. 433–436.

Levinson, S. E. (1985). Structural methods in automatic speech recognition. *Proc. IEEE 73*(11):1625.

Lowerre, B. (1976). The HARPY speech recognition system. Ph.D. thesis, Carnegie-Mellon University.

Matsunaga, S., and Kohda, M. (1988). Linguistic processing using a dependency structure grammar for speech recognition and understanding. *Proc. COLING '88*, pp. 402–407.

Myers, C., and Rabiner, L. R. (1981). A level building dynamic programming algorithm for connected word recognition. *IEEE Trans. Acoust. Speech Signal Process. ASSP-29*:284.

Myers, C., Rabiner, L. R., and Rosenberg, A. E. (1980). Performance tradeoffs in dynamic time warping algorithms for isolated word recognition. *IEEE Trans. Acoust. Speech Signal Process. ASSP-28*:623.

Nakagawa, S. (1987). Spoken sentence recognition by time-synchronous parsing algorithm of context free-grammar. *Proc. ICASSP '87*. pp. 829–832.

Ney, H. (1984). The use of one-stage dynamic programming algorithm for connected word recognition. *IEEE Trans. Acoust. Speech Signal Process. ASSP-32*:263.

Ney, H. (1987). Dynamic programming speech recognition using a context-free grammar. *Proc. ICASSP '87*, pp. 69–72.

Ney, H., Margel, D., Noll, A., Paeseler, A. (1987). A data-driven organization of the dynamic programming beam search for continuous speech recognition. *Proc. ICASSP '87*, pp. 833–836.

Ozeki, K. (1987). A multi-stage decision algorithm to select optimum bunsetsu sequence based on degree of kakariuke-dependency. *Trans. IEICE Jpn. J70-D*(3): 601. (in Japanese)

Rabiner, L. R., Rosenberg, A. E., and Levinson, S. E. (1978). Considerations in dynamic time warping for discrete word recognition. *IEEE Trans. Acoust. Speech Signal Process. ASSP-26*:575.
Rumelhart, D. E., and McClelland, J. L. (1986). *Parallel Distributed Processing*, vol.1. MIT Press, Cambridge, Mass.
Sakoe, H. (1979). Two-level DP-matching—a dynamic programming based pattern matching algorithm for connected word recognition. *IEEE Trans. Acoust. Speech Signal Process. ASSP-27*:588.
Sakoe, H. (1982). A generalized two-level DP-matching algorithm for continuous speech recognition. *Proc. IECE Jpn. E-65*:649.
Sakoe, H., and Chiba, S. (1971). Dynamic programming approach to continuous speech recognition. Proc. 7th ICA, 20C13.
Sakoe, H., and Chiba, S. (1978). A dynamic programming algorithm optimization for spoken word recognition. *IEEE Trans. Acoust. Speech Signal Process. ASSP-26*:43.
Sakoe, H., and Watari, M. (1981). Clockwise propagating DP-matching algorithm for connected word recognition. *Trans. Comm. Speech Res. Acoust. Soc. Jpn. S81-65*, pp. 517–524. (in Japanese)
Sakoe, H., Fujii, H., Yoshida, K., and Watari, M. (1988). A high-speed DP-matching algorithm based on frame synchronization, beam search and vector quantization. *Trans. IEICE Jpn. J71-D*:1650 (in Japanese).
Sakoe, H., Isotani, R., Yoshida, K., Iso, K., and Watanabe, T. (1989). Speaker-independent word recognition using dynamic programming neural networks. *Proc. ICASSP '89*, pp. 29–32.
Slutsker, G. S. (1968). Non-linear method of analysis of speech signal. Trudy N.I.I.R. (in Russian).
Sugamura, N., Shikano, K., and Furui, S. (1983). Isolated word recognition using phoneme-like templates. *Proc. ICASSP '83*, pp. 723–726.
Tebelskis, J. and Waibel, J. (1990). Large vocabulary recognition using linked predictive neural networks. *Proc. ICASSP '90*, pp. 437–440.
Velichko, V. M., and Zagoruyko, N. G. (1970). Automatic recognition of 200 words, *Int. J. Man-Machine Studies 2*:223.
Vintsyuk, T. K. (1968). Speech recognition by dynamic programming. *Kibernetika 4*: 81.
Vintsyuk, T. K. (1971a). Element-wise recognition of continuous speech consisting of given vocabulary. *Kibernetika 2*:133–143 (in Russian).
Vintsyuk, T. K. (1971b). Element-wise recognition of continuous speech consisting of given vocabulary. *Cybernetics 7*:361–372.
Waibel, A., Hanazawa, T., Hinton, G., Shikano, K., and Lang, K. J. (1989). Phoneme recognition using time-delay neural networks. *IEEE Trans. Acoust. Speech Signal Process. 37*:328.
West, N., Burr, D. J., and Ackland, B. D. (1983). Dynamic time warp pattern matching using an integrated multiprocessing array. *IEEE Trans. Comput. C-32*(8): 731.

17

Issues in Using Hidden Markov Models for Speech Recognition

BIING-HWANG JUANG and LAWRENCE R. RABINER AT&T
Bell Laboratories, Murray Hill, New Jersey

1. INTRODUCTION

The use of hidden Markov models (HMMs) for speech recognition has become increasingly popular in the past few years. Although the number of reported recognition systems based on HMMs is too large to discuss in detail here, it is worthwhile pointing out some of the most successful of these systems. These include the early work of Baker (1975) on the Dragon System at Carnegie Mellon University; the long-standing effort of Jelinek and colleagues at IBM on a voice dictation system (Jelinek, 1976; Bahl et al., 1983; Averbuch et al., 1987); the work at AT&T Bell Laboratories, IDA, MIT Lincoln Labs, and Philips on whole-word recognition using HMMs (Rabiner et al., 1983, 1986a, 1989b; Lee et al., 1988; Poritz and Richter, 1986; Lippmann et al., 1987; Bourlard et al., 1985); the DARPA Resource Management task (Chow et al., 1987; Lee, 1989); and other related efforts (Gupta et al., 1987; Derouault, 1987). The widespread popularity of the HMM framework can be attributed to its simple algorithmic structure, which is straightforward to implement, and to its clear performance superiority over alternative recognition structures.

Although hidden Markov modeling has significantly advanced the area of automatic speech recognition, the general problem of completely fluent, speaker-independent speech recognition is still far from being solved. For example, there is no system that is capable of reliably recognizing unconstrained conversational speech, nor does there exist a good way to infer statistically the language structure from a limited corpus of spoken sentences. In order to meet these broader challenges, it is worthwhile to put the method of hidden Markov modeling in the proper perspective and reexamine its strengths as well as some areas for further refinements and/or improvements. This is essentially the purpose of this chapter.

The strength of the HMM method lies in two major categories, namely the mathematical framework (theory) and the implementational structure (practice). Therefore, we will first reexamine these strengths with the hope of not only enhancing our understanding of the method but also gaining new insights which may lead to a better utilization of this rich mathematical tool. Next, we will discuss some aspects of the HMM method for which there are still open questions as to how best to apply them to speech recognition applications. These include hidden Markov model training criteria, modeling for discrimination, variance reduction methods for model parameter estimates, integration of features in different categories, explicit modeling of state duration, model clustering, and the use of prior knowledge.

2. HIDDEN MARKOV MODELING OF SPEECH

Speech is essentially a nonstationary signal. When we speak, our articulatory apparatus modulates the air pressure and flow to produce an audible sequence of sounds. Whereas the spectral content of any particular sound may include frequencies up to several thousand Hertz, our articulatory configuration (vocal tract shape, tongue movement, etc.) often does not undergo dramatic changes more than 10 times per second. Speech modeling thus involves the following two aspects: analysis of the short-time spectral properties of individual sounds, sampled at an interval on the order of 10 ms, and characterization of the long-time development of sound sequences on the order of 100 ms, due to articulatory configuration changes.

The short-time properties of an individual sound can be conveniently represented by a spectral measurement vector. Many spectral analysis methods have been proposed for speech signal modeling. These include such standard methods as measurement of the discrete (fast) Fourier transform (FFT), linear prediction (LPC), and autoregressive/moving average (ARMA) models (Allen and Rabiner, 1977; Schafer and Rabiner, 1971; Atal and Hanauer,

1971; Markel and Gray, 1976; Makhoul, 1975; Cadzow, 1982). Even the more traditional filter bank method of spectral analysis is still used in some systems (Dautrich et al., 1983), particularly in hardware implementations. To emphasize spectral properties which are known to be important to a human listener, auditory models can be incorporated in the overall spectral representation (Cohen, 1985; Ghitza, 1986). In speech modeling, we often call this short-time spectral vector an observation. Each observation is obtained regularly, say one every 10 to 30 ms.

There are many ways to characterize the temporal sequence of sounds, i.e., running speech, as represented by a sequence of spectral observations. The most direct way is to register the spectral sequence directly, without further modeling. Another way is to model the temporal sequence of spectra in terms of a Markov chain to describe the way one sound changes to another. As will be explained later, the two modeling phases, namely the short-time spectra of individual sounds and the probabilistic description of the sound changes, can be described in a mathematically consistent framework, thereby offering analytic solutions to speech problems.

For historical reasons, the direct representation of a sequence of speech spectra, without explicit probabilistic modeling, is called a "deterministic" template. If we denote the spectral vector at time t by $\mathbf{O}_t$ and the observed spectral sequence lasts from $t = 1$ to $t = T$, such a deterministic sequence is then represented by $\{\mathbf{O}_t\}_{t=1}^{T} = (\mathbf{O}_1, \mathbf{O}_2, \ldots, \mathbf{O}_T)$. In contrast, when an explicit probabilistic structure is imposed on the representation of the evolutional sequence, the resulting model is often classified as "stochastic." Such a dichotomy of views is not really necessary. Some perspectives on how the two seemingly different methodologies relate to each other have been given in Juang (1984) and Bridle (1984). In this chapter we discuss only the typical stochastic models which can be generalized without great difficulty.

Based on the above discussion, we are now ready to define formally a hidden Markov model for speech recognition. Consider a first-order N-state Markov chain as illustrated for $N = 3$ in Fig. 1. The system can be described as being in one of the N distinct states, $S_1, S_2, \ldots, S_N$, at any discrete time instant t. We use the state variable q_t, as the state of the system at discrete time t. The Markov chain is then described by a state transition probability matrix $A = [a_{ij}]$ where

$$a_{ij} = P(q_t = S_j \mid q_{t-1} = S_i), \qquad 1 \leq i, \quad j \leq N \tag{1}$$

Note that in (1) we have assumed homogeneity of the Markov chain so that the transition probabilities do not depend on time. These transition

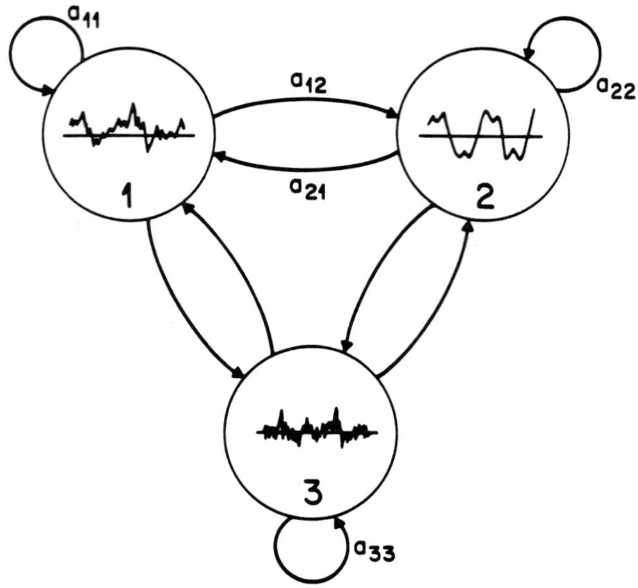

Figure 1 A first-order, three-state hidden Markov model.

probabilities satisfy the following axiomatic constraints:

$$a_{ij} \geq 0 \tag{2}$$

and

$$\sum_{j=1}^{N} a_{ij} = 1 \qquad \text{for all } i \tag{3}$$

Assume that at $t = 0$ the state of the system, q_0 is specified by an initial state probability vector $\pi' = [\pi_1, \pi_2, \ldots, \pi_N]$ where the prime denotes matrix transpose. Then for any state sequence $\mathbf{q} = (q_0, q_1, q_2, \ldots, q_T)$, where $q_t \in \{S_1, S_2, \ldots, S_N\}$, the probability of $\mathbf{q}$ being generated by the Markov chain is

$$P(\mathbf{q} \mid A, \pi) = \pi_{q_0}, a_{q_0 q_1} a_{q_1 q_2} \cdots a_{q_{T-1} q_T} \tag{4}$$

Suppose now that the state sequence $\mathbf{q}$ cannot be readily observed. Instead, we envision each observation $\mathbf{O}_t$ as being produced with the system in state q_t (where $q_t \in \{S_1, S_2, \ldots, S_N\}$). We assume the production of $\mathbf{O}_t$ in each possible state S_i, $i = 1, 2, \ldots, N$ is stochastic and is characterized by a set of observation probability measures $B = \{b_i(\mathbf{O}_t)\}_{i=1}^N$ where

$$b_i(\mathbf{O}_t) = P(\mathbf{O}_t \mid q_t = S_i) \tag{5}$$

If the state sequence $\mathbf{q}$ that led to the observation sequence $\mathbf{O} = (\mathbf{O}_1, \mathbf{O}_2, \ldots, \mathbf{O}_T)$ is known, the probability of $\mathbf{O}$ being generated by the system is assumed to be

$$P(\mathbf{O} \mid \mathbf{q}, B) = b_{q_1}(\mathbf{O}_1) b_{q_2}(\mathbf{O}_2) \cdots b_{q_T}(\mathbf{O}_T) \tag{6}$$

Therefore, the joint probability of $\mathbf{O}$ and $\mathbf{q}$ being produced by the system is simply the product of (4) and (6), written as

$$P(\mathbf{O}, \mathbf{q} \mid \pi, A, b) = \pi_{q_b} \prod_{t=1}^{T} a_{q_{t-1}q_t} b_{q_t}(\mathbf{O}_t) \tag{7}$$

It then follows that the stochastic process, represented by the observation sequence $\mathbf{O}$, is characterized by

$$\begin{aligned} P(\mathbf{O} \mid \pi, A, B) &= \sum_{\mathbf{q}} P(\mathbf{O}, \mathbf{q} \mid \pi, A, B) \\ &= \sum_{\mathbf{q}} \pi_{q_0} \prod_{t=1}^{T} a_{q_{t-1}q_t} b_{q_t}(\mathbf{O}_t) \end{aligned} \tag{8}$$

which describes the probability of $\mathbf{O}$ being produced by the system without assuming knowledge of the state sequence in which it was generated. The triple (π, A, B) defines a hidden Markov model and (8) is the probability measure defined by the model. The model parameter set is often denoted as $\lambda = (\pi, A, B)$.

In the development of the hidden Markov model methodology, the following problems are of particular interest. First, given the observation sequence $\mathbf{O}$ and a model λ, how do we efficiently evaluate the probability of $\mathbf{O}$ being produced by the source model λ, i.e., $P(\mathbf{O} \mid \lambda)$? Second, given the observation $\mathbf{O}$, how do we solve the inverse problem of estimating the parameters in λ? Although (8) defines the probability measure associated with model λ, knowledge of the most likely state sequence $\mathbf{q}$ that led to the observation $\mathbf{O}$ is desirable in many applications. The third problem, then, is how do we deduce from $\mathbf{O}$ the most likely state sequence $\mathbf{q}$ in a meaningful manner? According to convention (Ferguson, 1980), we call these three problems the evaluation problem, the estimation problem, and the decoding problem, respectively. In the following sections we describe several conventional solutions to these three standard problems.

2.1. The Evaluation Problem

The main concern in the evaluation problem is computational efficiency. Without complexity constraints, one can simply evaluate $P(\mathbf{O} \mid \lambda)$ directly from the definition of (8). Since the summation in (8) involves N^{T+1} possible $\mathbf{q}$ sequences, the total computational requirements are on the order of $2T \cdot N^{T+1}$ operations. The need to compute (8) without exponential growth of the computation, as a function of the sequence length T, is the first challenge for implementation of the HMM technique. Fortunately, using the well-known forward-backward procedure (Baum, 1972), this exorbitant computational requirement of the direct summation can easily be alleviated. By visualizing the Markov chain as a trellis structure, a forward induction procedure allows evaluation of the probability $P(\mathbf{O} \mid \lambda)$ to be carried out with only a computational requirement linear in the sequence length T and quadratic in the number of states N. Such a tremendous reduction in computation makes the HMM method attractive and viable for speech recognizer designs because the evaluation problem can be viewed as one of scoring how well an unknown observation sequence (corresponding to the speech to be recognized) matches a given model (or sequence of models) source, thus providing a mechanism for classification. The forward induction procedure is summarized in Appendix A for reference.

2.2. The Estimation Problem

Given an observation sequence (or a set of sequences) $\mathbf{O}$, the estimation problem involves finding the "right" model parameter values that specify a model most likely to produce the given sequence. In speech recognition, this is often called "training," and the given sequence, based on which we obtain the model parameters, is called the training sequence, even though the formulation here is statistical.

In solving the estimation problem, we often follow the method of maximum likelihood; that is, we choose model λ such that $P(\mathbf{O} \mid \lambda)$, as defined by (8), is maximized for the given training sequence $\mathbf{O}$. The Baum–Welch algorithm (Baum and Petrie, 1966; Baum and Egon, 1967; Baum and Sell, 1968; Baum et al., 1970) (often blended with the forward-backward algorithm because of its interpretation as an extension of the forward induction procedure to the evaluation problem) cleverly accomplishes this maximization objective in a two-step procedure. Based on an existing model λ' (possibly obtained randomly), the first step transforms the objective function $P(\mathbf{O} \mid \lambda)$ into a new function $Q(\lambda', \lambda)$ that essentially measures the divergence between the initial model λ' and an updated model λ. The Q function is defined, for the simplest case, as

$$Q(\lambda', \lambda) = \sum_{\mathbf{q}} P(\mathbf{O}, \mathbf{q} \mid \lambda') \log P(\mathbf{O}, \mathbf{q} \mid \lambda) \tag{9}$$

where $P(\mathbf{O}, \mathbf{q} \mid \lambda)$ is given in (7). Because $Q(\lambda', \lambda) \geq Q(\lambda', \lambda')$ implies $P(\mathbf{O} \mid \lambda) \geq P(\mathbf{O} \mid \lambda')$, we can then simply maximize the function $Q(\lambda', \lambda)$ over λ to improve λ' in the sense of increasing the likelihood $P(\mathbf{O} \mid \lambda)$. This maximization of the Q function over λ is the second step of the algorithm The algorithm continues by replacing λ' with λ and repeating the two steps until some stopping criterion is met. The algorithm is of a general hill-climbing type and is only guaranteed to produce fixed-point solutions, although in practice the lack of global optimality does not seem to cause serious problems in recognition performance (Paul, 1985). The Baum–Welch algorithm is also summarized in Appendix A, together with the aforementioned forward induction procedure, for reference.

The method of maximum likelihood is not the only possible choice for solving the estimation problem. As will be discussed later, other alternatives are attractive and offer different modeling perspectives.

2.3. The Decoding Problem

As noted above, we often are interested in uncovering the most likely state sequence that led to the observation sequence $\mathbf{O}$. Although the probability measure defined by a hidden Markov model, by definition, does not explicitly involve the state sequence, it is important in many applications to have knowledge of the most likely state sequence for several reasons. As an example, if we use the states of a word model to represent the sounds in the word, it may be desirable to know the correspondence between the speech segments and the sounds of the word, as the duration of the individual speech segments provides useful information for speech recognition purposes.

As with the second problem, there are several ways to define the decoding objective. The most trivial choice is, following the Bayesian framework, to maximize the (instantaneous) a posteriori probability

$$\gamma_t(i) = P(q_t = S_i \mid \mathbf{O}, \lambda) \tag{10}$$

that is, we decode the state at time t by choosing q_t to be

$$q_t = \arg \max_{1 \leq i \leq N} \gamma_t(i) \tag{11}$$

It is also possible to extend the definition of (10) to the cases of pairs of states or triples of states and so on. For example, the rule

$$(q_t, q_{t+1}) = \arg \max_{1 \leq i,j \leq N} P(q_t = S_i, q_{t+1} = S_j \mid \mathbf{O}, \lambda) \tag{12}$$

will produce the maximum a posteriori (MAP) result of the minimum number of incorrectly decoded state pairs, given the observation sequence $\mathbf{O}$.

While the above discussion shows the flexibility of possible localized decoding, we often choose to work on the entire state sequence $\mathbf{q}$ by maximizing $P(\mathbf{q} \mid \mathbf{O}, \lambda)$ for three reasons: (1) it is optimal for the unknown observation $\mathbf{O}$ in the MAP sense; (2) speech utterances are usually not so prohibitively long as to require locally (rather than globally) optimal decoding; and (3) it is possible to formulate the maximization of $P(\mathbf{q} \mid \mathbf{O}, \lambda)$ in a sequential manner so that it can be solved by dynamic programming methods such as the Viterbi algorithm (Forney, 1973).

Maximization of $P(\mathbf{q} \mid \mathbf{O}, \lambda)$ is equivalent to maximization of $P(\mathbf{q}, \mathbf{O} \mid \lambda)$ because $P(\mathbf{O} \mid \lambda)$ is not involved in the optimization process. From (7), we see that

$$\begin{aligned} &P(q_1, q_2, \ldots, q_t, \mathbf{O}_1, \mathbf{O}_2, \ldots, \mathbf{O}_t \mid \lambda) \\ &\quad = P(q_1, q_2, \ldots, q_{t-1}, \mathbf{O}_1, \mathbf{O}_2, \ldots, \mathbf{O}_{t-1} \mid \lambda) \cdot a_{q_{t-1}q_t} b_{q_t}(\mathbf{O}_t) \end{aligned} \tag{13}$$

If we define

$$\delta_t(i) \triangleq \max_{q_1, q_2, \ldots, q_{t-1}} P(q_1, q_2, \ldots, q_t = S_i, \mathbf{O}_1, \mathbf{O}_2, \ldots, \mathbf{O}_t \mid \lambda) \tag{14}$$

then the following recursion is true:

$$\delta_{t+1}(j) = \left[\max_i \delta_t(i) a_{ij} \right] b_j(\mathbf{O}_{t+1}) \tag{15}$$

The optimal state sequence is thus the one that leads to $\delta_T(q_T)$. This recursion is in a form suitable for application of the Viterbi algorithm, a detailed procedure for which is provided in Appendix B.

2.4. Speech Recognition Using Hidden Markov Models

The typical use of hidden Markov models in speech recognition is not very different from the traditional pattern-matching paradigm (Duda and Hart, 1973, Part I). Successful application of HMM methods usually involves the following steps: (1) define a set of L sound classes for modeling, such as phonemes or words; call the sound classes $V = \{v_1, v_2, \ldots, v_L\}$; (2) for each class, collect a sizable set (the training set) of labeled utterances that are known to be in the class; (3) based on each training set, solve the estimation problem to obtain a "best" model λ_i for each class v_i, $i = 1, 2, \ldots, L$; (4) during recognition, evaluate $P(\mathbf{O} \mid \lambda_i)$, $i = 1, 2, \ldots, L$, for the unknown utterance $\mathbf{O}$ and identify the speech that produced $\mathbf{O}$ as class v_j if

$$P(\mathbf{O} \mid \lambda_j) = \max_{1 \le i \le L} P(\mathbf{O} \mid \lambda_i) \tag{16}$$

Since the details of implementing an HMM recognizer are not essential to this chapter, we will omit them here. Interested readers should consult Jelinek et al. (1975) and Levinson et al. (1983) for more specifics related to individual applications. We will instead limit our discussion to several general issues related to hidden Markov model theory and implementation.

3. STRENGTHS OF THE METHOD OF HIDDEN MARKOV MODELS

As mentioned in the Introduction, the strengths of the hidden Markov model method lie in two broad areas: its mathematical framework and its implementational structure. In terms of the mathematical framework we discuss the method's consistent statistical framework and the way it provides straightforward solutions to related problems. In terms of the implementational structure, we discuss the inherent flexibility the method provides in dealing with various sophisticated speech recognition tasks and the ease of implementation, which is one of the crucial considerations in many practical engineering systems.

3.1. The Consistent Statistical Framework of the HMM Methodology

The foundation of the hidden Markov model methodology is built on the well-established field of statistics and probability theory. That is, the development of the methodology follows a tractable mathematical structure that can be examined and studied analytically.

The basic theoretical strength of the hidden Markov model is that it combines modeling of stationary stochastic processes (for the *short-time* spectra) and the temporal relationship among the processes (via a Markov chain) together in a well-defined probability space. The measure of such a probability space is defined by (8). This combination allows us to study these two separate aspects of modeling a dynamic process (like speech) using one consistent framework.

In addition, this combination of short-time static characterization of the spectrum within a state and the dynamics of change across states is rather elegant because the measure of (8) can be decomposed simply into a summation of the joint probability of **O**, the observation, and **q**, the state sequence, as defined by (7). The decomposition permits independent study and analysis of the behavior of the short-time processes and the long-term characteristic transitions. Since decoding and recognition are our main concerns, this also provides an intermediate level of decision that can be used to choose among alternate configurations of the models for the recognition task. This kind of flexibility with consistency is particularly useful for converting a time-varying signal such as speech, without clear anchor points that mark each sound change, into a sequence of (sound) codes.

3.2. The Training Algorithm for HMMs

Another attractive feature of hidden Markov models comes from the fact that it is relatively easy and straightforward to train a model from a given set of labeled training data (one or more sequences of observations).

When the maximum likelihood criterion is chosen as the estimation objective, i.e., maximization of $P(\mathbf{O} \mid \lambda)$ over λ, the well-known Baum–Welch algorithm is an iterative hill-climbing procedure that leads to at least a fixed-point solution, as explained in Section 2.2. If we choose the state-optimized (or decoded) likelihood defined by

$$L_\lambda(\bar{\mathbf{q}}) = \max_{\mathbf{q}} P(\mathbf{O}, \mathbf{q} \mid \lambda) \tag{17}$$

where

$$\bar{\mathbf{q}} = \arg\max_{\mathbf{q}} P(\mathbf{O}, \mathbf{q} \mid \lambda) \tag{18}$$

as the optimization criterion, the segmental k-means algorithm (Rabiner et al., 1986b; Juang and Rabiner, 1990), which is an extended version of the Viterbi training/segmentation algorithm (Jelinek, 1976), can be conveniently used to accomplish the parameter training task.

The segmental k-means algorithm, as can be seen from the objective function of (17), involves two optimization steps, namely, the segmentation step and the optimization step. In the segmentation step, we find a state sequence $\bar{\mathbf{q}}$ such that (17) is obtained for a given model λ and an observation sequence $\mathbf{O}$. Then, given a state sequence $\bar{\mathbf{q}}$ and the observation $\mathbf{O}$, the optimization step finds a new set of model parameters $\bar{\lambda}$ so as to maximize (17), i.e.,

$$\bar{\lambda} = \arg\max_{\lambda} \left\{ \max_{\mathbf{q}} P(\mathbf{O}, \mathbf{q} \mid \lambda) \right\} \tag{19}$$

Equation (19) can be rewritten as

$$\bar{\lambda} = \arg\max_{\lambda} \left\{ \max_{\mathbf{q}} [\log P(\mathbf{O} \mid \mathbf{q}, \lambda) + \log P(\mathbf{q} \mid \lambda)] \right\} \tag{20}$$

Note that $\max_{\lambda} [\log P(\mathbf{O} \mid \bar{\mathbf{q}}, \lambda) + \log P(\bar{\mathbf{q}} \mid \lambda)]$ consists of two terms that can be separately optimized, since $\log P(\bar{\mathbf{q}} \mid \lambda)$ is a function of only A, the state transition probability matrix, and $\log P(\mathbf{O} \mid \bar{\mathbf{q}}, \lambda)$ is a function of only B, the family of (intrastate) observation distributions. (We neglect the initial state probability for simplicity in presentation.) This separate optimization is the main distinction between the Baum–Welch algorithm and the segmental k-means algorithm. For a more detailed discussion of the segmental k-means algorithm, see Juang and Rabiner (1990).

These two training algorithms (Baum–Welch and segmental k-means) both result in well-formulation and well-behaved solutions. [For a theoretical comparison of the two methods in terms of likelihood differences and the state posteriori probability deviations, readers can consult Merhav and Ephraim (1991).] The segmental k-means algorithm, however, due to the separate optimization of the components of the model parameter sets, leads to a more straightforward (simpler with less computation and numerical difficulties) implementation, which will be discussed shortly.

The case of HMM training also extends to the choice of observation distributions. It is known (Liporace, 1982; Juang, 1985; Juang and Rabiner, 1985) that these algorithms can accommodate observation densities that are (1) strictly log-concave densities; (2) elliptically symmetric densities; (3) mixtures of distributions of the above two categories; and (4) discrete distributions. These choices of observation distribution in each state of the model allow accurate modeling of virtually unlimited types of data.

3.3. Modeling Flexibility

The flexibility of the basic HMM is manifested in three aspects of the model, namely model topology, observation distributions, and decoding hierarchy.

Many topological structures for HMMs have been studied for speech modeling. For modeling isolated utterances (i.e., whole words or phrases) we often use left-to-right models (Rabiner et al., 1983; Bakis, 1976) since the utterance begins and ends at well-identified time instants (except in the case of very noisy or corrupted speech) and the sequential behavior of the speech is well represented by a sequential HMM. For other speech modeling tasks, the use of ergodic models (Levinson, 1987) is often more appropriate. The choice of topological configuration and the number of states in the model often is a reflection of the a priori knowledge of the particular speech source to be modeled and is not in any way related to the mathematical tractability or implementational considerations.

In Section 3.2 we pointed out that the range of observation distributions that can be accommodated by well-developed training algorithms is rather large. There are no real analytical problems that make the use of any of this rather rich class of distributions impractical. Since speech has been shown to display quite irregular probability distributions (Jayant and Noll, 1984; Juang et al., 1985) in both waveform and spectral parameters, one indeed needs the freedom to choose an appropriate distribution model that fits the observation well and yet is easy to obtain.

In modeling spectral observations, we have found the use of mixture densities (Juang, 1985; Juang and Rabiner, 1985) beneficial. With $f_i(\cdot)$ denoting the kernel density function, the mixture density assumes the form

$$b(\mathbf{O}) = \sum_{i=1}^{M} c_i f_i(\mathbf{O}) \tag{21}$$

where c_i is the mixture component weights, $\sum_{i=1}^{M} c_i = 1$, and M is the number of mixture components. This mixture distribution function is used to characterize the distribution of the observations in each state. Such a density function has several interesting and important interpretations.

Figure 2 depicts graphically the result of mixing probability densities to represent the supported space. By varying the number of mixture components M, it can be shown that it is possible to approximate densities of virtually any shape (unimodal, multimodal, heavy-tail, etc.). Furthermore, if we approximate $b(\mathbf{O})$ by the principal value in the mixture, i.e.,

$$b'(\mathbf{O}) = \max_i c_i f_i(\mathbf{O}) \tag{22}$$

with proper scaling such that it is a legitimate density function, then the density can be considered as packed cells (probability in a multidimensional space). These two alternatives have been used in practical recognizer designs (Juang and Rabiner, 1985).

The expression of (22) can be further modified to take the form

$$b(\mathbf{O}) = \sum_i c_i I(\mathbf{O}, i) \tag{23}$$

where $I(\mathbf{O}, i)$ is the indicator function

$$I(\mathbf{O}, i) = \begin{cases} 1, & i = \arg\max_j f_j(\mathbf{O}) \\ 0, & \text{otherwise} \end{cases} \tag{24}$$

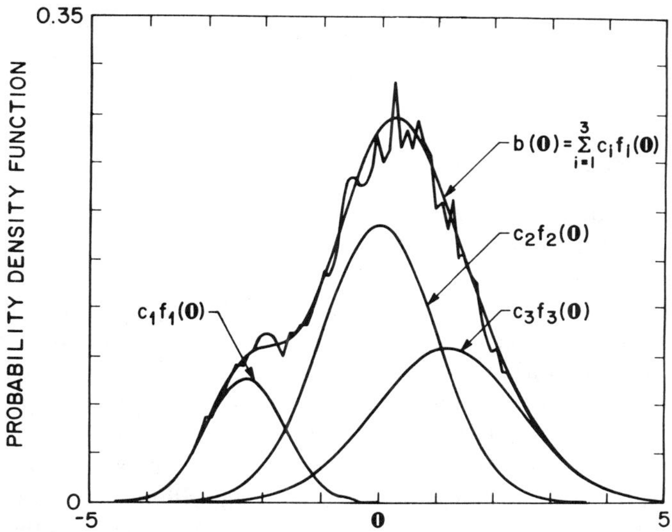

Figure 2 A mixture density function.

The density function $b(\mathbf{O})$ then becomes a discrete distribution in (23) and the resulting model is often called a discrete HMM. The mixture density thus allows us to bridge the gap between parametric and nonparametric distribution models.

Although we have focused on the choice of observation distributions within individual states of the model, one can apply the same ideas to the modeling of the *entire observation space* (lumped from all the classes to be modeled), as if it were a single state HMM source, before further modeling for classification. This flexibility of configuration gives rise to the so-called vector quantizer hidden Markov models (VQ-HMM) (Rabiner et al., 1083) semicontinuous HMM (Huang and Jock, 1989) and continuous HMM (Rabiner et al., 1984a; Poritz and Richter, 1986; Bahl et al., 1988a), all of which share the mathematical foundation described above.

It is important to stress that the choice of observation distributions also extends to the case of the hidden Markov model itself. One can form a hidden Markov model with each state characterized by another hidden Markov model; that is, $b_i(\mathbf{O})$ of each state, $i = 1, 2, \ldots, N$, can assume the form of an HMM probability measure as defined by (8). This principle is the basis of many of the subword unit–based speech recognition algorithms (Lee et al., 1988, 1989). This capability enhances the HMM flexibility even further since it allows manipulation of the decoding hierarchy, as well as modeling at various uncertainty levels inherent in the speech signal, as a result of proper use of the probability defined by (7) and (8).

3.4. Ease of Implementation

Two areas of concern in the implementation of any algorithm are the potential for numerical difficulties and the computational complexity. The hidden Markov model is no exception.

The potential numerical difficulties in implementing HMM systems come from the fact that the terms in the HMM probability measure of (7) and (8) are multiplicative. A direct outcome of the multiplicative chain is the need for excessive dynamic range in numerical values in order to prevent underflow or overflow problems in digital implementations.

There are two reasonable ways to avoid such numerical problems: numerical scaling and interpolation. The scaling algorithm, well documented in Levinson et al. (1983) and Juang and Rabiner (1985), alleviates the dynamic range problem by normalizing the forward and backward probabilities (Appendix A) at each time instant before they cause overflow or underflow. The scaling algorithm is naturally blended in the forward-backward procedure, which is the core of the solution to the evaluation problem and the estimation problem above. However, normalization alone does not entirely solve the numeric problems which result from insufficient data support. Insufficient data support can cause spurious singularities in the model parameter space. One may resort to parameter smoothing and interpolation to alleviate such numerical singularity problems. Although the topic of model parameter smoothing and interpolation is still one of the open issues in HMM theory, some trivial measures such as setting a numeric floor to prevent singularity are often found beneficial and are straightforward to implement (Rabiner et al., 1986a; Lee et al., 1991a).

With the understanding of the relationship between the trellis structure (Juang, 1984; Rabiner and Juang, 1986) and the decoding structure of the hidden Markov model, as discussed in the evaluation problem, we are able to apply the HMM to many complicated problems without much concern as to computational complexity. An analysis of the computational requirements in a typical speech recognition application may help to illustrate this point. For isolated word modeling, with a vocabulary of $V = 100$ words, a typical configuration is to use $N = 5$-state, $M = 5$-component mixture density HMMs for each word model. The observation vector is usually on the order of $D = 20$ elements. For a word utterance of duration $T = 50$ frames, the computation for decoding amounts to $V \cdot NT$ computations of local observation density and VN^2T combinatorics (typically a comparison and an addition). Since the computation of local observation density is usually the most expensive, we can approximate the total computation as $VNT(2DM)$, where we assume we can evaluate (21) by $2DM$ operations in the case where $f_i(\mathbf{O})$ is a Gaussian density with diagonal covariance. Thus the total computation is about $2VNTDM$ operations (where an operation is a multiplication and an addition). Hence for the values given above for V, N, T, D, and M we need about 5 million multiplication-addition operations to match the utterance to each of $V = 100$ word models. Since we assume the input is 50 frames long with 10 ms per frame, this implies that we need about 10 million multiply-add operations per second to achieve real-time performance. A typical off-the-shelf DSP processor of 10 million floating-point operations per second (MFLOPS) would thus be able to support a 100-word recognition vocabulary. Even as recognition vocabularies increase to size 1000 or more, the required processing often remains somewhere in the same range as above because of grammatical constraints which limit the average number of words following a given word to somewhere on the order of 100 words. [This effect is called the word average branching factor or perplexity and has been shown to be on the order of 100 for several large-vocabulary recognition tasks (Chow et al., 1987; Bahl et al., 1980).]

4. HIDDEN MARKOV MODEL ISSUES FOR FURTHER CONSIDERATION

The basic theory of hidden Markov modeling has been developed over the last two decades. As explained above, the theory includes the definition of the probability measure, the evaluation algorithm, the decoding algorithm, several parameter estimation procedures which are readily interpreted in terms of signal modeling, and a rather rigorous treatment of the HMM measure space (Petrie, 1969). However, when applied to speech recognition, there are some remaining issues to be resolved. We begin with a discussion of parameter estimation criteria as applied to optimal decoding of the observation sequence.

4.1. Parameter Estimation Criteria

The original HMM parameter estimation was formulated as an inverse problem: given an observation sequence $\mathbf{O}$ and an assumed source model, estimate the (source) parameter set λ which maximizes the probability that $\mathbf{O}$ was produced by the source. The method of maximum likelihood (ML), which seeks to maximize $P(\mathbf{O} \mid \lambda)$, is optimal according to this criterion. The Baum–Welch reestimation algorithm, as described previously, is a convenient straightforwardly implementable solution to the maximum likelihood HMM estimation problem.

The method of maximum likelihood, however, need not be optimal in terms of minimizing classification error rate in recognition tasks in which the observation $\mathbf{O}$ is said to be produced by one of the many (say L) source classes, $\{C_i\}_{i=1}^{L}$. This is the classic problem in isolated and connected word recognition tasks. To achieve the minimum classification error rate, the classical Bayes rule (Duda and Hart, 1983) requires that

$$C^*(\mathbf{O}) = C_i \quad \text{if } C_i = \arg\max_j P(C_j \mid \mathbf{O}) \tag{25}$$

where $\mathbf{O}$ is the unknown observation, which is to be classified into (recognized as) one of the L classes; $C^*(\cdot)$ denotes the decoded class of $\mathbf{O}$; and $P(C_j \mid \mathbf{O})$ is the (true) a posteriori probability of C_j given the observation $\mathbf{O}$. The decision rule of (25) is the well-known MAP decoder. Since

$$P(C_j \mid \mathbf{O}) = \frac{P(\mathbf{O} \mid C_j)P(C_j)}{P(\mathbf{O})} \tag{26}$$

and since $P(\mathbf{O})$ is a constant in this decoding context, the decision rule of (25) is often written as

$$C^*(\mathbf{O}) = C_i \quad \text{if } C_i = \arg\max_j P(\mathbf{O} \mid C_j)P(C_j) \tag{27}$$

in terms of the class prior $P(C_j)$ and the conditional probability $P(\mathbf{O} \mid C_j)$. This factorization, i.e., writing the joint probability as a product of an a priori probability and a conditional probability, is particularly appropriate in speech recognition, as will be explained shortly.

It is clear that the difficulty in minimizing misclassification rate stems from the fact that both the prior distribution $P(C_i)$ and the conditional distribution $P(\mathbf{O} \mid C_i)$ are generally unknown and have to be estimated from a given, finite, training set. For simple tasks, such as isolated word recognition where each word is a class and each word is represented by a unique model, the difficulties in implementing the MAP rule are minimal. However, for recognition of naturally spoken sentences from a large vocabulary (e.g., 1000 or more words) there are certain practical constraints that make MAP estimation difficult. These constraints have to do with the amount of available training data and how it is segmented and labeled, as well as with the assumptions made about how the actual speech source fits the model and the assumptions made in optimizing model parameters.

First, in order to obtain reliable estimates of $P(C_i)$ and $P(\mathbf{O} \mid C_i)$, we generally need a training set of sufficient size (i.e., large enough to sample adequately all relevant class interactions). When the vocabulary is large, this is difficult if not impossible to achieve. For example, suppose the vocabulary has 10,000 words, each of which represents a class. If we assume that 10 occurrences each, on average, are needed for reliable estimates of both $P(C_i)$ and $P(\mathbf{O} \mid C_i)$, this amounts to a total of $10{,}000 \times 10 = 10^5$ word utterances. (This is equivalent to $\sim$ 14 hours of speech, assuming two words per second.) Therefore, some other strategy is required to estimate $P(C_i)$ and $P(\mathbf{O} \mid C_j)$ reliably from a smaller training set. One possibility is to choose a set of classes to represent, instead of words, a reduced set of subword units, e.g., phonemes. This greatly reduces the requirements on the amount of training data since the number of subword classes essentially does not grow with the size of vocabulary. A consequence of using subword unit classes to represent the basic set of speech sounds is that an estimate of class probability, $P(C_i)$, can be obtained directly from a lexical description of the words, independent of the *spoken* training set. Furthermore, we can estimate $P(\mathbf{O} \mid C_i)$ for words from each of the subword units in the lexical entry for the words. This type of decomposition, namely breaking large sound classes like words and phrases into smaller ones like subword units, leads to one particular problem in HMM speech recognition: given an independent (and possibly incorrect) estimate of word probability, $P^*(C_i)$, how do we estimate $P^*(\mathbf{O} \mid C_i)$ such that the Bayes minimum error rate is attained? (Although we have used the example of decomposing words into subword classes, the same concept applies to higher levels, i.e., decomposing phrases into words.) Note that in this discussion the association between the training data $\mathbf{O}$ and the (subword) class is assumed to be known a priori (often as a result of hand labeling). We call this case the complete label case. Typical examples of complete label systems include most isolated word and connected word tasks and the case of hand-segmented and hand-labeled continuous speech recognition.

The decomposition described above (namely from typed text to words or subword units), while circumventing some of the training problems, leads to other problems, e.g., the need for detailed segmentation and labeling of the speech to correspond to the basic speech classes. For large-vocabulary continuous speech recognition, where the training data are extensive, this labor-intensive task cannot realistically be accomplished. Instead, we often have to rely on only partial knowledge of the data. For example, we generally know or assume we know the phoneme sequence of the words in the string, but not the direct relationship between each phoneme and the corresponding segment of speech. In these cases, we need some other criterion to guide the parameter estimation. We call this case the incomplete label case, and typical examples are the problems of estimating models of subword speech units from continuous speech and those of words in connected word tasks without prior word segmentation.

Another problem with the estimation procedure arises when the distribution, particularly the conditional probability $P(\mathbf{O} \mid C_i)$ is postulated to be the same as $P(\mathbf{O} \mid \lambda_i)$, where λ_i is, in the present case, a hidden Markov model to be estimated. Since we generally choose the *form* of the observation distribution before we have any solid knowledge of the characteristics of the source in each HMM state, there is the risk of a serious mismatch between the chosen observation distribution model and the actual data source. A similar mismatch potential exists in the Markovian structure of the model; that is, speech signals need not be Markovian. We therefore need to provide a mechanism for compensating such potential errors in modeling. We call this problem the model mismatch case.

It is important to note that if the chosen class model (the conditional as well as the a priori probability) is indeed the correct one, the method of maximum likelihood will lead to the asymptotically best recognition performance (in terms of highest correct classification rate) while allowing classes to be added or removed from the system specification without the need for complete retraining of all class reference models. These assumptions, however, are rarely true in practice. The above examples attempt to address these unrealistic assumptions from the viewpoint of modeling for recognition, aiming at the minimum Bayes error rate. The idea is to either obtain extremely accurate reference models or allow compensation for incorrect a priori probabilities in the overall modeling. Another alternative, called *corrective training*, tries to minimize the recognition error rate directly by identifying, during training, sources that lead to recognition errors (or near errors) and applying some type of correction rule to the parameter estimation to reduce the probability of these errors or near misses. This type of corrective training procedure intentionally seeks model parameter estimates which give suboptimal likelihood scores in order to reduce recognition error rate on the training set. Unlike the above three cases, corrective training uses the HMM as a form of *discriminant function* and is not concerned about modeling accuracy per se but more with recognition errors as they appear during training.

We now provide a more complete discussion of the complete label case, the incomplete label case, and the model mismatch case.

4.1.1. The Complete Label Case

In the complete label case, the association between the training data $\mathbf{O}$ and the class C_i is known a priori during training. This is a typical situation in classical pattern classification theory, and all the concerns about supervised learning (Duda and Hart, 1973) apply. However, what is unique in the current speech recognition problem is the use of a prescribed class prior model $P(C_i)$.

A detailed account of the issues involved in the complete label case is given in [54]. Consider the set of L HMM class models, $\Lambda = \{\lambda_i\}_{i=1}^{L}$, which are used to model the classes $\{C_i\}_{i=1}^{L}$, respectively. The complete set of models, Λ, defines a probability measure

$$P_\Lambda(\mathbf{O}) = \sum_i P_\Lambda(\mathbf{O} \mid C_i) P_\Lambda(C_i) \tag{28}$$

The notation of (28) is slightly different from that used earlier because we are explicitly using both classes and associated models simultaneously for classification purposes. Assume the prior $P_\Lambda^*(C_i) = P_0(C_i)$ is given or obtained independently of the *spoken* training set $\{\mathbf{O}^{(i)}\}$ where $\mathbf{O}^{(i)}$ are the portion of the training data that are labeled as C_i. For this case, Nadas et al. (1988) proposed the use of a slightly different training measure, namely the conditional maximum likelihood estimation (CMLE), obtained by

$$\Lambda_{\text{CMLE}}^* = \arg\max_\Lambda \prod_i P_\Lambda(C_i \mid \mathbf{O}^{(i)}) \tag{29}$$

The motivation for choosing this estimator is that the assumed model, $P_\Lambda(\mathbf{O}^{(i)} \mid C_i)$ or $P(\mathbf{O}^{(i)} \mid \lambda_i)$, *may be inappropriate for* $P(\mathbf{O}^{(i)} \mid C_i)$, so that the asymptotic optimality of the standard ML estimator is no longer guaranteed. Furthermore, the prior model $P_0(C_i)$ could be incorrect (poor estimate etc.), so that the true MAP decoder of (25) is virtually impossible to implement.

When the class a priori probability $P_0(C_i)$ is obtained independently of $\{\mathbf{O}\}$ and is not part of the model Λ to be optimized, then

$$\begin{aligned} L_{\text{CMLE}}^* &= \arg\max_\Lambda \prod_i P_\Lambda(C_i \mid \mathbf{O}^{(i)}) \\ &= \arg\max_\Lambda \sum_i \log \frac{P_\Lambda\left(\mathbf{O}^{(i)}, C_i\right)}{P_\Lambda\left(\mathbf{O}^{(i)}\right) P_0(C_i)} \\ &= \Lambda_{\text{MMI}}^* \end{aligned} \tag{30}$$

which is the well-known maximum mutual information (MMI) estimator. [Note that the second equality above comes from the fact that the logarithm is monotonic and the additive term $\log P_0(C_i)$ does not affect the maximum result.]

The effect of conditional maximum likelihood estimation, in terms of class prior robustness, i.e., uncertain or incorrect $P_\Lambda(C_i)$, can best be illustrated by the following example from Nadas et al. (1988). Suppose in the training data there are N_i occurrences of the class C_i and among these occurrences, the (discrete) observation **O** occurs jointly with class C_i $N_{i,\mathbf{O}}$ times. Then the maximum likelihood estimates of the a priori and the conditional probabilities are, respectively,

$$P^*_{\mathrm{ML}}(C_i) = \frac{N_i}{\sum_{j=1}^{L} N_j} \tag{31}$$

$$P^*_{\mathrm{ML}}(\mathbf{O} \mid C_i) = \frac{N_{i,\mathbf{O}}}{N_i} \tag{32}$$

The decoder that uses these estimates then decides that an observation **O** belongs to class C_i if

$$\frac{N_{i,\mathbf{O}}}{N_i} \frac{N_i}{\sum_{j=1}^{L} N_j} = \frac{N_{i,\mathbf{O}}}{\sum_{j=1}^{L} N_j} = \max_k \frac{N_{k,\mathbf{O}}}{\sum_{j=1}^{L} N_j} \tag{33}$$

which is simply

$$N_{i,\mathbf{O}} = \max_k N_{k,\mathbf{O}} \tag{34}$$

This is optimal when the total number of observations approaches infinity. However, when the a priori probability $P(C_i)$ is prescribed as $P_0(C_i)$, rather than $P^*_{\mathrm{ML}}(C_i)$, the dependence of the decision rule on the a priori probability becomes obvious if we continue to use $P^*_{\mathrm{ML}}(\mathbf{O} \mid C_i)$; that is, (33) becomes

$$\frac{N_{i,\mathbf{O}}}{N_i} \cdot P_0(C_i) = \max_j \frac{N_{j,\mathbf{O}}}{N_j} P_0(C_j) \tag{35}$$

The CMLE criterion of (29), on the other hand, leads to a set of equations

$$P^*_{\mathrm{CMLE}}(\mathbf{O} \mid C_i) = \frac{N_{i,\mathbf{O}}}{N_{\mathbf{O}}} \frac{\sum_{k=1}^{L} P^*_{\mathrm{CMLE}}(\mathbf{O} \mid C_k) P_0(C_k)}{P_0(C_i)} \tag{36}$$

where $N_{\mathbf{O}} = \sum_{j=1}^{L} N_{j,\mathbf{O}}$. This is obtained with the variational method by defining the Lagrangian for maximization as

$$\sum_{i=1}^{L} \sum_{\mathbf{O}} N_{i,\mathbf{O}} \log \frac{P(\mathbf{O} \mid C_i) P_0(C_i)}{\sum_{k=1}^{L} P(\mathbf{O} \mid C_k) P_0(C_k)} + \sum_{i=1}^{L} \theta_i \left(1 - \sum_{\mathbf{O}} P(\mathbf{O} \mid C_i) \right)$$

where θ_i, $i = 1, 2, \ldots, L$ are the Lagrange multipliers. Setting the derivative of the Lagrangian with respect to the probability parameters equal to zero, we find that $\sum_i \theta_i P(\mathbf{O} \mid C_i) = 0$ and that if $\theta_i \equiv 0$ is a solution, Eq. (36) follows as a result of unconstrained maximization of the first term in the Lagrangian.

That is, let

$$\Omega = \sum_{i=1}^{L}\sum_{\mathbf{O}} N_{i,\mathbf{O}} \log \frac{P(\mathbf{O} \mid C_i)P_0(C_i)}{\sum_{k=1}^{L} P(\mathbf{O} \mid C_k)P_0(C_k)}$$

$$= \sum_{i=1}^{L}\sum_{\mathbf{O}} N_{i,\mathbf{O}} \log P(\mathbf{O} \mid C_i) + \sum_{i=1}^{L}\sum_{\mathbf{O}} N_{i,\mathbf{O}} \log P_0(C_i)$$

$$- \sum_{\mathbf{O}} N_{\mathbf{O}} \log \sum_{k=1}^{L} P(\mathbf{O} \mid C_k)P_0(C_k)$$

Then

$$\frac{d\Omega}{dP(\mathbf{O} \mid C_i)} = \frac{N_{i,\mathbf{O}}}{P(\mathbf{O} \mid C_i)} - \frac{N_{\mathbf{O}} P_0(C_i)}{\sum_{k=1}^{L} P(\mathbf{O} \mid C_k)P_0(C_k)}$$

which is set to zero for $P(\mathbf{O} \mid C_i) = P^*_{\text{CMLE}}(\mathbf{O} \mid C_i)$, resulting in (36). Plugging (36) into the MAP rule of (25), we decide an unknown $\mathbf{O}$ to be from class C_i if

$$\frac{N_{i,\mathbf{O}}}{N_{\mathbf{O}}} \frac{\sum_{k=1}^{L} P^*_{\Lambda}(\mathbf{O} \mid C_k)P_0(C_k)}{P_0(C_i)} \cdot P_0(C_i)$$

$$= \max_j \frac{N_{j,\mathbf{O}}}{N_{\mathbf{O}}} \frac{\sum_{k=1}^{L} P^*_{\Lambda}(\mathbf{O} \mid C_k)P_0(C_k)}{P_0(C_j)} \cdot P_0(C_j) \tag{37}$$

which can be reduced to

$$N_{i,\mathbf{O}} = \max_j N_{j,\mathbf{O}} \tag{38}$$

which is independent of the a priori probability $P_0(C_i)$. In this sense, it was concluded by Nadas et al. (1988) that if $P_0(C_i)$ is not the *true* prior distribution (because of bad assumptions or estimation errors), the MLE will implement a suboptimal decoder (35) while the CMLE of (29) will lead to the correct MAP decoding results (asymptotically) because of the compensation built into the estimate of $P(\mathbf{O} \mid C_i)$.

Although the criterion of conditional maximum likelihood or maximum mutual information (for training) is attractive in terms of compensation problems associated with MAP decoding, some important concerns remain unanswered. The most immediate concern in using this criterion is the lack of a convenient and robust algorithm to obtain the estimate $P^*_{\text{CMLE}}(\mathbf{O} \mid C_i)$. In many practical situations, the procedure for obtaining the solution may be far more complicated than (36) would imply, particularly when HMMs are involved and the observation distribution is not of a discrete type. Previous attempts (Bahl et al., 1986; Brown, 1987) at using the MMI criterion have not produced an estimation procedure that is guaranteed to converge to an optimal solution either. Also, even though the above example demonstrates the robustness of CMLE against errors in the class prior, it is still not clear if CMLE is more robust than MLE when the form of the model (i.e., HMM) is incorrect for the speech source. Another problem is that the conditional maximum likelihood estimate has a larger variance than the maximum likelihood estimate. This therefore undermines the potential gain in offsetting the sensitivity due to the independently obtain $P_0(C_i)$ possibly being inaccurate when the decoder based on finite training data is used on testing data not included in the training set. In the case of insufficient training data (as is almost always the situation), there are other problems with practical implementations of the procedure. Although the motivation to have a criterion specifically aimed at minimizing the recognition error is a good one, the goal has not yet been achieved by CMLE methods.

4.1.2. The Incomplete Label Case

The case of incomplete labeling because of (1) practical difficulties in labeling and segmenting any large continuous speech data base and (2) the inherent ambiguity among different sound classes in terms of both class definition (i.e., inherent similarities in sound classes) and time uncertainty as realized in speech signals (i.e., it is not clear that exact segmentation boundaries between adjacent sounds universally exist). For the case of decomposing an isolated word into a prescribed phoneme sequence, we usually have a lexical description of the word in terms of the phoneme sequence (as described in a dictionary) and the spoken version of the word without any explicit time segmentation into corresponding phonemes. Under these conditions (which are typical for speech recognition), training of the prescribed subword is rather difficult due to the lack of a definitive labeling relating subword classes to specific intervals of speech. After all, if we do not know for sure that a training token $\mathbf{O}$ is in sound class C_i, the likelihood function $P(\mathbf{O} \mid C_i)$ cannot be defined, not to mention optimized.

There are several ways to handle the problem of incomplete labeling based on the idea of embedded decoding. One way is to retain the constraints of the known class sequence (in the previous example, the phoneme sequence) and solve for the "optimal" set of class models sequentially. Another alternative is to solve for the models of the sound classes simultaneously with the decoding into classes.

Consider first the case where we have partial knowledge; that is, a given training token $\mathbf{O}$ is known to correspond to a sequence of u class labels $h = (h_1, h_2, \ldots, h_u)$ (as determined from dictionary lookup of the words realized by **O** (where $h_j \in \{C_i\}_{i=1}^{L}$. The goal is to obtain the L models $\Lambda = \{\lambda_i\}_{i=1}^{L}$ corresponding to the L sound classes $C = \{C_i\}_{i=1}^{L}$, using the number of segments u and the class labels h as hard constraints in the decoding. The procedure begins by assuming a uniform segmentation $\mathbf{X} = (x_1, x_2, \ldots, x_u)$ of **O** into u speech intervals with the ith interval, x_i, corresponding to the sound class h_i. (Note that each interval is a sequence of spectral vectors.) Based on this initial segmentation, the likelihood functions are defined and individual class models are obtained by maximum likelihood (via the forward-backward procedure). For example, if $h_i = C_j$, then the segment x_i is used to define a likelihood function $P(x_i \mid C_j)$ for maximization. As a result, a set of sound unit models are created. With the new set of unit models, we further refine the segmentation of **O** into **X** (again assuming exactly u segments) by optimally decoding **O** using the Viterbi algorithm. This leads to an improved segmentation of **O**, which can then be used to give a refined set of sound models. This process is iterated until a reasonably stationary segmentation of **O** into intervals **X** is obtained.

This constrained decoding approach to the incomplete label case has been used in explicit acoustic modeling of phonemic classes (Lee et al., 1989, 1990) with good success. There are, however, some theoretical shortcomings to the method. One problem is that the segmentation/decoding results will be different for different numbers of segments, u, in the given string. Thus even the simple expedient of having multiple dictionary definitions of words can lead to inconsistent segmentation into sound classes. Since the criterion for decoding is conditioned on the "suboptimal" segmentation (due to the constraint of a designated number of segments), the overall model parameter estimation does not obey any global optimality principle similar to that of maximum likelihood. Although the procedure is practical, there appears to be room for theoretical improvement, which in turn may prove beneficial in practical implementations.

An alternative and probably more thorough way of handling the incomplete label problem is to combine the ideas of segmentation, decoding, and modeling and try to solve a large network for both the class models and the segmentations simultaneously, without any prescribed label sequence constraint. In this approach, the above iterative procedure, i.e., recursively and interleavingly improving data segmentation and model estimation, is again used. The key difference, in contrast to the constrained decoding approach discussed previously, is that the label sequence constraint is no longer retained in each iteration. This allows globally optimal decoding of **O**, given a set of sound unit models. This advantage, however, comes at the price of giving up the convenient and readily available lexical representation, h, from the dictionary. The architectural considerations of this approach are as follows.

As previously pointed out, the goal of signal modeling is to come up with a parsimonious, consistent representation of the source that displays a certain kind of variation in the observed output. Speech signals are known to have inherent time uncertainty due to speaking style and individual speaker characteristics, etc. This justifies the use of HMM for an individual sound class. The reason is that the incomplete data problem formulation (Juang, 1984; Dempster et al., 1977; Rabiner and Juang, 1986), based on which the fundamental HMM probability measure of (8) is defined, is particularly appropriate when explicit knowledge of the exact position (labeling) of the particular sound in the middle of an utterance is lacking. On the other hand, speech is a linear code (Chao, 1968) in the sense that decoded symbols come out one after another and no two symbols can appear at the same time. Therefore, the objective of optimizing the quantity defined by (7) will have to be accomplished during the recognition process. Thus, the problem in this alternative approach is twofold: (1) how to break the observation sequences into proper sound segments which can then be accurately modeled by HMMs, and (2) how to construct a global model, in lieu of the convenient phoneme set, such that decoding of an utterance in terms of the new sound unit models can be consistently achieved. This problem can be solved by the acoustic segment method (Lee et al., 1988, 1989). To summarize, let Λ be a super HMM, $\Lambda = (A_\Lambda, \{\lambda_i\}, \pi_\Lambda)$, where the set of sound unit HMMs $\{\lambda\}$ is used to characterize the observations in each state of Λ, respectively. The state-optimized likelihood is

$$L_\Lambda(\bar{\mathbf{q}}) = \max_{\mathbf{q}} P(\mathbf{O}, \mathbf{q} \mid \Lambda)$$

and the overall optimization criterion involves the two maximization steps:

$$\lambda^* = \arg\max_{\Lambda} L_\Lambda(\bar{\mathbf{q}}) = \arg\max_{\Lambda} \left\{ \max_{\mathbf{q}} P(\mathbf{O}, \mathbf{q} \mid \Lambda) \right\} \tag{39}$$

The segmental k-means algorithm (Rabiner et al., 1986b, Juang and Rabiner, 1990) can now be generalized to handle the task of finding λ^*, thereby providing the set $\{\lambda_i^*\}$. This new set $\{\lambda_i^*\}$ and the resulting state sequence $\bar{\mathbf{q}}$, after the segmental k-means algorithm converges, provide both the set of sound unit models and the string segmentation simultaneously and fully automatically.

4.1.3. Model Mismatch Issues

In speech modeling, we often choose (assume) the form of model before we actually know enough about the characteristics of the source. When we choose $P(\mathbf{O} \mid \lambda)$ to model $P(\mathbf{O} \mid C)$ for class C and perform maximum likelihood estimation, the optimality in decoding is meaningful only when **O** is indeed generated by the source λ. When the actual source is inconsistent with λ, we need to either improve $P(\mathbf{O} \mid \lambda)$ based on **O** or revise the decoding rule in some way. Here we discuss the first possibility.

Consider two hypotheses, H_1 and H_2, that observation **O** is from one of two statistical populations with probability measure μ_1 and μ_2, respectively. Assume μ_1 has a generalized probability density function f_i, i.e.,

$$d\mu_i(\mathbf{O}) = f_i(\mathbf{O})d\zeta(\mathbf{O}), \qquad 1 \leq i \leq 2 \tag{40}$$

where $\zeta(\mathbf{O})$ is the kernel [61]. Then

$$\log \frac{f_1(\mathbf{O})}{f_2(\mathbf{O})} = \log \frac{P(H_1 \mid (\mathbf{O})}{P(H_2 \mid \mathbf{O})} - \log \frac{P(H_1)}{P(H_2)}$$

is the information in **O** for discrimination in favor of H_1 against H_2. If $\{\mathbf{O}\}$ is a training set, the mean information for discrimination in favor of H_1 against H_2 for μ_1 given $\{\mathbf{O}\}$ is then

$$I(1:2;\{\mathbf{O}\}) = \frac{1}{\mu_1(\{\mathbf{O}\})} \int_{\{\mathbf{O}\}} f_1(\mathbf{O}) \log \frac{f_1(\mathbf{O})}{f_2(\mathbf{O})} d\zeta(\mathbf{O}) \tag{41}$$

which is sometimes called the Kullback–Leibler number, cross entropy, I-divergence, or *discrimination information* (Good, 1963; Hobson and Cheng, 1973; Johnson, 1979). When $\{\mathbf{O}\}$ is the entire sample space, we have

$$I(f_1 : f_2) = I(1:2) = \int f_1(\mathbf{O}) \log \frac{f_1(\mathbf{O})}{f_2(\mathbf{O})} d\zeta(\mathbf{O}) \tag{42}$$

We can use discrimination information in HMMs as follows (Ephraim et al., 1989). Let $R = (R_1, R_2, \ldots, R_T)$ be a sequence of constraints associated with the observation sequence $\mathbf{O} = (\mathbf{O}_1, \mathbf{O}_2, \ldots, \mathbf{O}_T)$ which we attempt to model. For example, R could be considered as the autocorrelation of the actual speech data. Let $\Omega(R)$ be the set of all models (probability distributions) which satisfy the constraints R. The minimum discrimination information (MDI) approach to HMM modeling is to find the HMM parameter set λ which minimizes the minimum discrimination information defined by

$$\nu(R,P) \triangleq \inf_{Q \in \Omega(R)} I(Q:P) = \inf_{Q \in \Omega(R)} \int q(\mathbf{O}) \log \frac{q(\mathbf{O})}{p(\mathbf{O} \mid \lambda)} d\mathbf{O} \tag{43}$$

The estimation criterion is thus the minimum over λ of the MDI of (43). Note that in (43) we have assumed both distributions P and Q have generalized probability density functions, denoted by $p(\cdot\lambda)$ and $q(\cdot)$, respectively, Clearly, the idea of (minimum) MDI model design is to allow a search into a general set of models (distributions) Ω with the HMM measure $P(\mathbf{O} \mid \lambda)$ as a contrasting reference so that the form of the chosen model $P(\cdot \mid \lambda)$ and what is revealed in the given data $\mathbf{O}$ can be better matched under the MDI measure.

MDI modeling involves two steps. First, given an HMM model $p(\cdot \mid \lambda)$, we find the solution, q, to the minimum discrimination information problem of (43). Then, with q given, we optimize the parameter values λ such that $\nu(R,P)$ is minimized.

The MDI modeling criterion carries some information theoretic justification and can be related to the MMI approach (Ephraim and Rabiner, 1988). However, there remain several unresolved issues with this method. The objective of improving the match between the observations $\mathbf{O}$ and the model λ is embedded in $\nu(R,P)$, which is computed for all $Q \in \Omega(R)$ with respect to the *chosen* model $P(\cdot \mid \lambda)$. Since $P(\cdot \mid \lambda)$ is often not a good measure of how well the model matches the data because of model mismatch, the goodness of the MDI probability density function (pdf), q, which minimizes $\nu(R,P)$, is difficult to measure in recognition tasks, not to mention the effects on classification errors. Furthermore, the nice computational structure of the HMM that makes possible the modeling of long training sequences is not preserved in the MDI pdf, q, resulting in exceptionally high computational complexity. In practice, some approximations are made to reduce the complexity problem (Ephraim et al., 1989). However, the problem of providing a better match between the data and the model through the information theoretic MDI measure is not completely solved. To date, it has not been demonstrated that the MDI model design procedure can bring about significant improvements in recognition performance (Ephraim et al., 1989). It is also not clear whether MDI (for model correctness) and CMLE/MMIE (for robust decoding) can be easily jointly formulated for an improved recognizer design.

4.1.4. *Corrective Training*

As explained earlier, the minimum Bayes risk or error rate is the theoretical recognizer performance bound, conditioned on the exact knowledge of both the prior and the conditional distributions. When both distributions are not known exactly, and the classifier needs to be designed based on a finite training set, there are several ways to try to reduce the error rate. One method is based on the theoretical link between discriminant analysis and distribution estimation (Duda and Hart, 1973). The idea here is to design a classifier (discriminant function) such that the minimum classification error rate is attained on the training set. In particular, we wish to design a classifier that uses estimates of $P(C_i)$ and $P(\mathbf{O} \mid C_i)$ and that achieves a minimum error rate for the training set in the same way a discriminant function is designed. The reason for using the HMM statistical framework in $P(\mathbf{O} \mid \lambda_i)$ for modeling $P(\mathbf{O} \mid C_i)$, as opposed to other discriminant functions, is to exploit the strengths of the HMMs, namely consistency, flexibility and computational ease, as well as their ability to generalize classifier performance to independent (open) data sets. The generalization capability of HMMs as discriminant functions is somewhat beyond the scope of this chapter and will not be discussed here. In the following, we focus on the issue of treating the estimation of the distributions $P(C_i)$ and $P(\mathbf{O} \mid \lambda_i)$ as a discriminant function design to attain the minimum error rate.

Bahl et al. (1988b) were the first to propose an error-correction strategy, which they named corrective training, to deal specifically with the misclassification problem. Their training algorithm was motivated by analogy with an error-correction training procedure for linear classifiers (Nilsson, 1965). In their proposed method, the observation distribution is of a discrete type, $B = [b_{ij}]$, where b_{ij} is the probability of observing a vector quantization code index (acoustic label) j when the HMM source is in state i. Each b_{ij} is obtained via the forward-backward algorithm as the weighted frequency of occurrence of the code index (Rabiner et al., 1983). The corrective training algorithm (Bahl et al., 1988b) works as follows. First, use a labeled training set to estimate the parameters of the HMMs $\Lambda = \{\lambda_i\}$ with the forward-backward algorithm. For each utterance $\mathbf{O}$, labeled as C_k, for example, evaluate $P(\mathbf{O} \mid \lambda_k)$ for the correct class C_k and $P(\mathbf{O} \mid \lambda_l)$ for each incorrect class C_l. (The evaluation of likelihood for the incorrect classes need not be exhaustive.) For every utterance where $\log P(\mathbf{O} \mid \lambda_l) > \log P(\mathbf{O} \mid \lambda_k) - \delta$, where δ is a prescribed threshold, modify λ_k and λ_l according to the following mechanism: (1) apply the forward-backward algorithm to obtain estimates b'_{ij} and b''_{ij}, using the labeled utterance $\mathbf{O}$ only, for the correct class C_k and incorrect class C_l, respectively; (2) modify the original b_{ij} in λ_k to $b_{ij} + \gamma b'_{ij}$ and the b_{ij} in λ_l to $b_{ij} - \gamma b''_{ij}$. When the state labels are tied for certain models, this procedure is equivalent to replacing the original b_{ij} by $b_{ij} + \gamma(b'_{ij} - b''_{ij})$. The prescribed adaptation parameter γ controls the "rate of convergence," and the threshold δ defines the "near-miss" cases. This corrective training algorithm therefore focuses on the parts of the model that are most important for word discrimination, a clear difference from the maximum likelihood principle.

Although Bahl et al. reported that the corrective training procedure worked better (in isolated word recognition tasks) than models obtained using the maximum mutual information or the conditional maximum likelihood criterion, the lack of a rigorous analytical foundation for the algorithm is one problem. Without a better theoretical understanding of the algorithm, the appeal of the method is primarily experimental. Other attempts to design

HMMs for minimum error rate or some form of class separation include the work by Ljolje et al. (1990) and by Sondhi and Roe (unpublished).

Several other forms of discriminative training were also proposed by Katagiri et al. (in preparation) who combined the adaptive learning concept in learning vector quantizer (LVQ) design (Kohonen, 1986) and the corrective training method described above, leading to a framework for the analysis of related training/learning ideas. Although this work is still in progress, the key result is that these adaptive learning methods can be formulated as a general risk minimization procedure based on a probabilistic descent algorithm, ensuring a stochastic convergence result. The generalized risks considered by Katagiri et al. (in preparation) include the regular classification error, the mean square error, nonlinear functions (e.g., sigmoid) of the likelihood, and several other general measures. The corrective training algorithm of Bahl et al. (1988) is just one possible choice for the minimization of a prescribed risk function.

4.2. Integration of Nonspectral and Spectral Features

The use of HMMs in speech recognition has been mostly limited to the modeling of short-time speech spectral information. That is, the observation **O** typically represents a smoothed representation of the spectrum of the speech signal at a given time, be it in terms of the traditional Fourier transform coefficients, linear prediction coefficients, or any other transformations of the estimated (power) spectrum. The spectral feature vector has proved extremely useful and has led to a wide variety of successful recognizer designs. This success can be attributed both to the range of spectral analysis techniques developed in the past three decades and to our understanding of the perceptual importance of the speech spectrum to the recognition of sounds. The success of spectral parameters for characterizing speech was further augmented by the introduction of the so-called delta-cepstrum (Furui, 1986), which attempts to model the differential speech spectrum much as the cepstrum models the instantaneous speech spectrum.

Besides spectral parameters, there are other speech features that are believed to contribute to the recognition and understanding of speech by humans. One such category of nonspectral speech features is prosody as is manifested on both the segmental and the suprasegmental levels (Lea, 1980). Physical manifestations of prosody in the speech signal include signal energy (suitably normalized), differential energy, and signal pitch (fundamental frequency). There are at least two issues of concern in integrating nonspectral with spectral features in a statistical model of speech, namely whether such features contribute to the performance of the statistical model for actual recognition tasks, and how the features should be integrated so as to be consistent with their physical manifestations in the production and perception of speech. The first issue is relatively easy to resolve based on experimental results. Several HMM-based recognition systems have incorporated log energy (and differential log energy) either directly in the feature vector or as an additional feature whose probability (or likelihood) is factored into the likelihood calculation (Bahl et al., 1983; Rabiner, 1984; Shikano, 1985). These systems have had moderate success (i.e., higher recognition performance). However, the level of performance improvement is considerably smaller than one might anticipate based on the importance of prosody in speech.

The problem with combining spectral and nonspectral features in a statistical framework is one of temporal rate of change. To attain adequate time resolution for the spectral parameters that characterize the varying vocal tract, we need to sample the spectral observations at a rate on the order of 100 times per second (10 ms frame update). The prosodic features of speech characterize stress and intonation, and these occur at a syllabic rate or about 10 times per second. [Of course, we can always oversample the features associated with the prosodic parameters to keep the rate the same as that of the spectral parameters—this in fact is what is currently done in most systems (e.g., Rabiner, 1984; Shikano, 1985).] Furthermore, the sequential characteristic change of different features may be too different to warrant a single, unified Markov chain. Thus one key question is how to combine two feature sets with fundamentally different time scales and possibly different sequential characteristics so as to perform optimum modeling and decoding.

A second question in combining fundamentally different (in nature) features concerns their statistical characterization. To be technically correct we need to know the joint distribution of the two feature sets. For statistically independent (or often just uncorrelated) feature sets we can represent the joint density as a product of the individual densities. However, in practice, there is usually some correlation between any pair of speech feature sets; hence some correction for the correlation is usually required. One proposed method of handling this problem is to perform a principal component analysis on the joint feature set before hidden Markov modeling is performed (Bocchieri and Doddington, 1986). Although this alleviates the difficulties somewhat, it is not a totally satisfying solution because the resulting feature set usually has no straightforward physical significance. Furthermore, the set of principal components is a function of the data and hence need not be optimal for unseen data (open-set problem).

The key points raised here concern structural issues for combining traditional spectral features with other features such as pitch and signal energy. Although the features can be simply combined in traditional ways, there is probably a much better way of integrating these fundamentally different features into a consistent statistical structure which achieves the minimum decoding error.

4.3. Duration Modeling in HMMs

One of the inherent limitations of the HMM approach is its treatment of temporal duration—both within a model and within units composed of one or more models (e.g., a word made up from subword models). Inherently, within a state of an HMM, the probability distribution of state duration is exponential, of the form

$$P_i(d) = (a_{ii})^{d-1}(1 - a_{ii})$$
$$= \text{probability of staying in state } S_i \text{ for exactly } d \text{ observations} \quad (44)$$

where a_{ii} is the self-loop state probability. This exponential duration model is inappropriate for almost any speech event.

Several alternatives for implementing different state duration models have been proposed. The most straightforward approach is to allow explicit state duration distributions, $P_{q_i}(d_i)$, to be associated with each state, S_i, using the concept of a semi-Markov chain (Ferguson, 1980; Russell and Moore, 1985) in which state transitions do not occur at regular time intervals. More formally we assume that for a given state sequence $\mathbf{q}$ in which there are $r-1$ state transitions such that the states visited are $q_1, q_2, \ldots, q_r$ with associated state durations of $d_1, d_2, \ldots, d_r$ (frames), the joint probability of the observation sequence $\mathbf{O}$ and the state sequence $\mathbf{q}$ given the model λ becomes

$$\begin{aligned} P(\mathbf{O}, \mathbf{q} \mid \lambda) = \; & \pi_{q_1} P_{q_1}(d_1) P(\mathbf{O}_1, \mathbf{O}_2, \ldots, \mathbf{O}_{d_1} \mid q_1) \\ & a_{q_1 q_2} P_{q_2}(d_2) P(\mathbf{O}_{d_1+1} \cdots \mathbf{O}_{d_1+d_2} \mid q_2) \\ & \vdots \\ & a_{q_{r-1} q_r} P_{q_r}(d_r) P(\mathbf{O}_{d_1+d_2+\cdots+d_{r-1}+1} \cdots \mathbf{O}_T \mid q_r) \end{aligned} \quad (45)$$

where $P_{q_i}(*)$ are the state duration probabilities. The probability measure for such a Markov model is, accordingly,

$$P(\mathbf{O} \mid \lambda) = \sum_q P(\mathbf{O}, \mathbf{q} \mid \lambda) \quad (46)$$

Based on these definitions, modeling of the source, including the duration model $P_i(d_i)$, can be implemented using a hill-climbing reestimation procedure of the type used previously (Levinson, 1986). Typically, $P_i(d_i)$ is treated as a discrete distribution over the range $1 \leq d_i \leq D_{max}$, where D_{max} represents the longest possible dwell in any state.

Although the above formulation handles the duration model *simultaneously* with the Markovian transition and the local (state) distribution models and can lead to analytical solutions, there are drawbacks with such a duration model for speech recognition. One such drawback is the greatly increased computational complexity due to the loss of regularity in transition timing. In the traditional HMM, transitions are allowed to occur at every time frame, since the model topology includes a return transition back to the original state. In the semi-Markov model, transitions do not occur at regular time [the return transition is part of the duration model $P_i(d_i)$] and this leads to a significantly more complicated lattice for decoding an input string. In Rabiner (1989) it was estimated that the semi-Markov model incurs a factor of 300 increase in computational complexity for a value of $D_{max} = 25$. The much increased complication in decoding lattice often renders many search algorithms, such as the beam search (Lowerre and Reddy, 1980) and the stack algorithm (Jelinek, 1969) for handling large problems, extremely difficult to implement. Another problem with the semi-Markov model is the large number of parameters (D_{max}) associated with each duration model that must be estimated in addition to the usual HMM parameters. Finally, it is not clear how accurate or how robust the estimated $P_i(d_i)$ needs to be in order to be beneficial in speech recognition.

One proposal (Levinson, 1986) to alleviate some of these problems is to use a parametric state duration distribution model instead of the nonparametric ones used above. Several parametric distributions have been considered, including the Gaussian family (with constraints to avoid negative durations) and the Gamma family (Levinson, 1986). These families represent the most rigorous attempt to utilize the duration information explicitly and analytically in speech modeling and recognition to date. Other less involved attempts include modeling the state duration with a uniform distribution, requiring only an estimate of the duration range and minimum and maximum duration allowed in a particular state. This simple duration model has been applied with good success (Lowerre and Reddy, 1980) for some tasks.

The major difficulty in modeling the duration information is that it is much more sparse than spectral information; i.e., there is only one duration per state. Hence we either alter the structure of the HMM (e.g., to a semi-Markov model), thereby losing many of the regularity properties of the original HMM formulation, or seek alternative implementation structures, as in the case of prosodic information. For duration information a simple-minded approach is to treat the spectral modeling and the duration modeling as two separate, loosely connected problems. Hence the regular

HMM estimation (spectral) is performed on the given observation sequence $\mathbf{O}$. Then the best state sequence $\bar{\mathbf{q}}$ which maximizes $P(\mathbf{O}, \mathbf{q} \mid \lambda)$ is found using the Viterbi algorithm. Finally, estimates of $P_i(d_i)$ are obtained based on the optimal state sequence $\bar{\mathbf{q}}$ either by the maximum likelihood method or from simple frequency of occurrence counts (Rabiner et al., 1986a). Often the duration d_i for state i is normalized by the overall duration T in order to account for the inherent variation in speaking rate. This approach is usually called the postprocessor duration model because the standard decoding is performed first and the duration information is available only after the initial processing is finished. Although the postprocessor duration model has had some success (Rabiner et al., 1986a), the questions of optimality of the estimate, robustness of the solution, and other criteria for successful utilization of duration information, especially as applied to speech recognition, remain unanswered.

4.4. Other Issues

Although we have discussed several of the most important open issues in HMM systems, especially as applied to speech recognition, a few more areas of concern are worth noting. These include the issue of model clustering and model splitting for handling inhomogeneous data, the assumption that observations within a state of a semi-Markov model are statistically independent, and the idea of using higher-order models. We briefly discuss each of these issues here.

4.4.1. Model Clustering and Splitting

One of the basic assumptions in statistical modeling is that the variability in the observations from an information source can be modeled by statistical distributions. For speech recognition, the source could be a single word, a subword unit like a phoneme, or a word sequence. Because of variability in the source production (e.g., accents, speed of talking) or the source processing (e.g., transmission distortions, noise), it is often expedient to consider using more than a single HMM to characterize the source. There are two motivations behind this multiple-HMM approach. First, lumping all the variability together from inhomogeneous data sources leads to unnecessarily complex models, often yielding lower modeling accuracy. Second, some of the variability, or rather the inhomogeneity in the source data, may be known a priori, thus warranting separate modeling of the source data sets. Here, our main concern is the first case, i.e., automatic modeling of an inhomogeneous source with multiple HMMs, as the latter (manual) case is basically straightforward.

There exist a number of generalized clustering algorithms, such as the k-means clustering algorithm, the generalized Lloyd algorithm as is widely used in vector quantizer designs (Linde et al., 1980), or the greedy growing algorithm found in set partition or decision tree designs (Breiman et al., 1984), all of which are suitable for the purpose of separating inconsistent training data so that each divided subgroup becomes more homogeneous and therefore is better modeled by a single HMM. The nearest-neighbor rule required in these clustering algorithms is simply to assign an observation sequence $\mathbf{O}$ to cluster i if

$$P(\mathbf{O} \mid \lambda_i) = \max_j P(\mathbf{O} \mid \lambda_j)$$

where λ_j's denote the models of the clusters. Successful application of the model clustering algorithms to the speech recognition problem, using the straightforward maximum likelihood criterion, has been reported (Rabiner et al., 1989a). However, when other estimation criteria are used, the interaction between multiple-HMM modeling and the Bayes minimum error classifier design remains an open question in need of further study.

An alternative to model clustering is to arbitrarily subdivide a given speech source into a large number of subclasses with specialized characteristics and then consider a generalized procedure for model merging based on source likelihood considerations. By way of example, for large vocabulary speech recognition we often try to build specialized units (context sensitive) for recognition. For example, we could consider building units which are a function of the sound immediately preceding the unit (left context) and the sound immediately following the unit (right context). There are on the order of 10,000 such units in English. Many of the units are functionally almost identical. The problem is how to determine which pairs of units should be merged (so that the number of model units is manageable and the variance of the parameter estimate is reduced). To set ideas, consider two unit models, λ_a and λ_b, corresponding to training observation sets $\mathbf{O}_a$ and $\mathbf{O}_b$, and the merged model λ_{a+b}, corresponding to the merged observation sets $\{\mathbf{O}_a, \mathbf{O}_b\}$. We can then compute the change in entropy (i.e., loss of information) resulting from the merged model as

$$\Delta H_{ab} = H_a + H_b - H_{a+b} \tag{47}$$

$$= -P(\mathbf{O}_a \mid \lambda_a)\log P(\mathbf{O}_a \mid \lambda_a) - P(\mathbf{O}_b \mid \lambda_b)\log P(\mathbf{O}_b \mid \lambda_b) + P(\{\mathbf{O}_a, \mathbf{O}_b\} \mid \lambda_{a+b})\log P(\{\mathbf{O}_a, \mathbf{O}_b\} \mid \lambda_{a+b}) \tag{48}$$

Whenever ΔH_{ab} is small enough, it means that the change in entropy resulting from merging the models will not affect system performance (at least on the training set) and the models can be merged. The question of how small is acceptable is dependent on specific applications. Other practical questions of a similar nature also remain. However, this model merging technique has been used successfully by Lee (1989) to create a generalized set of triphone models for large-vocabulary speech recognition.

4.4.2. Parameter Significance and Statistical Independence

Consider an observation sequence $\mathbf{O} = \{\mathbf{O}_1, \mathbf{O}_2, \ldots, \mathbf{O}_T\}$. If the state sequence that led to the production of $\mathbf{O}$ is known, then the conditional probability of $\mathbf{O}$ given the source model and the state sequence is expressed in (6), where the product form implies statistical independence (M. Sondhi, S. Levinson, and P. de la Noue, unpublished; Wellekens, 1987). The partial HMM measure of (7) shows that the contribution of the underlying Markov chain [as expressed in (4)] is to multiply the result of (6), a result equivalent to assuming that the observations within a state are independent. We have already discussed the structural advantages of the Markov chain assumption. However, the significance of the Markov chain contribution, namely the product of state transitions, in terms of speech recognition performance is still not well understood.

Although, in theory, the importance of the state transition coefficients a_{ij} is the same as that of the state observation density $b_j(\mathbf{O})$ [in training, a_{ij} affects the parameter estimate of $b_j(\cdot)$ and vice versa], in normal practice of speech recognition this is not usually the case. This paradox is due to the discrimination capability of the a_{ij}'s relative to that of the $b_j(\mathbf{O})$'s. It should be clear that since a_{ij} is the relative frequency of state transition from state i to state j, its dynamic range (especially in a left-to-right model) is severely constrained. For example, the self-loop coefficient, a_{ii}, which can be interpreted roughly as $(\bar{d}_i - 1)/\bar{d}_i$ where $\bar{d}_i$ is the average duration in state i, typically is constrained between 0.5 (for $\bar{d}_i = 2$) and 0.95 (for $\bar{d}_i = 20$); hence the dynamic range over all states and models is less than 2 to 1. Similarly, the jump coefficient, $a_{i,i+1} = 1 - a_{ii}$, is constrained to be between 0.05 and 0.5, a somewhat larger dynamic range. However, there is only one jump coefficient for every $\bar{d}_i$ self-transitions, again effectively reducing the dynamic range of the product of the a_{ij} coefficients.

Consider now the $b_j(\mathbf{O})$ densities. These often involve the product of D local densities, each of which has almost unlimited dynamic range. Furthermore, each of the state densities is highly localized in the acoustic space, thereby leading to the excellent classification/discrimination capability of HMMs.

When we combine the a_{ij}'s and the $b_j(\mathbf{O})$'s to give the probability distribution of the HMM, we find in practice that, for a left-to-right model, the a_{ij}'s can be neglected entirely (i.e., set all $a_{ii} = a_{i,i+1} = 0.5$) with no effect on recognition performance. This effect is due to the lack of discrimination capability of the a_{ij}'s, as described above. There seems to be no simple way of changing this situation, nor is it entirely clear that it is important to do so.

The above analysis points out the existence of the unbalanced numerical significance of the a's and the b's in the likelihood calculation of the HMM and should not be taken to discredit totally the usefulness of the Markov chain contribution. In fact, the transition probability still plays an important role in parameter estimation. For some specific HMMs, e.g., large ergodic models with a large number of states, accurate calculation of the state transition probabilities is vital in obtaining a good model, although the trend of unbalanced significance between the state transition probability and the observation density remains.

The desire to change the significance of the Markov chain contribution, via state transition probabilities, is further reflected in the formulation of semi-Markov models as discussed in Section 4.7. The probability of a Markovian sequence of (4) is revised to a general form of product $P_{q_1}(d_1)P_{q_2}(d_2)\cdots$ as in (45). This more general form of Markov chain dependence allows introduction of a singularity in $P_q(d)$, thereby increasing its numerical significance under certain conditions. [For example, if we specify $P_q(d) = 0$ for $d \leq d_0$, the HMM system has to undergo at least d_0 frames in state q before it moves out of that state, regardless of the value of $b_q(\mathbf{O}_t)$ during that period of time.] Such a provision for introducing singularities could have a major impact on the recognition performance (Lowerre and Reddy, 1980).

When semi-Markov models are chosen, we often use an independence (within each state) assumption for simplicity. In the formulation of (45), therefore, the joint probability $P(\mathbf{O}_{d_{i-1}+1}, \ldots, \mathbf{O}_{d_i+d_{i+1}} \mid q_i)$ becomes $\prod_j P(\mathbf{O}_{d_{i-1}+j} \mid q_i)$. This greatly simplified treatment may be an area for improvement. The argument is that within a state (especially one representing a stationary sound like a vowel) the observations are highly correlated. Thus, assuming independence gives far too much emphasis to the match within such stationary states. The problem is one of form as well as substance. It is difficult to choose an appropriate form for $P(\cdot \mid q, \lambda)$, and it is even more difficult to estimate the parameters of the chosen form from any training set of reasonable size (Sondhi et al., unpublished). The problems are similar to those of any reduced rate measurement, namely one of measurement of $P(\cdot \mid q, \lambda)$

for each state, rather than several measurements per state. Hence we are trying to estimate parameters of a much more complicated representation from far less data than in the simple, uncorrelated observations, case. As such, this problem is one for which there may never be a good solution.

4.4.3. Higher-Order HMM

Until now, almost all HMM formulations for speech recognition are based on a simple first-order Markov chain. At the acoustic signal processing level, this low-order modeling may be acceptable because the time scale of processing for each frame of signal is kept on the order of 1- ms. When an HMM is used in higher levels of a recognition system, such as syntactic or semantic processing, the first-order formulation often turns out to be inadequate. This is because the grammatical structure of the English language (and perhaps any other language) obviously cannot be properly modeled by a first-order Markov chain. Although the structural simplicity of a first-order model makes the computation simple and straightforward, there may be a need to complete the analytical framework of higher-order models. Also, for such higher-order models to be practically useful, many of the implementational advantages of the first-order case may have to be formulated in an appropriate manner.

5. SUMMARY

In this chapter we have reviewed the statistical method of hidden Markov models (HMMs). We showed that the strengths of the method lie in the consistent statistical framework that is flexible and versatile, particularly for speech applications, and the case of implementation that makes the method practically attractive. We also pointed out some areas of the general HMM method that deserve more attention with the hope that increased understanding will lead to performance improvements for many applications. These areas include the modeling criteria, particularly the problem of minimum classification error, incorporation of new (nonspectral) features as well as prior linguistic knowledge, and the modeling of state durations and its use in speech recognition. With our current understanding, HMM systems have been shown capable of achieving recognition rates of upward of 95% word accuracy in certain speaker-independent tasks with vocabularies on the order of 1000 words. With further progress, it is not difficult to foresee HMM-based recognition systems that perform well enough to make the technology usable in everyday life.

ACKNOWLEDGMENT

The authors acknowledge the excellent comments and criticisms provided by Yariv Ephraim and C. H. Lee on earlier drafts of this chapter.

Appendix A. BAUM–WELCH REESTIMATION WITH FORWARD-BACKWARD ALGORITHM

In the Baum–Welch reestimation method, the objective function $P(\mathbf{O} \mid \lambda)$ is first transformed into an auxiliary function $Q(\lambda', \lambda)$ according to (9). The function $Q(\lambda', \lambda)$ is then maximized over λ, resulting in a new model $\bar{\lambda}$, at which max $Q(\lambda', \lambda)$ is attained.

Note that

$$\log P(\mathbf{O}, \mathbf{q} \mid \lambda) = \log \pi_{q_0} + \sum_{i=1}^{T} \log a_{q_{t-1}q_t} + \sum_{i=1}^{T} \log b_{q_i}(\mathbf{O}_i)$$

The auxiliary function can then be decomposed as

$$\begin{aligned} Q(\lambda', \lambda) &= \sum_{q} P(\mathbf{O}, \mathbf{q} \mid \lambda') \log P(\mathbf{O}, \mathbf{q} \mid \lambda) \\ &= Q_{\pi}(\lambda', \pi) + \sum_{i=1}^{N} Q_{a_i}(\lambda', \mathbf{a}_i) \\ &\quad + \sum_{i=1}^{N} Q_{b_i}(\lambda', \mathbf{b}_i) \end{aligned}$$

where $\pi = [\pi_1, \pi_2, \ldots \pi_N]$, $\mathbf{a}_i = [a_{i1}, a_{i2} \ldots a_{iN}]$, $\mathbf{b}_i$ is the parameter vector that defines $b_i(\cdot)$,

$$Q_{\pi}(\lambda', \pi) = \sum_{i=1}^{N} P(\mathbf{O}, q_0 = S_i \mid \lambda') \log \pi_i$$

$$Q_{a_i}(\lambda', \mathbf{a}_i) = \sum_{j=1}^{N} \sum_{t=1}^{T} P(\mathbf{O}, q_{t-1} = S_i, q_t = S_j \mid \lambda') \log a_{ij}$$

and

$$Q_{b_i}(\lambda', \mathbf{b}_i) = \sum_{t=1}^{T} P(\mathbf{O}, q_t = S_i \mid \lambda') \log b_i(\mathbf{O}_t)$$

Therefore, maximization of $Q(\lambda', \lambda)$ over λ can be accomplished by maximizing the individual terms separately. Maximization of these individual terms, Q_π, Q_{a_i}, and Q_{b_i}, for $i = 1, 2, \ldots, N$, subject to the constraints

$$\sum_{j=1}^{N} \pi_j = 1,$$

$$\sum_{j=1}^{N} a_{ij} = 1 \quad \text{for all } i$$

and, if $b_i(\mathbf{O}_i) = b_{ik}$ for discrete $\mathbf{O}_t$

$$\sum_{k=1}^{K} b_{ik} = 1 \quad \text{for all } i$$

is well known. (K is the number of distinct symbols that the observation $\mathbf{O}_t$ can assume in this case.) These individual auxiliary functions have the same form $\sum_{j=1}^{N} w_j \log y_j$, which as a function of $\{y_j\}_{j=1}^{N}$, subject to the constraints $\sum_{j=1}^{N} y_j = 1$ and $y_j \geq 0$, attains a global maximum at the single point

$$y_j = \frac{w_j}{\sum_{i=1}^{N} w_i}, \qquad j = 1, 2, \ldots, N$$

This results in the following reestimation transformation for obtaining $\bar{\lambda} = [\bar{\pi}, \bar{A}, \bar{B}]$:

$$\bar{\pi} = \frac{P(\mathbf{O}, q_0 = S_i \mid \lambda)}{P(\mathbf{O} \mid \lambda)}$$

$$\bar{a}_{ij} = \frac{\sum_{t=1}^{T} P(\mathbf{O}, q_{t-1} = S_i, q_t = S_j \mid \lambda)}{\sum_{t=1}^{T} P(\mathbf{O}, q_{t-1} = S_i \mid \lambda)}$$

and

$$\bar{b}_{ik} = \frac{\sum_{t=1}^{T} P(\mathbf{O}, q_t = S_i \mid \lambda)\delta(\mathbf{O}_t, k)}{\sum_{t=1}^{T} P(\mathbf{O}, q_t = S_i \mid \lambda)}$$

where

$$\delta(\mathbf{O}_t, k) = 1 \quad \text{if } \mathbf{O}_t \text{ is the } k\text{th symbol}$$
$$= 0 \quad \text{otherwise}$$

These transformations can be easily calculated by the following forward-backward procedure.

Define the forward variable, $\alpha_t(i)$, as

$$\alpha_t(i) = P(\mathbf{O}_1, \ldots, \mathbf{O}_t, q_t = S_i \mid \lambda)$$

i.e., the probability of the partial observation sequence up to time t and state $q_t = S_i$ at time t. The forward variable can be calculated inductively, starting from $t = 0$, as follows:

1. Initialization: $\alpha_0(i) = \pi_i, 1 \le i \le N$
2. Induction: for $t = 1, 2, \ldots, T$ and $1 \le j \le N$

$$\alpha_t(j) = \left[\sum_{i=1}^{N} \alpha_{t-1}(i) a_{ij}\right] b_j(\mathbf{O}_t)$$

3. Final calculation:

$$P(\mathbf{O}_1, \mathbf{O}_2, \ldots, \mathbf{O}_T \mid \lambda) = \sum_{i=1}^{N} P(\mathbf{O}_1, \mathbf{O}_2, \ldots, \mathbf{O}_T, q_T = S_i \mid \lambda)$$
$$= \sum_{i=1}^{N} \alpha_T(i)$$

The final calculation above produces the result of $P(\mathbf{O} \mid \lambda)$. It is easy to see that this forward induction procedure requires on the order of N^2T calculations, rather than $2TN^T$ as required by the direct calculation of Eq. (8).

In a similar manner, we can define a backward variable $\beta_t(i)$, as

$$\beta_t(i) = P(O_{t+1}, \ldots, O_T \mid q_t = S_i, \lambda)$$

i.e., the probability of the partial observation sequence from $t + 1$ to the end, given state $q_t = S_i$ at time t. Again, $\beta_t(i)$ can be calculated inductively:

1. Initialization: $\beta_t(i) = 1, 1 \le i \le N$
2. Induction: for $t = T - 1, T - 2, \ldots, 1, 1 \le i \le N$,

$$\beta_t(i) = \sum_{j=1}^{N} a_{ij} b_j(\mathbf{O}_{t+1}) \beta_{t+1}(j)$$

Many quantities required in the reestimation transformation can now be expressed in terms of the forward and backward probabilities. In particular,

$$P(\mathbf{O}, q_t = S_i \mid \lambda) = \alpha_t(i)\beta_t(i)$$

$$P(\mathbf{O} \mid \lambda) = \sum_{i=1}^{N} \alpha_t(i)\beta_t(i)$$

$$= \sum_{i=1}^{N} \alpha_T(i)$$

and

$$P(\mathbf{O}, q_{t-1} = S_i, q_t = S_j \mid \lambda)$$
$$= \alpha_{t-1}(i)\alpha_{ij} b_j(\mathbf{O}_t)\beta_t(j)$$

This leads to the reestimates in the forward-backward procedure:

$$\bar{\pi} = \frac{\alpha_0(i)\beta_0(i)}{\sum_{j=1}^{N} \alpha_T(j)}$$

$$\bar{a}_{ij} = \frac{\sum_{t=1}^{T} \alpha_{t-1}(i)\alpha_{ij} b_j(\mathbf{O}_t)\beta_t(j)}{\sum_{t=1}^{T} \alpha_{t-1}(i)\beta_{t-1}(i)}$$

and

$$\bar{b}_{ik} = \frac{\sum_{t-1}^{T} \alpha_t(i)\beta_t(i)\delta(\mathbf{O}_t, k)}{\sum_{t=1}^{T} \alpha_t(i)\beta_t(i)}$$

Appendix B. VITERBI ALGORITHM

The formal steps in the Viterbi algorithm for finding the best state sequence are as follows:

1. Initialization:

$$\delta_0(i) = \pi_i, \qquad 1 \le N$$

2. Recursion: for $1 \le t \le T$ and $1 \le j \le N$

$$\delta_t(j) = \max_{1 \le i \le N} [\delta_{t-1}(i)\alpha_{ij}] b_j(\mathbf{O}_t)$$

$$\psi_i(j) = \arg \max_{1 \le i \le N} [\delta_{t-1}(i) a_{ij}]$$

3. Termination:

$$P^* = \max_{1 \le i \le N} \delta_T(i)$$

$$i_T^* = \arg \max_{1 \le i \le N} \delta_T(i)$$

4. Path (state sequence) backtracking: for $t = T - 1, T - 2, \ldots, 1,$

$$i_t^* = \psi_{t+1}(i_{t+1}^*)$$

REFERENCES

Allen, J. B., and Rabiner, L. R. (1977). A unified theory of short-term spectrum analysis and synthesis. *Proc. IEEE 65*(11):1558–1564.

Atal, B. S., and Hanauer, S. L. (1971). Speech analysis and synthesis by linear prediction of the speech wave. *J. Acoust. Soc. Am. 50*(2):637–655.

Averbuch, A., et al. (1987). Experiments with the TANGORA 20,000 word speech recognizer. *Proc. ICASSP-87*, Dallas, Texas, April, pp. 701–704.

Bahl, L. R., Bakis, R., Cohen, P. S., Cole, A. G., Jelinek, F., Lewis, B. L., and Mercer, R. L. (1980). Further results on the recognition of a continuously read natural corpus. *Proc. ICASSP-80*, Denver, Colorado, April, pp. 872–875.

Bahl, L. R., Jelinek, F., and Mercer, R. L. (1983). A maximum likelihood approach to continuous speech recognition. *IEEE Trans. Pattern Anal. Machine Intell. PAMI-5*: 179–190.

Bahl, L. R., Brown, P. F., de Souza, P. V., and Mercer, R. L. (1986). Maximum mutual information estimation of hidden Markov model parameters for speech recognition. *Proc. ICASSP-86*, Tokyo, April, pp. 49–52.

Bahl, L. R., Brown, P. F., de Souza, P. V., and Mercer, R. L. (1988a). Speech recognition with continuous parameter hidden Markov models. *Proc. ICASSP-88*, New York, April, pp. 40–43.

Bahl, L. R., Brown, P. F., de Souza, P. V., and Mercer, R. L. (1988b). A new algorithm for the estimation of hidden Markov model parameters. *Proc. ICASSP-88*, New York, April, pp. 493–496.

Baker, J. K. (1975). The DRAGON system—an overview. *IEEE Trans. Acoust. Speech Signal Process. ASSP-23*(1):24–29.

Bakis, R. (1976). Continuous speech word recognition via centisecond acoustic states. *Proc. ASA Meeting*, Washington, D.C., April.

Baum, L. E. (1972). An inequality and associated maximum technique in statistical estimation for probabilistic functions of Markov processes. *Inequalities 3*:1–8.

Baum, L. E., and Egon, J. A. (1967). An inequality with applications to statistical estimation for probabilistic functions of a Markov process and to a model for ecology. *Bull. Am. Meteorol. Soc. 73*:360–363.

Baum, L. E., and Petrie, T. (1966). Statistical inference for probabilistic functions of finite state Markov chains. *Ann. Math. Stat. 37*:1554–1563.

Baum, L. E., and Sell, G. R. (1968). Growth functions for transformations on manifolds. *Pac. J. Math 27*(2):211–227.

Baum, L. E., Petrie, T., Soules, G., and Weiss, N. (1970). A maximization technique occurring in the statistical analysis of probabilistic functions of Markov chains. *Ann. Math. Stat. 41*(1):164–171.

Bocchieri, E. L., and Doddington, G. R. (1986). Frame-specific statistical features for speaker independent speech recognition. *IEEE Trans. Acoust. Speech Signal Process. ASSP-34*(4):755–764.

Bourlard, H., Kamp, Y., Ney, H., and Wellekens, C. J. (1985). Speaker-dependent connected speech recognition via dynamic programming and statistical methods.

In *Speech and Speaker Recognition*, M. R. Schroeder, Ed. Karger, Basel, pp. 115–148.

Breiman, L. (1984). *Classification and Regression Trees*. Wadsworth, Monterey, Calif., 1984.

Bridle, J. S. (1984). Stochastic models and template matching: Some important relationships between two apparently different techniques for automatic speech recognition. *Proc. Inst. Acoust. Autumn Conf.*, November, pp. 1–8.

Brown, P. F. (1987). *The acoustic-modeling problem in automatic speech recognition*. Ph.D. thesis, Dept. of Computer Science, Carnegie Mellon University.

Cadzow, J. A. (1982). ARMA modeling of time series. *IEEE Trans. Pattern Anal. Machine Intell. PAMI-4*:124–128.

Chao, Y. R. (1968). *Language and Symbolic Systems*. Cambridge Univ. Press, Cambridge, UK.

Chow, Y. L., et al. (1987). BYBLOS: The BBN continuous speech recognition system, *Proc. ICASSP '87* Dallas, April, Paper 3.7.1, pp. 89–92.

Cohen, J. (1985). Application of an adaptive auditory model to speech recognition. Presented at the 110th Meeting of Acoustical Society of America, Nashville, Tenn., November.

Dautrich, B. A., Rabiner, L. R., and Martin, T. B. (1983). On the effects of varying filter bank parameters on isolated word recognition. *IEEE Trans. Acoust. Speech Signal Process. ASSP-31*:793–806.

Dempster, A. P., Laird, N. M., and Rubin, D. B. (1977). Maximum likelihood from incomplete data via the EM algorithm. *J. Roy. Stat. Soc. 39*(1):1–38.

Derouault, A. M. (1987). Context dependent phonetic Markov models for large vocabulary speech recognition. *Proc. ICASSP '87*, April, Dallas, Paper 10.1.1, pp. 360–363.

Duda, R. O., and Hart, P. E. (1973). *Pattern Classification and Scene Analysis*. Wiley, New York.

Ephraim, Y., and Rabiner, L. R. (1988). On the relations between modeling approaches for information sources. *Proc. ICASSP-88*, New York, April, pp. 24–27.

Ephraim, Y., Dembo, A., and Rabiner, L. R. (1989). A minimum discrimination information approach for hidden Markov modeling. *IEEE Trans. Inform. Theory 35*(5):1001–1013.

Ferguson, J. D. (1980). Hidden Markov analysis: An introduction. In *Hidden Markov Models for Speech*. Institute for Defense Analysis, Princeton, N.J.

Forney, G. D. (1973). The Viterbi algorithm. *Proc. IEEE 61* 61:268–278.

Furui, S. (1986). Speaker independent isolated word recognition based on dynamics emphasized cepstrum. *Trans. IECE Jpn. 69*(12):1310–1317.

Ghitza, O. (1986). Auditory nerve representation as a front-end for speech recognition in a noisy environment. *Comput. Speech and Language 1*(2):109.

Good, I. J. (1963). Maximum entropy for hypothesis formulation, especially for multidimensional contingency tables. *Ann. Math. Stat. 34*:911–934.

Gupta, V. N., Lennig, M., and Mermelstein, P. (1987). Integration of acoustic information in a large vocabulary word recognizer. *Conf. Proc. IEEE Acoustics, Speech, and Signal Processing*, April, pp. 697–700.

Hobson, A., and Cheng, B. (1973). A comparison of the Shannon and Kullback information measures. *J. Stat. Phys.* *7*(4):301–310.

Huang, X., and Jack, M. A. (1989). Unified techniques for vector quantization and hidden Markov models using semi-continuous models. *Proc. ICASSP-89*, Glasgow, May, pp. 639–642.

Jayant, N. S., and Noll, P. (1984). *Digital Coding of Waveforms*. Prentice-Hall, Englewood Cliffs.

Jelinek, F. (1969). A fast sequential decoding algorithm using a stack. *IBM J. Res. Dev.* *13*:675–685.

Jelinek, F. (1976). Continuous speech recognition by statistical methods. *Proc. IEEE* *64*:532–536.

Jelinek, F., Bahl, L. R., and Mercer, R. L. (1975). Design of a linguistic statistical decoder for the recognition of continuous speech. *IEEE Trans. Inform. Theory* *IT-21*:250–256.

Johnson, R. W. (1979). Determining probability distributions by maximum entropy and minimum cross-entropy. *Proc. APL79* (ACM 0-89791-005), May, pp. 24–29.

Juang, B. H. (1984). On the hidden Markov model and dynamic time warping for speech recognition—a unified view. *AT&T Tech. J.* *63*(7):1213–1243.

Juang, B. H. (1985). Maximum likelihood estimation for mixture multivariate stochastic observations of Markov chains. *AT&T Tech. J.* *64*(6):1235–1249.

Juang, B. H., and Rabiner, L. R. (1985). Mixture autoregressive hidden Markov models for speech signals. *IEEE Trans. Acoust. Speech Signal Process. ASSP-33*(6): 1404–1413.

Juang, B. H., and Rabiner, L. R. (1990). The segmental k-means algorithm for estimating parameters of hidden Markov models. *IEEE Trans. Acoust. Speech Signal Process*, *AS-38*(*9*):1639–1641.

Juang, B. H., Rabiner, L. R., Levinson, S. E., and Sondhi, M. M. (1985). Recent developments in the application of hidden Markov models to speaker-independent isolated word recognition. *Proc. ICASSP-85*, Tampa, March, pp. 9–12.

Katagiri, S., Lee, C. H., and Juang, B. H. (1990). A generalized probabilistic descent method, *Proc. of Fall Meeting of Acoustic Soc. of Japan*, September:141–142.

Kohonen, T. (1986). Learning vector quantization for pattern recognition. Helsinki Univ. of Technology, Report TKK-F-A6-1, November.

Kullback, S. (1958). *Information Theory and Statistics*. Wiley, New York.

Lea, W. A. (1980). Prosodic aids to speech recognition. In *Trends in Speech Recognition*, W. A. Lea (Ed.). Prentice-Hall, Englewood Cliffs, N.J.

Lee, C. H., Soong, F. K., and Juang, B. H. (1988). A segment model based approach to speech recognition. *Proc. ICASSP-88*, New York, April, pp. 501–504.

Lee, C. H., Juang, B H., Soong, F. K., and Rabiner, L. R. (1989). Word recognition using whole word and subword models. *Proc. ICASSP-89*, Glasgow, May, pp. 683–686.

Lee, C. H., Lin, C. H., and Juang, B. H. (1991a). A study on speaker adaptation of the parameters of continuous density hidden Markov models. *IEEE Trans. Acoust. Speech Signal Process. AS-39*(*4*):806–814.

Lee, C. H., Rabiner, L. R., Pieraccini, R., and Wilpon, J. G. (1990). Acoustic modeling for large vocabulary speech recognition. *Comput. Speech and Language*, (*4*): 127–165.

Lee, K. F. (1989). *Automatic Speech Recognition, the Development of the SPHINX System*, Kluwer, Boston.

Levinson, S. E. (1986). Continuously variable duration hidden Markov models for automatic speech recognition. *Comput. Speech and Language 1*(1):29–45.

Levinson, S. E. (1987). Continuous speech recognition by means of acoustic-phonetic classification obtained from a hidden Markov model. *Proc. ICASSP '87*, Dallas, April: pp. 93–96.

Levinson, S. E., Rabiner, L. R., and Sondhi, M. M. (1983). An introduction to the application of the theory of probabilistic functions of a Markov process to automatic speech recognition. *Bell Syst. Tech. J. 62*(4):1035–1074.

Linde, Y., Buzo, A., and Gray, R. M. (1980). An algorithm for vector quantizer design. *IEEE Trans. Commun. COM-28*:84–95.

Liporace, L. A. (1982). Maximum likelihood estimation for multivariate observations of Markov sources. *IEEE Trans. Inform. Theory IT-28*(5):729–734.

Lippmann, R. P., Martin, E. A., and Paul, D. B. (1987). Multistyle training for robust isolated word speech recognition. *Proc. ICASSP '87*, Dallas, April, pp. 705–708.

Ljolje, A., Ephraim, Y., and Rabiner, L. R. (1990). Estimation of hidden Markov model parameters by minimizing empirical error rate. *Proc. ICASSP*, Albuquerque, New Mexico, April, pp. 709–712.

Lowerre, B., and Reddy, R. (1980). The HARPY speech understanding system. In *Trends in Speech Recognition*, W. Lea (Ed.). Prentice–Hall, Englewood Cliffs, pp. 340–346.

Makhoul, L. (1975). Linear prediction: A tutorial review. *Proc. IEEE 63*:561–580.

Markel, J. D., and Gray, A. H. Jr. (1976). *Linear Prediction of Speech*. Springer–Verlag, New York.

Merhav, N., and Ephraim, Y. (1991). Maximum likelihood hidden Markov modeling using a dominant sequence of states. *IEEE Trans. Acoust. Speech Signal Process.*, in press.

Nadas, A., Nahamoo, D., and Picheny, M. A. (1988). On a model-robust training method for speech recognition. *IEEE Trans. Acoust. Speech Signal Process. ASSP-36*(9):1432–1436.

Nilsson, N. J. (1965). *Learning Machines*. McGraw Hill, New York.

Paul, D. B. (1985). Training of HMM recognizers by simulated annealing. *Proc. ICASSP-85*, Tampa, Florida, March, pp. 13–16.

Petrie, T. (1969). Probabilistic functions of finite state Markov chains. *Ann. Math. Stat. 40*(1):97–115.

Poritz, A. B., and Richter, A. G. (1986). On hidden Markov models in isolated word recognition. *Proc. ICASSP '86*, Tokyo, April, pp. 705–708.

Rabiner, L. R. (1984). On the application of energy contours to the recognition of connected word sequences. *AT&T Bell Labs. Tech. J. 63*:1981–1995.
Rabiner, L. R. (1989). A tutorial on hidden Markov models and selected applications in speech recognition. *Proc. IEEE 77*(2):257–286.
Rabiner, L. R., and Juang, B. H. (1986). An introduction to hidden Markov models. *IEEE ASSP Mag. 4*(1):4–16.
Rabiner, L. R., Levinson, S. E., and Sondhi, M. M. (1983). On the application of vector quantization and hidden Markov models to speaker-independent isolated word recognition. *Bell Syst. Tech. J. 62*(4):1075–1105.
Rabiner, L. R., Juang, B. H., Levinson, S. E., and Sondhi, M. M. (1986a). Recognition of isolated digits using hidden Markov models with continuous mixture densities. *AT&T Tech. J. 64*(6):1211–1222.
Rabiner, L. R., Wilpon, J. G., and Juang, B. H. (1986b). A segmental *k*-means training procedure for connected word recognition. *AT&T Tech. J. 65*(3):21–31.
Rabiner, L. R., Lee, C. H., Juang, B. H., and Wilpon, J. H. (1989a). HMM clustering for connected word recognition. *Proc. ICASSP-89*, Glasgow, May, pp. 405–408.
Rabiner, L. R., Wilpon, J. G., and Soong, F. K. (1989b). High performance connected digit recognition using hidden Markov models. *IEEE Trans. Acoust. Speech Signal Process. ASSP-37*(8):1214–1225.
Russell, M. J., and Moore, R. K. (1985). Explicit modeling of state occupancy in hidden Markov models for automatic speech recognition. *Proc. ICASSP '85*, Tampa, Fla., May, pp. 5–8.
Schafer, R. W., and Rabiner, L. R. (1971). Design of digital filter banks for speech analysis. *Bell. Syst. Tech. J. 50*(10):3097–3115.
Shikano, K. (1985). Evaluation of LPC spectral matching measures for phonetic unit recognition. Technical Report, Computer Sciences Dept., Carnegie Mellon Univ., May.
Wellekens, C. (1989). Explicit time correlation in hidden Markov models for speech recognition. *Proc. ICASSP*, pp. 384–386.

18

Neural Network Approaches for Speech Recognition

ALEXANDER WAIBEL Carnegie-Mellon University, Pittsburgh, Pennsylvania

1. INTRODUCTION

In the light of three decades of activity aimed at systems capable of understanding natural human speech, one might wonder if the problem has been solved and, if not, what might be required to achieve this challenging goal. To be sure, the field has seen impressive advances and several very important lessons have been learned. Despite these inroads, however, present-day systems still fall short of the ease and reliability with which humans can communicate with each other by speech. This gap has indeed led to a continuing search for new models and techniques that might bring us closer to machines that have this ability, but also to a better understanding of cognitive processes such as speech perception and understanding.

Neural networks are the most recent development in this search for new models of speech understanding. Most of what will be described is relatively recent work and only experiments on parts of the problem have been completed to date. Although very promising work in this direction has begun, no fully integrated speaker-independent large-vocabulary speech understanding systems based on neural networks alone yet exist. What fuels the excitement and what is the promise behind "neural networks," "connectionism," or "parallel distributed processing," as these models have often been termed interchangeably?

A partial list of attractive properties of connectionism is given in the following. Note that some of these properties are not necessarily unique to the connectionist approach. Indeed, some of the important ones are shared with other recognition schemes and are partially responsible for their success.

- *Massive parallelism*: A connectionist net is composed of many simple computing units, and computation is performed in parallel and in a distributed fashion by many interconnected computing elements. Potential advantages resulting from this are speed, regularity (for hardware implementation), and fault tolerance.
- *Constraint satisfaction*: Processing in connectionist nets is not performed sequentially and does not depend on the performance of any one single computing element, but on many processing elements' joint evaluation of numerous interrelated constraints.
- *Learning*: Massively parallel architectures cannot be easily programmed and parallel distributed processing models must rely on automatic learning algorithms. A number of such algorithms now exist, including training techniques for multilayer perceptrons (back propagation), Boltzmann machines, learning vector quantization, and associative nets (for reviews see Lippmann, 1987; Rumelhart and McClelland, 1986; Hinton, 1987; Kohonen, 1988). These learning algorithms optimize local computing elements to jointly improve some more global overall objective.
- *Stochastic modeling, uncertainty, variability, fuzziness*: Connectionist models deal with variability and noise by finding suitable probabilistic generalizations. Probabilities are encoded in such a network predominantly as patterns of activity across its elements, instead of single scalar values. Connectionist networks do not assume any particular statistical distributions, and hence no parametric assumptions need to be made.
- *Nonlinear modeling*: Connectionist networks are nonlinear models that can implement nonlinear classifiers and mapping functions and represent multimodel distributions and complex relationships. This may lead to better performance than linear models in various classification, mapping, and interpolation tasks.
- *Discovery of "hidden" knowledge*: Connectionist networks generate hidden knowledge, abstractions, and generalizations in the process of solving a more complex problem. In multilayer perceptrons, this hidden knowledge is often encoded in the connection weights learned

by so-called hidden units (Rumelhart and McClelland, 1986). If knowledge can be extracted from or encoded into these networks effectively, this may provide mechanisms for bridging the gap between knowledge-based approaches and stochastic models.

- *Uniformity*: Computation in connectionist nets is performed by simple underlying computing elements and the interaction between them. The computing steps performed by a particular unit (usually simple multiplications and additions) are generally independent of the task that the network is trying to solve. This is very useful for hardware implementations as the units are simple (cheap) and the same units can be used for a variety of tasks. Uniformity is also attractive as a means of achieving sensory fusion, i.e., the potential combination of different signals or input information (in speech, for example, phonetic and visual cues, or syntactic, semantic, and prosodic cues, etc.), possibly at varying levels of processing.
- *Speed—learning vs. recognition*: By virtue of massive parallel computation, connectionist nets can run very efficiently. Some connectionist models, however, do require considerable training to be performed.
- *Brain-style computation (?)*: Connectionist models attempt to simulate the style of computation as it is performed in the nervous system. Although similarities do exist, this analogy should not be carried too far. Our understanding of "brain-style computation" is still too fragmentary and our present engineering efforts too limited to warrant such comparisons. It is also necessary to replicate the brain accurately to build useful computer speech understanding systems. The human brain is, however, an existence proof that intelligent communication via speech is possible and insights and intuitions gleaned from its computational mechanisms could* usefully inspire new models and ideas for practical design.

Our goal in the following is to present a review of connectionist speech recognition models to date. We will omit a general introduction and assume that the reader is familiar with some of the more common neural network models. For more detailed introductions to neural networks in general, we recommend tutorial papers by Lippmann (1987) and Hinton (1987). Fundamentals of connectionist models and learning algorithms (e.g., back propagation, Boltzmann machines, associative nets, LVQ) will also not

*Along with many other scientific disciplines, of course.

be covered here. Important background material may be found in Rumelhart and McClelland (1986), Rumelhart et al. (1986), Hinton and Sejnowski (1986), and Kohonen (1988). We will limit our discussion to a review of connectionist advances specifically applied to speech recognition and understanding. So much has been written on this subject recently that we cannot hope to review all the activity in this area. We will limit ourselves therefore to a representative subset of important current activity, along with presentation of some of our own recent work. Three important levels of the speech understanding problem will be addressed:

- The phonemic level
  - Temporally static networks
  - Temporally dynamic networks
  - Modularity, scaling
  - Phoneme spotting
  - Speaker independence
- The word level
  - Static full-word models
  - Dynamic full-word models
  - Dynamic large-vocabulary models
  - Word model enhancements
- The language level
  - Word category prediction and disambiguation
  - Recurrent networks
  - Parsing

2. THE PHONEMIC LEVEL

Much of the earliest work in connectionist speech recognition has focused on phoneme recognition. This choice is motivated largely by the fact that phoneme recognition is (compared to the full-speech understanding problem) a tractable subproblem, difficult enough to be interesting, and also a useful focus, as large-vocabulary systems need to make use of atomic subunits or submodels of speech. Phoneme classification networks can be divided into two groups: (1) those that require precise temporal alignment of input tokens for accurate recognition performance (making them temporally static classifiers) and (2) those that do not (making them temporally dynamic or shift invariant).

2.1. Temporally Static Networks

In a static model, a speech pattern or feature vector centered around some boundary or target point is used as input to a neural network used as a classifier with outputs representing each of the phoneme categories in a language.

Figure 1 serves as an illustrative example of an early static application of neural networks to phoneme recognition tasks. Here, Huang and Lippmann (1988; Lippmann and Gold, 1987), apply a three-layer back-propagation network to vowel classification. The back-propagation training procedure was used and successfully learned to form nonlinear classification regions around vowel classes. For input features formant frequencies F_1 and F_2, as measured by Peterson and Barney (1952) in studies of adult and child male and female subjects, were used. The network has 50 units in its hidden layer and was trained for 50,000 trials, resulting in interformant boundaries comparable to those one would draw by hand and those formed by more traditional nonlinear classification techniques such as k-nearest neighbor classifiers (Duda and Hart, 1973; Makhoul et al., 1985). No parametric assumptions needed to be made, and the network provided a good and early demonstration of the usefulness of such networks as nonparametric nonlinear classifiers. The input features, however, were obtained by a human experimenter measuring formants from actual speech spectrograms, indeed a cognitive task in itself.

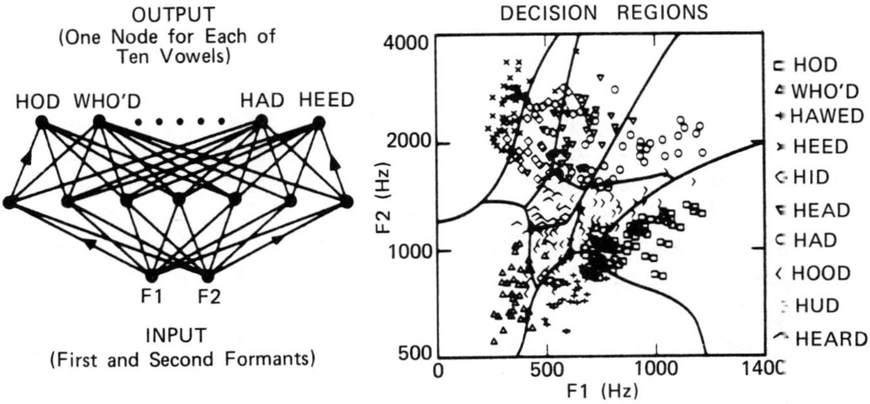

Figure 1 A three-layer back-propagation network used to form classification boundaries on the formants F_1 and F_2 for vowels.

Elman and Zipser (1987) reported phoneme classification based on experiments using actual speech patterns for the voiced-stop consonants /b, d, g/ (followed by the vowels /a, i, u/). 505 tokens of the nine discrete voiced-stop syllables were parsed from recordings of a single male speaker using a 10-kHz sampling rate applied to 3.5-kHz low-pass filtered speech. Twenty 16-coefficient discrete Fourier transforms (DFTs) are computed at overlapping 3.2-ms intervals to form the input of a three-layer back-propagation network. In a series of experiments, hidden layer and output layer node counts were varied. In one case, nine output nodes corresponding to the nine possible syllables were used; in two other cases an output node count of three corresponding to the three voiced-stop phonemes /b, d, g/ was used. More than 100,000 training passes were run for each experiment, using approximately half of the tokens as training exemplars. Recognition rates for the disjoint test data set were 84% for whole syllables, 98.5% for vowels, and 92.1% for voiced-stop consonants. Elman and Zipser found that introducing uniformly distributed white noise to training tokens at the input layer improved recognition rates to 90%, 99.7%, and 95%, respectively. The noise source tended to obscure idiosyncrasies of the training data and improve the networks' ability to generalize to unseen test data.

Both of these models used the back-propagation training procedure, but other methods have been proposed with similar success. Among them, good results were reported by Niranjan and Fallside (1988) using radial basis functions, by Kohonen and colleagues using phonotopic maps and learning vector quantization (LVQ) (Kohonen, 1988; Kohonen et al., 1988), and by Prager et al. (1986) using Boltzmann machines.

2.2. Temporally Dynamic Networks

The experiments reported above achieved successful classification for patterns that were presegmented and extracted from the signal by hand or by some presegmentation procedure. In a full system design, however, such segmentation would eventually have to be performed automatically, and even the best automatic segmentation schemes are prone to make errors. These errors, in turn, generally result in higher error rates further along in the recognition process. A robust speech recognition system should therefore *scan* the speech signal for useful cues *without* relying on presegmentation, basing its overall recognition decision on the sequence and co-occurrence of a sufficient set of those cues. Neural networks, therefore, should be temporally dynamic or shift invariant (i.e., classification that is unaffected by temporal shifts of the input speech train). The experiments reported in the following all employed techniques aimed at yielding shift-invariant phoneme recognition.

2.2.1. *The Time Delay Neural Network (TDNN)*

One of the earlier models that demonstrated successfully that this can be done in a connectionist framework was reported by Waibel (1989), Waibel et al. (1989a,b), and Lang et al. (1990). Their time delay neural network (TDNN) architecture was aimed at high-accuracy phoneme recognition under varying conditions of phoneme duration and temporal location within the speech signal. Figure 2 illustrates the TDNN architecture. It consists of neural units that use time-delayed connections at each layer to capture varying amounts of contexts at the layer below. The sizes of the resulting temporal input windows increase with increasing layers to learn increasingly coarser abstractions at progressively higher layers. As lower layers produce firing patterns, higher layers observe the resulting patterns of activations. During training shift invariance was achieved by making time-shifted copies of the net and linking their corresponding weights. Thus, knowledge of position in time was removed and the network had to spot relevant features anywhere in the input range to assemble sufficient evidence in favor of one of the output phoneme classes. A set of /b, d, g/ tokens was extracted independent of context from a large-vocabulary data base of 5240 Japanese words (Sagisaka et al., 1987) spoken by three male speakers, resulting in approximately 200 training and 200 test tokens per speaker. Recognition results on test data yielded an average recognition rate of 98.5% across all speakers.* Cursory studies of the effects of temporal shifts of the test data input spectra with respect to vowel onset suggested that nominal shifts had little appreciable effect on recognition rates. Comparisons to a variety of hidden Markov models (HMMs) also showed that significantly higher recognition rates could by obtained using the TDNN for this discrimination task (Waibel et al., 1987, 1989a). Using the same speech data, the HMMs tried achieved an average recognition rate of 92.7%. Detailed analysis of the internal representations formed by the network also showed that a number of interesting, linguistically plausible features were "discovered." Hidden layer activations showed specific response to acoustic-phonetic features such as detectors for unvoiced speech, vowel onsets, and rising or falling formants.

2.2.2. *Recurrent Nets*

A recurrent architecture aimed at shift-invariant phoneme recognition was developed by Watrous (1988). His temporal flow model was evaluated on

*Higher recognition rates than those reported here have been observed in later training runs.

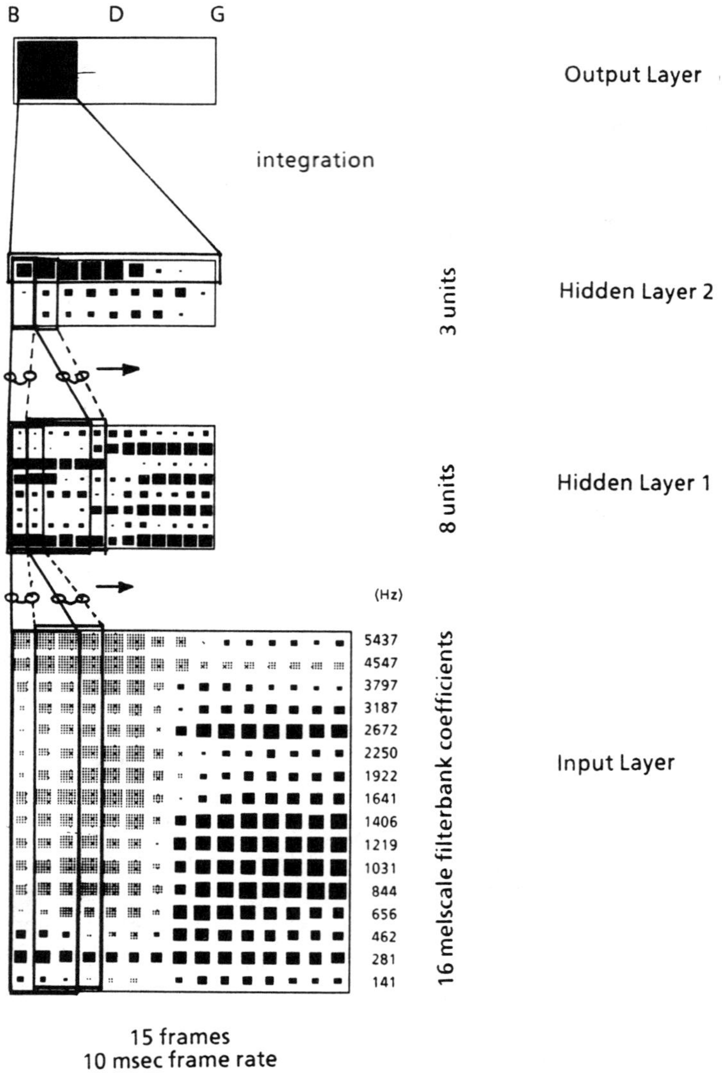

Figure 2 A time-delay neural network (TDNN) for stop consonants /b, d, g/. Eight hidden units in hidden layer 1 are fully interconnected with a set of 16 spectral coefficients and two delayed versions illustrated by the window over 48 input units. Each of these eight units in hidden layer 1 produces patterns of activation as the window moves through input speech. A five-frame window scanning these activation patterns over time then activates each of three units in hidden layer 2. These activations over time in turn are integrated into one single output decision. Note that the final decision is based on the combined acoustic evidence, independent of *where* in the given input interval (15 frames or 150 ms) the /b, d, or g/ actually occurred.

consonant (/b, d, g/) and vowel (/i, a, u/) tasks and achieved similarly good performance figures. It employed recurrent connections, nonbinary output targets (the network is trained to produce a Gaussian-distributed activation across its output nodes), and time-delayed connections. The model was applied to hand-segmented speech from a single male speaker, yielding recognition rates of 99.2% for the /b, d, g/ task and 100% for the /i, a, u/ task. Extensions of these ideas to connected multispeaker recognition of four English letters, B, D, E, and V, have also been successfully carried out by Kuhn et al. (1989).

2.2.3. Learning Vector Quantization (LVQ)

McDermott and Katagiri (1989; McDermott et al., 1990) applied LVQ, proposed by Kohonen and collaborators (Kohonen 1988; Kohonen et al., 1988b), to the same Japanese consonant recognition tasks used by Waibel et al. The input layer structure and the shift-tolerant design were motivated by experience with the TDNN and allowed for comparisons (Waibel et al., 1989a). Figure 3 illustrates the moving spectral window used as input features to the LVQ classifier and its extension LVQ2. Hidden layer connections were initialized using a traditional k-means clustering algorithm (Makhoul et al., 1985). The network achieved a 99.2% recognition rate for the voiced-stop consonants /b, d, g/ excised from spoken isolated word utterances from a single male speaker. Performance for all stops, fricatives, and affricates for the same speaker, using a larger LVQ2 network, was 97.1 and 97.7%.* Network training time was somewhat less than training time for a comparable TDNN on the same task, at the cost of a three-fold increase in the total number of connections required in the network and an increase in the time required for posttraining recognition. Unlike feedforward multilayer networks, LVQ networks do not readily generate hidden abstractions of knowledge. On the other side, vectors are produced that provide an efficient encoding of speech that can be incorporated into traditional stochastic models. The potential advantage is that LVQ is a supervised training procedure and LVQ-trained vectors provide an encoding that possibly better represents phonetically relevant featural distinctions. The resulting encoding could therefore replace traditional (unsupervised) vector quantization and yield a crisper representation of speech for improved hidden Markov modeling. Work in this direction is in progress. (Iwamida et al., 1990; McDermott et al., 1990).

*Depending on the amount of training data used.

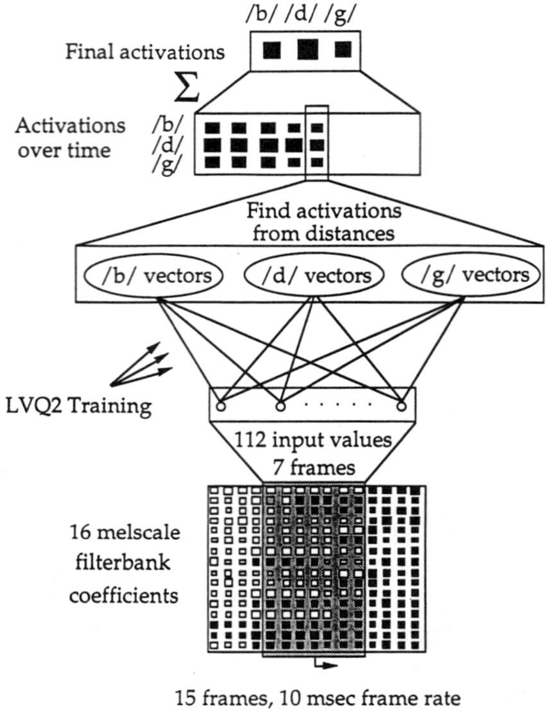

Figure 3 An LVQ network used for the /b, d, g/ recognition task.

2.3. Extensions

Several remaining problems, however, need to be addressed. The first is the problem of scaling and incremental learning, which plagues many neural network discriminant classifiers and can lead to prohibitively large amounts of training time and inflexible classifiers. It would be preferable if larger networks could be gradually built from smaller previously trained subcomponents. Second, can such networks be used to reliably *spot* phonemes in running speech, rather than to discriminate among them? Finally, how do these networks perform when faced with multispeaker or speaker-independent tasks? A number of extensions have been studied to answer these questions.

2.3.1. *Scaling*

Back-propagation training of large phonemic networks is an example of a training task that can require considerable computational resources. One approach to alleviating this training problem is the introduction of specially designed fast simulators [Haffner 88, Haffner 89]. Beyond speed improvements, however, it is clear that the full complexity of cognitive tasks requires incremental learning of new information and categories. Modular, incremental training was explored based on TDNNs to allow for more flexible, efficient, structured design of large neural network–based systems (Waibel, 1988, 1989; Waibel et al., 1989b). In doing so, one would certainly not want to lose the distributed, globally optimizing nature of connectionist learning. The underlying idea here is a compromise between these two extremes, namely to reuse the featural abstractions encoded in the *hidden* units. Hidden units from networks trained to perform smaller subtasks are linked into larger nets aiming at more complex tasks. A number of architectural schemes aimed at increasing the scale of phoneme recognition networks through an arrangement of interconnected modules were evaluated and found to yield performance as good as or better than that of monolithically trained nets. Figure 4 is an illustration of a modular stop-consonant TDNN that was trained by merging two submodules using "connectionist glue." Here connections to hidden units at the first hidden layer of a /b, d, g/-discriminating net and to the first hidden layer of a /p, t, k/ are frozen, while connections from these units to common higher layers are trained to join the two nets. The "connectionist glue" units are additional units that are free to learn missing features when these separate networks are merged (Waibel et al., 1989b). Figure 5 shows a network that was constructed in a modular fashion from subnetworks to recognize all phonemes in a Japanese large-vocabulary isolated word data base. Consonant discrimination results well in excess of 96% were achieved using such networks.

2.3.2. *Phoneme Spotting*

A related problem here is whether a network is able to spot phonemes in running speech rather than discriminate among them if a range of suitable input data is provided. Various experiments that enhance a TDNN's ability to model shift-invariant features of speech have shown that this can indeed be done. By introducing training patterns with varying time shifts around a particular phoneme boundary and counterexamples, Lang (1989) and Miyatake et al. (1990) showed that excellent letter- and phoneme-spotting performance can be achieved. The latter study reported 98% phoneme-spotting performance over running Japanese isolated word utterances. Word recognition

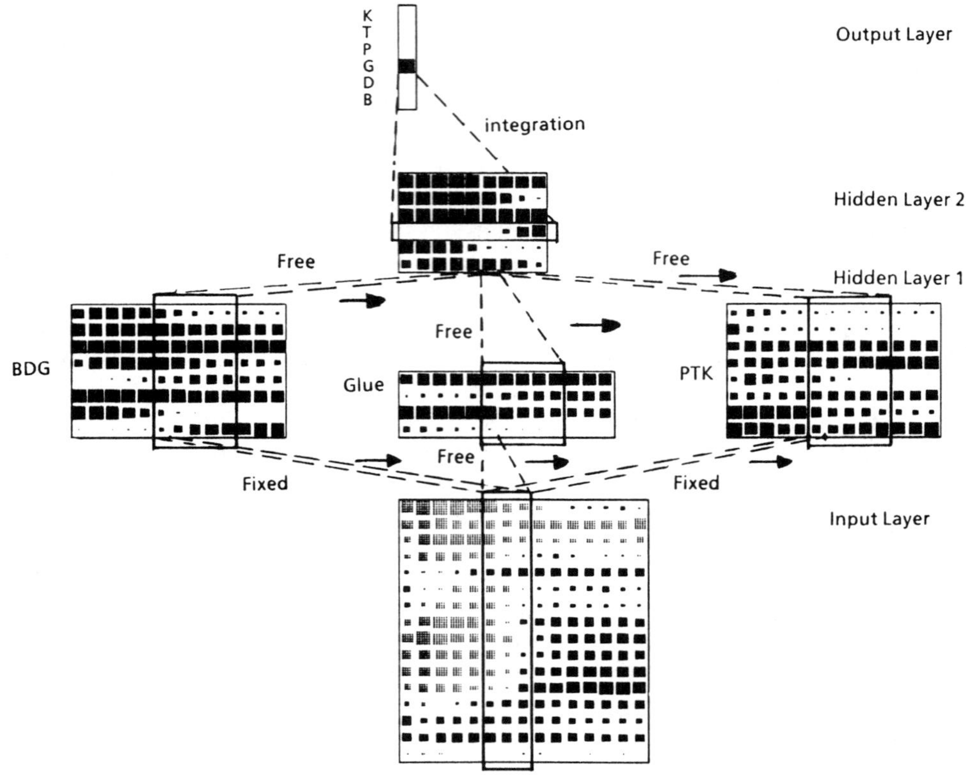

Figure 4 Incremental training: from /b, d, g/ to /b, d, g, p, t, k/.

based on these results can indeed be carried out, as we shall see in the next section.

2.3.3. *Speaker-Independent/Multispeaker Recognition*

Multispeaker and speaker-independent recognition has also been attempted by a number of investigators. As with other recognition techniques, speaker independence can be achieved by training a connectionist network using training data from many different speakers. This has been demonstrated by Leung (1988) and Leung and Zue (1988) for phoneme recognition in continuous English speech from the TIMIT speech data base. Phoneme recognition performance of up to 64% was achieved for a 16-vowel discrimination task, given excised frames of speech. If other sources of information (both numeric

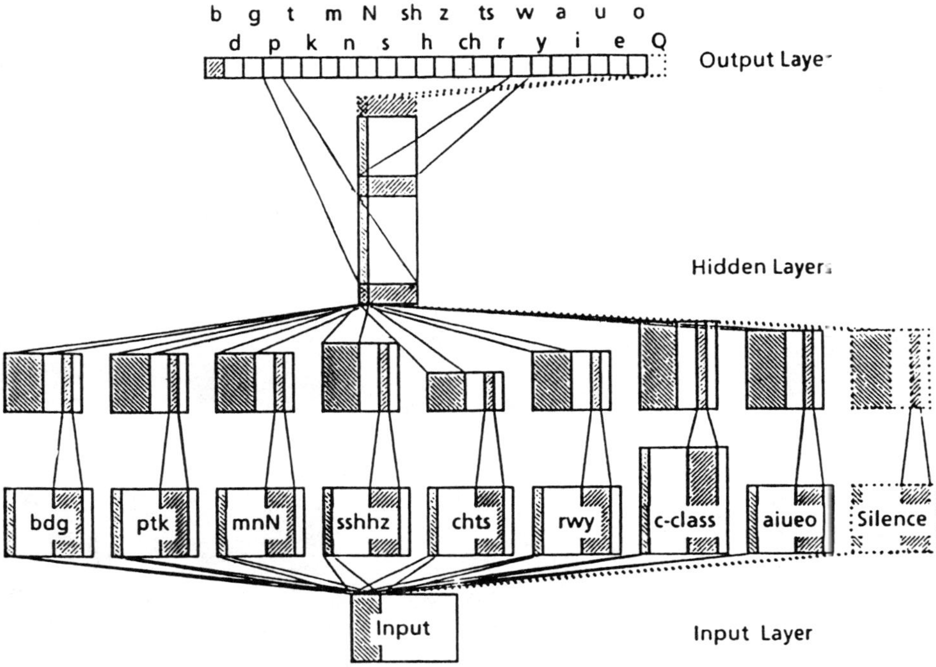

Figure 5 Modular all-phoneme TDNN.

and symbolic), such as durational and contextual information, are provided to the network, performance improves to up to 77%. The study illustrates how connectionist models lend themselves to fusing heterogeneous sources of information and knowledge gracefully.

Another model aimed at multispeaker and speaker-independent recognition was proposed by Hampshire and Waibel. It was motivated by the desire to mimic humans' ability to adapt rapidly within a syllable or two (Kato and Kakehi, 1987) to a speaker's voice. Their model seeks to represent speaker differences explicitly and provide the networks with a mechanism that rapidly focuses on a suitable speaker or set of speakers. The Meta-Pi architecture is the mechanism by which this was done. It is a hierarchical connectionist phoneme classifier that performs multispeaker phoneme discrimination at speaker-*de*pendent rates. The overall network is composed of speaker-dependent submodules (i.e., TDNNs trained to classify the phonemes of a particular speaker) that are linked together by an integrating superstructure. The Meta-Pi superstructure is itself a TDNN. The outputs of the superstructure act as connection weights, gating the speaker-dependent classification decisions of all the modules to a global classification decision. Thus, the global classification decision is a linear combination of the constituent speaker-dependent classification decisions. The connections of the Meta-Pi superstructure are trained so that the resulting combination of speaker-conditional module outputs produces a correct global classification. Individual speaker-dependent nets are trained individually, as before, as regular TDNN's. Figure 6 shows how the error signal obtained from the global classification output then back-propagates through multiplicative con-

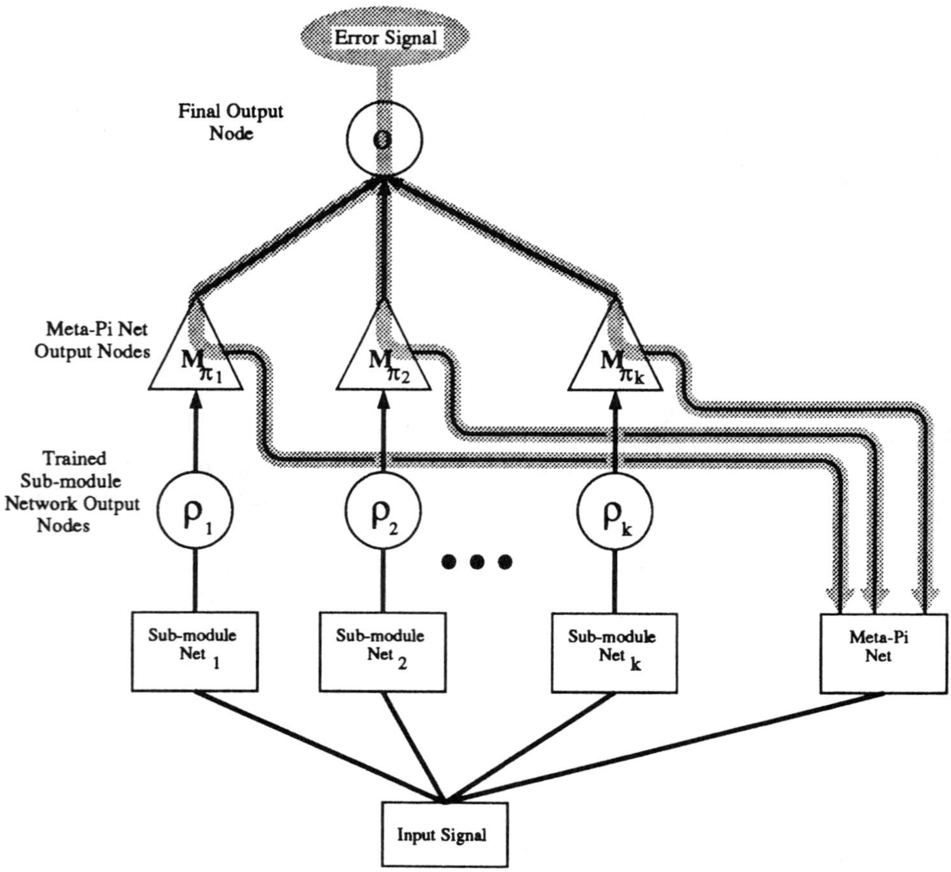

Figure 6 Error back propagation in a Meta-Pi architecture.

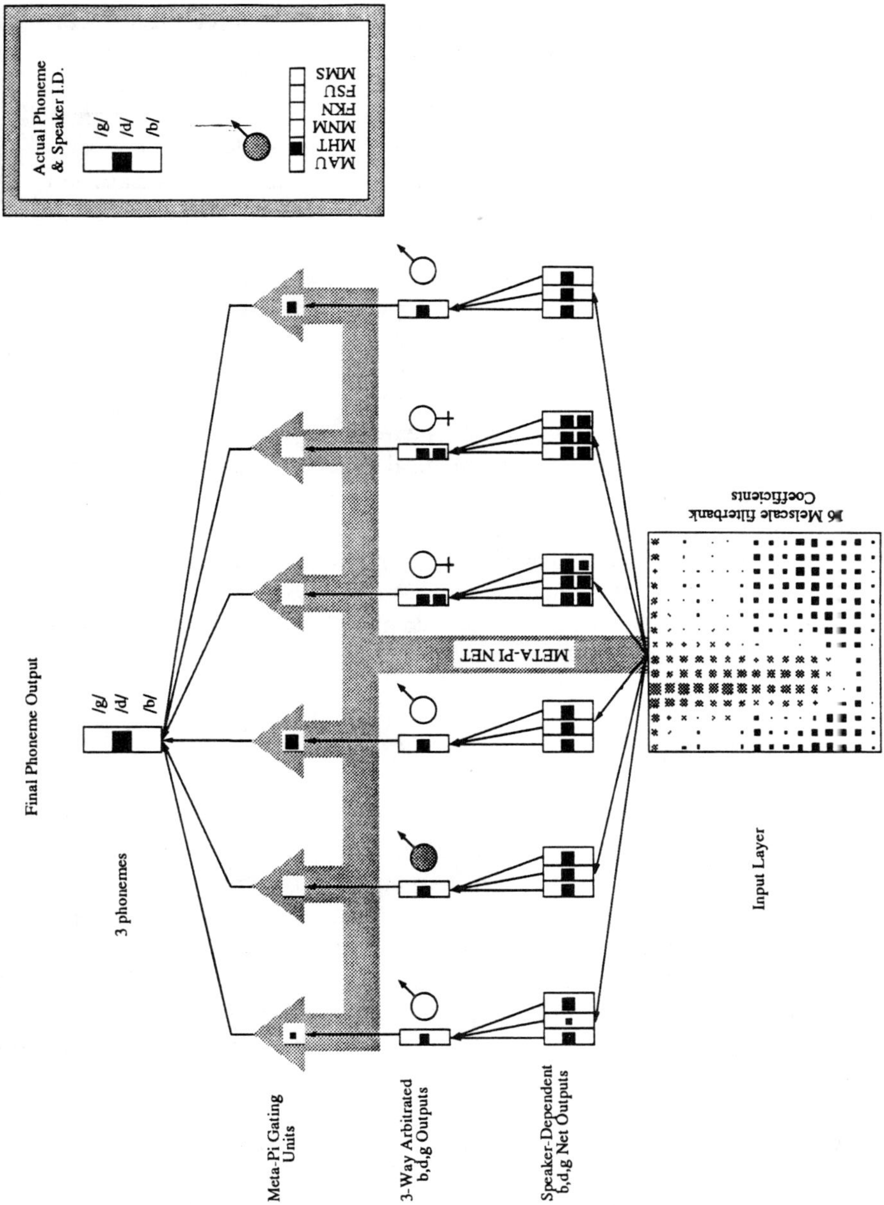

Figure 7 Typical activations in a Meta-Pi network for multispeaker phoneme discrimination.

nections into the Meta-Pi superstructure to optimize the integrating Meta-Pi net. Figure 7 illustrates the fully trained Meta-Pi architecture performing the /b, d, g/ recognition task on speech from six speakers (four male, two female). Here a /d/ token from speaker MHT is input. The Meta-Pi structure gates a mixture of classification decisions from three male speakers' subnetworks to produce the correct overall classification decision. A Bayesian analysis of the Meta-Pi architecture indicates that the classifier performs robust multispeaker phoneme discrimination (98.4%) by learning to identify and separate relevant speaker types associated with particular signal patterns. This result approaches the mean speaker-dependent rate of 98.7% and significantly improves the discrimination rate of 95.9% obtained for a single TDNN trained on all six speakers' training data. This unconstrained form of rapid speaker adaptation could be extensible to speaker independence, where speaker-specific subnetworks could be replaced by nets responsible for groups of speakers or where new prototypical speaker (group) specific nets are incrementally added only if this should help to improve performance.

Many other improvements in phoneme level recognition, particularly in view of speaker independence, have been attempted, including the use of improved objective functions (Hampshire and Waibel, 1990), phonetic features motivated by speech knowledge (Bengio et al., 1989), and different input representations (Kamm, 1989).

3. THE WORD LEVEL

Although good results have been achieved using neural networks for phonemic patterns, the question remains whether this technology can be used effectively for word recognition as well. An early set of experiments simply extended the classification capabilities of these networks by applying an entire word's coefficient matrix to the inputs of static full-word networks with output units for each word to be classified. Good results were achieved, but time alignment and word end point detection are problems that limit this approach. Similarly limiting is the fact that only small vocabularies can be handled in this fashion, because network size and training time become prohibitively large with increasing vocabulary size. To overcome the former limitations, networks that model time alignment and/or shift invariance internally have been developed for small-vocabulary recognition. For large-vocabulary recognition, subword units such as phonemes or syllables must be employed. A number of novel techniques are emerging that attempt integration of connectionist subword models into words and sentences. A majority of them could be characterized as hybrid techniques, that is, techniques that seek to combine the perceived strengths of neural networks at the pattern recognition level with the strengths at modeling stochastic sequences of conventional methods such as hidden Markov models, Viterbi decoding, or dynamic programming.

3.1. Static Full-Word Models

Among the early static full-word models, Lippmann and Gold studied a number of back-propagation network architectures applied to the task of isolated digit recognition (Lippmann and Gold, 1987). Seven isolated monosyllabic digits were obtained from the TI Isolated Word Database representing speech from 16 different speakers. The speech data were sampled at 12 kHz, windowed, and a discrete Fourier transform performed; preprocessing produced 15-coefficient Mel-scale spectra at 10-ms intervals. These spectra were used to develop two 11-point cepstra offset by 30 ms in time; the cepstrum was taken from the maximum acoustic energy segment of each digit. These cepstra served as input to a series of networks all having 22 input layer nodes and 7 output layer nodes (corresponding to the 7 digits). Seventy training and 112 testing tokens were obtained for each speaker, and networks were trained and tested for single speakers only. A three-layer network yielded the best connectionist recognition performance of 92.3%, averaged over all 16 speakers.

Peeling and Moore (1987) also ran experiments with isolated digit recognition. They used a three-layer network with 50 hidden-layer nodes. Sixty 19-coefficient spectra taken at 20-ms intervals were used as input in order to capture the longest duration utterances. Shorter utterances were zero padded and time shifted randomly in the network input "window." Isolated digit speech data were taken from the 40-speaker Royal Speech and Radar Establishment (RSRE) data base. Speaker-dependent recognition under these conditions was 99.7%.

Burr (1988a) conducted a series of experiments in isolated E-set and polysyllabic word recognition using a single-layer perceptron. The network input comprised twenty 64-coefficient spectra; in separate experiments these spectra were computed using smoothed DFT and linear predictive coding (LPC) techniques. Input tokens were temporally aligned in the spectral "window" using a DP time alignment procedure. Five tokens of 20 polysyllabic words containing three to five syllables were recorded from a single male speaker. Training tokens were also used as testing tokens in this experiment—under these conditions, recognition rates were, not surprisingly, nearly 100%. Burr also ran recognition experiments on single-syllable words recorded from a single male speaker. Twenty tokens of each of the nine single-syllable E-set words were obtained. Half of the tokens were reserved for training and half for testing. Recognition accuracy under these conditions was 91.4%. Word recognition was increased to 98.2% following modifications to the network's input layer structure and spectral estimation methods; these modifications focused network activity on the first 40% of each word.

3.2. Dynamic Full-Word Models

Word recognition of static classifiers is sensitive to time alignment and needs to rely on end point detection, as each connection represents a specific portion of an utterance. During fast and slow speech the relative position of acoustic features will generally be distorted, and a static net would only be able to blur its internal representation to compensate for such distortions. Word end points cannot always be determined reliably (particularly in noise and in continuous speech). Several connectionist models described here seek to overcome some of these problems.

3.2.1. A Word Level TDNN

Bottou (1988) used a large TDNN for recognition of small vocabularies and a novel time-warping approach to increase the temporal variance of isolated words and achieve shift-invariant speaker-independent word recognition on five consonant-vowel French words (Bottou et al., 1989). Single exemplars of each word were obtained from six speakers. Speech from four speakers was used for training and speech from the remaining two speakers constituted testing data. The data were sampled at 10 kHz and used to compute 256-point DFTs at 12.8-ms intervals. These spectra were reduced to 16 spectral mel-scale coefficients covering a frequency range of 100 Hz to 5 kHz. These formed the input to a 65-time-frame TDNN input layer. The TDNN architecture was in principle similar to the one described above except that the higher-layer units undersampled (skipped every other) activations from lower-layer outputs for greater efficiency. Bottou took the original 20-token training set and created a total of 400 additional training tokens by time warping the original set independent of phonetic structure. The extent of warping ranged from warping 80% of the word into 50% of the TDNN input spectra to warping 50% of the word into 80% of the input spectra. Occasionally, warping was so extreme that it eliminated consonant portions of words. The TDNN was trained on the original 20 tokens, plus these 400 "synthesized" versions. After training, Bottou achieved 100% recognition on all 20 original training tokens and 94% recognition on the 400 warped tokens.* The recognition rate on test data was 90% using this technique of artificially expanding the training set by means of temporal warping. In a separate experiment involving word recognition on the TI 20-word data base, Kammerer and Kupper (1988) realized a 30% reduction in the number of classification errors on test data by using a similar time-warping technique to artificially increase the size and variance of their training token set. Their recognition results were 99.6% for speaker-dependent experiments and 97.3% for a speaker-independent trial.

3.2.2. *Tank–Hopfield Time Delays*

Tank and Hopfield (1987) developed an analog neural network model for recognizing particular stimulus sequences (comprising letters of a word) that were slightly distorted and embedded in larger sequences. The network employed a series of detectors $D_A \ldots D_X$ for single elements of a stimulus sequence; each of these detectors was replicated over a series of time delays, allowing the network to detect a single element of the sequence of interest across a range of time segments $f_1(t) \ldots f_k(t)$. Appropriate combinations of these time-shifted detectors fed a recognition unit V corresponding to the precise sequence to be detected. Inhibitory connections between recognition units minimized network output for stimulus sequences not closely matching the desired sequence. The network was very effective in locating distorted letter sequences embedded in larger sequences. In follow-on work, Unikrishnan et al. (1988) used this same network paradigm to achieve a 99.3% recognition rate on random sequences of digits.

3.2.3. *The Dynamic Neural Net*

Sakoe (1987) and Sakoe et al. (1989) developed a dynamic programming neural network (DNN) for speaker-independent word recognition. This network employed a three-layer back-propagation architecture capable of dynamically warping its input. The input layer consisted of a series of 10-coefficient Mel-scale spectra taken at 16-ms intervals. These spectra were linked in groups of two frames to single groups of four hidden units; each hidden unit group represented a temporal shift from its predecessor. All hidden layer unit groups were fully connected to a decision output unit corresponding to one of ten spoken digits. Speech from 50 speakers was used to train the networks in

*The relatively low rate for the warped training set was due to the extreme warping performed on a small number of those tokens.

two ways. In a temporally prewarped training method called "fixed time alignment," all training tokens for a particular word were time warped to a standard pattern prior to training. In an alternative training procedure called "adaptive time alignment," each token of a word was interactively warped in order to produce the maximum output activation of the network. Once the adaptive alignment was complete, the back-propagation iteration for that token was performed. Recognition performance was tested on tokens obtained from 57 speakers (none of whom were used for training). Recognition rates were 97.5% for networks trained with the fixed time alignment procedure and 99.3% for networks trained using the adaptive time alignment procedure. The added computational cost of the recognition improvement afforded by the adaptive time alignment training procedure was substantial.

3.2.4. Predictive Neural Nets

Most connectionist models that we have discussed so far apply neural nets as classifiers of either word patterns or subpatterns. For classification, the input usually consists of a coefficient matrix and the output approximates a bit pattern representing the classification results. In addition to learning discrete classifications, however, neural networks can also learn nonlinear mapping functions between real-valued inputs and outputs. This can be exploited in speech for various signal mapping and coding applications, including noise suppression (Tamura and Waibel, 1988) and nonlinear predictive coding (Lapedes and Farber, 1987). The neural networks have been used successfully in the neural prediction model proposed by Iso and Watanabe (1990) and the hidden control neural network proposed by Levin (1990). Both of these models have so far been limited to small-vocabulary recognition tasks (i.e., digits), but they appear to yield high speaker-independent recognition performance. Extensions are also possible, as we shall see later.

The basic idea is illustrated in Fig. 8. A two-frame window of input coefficients is input into a multilayer feedforward net trained to produce at its output a frame of coefficients that is as close as possible to the next (future) speech frame. The distance between this predicted frame and the actual next speech can be measured as a prediction error or distortion, and this distortion is used as an error criterion for back-propagation training. Given a set of predictor networks, one can imagine training each predictor for a separate region of an utterance. Each predictor net becomes specialized to best predict this portion of an utterance, so that the prediction error is likely to be lowest in these regions. A word is then represented by the *sequence* of predictor nets that best predicts the actual observed speech. Dynamic programming is used as a mechanism to apply each predictor sequentially over time to best ap-

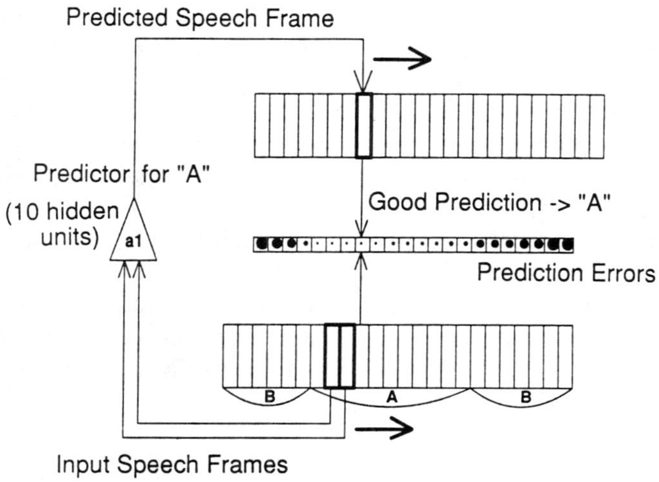

Figure 8 Modeling a phoneme by signal prediction.

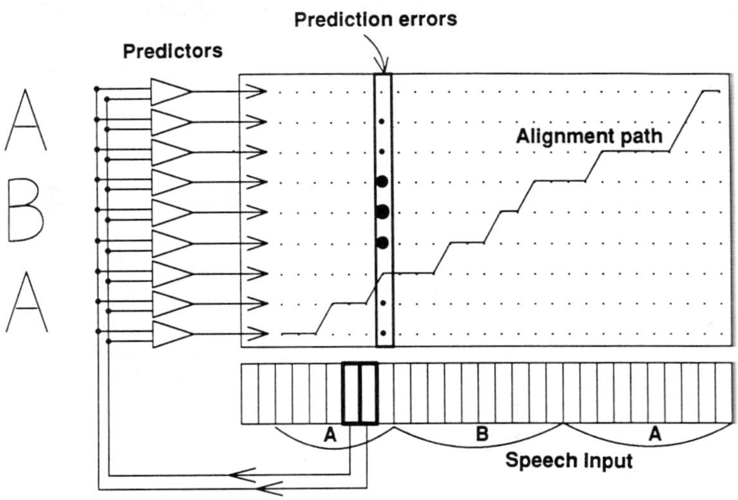

Figure 9 A neural prediction model.

proximate the actual signal. Figure 9 shows this alignment step based on the matrix of distances between actual speech frames and predicted frames. During training an alignment path is determined by dynamic programming. Each predictor is then trained to minimize the error between its output and the speech frames that it was assigned to predict by the DP alignment path. During recognition the word whose sequence of predictors minimizes the error between predicted frames and actual signal frames is chosen. Iso and Watanabe (1990) used 10-mel-scale cepstral coefficients and amplitude change as inputs to their networks. The number of predictors used depended on the utterance and ranged (for Japanese digits) between 9 and 14. Each predictor net has three layers, an input layer of two 11-coefficient frames, 9 hidden units, and 11 predicted output coefficients. Excellent performance (0.2% error) was reported for a Japanese speaker-independent isolated digit recognition task uttered over telephone lines. This result compared favorably with other techniques—0.7% for the DNN (Sakoe, 1987; Sakoe et al., 1989) and 1.1% for DP matching (Sakoe and Chiba, 1978) tested on the same data.

The model proposed by Levin is similar to the one described above and is illustrated in Fig. 10. As before, it uses nonlinear prediction by neural nets to measure a model's fit to the input data. Unlike the neural prediction model, however, it uses only one single predictor for an entire word and a sequence of varying input flags or "control units" that switch the predictor into alternate

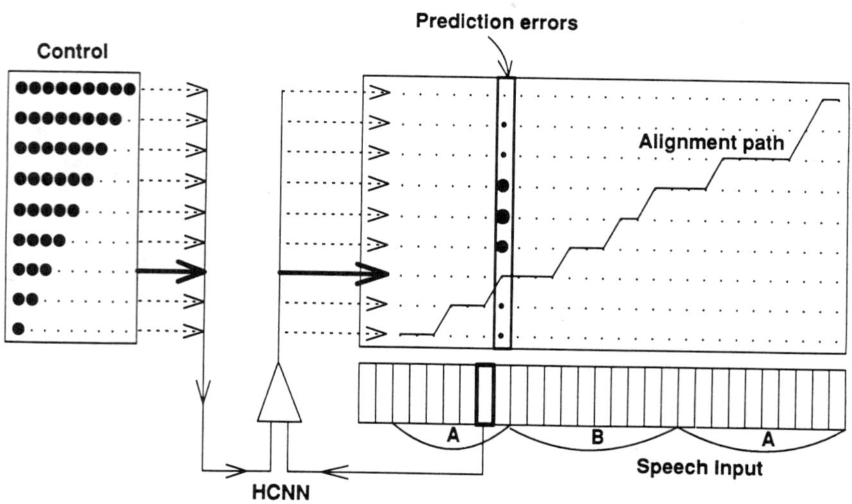

Figure 10 The hidden control neural network.

modes of operation as time progresses. Similar to "counter nodes"* (Kukich, 1988), these units are used to control the sequential state of the network. The predictor network used by Levin had 24 speech inputs (12 cepstral and 12 deltacepstal parameters), 30 hidden units, 24 predicted outputs, and 8 control input units. The control units turn on sequentially when appropriate and remain on as additional bits are activated ("thermometer" representation). The correspondence between an input speech frame and a control transition (when a new bit is turned on) is determined by Viterbi alignment. During training, the Viterbi algorithm determines the state of control unit settings for each speech input frame given a trained predictor net. Based on this state, then, the prediction error produced by the predictor net is reduced by back-propagation learning, given each frame of input speech and its corresponding control unit setting. The network was tested on connected digits from the TI-digit data base (using male speakers only). Using independent test data but from the same speakers used in training, a word recognition rate of 99.3% was achieved.

3.3. Dynamic Large-Vocabulary Models

Given the encouraging results for the connectionist models reviewed in the previous section, we would now like to explore possible extensions to large-vocabulary recognition. The most significant difference here is that large-vocabulary word recognition models must rely on units smaller than the word (such as syllables or phonemes) to decompose the large number of words into a limited set of atomic subunits that can be trained and optimized for use in (ideally) any vocabulary. How can connectionist models contribute usefully to this problem?

Most popular at present are so-called hybrid models, which seek to combine the perceived strengths of connectionist models with those of more classical recognition techniques such as dynamic programming or hidden Markov modeling. In this approach, connectionist models are viewed as high-performance nonlinear classifiers or predictors that could replace distance metrics, or vector quantization steps commonly found at the front end of most typical recognizers. Dynamic programming, Viterbi alignment, and/or hidden Markov models are then viewed as mechanisms for providing the additional constraints that phonemes must be of acceptable order, duration, and likelihood to produce a legal word.

---

*Proposed to control state information in networks used for spelling correction.

3.3.1. TDNN-LR-DP

Based on the Japanese large-vocabulary isolated word data base described before (Sagisaka et al., 1987; Waibel et al., 1987, 1989a), a number of experiments were carried out to extend TDNNs to large-vocabulary recognition. In this approach a TDNN is trained as before to classify input speech into one of several phoneme output categories over running speech (in this case entire words spoken in isolation). As the original TDNNs were trained on excised phoneme tokens only, several enhancements were introduced. First, the original excised phoneme training patterns were artificially misaligned in time by various offsets. This was particularly effective for phoneme *spotting* as opposed to discrimination, as it enforced shift-invariant phoneme classification even in transitory regions between phonemes. In doing so, the phoneme spotting rate was improved from 95.8 to 98.0% and, more important, the false alarm rate* decreased from 62.2 to 23.2%.† For word recognition a silence category was necessary, which was added by modular design to the existing net (Miyatake et al., 1990).

A total of 24 phonemes (5 vowels, 18 consonants, and silence) were spotted in this fashion by TDNNs shifted across time providing the front end for phoneme-based word recognition (Fig. 11 shows typical phoneme output unit firings over time). To do so, the output categories were considered to be output probability estimates of the likelihood of each phoneme occurring in a particular position in time. An LR parser provided top-down predictions of the set of phonemes that are legal under a given dictionary. The likelihood of a given phoneme predicted by the LR parser is evaluated at the outputs of the TDNNs at each time frame. Dynamic time warning (DTW) is then applied to find an optimal alignment between TDNN outputs and the predicted phoneme states. For duration control, each phoneme state was expanded to the average number of frames of that phoneme before DTW was carried out. Recognition experiments on various vocabularies were carried out with this system. All experiments were independent of vocabulary‡ and were performed on independent test data (phonemes not used for training). For a 500-word test vocabulary, first choice accuracy of 98% was achieved. For a large vocabulary of 5000 words, recognition rates as high as 92.6% were obtained. Second and fifth choice rates for the latter vocabulary were 97.6 and

*Presumably due to previously undefined transitory regions.
†All recognition tests were run on independent test data from the same speaker.
‡The phonemes used for training were extracted from words of a different vocabulary than the one used for testing.

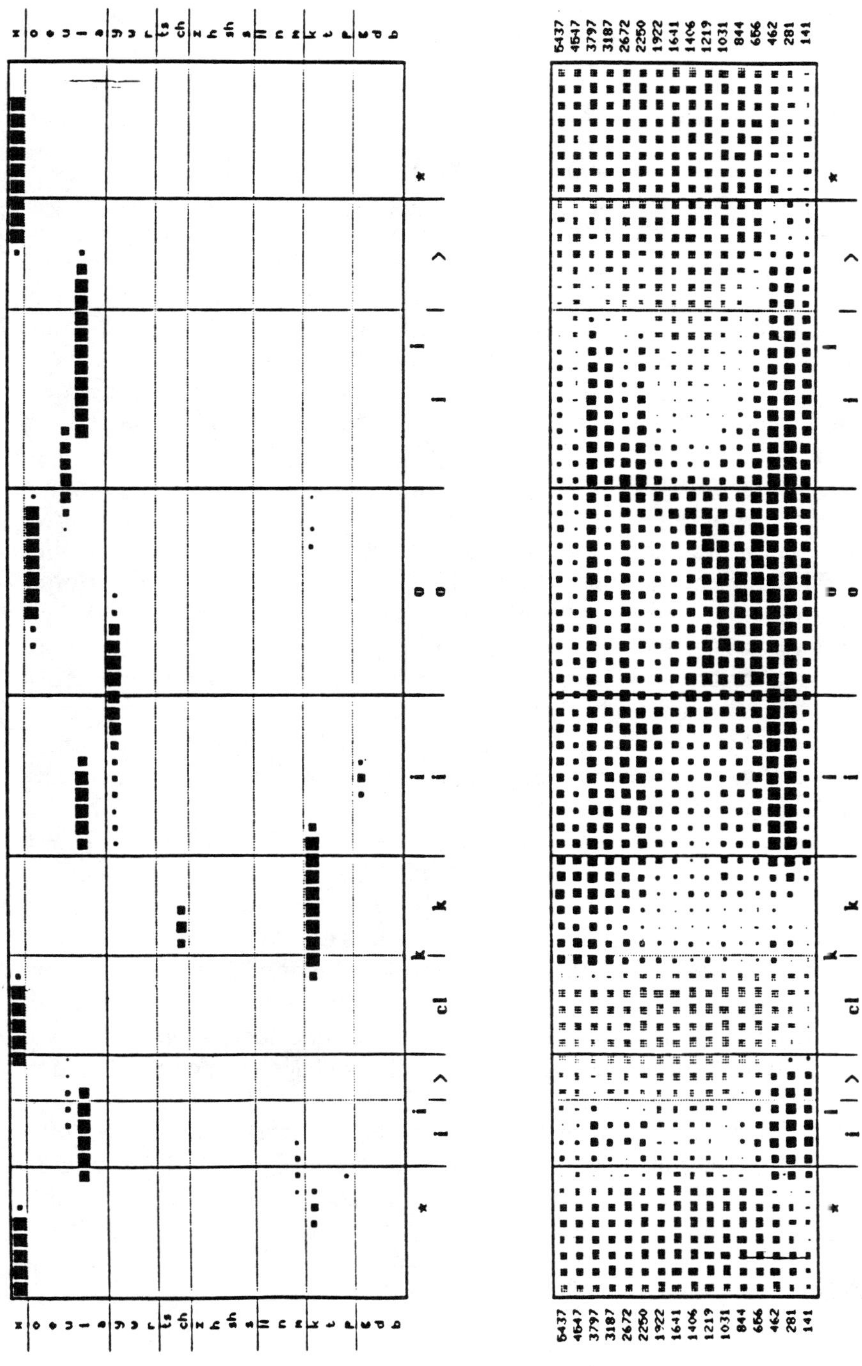

Figure 11 TDNN phoneme spotting output for word "ikioi."

99.1%, respectively, indicating that most confusions occurred among a small group of acoustically similar words (e.g., "itai" → "ittai").

3.3.2. Neural Nets and Hidden Markov Models

Some of the earliest attempts to combine the strengths of neural net classifiers with traditional word modeling techniques were developed by Bourlard and Wellekens (1988, 1989) and Morgan and Bourlard (1990). In theoretical and experimental work they have shown that the outputs of a multilayer perceptron (feedforward network) trained by back propagation of a mean square error may be considered to be estimates of the maximum a posteriori probabilities of a corresponding class. They have since built on this notion to construct hidden Markov model chains where the output activations of a local multilayer perceptron (MLP) are used as output probabilities for the states in a traditional HMM. Viterbi alignment is performed to assign the framewise MLP firings to corresponding states and to compute an overall word output probability. Not unlike HMM systems, an iterative optimization procedure was then introduced [Bourlard 90] that combines Viterbi alignment with back-propagation learning. This procedure performs a forward pass on the MLPs over time and assigns these outputs to their respective best matching states by way of Viterbi alignment. Back propagation of errors is then performed to improve the outputs of the MLPs given this assignment and the procedure iterates. In many of their experiments the MLP consisted of a three-layer feedforward network with nine binary 132-bit input vectors encoding the input spectrum over nine frames, and 50 outputs representing each phoneme. A variable number of hidden units was used. Several additional modifications have been added to achieve good performance (Morgan and Bourlard, 1990). First, the output a posteriori probability estimates produced by the MLPs were normalized by their respective prior probabilities to eliminate a bias due to uneven distributions in number of tokens per phoneme. Second, extra word transition penalties were introduced to reduce insertion errors. Third, a cross-validation learning scheme was applied that improved generalization considerably. Here the performance of the trained networks was checked against a cross-validation "test" set during each iteration to determine at what point test performance started to degrade (i.e., overfitting to the training data). Training was then halted. In speaker-dependent German large-vocabulary (918 words) continuous speech recognition experiments, word recognition performance currently approaches 60%. In the absence of a language model, this compared favorably with an HMM evaluated over the same data under similar conditions (Morgan and Bourlard, 1990).

Franzini et al. (1989) have investigated two similar methods for combining neural nets with hidden Markov models to perform continuous speech recognition. These methods were applied to speaker-independent recognition of continuous digit strings. In the first method, a network is trained to identify the word and phone to which a given frame of speech belongs. A four-layer recurrent network is used with LPC cepstrum coefficients over 70 ms of speech as input and one output unit corresponding to each word and each phone. In order to generate word and phone labels to use in training the network, a conventional discrete HMM is trained on the task, and a forced Viterbi alignment is performed with the HMM. The network is then trained on the labels generated by the alignment. During recognition, the network's input window is shifted, frame-by-frame, across an entire sentence, generating a vector of outputs for each input frame. Subsequently, an HMM is used to combine these output values into a final sentence hypothesis. The output values are treated as output probabilities associated with transitions in the HMM, and the recognition is performed as in conventional HMM systems, using a Viterbi search.

The second method (Franzini et al., 1990) also used a four-layer neural net with HMM postprocessing, but optimization of the two components is more tightly coupled. This method, connectionist Viterbi training (CVT), is a variant of the Viterbi training (or segmental k-means) method for training HMMs. As before, a discrete HMM is trained and an initial force alignment performed using this HMM. The network is then iteratively optimized by repeated application of back-propagation training, Viterbi alignment, and reestimation of transition probabilities. A cross-validation set was also used as a halting criterion. A second important difference from the previous method is that the network's output nodes model output probabilities corresponding to states in an HMM phone model. Such phone models can therefore be used to construct different words, and the system is extensible to large-vocabulary recognition. Both of the methods described, although still being investigated, have produced good results. The first method achieved 97% word accuracy on the Texas Instruments continuous digits data base, and the CVT procedure has reached 99% word accuracy on the TI digits.

3.3.3. *Linked Predictive Neural Networks*

Another word level model is an extension of work discussed in the previous section for small-vocabulary recognition, i.e., the use of neural nets as nonlinear predictors of speech. For use in large-vocabulary recognition, words must here again be decomposed into subword units such as phonemes or syllables and an optimal model for these units must be trained. Work by Tebelskis and

Waibel (1990) has demonstrated that this can be done without the need for segmentation by jointly optimizing time alignment and connection weights and by linking the weights (as in Waibel et al., 1989a) of sets of network predictors corresponding to the same phoneme symbols. Experiments with the linked predictive neural network (LPNN) resulted in 94% recognition performance for speaker-dependent isolated word recognition over a data base of 234 confusable Japanese words and 90% over a confusable 1000-word vocabulary.*

The operation and training of the LPNN are shown in Fig. 12. As in the neural prediction model, a set of predictors are assigned to different portions of a word. Here these portions are defined to be phonemes and each oc-

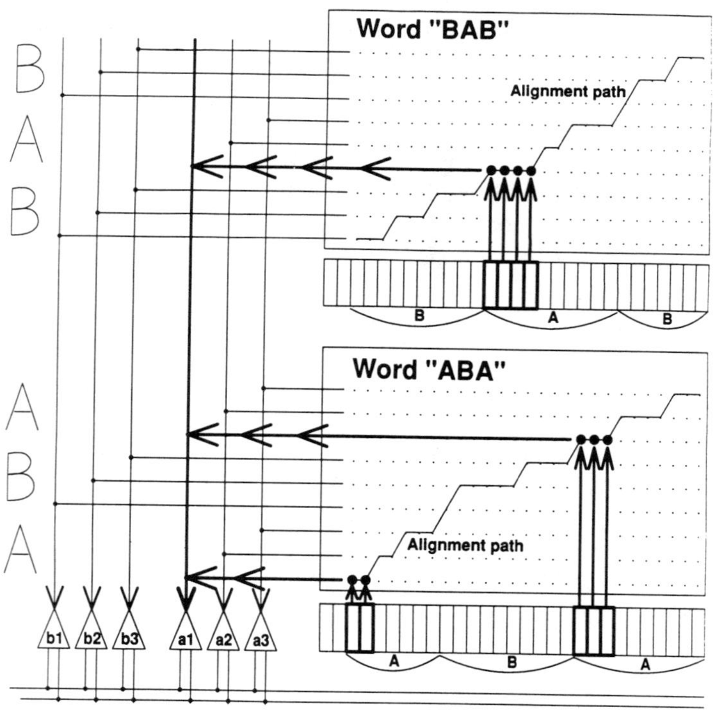

Figure 12 LPNN training.

*The data in this evaluation were a confusable subset from the Japanese large-vocabulary isolated word data base used in other experiments discussed above.

currence of the same phoneme is modeled by the same set of predictors. In Fig. 12, for example, two words "BAB" and "ABA" may consist of the same phonemes in different order and position. Time alignment over the sequence of predictors is done as before, but all prediction errors assigned to the same phoneme (or portion thereof) train the same predictor net by way of a linkage pattern that defines the legal phoneme sequence of a word. A number of enhancements of this basic scheme have been found effective. A set of parallel predictors was added to each phoneme model to allow the LPNN to better represent alternate pronunciations and context dependences. An assignment of each alternate was not predetermined, but the system decided itself which alternate to use based on the prediction errors produced by each alternate. Significant improvements were also obtained when phoneme pairs that are distinguishable only on the basis of duration (e.g., in Japanese: "k" vs. "kk") were represented by different sets of predictors. Figure 13 shows an example of processing in the LPNN. Here the prediction errors obtained by each phoneme state predictor during the word "shikisai" and the corresponding alignment path are shown.

3.4. Connectionist Word Level Enhancements

In addition to the connectionist word models described so far, other connectionist solutions to the word recognition problem and enhancements have been reported. These include connectionist alignment strategies and connectionist postprocessors (to enhance discrimination performance at the word level). Further advances are likely to emerge from theoretical work (in progress) aiming at finding unified formulations for connectionist and stochastic models of speech (Bridle, 1990; Young, 1990).

3.4.1. Viterbi Net

A connectionist solution to the problem of alignment and sequential control was proposed by Lippmann and Gold (1989) and is called the Viterbi net. Figure 14 shows this network. The triangular nodes of the network corresponded to single nodes in an HMM word model; each of these nodes performed a thresholding and time delay function. Input layer nodes accepted mel and differential mel cepstra updated at 10-ms intervals. Connection strengths between input and HMM nodes were set to values obtained by conventional HMMs. The small subnetworks feeding input to the HMM nodes were used to select the maximum of two competing inputs. This network achieved a 99.4% word recognition rate—virtually identical to that achieved by nonconnectionist HMM recognizers. The network might, however, be of

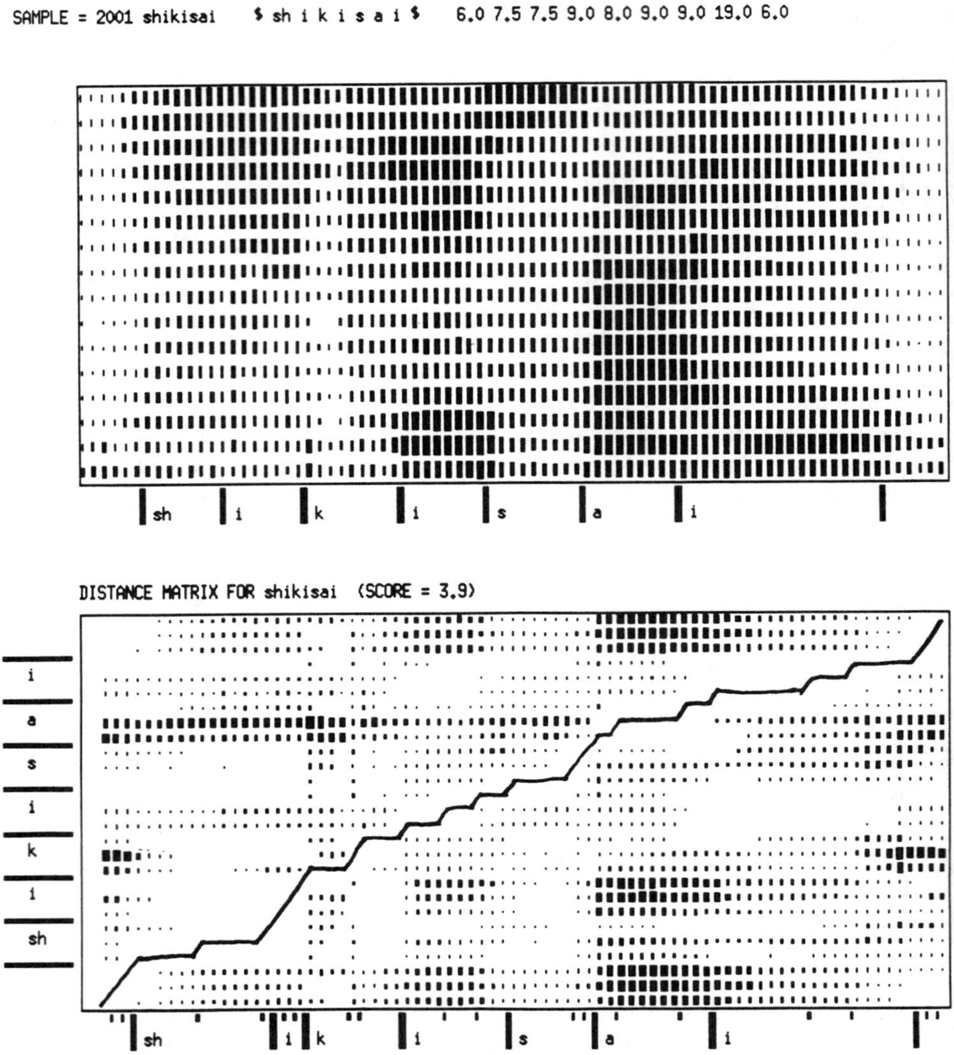

Figure 13 Prediction errors for different phoneme predictors during the word "shikisai."

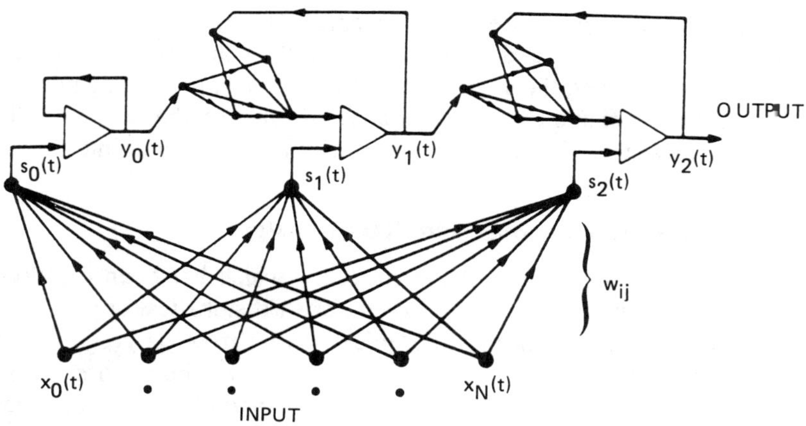

Figure 14 Lippmann and Gold's Viterbi net.

interest for hardware implementations as well as provide a basis for trainable connectionist extensions of Viterbi decoding.

3.4.2. *Connectionist Postprocessing*

Another enhancement employed successfully by several investigators is to use connectionist models as postprocessors. In this approach recognition is first carried out by DTW or HMMs. A connectionist classifier is then applied to discriminate between potentially confusable pairs before a final recognition decision is made. The study previously reported by Burr (1988b) is one example. Prealignment was done by DTW before a connectionist classifier was applied. In a similar vein, (sometimes considerable) improvements were achieved by connectionist postprocessors (multilayer perceptrons or LVQs) following a hidden Markov model-based recognition pass (Huang et al., 1988; Katagiri and Lee, 1990).

4. THE LANGUAGE LEVEL

Beyond recognition of words, connectionist models have also been applied to language models and natural language processing. The attempts are driven by the desire to develop more robust and perhaps cognitively plausible models of language. Indeed, some recent work that we review here suggests productive uses for the development of spoken language systems. Of interest in this regard are the possibility of representing complex syntactic and semantic relationships stochastically and the hope to learn them automatically from text (or speech). Moreover, the uniformity of connectionist processing may allow for complex codes that include nonsyntactic information that may be difficult to incorporate by traditional means (e.g., pragmatic or prosodic information).

4.1. Word Prediction, Coding, and Disambiguation

A very direct approach to language modeling which has been explored is connectionist N-gram modeling. Straightforward statistical approaches to N-gram models become intractable quickly as N grows large. The number of parameters to estimate grows exponentially, and the requirements for sample data grow excessive. Nakamura and Shikano (1989) have proposed the NETgram, a connectionist network architecture which learns N-gram models efficiently. The basic bigram network had a localist input-output representation for word category, and the network was trained to predict the category of the next word given the current word by using error back propagation. A NETgram architecture for a particular value of N words is constructed by augmenting a trained NETgram for $N - 1$ words. In tests on the Brown corpus, the architecture had comparable performance to the traditional statistical method, but the number of parameters in a NETgram does not increase exponentially with N.

Rather then predicting the next word given a sequence of previous words (or word categories), networks have also been used as autoassociators or disambiguators to learn suitable codes for language. An example of the former is PARSNIP (Hanson and Kegl, 1987), a three-layer feedforward back-propagation network that was trained to reproduce an input word category sequence at its outputs by way of a set of hidden units. The units develop during training a code for language that can usefully incorporate information that extends beyond the first one or two previous words. An extension of this idea that could perhaps be applied more directly to speech was proposed by Benello et al. (1989). Here a multilayer feedforward net was trained to produce an unambiguous word category as output, given an input that consisted of current ambiguous word categories and several unambiguous (known) preceding word categories. After training, the network could correctly disambiguate 95% of the words in previously unanalyzed text.

4.2. Recurrent Networks

Recursive network architectures have also been shown to learn fairly difficult syntactic relationships. The explicit purpose for their recurrent connections was to provide a network with state sequence information. Early connectionist research focusing on the design and training of recurrent networks to do this was done by Jordan (1986) (see Fig. 15), and his work spawned several other papers on the subject. Various forms of recurrence were used to achieve temporal sequencing and shift invariance for time-varying signals including robotic control, speech production, and phoneme and word recognition, but several applications to modeling syntax now also exist. Elman and Zipser (1988) developed a three-layer network with "context" units that formed a feedback mechanism between the hidden and input layers of the network. Using this structure (very similar to that illustrated in Fig. 16), they ran a series of experiments to assess the network's ability to represent temporally sequential relationships in the input data. Network performance was judged on its ability to predict future input states, given present input state and former internal (hidden) state. In effect, the network was tasked with learning discrete state-space trajectories. They successfully trained the network to predict follow-on states for a set of three discrete trajectories in one experiment. In a more complex task they trained a similar network with 200 variable-length sentences generated from a 15-word lexicon. The training was conducted with the objective of correctly predicting the next letter of the sequence representing a given word in the lexicon. The trained network performed the task consistently; prediction errors were typically high for the first letter of each word and dropped rapidly (indicating high-confidence predictions) as the letter stream corresponding to the word was processed.

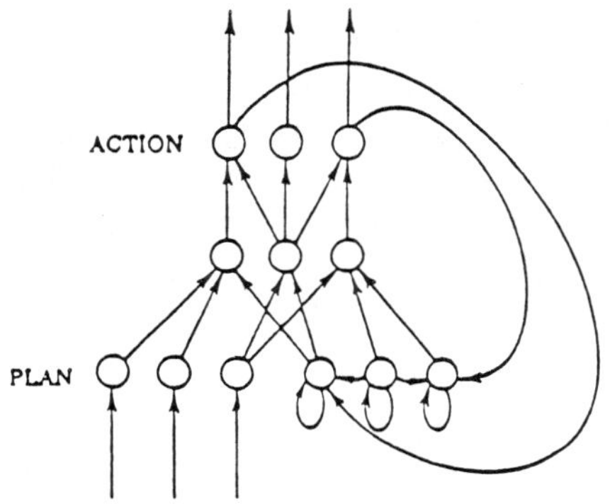

Figure 15 Recurrent network used by Jordan (1986) to generate an unfolding output sequence given a static input plan.

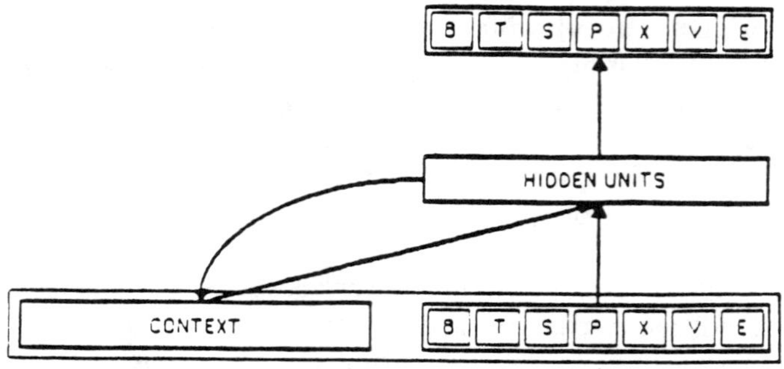

Figure 16 The recurrent network form used by Elman and Servan-Schreiber in their word recognition experiments.

Servan-Schreiber et al. (1988a,b) expanded on Elman's work using the same recurrent connection paradigm. They trained a recurrent network with 200,000 strings of varying length drawn from a finite-state grammar. After training, the network was tested with 20,000 strings drawn randomly from the 200,000-string training set. Since substrings of different full strings could be identical—thereby leading to different predictions for next state—performance measures accounted for multiple predictions of follow-on states. Under these criteria, the network predicted next states flawlessly for all 20,000 "test" strings. When tested with 130,000 strings, only 0.2% of which were consistent with the finite-state grammar, the network rejected all 99.8% nongrammatical strings while it correctly processed all grammatical strings.

The results of Elman and Zipser and Servan-Schreiber et al. illustrated the effectiveness of capturing temporal context with representations of sequential state. Extensions to continuous outputs have also been proposed for control systems application (Pearlmutter, 1988). These networks follow continuous state space trajectories, in contrast to sequences of discrete states, and might be useful for speech applications as well.

4.3. Parsing

Another approach to the problem of temporal context has been taken by Jain (1989) and Jain and Waibel (1990a) in work on connectionist parsing. Instead of requiring networks to learn to capture arbitrarily complex and distant temporal context information through recurrence, the process of capturing context is explicitly built into the task to be learned. This work extends the standard back-propagation paradigm to allow the construction of well-behaved storage buffers within networks which operate on sequential input. The parsing networks are constructed in a hierarchical fashion from separately trained modules. Each module performs some transformation of either input data or substructures built by other modules. Arbitrarily complex information structures can be built up over time in this manner. Jain and Waibel have successfully trained a network to parse grammatically complex sentences including passive constructions and center embedded clauses. The dynamic parsing behavior is predictive; the trained network produces hypotheses about sentence structure at every moment in time and confirms or revises hypotheses as input words are processed. In Fig. 17, the temporal activation patterns corresponding to the semantic role description for a single clause unit are plotted for a passive sentence. The initial hypothesis of

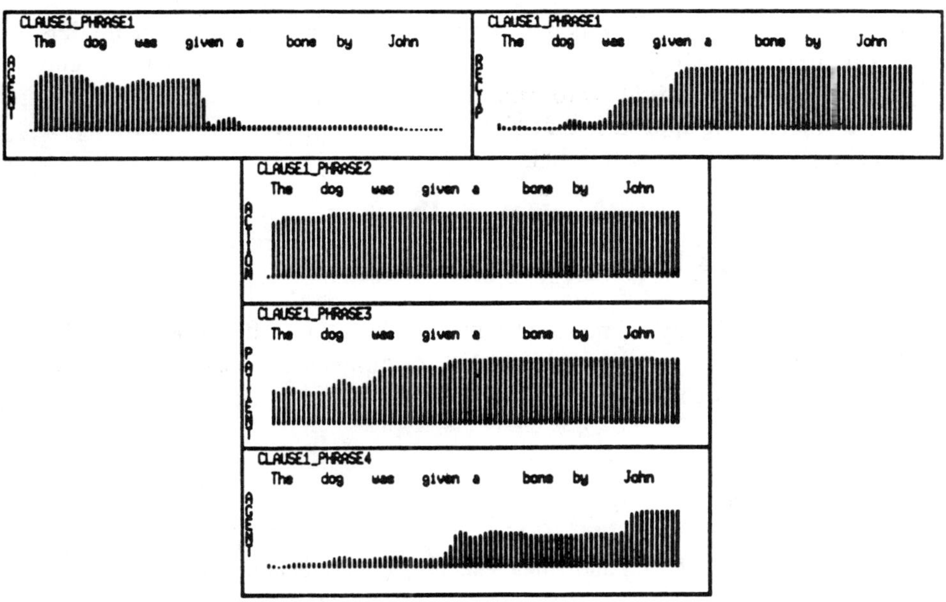

Figure 17 Dynamic role assignment behavior for "The dog was given a bone by John." The phrase numbers correspond to the following phrases: "[The dog] [was given] [a bone] [by John]." The network begins with an agent/action/patient assignment and finishes with the correct recipient/action/patient/agent assignment. All roles are predicted before their respective phrases are processed.

the common agent/action/patient role structure is quickly revised when the passive construction is detected.

The parser has the advantage that its structure is learned automatically from text data and not programmed. This should allow extensions to parsers that compensate for word recognizer confusions and syntactically ill-formed spontaneous speech. In preliminary evaluations the present version of the parser was found to be tolerant of ungrammatical sentences and various other types of degradations (Jain and Waibel, 1990a,b). A trainable parser may also easily incorporate other nonsymbolic information, such as prosodic cues, intonation, stress, intensity, and rhythm, that have so far been ignored in most language models of speech.

5. CONCLUSION

In this chapter we have provided a review of recent research on applying neural networks to speech recognition. At this writing, research in the field is rapidly expanding and new models for phoneme, word, and language modeling continue to emerge. This creative search for novel solutions to the speech recognition problem is likely to stimulate new insights and intuitions beyond the connectionist approach that should lead to a better understanding of speech recognition system design.

ACKNOWLEDGMENT

The author gratefully acknowledges his collaborators, Mike Franzini, John Hampshire, Ajay Jain, and Joe Tebelskis, for numerous discussions and help in preparing this chapter.

REFERENCES

Benello, J., Mackie, A. W., and Anderson, J. A. (1989). Syntactic category disambiguation with neural networks. *Comput. Speech Language 3*:203–217.

Bengio, Y., Cardin, R., Cosi, P., and DeMori, R. (1989). Speech coding with multilayer networks. *IEEE Int. Conf. Acoustics, Speech, and Signal Processing*, May, pp. 164–167.

Bottou, L.-Y. (1988). Reconnaissance de la Parole par Reseaux multi-couches. *Proc. Neuro-Nimes 88*, November.

Bottou, L., Fogelman-Soulie, F., Blanchet, P., and Lienard, J. S. (1989). Experiments with time-delay networks and dynamic time warping for speaker independent isolated digits recognition. *Proc. Eurospeech*, September.

Bourlard, H., and Wellekens, C. J. (1989). Links between Markov models and multilayer perceptrons. In *Advances in Neural Network Information Processing Systems*. D. S. Touretzky (ed.) Morgan Kaufman, San Mateo, pp. 502–510.

Bourlard, H., and Wellekens, C. J. (1989). Speech pattern discrimination and multilayer perceptrons. *Comput. Speech Language 3*:1–19.

Bridle, J. S. (1990). Alpha-nets: A recurrent neural network architecture with a hidden Markov model interpretation. *Speech Commun.*, in press.

Burr, D. J. (1988a). Speech recognition experiments with perceptrons. In *Advances in Neural Information Processing Systems*. Morgan Kaufmann.

Burr, D. J. (1988b). Experiments on neural net recognition of spoken and written text. *IEEE Trans. Acoust. Speech Signal Process 36*:1162–1168.

Duda, R. O., and Hart, P. E. (1973). *Pattern Classification and Scene Analysis* Wiley, New York.

Elman, J. L. (1988). Finding structure in time. Technical Report CRL 8801, University of California, San Diego.

Elman, J. L., and Zipser, D. (1987). Learning the hidden structure of speech. Technical Report, University of California, San Diego, February.

Franzini, M., Witbrock, M., and Lee, K. F. (1989). A connectionist approach to continuous speech recognition. *IEEE Int. Conf. Acoustics, Speech, and Signal Processing*, May.

Franzini, M. A., Lee, K. F. and Waibel, A. H. (1990). Connectionist Viterbi training: A new hybrid method for continuous speech recognition. *IEEE Int. Conf. Acoustics, Speech, and Signal Processing*, April.

Haffner, P., Waibel, A., Sawai, H., and Shikano, K. (1988). Fast back-propagation learning methods for neural networks in speech. Technical Report TR-I-0058, ATR Interpreting Telephony Research Laboratories, November.

Haffner, P., Waibel, A., Sawai, H., and Shikano, K. (1989). Fast back-propagation learning methods for large phonemic neural networks. *Proc. Eurospeech*, September, pp. 553–556.

Hampshire, J., and Waibel, A. (1990). A novel objective function for improved phoneme recognition using time delay neural networks. *IEEE Trans. Neural Networks*, June, *1*:216–228.

Hanson, S. J., and Kegl, J. (1987). PARSNIP: A connectionist network that learns natural language grammar from exposure to natural language sentences. *Proc. Ninth Annual Conf. Cognitive Science Society*, pp. 106–119.

Hinton, G. E. (1989). Connectionist learning procedures. *Artif. Intell. 40*:185–234.

Hinton, G. E., and Sejnowski, T. J. (1986). Learning and relearning in Boltzmann machines. In *Parallel Distributed Processing; Explorations in the Microstructure of Cognition*, J. L. McClelland and D. E. Rumelhart (Eds.). MIT Press, Cambridge, pp. 282–317.

Huang, W., and Lippmann, R. (1988). Neural net and traditional classifiers. In *Neural Information Processing Systems*, D. Anderson (Ed.). American Institute of Physics, New York, pp. 387–396.

Huang, W. M., Lippmann, R. P., and Nguyen, T. (1988). Neural nets for speech recognition. *Conf. Acoustical Society of America*, May.

Iso, K., and Watanabe, T. (1990). Speaker-independent word recognition using a neural prediction model. *IEEE Int. Conf. Acoustics, Speech, and Signal Processing*, April.

Iwamida, H., Katagiri, S., McDermott, E., and Tohkura, Y. (1990). A hybrid speech recognition system using HMMs with an LVQ-trained codebook. *IEEE Int. Conf. Acoustics, Speech, and Signal Processing*, April, p. 35.S10.1.

Jain, A. (1989). A connectionist architecture for sequential symbolic domains. Technical Report CMU-CS-89-187, Carnegie Mellon University, December.

Jain, A., and Waibel, A. (1990a). Robust connectionist parsing of spoken language. *IEEE Int. Conf. Acoustics, Speech, and Signal Processing*, April.

Jain, A., and Waibel, A. (1990b). Incremental parsing by modular recurrent connectionist networks. In *Advances in Neural Information Processing Systems*, D. S. Tovretzky (ed.) Morgan Kaufman, San Mateo, pp. 364–371.

Jordan, M. (1986). Attractor dynamics and parallelism in a connectionist sequential machine. In *Proceedings of the 1986 Cognitive Science Conference*. L. Erlbaum, Hillsdale, N.J.

Kamm, C. A. (1989). Effects of input representation on speaker-independent digit recognition using neural networks. in *Proceedings of Speech Tech '89*, Media Dimensions, May, pp. 301–303.

Kammerer, B., and Kupper, W. (1988). Experiments for isolated word recognition with single and multi-layer perceptrons. In *Proceedings of the First International Conference of the Neural Network Society*. INNS, Boston, p. 302.

Katagiri, S., and Lee, C. H. (1990). A new HMM/LVQ hybrid algorithm for speech recognition. *Proc. GlobeCom '90*, November.

Kato, K., and Kakehi, K. (1987). Listener adaptability to individual speaker differences. *J. Acoust. Soc. Jpn.*, in press.

Kohonen, T. (1988). *Self-Organization and Associative Memory*, 2nd ed. Springer, Berlin.

Kohonen, T., Torkkola, K., Shozakai, M., Kangas, J. and Venta, O. (1988a). Phonetic typewriter for Finnish and Japanese. *IEEE Int. Conf. Acoustics, Speech, and Signal Processing*, April, pp. 607–610.

Kohonen, T., Makisara, K., and Saramaki, K. (1988b). Statistical pattern recognition with neural networks: Benchmarking studies. *IEEE Proc. 2nd Annual Int. Conf. Neural Networks*, July.

Kuhn, G., Watrous, R. L., and Ladendorf, B. (1989). Connected recognition with a recurrent network. *Proc. Neurospeech '89*, May.

Kukich, K. (1988). Back-propagation topologies for sequence generation. *IEEE Int. Conf. Neural Networks*, pp. 301–308.

Lang, K. (1989). A time delay neural network architecture for speech recognition. Ph.D. thesis, Carnegie Mellon University.

Lang, K. J., Hinton, G. E., and Waibel, A. H. (1990). A time-delay neural network architecture for isolated word recognition. *Neural Networks J. 3*:23–44.

Lapedes, A., and Farber, R. (1987). Nonlinear signal processing using neural networks; prediction and system modeling. Technical report LA-UR-87-2662, Los Alamos National Laboratory.

Leung, H. C. (1988). Some phonetic recognition experiments using artificial neural nets. *IEEE Int. Conf. Acoustics, Speech, and Signal Processing*, April.

Leung, H. C and Zue, V. W. (1989). Applications of error back-propagation to phonetic classification In Advances in Neural Information Processing Systems, D. S. Touretzky (ed.) Morgan Kaufman, San Mateo, pp. 206–214.

Levin, E. (1990). Speech recognition using hidden control neural network architecture. *Proc. Int. Conf. Acoustics, Speech, and Signal Processing*, April.

Lippmann, R. P. (1987). An introduction to computing with neural nets. *IEEE ASSP Mag.* April, pp. 4–22.

Lippmann, R. P., and Gold, B. (1987). Neural-net classifiers useful for speech recognition. *IEEE Int. Conf. Neural Networks*, June.

Makhoul, J. I., Roucos, S., and Gish, H. (1985). Vector quantization in speech coding. *IEEE Proc. 73*:1551–1588, November.

McDermott, E., and Katagiri, S. (1989). Shift-invariant, multi-category phoneme recognition using Kohonen's LVQ2. *IEEE Int. Conf. Acoustics, Speech, and Signal Processing*, May, p. 9.S3.1.

McDermott, E., Iwamida, H., Katagiri, S., and Tohkura, Y. (1990). Shift-tolerant LVQ and hybrid LVQ-HMM for phoneme recognition. In *Readings in Speech Recognition*, A. Waibel and K. F. Lee (eds.) Morgan Kaufmann, San Mateo, pp. 425–438.

Morgan, N., and Bourlard, H. (1990). Continuous speech recognition using multilayer perceptrons with hidden Markov models. *IEEE Int. Conf. Acoustics, Speech, and Signal Processing*, 26.S8.1.

Nakamura, M., and Shikano, K. (1989). A study of English word category prediction based on neural networks. *IEEE Int. Conf. Acoustics, Speech, and Signal Processing*, May, p. 731–734.

Niranjan, M., and Fallside, F. (1988). Neural networks and radial basis functions in classifying static speech patterns. Technical report CUED/F-INFENG/TR-22, Cambridge University Engineering Department.

Pearlmutter, B. (1988). Learning state space trajectories in recurrent neural networks. Technical Report CMU-CS-88-191, Carnegie Mellon University, December.

Peeling, S., and Moore, R. (1987). Experiments in isolated digit recognition using the multi-layer perceptron. Technical Report 4073, Royal Speech and Radar Establishment (RSRE), December.

Peterson, G. and Barney, H. (1952). Control methods used in a study of vowels. *J. Acoust. Soc. Am. 24*:175–184, March.

Prager, R. W., Harrison, T. D., and Fallside, F. (1986). Boltzmann machines for speech recognition. *Comput. Speech Language 1*:3–27, March.

Rumelhart, D. E., and McClelland, J. L. (1986). *Parallel Distributed Processing; Explorations in the Microstructure of Cognition*. MIT Press, Cambridge, Mass.

Rumelhart, D. E., Hinton, G. E., and Williams, R. J. (1986). Learning representations by back-propagating errors. *Nature 323*:533–536.

Sagisaka, Y., Takeda, K., Katagiri, S., and Kuwabara, H. (1987). Japanese speech database with fine acoustic-phonetic transcriptions. Technical Report, ATR Interpreting Telephony Research Laboratories, May.

Sakoe, H. (1987). Dynamic neural network—A new speech recognition model based on dynamic programming and neural network. IEICE Technical Report, December.

Sakoe, H. and Chiba, A. (1978). Dynamic programming optimization for spoken word recognition. *IEEE Trans. Acoust. Speech Signal Process. ASSP-26*(1):43–49.

Sakoe, H., Isotani, R., Yoshida, K., Iso, K., and Watanabe, T. (1989). Speaker-independent word recognition using dynamic programming neural networks. *IEEE Int. Conf. Acoustics, Speech, and Signal Processing*, May., pp. 29–32.

Miyatake, M., Sawai, H., and Shikano, K. (1990). Integrated training for spotting Japanese phonemes using large phonemic time-delay neural networks. *IEEE Int. Conf. Acoustics, Speech, and Signal Processing*, May.

Servan-Schreiber, D., Cleeremans, A., and McClelland, J. L. (1988a). Learning sequential structure in simple recurrent networks. *Proc. Second IEEE Conf. Neural Information Processing Systems.*

Servan-Schreiber, D., Cleeremans, A., and McClelland, J. L. (1988b). Encoding sequential structure in simple recurrent networks. Technical Report CMU-CS-88-183, Carnegie Mellon University, November.

Tamura, S., and Waibel, A. (1988). Noise reduction using connectionist models. *IEEE Int. Conf. Acoustics, Speech, and Signal Processing*, April, p. S12.7.

Tank, D. W., and Hopfield, J. J. (1987). Neural computation by concentrating information in time. *Proc. Natl. Acad. Sci. USA* 1896–1900. April.

Tebelskis, J., and Waibel, A. (1990). Large vocabulary recognition using linked predictive neural networks. *IEEE Int. Conf. Acoustics, Speech, and Signal Processing*, April.

Unikrishnan, K., Hopfield, J., and Tank, D. (1988). Learning time delayed connections in a speech recognition circuit. Snowbird Conference, March.

Waibel, A. (1989). Consonant recognition by modular construction of large phonemic time-delay neural networks. In *Advances in Neural Network Information Processing Systems*. D. S. Touretzky (ed.) Morgan Kaufmann, San Mateo, pp. 215–223.

Waibel, A. (1989). Modular construction of time-delay neural networks for speech recognition. *Neural Computation 1*:39–46.

Waibel, A., Hanazawa, T., Hinton, G., Shikano, K., and Lang K. (1987). Phoneme recognition using time-delay neural networks. Technical Report TR-1-0006, ATR Interpreting Telephony Research Laboratories, October.

Waibel, A., Hanazawa, T., Hinton, G., Shikano, K., and Lang K. (1989a). Phoneme recognition using time-delay neural networks. *IEEE Trans. Acoust. Speech Signal Process. 37*:328–339, March.

Waibel, A., Sawai, H., and Shikano, K. (1989b). Modularity and scaling in large phonemic neural networks. *IEEE Trans. Acoust. Speech Signal Process. 37*: 1888–1898, December.

Watrous, R. (1988). Speech recognition using connectionist networks. Ph.D thesis, University of Pennsylvania.

Young, S. J. (1990). Competitive training: A connectionist approach to the discriminative training of hidden Markov models. Technical Report CUED/F-INFENG/TR.41, Cambridge University, March.

19

Speaker-Independent and Speaker-Adaptive Recognition Techniques

SADAOKI FURUI NTT Human Interface Laboratories, Tokyo, Japan

1. INDIVIDUAL INFORMATION IN SPEECH WAVES

Voice individuality can be attributed to anatomical differences between speakers, as in the vocal tract length and the size of the nasal cavity, and speaker-dependent speaking habits, such as accent, speech rate, and loudness. These variations are combined together in the speech waves The average characteristics of variations in vowel formant frequencies due to gender and age can be modeled by the shift in the logarithmic frequency scale maintaining the relative location of the vowels. Individual characteristics are extremely complicated, however, and it has also been observed that mutual effects exist between phonetic and individual information, specifically, the differences in individual information depending on the phonemes (Furui, 1986b).

The most reliable way to tackle the individuality problem and achieve high recognition performance is to "train" the system using the new speaker's voice. This method, called the speaker-dependent method, has several problems, however (Lee, 1989):

1. The training session is inconvenient for users.

2. A large amount of processing is necessary before the system becomes available for use. Sometimes users must wait for several hours after uttering training speech for the processing to be completed.
3. There are several applications in which training cannot be performed, such as telephone directory assistance and banking services.
4. In some applications, such as dictations of conferences and trials, more than one speaker is present.
5. Storing the parameters of all speakers individually requires an extremely large amount of storage in some cases.
6. Each speaker's voice is essentially variable under the influence of stress, tiredness, illness, differences in location and characteristics of telephone sets and microphones, variations in background noise, room acoustic characteristics, and frequency band limitation, and so on. Voices also change over time.

Methods for recognizing speech in a speaker-independent fashion have therefore been investigated by many researchers. By using a reference template obtained by averaging the spectral patterns of many speakers for each word, high recognition accuracy can be achieved only with a small vocabulary and no similar word pairs in the spectral domain. If it is necessary to recognize accurately a large vocabulary uttered by unrestricted speakers, the recognition system must incorporate more sophisticated functions for coping with voice individuality.

2. SPEAKER-INDEPENDENT RECOGNITION METHODS

2.1. Classification of Speaker-Independent Recognition Methods

A number of approaches have been attempted in an effort to build speaker-independent recognition systems. These include the signal processing approach, information theoretic approach, pattern recognition approach, artificial intelligence approach, statistical approach, and neural network approach. Each of these is briefly outlined in the following.

2.2. Signal Processing Approach–Invariant Features

The most basic and desirable approach to speaker independence is to find feature parameters that are invariant between speakers, that is, commonly extracted from the same phonemes or words uttered by all speakers and effective for separating different phonemes and words. This approach is actually very ambitious, since there is no evidence that common physical features exist in the same words uttered by different speakers even if they can be clearly recognized by humans (Furui, 1989a).

One of the successful feature parameters which have been investigated and widely used is the dynamic spectral feature set, namely differential spectral parameters (Furui, 1986a). The set includes the Δ cepstrum and Δ power. They are defined as first-order regression coefficients and logarithmic energy, respectively. The length of the parameter sequences is usually set empirically between 40 and 100 ms. These parameters are usually used in combination with conventional cepstrum coefficients and logarithmic energy. This method has been used in a large number of recognition systems, and its effectiveness in improving recognition accuracy, especially in speaker-independent cases, has been confirmed.

2.3. Information Theoretic Approach—Discriminant Methods

Discriminant features and distance measures effective for discriminating phonemes uttered by unspecified speakers have been created based on the information theoretic approach. Recently, this approach has been actively investigated in the framework of hidden Markov model (HMM) recognition systems. In this approach, the maximum mutual information method and various discrimination methods, including corrective training and sophisticated decision boundary adjustment techniques, have been tried (Mariani, 1989).

The corrective training (Bahl, 1988) is similar to the well-known error correcting training procedure for linear classifiers and works by iteratively adjusting the parameter values so as to make correct words more probable and incorrect words less probable. When a recognition error does occur, or a wrong candidate becomes difficult to separate from the right one, the initial model is modified to lower the probability of the labels responsible for the mistake or the "near miss." The process is repeated with the modified parameters and is stopped when no more improvements are obtained. Acoustically confusable words can be used to reduce the duration of the process. Although convergence of the algorithm has not been proved, experimental evidence suggests that it does converge and that the algorithm leads to fewer recognition errors than maximum likelihood estimation.

The maximum mutual information method (Bahl et al., 1986; Merialdo, 1988) is similar to the corrective training method but is more formalized. This method determines the model parameters which maximize the probability of generating the acoustic data given the right word sequence and, at the same time, minimizes the probability of generating any wrong word sequence, especially the most frequent ones.

Phonetically sensitive transformation of speech features, which is designed to discriminate between in-class and out-of-class confusion data and which is a function of the phonetic state, has yielded improvement in speech recognition performance (Doddington, 1989). The basic concept of this method is similar to that of the corrective training approach.

A spectral representation, called IMELDA (Integrated MEL-scale representation using Linear Discriminant Analysis) has been tested in speaker-independent connected-word and isolated-word recognition with undegraded speech and with speech degraded by adding white noise and by applying a 6 dB/octave spectral tilt (Hunt and Lefebvre, 1989). The representation consists of linear discriminant functions derived from the filter-bank output, and under all conditions it outperformed all other representations, including an auditory model, weighted mel-cepstrum coefficients, and cepstrum coefficients augmented with dynamic features.

One of the problems associated with these discriminant methods is that they are very sensitive to the difference between training and testing samples, since the recognition strategies are elaborately tuned to the training samples.

2.4. Pattern Recognition Approach—Multiple-Template Method

The multiple-template method, in which multiple templates representing individual variabilities are designed by clustering techniques to cover the range of individual variation, is one of the successful methods for speaker-independent recognition. Each word in the vocabulary is uttered by many speakers, and these utterances are then divided into clusters. Several algorithms are used in combination to achieve clustering. The speech sample at the center of each cluster or the mean value for the speech data associated with each cluster is stored as a reference template (prototype).

In the recognition phase, distances (or similarities) between an input utterance and all multiple templates of all vocabulary are calculated, and the word with the smallest distance is selected as the word spoken. In order to increase the reliability, the KNN (K-nearest neighbor) technique is often used for the decision (Levinson et al., 1979; Rabiner and Wilpon, 1981). Here, the K reference templates with the smallest distances from the input speech are selected from the multiple-reference template set for each word, with the mean value for these K templates being calculated for each word. The word with the smallest mean value is then selected as the recognition result.

Figure 1 is a block diagram of the multiple template–based recognition system (Rabiner et al., 1979; Rabiner and Wilpon, 1979; Wilpon et al., 1982). The effectiveness of this method has been experimentally confirmed in word recognition and connected digit recognition. With 12 templates for each

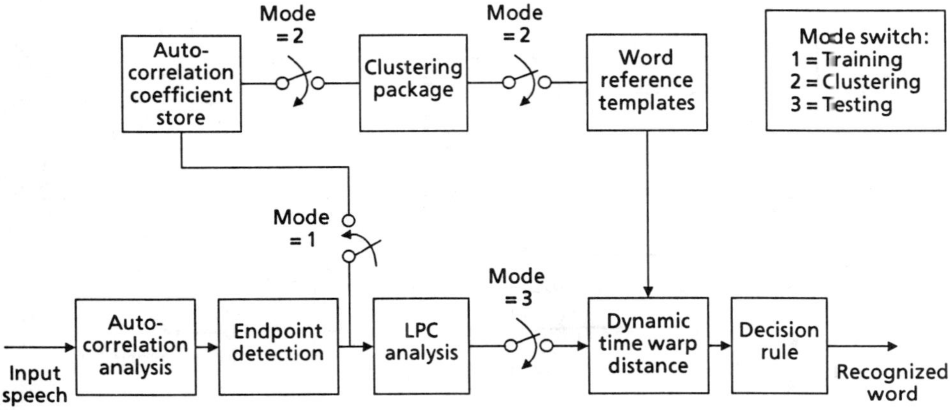

Figure 1 Block diagram of the multiple template–based speaker-independent word recognition system.

word, the recognition accuracy for $K = 2$ to 3 was revealed to be higher than for $K = 1$. Accuracies of 98.2 and 91.0% for an English digit task and for a 129-isolated-word task, respectively, were reported.

High recognition accuracy can be expected if a sufficiently large number of reference templates are created based on training utterances by a sufficiently large number of speakers. Three research questions are associated with this approach: How many speakers should be used for training? How many templates should be created? How should the multiple templates be created?

When the vocabulary size is large, the multiple-template-approach has the problems of requiring huge storage for reference templates and extensive computation, since multiple templates equivalently increase the vocabulary size. The SPLIT method, in which each word is represented by a sequence of VQ codebook elements (phoneme-like templates), was proposed to cope with these problems (Sugamura et al., 1983). The SPLIT method is block diagramed in Fig. 2. In this method, a universal codebook constructed based on a multispeaker, multiword data base is commonly used for every speaker, and multiple templates are represented by sequences of the universal codebook indices. Since spectral distances are calculated between input speech and universal codebook elements, the amount of computation and the storage for reference templates are largely reduced.

To further reduce the amount of computation, a VQ-based preprocessor has been combined with the SPLIT method as shown in Fig. 3 (Furui,

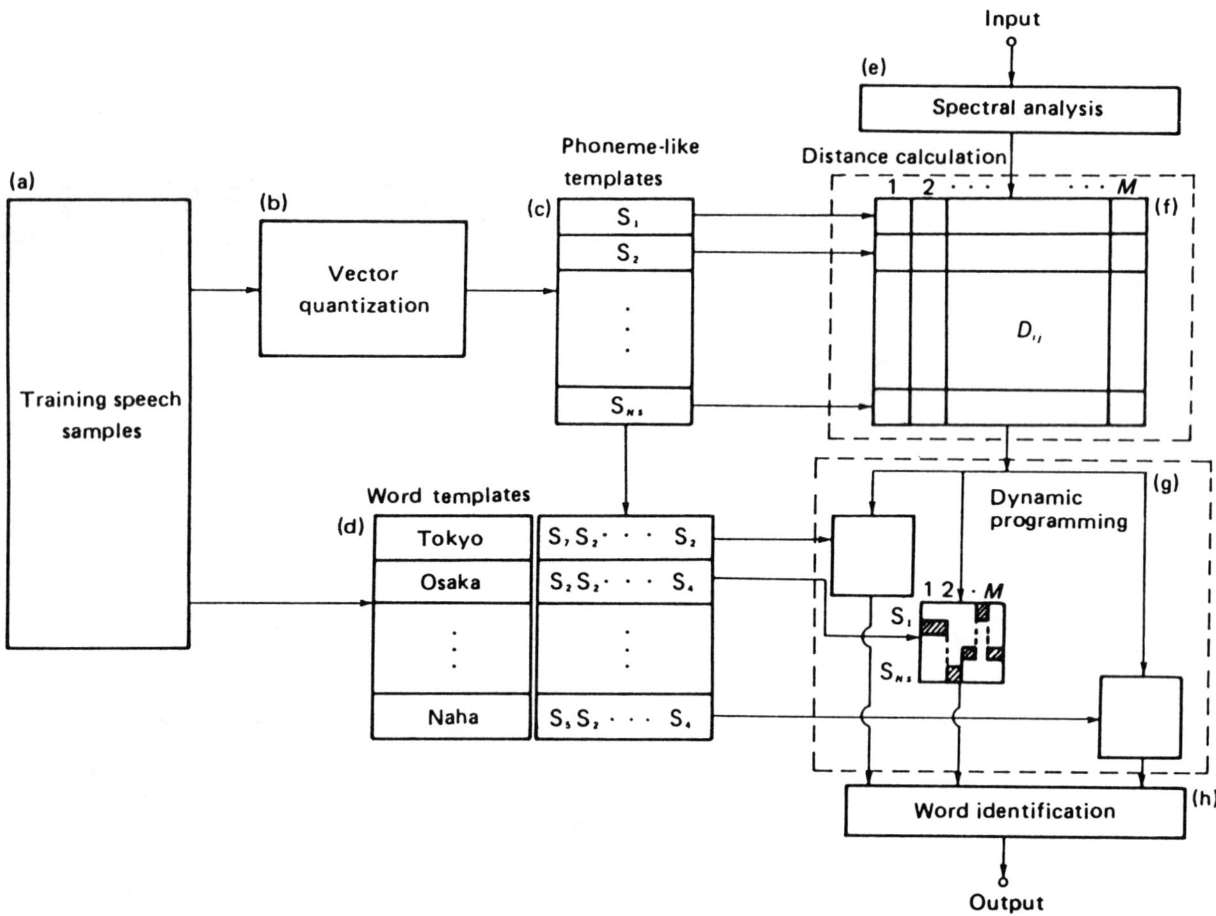

Figure 2 SPLIT isolated word recognition system in which each word is represented by a sequence of VQ codebook elements (phoneme-like templates).

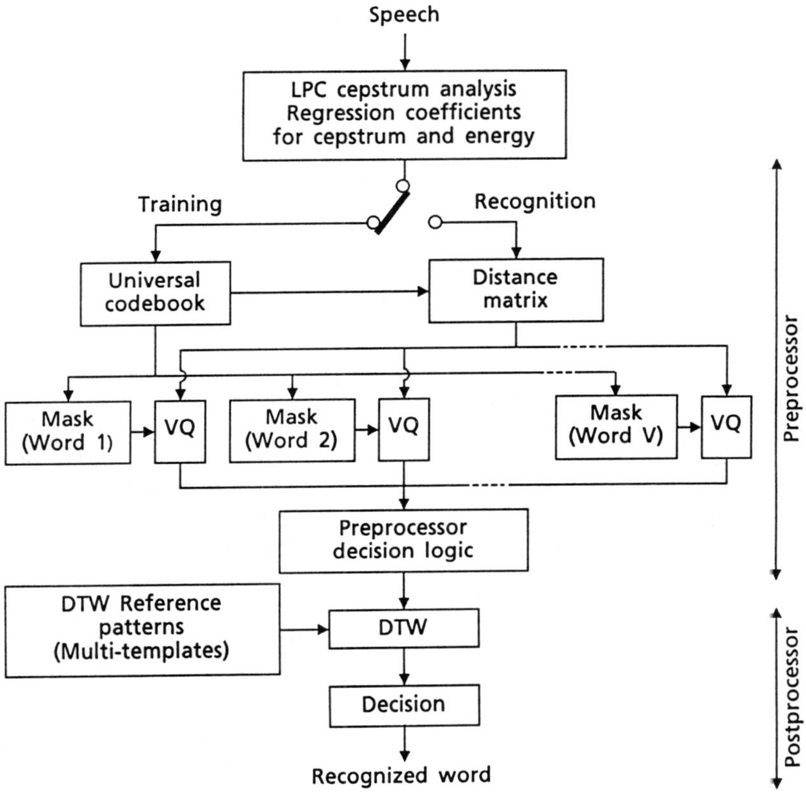

Figure 3 Combination of VQ-based preprocessor and the SPLIT recognition system.

1988, 1990). Here, a speech wave is analyzed by the time functions of instantaneous cepstrum coefficients, Δ cepstrum coefficients, and Δ power. A separate codebook for each word in the vocabulary is designed as a subset of the universal codebook. These word-specific codebooks are used for front-end processing to eliminate word candidates with large distance (distortion) scores. The SPLIT method subsequently resolves the choice between the remaining word candidates. This method reduces the number of calculations to one-tenth that of the SPLIT method or less while maintaining recognition accuracy.

2.5. Artificial Intelligence Approach—Knowledge Engineering

Phonetic features or representation rules that are robust against speaker-to-speaker variation have been investigated using knowledge engineering approaches based on spectrogram-reading experiments. In the ANGEL system developed at Carnegie-Mellon University, recognition is achieved by using phoneme location modules and classification modules, which have the tasks of segmenting and identifying the corresponding segments (Adams and Bisiani, 1986). Examples of such modules are the stop, fricative, closure, and sonorant modules. A 41% word recognition accuracy was obtained using this approach for the DARPA resource management task having a perplexity of 34.

Segmentation and labeling using a knowledge-based approach has been investigated in the SUMMIT phonetically based spoken language understanding system created at MIT (Zue et al., 1989). In contrast to many of the past efforts that made use of heuristic rules whose development requires intense knowledge engineering, this approach attempts to express the speech knowledge within a formal framework using well-defined mathematical tools. In the system, features and decision strategies are discovered and trained automatically using a large data base.

The segmentation process produces a multilevel representation called a "dendrogram." The speech spectrogram is segmented into units of different levels of description, from fine to coarse, with the last segment being the whole sentence. This process is based on the computation of a similarity measure between adjacent segments using the Euclidean distance for the average spectral vectors of each region previously delimited, and on the merging of similar ones. A phoneme lattice is then obtained by using a statistical classifier. The top-choice phonetic classification accuracy is about 70% and within the top three over 90% of the time.

2.6. Statistical Approach—HMM Methods

Variations of phonetic features and transitions between them can be represented efficiently and flexibly in a probabilistic manner by hidden Markov models (HMMs), which are trained using utterances by many speakers. The HMM method has another advantage in that it uses a relatively small amount of computation for recognition, since each word or phoneme is represented by a sequence of a small number of states.

In a small vocabulary-isolated French word recognition task, 85% and 89% recognition accuracies were obtained for digits and the signs of the zodiac, respectively, using utterances collected over the public telephone analog network (Dutoit, 1987). For English digits recorded over a standard dialed-up telephone line, a 96.3% recognition accuracy was obtained using discrete HMM (Rabiner et al., 1983).

It was also reported that the performance can be significantly improved by using continuous Gaussian mixture densities HMM (Rabiner et al., 1985). In a connected digit recognition experiment using continuous HMM, multiple models, and mixture densities, a 97.1% sentence (string) recognition rate was obtained (Rabiner et al., 1988).

A speaker-independent word-spotting algorithm based on HMMs, which can recognize a predefined set of vocabulary items embedded in fluent speech, has also been investigated (Wilpon et al., 1989). The task researched was automatic call handling of command phrases for making "operator-assisted calls." In the experiment, it was assumed that only a single key word was spoken for each spoken input, which consisted of a background signal period, a speech period, and another background signal period. The experimental results show that for a vocabulary of five words 87.1% of key words are correctly recognized when they are spoken over the long-distance telephone network. Several robust training methods as well as detailed analyses of state-level likelihood and duration scores were used to achieve this performance.

One of the most successful examples of HMM-based speaker-independent continuous speech recognition systems is the SPHINX large-vocabulary system (Lee and Hon, 1988; Lee, 1989). This system achieved speaker-independent word accuracies of 70.6%, 93.7%, and 95.8% on the 997-word DARPA resource management task with grammars of perplexity 997, 60, and 20, respectively. The results are almost the same as those obtained by speaker-dependent systems. The principal techniques used in the SPHINX system are:

1. VQ-based discrete HMM
2. Training by a large multispeaker data base
3. Interpolation of robust models and context-dependent as well as speaker-dependent detailed models by a deleted interpolation technique
4. Multiple codebook including various feature parameters, such as cepstrum, Δ cepstrum, power, and Δ power
5. Function word–dependent phone modeling
6. Generalized triphone modeling

The details of these techniques are described by Lee and Alleva in Chapter 20. The experimental results indicate that better performance did not simply materialize from adding training data but that the modeling techniques had to be improved to take advantage of the ample training data.

The same task has been performed at MIT Lincoln Labs with slightly worse results, using word-context-free triphone models with tied continuous Gaussian mixture HMMs (Paul, 1989). Other HMM-based systems include the TI system using continuous density word HMM (TI, 1987) and the SRI system using discrete density phoneme HMM (Murveit and Weintraub, 1988). They achieved word recognition accuracies of 44% and 40%, respectively, for the resource management task under the no-grammar conditions.

The HMM clustering procedure, the combination of multiple-template and HMM methods, has been proposed and applied to connected word systems and to large-vocabulary recognition based on phonelike units (Rabiner et al., 1989). In this method, a splitting procedure initializes each new cluster (statistical model) by splitting off all tokens in the training set which are poorly represented (in the likelihood sense) by the current set of models. It was reported that, for speaker-independent connected digit recognition on the TI data base, the recognition performance using two HMM-clustered models was as good as or better than previous results using 4 to 6 models/digit obtained from template-based clustering.

2.7. Neural Network Approach

Since neural networks have the capabilities of sophisticated nonlinear discrimination, they have recently been used for speaker-independent recognition. The dynamic programming neural network (DNN), based on the integration of DP (DTW) and multilayer neural networks, has been proposed as a method for speaker-independent word recognition (Sakoe et al., 1989). The detailed structure of DNN is explained in Chapter 16. A high recognition accuracy of 99.3% was obtained for isolated Japanese digit recognition using this method.

A four-layer neural network with recurrent connections, as shown in Fig. 4, has been proposed for the task of speaker-independent continuous speech recognition (Franzini et al., 1989). In the network, the representation on the first hidden layer is copied through time to provide a history mechanism. The network generates and refines hypotheses concerning the identity of an utterance over successive intervals. The hypotheses are used as inputs to a Markov chain–based Viterbi recognizer, which produces a final identification of the entire utterance. The combination of neural network and Markov modeling techniques allows the system to take advantage of the superior static pattern classification ability of neural networks as well as the well-known tem-

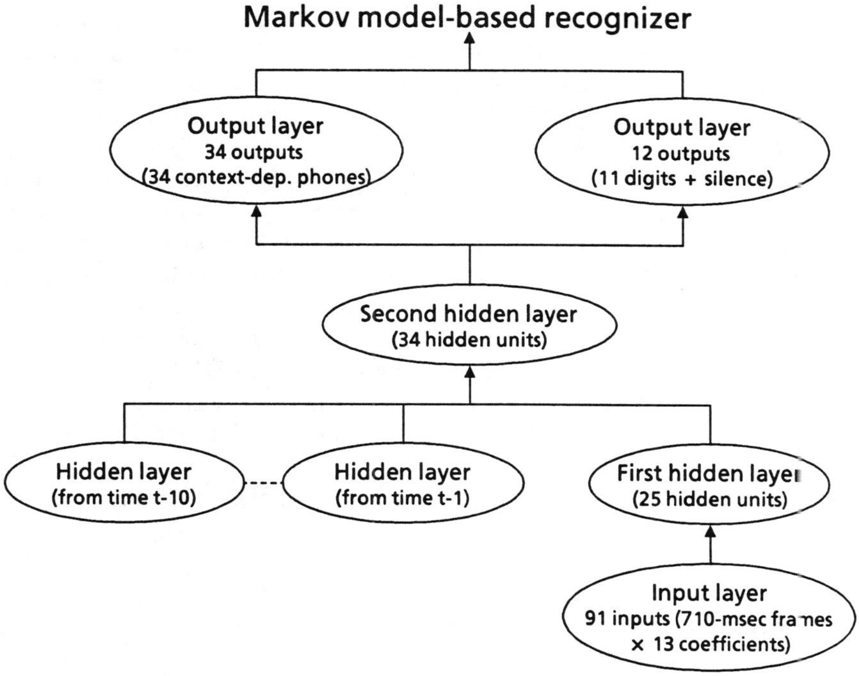

Figure 4 Neural network used in the speaker-independent connected speech recognition system.

poral processing capability of Markov models. This method achieved 97.0% word accuracy and 91.0% sentence accuracy in preliminary tests on the TI connected-digits data base.

2.8. Advantages and Limits of Speaker-Independent Methods

The speaker-independent approach has the advantage that a large data base for building the recognition system can easily be obtained in comparison with the speaker-dependent approach. The larger the data base, the more sophisticated the model can be designed and the better the performance becomes. In the process of introducing sophisticated models, however, it is important to use speech knowledge properly and to analyze the task and error characteristics carefully.

One of the disadvantages of the speaker-independent approach is that it neglects various useful characteristics of the speaker in spite of the fact that they can be learned after recognition of several words or sentences. If these characteristics can be properly used, the recognition process is expected to be accelerated due to the narrowing of the search space. Another disadvantage is that when the distributions of feature parameters are very broad or multimodal, as in the cases of combinations of male and female voices as well as various dialects, it is difficult to separate phonemes using speaker-independent methods. In these situations, nonsense matches are likely to be produced. For example, the first and the second halves of a word may be matched against the male and female halves of a bimodal distribution, respectively.

Another practical difficulty in generating speaker-independent recognition systems is that model specificity and model trainability are two incompatible goals when only a fixed amount of training data is given. More specificity usually reduces trainability, and increased trainability usually results in overgenerality.

To cope with these problems, it is essential to introduce speaker-adaptation techniques.

3. SPEAKER ADAPTATION AND NORMALIZATION

3.1. Classification of Speaker Adaptation and Normalization Methods

Speaker adaptation or normalization ("speaker adaptation" will indicate both adaptation and normalization hereafter) is the method of automatically adapting reference templates to each new speaker or normalizing (reducing) interspeaker variation in each input speech based on the transformation rules obtained using a few training words or short sentences. In large-vocabulary recognition systems, training by the utterances of all the vocabulary words is too troublesome for the users and consequently unrealistic. Therefore, training by a few words or short sentences is a practical and realistic solution, although it is less convenient than the speaker-independent systems, which have no need for training.

Speaker-adaptation methods are generally classified into supervised (text-dependent) methods, in which training words or sentences are known, and unsupervised methods, in which arbitrary utterances can be used. Both methods can also be classified into offline methods, in which training words or sentences must be uttered before the recognition, and online methods, in which utterances for recognition are, at the same time, used for training.

Ideally for users, the system should work as if it were a speaker-independent system which requests no additional training utterance of each speaker. Also, the system should actually adapt to the speaker's voice automatically using utterances for recognition. Such a system can be realized by the unsupervised, online adaptation mechanism. Humans have been found to have a mechanism of unsupervised, online speaker adaptation (Kato and Furui, 1985). Specifically, although the first several words uttered by a speaker new to the listener may be somewhat unintelligible, the latter quickly becomes accustomed to the former's voice. Thus, the intelligibility of the speaker's voice increases particularly after the listener hears several words.

Since individual information (individuality) varies depending on the phoneme classes, it is effective to introduce the mechanism of explicitly or implicitly recognizing the phonemes included in the training utterances into the unsupervised adaptation process. By recognizing the phonemes and utilizing the phoneme-dependent individual information, adaptation performances can be improved. In such processes, it is crucially important that phoneme decision errors do not cause inappropriate adaptation. When supervised methods are used for online adaptation, it is necessary to inform the adaptation system of the correct answer of each input utterance.

The following subsections summarize the major speaker-adaptation methods which have recently been investigated.

3.2. Speaker Cluster Selection Methods

One of the basic speaker-adaptation methods is the speaker cluster selection method. In this method, it is assumed that speakers can be divided into clusters, within which the speakers are similar. From many sets of phoneme template clusters representing speaker variability, the most suitable set for the new speaker is automatically selected.

An unsupervised speaker-adaptation method based on this principle has been applied to vowel recognition in continuous speech (Sugiyama, 1985). In this investigation, many sets of the five Japanese vowels are stored, and the vowel set which produces the smallest spectral distortion is selected for each new speaker. A method of adjusting the selected vowel template set to the speaker, using spectral parameters extracted from the speech period having high reliability of vowel recognition, has also been investigated.

In an HMM-based supervised method, a group of speakers is divided into speaker clusters, and a codebook and HMMs are generated for each cluster (Lee, 1989). Cluster-specific HMM parameters are derived by a training process using speakers in the cluster, or by a training process using all speakers and then converting the parameters to cluster-specific parameters through probabilistic mapping. Cluster-specific parameters for the cluster selected to be the most suitable for the speaker are used in the recognition stage as illustrated in Fig. 5. It has been observed that when the size of training data

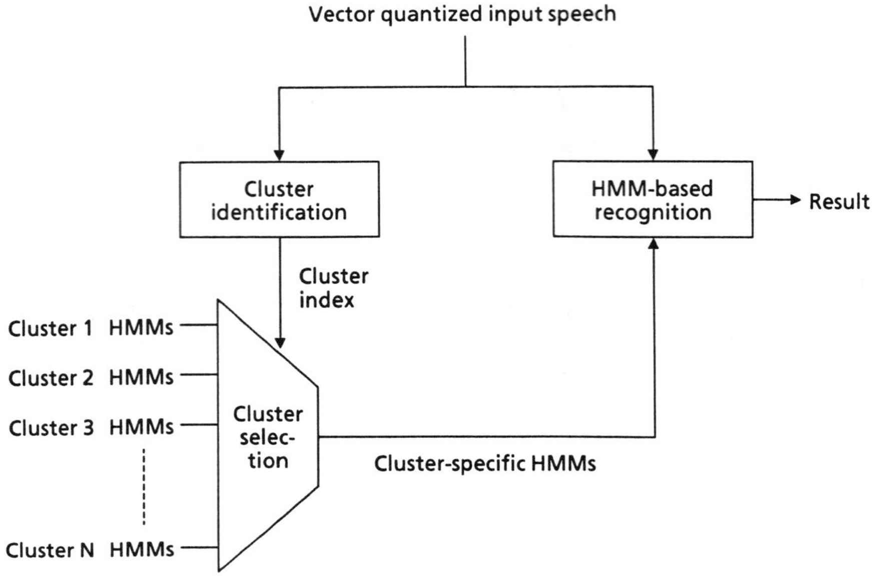

Figure 5 Block diagram of speaker adaptation using the cluster selection method.

is restricted, the possible number of clusters is limited, and therefore the adaptation performance is limited.

3.3. Interpolated Reestimation Algorithm

The deleted interpolation technique (Jelinek and Mercer, 1980) has recently been investigated for generating speaker-adaptive parameters by interpolating speaker-independent, or cluster-independent, fixed existing parameters and speaker-dependent parameters created from a small number of training sentences. The speaker-independent parameters are well trained but less appropriate for the individual speaker. On the other hand, the speaker-dependent parameters are appropriate for the individual speaker but poorly trained because of limited training data. The deleted interpolation technique can be used to combine the two parameter sets into estimates that are more suitable than speaker-independent parameters, yet more robust than speaker-dependent ones.

In a trial within the framework of the SPHINX system (Lee, 1989), five measures with various degrees of reliability and robustness were used: (1) speaker-dependent parameters, (2) speaker-independent parameters,

(3) speaker-dependent parameters with tied phone states, (4) co-occurrence-smoothed speaker-dependent parameters, and (5) similar speaker parameters from the most similar training speakers. Speaker-dependent parameters were derived by running the forward-backward algorithm on the training sentences. Figure 6 is a block diagram of the interpolated reestimation learning in SPHINX. With 30 adaptation sentences, the word error rate was reported to be reduced by about 5 to 10% using this method when no grammar was used. When a grammar was used, the improvement was smaller.

3.4. Codebook Adaptation/Normalization Algorithm

Instead of interpolating speaker-independent and speaker-dependent parameters, speaker-adaptive parameters can be estimated from speaker-independent parameters based on mapping rules. The mapping rules are estimated from the relationship between speaker-independent and speaker-dependent parameters. Within the framework of VQ-based speech recognition, both supervised and unsupervised methods of adapting the speaker-independent codebook to a new speaker or normalizing (adjusting) the input

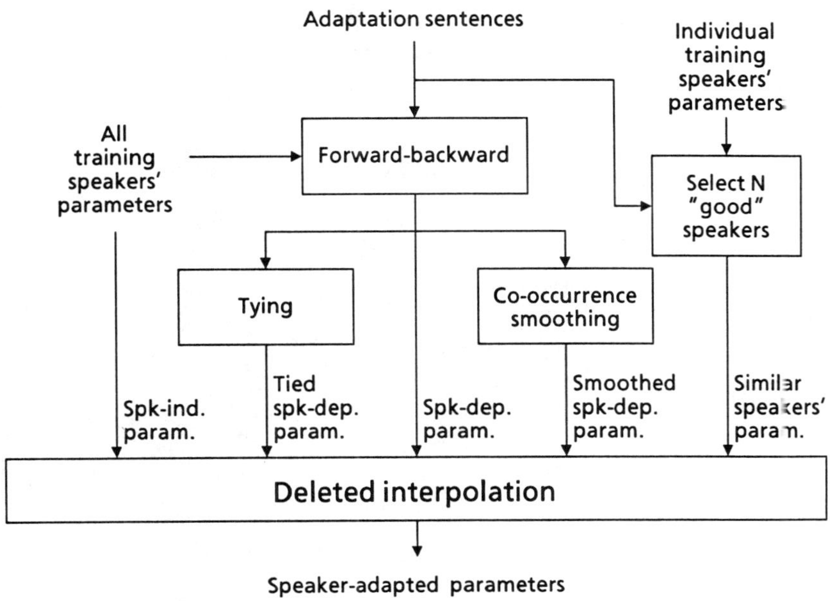

Figure 6 Block diagram of interpolated reestimation learning in SPHINX.

speech to the codebook have been proposed. Each word is represented in the word dictionary as single or multiple sequences of codebook entries. In the latter case, individual variations on how a word is uttered are modeled by multiple code sequences. The code sequences are not changed during the adaptation and are universally used for all speakers.

3.4.1. *Supervised Adaptation*

In the case of supervised adaptation, the mapping rules are obtained through DTW or a forward-backward algorithm. Figure 7 is a block diagram of a supervised adaptation method (Shikano et al., 1986). The utterances by a reference speaker are initially used to create an initial codebook and these utterances are vector-quantized, that is, converted into sequences of codebook entries. In the training stage, training utterances by a new speaker are converted into code sequences and time-aligned with the same words or sentences uttered by the reference speaker, using the DTW technique. The spectral mapping function between the codebook elements of these two speakers is obtained based on the alignment functions, that is, the correspondence between the time axes.

Each codebook element exists in various words, and each codebook element of the reference speaker corresponds to various elements of the new speaker. Thus, a histogram of correspondences between codebook elements of the reference speaker and the new speaker—that is, a histogram of co-

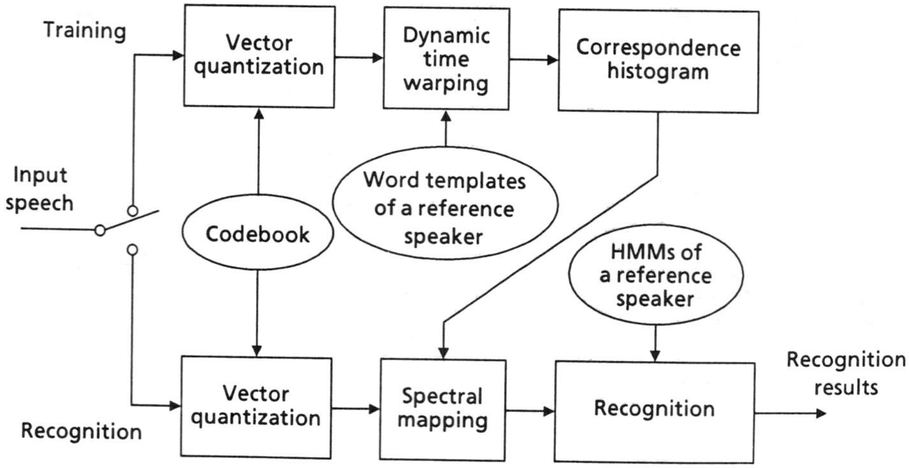

Figure 7 Block diagram of speaker adaptation by spectral mapping.

occurrences of codebook elements—is calculated using the alignment results for all training words or sentences. The mapping function is weighted by the histogram to find the best correspondence rule.

In the recognition stage, input speech is vector-quantized and mapped (normalized) to the reference speaker's spectrum at every frame using the mapping rules. The similarity between normalized input speech and each word of the reference speaker is then calculated and used for the recognition decision. Experimental results show that recognition accuracies for 92-word and 422-word recognition are improved from 64 to 83% and from 48 to 81%, respectively, by this method. Fuzzy VQ has also been introduced to decrease the quantization error without increasing the codebook size. The number of training words was confirmed to be reducible to one-fourth of that without fuzzy VQ.

In HMM-based recognition, HMM speech models derived from one "prototype" speaker are transformed so that they model the speech of a new speaker (Schwartz et al., 1987; Nishimura and Sugawara, 1988). This transformation is accomplished by probabilistic spectral mapping from one speaker's spectral space to another's. The transformation matrix, which represents the conditional probability of a quantized spectrum of the new speaker, given a quantized spectrum of the reference speaker, is computed by applying a modified forward-backward algorithm to training utterances.

In the adaptive mode of the BYBLOS system developed at BBN Laboratory, a 97% word recognition rate was obtained for a 350-word continuous task with perplexity ranging from 30 to 60 (Chow et al., 1987). In this method, adaptation is performed from a reference speaker to a new speaker using 15 s of training speech. It was reported that the results are close to those obtained with speaker-dependent training with 15 min of speech. Using an improved version of the adaptation method, in which the phoneme-dependent mapping procedure is implemented and more precise time alignment is introduced, a 90.4% recognition rate was obtained for the standard DARPA data base with a grammar of perplexity 60 (Feng, 1989). In this case training utterances having a 2-min duration were used. The speaker-dependent performance with 28 min of training speech was 92.9%.

An alternative approach to speaker adaptation for the IBM large-vocabulary HMM-based speech recognition system has also been investigated (Rigoll, 1989). The approach is based on the use of a stochastic model, called the "speaker Markov model," representing the different properties of the new speaker and the reference speaker. The speaker Markov model indicates which prototype of the new speaker is likely to occur and what acoustic parameters are generated by the reference speaker if at the same time a certain prototype is generated by the new speaker, as shown in Fig. 8. It is possible

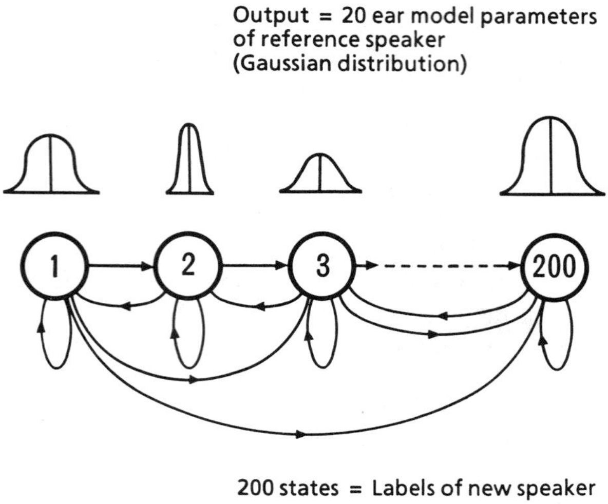

Figure 8 Speaker Markov model describing the spectral differences of two speakers.

to interpret the principle used in this approach as a consequent application of the information theory principle that is the basis of the entire IBM speech recognition system. This is illustrated in Fig. 9. The average recognition rate dropped only from 96.4% to 95.2% for a 5000-word vocabulary task when the duration of training utterances was reduced from the usual 20 min to 5 min.

3.4.2. Supervised Adaptation Using Neural Networks

An adaptation method using neural networks, as shown in Fig. 10, has also been investigated (Iso et al., 1989; Montacie et al., 1989). In this method, nonlinear continuous mapping between the reference spectral space and the spectral space of a new speaker is represented by a multilayer neural network. The network is used for adapting reference templates to a new speaker or adapting input speech spectra to the reference templates. Figure 10 shows the latter case.

The network training is executed using the back-propagation method so that the network generates reference template spectra at the output layer when corresponding training utterance spectra are given to the input layer. Corresponding data sets of reference templates and training utterance spectra are obtained using the DTW technique in the same way as the supervised adaptation method described in the previous subsection. One difference between the previous VQ-based method and this method is that in the latter

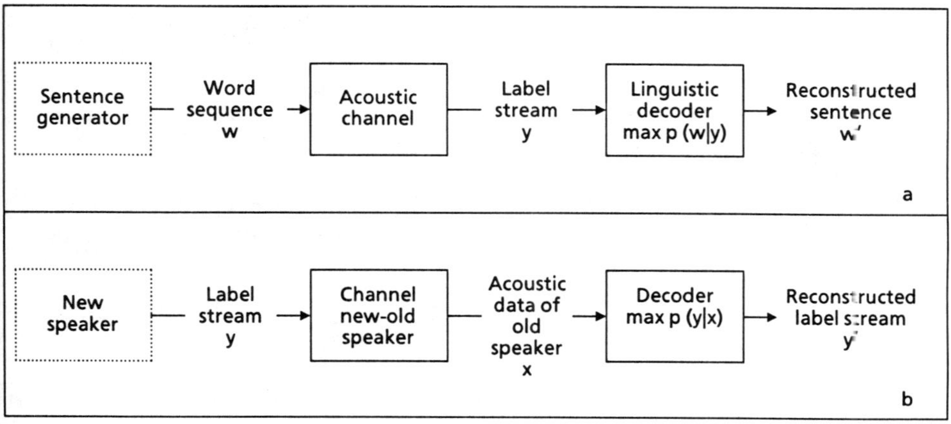

Figure 9 Comparison of the information theory principle used in the IBM speech recognition system (a) and used for speaker adaptation (b).

VQ-distortion of training utterance spectra can be avoided and nonlinear mapping can be realized.

This method was evaluated using an HMM-based 100-Japanese-word recognition system (Iso et al., 1989). When 202 difference vocabulary words

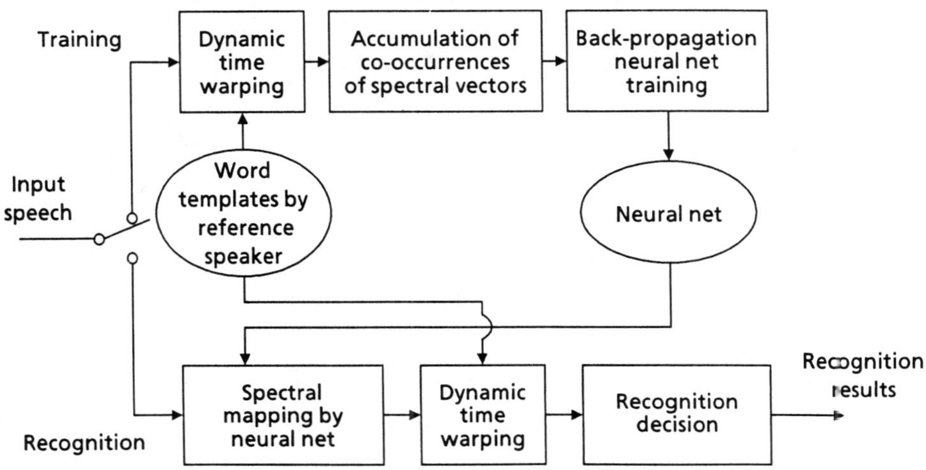

Figure 10 Block diagram of speaker adaptation by neural network.

were used for neural net training, the recognition rate was shown to be improved from 82.0 to 99.0%. The problem at this moment is that the method requires a large number of training words.

3.4.3. Unsupervised Adaptation

Figure 11 is a block diagram of an unsupervised codebook adaptation method (Furui, 1989b, 1989c). This method is based on an adaptation algorithm for

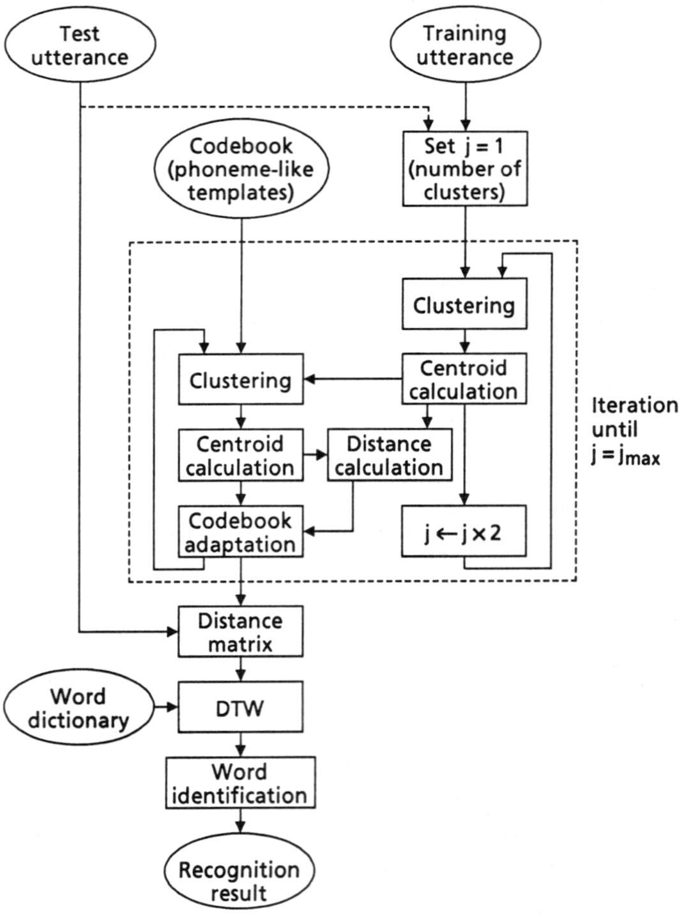

Figure 11 Principal structure of a word recognition system with the unsupervised codebook adaptation mechanism.

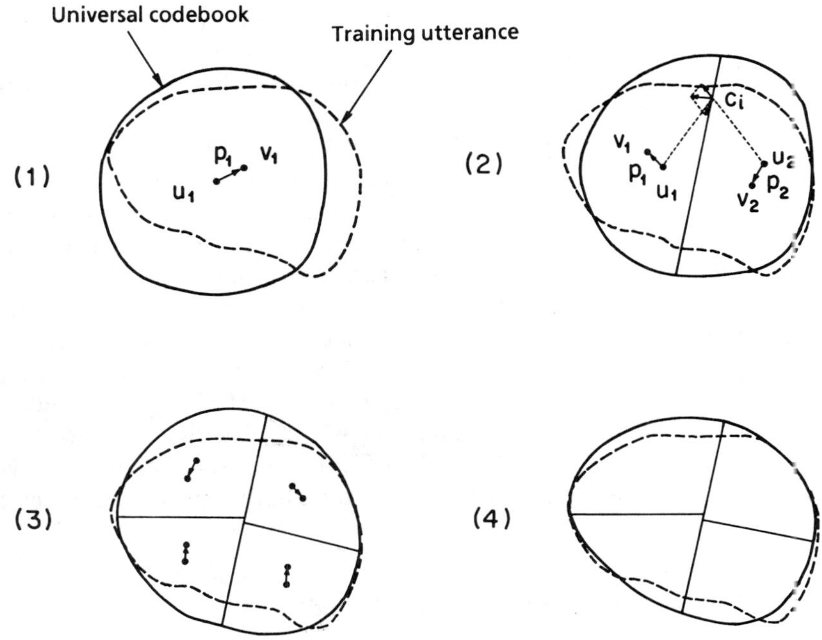

Figure 12 Hierarchical codebook adaptation algorithm maintaining continuity between adjacent clusters. v_j; centroid of jth training spectra cluster; u_j, centroid of corresponding codebook cluster; p_j, deviation vector; c_i, shifting vector for ith codebook element.

a segment vocoder (Shiraki and Honda, 1990). First, an initial codebook and a VQ-indexed word dictionary are prepared. The initial codebook is produced by clustering the voices of multiple speakers and commonly serves as the initial condition for every new speaker.

In the adaptation process, a set of spectra form the training utterances of a new speaker and the reference codebook elements are clustered hierarchically in an increasing number of clusters. Using deviation vectors between centroids of the training spectra clusters and the corresponding codebook clusters, either codebook elements or input frame spectra are shifted so that the corresponding centroids coincide. Continuity between adjacent clusters is maintained by determining the shifting vectors as the weighted sum of the deviation vectors of adjacent clusters (see Chapter 7). Adaptation is thus performed hierarchically from global to local individuality.

Figure 12 illustrates the hierarchical adaptation procedures of shifting the codebook elements from the beginning to the four-cluster stage. The size of the codebook is maintained throughout the adaptation process. Using the hierarchical technique, phonetic correspondence between training utterance spectra and codebook elements is roughly extracted and maintained.

This method was evaluated by a cepstrum-based 100-Japanese-word recognition system in which each word is represented by multiple sequences of codebook entries. Four speakers representing interspeaker variability are selected and the voices of these speakers are used to create an initial universal codebook and word dictionary. The universal codebook has 1024 entries. The multiple template method ensures moderate recognition performance for new speakers at the beginning of the adaptation process. Results of experiments show that the adaptation using 10 arbitrary training words reduces the mean word recognition error rate from 4.9 to 2.9%. Since the error rate for speaker-dependent recognition is 2.2%, this method is highly effective. Several modifications of the adaptation method have also been investigated (Furui, 1989b).

3.5. Reference Templates Generation Method

A supervised adaptation procedure has been devised to generate phoneme templates adapted to each speaker using a set of transformation rules and the templates extracted from training words (Furui, 1980). Each new speaker is requested to provide utterances of only a fraction of the entire vocabulary as a training set. The transformation rules indicate the relationship between the templates for the entire vocabulary and those extracted from the training words. The rules are obtained through multiple regression analysis in a pretraining stage in which a group of speakers provides utterances of the entire vocabulary. The rules are universally used for all new speakers. The generated templates and speaker-independent templates are interpolated to create templates to be used in the recognition stage.

Results of recognition experiments using 67 Japanese airport names uttered by 30 male speakers have ascertained the effectiveness of this training procedure. A mean recognition accuracy of 98.2% was obtained after a 12-word training procedure.

4. CONCLUSION

Speaker-adaptation or speaker-normalization methods, which need training utterances, are generally less convenient than speaker-independent methods.

If unsupervised online adaptation using arbitrary words or sentences is possible, however, the recognition system can be used like a speaker-independent system. Such a recognition system automatically adapts to a new speaker's voice according to the current speech characteristics in an incremental and nonintrusive manner. The system improves its performance as the number of recognition utterances increases, without requiring any additional training utterances from the speaker.

Unsupervised methods are considered to be applicable also to normalization of other speech variations due to additive noise, the Lombard effect, room acoustics, telephone/microphone characteristics, transmission characteristics, and so on.

From the viewpoint of convenience for the users and present speaker-independent recognition techniques, it is desirable to combine speaker-independent and speaker-adaptive techniques. In other words, it is ideal to begin with a speaker-independent system and to adapt the parameters to better suit the input speaker. This method requires no delay in enrolling the speaker and realizes relatively high performance even from the beginning. In experimental systems, adaptation from parameters of a single difference speaker is convenient for easily showing the improvement of performance, but it requires a large number of adaptation sentences or words from the new speaker as well as considerable time to perform the adaptation.

We should pursue improvement of the speaker-independent recognition methods as far as possible and create highly effective speaker-adaptation methods, which can be combined with the speaker-independent methods, to further enhance performance. For this purpose, it will be necessary to advance basic research on elucidating the mechanism of individuality in speech spectra and on inventing a sophisticated yet tractable model of speech variation based on a sufficiently large data base. Clarifying the mechanism of human speech perception and lexical access will also be essential as long-term research.

REFERENCES

Adams, D. A., and Bisiani, R. (1986). The Carnegie Mellon University distributed speech recognition system. *Speech Technol.* March/April, pp. 14–23.

Bahl, L. R., Brown, P. F., de Souza, P. V., and Mercer, R. L. (1986). Maximum mutual information estimation of hidden Markov model parameters for speech recognition. *Proc. IEEE Int. Conf. Acoustics, Speech, and Signal Processing*, Tokyo, 2.3.

Bahl, L. R., Brown, P. F., de Souza, P. V., and Mercer, R. L. (1988). A new algorithm for the estimation of hidden Markov model parameters. *Proc. IEEE Int. Conf. Acoustics, Speech and Signal Processing*, New York, S11.2.

Chow, Y. L., Dunham, M. O., Kimball, O. A., Krasner, M., A., Kubala, G. F., Makhoul, J., Roucos, S., and Schwartz, R. (1987). BYBLOS: The BBN continuous speech recognition system. *Proc. IEEE Int. Conf. Acoustics Speech, and Signal Processing*, Dallas, 3.7.

Doddington, G. R. (1989). Phonetically sensitive discriminants for improved speech recognition. *Proc. IEEE Int. Conf. Acoustics, Speech, and Signal Processing*, Glasgow, S10b.11.

Dutoit, D. (1989). Evaluation of speaker-independent isolated-word recognition systems over telephone network. *European Conf. on Speech Technology*, Edinburgh, pp. 241–244.

Feng, M. W. (1989). Iterative normalizing for speaker-adaptive training in continuous speech recognition. *Proc. IEEE Int. Conf. Acoustics, Speech, and Signal Processing*, Glasgow, S12.4.

Franzini, M. A., Witbrock, M. J., and Lee, K.-F. (1989). A connectionist approach to continuous speech recognition. *Proc. IEEE Int. Conf. Acoust., Speech, Signal Processing*, Glasgow, S8.10.

Furui, S. (1980). A training procedure for isolated word recognition systems. *IEEE Trans. Acoust. Speech Signal Process. ASSP-28*:129–136.

Furui, S. (1986a). Speaker-independent isolated word recognition using dynamic features of speech spectrum. *IEEE Trans. Acoust. Speech Signal Process. ASSP-34*: 52–59.

Furui, S. (1986b). Research on individuality features in speech waves and automatic speaker recognition techniques. *Speech Commun. 5*:183–197.

Furui, S. (1988). A VQ-based preprocessor using cepstral dynamic features for speaker-independent large vocabulary word recognition. *IEEE Trans. Acoust. Speech Signal Process. ASSP-36*:980–987.

Furui, S. (1989a). *Digital Speech Processing, Synthesis, and Recognition*. Marcel Dekker, New York.

Furui, S. (1989b). Unsupervised speaker adaptation method based on hierarchical spectral clustering. *Proc. IEEE Int. Conf. Acoustics, Speech, and Signal Processing*, Glasgow, S6.9.

Furui S. (1989c). Unsupervised speaker adaptation based on hierarchical spectral clustering. *IEEE Trans. Acoust. Speech Signal Process. ASSP-37*:1923–1930.

Furui, S. (1990). On the use of hierarchical spectral dynamics in speech recognition. *Proc. IEEE Int. Conf. Acoustics, Speech, and Signal Processing*, Albuquerque, S15a.10.

Hunt, M. J., and Lefebvre, C. (1989). A comparison of several acoustic representations for speech recognition with degraded and undegraded speech. *Proc. IEEE Int. Conf. Acoustics, Speech, and Signal Processing*, Glasgow, S6.3.

Iso, K., Asogawa, M., Yoshida, K., and Watanabe, T. (1989). Speaker adaptation using neural network. *Proc. Spring Meeting Acoust. Soc. Jpn.*, 1-6-16.

Jelinek, F., and Mercer, R. L. (1980). Interpolated estimation of Markov source parameters from sparse data. In *Pattern Recognition In Practice*, E. S. Gelsema and L. N. Kanal (Eds.), North–Holland, Amsterdam, pp. 381–397.

Kato, K., and Furui, S. (1985). Listener adaptability for individual voice in speech perception. *Trans. Committee Hearing Res. Acoust. Soc. Jpn.*, H-85-5.

Lee, K.-F., and Hon, H.-W. (1988). Large-vocabulary speaker-independent continuous speech recognition using HMM. *Proc. IEEE Int. Conf. Acoustics, Speech, and Signal Processing*, New York, S3.7.

Lee, K.-F. (1989). *Automatic Speech Recognition, the Development of the SPHINX System*, Kluwer, Boston.

Levinson, S. E., Rabiner, L. R., Rosenberg, A. E., and Wilpon, J. G. (1979). Interactive clustering techniques for selecting speaker independent reference templates for isolated word recognition. *IEEE Trans. Acoust. Speech Signal Process. ASSP-27*:134–141.

Mariani, J. (1989). Recent advances in speech processing. *Proc. IEEE Int. Conf. Acoustics, Speech, and Signal Processing*, Glasgow, S9.1.

Merialdo, B. (1988). Phonetic recognition using hidden Markov models and maximum mutual information training. *Proc. IEEE Int. Conf. Acoustics, Speech, and Signal Processing*, New York, S3.4.

Montacie, C., Choukri, K., and Chollet, G. (1989). Speech recognition using temporal decomposition and multi-layer feed-forward automata. *Proc. IEEE Int. Conf. Acoustics, Speech, and Signal Processing*, Glasgow, S8.6.

Murveit, H., and Weintraub, M. (1988). 1000-word speaker-independent continuous-speech recognition using hidden Markov models. *Proc. IEEE Int. Conf. Acoustics, Speech, and Signal Processing*, New York, S3.5.

Nishimura, M., and Sugawara, K. (1988). Speaker adaptation method for HMM-based speech recognition. *Proc. IEEE Int. Conf. Acoustics, Speech, and Signal Processing*, New York, S5.7.

Paul, D. (1989). The Lincoln robust continuous speech recognizer. *Proc. IEEE Int. Conf. Acoustics, Speech, and Signal Processing*, Glasgow, S9.4.

Rabiner, L. R., Levinson, S. E., Rosenberg, A. E., and Wilpon, J. G. (1979). Speaker-independent recognition of isolated words using clustering techniques *IEEE Trans. Acoust. Speech Signal Process ASSP-27*:336–349.

Rabiner, L. R., and Wilpon, J. G. (1979). Speaker-independent isolated word recognition for a moderate size (54 word) vocabulary. *IEEE Trans. Acoust. Speech Signal Process. ASSP-27*:583–587.

Rabiner, L. R., and Wilpon, J. G. (1981). Isolated word recognition using a two-pass pattern recognition approach. *Proc. IEEE Int. Conf. Acoustics, Speech, and Signal Processing*, Atlanta, 19.2.

Rabiner, L. R., Levinson S. E., and Sondhi, M. M. (1983). On the application of vector quantization and hidden Markov models to speaker-independent, isolated word recognition. *Bell Syst. Tech. J. 62*(4):1075–1105.

Rabiner, L. R., Juang, B. H., Levinson, S. E., and Sondhi, M. M. (1985). Recognition of isolated digits using hidden Markov models with continuous mixture densities. *AT&T Tech. J. 64*(6):1211–1234.

Rabiner, L. R., Wilpon, J. G., and Soong, F. K. (1988). High performance connected digit recognition using hidden Markov models. *Proc. IEEE Int. Conf. Acoustics, Speech, and Signal Processing*, New York, S3.6.

Rabiner, L. R., Lee, C. H., Juang, B. H., and Wilpon, J. G. (1989). HMM clustering for connected word recognition. *Proc. IEEE Int. Conf. Acoustics, Speech, and Signal Processing*, Glasgow, S8.5.

Rigoll, G. (1989). Speaker adaptation for large vocabulary speech recognition systems using 'speaker Markov models'. *Proc. IEEE Int. Conf. Acoustics, Speech, and Signal Processing*, Glasgow, S1.2.

Sakoe, H., Isotani, R., Yoshida, K., Iso, K., and Watanabe, T. (1989). Speaker-independent word recognition using dynamic programming neural networks. *Proc. IEEE Int. Conf. Acoustics, Speech, and Signal Processing*, Glasgow, S1.8.

Schwartz, R., Chow, Y.-L., and Kubala, F. (1987). Rapid speaker adaptation using a probabilistic spectral mapping. *Proc. IEEE Int. Conf. Acoustics, Speech, and Signal Processing*, Dallas, 15.3.

Shikano, K., Lee, K.-F., and Reddy R. (1986). Speaker adaptation through vector quantization. *Proc. IEEE Int. Conf. Acoustics, Speech, and Signal Processing*, Tokyo, 49.5.

Shiraki, Y., and Honda, M. (1990). Speaker adaptation algorithms based on piecewise moving adaptive segment quantization method. *Proc. IEEE Int. Conf. Acoustics, Speech, and Signal Processing*, Albuquerque, S12.5.

Sugamura, N., Shikano, K., and Furui, S. (1983). SPLIT isolated word recognition using phoneme-like templates. *Proc. IEEE Int. Conf. Acoustics, Speech, and Signal Processing*, Boston, 16.3.

Sugiyama, M. (1986). Unsupervised speaker adaptation methods for vowel templates. *Proc. IEEE Int. Conf. Acoustics, Speech, and Signal Processing*, Tokyo, 49.4.

TI Speech Recognition Group (1987). TI speech recognition technology development. DARPA Speech Recognition Workshop, October.

Wilpon, J. G., Rabiner, L. R., and Bergh, A. (1982). Speaker-independent isolated word recognition using a 129-word airline vocabulary. *J. Acoust. Soc. Am. 72*:390–396.

Wilpon, J. G., Lee, C. H., and Rabiner, L. R. (1989). Application of hidden Markov models for recognition of a limited set of words in unconstrained speech. *Proc. IEEE Int. Conf. Acoustics, Speech, and Signal Processing*, Glasgow, S6.1.

Zue, V., Glass, J., Phillips, M., and Seneff, S. (1989). Acoustic segmentation and phonetic classification in the SUMMIT system. *Proc. IEEE Int. Conf. Acoustics, Speech, and Signal Processing*, Glasgow, S8.1.

20

Continuous Speech Recognition

KAI-FU LEE Apple Computer, Inc., Cupertino, California

FIL ALLEVA Carnegie-Mellon University, Pittsburgh, Pennsylvania

1. INTRODUCTION

Continuous speech recognition (CSR) is an important area of speech research. While isolated-word recognition (IWR) systems have some applications, they are slow and awkward for larger realistic tasks. Although CSR is inherently much more difficult than IWR, considerable progress has been made in recent years. This is the focus of this chapter.

Recognition of continuously spoken speech is significantly more difficult than that of isolated words. Its complexity is caused by three properties of continuous speech. First, word boundaries are unclear in continuous speech. In isolated-word recognition, word boundaries are known and can be used to improve the accuracy and limit the search. In continuous speech, however, word boundaries are usually difficult to find. For example, in the phrase *this ship*, the /s/ in *this* is often omitted. In a sentence like "*we were away a year*," the whole sentence is one long vocalic segment, and word boundaries are difficult to locate. Second, *coarticulatory effects* are much stronger in continuous speech. Although we may try to pronounce words as concatenated sequences of phones, our articulators cannot move instantaneously to produce unaffected phones. As a result, a phone is strongly influenced by the context in which it appears. In continuous speech, this effect occurs between words and is much harder to predict. Moreover, with the faster speaking rate in continuous speech, within-word coarticulation is also accentuated. Third, *content words* (nouns, verbs, adjectives, etc.) are often emphasized, while *function words* (articles, prepositions, pronouns, short verbs, etc.) are poorly articulated. In particular, the phones in function words are often shortened, skipped, or distorted.

As a result, error rates increase drastically from isolated-word to continuous speech. Bahl et al. (1981) reported error rates of 3.1 and 8.7% for IWR and CSR, respectively, for a constrained 1000-word vocabulary. On a 5000-word natural language task, a more advanced system (Bahl et al., 1989a) resulted in 3.3 and 11.0% error rates for IWR and CSR, respectively.

In addition to degraded accuracy, another difficulty with CSR is the organization of the search. With isolated words, search can be simply organized as a process that matches every word against an endpointed segment of speech, and returns the best-matching model as the recognized word. With continuous speech, word boundaries are unknown, so many boundaries need to be hypothesized and verified for each word. This makes the search problem potentially exponential. In other words, IWR matches every word against a segment of speech, whereas CSR matches every sentence. In any given language, the number of possible sentences is much larger than the number of words. With IWR, exhaustive search is often feasible, so that a poorly organized search is acceptable. However, with CSR, an exhaustive search is infeasible. Thus, the area of search organization and heuristic pruning is important for CSR.

The most successful continuous speech recognition systems (Bahl et al., 1983; Chow et al., 1987; Rabiner et al., 1989a; Lee et al., 1990a) today are based on hidden Markov models (HMMs). This chapter will focus on HMM-based continuous recognition algorithms. It should be noted that many of these algorithms are similar to other techniques [such as dynamic programming—see Myers and Rabiner (1981)], and can be applied to other techniques [such as neural networks (Franzini et al., 1989) or knowledge-based systems (Zue et al., 1990)].

Since the fundamentals of HMMs have already been discussed in a previous chapter, we will proceed directly to the techniques and algorithms used in HMM-based continuous speech recognition. There are three problems that must be resolved to extend speech recognition to continuous speech: (1) the *training* problem, or how to train HMMs for continuous speech, (2) the *modeling* problem, or how to enhance acoustic-phonetic modeling for continuous speech, and (3) the *recognition* problem, or how to organize the search and prune the search space to deal with the combinatorial explosion in continuous speech. These three issues will be addressed in Sections 2, 3, and 4, respectively. Section 5 closes with some concluding remarks.

2. THE TRAINING PROBLEM

Many speech recognition techniques face problems in training for continuous speech because word boundaries are not automatically detectable. Often, hand marking is needed, which is tedious and may be suboptimal. HMMs, however, do not have this problem. In fact, training HMMs on continuous speech is not much different from training on isolated words.

Since the state sequence is hidden in HMMs, *it doesn't matter where the word boundaries are.* We simply pretend that a sentence is a word. For each

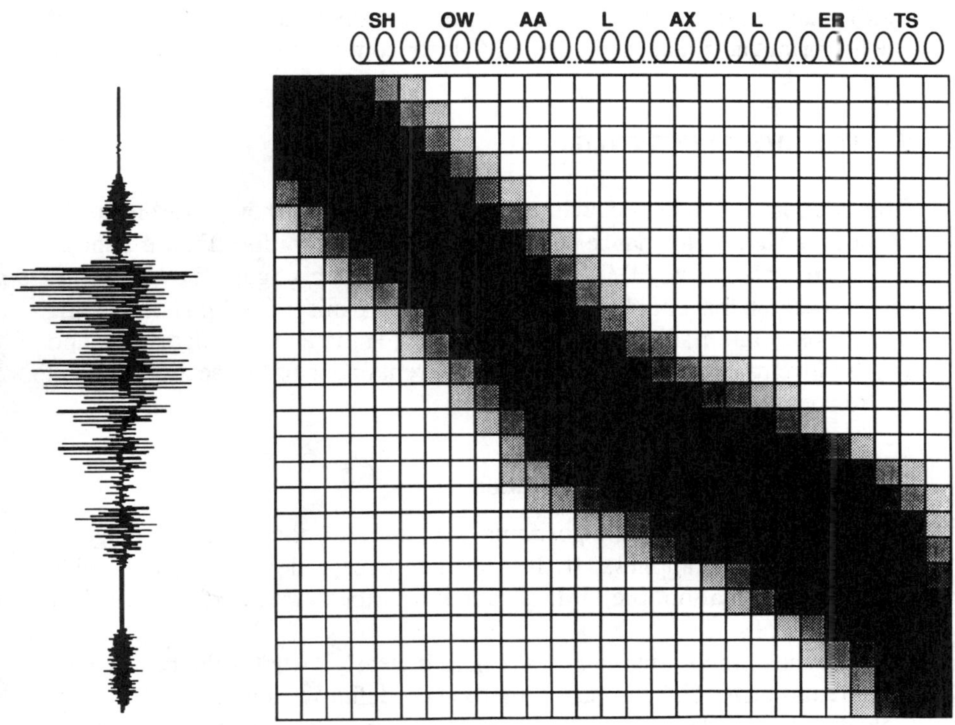

Figure 1 The alignment of states vs. speech using the forward-backward algorithm. Darker squares correspond to large α and represent more plausible alignments.

sentence, we need only the orthographic transcription. First, each word is instantiated with its HMM (which may, in turn, be a concatenation of subword models). Next, the words in the sentence are concatenated with optional silence models between words, as well as at the beginning and the end of the sentence. This large concatenated sentence HMM is then trained on the entire sentence using the forward-backward algorithm (Bahl et al., 1983). Figure 1 shows the concatenation of phone models into a sentence model and the alignment of states with speech.

Alternatively, the segmental K-means algorithm (Rabiner et al., 1986) could be used. This algorithm first aligns each frame of speech with a state and then reestimates all the parameters given the alignments. Iterative execution of this algorithm improves the alignments, as well as the word boundaries. It yields comparable results to the forward-backward algorithm for continuous density HMMs.

For both algorithms, training on continuous speech requires almost no change in algorithm or code. The only difference is the creation of a sentence HMM that considers multiple ways a sentence might be uttered. This is one of the important advantages in using HMMs for continuous speech recognition.

3. THE MODELING PROBLEM

This section is concerned with enhanced hidden Markov modeling techniques for high performance continuous speech recognition. Figure 2 shows a block diagram of an HMM-based CSR system. This figure shows four areas of research that could improve system performance: (1) signal processing and representation, (2) the HMM structure, (3) the learning algorithm, and (4) the inventory of subword units. In the remainder of this section, we will address each of these four areas.

3.1. Improved Signal Representation

Although HMM learning algorithms are powerful, the cliché "garbage in garbage out" still applies. If the signal representation does not provide sufficient information, the HMMs cannot be expected to perform the right discriminations.

The most important issue to ask here is clearly, "what is the best speech representation?" Unfortunately, there is no definitive answer to this question, due to a shortage of work on comparative studies. Two of the exceptions are Davis and Mermelstein (1980) and Rabiner et al. (1985). Davis and

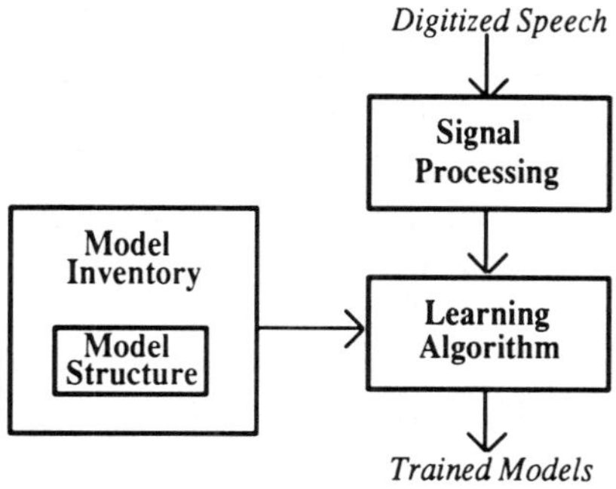

Figure 2 An HMM-based CSR architecture.

Mermelstein (1980) found that mel-frequency cepstrum gave the best performance, while Rabiner et al. (1985) demonstrated that LPC cepstrum outperformed four other parameters. Both of these representations have been used by a number of successful speech recognition systems (Rabiner et al., 1988; Lee, 1989; Chow et al., 1987; Paul and Martin, 1988). More recently, auditory models (Seneff, 1988; Cohen, 1989) based on human hearing have been proposed, but no overwhelming performance improvements have been shown yet.

For discrete HMM systems, another opportunity to improve the signal representation is in the vector quantization (VQ) process. VQ is a preprocessing stage that replaces each frame of speech with the index of the nearest matching centroid vector. The centroid vectors were precomputed using a clustering algorithm. The distance metric used in both clustering and quantization is insensitive to the type of speech events being compared. Moreover, it is detached from HMM learning, and errors in this stage may be harmful for HMM learning later. In view of this problem, techniques have been proposed to inject more phonetic information into the HMMs. One method is Kohonen's learning vector quantization (LVQ) algorithm (Iwamida et al., 1990) to organize multiple centroids for each phoneme to improve discrimination. Another is supervised vector quantization (SVQ) (Bahl et al., 1989a). SVQ iteratively improves the SVQ centroids by using phonetically labeled training frames to improve phonetic consistency of the centroids. Both VQ algorithms aimed to increase the phonetic information in the VQ stage.

One of the greatest improvements in CSR systems involves the incorporation of temporal changes in the spectra. One of the first efforts that incorporated dynamic changes was Furui's use of *delta cepstrum* (Furui, 1986), or a slope measure computed over a window of about five frames (50 ms). Simpler measures such as differenced coefficients (Shikano et al., 1986) have also been used with success. These differential parameters not only provide additional information about spectral movement but also broaden the scope of a single frame that is presented to the HMM. An alternative to these differential coefficients is to concatenate frames and then reduce the dimensionality (Brown, 1987). The incorporation of transitional information has reduced errors by as much as 50% (Rabiner et al., 1988; Lee, 1989).

The combination of heterogeneous parameters has also been found to be helpful. Power and differenced power (Lee 1989) have been used successfully in a LPC-based system, which does not have an adequate power model. A more ambitious attempt integrated many different features, including speech level, rate of spectral transition, frame and filter energy, and differential frame and filter energy (Doddington, 1989). Although these measures are likely to be correlated, the use of dimensionality reduction techniques, such as principal component analysis, effectively decorrelates them.

3.2. Improved HMM Structures

The previous section was concerned with how to condition the speech parameters for hidden Markov modeling. In this section, we will discuss several techniques that improve hidden Markov modeling for better continuous speech recognition. The "improved" HMM should be a more "correct" model of speech. For example, if a phoneme undergoes many changes temporally, sufficient states must be allocated for it. Or if a model must absorb male and female voices, sufficient codewords (for discrete HMM) or mixtures (for continuous HMM) must be assigned to model a bimodal distribution. Having a correct model is very important, since the maximum likelihood estimation (MLE) technique used in HMM learning relies on correct model assumptions (Brown, 1987).

Given sufficient data, hidden Markov models will yield superior performance with finer models. But how does one create finer models? One obvious way is to increase the number of states. More states can model finer within-unit variations, which improves the "correctness" of the model. Deng et al. (1989) found that tailoring model length to the average vowel duration improves performance. In that study, Deng et al. used the Bakis model, where each state has a *stay* transition, a *progressive* transition, and a *skip* transition. An example of the Bakis model is shown in Fig. 3A. Doddington (1989) also used the Bakis model and found that best results were obtained when the number of states corresponded to the average number of frames for the units

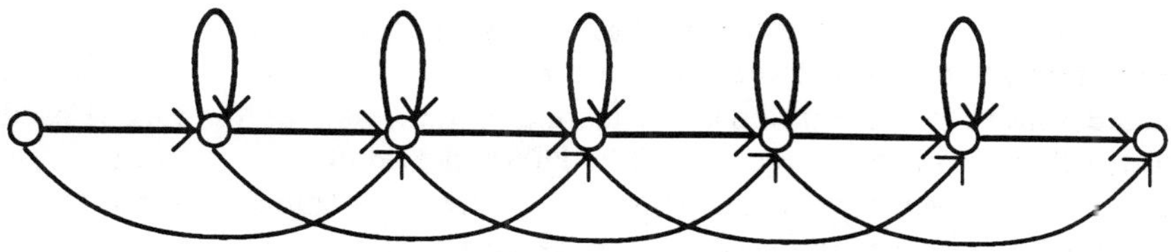

(A)

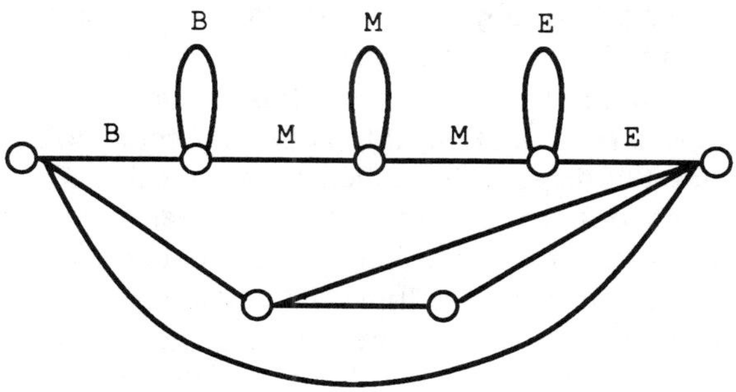

(B)

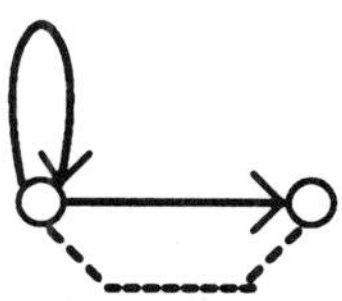

(C)

Figure 3 Three different types of HMM topologies.

to be modeled.* Moreover, detailed modeling was enforced by allowing a maximum state duration of three frames.

One problem with the Bakis (or any linear) model is that it will not be able to recognize tokens shorter than the shortest skip path. A possible solution is the model structure shown in Fig. 3B. This structure has explicit paths for tokens shorter than four frames and has yielded good results for phonetic

*Note that sufficient data must exist to model this level of detail. For Doddington's study (1989) over 8000 sentences were available to train 29 models.

modeling (Lee, 1989). Alternatively, skip arcs, or arcs that enable movement from one state to another without consuming an input frame, can be used to enable a more expedient departure from a model. Bahl et al. (1988a) used models that could be completely skipped to model fenones (very short acoustic events). However, while fenones have been successfully applied to IWR, they have not shown comparable promise for CSR.

Another area of finer modeling is at the output probability density level. For discrete hidden Markov models, a larger codebook increases precision in acoustic modeling. Typically, about 200–300 codewords have been used. However, for most tasks, models become untrainable if we increase the number of codewords substantially. In view of this problem, *multiple codebooks* have been proposed (Gupta et al., 1987). Several codebooks, each modeling a disjoint subset of the speech coefficients, are created. By assuming that the multiple codebooks are independent, the forward-backward algorithm can be modified to train these multiple probability density functions (pdfs) simultaneously. The concept of multiple codebooks could be combined with transitional and other parameters. For example, one codebook might be used for instantaneous parameters, another for differential parameters, and yet a third for power (Lee, 1989).

For mixture Gaussian continuous density HMMs, the number of mixtures per state can be increased to model acoustics in more detail. Rabiner et al. (1988) used up to nine Gaussian mixtures per state in a continuous digit recognizer. However, adding more mixtures per state increases the number of parameters very quickly. One way to control the parameter growth is to use *semicontinuous hidden Markov models* (SCHMMs) (Huang and Jack, 1989; Bellegarda and Nahamoo, 1989), a compromise between continuous HMMs and discrete HMMs. The SCHMM is a continuous density HMM, where the mixtures are shared by all the states. If we view these mixtures as a codebook, then the SCHMM is also similar to the discrete HMM, except that continuous density codebooks are used and that many codewords can be matched at each state. Compared to the continuous density HMM, the SCHMM could have a finer acoustic model because each state could use a large number of mixtures. At the same time, the SCHMM is more trainable because the number of total mixtures is likely to be smaller than the number in standard state-dependent continuous density HMMs.

3.3. Improved Learning and Discrimination

For learning, most HMM-based systems use the forward-backward algorithm (Baum, 1972; Jelinek, 1976; Bahl et al., 1983), which adjusts the parameters to obtain an approximation to the maximum-likelihood estimates of the HMM parameters. Maximum likelihood estimators have many desirable properties, and many successful systems (Jelinek, et al., 1985; Chow et al., 1987; Lee, 1989; Rabiner et al., 1988) are based on MLEs. However, maximum likelihood estimation has one serious flaw: it assumes that the underlying models are correct. In reality, however, typical HMMs make extremely inaccurate assumptions about the speech production process.

In view of this problem, maximum mutual information estimation (MMIE) (Brown, 1987) was proposed. MMIE minimizes information needed to specify the word sequence when the acoustic sequence is known. Given an acoustic sequence Y and the corresponding word sequence W, MMIE not only tries to increase $\Pr(Y \mid W)$ (as does MLE), but also tries to decrease $\Pr(Y \mid W')$ for incorrect words W'. Training against negative exemplars increases the importance of a correct recognition.

Another alternative, the corrective training algorithm (Bahl et al., 1988b) attempts to maximize the recognition rate on the training data. It does so by modifying the HMM parameters to not only increase the probability for the correct training data but also decrease the probability for the misrecognitions and near misses. Although corrective training does not have the theoretical justifications of MMIE, it has performed well for isolated word recognition (Bahl et al., 1988b), as well as continuous speech recognition (Lee and Mahajan, 1990).

Another discriminant learning approach is based on *linear discriminations* (Brown, 1987; Doddington, 1989). This approach is similar to corrective training in that both correct and near-miss data are used to reinforce discrimination. A transformation matrix is computed so that the variance for the correct data is minimized, while the variance between the correct and the near-miss classes is maximized. Brown (1987) successfully used a single linear discriminant for all speech frames, essentially as a postprocessor. In a more complex application, Doddington (1989) created state-dependent linear discriminants. First, correct and near-miss samples are collected to compute a within-class and a between-class covariance matrix for each state. These matrices are then used to transform the speech data to maximize the between-class variance and minimize the within-class variance for each state. Both studies showed dramatic error rate reductions.

These algorithms add a discriminatory capability to CSR. Note that, in general, the best results were obtained not by discriminating all words or all phones but by *selectively* discriminating words and phones that may cause confusion. Given this goal, their performance invariably improves for specific tasks and grammars; however, they are also likely to be less robust, making task switching or adding new words difficult.

3.4. Improved Subword Inventory

The previous improvements assume the existence of a fixed subword inventory, such as phonemes, triphones, or syllables. In this section, we examine how the inventory of these units can be enhanced.

For connected speech recognition of small vocabularies, words are viable models. Indeed, when sufficient training exists, detailed word models have led to the best results (Rabiner et al., 1988; Doddington, 1989). However, for large-vocabulary tasks, word models are no longer feasible due to the large number of parameters involved and the lack of sufficient data per word.

Phonemes are natural units of speech. However, because phonemes are highly affected by context, phoneme models have very broad distributions and are not adequate for high-performance speech recognition. The most prominent contextual effect is that of the immediate phonetic neighbor. That is, the realization of a phone is highly affected by its left and right neighbors. This effect is accentuated in continuous speech, where phones are shorter. In view of this, triphones (Schwartz et al., 1985) were introduced to model every phone in the context of its left and right neighbors. Bahl et al. (1989a) also found that while detailed word modeling is the most effective technique for IWR, for CSR contextual modeling gave the best results. BYBLOS (Kubala et al., 1988), a speech recognition system based on triphones, has produced very good results. Many other successful systems (Paul and Martin, 1988; Lee, 1989; Bernstein et al., 1989; Lee et al., 1990b; Bahl, et al., 1989a) are based on triphones or triphone-like units.

For HMM systems, triphones are implemented by replacing each phone with its corresponding triphone. But what about word-boundary triphones? The first phone of a word is clearly affected by the previous word, and the last phone of a word is affected by the next word. Implementation of this effect involves modification of the search algorithm to use dynamically the right word-boundary models. One procedure for cross-word triphone modeling has been documented in Hwang et al. (1989).

Another problem with triphone units is that there are a great many of them. With about 50 phones, there are potentially 125,000 triphones. Even after discarding impossible triphones, there are still tens of thousands of triphones. Moreover, if we wish to model other allophonic variations (such as stress, word boundary, syllable boundary, or quad-phones), the number of models grows geometrically. One approach to controlling this proliferation of models is to combine similar triphones (or allophones). This is phonetically plausible, because many triphones are similar. Lee (1990) used an agglomerative clustering to reduce triphones into generalized triphones. A maximum likelihood distance was used in order to remain true to the training criterion. Generalized triphones have also led to high-performance recognition (Lee, 1989).

If the amount of training is too limited to even perform triphone/allophone clustering, human knowledge can be used to reduce the number of contextual variations. For example, in view of poor recognition for plosive sounds, Derouault (1987) used context-dependent plosive models that depended only on the place of articulation of the neighboring vowel. These models substantially improved performance on plosives. As another example, Deng et al. (1989) created multiple context-dependent vowel models that shared the same output probabilities but had their own transition probabilities. This type of context-dependent modeling is particularly appropriate when there is an insufficient training data to support training all parameters and when the spectral properties are invariant across context while the durational properties are variable.

Another difficulty with continuous speech is the poor articulation of function words. To deal with function words, *function word–dependent phone models* (Lee, 1990) were proposed. These models focus on the most confusable subvocabulary. Yet they can be well trained because function words occur frequently. To ensure robust modeling, function word–dependent phone models (like other context-dependent phone models) can also be interpolated with context-independent phone models.

All of the foregoing techniques create special models for specific contexts or words. Another approach is to create multiple models for the same event. For example, one model could be used for male speakers and another for female speakers. For recognition, both gender spaces are searched while gender consistency in each hypothesized sentence is maintained (Doddington, 1989). Gender clustering is a simple rule-derived clustering. Alternatively, automatic clustering techniques could be used to create multiple models at the word or phoneme level. Rabiner et al. (1989b) found that creating clusters for outliers significantly improves recognition accuracy.

Finally, multiple pronunciations of words could be used to accommodate different ways a word may be spoken. This is useful for modeling words with multiple pronunciations (such as *the*), or even words that might be pronounced differently because of dialect or habit (such as *tomato*). The combination of phonological rules and HMMs was explored by Bernstein et al. (1989). It was found that full phonological expansion increased the number of parameters excessively. The best combination of phonology and HMM learning is to allow a small number of speaking variations (about 1.3 pronunciations per word).

3.5. Summary

We have described a number of methods that improved acoustic modeling using HMMs. These techniques are inspired by knowledge of acoustics-phonetics, as well as an understanding of the strengths and weaknesses of the HMM. These techniques enhance the detail modeled by HMMs, while keeping the parameters trainable. The combination of improved parameters, improved HMMs, improved learning, and improved subword inventory can reduce the error rate of a CSR system by an order of magnitude (Lee, 1989).

4. THE RECOGNITION PROBLEM

4.1. Overview of the Problem

A search paradigm is employed to select from a number of alternatives a solution that satisfies some goal criteria. The ultimate goal of continuous speech recognition is to determine a *correct* decoding (in terms of a sequence of words) of an observed acoustic sequence. Unfortunately, constructing an evaluation that can always identify the correct sequence of words is impossible. Therefore, we modify the goal of the search paradigm to select the sequence of words that is most likely for our acoustic models given the observed acoustic sequence. This leads to the following definition of CSR.

> Given an acoustic sequence y, derive a word sequence w' such that $P(w' \mid y) = \max_w P(w \mid y)$.

By applying Bayes rule to this definition of CSR we derive Eq. (1), or the fundamental equation of speech recognition.

$$P(w \mid y) = \frac{P(w) \times P(y \mid w)}{P(y)} \tag{1}$$

Fundamental equation of speech recognition

Equation (1) allows us to express $P(w \mid y)$ in terms we can compute. $P(w)$ is the a priori probability of the word sequence w. $P(w)$ is computed from the language model. $P(y \mid w)$ is the conditional probability of the acoustic sequence y given the word sequence w. $P(y \mid w)$ is computed from the acoustic model. Finally, $P(y)$ is the probability of the acoustic sequence y. Determining a reasonable model for $P(y)$ is difficult and often not necessary, since

$P(y)$ is equivalent for all complete decodings of y. Thus, we discard $P(y)$ and rewrite Eq. (1) as

$$P(w,y) = P(w) \times P(y \mid w) \tag{2}$$

4.1.1. *The Language Model*

Although the development of language models is covered elsewhere in this volume, the choice of language model (or $P(w)$ in Eq. 2) has a significant impact on the recognition process. In particular, the constraint provided by a language model can substantially improve a system's performance, and the size of the search space generated by a language model often determines the best search algorithm. Thus, we provide a brief overview here.

Language models used by CSR systems fall into numerous categories.

- *Uniform* language models, where every word in any sentence is equally probable. The size of the language model is 0 but so is the constraint.
- *Stochastic* language models such as trigram, bigram, and unigram models (Jelinek et al., 1985), where the probability of a word in a sequence is estimated based on the probability that it belongs to an equivalence class defined by the predecessor words of that sequence. The language probability of the word sequence is then given by the joint probability of all the words in the sequence.
- *Finite state* languages (Baker, 1975; Bahl et al., 1978) are simple artificial languages that model all legal sentences using a single network.
- *Other* possible language models include context-free (Ney, 1987), unification (Hemphill and Picone, 1989), statistical tree-based (Bahl et al., 1989b), and case frame grammars.

These language models provide various levels of constraint to the recognizer. Finite state grammars tend to be constraining; however, they are not realistic grammars for most tasks. On the other hand, context-free grammars and unification grammars are more realistic, but have too many states to enumerate. Statistical grammars can be small (unigram, bigram), but as the number of previous words modeled increases to trigrams, the number of states grows very quickly.

4.1.2. *The Acoustic Model*

The second component of Eq. (2), $P(y \mid w)$, is estimated by the acoustic model. In this chapter, this is assumed to be estimated by hidden Markov models. To compute $P(y \mid w)$ for a given y and w, the model corresponding to the sentence w is created by concatenating word or subword models. Then, the forward algorithm gives $P(y \mid w)$.

4.1.3. *Combining Language and Acoustic Models*

Although probability theory suggests that the acoustic and language probabilities can be combined through multiplication, in practice some weighting is necessary. This is because the HMM acoustic model makes certain inaccurate independence assumptions that result in an underestimated acoustic model probability. Consequently, the two quantities have vastly different dynamic ranges. One way to balance the two is to replace the term $P(w)$ with the term $P(w)^l$ in Eqs. (1) and (2) (Bahl et al., 1980). The exponent l is referred to as the *language weight* and is determined experimentally in order to optimize the recognition performance of the CSR system. Since the acoustics are underestimated, $l > 1$. Typical values of l are between 2 and 5.

4.1.4. *Unrestricted Search*

Having described the two components of a speech recognizer and how they are combined, we now proceed to the problem of finding the most likely word sequence given these components and a spoken utterance.

In order to illuminate the problems of continuous speech recognition, we first examine a basic form of search, simple enumeration. In CSR, a search that enumerates all possibilities is guaranteed to find the optimal sequence of words w, given an observed acoustic sequence y. The following algorithm describes the enumerative search.

Enumerative Search

1. For each possible sequence length, N,
    1. Enumerate each possible word sequence, w, of length N,
        a. Evaluate $P(w \mid y)$.
        b. Select most probable w.

As described before, $P(w \mid y)$ is determined by combining the HMM acoustic model [giving $P(y \mid w)$] and the language model [giving $P(w)$].

4.1.5. *Search Enhancements*

Unfortunately, except for the smallest of problems, the enumerative search will fail to terminate in a practical amount of time, since the number of word sequences grows exponentially with the length of the word sequence N. Nonetheless, all CSR search algorithms represent a restructuring of this simpleminded algorithm in order to find their solutions in a *small* finite period of time. This restructuring involves applying one or more of the following techniques to the unrestricted search procedure outlined above:

- Optimized hypothesis generation
- Problem space reduction
- Search reduction
- Knowledge application

Optimized hypothesis generation is accomplished by merging common partial hypotheses, which is the basis of a tree search. The set of partial hypotheses is constructed incrementally so that each unique partial solution is evaluated exactly once. For example, in Fig. 4A, two sentences are generated sequentially in their entirety. Recognition using this representation requires two complete sequential searches. On the other hand, in Fig. 4B, a tree was generated by merging common portions of the two sentences. Using the tree in 4B for recognition, the partial solution (ab) is evaluated only once. By evaluating common partial solutions, significant savings can be realized in CSR systems.

Problem space reduction involves transforming the state space of the problem to make the search more efficient. This involves creating a graph from the search tree such that no state of the search tree is duplicated. Thus, each unique state is visited exactly once, resulting in major reductions in the search space. To illustrate this, consider the search tree and the search graph in Fig. 5. By recombining the paths at state d the number of nodes that must be visited can be reduced. In CSR systems, paths are generally merged when they represent identical word sequences. Paths can also be merged at *juncture states*, such as state d in Fig. 5, in the grammar network.

Once it has been determined that two paths should be merged, one must decide how to combine their scores. If these paths represent identical word

Figure 4 Merging common partial solutions.

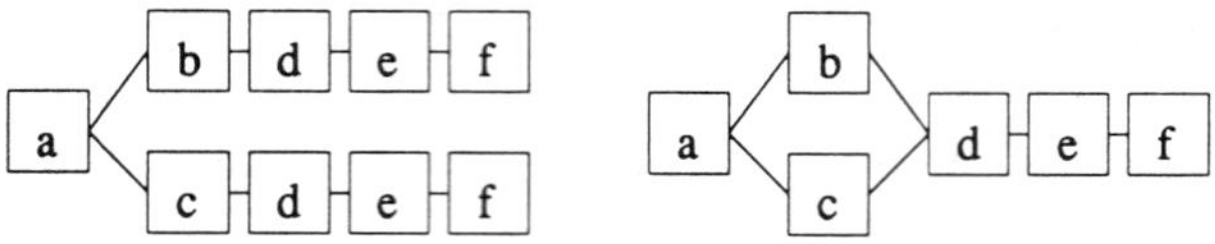

Figure 5 Merging common states.

sequences, it is consistent with the forward-backward training to add the probabilities. Some CSR systems may choose to take a maximum for efficiency.* If the paths have different histories, a maximum can be taken when they are merged. This still guarantees optimal top choice, since the inferior path at that point could not possibly exceed the superior one. However, this technique can produce no more than the top candidate.

Search reduction involves pruning hypotheses from further consideration. The eliminated hypotheses fall into two categories:

1. *Probably* inferior hypothesis, whose partial evaluation is worse than an already completed evaluation.
2. *Probably* inferior hypothesis, whose partial evaluation is worse than some heuristically determined threshold. Depending on the heuristics used, this elimination may lead to a search error.†

Knowledge application involves applying expert knowledge in the CSR problem domain in order to improve the efficiency of the search. Expert knowledge is most effective in the construction of the knowledge base that constrains the search in the acoustic and language modeling domains. Tighter constraints in the knowledge base translate directly into smaller search spaces. Another area of knowledge application is in the design of heuristic evaluation functions. Heuristic evaluation functions are used to determine which partial hypotheses should be pursued and which ones should be eliminated from consideration.

In the next two sections we will look at how two algorithms, stack decoding and Viterbi search, have applied the techniques of optimized hypothesis

*CSR systems often use log probability representations. A search algorithm that always takes the maximum would require only summations of log probabilities. This has the advantage of avoiding costly multiples and floating-point underflows.

†A search error means the search terminated with a suboptimal solution.

generation, problem space reduction, and heuristic pruning to the CSR problem.

4.2. Stack Decoding

Stack decoding was first used by IBM (Bahl et al., 1983) and is used in their large-vocabulary CSR systems (Bahl et al., 1989a). The stack decoding algorithm is similar to the A^* search (Nilsson, 1971) used in artificial intelligence. In the following paragraphs we will look at the basic stack decoding algorithm and discuss the features that make stack decoding appropriate for CSR. We will also examine the problems that stack decoding presents.

4.2.1. An Optimal Stack Decoding Algorithm

Stack decoding constructs a search tree from a state graph S that is described by the language model. The states in the graph represent the abstract states in language, and the branches in the search tree are words that cause transition from one language state to another. We first give a stack decoding algorithm that guarantees the optimal solution:

Optimal Stack Search

1. Add initial states of the language to the queue;
   evaluate these states.
2. Remove from the queue the state s, with the best evaluation.
3. For each word w that transits from state s to s':
   1. Consider *all* possible beginning and ending times for the word w corresponding to s'.
   2. Evaluate s' as the product of the probability at s and the probability that w matches the input.
   3. Add s' to the queue.
      If s' is already in the queue, merge the two paths.
4. Terminate if the best state in the queue is also a complete path.
5. Go to step 2.

An important advantage of stack decoding is its consistency with the forward-backward training algorithm. In the forward-backward algorithm, the words w and the speech y are aligned as shown in Fig. 6. In the forward-backward alignment, all paths that reaches a node $d_{n,m}$ are summed. Moreover, all nonzero (dark area in Fig. 6) beginning and ending times are considered. Both of these properties are faithfully preserved in the stack decoding algorithm. In step 3.2, the probabilities are summed, and in step 3.1, multiple beginning and ending times are considered.

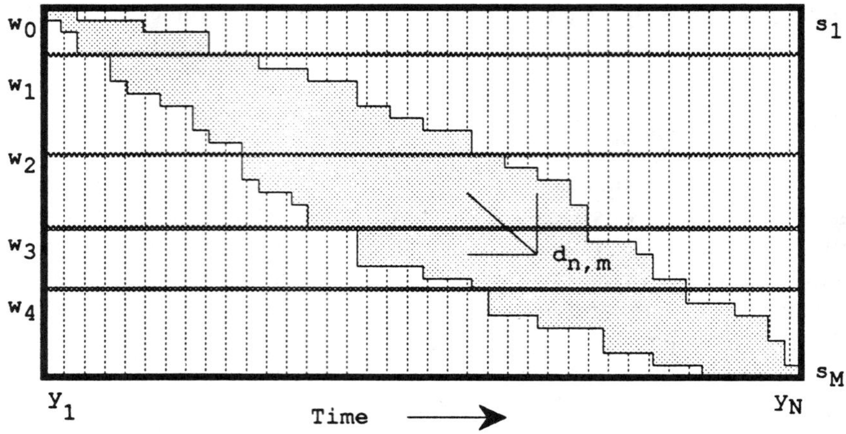

Figure 6 The forward-backward alignment space. Each grid point d in the alignment space corresponds to $P(s_1^m \mid y_1^n)$. s_1^M is the sequence of Markov states obtained by concatenating the Markov states of w_{0-4}. The dark area represents the nonzero values in the alignment space.

In addition to the extra computation, stack decoding requires an evaluation function to compare hypotheses of different lengths. The most obvious evaluation function is perhaps

$$\sum_{n=0}^{N} P(w,y_1^n) = P(w) \times \sum_{n=0}^{N} P(y_1^n \mid w) \tag{3}$$

This evaluation function simply uses the forward algorithm to evaluate partial paths. However, it is unsuitable because it fails to normalize the paths that have different lengths. This will cause the search always to prefer shorter paths, resulting in a breadth-first search. The ideal evaluation function should have the following three properties:

1. *Normalizing.* The evaluation must allow comparison of paths that account for different portions of the acoustic sequence.
2. *Discriminating.* For the optimal path w, the evaluation tends to increase over time. For all other paths, it tends to decrease.
3. *Near-optimal.* If the evaluation of a complete path a exceeds that of all partial paths, path a is likely to be the optimal solution.

Equation (4) is an evaluation function that better satisfies the above criteria.

$$P(w) \times \sum_{n=0}^{N} \frac{P(y_1^n \mid w)}{e_n^*} \tag{4}$$

where e_n^* is an estimate of $P(y_1^n)$. Recall that when the fundamental equation of speech recognition (Eq. 1) was introduced, we stated that $P(y)$ can be eliminated because it is a constant for complete hypotheses. In the case of stack decoding, this no longer holds, as hypotheses can have different lengths and need to be normalized. Thus, e_i^*, an estimate of $P(y)$, is used to that end.

Although we cannot compute e_i^* perfectly, it could be estimated. For example, if discrete symbols (vector quantization) are used to represent the acoustic sequence, the following Nth-order estimates can be computed:

$e^* = \sqrt[N]{e_N^*}$	average probability per frame
$e_i^* = (e^*)^i$	zeroth-order estimate
e_i^* for $y_1^i = e^*(y_1) \cdot e^*(y_2) \cdots e^*(y_i)$	first-order estimate
e_i^* for $y_1^i = e^*(y_1) \cdot e^*(y_2 \mid y_1) \cdots e^*(y_i \mid y_{i-1})$	second-order estimate

Equation (4) enables comparison of paths with different lengths. However, if e_i^* excessively overestimates $P(y_1^n)$, the search degenerates to a breadth-first search, which is too inefficient. On the other hand, if e_i^* excessively underestimates $P(y_1^n)$, then the search degenerates into an inadmissible best-first search, which may yield an inferior answer. In theory, if $e_i^* \geq P(y_1^n)$, the search will find an optimal solution. In practice, an optimal search is too inefficient, and a normalizing parameter α is used to balance the two extremes, as shown in Eq. (5).

$$P(w) \times \sum_{n=0}^{N} \frac{P(y_1^n \mid w)}{e_n^*} \alpha^{N-n} \tag{5}$$

In addition to using a small α to make the search more efficient, *pruning* can be introduced in several ways, as illustrated in the following stack decoding algorithm with pruning:

Stack Search with Pruning

1. Add initial states of the language to the queue;
   evaluate these states.
2. Remove from the queue the state s with the best evaluation.
3. For each word w that transits from states s to s':

1. Consider *all* possible beginning and ending times for the word w corresponding to s'.
2. Evaluate s' according to Eq. (5).
3. Add s' to the queue.
   If s' is already in the queue, merge the two paths.

4. [Δ1] Terminate if the best complete path in the queue is within τ of the best-scoring path in the queue.
5. [Δ2] Remove states from the queue that are more than ϵ from the best score in the queue.
6. [Δ3] Further prune the queue to keep only the S best paths.
7. Go to step 2.

For a real-time system, waiting for all hypotheses in the queues to be proved to be worse than the best complete path might be intolerable. Thus, the first change, marked by [Δ1], terminates the search when the complete theory appears "good enough." [Δ2] uses a predetermined threshold ϵ to eliminate a newly generated hypothesis, if its score is much worse than the best hypothesis. Both of these are applications of knowledge-guided heuristics. [Δ3] in step 6, however, is a blind heuristic that refrains from exceeding a maximum queue size by pruning hypotheses. By increasing S and τ, and by reducing ϵ, the number of states considered can be reduced, which may introduce search errors.

4.3. Viterbi Beam Search

Viterbi beam search has successfully been applied in the Harpy (Lowerre, 1976), Sphinx (Lee and Hon, 1988; Lu, 1988; Lee et al., 1990a), SPICOS (Ney et al., 1987), and BYBLOS (Chow et al., 1987) CSR systems. We will now describe the basic Viterbi search algorithm and the application of heuristic pruning that yields the Viterbi beam search. We will also discuss the features that make Viterbi search appropriate for CSR and examine the problems that Viterbi search presents.

4.3.1. Optimal Viterbi Search

The Viterbi search constructs its search tree from a state graph S that is specified by the grammar network and the word models. The word models may be represented directly as Markov nets, or they may be further specified by a network whose states represent subword units. Figure 7 illustrates this expansion process.

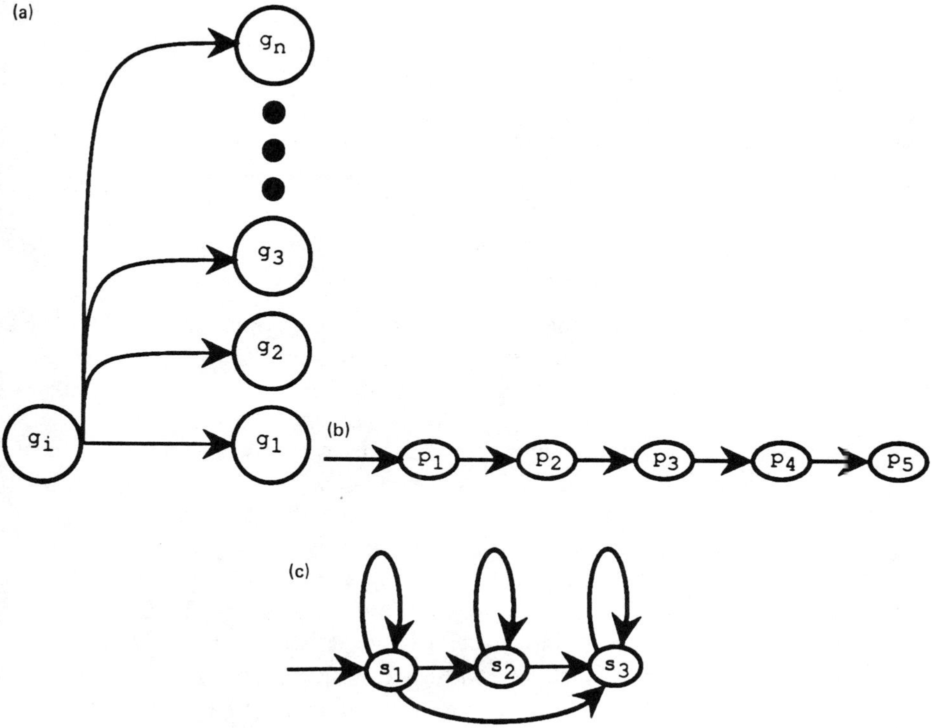

Figure 7 (a) A small portion of the grammar net G. Each grammar state g_i corresponds to a word model $W = Wm(G_i)$ where the function Wm may be many to one. (b) A word W expanded to its network of sub-word units P. (c) Each p_i is expanded to its Markov net.

Optimal Viterbi Search

1. Create N state lists SL, where N is the length of the acoustic sequence, and where each list contains one entry for each Markov state in the state graph S.
2. Initialize $SL[t]$ for $t = 0$ by setting the probability of the initial state equal to 1.0 and all others to 0.0.
3. Clear $SL[t + 1]$.
4. For each state s in $SL[t]$
    1. For each transition from s specified by S
        a. Compute the transition probability.

   b. If the destination state in $SL[t + 1]$ is uninitialized, then update it with a score and backpointer to *state*.
   c. If the destination state in $SL[t + 1]$ is already initialized, then check if this transition is better. If it is then update it with a new score and backpointer to *state*.
5. If $t = N$, backtrack and return.
6. Else $t = t + 1$ and go to step 3.

Unlike stack decoding, the Viterbi search is *time synchronous*. That is, at any given time, each partial solution w' accounts for the same portion of the acoustic sequence, namely y_1^i. Thus, partial solutions may be directly compared without elaborate evaluation functions.

However, in step 4 of the Viterbi search, both within-word and between-word transitions are considered in a unified framework. This has the advantage of efficiently evaluating partial solutions due to its dynamic programing properties. A negative consequence of this is that partial solutions are recombined using a *max* function instead of a *sum* function. This is computationally efficient because nonmaximum paths can simply be discarded, and *max* is faster than *sum*. However, it is inconsistent with the summing operation used by the forward-backward algorithm. For an input y and a hypothesized sentence w, the stack decoding reports $P(y \mid w)$, whereas the Viterbi search reports $P(y \mid w,s)$, where s is the most likely state sequence.

4.3.2. *The Viterbi Beam Search*

A Viterbi search takes advantage of dynamic programming techniques that reduce the problem size to $O(\textit{number of states} \times \textit{length of input})$. However, the problem is still too large when complex finite state or stochastic grammars are used, since each state must be enumerated. By observing that most states in $SL[t]$ have zero or near-zero probability, we can modify the Viterbi search to consider only states that have probabilities that are within ϵ of the best state in $SL[t]$, where ϵ defines the *beam* width of the search. The Viterbi beam search is now defined as follows.

Viterbi Beam Search

1. Create N heaps B for the dynamic state beams. N is the length of the acoustic sequence.
2. Add the initial state to $B[0]$ with probability 1.0. Set $t = 0$.
3. Clear $B[t + 1]$.
4. For each s in $B[t]$.

1. For each transition from *state* specified by S.
   a. Compute the transition probability.
   b. If the destination state is not in $B[t+1]$ then add it to $B[t+1]$ with its score and backpointer set to *state*.
   c. If the destination state is already in $B[t+1]$ then check if this transition is better. If it is then update the state with the new score and backpointer to s.

5. If $t = N$, backtrace and terminate.
6. Find best state in $B[t+1]$ and discard states that are more than ϵ worse than the best state.
7. $t = t + 1$ and go to step 3.

The clear advantage of the Viterbi beam search is that the problem size is reduced to O(*ave beam size* $\times$ *length of input*) and the path-merging function *max* has a significant computational savings over the *add* function when applied to log probabilities. Although the best *ave-beam-size* cannot be determined a priori, it can be determined empirically. The advantage of a dynamic beam heuristic is that it will consider only *good* hypothesis relative to the best hypothesis. The beam heuristic therefore allows the search to consider many hypotheses when there is no clear best hypothesis. Conversely, when there is a clear best hypothesis only a few alternatives need to be maintained by the search.

We now address some implementation issues for the beam search. If we want to keep the problem complexity at O(*ave-beam-size* $\times$ *length-of-input*) instead of O(*number-of-states* $\times$ *length-of-input*), we cannot enumerate all the states. Yet, without enumeration the beam cells cannot be randomly accessed. This can lead to large constant factors in the innermost loop of the search. Typically this problem has been solved by creating a *from-state-list* and a *to-state-list* that enumerate the active states in S. The heaps B then become pointers to the active cells in these two lists. When a transition is performed it can easily be determined if a cell is active. If the cell is not active, it can be added to the heap. If the cell is already active, it will be updated if appropriate. No change needs to be made to the heap. Backpointers can be handled either as part of the heap or as separate data structures.

But what if it is impractical to enumerate the state graph S? One solution, of course, would be the use of a hashing function using a key based on the grammar state, word, and HMM state. Another solution used by the SPICOS (Ney et al., 1987b) system is to dynamically create a hierarchical structure from the active grammar states, active words, and active HMM states. The dynamic grammar traversal requires additional computation, but has the advantage of saving substantial storage.

4.3.3. *Viterbi Beam Search Variants*

Two variants of the Viterbi beam search have been implemented at BBN. The first variant (Chow et al., 1987) differs from the Viterbi search in that it *adds* probabilities when paths merge within words and takes the *maximum* when paths merge at word boundaries. The chief advantages are that this algorithm is more efficient than stack decoding and brings the decoding process *closer* to the training condition. The algorithm also preserves the attribute that all paths are the same length to keep pruning simple. The problems with this approach are that (1) it is slower than the Viterbi search, (2) it still considers only one boundary point at word exits, and (3) paths are combined (and their probabilities added) even when they have different histories.

The second variant on the Viterbi beam search is the N-Best search (Schwartz and Chow, 1990), which provides the N best matching sentences. The N-Best search maintains a separate beam for each state. (This is in contrast to the Viterbi search, which uses a global beam with one path per active state.) When paths merge at a state s the N best alternative paths to s are kept in the local beam. As a result, this allows the search to report on the top N matching sentences. Note that this search is actually a hybrid that is similar to both the Viterbi beam search and the stack decoding.

4.4. Summary

In this section, we have provided a tutorial overview of the search problem in CSR. Because the search space in CSR problems is often enormous, the study of efficient search algorithms is important. We have described efficient search organizations, as well as heuristic pruning and searching strategies that produce near-optimal results in polynomial time.

Two search algorithms were described in detail. The Viterbi beam search sacrifices optimality early in exchange for efficiency and ease of evaluation, while stack search initially aims for optimality but then violates it higher levels. On the one hand, the Viterbi beam search can be implemented easily and efficiently if the state graph can be fully expanded into Markov states. On the other hand, full expansion is impossible for large language models, for which stack decoding is more suitable. However, the implementation of a stack decoder requires the maintenance of ordered lists of hypotheses and the use of a heuristic evaluation function.

5. CONCLUSION

In this chapter, we have described the problem of continuous speech recognition within the framework of hidden Markov modeling. CSR is far more difficult than IWR because of the greater confusability, the lack of word boundaries, and the less clear articulation.

We identified three major problems in continuous speech recognition: training, modeling, and searching. Training is the simplest problem, because with HMMs, CSR training is essentially the same as IWR training. The second problem, improved modeling, involves finding better features, more correct model structures, a better set of models, and better learning algorithms. We have described a number of ideas in each area. Finally, for the search problem, we have described two search algorithms for CSR that use efficient algorithms and heuristic pruning to find good approximate solutions to the exponential CSR problem in polynomial time.

Although the best current CSR systems still have much higher error rates than the best IWR systems, it may be largely due to the inherent difficulty of CSR. The various techniques presented in this chapter have already substantially narrowed the gap between IWR and CSR. However, the best IWR and CSR systems are still an order of magnitude worse than human performance (Brown, 1987; Doddington, 1989). Clearly, much work is needed in both areas.

REFERENCES

Bahl, L. R., Baker, J. K., Cohen, P. S., Cole, A. G., Jelinek, F., Lewis, B., L., and Mercer, R. L. (1978). Automatic recognition of continuously spoken sentences from a finite state grammar. *IEEE Int. Conf. Acoustics, Speech, and Signal Processing*, April, pp. 418–421.

Bahl, L. R., Bakis, R., Jelinek, F., and Mercer, R. L. (1980). Language-model/acoustic channel balance mechanism. *IBM Technical Disclosure Bull.* *23*(7B):3464–3465.

Bahl, L. R., Bakis, R., Cohen, P. S., Cole, A. G., Jelinek, F., Lewis, B. L., and Mercer, R. L. (1981). Speech recognition of a natural text read as isolated words. *IEEE Int. Conf. Acoustics, Speech, and Signal Processing*, April, pp. 1168–1171.

Bahl, L. R., Jelinek, F., and Mercer, R. (1983). A maximum likelihood approach to continuous speech recognition. *IEEE Trans. Pattern Anal. Machine Intell. PAMI-5*(2):179–190.

Bahl, L. R., Brown, P. F., De Souza, P. V., and Mercer, R. L. (1988a). Acoustic Markov models used in the Tangora speech recognition system. *IEEE Int. Conf. Acoustics, Speech, and Signal Processing*, April, pp. 497–500.

Bahl, L. R., Brown, P. F., De Souza, P. V., and Mercer, R. L. (1988b). A new algorithm for the estimation of hidden Markov model parameters. *IEEE Int. Conf. Acoustics, Speech, and Signal Processing*, April, pp. 493–496.

Bahl, L. R., et al. (1989a). Large vocabulary natural language continuous speech recognition. *IEEE Int. Conf. Acoustics, Speech, and Signal Processing*, May, pp. 465–467.

Bahl, L. R., Brown, P. F., De Souza, P. V., and Mercer R. L. (1989b). A tree-based statistical language model for natural language speech recognition. *IEEE Trans. Acoust. Speech Signal Process. ASSP-37*(7):1001–1008.

Baker, J. K. (1975). Stochastic modeling as a means of automatic speech recognition. Ph.D. thesis, Computer Science Department, Carnegie Mellon University, April.

Baum, L. E. (1972). An inequality and associated maximization technique in statistical estimation of probabilistic functions of Markov processes. *Inequalities 3*: 1–8.

Bellegarda, J. R. and Nahamoo, D. (1989). Tied mixture continuous parameter models for large vocabulary isolated speech recognition. *IEEE Int. Conf. Acoustics, Speech, and Signal Processing*, May, pp. 13–16.

Bernstein, J. Cohen, M., Murveit, H., and Weintraub, M. (1989). Linguistic constraints in hidden Markov model based speech recognition. *IEEE Int. Conf. Acoustics, Speech, and Signal Processing*, May, pp. 699–702.

Brown, P. (1987). The acoustic-modeling problem in automatic speech recognition. Ph.D. thesis, Computer Science Department, Carnegie Mellon University, May.

Chow, Y. L., Dunham, M. O., Kimball, O. A., Krasner, M. A., Kubala, G. F., Makhoul, J., Roucos, S., and Schwartz, R. M. (1987). BYBLOS: The BBN continuous speech recognition system. *IEEE Int. Conf. Acoustics, Speech, and Signal Processing*, April, pp. 89–92.

Cohen, J. R. (1989). Application of an auditory model to speech recognition. *J. Acoust. Soc. Amer. 85*(6):2623–2629.

Davis, S. B., and Mermelstein, P. (1980). Comparison of parametric representations of monosyllabic word recognition in continuously spoken sentences, *IEEE Trans. Acoust. Speech, Signal Process. ASSP-28*(4):357–366.

Deng, L., Lennig, M., and Mermelstein, P. (1989). Use of vowel duration information in a large vocabulary word recognizer. *J. Acoust. Soc. Am.* 86(2):540–548.

Derouault, A.-M. (1987). Context-dependent phonetic Markov models for large vocabulary speech recognition. *IEEE Int. Conf. Acoustics, Speech, and Signal Processing*, April, pp. 360–363.

Doddington, G. R. (1989). Phonetically sensitive discriminants for improved speech recognition. *IEEE Int. Conf. Acoustics, Speech, and Signal Processing*, May.

Franzini, M., Witbrock, M., and Lee, K. F. (1989). A Connectionist Approach to Continuous Speech Recognition. *IEEE Int. Conf. Acoustics, Speech, and Signal Processing*, May, pp. 425–428.

Furui, S. (1986). Speaker-independent isolated word recognition using dynamic features of speech spectrum. *IEEE Trans. Acoust. Speech Signal Process. ASSP-34*(1): 52–59.

Gupta, V. N., Lennig, M., and Mermelstein, P. (1987). Integration of acoustic information in a large vocabulary word recognizer. *IEEE Int. Conf. Acoustics, Speech, and Signal Processing*, April, pp. 697–700.

Hemphill, C., and Picone, J. (1989). Robust speech recognition in a unification grammar framework. *IEEE Int. Conf. Acoustics, Speech, and Signal Processing*, May, pp. 723–726.

Huang, X. D., and Jack, M. A. (1989). Semi-continuous hidden Markov models for speech recognition. *Comput. Speech Language 3*(3):239–252.

Hwang, M. Y., Hon, H. W., and Lee, K. F. (1989). Modeling between-word coarticulation in continuous speech recognition. *Proc. Eurospeech*, September.

Iwamida, H., Katagiri, S., McDermott, E., and Tohkura, Y. (1990). A hybrid speech recognition system using HMMs with an LVQ-trained codebook. *IEEE Int. Conf. Acoustics, Speech, and Signal Processing*, April, pp. 489–492.

Jelinek, F. (1976). Continuous speech recognition by statistical methods. *Proc. IEEE 64*(4):532–556.

Jelinek, F., et al. (1985). A real-time, isolated-word, speech recognition system for dictation transcription. *IEEE Int. Conf. Acoustics, Speech, and Signal Processing*, March, pp. 858–861.

Kubala, G. F., Chow, Y., Derr, A., Feng, M., Kimball, O., Makhoul, J., Price, P., Rohlicek, J., Roucos, S., Schwartz, R., and Vendegrift, J. (1988). Continuous speech recognition results of the BYBLOS system on the DARPA 1000-word resource management database. *IEEE Int. Conf. Acoustics, Speech, and Signal Processing*, April, pp. 291–294.

Lee, C. H., Rabiner, L. R., Pieraccini, R., and Wilpon, J. G. (1990b). Acoustic modeling for large vocabulary speech recognition. *Comput. Speech Language*, in press.

Lee, K. F. (1988). Large-vocabulary speaker-independent continuous speech recognition: The SPHINX system. Ph.D. thesis, Computer Science Department, Carnegie Mellon University, April.

Lee, K. F. (1989). *Automatic Speech Recognition: The Development of the SPHINX System*. Kluwer Academic Publishers, Boston.

Lee, K. F. (1990.) Context-dependent phonetic hidden Markov models for continuous speech recognition. *IEEE Trans. Acoust. Speech Signal Process.*, April, pp. 599–609.

Lee, K. F., and Hon, H. W. (1988). Large-vocabulary speaker-independent continuous speech recognition. *IEEE Int. Conf. Acoustics, Speech, and Signal Processing*, April, pp. 123–126.

Lee, K. F., and Mahajan, S. (1990). Corrective and reinforcement learning for speaker-independent continuous speech recognition. *Comput. Speech Language*, April.

Lee, K. F., Hon. H. W., and Reddy, R. (1990a). An overview of the SPHINX speech recognition system. *IEEE Trans. Acoust. Speech Signal Process.*, January, pp. 35–45.

Lowerre, B. T. (1976). The HARPY speech recognition system. Ph.D. thesis, Computer Science Department, Carnegie Mellon University, April.

Myers, C. S., Rabiner, L. R. (1981). Connected digit recognition using a level building DTW algorithm. *IEEE Trans. Acoust. Speech Signal Process. ASSP-29*(3): 351–363.

Ney, H. (1987). Dynamic programming speech recognition using a context-free grammar. *IEEE Int. Conf. Acoustics, Speech, and Signal Processing*, April, pp. 3.2.1–3.2.4.

Ney, H., Mergel, D., Noll, A., and Paeseler, A. (1987). A data-driven organization of the dynamic programming beam search for continuous speech recognition. *IEEE Int. Conf. Acoustics, Speech, and Signal Processing*, April, pp. 833–836.

Nilsson, N. J. (1971). *Problem-Solving Methods in Artificial Intelligence*. McGraw-Hill, New York.

Paul, D. B., and Martin, E. A. (1988). Speaker stress-resistant continuous speech recognition. *IEEE Int. Conf. Acoustics, Speech, and Signal Processing*, April, pp. 283–286.

Rabiner, L. R., Juang, B. H., Levinson, S. E., and Sondhi, M. M. (1985). Recognition of isolated digits using hidden Markov models with continuous mixture densities. *AT&T Tech. J. 64*(6):1211–1233.

Rabiner, L. R., Wilpon, J. G., and Juang, B. H. (1986). A segmental *k*-means training procedure for connected word recognition based on whole word reference patterns. *AT&T Tech. J. 65*(3):21–31.

Rabiner, L. R., Wilpon, J. G., and Soong, F. K. (1988). High performance connected digit recognition using hidden Markov models. *IEEE Int. Conf. Acoustics, Speech, and Signal Processing*, April, pp. 119–122.

Rabiner, L. R., Wilpon, J. G., and Soong, F. K. (1989a). High performance connected digit recognition using hidden Markov models. *IEEE Trans. Acoust. Speech Signal Process. 37*(8):1214–1225.

Rabiner, L. R., Lee, C. H., Juang, B. H., and Wilpon, J. G. (1989b). HMM clustering for connected word recognition. *IEEE Int. Conf. Acoustics, Speech, and Signal Processing*, April, pp. 405–408.

Schwartz, R., and Chow, Y. L. (1990). The optimal *N*-best algorithm: An efficient procedure for finding multiple sentence hypotheses. *IEEE Int. Conf. Acoustics, Speech, and Signal Processing*, April, pp. 81–84.

Schwartz, R., Chow, Y., Kimball, O., Roucos, S., Krasner, M., Makhoul, J. (1985). Context-dependent modeling for acoustic-phonetic recognition of continuous speech. *IEEE Int. Conf. Acoustics, Speech, and Signal Processing*, April, pp. 1205–1208.

Seneff, S. (1988). A joint synchrony/mean-rate model of auditory speech processing. *J. Phonet. 16*(1):55–76.

Shikano, K., Lee. K, and Reddy, D. R. (1986). Speaker adaptation through vector quantization. *IEEE Int. Conf. Acoustics, Speech, and Signal Processing*, April, pp. 2643–2646.

Zue, V., Glass, M., Phillips, M., and Seneff, S. (1990). The summit speech recognition system: Phonological modelling and lexical access. *IEEE Int. Conf. Acoustics, Speech, and Signal Processing*, April, pp. 49–52.

21

Principles of Lexical Language Modeling for Speech Recognition

FREDERICK JELINEK, ROBERT L. MERCER, and SALIM ROUKOS IBM Research Division, Yorktown Heights, New York

1. INTRODUCTION

A speech recognizer is a device which automatically transcribes speech into text. The recognizer is usually based on some finite vocabulary that restricts the words that can be "printed" out. For the purpose of this chapter, a word is defined by its spelling. Two differently spelled inflections or derivations of the same stem are considered different words. Homographs having different parts or speech or meanings constitute the same word.

A language model assigns a probability value to every string of words w_1, $w_2, \ldots, w_n$ taken from the prescribed vocabulary. We will now outline its role in speech recognition.

Let

$$W = w_1, w_2, \ldots, w_n \tag{1}$$

denote a string of n words, and let A denote the acoustic evidence (data) on the basis of which the recognizer will make its final decision about which words were spoken. If $P(W \mid A)$ denotes the probability that the words W were spoken, given that the evidence A was observed, then the recognizer should decide in favor of a word string $\hat{W}$ satisfying

$$P(\hat{W} \mid A) = \max_{W} P(W \mid A) \tag{2}$$

Thus the recognizer will pick the most likely word string given the observed acoustic evidence.

The well-known Bayes formula allows us to rewrite the right-hand side probability of (2) as

$$P(W \mid A) = \frac{P(W)P(A \mid W)}{P(A)} \tag{3}$$

where $P(W)$ is the probability that the word string W will be uttered, $P(A \mid W)$ is the probability that when the speaker says W the acoustic evidence A will be observed, and $P(A)$ is the average probability that A will be observed.

Since the maximization in (2) is carried out with the variable A fixed, it follows from (2) and (3) that the aim of the recognizer is to find the word string $\hat{W}$ that maximizes the product $P(W)P(A \mid W)$, i.e.,

$$\hat{W} = \arg\max_{W} P(W)P(A \mid W) \tag{4}$$

The a priori probabilities $P(W)$ whose values are given by the language model are thus central to the speech recognition process.

The purpose of this chapter* is to discuss the problems of lexical language modeling† and to suggest some lines of possible solution. The next section describes in some detail what a statistical language model is. Section 3 introduces perplexity, a measure of language model quality based on ideas of information theory, and defines the concept of average mutual information. Sections 4 and 5 are devoted to problems of language model construction from word n-grams extracted from relatively sparse text data bases. Section 6 concerns attempts to incorporate into the language model information contained in the more distant past of the recognized text. In Section 7 we face the problem of selecting the recognizer's vocabulary and tailoring it to particular users. Finally, Section 8 discusses the construction of language models that adjust to changes of the vocabulary or discourse domain.

The appendices of this chapter are organized as follows. Appendix A introduces the hidden Markov model concept that is useful in the formulation

---

*This chapter is an updated version of the authors' widely distributed 1985 report "Self-Organized Language Modeling for Speech Recognition."

†The adjective "lexical" is intended to specify that the language models in question will depend on the statistical characteristics of the words as spelled and that grammatical or semantic properties will not be taken advantage of directly.

and optimization of language models. Appendix B is devoted to automatic estimation of language model parameters. Finally, Appendix C is devoted to the discovery of statistically significant word classes.

2. PRACTICAL ASPECTS OF LANGUAGE MODELING

Using elementary rules of probability theory, the language model's probability $P(W)$ can be formally decomposed as

$$P(W) = \prod_{i=1}^{n} P(w_i \mid w_1, \ldots, w_{i-1}) \tag{5}$$

where $P(w_i \mid w_1, \ldots w_{i-1})$ is the probability that w_i will be spoken given that words $w_1, \ldots, w_{i-1}$ were said previously.

The sequential decomposition (5) is appropriate for speech recognition because it leads to the development of natural intermediate decision criteria allowing a minimal delay in the response of the recognizer to the progressing dictation. In reality, of course, the probabilities $P(w_i \mid w_1, \ldots, w_{i-1})$ would be impossible to estimate for even moderate values of i: most histories $w_1, \ldots, w_{i-1}$ would be unique or would have occurred only a very few times. Indeed, for a vocabulary of size L there are $L^{(i-1)}$ different histories, and so to specify $P(w_i \mid w_1, \ldots, w_{i-1})$ completely, L^i values would have to be estimated. This is an astronomically large number for practical values of L (for $L = 20000$ and $i = 3$, L^i is equal to 80 trillion!), and thus the probabilities $P(w_i \mid w_1, \ldots, w_{i-1})$ could be neither stored nor retrieved when needed by the recognizer.

It follows that the various possible conditioning histories $w_1, \ldots, w_{i-1}$ must be considered to belong to only a manageable number M of different equivalence classes. It is intuitively obvious that thinking of many of the histories as equivalent will not appreciably weaken the ability to predict the new word w_i. In fact, for practical vocabulary sizes and $i > 3$, most of the N^i word strings $w_1, \ldots, w_i$ will never occur in English.

Let \$ be a (many to one) mapping of histories into some (perhaps large) number of equivalence classes. If $\$[w_1, \ldots, w_{i-1}]$ denotes the equivalence class of the string $w_1, \ldots, w_{i-1}$, then the probability $P(W)$ may be approximated by

$$P(W) = \prod_{i=1}^{n} P\left(w_i \mid \$[w_1, \ldots, w_{i-1}]\right) \tag{6}$$

A very practical method of language modeling of natural text is based on a very simple equivalence classification: histories are equivalent if they end in the same two words. Thus

$$P(W) = \prod_{i=1}^{n} P(w_i \mid w_{i-2}, w_{i-1}) \tag{7}$$

Crucial to the trigram language model (7) is the estimation of the basic probabilities $P(w_3 \mid w_1, w_2)$. One's first impulse is to do so by the simple relative frequency approach.

$$P(w_3 \mid w_1, w_2) = f(w_3 \mid w_1, w_2) = \frac{C(w_1, w_2, w_3)}{C(w_1, w_2)} \tag{8}$$

where the function C counts the number of times the string in its argument occurred in a training text. But some thought will reveal that (7) is an inadequate formula, since many possible English word trigrams w_1, w_2, w_3 never actually take place even in very large corpora. A language model based on (8) would assert the impossibility of such trigrams, assign $P(W) = 0$ to strings W containing them, and would thus force a large number of word errors in a recognizer operating under the statistical decision criterion

$$\hat{W} = \arg\max_{W} P(W)P(A \mid W) \tag{4}$$

In fact, when the IBM group divided a corpus of patent descriptions (limited to a 1000-word vocabulary) into test and training subsets (of size 300,000 and 1,500,000 words, respectively), they found that 23% of trigrams appearing in the test subset never took place in the training subset!

It is therefore necessary to "smooth" the trigram frequencies. This can be done by interpolating trigram, bigram, and unigram relative frequencies,

$$P(w_3 \mid w_1, w_2) = q_3 f(w_3 \mid w_1, w_2) + q_2 f(w_3 \mid w_2) + q_1 f(w_3) \tag{9}$$

where the nonnegative weights satisfy $q_1 + q_2 + q_3 = 1$. Since $f(w_3 \mid w_1, w_2)$ approximates $P(w_3 \mid w_1, w_2)$ better if it is based on a larger count $C(w_1, w_2)$, the estimation will be improved by letting the weights depend on the conditioning counts $C(w_1, w_2)$ and $C(w_2)$. In Appendix B we show how the principle of *cross-validation* (Stone, 1974; Jelinek and Mercer, 1980) can be used to choose the values q_1 to satisfy the maximum likelihood criterion* and in this sense achieve an optimal interpolation (9).

An intuitive feeling for the quality of the language model (7) may be gleaned from Table 1. The language model is used to predict the words of

*The maximization is carried out based on some "typical" large text W, different from the training text used to estimate the relative frequencies f. The weights q_i are adjusted to maximize the probability $P(W)$ when formula (9) is substituted into (7).

Table 1 Ranks of Words Predicted by a Trigram Language Model

1.	I	are	**to**	have	be	the	**the**	document
2.	this	will		know	this	**of**		question
3.	one	the		do	these			first
4.	two	would		get	problems			company
5.	a	also		make	a			data
6.	three	do		go	any			system
7.	please	**need**		use	some			other
8.	in			develop	those			equipment
9.	**we**			take	**all**			product
.				.				.
.				.				.
.				.				.
83.				create				art
84.				communicate				number
85.				**resolve**				major
.								.
.								.
.								.
655.								dates
656.								security
657.								**important**

1.	thing	.	**the**	company	page	weeks
2.	point	.		**next**	exhibit	years
3.	to	to			meeting	pages
4.	parts	of			day	months
5.	points	and			**two**	**days**
6.	factors	in				
7.	things	that				
.	.	.				
.	.	.				
.	.	.				
14.	**issues**	relative				
.		.				
.		.				
.		.				
71.		still				
72.		I				
73.		requiring				
74.		**within**				

the sentence “We need to resolve all of the important issues within the next two days.” The figure displays all the words that are predicted to be more likely than the actual word, given the language model’s perfect knowledge of the past (at the beginning of the sentence nothing but the fact that it begins is known). For instance, when the two preceding words are thought to be “all the”, then the most likely next word is “document,” and the words “question”, ..., “security” are all more likely than the actual word “important,” which is the 657th likeliest, given that particular past. We can observe that the language model (7) whose trigram probabilities are estimated by (9) is quite good at predicting function words but that it is uncertain about content words, just as human beings often are (no claim is made here that they wouldn’t do better; just that they too might have trouble).

Of course, one example proves nothing. We would like to have an objective measure of language model quality. It can be based on results from information theory, which concerns itself with sources of information. We will derive such a measure in the next section.

3. PERPLEXITY: MEASURES OF LANGUAGE MODEL QUALITY

In this section we will consider objective measures of language model quality. They will be based on results of information theory, which concerns itself with sources of information. In simplest terms, a source is a device which puts out symbols chosen from some finite set $\mathcal{V}$ that is known to the user. For the purposes of this chapter, the symbols in question are words, and $\mathcal{V}$ is the vocabulary. Underlying the source operations is a statistical law which governs the choice of output words. When putting out a word, a source provides information by removing the user’s uncertainty about the identity of that word. Thus a source contains more information if the uncertainty about the next output is greater. For a given size L of the set $\mathcal{V}$, uncertainty and therefore information is maximal if each of the possible words is chosen with equal probability and independently of previously chosen words. The information content of such a source is

$$I = \log L \tag{10}$$

The correctness of the logarithmic measure in (10) can be appreciated by noting that putting out two words from the source is equivalent to putting out one two-word symbol from a corresponding uniform source of set size $L' = L^2$ and that this latter act should provide twice as much information. But

$$I' = \log L' = \log L^2 = 2 \log L = 2I \tag{11}$$

so that (10) agrees with our intuition.

Let w denote a word put out by the source with probability $P(w)$. Then it can be shown in general that the proper measure of information is entropy H (in thermodynamics it measures disorder), which for sources that choose their words independently of each other is given by

$$H = -\left[\sum_{w=1}^{L} P(w) \log P(w)\right] \tag{12}$$

Note that if the source is uniform, i.e., if $P(w) = 1/L$, then (12) reduces to (10).

The logarithms in (12) are usually taken to the base 2, in which case the information is measured in units of binary symbols (0 or 1) called "bits" [for $L = 2$, (10) yields $I = 1$)]. In fact, the fundamental coding theorem of information theory (Gallager, 1968) states that on the average it takes H bits to represent a word put out by a source of entropy H (the idea is to encode source output sequences into binary sequences from which the original word sequences can be fully recovered).

It follows from (12) and (10) that a source of entropy H can be thought of as one that has as much information content as a source which chooses words equiprobably from an alphabet of size

$$L' = 2^H \tag{13}$$

By direct extension of (12) it follows that if w_i is the ith word put out, then for general sources

$$H = -\lim_{n\to\infty}\left(\frac{1}{n}\right)\left\{\sum P(w_1,\ldots,w_n)\log P(w_1,\ldots,w_n)\right\} \tag{14}$$

where the sum is over all sequences $w_1, \ldots, w_n$. It is easily checked that when the source generates words independently of each other, i.e.,

$$P(w_1,\ldots,w_n) = P(w_1)P(w_2)\cdots P(w_n)$$

then (14) reduces to (12).

If the source is "well-behaved" so that sufficiently long sequences of words are typical of it and can be used to deduce its statistical structure (this property is technically known as ergodicity), then it can be shown that (14) is equivalent to

$$H = -\lim_{n\to\infty}\left(\frac{1}{n}\right)\log P(w_1,w_2,\ldots,w_n) \tag{15}$$

where the sequence w_1, w_2, ..., w_n denotes the words of actual text. Thus, in the well-behaved case, entropy can be directly estimated from large text corpora.

We can now apply the above theory to language models. Language can be thought of as an information source whose outputs are words w_i. The amount of information per word in some corpus W can then be estimated by

$$H = -\left(\frac{1}{n}\right)\log P(w_1, w_2, \ldots, w_n) \tag{16}$$

where n is the size of the corpus W (of course, the larger the text size n, the better the estimate). Because the information content is equal to the average number of bits it takes to specify a word, as pointed out in the paragraph following (12), (16) is also an estimate of recognition difficulty of speech generated by the same "mechanism" as gave rise to the corpus. In fact, it is exactly H bits of information that the recognizer must extract on the average from the acoustic data in order to determine a spoken word.

In order to determine (16), one would need to know the actual probabilities $P(w_1, w_2, \ldots, w_n)$ of strings of the language. These probabilities are in practice unknowable. At best we can have their estimates $\hat{P}(w_1, w_2, \ldots, w_n)$ which the language model provides to the recognizer. Thus from the point of view of the recognizer, the difficulty of recognition of speech generated by the same mechanism as the given (large) sample text is measured by its *logprob*

$$\mathrm{LP} = -\frac{1}{n}\log \hat{P}(w_1, w_2, \ldots, w_n) \tag{16}$$

It is easy to show that LP $\geq H$, assuming properly ergodic behavior of the text-generating source. Thus mathematics confirms our intuitive feeling that any misestimate of the underlying probabilities will only increase the difficulty of the recognition task. In the case of the trigram language model [see (8)]

$$\mathrm{LP} = -\frac{1}{n}\sum_{i=1}^{n}[\log \hat{P}(w_i/w_{i-1}, w_{i-2})] \tag{17}$$

Perhaps because it is felt to be intuitively more satisfying, it is customary in speech recognition to measure the difficulty of a recognition task relative to a given language model by the value of its perplexity PP (Bahl et al., 1977; Sondhi and Levinson, 1978) defined by [compare this with (13)]

$$\mathrm{PP} = 2^{\mathrm{LP}} = \hat{P}(w_1, w_2, \ldots, w_n)^{-1/n} \tag{18}$$

Thus to the first order of approximation (which disregards the acoustic distances between particular words) the task can be thought to be as difficult as would the recognition of a text that was generated by choosing words from a vocabulary of size PP, independently of each other and with uniform probability. Perplexity is therefore a measure of average branching of the text as seen from the point of view of the language model.

In practice, the value of perplexity depends on the sample test text on which it is measured, as well as on the training text from which the probabilities underlying the language model have been estimated. It is of course crucial that the two texts not overlap. If the two texts are matched, then the larger the training text, the lower the perplexity (because size of training text improves the accuracy of the estimates derived from it). Also, the larger the test text, the more accurately does PP measure the difficulty of recognition. For practical tasks of office correspondence dictation based on 20,000-word vocabularies, the perplexity of a trigram language model turns out to be between 100 and 200.

As pointed out above, the perplexity measure does not take into account the acoustical aspect of the problem. It treats words as abstract symbols and will have the same value whether the pronunciation of the words of the vocabulary is very similar (e.g., the words are short) or very distinct. In fact, it can be seen from (17) that LP will be relatively uninfluenced by the values of $P(w_3 \mid w_1, w_2)$ for rare words w_3, because such words will occur rarely in the text and their effect on LP will be made negligible by the divisor n. However, if such words were not to have distinctive pronunciations, then they would have the tendency to be decoded in error by the maximum a posteriori probability rule

$$\hat{W} = \arg\max_{W} P(W)P(A \mid W) \tag{4}$$

because their replacement by wrong words in the hypothesized string $w_1, w_2, \ldots, w_n$ may increase the overall value of $P(W)$ to more than compensate for the decrease in the factor $P(A \mid W)$.

To put acoustics into the evaluation function, we first observe that LP can be regarded as an estimate of the *average conditional entropy*

$$\begin{aligned} H(W_3 \mid W_1, W_2) &= -\sum_{w_1,w_2,w_3} P(w_1,w_2,w_3) \log P(w_3 \mid w_1,w_2) \\ &= -\sum_{w_1,w_2} P(w_1,w_2) \left[\sum_{w_3} P(w_3 \mid w_1,w_2) \log P(w_3 \mid w_1,w_2) \right] \end{aligned} \tag{19}$$

where the sum in the middle expression is over all strings w_1, w_2, w_3. In fact, as the last expression in (19) indicates, $H(W_3 \mid W_1, W_2)$ can be regarded as the average entropy of the next word w_3 when the relevant history w_1, w_2 is known.

Let a_i denote the acoustic string resulting when the word w_i was pronounced. Since it is approximately true that a_j for $j < i$ does not influence the recognition of w_i once w_j are known for $j < i$ (this observation is completely valid for isolated word recognition), and since a_i is known to the recognizer, then Ferretti et al. (1989) observe that an improved measure of recognition difficulty (and therefore of language model quality) would be the average entropy

$$H(W_3 \mid A_3, W_1, W_2) = -\sum P(w_1, w_2, w_3, a_3) \log P(w_3 \mid a_3, w_1, w_2) \quad (20)$$

where the sum is over all combinations of w_1, w_2, w_3, a_3. But $H(W_3 \mid A_3, W_1, W_2)$ can be estimated by the quantity

$$\hat{H} = -\frac{1}{n} \sum_{i=1}^{n} [\log \hat{P}(w_i \mid a_i, w_{i-1}, w_{i-2})] \quad (21)$$

where a_i is the acoustic sequence observed as a result of w_i being uttered.

To compute $\hat{H}$, we must evaluate the probability in its argument. The following decomposition is clearly valid for isolated word recognition and should be a good approximation in the general case:

$$\hat{P}(w_i, a_i, w_{i-1}, w_{i-2}) = P(a_i \mid w_i) \hat{P}(w_i \mid w_{i-2}, w_{i-1}) \quad (22)$$

Thus $\hat{P}(w_i \mid a_i, w_{i-1}, w_{i-2})$ is obtained by dividing (22) by

$$\sum P(a_i \mid w) \hat{P}(w \mid w_{i-2}, w_{i-1})$$

where the sum is over all words w in the vocabulary. Such a sum has 20,000 terms. However, to each acoustic string a and history $h_i = w_{i-2}, w_{i-1}$, there corresponds a *fast match list* $\mathrm{FM}(a, h_i)$ of words w for which the product $P(a \mid w)\hat{P}(w \mid h_i)$ is significant (Bahl et al., 1989). Such a list need have, on the average, no more than 30 members,* so the sum can be restricted to the list $\mathrm{FM}(a, h_i)$.

Finally, Ferretti et al. observe that is it well known that Markov models underestimate the values $P(a \mid w)$ because they do not take into account the correlation between contiguous acoustic frames. In practical recognition systems, this is compensated for by raising $P(a \mid w)$ to a small, empirically

*Given a and h_i, a fast match list is obtained by first determining in some simplified and therefore fast manner a set of words $\mathrm{FW}(a)$ for which $P(a \mid w)$ exceeds some threshold, and then eliminating from $\mathrm{FW}(a)$ all words w' for which $P(a \mid w')P(w' \mid w_{i-2}, w_{i-1})$ is too small (Bahl et al., 1989).

determined power which Brown (Brown, 1987) gives as 0.25. Thus, defining

$$S(w,a) = P(a \mid w)^{.25} \tag{23}$$

we get our final acoustic-sensitive measure of language model quality called *speech decoder logprob*,

$$\text{SDLP} = -\frac{1}{n}\sum_{i=1}^{n} \log \left[\frac{S(w_i,a_i)\hat{P}(w_i \mid w_{i-1},w_{i-2})}{\sum_{FM(a_i,h_i)} S(w,a_i)\hat{P}(w \mid w_{i-1},w_{i-2})} \right] \tag{24}$$

Ferretti et al. show that the SDLP measure correlates excellently with probability of error. Its weakness is that it depends through the function $S(w,a)$ on the particular recognizer being used.*

We conclude this section by introducing a measure of information provided by output symbols y of one source about outputs x of a related source. On this measure we will base several very useful self-organizing algorithms. For instance, we will be able to partition automatically the vocabulary into semantic/syntactic word classes (Appendix C) and to find automatically fixed phrases (Section 5) that act as a single word (e.g., New York).

Intuitively, the information provided by source 1 and source 2 should be equal to the difference between the uncertainties about the outputs of source 2 before and after the output of source 1 was observed.

Let $P(x \mid y)$ be the probability that the symbol x will be put out by source 2 given that y was observed as the output of source 1. Then since entropy is a measure of uncertainty, it follows from (12) that once y is known, the

*There are many ways to evaluate the function (24). The most accurate one is, of course, to have a subject utter the text $w_1, w_2, \ldots, w_n$, thus generating the acoustic strings $a_1, a_2, \ldots, a_n$. The recognizer can then provide the actual fast match lists $\text{FM}(a_i,h_i)$ corresponding to the acoustics and the text.

A less accurate method is for the subject to pronounce only once all the distinct words $v_1, v_2, \ldots, v_k$ found in the text $w_1, w_2, \ldots, w_n$, and use the resulting acoustic strings to create semiartificial acoustics $a_1, a_2, \ldots, a_n$.

In a final simplification, the fast match lists $\text{FM}(a_i,h_i)$ can be predetermined, independent of a_i, to depend only on the identify of the word w_i that gave rise to a_i, by simply finding the set of words which are expected to be phonetically similar to w_i.

It may even be possible to dispense with any pronunciation of the test text at all and to replace a_i by the phonetic baseform of the corresponding word w_i. However, experiments validating the resulting simplified measure of language model quality are yet to be carried out.

uncertainty about what x is going to be is equal to

$$H(X \mid y) = -\sum_{x=1}^{L} P(x \mid y) \log P(x \mid y) \tag{25}$$

Thus on the average, after the output of source 1 has been observed, the remaining uncertainty about source 2 is

$$\begin{aligned} H(X \mid Y) &= \sum_{y=1}^{L'} P(y) H(X \mid y) \\ &= -\sum_{y=1}^{L'} \sum_{x=1}^{L} P(y) P(x \mid y) \log P(x \mid y) \end{aligned} \tag{26}$$

where L' is the size of the output alphabet of source 2 and $P(y)$ is the probability that it will put out the symbol y. Consequently, the information that source 1 provides about source 2 is

$$\begin{aligned} I(X;Y) &= H(X) - H(X \mid Y) \\ &= \sum_{y=1}^{L'} \sum_{x=1}^{L} P(y) P(x \mid y) \log \frac{P(x \mid y)}{P(x)} \\ &= \sum_{y=1}^{L'} \sum_{x=1}^{L} P(x,y) \log \frac{P(x,y)}{P(x)P(y)} \end{aligned} \tag{27}$$

where $H(X)$ denotes the entropy of source 1 given by (12), the second equality follows directly from (12) and (26), and the last equality from straightforward algebraic manipulation based on the fact that $P(x,y) = P(y)P(x \mid y)$.

It is worth noting from the last expression of (27) that the function $I(X;Y)$ is totally symmetrical in the variables x and y and therefore source 2 supplies as much information about source 1 as does source 1 about source 2! Thus $I(X;Y)$ measures *mutual* information between the two sources. Furthermore, it can be seen from (27) that $I(X;Y)$ is an *average* of the quantity

$$I(x;y) = \log \frac{P(x,y)}{P(x)P(y)} \tag{28}$$

which is referred to as the *mutual information* between the particular symbols x and y. $I(X;Y)$ is called the *average mutual information function*.

While we have derived the form of $I(X;Y)$ by essentially intuitive reasoning, information theory (Gallager, 1968) shows its fundamental nature by basing on it precise theorems concerning transmission of information through noisy communication channels. That is, $I(X;Y)$ is related to the number of bits that can be reliably transmitted per symbol x to a receiver observing the symbols y when these two are related by a fixed probability distribution $P(y \mid x)$.

4. PROBLEMS WITH TRIGRAM MODEL PROBABILITY ESTIMATION

The interpolation formula

$$P(w_3 \mid w_1, w_2) = q_3 f(w_3 \mid w_1, w_2) + q_2 f(w_3 \mid w_2) + q_1 f(w_3) \qquad (9)$$

which estimates $P(w_3 \mid w_1, w_2)$ has many problems. If the dictation contains a trigram w_1, w_2, w_3 which never took place in the training text, then w_3 must be estimated from bigrams and unigrams. If the training text didn't even contain w_2, w_3, then the language model (9) will fall back on the unigram relative frequency $f(w_3)$. The difficulty is that $f(w_3)$ puts most weight on the most frequent words in the vocabulary, such as "the," "a," etc. (see Table 1 for the unnaturally high position of "the" in the prediction sequence). But since there must be a good linguistic reason why the sequence [w_2, "the"] never occurred in training, the language model should actually assign a lower probability to the sequence [w_2, "the"] than to unseen sequences [$w_2, w_{3''}$] where $w_{3''}$ is some low-frequency word (this argument applies to all high-frequency words w_3, not just to "the")!

Thus formula (9) should be improved. One possibility is the use of parts of speech (POS). Let $g(w_i) = g_i$ denote the classification of the word w_i reflecting its POS use in the text (for many words the classifications depend on context; e.g., "light" can be a noun, verb, adjective, or adverb). Then it is possible to predict the word w_3 in two steps: first its class and then a choice from the class. For example, if w_1, w_2 was never followed in the training text by a determiner, then any particular determiner should be predicted only with a small probability. Using a POS classification, the estimate in question would be based on the formula

$$kh(w_3 \mid w_1, w_2) = \sum_{g_3} k(w_3 \mid g_3) h(g_3 \mid w_1, w_2) \qquad (29)$$

which could be added as a fourth interpolation component to (9). Note that if the weights q_i are chosen optimally (as we claim we are doing), then adding a component can never make the language model worse. If one component is worthless compared to the other ones, it will simply get a very low weight q.

The probabilities k and h used in (29) can be estimated by a statistical procedure described in Appendix B. In particular, k can be derived by first annotating (automatically, see Appendix B) the words in the training text by their parts of speech g and then by computing the relative frequency of words w occurring in text positions annotated by g. The probability $h(g_3 \mid w_1, w_2)$ itself may best be obtained by interpolation of relative frequencies computed from annotated text, e.g.,

$$h(g_3 \mid w_1, w_2) = r_3 f(g_3 \mid w_1, w_2) + r_2 f(g_3 \mid w_2) + r_1 f(g_3) \qquad (30)$$

Of course, the interpolation (30) may even have components such as $f(g_3 \mid g(w_1), g(w_2))$, etc., that may be used instead of or even in addition to the other components. The disadvantage of components having functions $g(w_i)$ in the conditioning part is that during recognition the assignment of the POS $g(w_i)$ to the word w_i is in fact stochastic and depends on the context $w_1, \ldots, w_{i-1}, w_{i+1}, \ldots, w_k$. For a hypothesized history $w_1, \ldots, w_i$, the decoder would have to keep separate its various possible POS assignments $g(w_1), \ldots, g(w_i)$.

The trigram probability estimate

$$P(w_3 \mid w_1, w_2) = q_3 f(w_3 \mid w_1, w_2) + q_2 f(w_3 \mid w_2) + q_1 f(w_3) + q_4 + kh(w_3 \mid w_1, w_2) \qquad (31)$$

has, in fact, been observed to result in a real lowering of the perplexity of the language model (Bahl et al., 1986).

In our experiments we have based the POS classification on the 1972 Harvard Predictive Analyzer. We obtained 29 parts of speech and added 11 additional ones related to semantics, such as *date*, *number*, *name*, and *company*. However, our opinion is that the above POS classification is too crude and not necessarily suited to language modeling. Our POS set contains only one verb category (i.e., it does not distinguish between transitive, intransitive, or any other verbs) and nouns are divided only into plurals and singulars. It might further be useful to split small but frequent categories such as prepositions: "in" and "on" might turn out to predict different noun phrases as their objects. In any case, the approach using conventional parts of speech leads to classes of grossly unequal frequency and possibly dubious information value.

In Appendix C we show two methods of deriving automatically a prescribed number M of "nuclear" parts of speech (NPOS) which provide a more suitable equivalence classification than the natural ones. The first method (from which the name is derived) starts out by taking the M most frequent words of the vocabulary and assigning them to different NPOS classes. Each of the remaining $L - M$ nonnuclear words is originally allowed to have any of the M NPOS. Through a self-organized "clustering" process, the words gradually shed most of their class possibilities, until all words belong to a maximum of, say, 4 NPOS classes each. Most words actually end up belonging only to the class of that nuclear word to which they are most similar in their semantic/syntactic behavior. The clustering process is based directly on the formula

$$\sum_{g_3} k(w_3 \mid g_3) h(g_3 \mid g(w_1), g(w_2)) \tag{32}$$

i.e., on the language model component that will make use of the resulting NPOS classification. This approach when applied to the predictor (31) yields a 15% improvement in language model perplexity.

5. CONCLUDING REMARKS ON N-GRAM MODELING

The problems discussed in the previous section are related to "smoothing" of the relative frequency estimates $f(w_3 \mid w_1, w_2)$ to compensate for the (necessary) sparseness of data on which they are based. A direct way to achieve an improvement in language model quality is to derive it from a much larger training text. But this approach leads to further questions:

1. Where would one get a sufficiently large corpus (e.g., 1 trillion words for a vocabulary of 20,000 words)?
2. If the matrix $f(w_3 \mid w_1, w_2)$ were not sparse, how would one store and retrieve the information?
3. If a resulting trigram model turned out to be a good one, why not use an n-gram model with $n > 3$? For the language model to extract information as well as humans do, doesn't n have to be extremely large?
4. The resulting language model would still be bound to the area of discourse from which it was extracted. How about language models for other areas?

We have no answer to the first two questions. We will deal in Section 6 with information contained in the distant past, and in Section 8 with adaptation of language models to new subjects of discourse.

Formula (9) is not the only way to use bigrams and unigrams to smooth out trigram estimates. Another method is due to Katz (1987). It "backs off" from a trigram to a bigram to a unigram estimate, rather than interpolating between the three. The idea is incorporated in the approximate formula

$$P(w_3 \mid w_1, w_2) = \begin{cases} s_3 f(w_3 \mid w_1, w_2) & \text{if } C(w_1, w_2, w_3) > 0 \\ s_2 f(w_3 \mid w_2) & \text{if } C(w_1, w_2, w_3) = 0 \\ & \quad \text{and } C(w_2, w_3) > 0 \\ s_1 f(w_3) & \text{otherwise} \end{cases} \tag{33}$$

where s_i are factors that depend on the counts C and assure that the probability P when summed over all words w_3 adds up to 1.

The trigram model is, of course, too rigid. Since strings of different lengths are never put into the same equivalence class, advantage cannot be taken of the probably similar predictive power of phrases with the same heads (e.g., "big machine," "light machine," "machine"). At least some skipping of words would surely be in order. For instance, since determiners might have strong predictive qualities when they immediately precede the word w_i to be predicted, but weak ones when there is at least one word between them and w_i, then a better equivalence classification of the history $w_1, \ldots, w_{i-1}$ [see (6)] might be

$$P(w_i \mid w_1, \ldots, w_{i-1}) = \begin{cases} P(w_i \mid w_{i-3}, w_{i-1}) & \text{if } w_{i-2} = \text{determiner} \\ P(w_i \mid w_{i-2}, w_{i-1}) & \text{otherwise} \end{cases} \tag{34}$$

A word is a rather arbitrary building block for a language model. It can surely be argued that "Puddleby on the Marsh" or "nuclear magnetic resonance" should be basic elements of the vocabulary $\mathcal{V}$, rather than composites taking up all three positions of the trigram (and more) and considerably weakening the coupling between the past and the future. An automatic way of deciding which phrases should be included in the vocabulary was suggested by R. L. Mercer (personal communication) and by Choueka (1988). It is based on an iterative use of the concept of mutual information [see Section 3, Eq. (28)] between adjacent words

$$J(w_1, w_2) = \log \frac{C(w_1, w_2)n}{C(w_1)C(w_2)} \tag{35}$$

In (35), C denotes the count function and n is the size of the training corpus over which the counts were collected.

Since a large value of $J(w_1, w_2)$ indicates that the words w_1 and w_2 occur as a sequence much more frequently than can be expected from pure chance [namely, that $nC(w_1, w_2) > C(w_1)C(w_2)$], $J(w_1, w_2)$ can be used as a measure of desirability of including w_1, w_2 as a unit into the vocabulary. We therefore establish two thresholds, TJ and TC, and start out by augmenting the basic vocabulary $\mathcal{V}_0$ by all phrases w_1, w_2 for which simultaneously

$$J(w_1, w_2) > \text{TJ} \quad \text{and} \quad C(w_1, w_2) > \text{TC} \tag{36}$$

In this way a new vocabulary $\mathcal{V}_1$ is established. The second threshold is needed to make sure that the first comparison of (36) is reliably based on sufficient data.

The counts are next adjusted to reflect their new base vocabulary $\mathcal{V}_1$. That is, for every new element $w = w_1, w_2$ of $\mathcal{V}_1$, the old counts $C(w_1)$ and $C(w_2)$ are decremented by $C(w_1, w_2)$. The process (35) then resumes with $\mathcal{V}_1$ as its basis, leading to a new vocabulary $\mathcal{V}_2$. The augmentation terminates when no new phrases over the current vocabulary can be found satisfying (35). It is obvious that, in principle, the final vocabulary may include phrases of unlimited length.

In this and the previous sections, we have based estimates of trigram probabilities on relative frequency. This assumes that an n-gram which has been seen k times is k/m more probable than one which has been seen m times. This approach forces us to conclude that an n-gram that has been seen even once is infinitely more probable than an unobserved n-gram, and yet we know that a single appearance could have been a completely chance event! We will now try to escape from this old trap.

The interpolations (9) and (31) used weights estimated by the principle of cross-validation. But that principle can also be used in another way altogether (Jelinek and Mercer, 1985). Let us take the view that all n-grams seen r times in the corpus W_1 are equally probable (including all n-grams seen 0 times) and estimate the exact size of that probability from a (usually much smaller) corpus W_2. In fact, let $\mathcal{S}_r$ denote the set of different n-grams that have been seen exactly r times in corpus W_1, and let N_r be the size of the set $\mathcal{S}_r$. Then if x denotes an n-gram in $\mathcal{S}_r$, we will set its probability to

$$P(x) = \frac{q_r}{N_r} \tag{37}$$

and determine the value of q_r from corpus W_2.

The way to do that is to sort W_2 into n-grams and consider them to have been independently generated.* The values of q_r are then chosen to maximize the probability of the total observed n-gram collection arising out of W_2. Let that collection consist of M n-grams, and let t_r be the number of n-grams in the collection that belong to the set $\mathcal{S}_r$. Elementary calculus shows

---

*There is a problem here, of course. It is not clear if one should sort overlapping or nonoverlapping n-grams. What is worse is that W_2 certainly did not arise out of independent generation of n-grams! (*Footnote continues on page 668.*)

that the probability of the collection will be maximized by the choice

$$q_r = \frac{t_r}{M} \tag{38}$$

The above methodology can be used to replace relative frequency in all the formulas estimating probabilities, including (33), the basis of the backing-off approach.

Church and Gale (1991) suggested an interesting enhancement. They observe that it is not necessary to give the same probability to all n-grams occurring r times, if we have extra information. They suggest that all bigrams w_1, w_2 be divided into bins according to the value of the product $f(w_1)f(w_2)$ and that the estimation procedure be carried out separately for different bins. Here it is assumed that enough data exist so as to make $f(w)$ a good estimate of $P(w)$. By this procedure one gets an estimate $\hat{P}(w_2 \mid w_1)$. Bootstrapping this approach one then divides trigrams w_1, w_2, w_3 into bins according to the value of the product $\hat{P}(w_1, w_2)P(w_3 \mid w_2)$ and again estimates trigram probabilities on the basis of separate bin estimations, etc.

There are, of course, other ways of binning. A fruitful one may be based on part-of-speech assignment of bigrams. In such a case one would estimate separately the probabilities $\hat{P}(w_1, w_2 \mid g_1, g_2)$ of bigrams w_1, w_2 that have been annotated in both the original and cross-validation texts (W_1 and W_2) by the

---

Take the case of bigrams, for instance. If $W_2 = w_1, w_2, \cdots, w_{M+1}$ then an acceptable model of generation of W_2 would assign to W_2 the probability

$$P(W_2) = \frac{\prod_{i=1}^{M} P(w_i, w_{i+1})}{\prod_{i=2}^{M} P(w_i)} \tag{39}$$

that should be maximized by appropriate choice of q_r in (37). Unfortunately, assignment (38) maximizes only the numerator of the above expression. The problem is that

$$P(w_i) = \sum_{w} P(w_i, w)$$

is also a function of q_r and that rigorous maximization of (39) is too complex a procedure.

One way to preserve the approximate correctness of formula (38) is to base the estimation on bigrams that are separated in W_2 by adequate intervals of ignored text. The longer such intervals, the more independent will be the bigrams included in the collection. This approach is possible if W_2 is large, in which case we seem to be wasting training text. However, the ignored intervals can be included in W_1!

part-of-speech pair g_1, g_2. The final estimate would then be

$$\hat{P}(w_1, w_2) = \sum_{g_1 g_2} P(g_1, g_2)\hat{P}(w_1, w_2 \mid g_1, g_2) \tag{40}$$

Such an approach presupposes that the sets of pairs w_1, w_2 determined by g_1, g_2 have a sufficient number of members to make the estimation process involving (37) and (38) a reliable one, and that the estimate of the probabilities $P(g_1, g_2)$ itself be reliable. If these two conditions were met only by a subset $\mathcal{S}$ of pairs g_1, g_2, one could modify (40) and get

$$\hat{P}(w_1, w_2) = \sum_{\mathcal{S}} P(g_1, g_2)\hat{P}(w_1, w_2 \mid g_1, g_2) + (1 - P(\mathcal{S}))\hat{P}(w_1, w_2 \mid \bar{\mathcal{S}}) \tag{41}$$

where $\bar{\mathcal{S}}$ denotes a single bin containing bigrams not belonging to the set $\mathcal{S}$, and

$$P(\mathcal{S}) = \sum_{\mathcal{S}} P(g_1, g_2)$$

The bootstrap to trigram estimation via POS bins is perhaps more problematic, since a relatively smaller number of sets determined by POS trigrams would meet the above two reliability conditions. In any case, the corresponding set $\bar{\mathcal{S}}$ of "inconvenient" trigrams would be large and would have to be binned by a different principle (perhaps based on probabilities $\hat{P}(w_1, w_2)P(w_3 \mid w_2)$ as suggested by Church and Gale.)

6. INCORPORATING PAST INFORMATION INTO THE LANGUAGE MODEL

When predicting what will be said, human beings often use information in words whose position considerably precedes that of the last two. To come close to human performance (and it is not known what precise considerations account for it) prediction would have to be based on sophisticated syntactic and semantic analysis.

A simple method that uses past information was suggested by DeMori and Kuhn (1991). The basic idea is to add (by interpolation) a unigram *cache* component to the trigram language model (9). The probability PC(w) assigned by this component to the word w is made equal to its relative frequency of occurrence in a "window" viewing the last K words of the text.

There is no reason not to generalize this idea and construct a full trigram cache component. The following procedure gives excellent results (Jelinek et al., 1991). We intend to apply our method to the creation of documents, and accordingly the past window for predicting the $n+1$th word of the current document is of size

$$K = \max\{n, NP\} \tag{42}$$

where NP is the length of the previous document dictated by the user. Initially, the window views the previous document. At time $n < NP$ the window views the text consisting of all the past words of the current document and of the $NP - n$ last words of the previous document. At time $n \geq NP$ the window simply views the past text of the current document.

Relative frequencies of trigrams, bigrams, and unigrams appearing in the window are computed and denoted by $fc(w_3 \mid w_1, w_2)$, $fc(w_3 \mid w_2)$, and $fc(w_3)$, respectively. The trigram cache component is then given by

$$Pc(w_3 \mid w_1, w_2) = 0.25 fc(w_3 \mid w_1, w_2) + 0.25 fc(w_3 \mid w_2) + 0.5 fc(w_3) \tag{43}$$

It is obvious that the weights in (43) have to be constants, because the vast majority of bigrams and trigrams observed in windows are singletons. The interpolation in (43) has been observed to be quite insensitive to the values of the weights, and the ones used in (43) are as good as any.

It is possible to further augment the cache component. One can see from the perusal of texts that it is not simply words that repeat, but various forms of words, such as inflections or close derivations (e.g., absent, absentee, absence). To include this phenomenon, it is necessary to have a training text and proceed as follows:

Let $\mathcal{S}_w$ be the set of forms of a given word w. Let $N_r(x \mid w)$ denote the number of times the word x belonging to $\mathcal{S}_w$ is found in the training text in a window following the appearance of the word w. We will use the values $N_r(x \mid w)$ to construct a unigram probability distribution $q_n(w_n)$ for predicting the nth word of a text. We will let $N_r(x \mid w) = 0$ for all x not belonging to $\mathcal{S}_w$, and define

$$N_{rn}(x) = \max_{i=1}^{K} N_r(x \mid w_{n-i}) \tag{44}$$

Obviously, $N_{rn}(x) = 0$ for all words x that belong to no set $\mathcal{S}_{w_{n-i}}$. Letting further

$$N_{rn} = \sum_x N_{rn}(x)$$

we get the distribution

$$q_n(w) = \frac{N_{rn}(w)}{N_{rn}} \tag{45}$$

Note that $q_n(w)$ depends on the identity of words in the window preceding w_n and is thus dynamic.

One can similarly obtain dynamic bigram and trigram distributions $q_n(w_3 \mid w_2)$ and $q_n(w_3 \mid w_1, w_2)$. For instance, let $N_r(x \mid w_2; w)$ denote the number of times the word x belonging to $\mathcal{S}_w$ follows the word w_2 and is found in a window following the appearance of the word w in the training text. Let $N_r(x \mid w_2; w) = 0$ for all x not belong to $\mathcal{S}_w$, and define

$$N_{rn}(x \mid w_2) = \max_{i=1}^{K} N_r(x \mid w_2; w_{n-i}) \tag{46}$$

Letting further

$$N_{rn}(w_2) = \sum_x N_{rn}(x \mid w_2)$$

we get the distribution

$$q_n(w \mid w_2) = \frac{N_{rn}(w \mid w_2)}{N_{rn}(w_2)} \tag{47}$$

The definition and construction of $q_n(w_3 \mid w_2, w_1)$ is similar. We have thus produced a new dynamic component

$$P_r(w_3 \mid w_1, w_2) = r_3 q_n(w_3 \mid w_1, w_2) + r_2 q_n(w_3 \mid w_2) + r_1 q_n(w_3) \tag{48}$$

which can be added (interpolated) to the overall language model in addition to the direct trigram cache $P_c(w_3 \mid w_1, w_2)$ defined in (43).

It should be noted that if the training text on which the relative frequencies of (48) are based is very similar to the documents that will be created, then $P_r(w_3 \mid w_1, w_2)$ has the capacity to predict phrases, not just words. For instance, if "nuclear magnetic resonance" has occurred in the window as well as in training text, then $N_{rn}(x)$ will assign nonzero counts to "magnetic" as well as "resonance," and $N_{rn}(x \mid \text{nuclear})$ will assign nonzero count to "magnetic," etc. It might even be argued that the component (48) will be a stronger predictor than (43), since it is based on more data, but only if the training text mirrors the intended dictation.

What makes $P_c(w_3 \mid w_1, w_2)$ especially valuable is that it has the capacity to reflect the unforeseen nature of dictation: once a word or a phrase unknown to the vocabulary is spoken by the user (and described to the system, perhaps through correction of the consequent recognition error), the cache component incorporates it into the language model and it ceases to be unexpected in further dictation!

7. VOCABULARY SELECTION AND PERSONALIZATION FROM TEXT DATA BASES

Every speech recognizer is based on a vocabulary $\mathcal{V}$ of finite size L whose words are the only ones that can appear in the transcribed text. As a consequence, any attempt to recognize a word not belonging to $\mathcal{V}$ will result in an error. Such an error will force the language model into a wrong state, increasing the probability of an error in the next word. Since the complexity of a language model and the difficulty of constructing it grow with the vocabulary size, it is desirable to keep the size as small as possible. Thus the vocabulary must be chosen carefully so as to maximize the expected coverage of the dictated text, defined as the percentage of words in the running text that belong to $\mathcal{V}$.

Since the IBM recognizer is intended for dictation of office correspondence, its 20,000-word vocabulary is based on three collections of business memoranda and letters concerned with computer manufacture and usage. The vocabulary was determined in two steps. First, a basic subvocabulary of size M was constructed, consisting of words whose availability was thought necessary for the task, such as recognizer control words, numbers "thousand," "million," etc.), and dates ("Monday," "November," etc.). Then the remaining $20{,}000 - M$ words were chosen to be those that have the highest minimal count in our three data bases (OC1, OC2, and OC3) and do not belong to the subvocabulary. In fact, let $C_1(w)$, $C_2(w)$, and $C_3(w)$ be the counts of the word w in the data bases OC1, OC2, and OC3. Define the minimal count

$$C_m(w) = \min[C_1(w), C_2(w), C_3(w)] \tag{49}$$

and arrange the words in descending order of the value of the minimal count C_m. Then the $20{,}000 - M$ highest ordered words that were not in the basic subvocabulary are chosen.

Basing the selection method on the minimal count (49) results in omission from the vocabulary of words that have little universal use and are peculiar to a particular data base.

We have tested the coverage of our 20,000-word vocabulary on several additional data bases in our possession. We have noted that 97% of words in their running texts were contained in $\mathcal{V}$. The 3% of the text not covered constitutes a lower bound on the error rate of the recognizer. Greater coverage can be achieved by both increasing the vocabulary size and personalizing the vocabulary to the users of the system. Inspection of the American Heritage *Word Frequency Book* (Carroll et al., 1971) clearly indicates that simply increasing the vocabulary size will not be enough. In fact, quite familiar words such as "admonition" and "deluded" are not among the 86,000 most frequent words. Furthermore, the *Word Frequency* book will not contain technical words particular to an individual user's field! On the other hand, there are only 19,000 distinct words in the 1 million words large corpus OC1, and Henry Kucera (personal communication) observes that there are only 29,000 distinct words in all of Shakespeare's works. It should therefore be possible to achieve high coverage through appropriate personalization.

Using a slight modification of the vocabulary selection method described above, we constructed vocabularies of varying sizes. Table 2 shows their coverage on the corpus PDB (created by one person) when names and acronyms occurring in it were disregarded.

Our method of fixed-vocabulary construction may be satisfactory if the intended user belongs to some easily identifiable category (e.g., employee of the data processing industry) for which text data bases are available. Still, it is hardly acceptable that before purchasing a recognizer, a customer would have to identify its use so it could be equipped with an appropriate language model.

To keep its complexity down, a language model should probably be based on an economical set of building blocks (perhaps stems or morphemes) from which word inflections and derivations could be constructed. Kucera (personal communication) estimates that vocabulary compression by a factor of 4 may be attainable in this way even for English, a practically noninflecting language.

A crude method of personalization is achieved by maintaining during dictation a dynamically varying vocabulary consisting at any given moment of the last L different words used. The coverage is then defined as the asymptotic probability that the word spoken next already belongs to the vocabulary. Table 3 gives the result for the PDB data base. The second column is the good news: 99% coverage is achieved by a 15,000 word vocabulary. The third column is the bad: it takes 620,000 words of text before a set of 15,000 different words is assembled. This then is a slow way of vocabulary personalization. It may, however, be practical for an installation that would pool all the texts produced by all the resident users.

Table 2 Static Coverage of Sample Text as a Function of Vocabulary Size

Vocabulary Size	Text Coverage (%)
5,000	92.5
10,000	95.9
15,000	97.0
20,000	97.6

Table 3 Dynamic Coverage of Sample Test as a Function of Vocabulary Size

Vocabulary Size	Text Coverage (%)	Text Size Needed To Reach Coverage
5,000	95.5	52,000
10,000	98.2	230,000
15,000	99.0	620,000
20,000	99.5	1,300,000

The above dynamic method of vocabulary creation is useful provided it is accompanied by an algorithm capable of creating a corresponding language model. Fortunately, the cache language model (43) introduced in the preceding section gives hope of good performance.

For direct personalization it would be desirable for a request for a new word to bring into the vocabulary automatically a triggered subset of other words that collocate with it. We have not been very successful with this trigger language model idea. We think the problem is that the appropriate composition of triggered subsets varies greatly with the user as well as with the particular document being created. To be assured of coverage, one would have to use a union of such subsets. But that many words being added would push out from the fixed-size vocabulary other words that would turn out to be more useful. Therefore, we recommend that the triggered subsets pertaining to a word w be composed only of appropriate derivations of w, as was discussed in the previous section in connection with the cache language model component (48).

8. LANGUAGE MODELS FOR VARYING DISCOURSE DOMAINS

One of the IBM trigram language models was constructed from three combined text data bases containing a total of 25 million words. The extracted trigrams made up of words in the chosen vocabulary covered 84% of correspondingly restricted trigrams in the test data base PDB, while the coverage by trigrams collected solely from one of the three data bases (OC1, containing 1 million words) was only 55%.

These percentages are indicative of the fact that large training texts pertaining to specific discourse domains are required to generate language models of adequately low perplexity. Such texts will in many cases turn out to be hard to get.

We have two very related problems: (1) how to create a language model for a new domain and (2) how to create a language model appropriate to a particular user.* This section is concerned with their solution.

The trigram language model for a dynamically changing active (personalized) vocabulary cannot consist solely of a cache component constructed from the text created by the user. The user will never dictate enough. A straightforward way to proceed would be first to determine the appropriate initial vocabulary from the writings of the user, and then to extract the model from an already stored trigram collection appropriate to that vocabulary. The last column of Table 3 indicates that a vocabulary determined from a reasonable amount of text created by a single user (say, at most, 100,000 words) cannot be the final one: the vocabulary will change a lot as the user continues using the system.

Fortunately, there is another way to proceed that holds a lot of promise. We will first introduce a method for deriving trigram probabilities for relatively infrequent words that were added to the vocabulary during the dynamic process (see Section 7) when the basic language model is fixed (Jelinek et al., 1990).† We will then generalize the approach to the creation of a language model for a completely new domain.

Consider a fixed vocabulary $\mathcal{V}_0$ and a training corpus. Let us label the words w of the latter by the function $l(w)$ as follows:

$$l(w) = \begin{cases} w & \text{if } w \text{ is in } \mathcal{V}_0 \\ u & \text{otherwise} \end{cases} \tag{50}$$

We can consider the sequences of labels as a new text and derive in the usual way [see (9)] its language model $P(l_3 \mid l_1, l_2)$. We want to use the latter as the basis for a language model for an augmented vocabulary $\mathcal{V}_1$ which is an extension of $\mathcal{V}_0$ by some set $\mathcal{V}_R$. The idea is to define for each word w in $\mathcal{V}_1$ a set $\mathcal{S}_w$ of its synonyms in the vocabulary $\mathcal{V}_0$,

$$\mathcal{S}_w = \{z_1, z_2, \ldots, z_K\}, \qquad v_i \text{ in } \mathcal{V}_0 \tag{51}$$

*Comparisons of Tables 2 and 3 show the advantage of dynamic vocabulary personalization. Obviously, the advantages of language model personalization will turn out to be even more pronounced.

†The method presented below, and specified in (52), is capable of providing statistics for new words requested by the user. In this way it is complementary to the cache language component introduced in Section 6, which can be used for the same purpose.

where $K = 1$ and $z_1 = w$ if w belongs to $\mathcal{V}_0$. The language model for $\mathcal{V}_1$ will then be given by

$$Q(w_3 \mid w_1, w_2) = \sum_{z \text{ in } \mathcal{S}_{w_3}} k(w_3 \mid z) P(z \mid l(w_1), l(w_2)) \tag{52}$$

It remains to determine the sets $\mathcal{S}_w$ and the distributions $k(w \mid z)$ for w in $\mathcal{V}_1$. Let us assume that for each word w in $\mathcal{V}_R$ we have available contexts* $l(w_{1i})\,l(w_{2i})\,w\,l(w_{4i})\,l(w_{5i}), i = 1, 2, \ldots, I(w)$, in which w appears in a training text w_R. The probability of generating the sequence $w\,l(w_{4i})\,l(w_{5i})$, given the initial bigram $l(w_{1i})\,l(w_{2i})$, is then given by

$$\begin{aligned} P\,(w, l(w_{4i}), l(w_{5i}) \mid l(w_{1i}), l(w_{2i})) \\ = \sum_{z \text{ in } \mathcal{S}_w} k(w \mid z)\,[P\,(z \mid l(w_{1i}), l(w_{2i})) \\ \times\, P\,(l(w_{4i}) \mid l(w_{2i}), z) P\,(l(w_{5i}) \mid z, l(w_{4i}))] \end{aligned} \tag{53}$$

Since, all things being equal, we wish to assign large probabilities to contexts that actually took place, a heuristic way to determine the synonym sets $\mathcal{S}_w$ for words w in $\mathcal{V}_R$ is simply to find the K words z in $\mathcal{V}_0$ that give the maximum scores for the function

$$\sum_{i=1}^{I(w)} \log [P\,(z \mid l(w_{1i}), l(w_{2i}))\; P\,(l(w_{4i}) \mid l(w_{2i}), z)\; P\,(l(w_{5i}) \mid z, l(w_{4i}))] \tag{54}$$

This is reasonable because such words z are those that most "like" the contexts $l(w_{1i})\; l(w_{2i})\; w\; l(w_{4i})\; l(w_{5i})$, $i = 1, 2, \ldots, I(w)$, in which the word w appeared.

To determine the distributions $k(w \mid z)$ for each z in $\mathcal{V}_0$ one should choose them so as to maximize the probability of generating $\mathcal{W}_R$, the part of the training text in which the words of $\mathcal{V}_R$ were observed,† by the somewhat degenerate

---

*It is assumed here that the probabilities $P(l_3 \mid l_1, l_2)$ used in (52) are estimated from some large original training text. Then once the user starts generating text, the latter is collected until enough of it exists to form $\mathcal{W}_R$. At this time the user has utilized words of $\mathcal{V}_R$ that were not found in the fundamental vocabulary $\mathcal{V}_0$.

†After training, it is desirable to somewhat modify the original language model $P(l_3 \mid l_1, l_2)$, because for any words w' and w'' of the original vocabulary $\mathcal{V}_0$ we want to maintain the original relation between their probability values $P(w' \mid l_1, l_2)$ and $P(w'' \mid l_1, l_2)$. We also take the point of view that the inclusion of the vocabulary $\mathcal{V}_R$ should in no way decrease the probability $P(u \mid l_1, l_2)$ of the unknown word (one not

hidden Markov model implied by formula (52) (see Appendix A). This can be done without difficulty using a simple variant (Appendix B) of the Baum–Welch reestimation algorithm (Baum, 1972) provided the sets $\mathcal{S}_w$ are not too large.

The above version of the synonym language modeling method is appropriate when the trigram language model is adequately determined for some basic vocabulary to which we wish to add additional words.* We will now generalize the approach to the creation of a language model for a completely new domain. To do so we will make several assumptions:

1. An adequate trigram language model exists for some fundamental domain.
2. The fundamental domain is sufficiently general so that it exhibits all of the grammatical phenomena present in any domain.
3. Most words in the new domain have a function that is fulfilled by some words (that may or may not have the same spelling) present in the fundamental domain.

If the above assumptions hold, then it is reasonable to expect that a language model for the new domain may be adequately constructed by taking the view that all of its words are statistical synonyms of words in the fundamental domain in the sense of (52). The language model in the new domain will be a hidden Markov model (HMM; see Appendix A) whose state transition

---

belonging to either $\mathcal{V}_0$ or $\mathcal{V}_R$. But for w in $\mathcal{V}_0$,

$$Q(w \mid l_1, l_2) = k(w \mid z = w) P(w \mid l_1, l_2)$$

and it is therefore necessary to base the synonym language model on the renormalized probabilities

$$P'(w \mid l_1, l_2) = \frac{P(w \mid l_1, l_2)/k(w \mid z = w)}{\sum_{w'} P(w' \mid l_1, l_2)/k(w' \mid z = w')},$$

---

*This method should then be used in conjunction with the cache component of Section 6 to provide the language model with statistics about previously unknown words. While the cache component works instantaneously, the synonym component has a build-in delay in that several (about 10, in our experience) uses of the word w are needed to provide enough contexts $l(w_{1i})l(w_{2i})wl(w_{4i})l(w_{5i})$ for the determination of an accurate synonym list $\mathcal{S}_w$ and for a well-estimated generation distribution $k(w \mid s)$.

probabilities were determined by the trigram probabilities of the fundamental language model. In fact, the language model for the new domain will be formally the same as the POS language model (32), but its "parts of speech" will be the statistical synonyms in the fundamental domain of words in the new domain.

We proceed as follows. As before, to each word w in the new domain there will correspond a set of synonyms $\mathcal{S}_w$ of words z in the fundamental domain whose trigram language model is specified by $h(z_3 \mid z_1, z_2)$. The new language model has states specified by pairs $s_i = (z_{i-1}, z_i)$. The transition probability from state $s = (z_1, z_2)$ to state $s' = (z'_1, z'_2)$ is given by

$$q(s' \mid s) = \begin{cases} h(z'_2 \mid z_1, z_2) & \text{if } z'_1 = z_2 \\ 0 & \text{otherwise,} \end{cases} \tag{55}$$

To the states s is attached an output probability distribution

$$r(w \mid s) = k(w \mid z_2) \quad \text{where } s = (z_1, z_2) \tag{56}$$

where $k(w \mid z_2) = 0$ if z_2 does not belong to the synonym set $\mathcal{S}_w$.

The generation of words in the new domain is then modeled by the HMM specified in (55) and (56). That is, the model makes transitions between hidden states s and s' with probability $q(s' \mid s)$, and when state $s' = (z'_1, z'_2)$ is reached, it puts out a word w with probability $k(w \mid z'_2)$.

Since the transition function $h(z_3 \mid z_1, z_2)$ is specified by the trigram language model of the fundamental domain, it is necessary only to extract the synonym set $\mathcal{S}_w$ and the output functions $k(w \mid z)$ from the training text of the new domain. If $|\mathcal{V}_N|$ is the size of the new vocabulary $\mathcal{V}_N$, and K the maximal size of the synonym sets, then we need to determine at most $K^*|\mathcal{V}_N|$ parameters instead of $|\mathcal{V}_N|^3$ parameters necessary for a trigram language model. This is therefore a method for language model creation from relatively sparse data.

Obviously, once the synonym sets are known, the output function $k(w \mid z)$ can be determined by the Baum–Welch reestimation algorithm (see Appendix B). An adequate initialization is to let

$$k^*(z \mid w) = \begin{cases} \dfrac{1}{\mathcal{S}_w} & \text{if } z \text{ is in } \mathcal{S}_w \\ 0 & \text{otherwise} \end{cases} \tag{57}$$

(where $|\mathcal{S}_w|$ denotes the size of the set $\mathcal{S}_w$) and then define the initial distribution $k(w \mid z)$ by

$$k_0(w \mid z) = \frac{k^*(z \mid w) f(w)}{k^*(z)} \tag{58}$$

where $f(w)$ is the unigram frequency of w in the training set, and

$$k^*(z) = \sum_{w'} k^*(z \mid w')f(w') \tag{59}$$

The final problem is the determination of the synonym sets $\mathcal{S}_w$. In principle, given sufficient computing power, one could determine these incrementally, first letting $\mathcal{S}_w$ be the entire vocabulary for all w, and gradually restricting the sets $\mathcal{S}_w$ by keeping in them those synonyms z for which $k(w \mid z)h(z)$ is the largest, where $h(z)$ is the unigram distribution of the word z in the fundamental domain. But this process is probably too expensive for any reasonably sized vocabulary, so a heuristic is necessary.

One plausible heuristic is based on the assumption that, to the first order of approximation, if $\mathcal{V}^*$ is the set of words common to both domains, then all words in $\mathcal{V}^*$ function in the same way. We then proceed as follows:

1. Let $\mathcal{V}_F = \mathcal{V}^* \cup \mathcal{V}_R$ be the vocabulary of the fundamental domain. Let $\mathcal{V}_{F'} = \mathcal{V}^* \cup u$ be an auxiliary vocabulary, where u is a special word representing all the words in $\mathcal{V}_R$. In a straightforward way create a trigram model for the fundamental domain with probabilities $g(z_3 \mid l_1, l_2)$, where z_3 is in $\mathcal{V}_F$ and l_1 and l_2 are in $\mathcal{V}_{F'}$.
   Define

$$g_l(l \mid y_1, y_2) = \begin{cases} g(l \mid l_1, l_2) & \text{if } l \text{ is in } v^* \\ \sum_{z \text{ in } \mathcal{V}_R} g(z \mid l_1, l_2) & \text{if } l = u \end{cases} \tag{60}$$

2. Create a mapping of words from the new vocabulary v_N into the vocabulary $v_{F'}$ as follows:

$$l(w) = \begin{cases} w & \text{if } w \text{ is in } v^* \\ u & \text{otherwise} \end{cases} \tag{61}$$

3. For every word w in v_N, find its contexts $w_{1i}w_{2i}ww_{4i}w_{5i}$, $i = 1, 2, \ldots, I(w)$, in the training set.
4. For each w in v_N, define $\mathcal{S}_w$ to be the set of those K words z in v_F for which the value of

$$\sum_{i=1}^{I(w)} \log [g(z \mid l(w_{1i}), l(w_{2i}))g_1(l(w_{4i}) \mid l(w_{2i}), l(w))$$
$$\times\, g_{1i}(l(w_{5i}) \mid l(w), l(w_{4i}))] \tag{62}$$

is largest.*

The HMM structure of the synonym language model (55) and (56) has a run-time disadvantage. Each hypothesized past $w_1 w_2 \cdots w_k$ will imply a probability distribution $P(z_1', z_2' \mid w_1, w_2, \ldots, w_k)$ over the hidden states $s' = (z_1', z_2')$ which the recognizer will have to store and update. In fact, let

$$Q(z_1, z_2 \mid w_1, w_2, \ldots, w_k) = \sum_{z_1'} h(z_2 \mid z_1', z_1) P(z_1', z_1 \mid w_1, w_2, \ldots, w_k) \quad (63)$$

then

$$P(z_1', z_2' \mid w_1, w_2, \ldots, w_{k+1}) = \frac{k(w_{k+1} \mid z_2') Q(z_1', z_2' \mid w_1, w_2, \ldots, w_k)}{\sum_{z_1, z_2} k(w_{k+1} \mid z_2) Q(z_1, z_2 \mid w_1, w_2, \ldots, w_k)} \quad (64)$$

The sum in (63) is over all synonyms z_1' of w_{k-1}, and the sum in (64) is over all synonym pairs z_1, z_2, of w_k and w_{k+1}, respectively. The computations (64) are therefore doable, provided the synonym list is adequately restricted.

It may be desirable not to base the transition probability $h(z_3 \mid z_1, z_2)$ of (55) simply on the trigram language model for some fundamental domain, but to adjust it first to the purpose to which it is intended. We have in mind to extract it from many overlapping domains $W_1, W_2, \ldots, W_L$ with a common core vocabulary $\mathcal{V}^*$:

1. Start by extracting the probabilities $g(l_3 \mid l_1, l_2)$ and $g_1(l_3 \mid l_1, l_2)$ of (60) from the total training data $W_1 + W_2 + \cdots + W_L$.
2. For each domain W_i separately, find synonym sets $\mathcal{S}_{iw}$ for words in the corresponding vocabulary $\mathcal{V}_i$, using criterion (62).
3. Estimate separate output probability functions $k_i(w \mid z)$ (valid for training set W_i), $i = 1, 2, \ldots, L$, and a common transition function $h(z_3 \mid z_1, z_2)$ by running the Baum–Welch algorithm whose single iteration would consist of processing the entire training data W_1, W_2, $\ldots$, W_L.
4. The final transition function $h(z_3 \mid z_1, z_2)$ would then serve as the basis of the HMM language model for any new task domain. The statistical synonym sets $\mathcal{S}_w$ and the output function $k(w \mid z)$ of this language model would have to be estimated as described in the preceding paragraphs.

*Note that this procedure allows even words w in $\mathcal{V}^*$ to have K synonyms in $\mathcal{V}_F$ (of which, presumably, its homograph would be one), thus allowing some loosening of the original rigid assumption.

Appendix A. HIDDEN MARKOV CHAINS: A BASIS FOR LANGUAGE MODELING

This appendix is devoted to a discussion of the probabilistic concept of hidden Markov chains, which is the basis for most of the methods of language modeling introduced in the body of this chapter.

A Markov chain* is a random process that generates sequences $s_1 s_2 \cdots s_n \cdots$ of elements called *states* that are picked from some finite set $\mathcal{T}$.† For all pairs s, s', in $\mathcal{T}$ there is defined a transition probability $q(s' \mid s)$ of one-step state-to-state transitions $s \rightarrow s'$ that governs the process:

$$\sum_{s' \text{ in } \mathcal{T}} q(s' \mid s) = 1 \quad \text{for all } s \text{ in } \mathcal{T} \tag{A.1}$$

A Markov chain has finite memory, in that the probability of any sequence of states $s_1, s_2, \cdots s_n$ is given by

$$P(s_1, s_2, \ldots, s_n) = \prod_{i=1}^{n} q(s_i \mid s_{i-1}) \tag{A.2}$$

In (A.2) the first term of the product is $q(s_1 \mid s_0)$, and we can think of s_0 as an implied obligatory starting state.

All transitions $s \rightarrow s'$ such that $q(s' \mid s) > 0$ are said to be allowed; the rest are prohibited. An intuitive feeling about a given Markov chain may be gained by diagraming the allowed transitions between states. Figure 1 is an

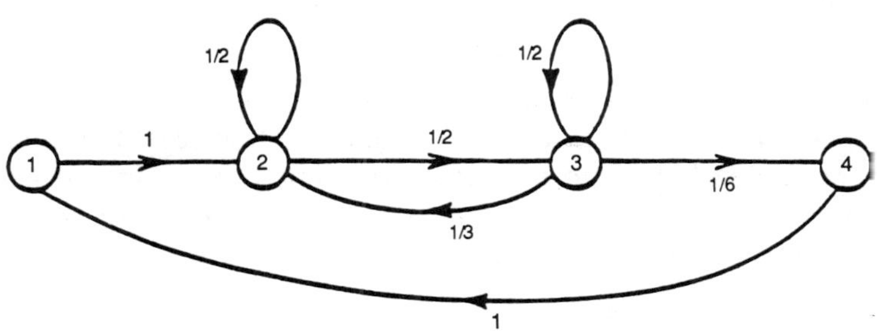

Figure 1 Example of a four-state Markov chain.

---

*The formulation is due to A. A. Markov, who actually used it to study texts (Markov, 1913). The reason for the terminology "chain" will become apparent below.

†We are introducing only the most rudimentary Markov chains, the so-called homogeneous finite state MCs that suffice for our purposes.

example of such a diagram for a four-state chain (its appearance suggests the terminology) which does not allow all the possible transitions (e.g., transitions from state 2 to states 1 and 4 are prohibited). Note that self-transitions are possible, such as those of states 2 and 3 to themselves. The numbers attached to the arrows specify the probability with which a transition is taken out of its initial state. A chain can be viewed as a mechanism that, starting operation in the initial state, changes its state at discrete time intervals along some allowed transition. A state sequence $s_1, s_2, \ldots$ has positive probability if and only if it can be traced in the diagram as a path leading out of the designated initial state s_0. Transition diagrams, although impractical for large state sets, define the chain fully.

A hidden Markov model (HMM) is a Markov chain that has an output distribution associated with each of its states.* There is a finite set $\mathcal{X}$ of outputs, and to each state s in $\mathcal{T}$ there corresponds a probability distribution $r(x \mid s)$ over the output set $\mathcal{X}$:

$$\sum_{x \text{ in } \mathcal{X}} r(x \mid s) = 1 \quad \text{for all } s \text{ in } \mathcal{T} \tag{A.3}$$

As the Markov chain underlying the HMM arrives at a state s, it chooses an output x with probability $r(x \mid s)$ and reveals it to the observer. The identity of s stays hidden, throughout.

The probability that a given output sequence $X = x_1, x_2, \ldots, x_n$ will be generated is

$$P(X) = \sum \left[\prod_{i=1}^{n} r(x_i \mid s_i) q(s_i \mid s_{i-1}) \right] \tag{A.4}$$

where the sum is over all sequences $S = s_1, s_2, \ldots, s_n$. Since there are potentially K^n of them (where K is the size of the state space $\mathcal{T}$), formula (A.4) seems prohibitively hard to evaluate. Fortunately, by taking advantage of the limited memory of Markov chains, it is possible to derive an algorithm computing (A.4) whose complexity grows only linearly with n. In fact,

$$\begin{aligned} P(x_1, \ldots, x_{k-1}, x_k, s_k = s') \\ = \sum_{s \text{ in } \mathcal{T}} P(x_i, \ldots, x_{k-1}, s_{k-1} = s) r(x_k \mid s') q(s' \mid s) \end{aligned} \tag{A.5}$$

*We use this formulation of HMMs because it is adequate for our purposes. Other formulations are possible. HMMs were originally called Markov sources by Shannon (1948).

and so the left-hand side of (A.5) can be computed recursively starting with the initial condition

$$P(x_1,s_1 = s') = r(x_1 \mid s')q(s' \mid s_0) \tag{A.6}$$

where s_0 is the obligatory initial state.

A question of interest when dealing with HMMs, is the following:

> Given an output sequence $x_1, x_2, \ldots, x_n$, what is the most likely state sequence $s_1, s_2, \ldots, s_n$ that could have caused the observed output?

The answer to this question is supplied by the Viterbi (1967) algorithm, and we will apply it in Appendix B to part-of-speech annotation of text.

What we want to do then is to find the sequence $s_1, s_2, \ldots, s_n$ maximizing the probability $P(s_1,s_2,\ldots,s_n,x_1,x_2,\ldots,x_n)$. Let us first find this maximum over all sequences whose $k + 1$th state equals some arbitrary state s. Now,

$$\begin{aligned}
P(s_1,s_2,\ldots,s_k,s_{k+1} &= s,x_1,x_2,\ldots,x_k,x_{k+1}) \\
&= P(s_{k+1} = s,x_{k+1} \mid s_k)P(s_1,s_2,\ldots,s_k,x_1,x_2,\ldots,x_k) \\
&= q(s \mid s_k)r(x_{k+1} \mid s)P(s_1,s_2,\ldots,s_k,x_1,x_2,\ldots,x_k)
\end{aligned}$$

and therefore

$$\max_{s[1,k]} P(s_1,s_2,\ldots,s_k,s_{k+1} = s,x_1,x_2,\ldots,x_k,x_{k+1})$$

$$= \max_{s'} \left[q(s \mid s')r(x_{k+1} \mid s) \max_{s[1,k-1]} P(s_1,s_2,\ldots,s_k = s',x_1,x_2,\ldots,x_k) \right] \tag{A.7}$$

where the max on the left-hand side is over all sequences $s[1,k] = s_1, s_2, \ldots, s_k$, and the second max is over all sequences $s[1,k-1] = s_1, s_2, \ldots, s_{k-1}$. Moreover, if the maximizing state of the right-hand side is $s' = s^*$, then s_k in the maximizing sequence of the left-hand side is given by the same state s^*. It follows by induction that the maximizing sequence $s[1,k-1]$ on the left- and right-hand sides of (A.7) are identical, and therefore that to any state $s_{k+1} = s$ there leads its own unique maximizing sequence $s_1(s_2), s_2(s_3), \ldots, s_k(s)$.

Therefore, we can get the desired maximizing state sequence of $P(s_1, s_2, \ldots, s_n, x_1, x_2, \ldots, x_n)$ in a step-by-step manner:

1. For all states s in $\mathcal{T}$ find the maximizing first state $s_1(s)$:

$$\max_{s_1} P(x_1,x_2,s_1,s_2 = s) = \max_{s'} \left[q(s \mid s')r(x_2 \mid s)q(s' \mid s_0)r(x_1 \mid s') \right]$$

2. For all states s in $\mathcal{T}$ find the maximizing second state $s_2(s)$:

$$\max_{s_2} P(x_1,x_2,x_3,s_1(s_2),s_2,s_3 = s)$$

$$= \max_{s'} \left[q(s \mid s') r(x_3 \mid s) P(x_1, x_2, s_1(s'), s_2 = s') \right]$$

$$\vdots$$

$n-1$. For all states s in $\mathcal{T}$ find the maximizing $n-1$th state $s_{n-1}(s)$:

$$\max_{s_{n-1}} P(x_1, x_2, \ldots, x_n, s_1(s_2), s_2(s_3), \ldots, s_{n-1}, s_n = s)$$
$$= \max_{s'} \left[q(s \mid s') r(x_n \mid s) \right.$$
$$\left. P(x_1, x_2, s_1(s_2), s_2(s_3), \ldots, s_{n-2}(s'), s_{n-1} = s') \right]$$

n. Find the maximizing last state $s_n = s^*$ and therefore the maximizing state sequence $s_1(s_2)$, $s_2(s_3)$, $\ldots$, $s_{n-1}(s_n)$, $s_n = s^*$:

$$\max_{s} P(x_1, x_2, \ldots, x_n, s_1(s_2), s_2(s_3), \ldots, s_{n-1}(s), s_n = s)$$

The above n steps are called the Viterbi (1967) algorithm.

Appendix B. ESTIMATION OF LANGUAGE MODEL PARAMETERS AND THE FORWARD-BACKWARD ALGORITHM

A hidden Markov model is a possible language model if its output alphabet $\mathcal{X}$ is equal to the vocabulary v. All the language models introduced in this chapter are HMMs, as we will now show.

The interpolated language model

$$P(w_3 \mid w_1, w_2) = q_3 f(w_3 \mid w_1, w_2) + q_2 f(w_3 \mid w_2) + q_1 f(w_3) \tag{9}$$

has four states: an initial state s_0 and states t_1, t_2, and t_3. The defining transition probability is

$$q(t_3 \mid s_0) = q_3; \quad q(t_2 \mid s_0) = q_2; \quad q(t_1 \mid s_0) = q_1; \tag{B.1}$$

and the defining output probabilities are

$$r(x \mid t_3) = f(x \mid w_1, w_2); \quad r(x \mid t_2) = f(x \mid w_2); \quad r(x \mid t_1) = f(x); \tag{B.2}$$

In this case, it is presumed that the output probabilities are known (they are computed from relative frequencies of word n-grams corresponding to some training text) and that the transition probabilities need to be determined. Later in this appendix we will show how this can be done.

In Section 4 we introduced the language model component

$$kh(w_3 \mid w_1, w_2) = \sum_{g_3} k(w_3 \mid g_3) h(g_3 \mid w_1, w_2) \tag{29}$$

where $h(g_3 \mid w_1, w_2)$ was defined by

$$h(g_3 \mid w_1, w_2) = r_3 f(g_3 \mid w_1, w_2) + r_2 f(g_3 \mid w_2) + r_1 f(g_3) \tag{30}$$

Formula (30) has exactly the same form as (9), except that the output alphabet $\mathcal{X}$ is not the vocabulary but the POS set. The output probabilities are again presumed known from POS annotation of training text.*

The language model (29) has $K + 1$ states, where K is the size of the POS alphabet. Therefore, $\mathcal{T} = \{s_0, t_1, \ldots, t_K\}$. The output alphabet $\mathcal{X}$ is the vocabulary v. The transition probabilities are

$$q(t_i \mid s_0) = h(t_i \mid w_1, w_2) \tag{B.3}$$

and the output probabilities are

$$r(x \mid t) = k(x \mid t) \tag{B.4}$$

The transition probabilities are presumed given, and the problem is to find the output probabilities.

In Section 8 we introduced the synonym language model

$$Q(w_3 \mid w_1, w_2) = \sum_{s \text{ in } \mathcal{S}_{w_3}} k(w_3 \mid s) P(s \mid l(w_1), l(w_2)) \tag{52}$$

In it, the set $\mathcal{S}_{w_3}$ of synonyms of word w_3 was determined on the basis of the contexts in which w_3 was found embedded in the training text. The problem is to determine the output probabilities $k(w \mid s)$ so as to maximize the probability

$$\prod_{i=1}^{n} Q(w_i \mid w_{i-2}, w_{i-1}) \tag{B.5}$$

of a training text $w_1 w_2 \cdots w_n$, where the probabilities $Q(w_i \mid w_{i-2}, w_{i-1})$ are given by formula (52). The problem then has exactly the same form as that connected with formula (29).

Next, in Section 8 we introduced a transformational language modeling method for transforming a language model determined for a fundamental domain into a language model suitable for a new domain. That method was formulated in terms of HMM terminology. We repeat it for completeness.

The fundamental domain determines a trigram probability distribution $h(l_3 \mid l_1, l_2)$, where l_i denotes words in its vocabulary. The language model

*The annotation is achieved with the help of a POS language model, as described later in this appendix.

of the new domain is an HMM whose states are bigrams, $t = (l_1, l_2)$, whose transition probabilities are

$$q(t' \mid t) = \begin{cases} h(l'_2 \mid l_1, l_2) & \text{if } l'_1 = l_2 \\ 0 & \text{otherwise} \end{cases} \tag{55}$$

and whose output probability distribution is

$$r(w \mid t) = k(w \mid l_2) \quad \text{where } t = (l_1, l_2) \tag{56}$$

where $k(w \mid l_2) = 0$ if l_2 does not belong to the synonym set $\mathcal{S}_w$. Assuming the synonym set $\mathcal{S}_w$ to have been determined, the problem is again to estimate the output probabilities $k(w \mid s)$. Note that the setup of the transformational model is different from that of the synonym language model, since in the latter the initial states $t = (l(w_1), l(w_2))$ of the HMM are fully determined from the text. Here, therefore, the estimation of $k(w \mid l)$ requires the application of the full-fledged forward-backward algorithm (see below).

Finally, let us consider the POS language model used to annotate text in order to create the language model component (29) and (30) discussed in Section 4. The states of this model are POS bigrams, $t = (g_1, g_2)$. The transition probabilities are defined by

$$q(t' \mid t) = \begin{cases} h(g'_2 \mid g_1, g_2) & \text{if } g'_1 = g_2 \\ 0 & \text{otherwise} \end{cases} \tag{B.6}$$

and the output probabilities by

$$r(w \mid t) = k(w \mid g_2) \tag{B.7}$$

This HMM therefore has the same form as that of the transformational model (55) and (56), except that the POS language model requires determination of both the state transition and the output probabilities.

The Viterbi algorithm can be used (see Appendix A) to find the most likely state sequence given an observed sentence $w_1 w_2 \cdots w_n$. The sentence can then be annotated by the unique POS sequence that corresponds to the found state sequence, thus facilitating the creation of the language model component (29) and (30).

Let us now consider the determination of the unknown probability parameters of the various HMMs of this chapter. If it were possible to observe the operation of the HMM, one would annotate the observed sequence $x_1 x_2 \cdots x_n$ by the state sequence $s_1 s_2 \cdots s_n$ that "caused" it, and estimate the required probabilities by relative frequencies, that is, if $C(t,x)$ denotes the number of times state t generated output x, and $C(t,t')$ is the number of times $s_i = t$ and $s_{i+1} = t'$, then we would estimate

$$q(t' \mid t) = \frac{C(t,t')}{C(t)}, \qquad r(x \mid t) = \frac{C(t,x)}{C(t)} \tag{B.8}$$

where

$$C(t) = \sum_{t' \text{ in } \mathcal{T}} C(t,t') \tag{B.9}$$

Now such annotation of the output sequence is not actually possible in any of our language models, except the last one, because (1) the text is not actually produced by any HMM, and (2) only in the POS language model do the hidden states have meaning accessible to experts (i.e., parts of speech of the text words). Before we discuss how to proceed in general, let us remark that hand annotation is truly appropriate in the POS case, where eventual automatic annotation of text is the aim.* It is feasible because the number of POS bigrams is not too large, and the annotation can itself be semiautomatized by use of the Viterbi algorithm.†

Since it is not possible to estimate the HMM probabilities by counting, we may try another idea. Suppose, for instance, given the observed data $X = x_1x_2\cdots x_n$ it were possible to estimate the probabilities $P_i(t \mid X)$ that when the HMM generated x_i it was in state t, as well as the probability $P_i(t,t' \mid X)$ that it was in state t when generating x_i and in state t' when generating x_{i+1}. Note that if $P_i(t,t' \mid X)$ were equal to 1 for the transition $t \to t'$ which actually took place and 0 for the other transitions, then

$$\sum_{i=1}^{n-1} P_i(t,t' \mid X) = C(t,t') \tag{B.10}$$

*The only method of totally automatic estimation is based on the maximum likelihood criterion. The problem is that it results in a model that has minimum perplexity on training data, but that does not necessarily annotate by the intuitively desirable POS labels. In fact, recent results by Merialdo (1991) show that even when the model is initialized by statistics obtained from hand annotation, the automatic annotation error rate of its final version is worse than that obtained from its initial version. On the other hand, in model (29) the POS parameters g_3 serve only to divide words w_3 into equivalence classes which need not correspond exactly to the parts of speech of linguistics. Therefore, maximum likelihood estimation can be used for that task.

†We can hand-annotate some part of text and estimate the probabilities (B.8). These can then be used in a Viterbi mode to annotate more text. In turn, this annotation can be hand-corrected and probabilities (B.8) reestimated on the basis of the total text. One can then Viterbi-annotate more text, etc.

where $C(t,t')$ is the count used in (B.8). Similarly, if $I(w,w')$ denotes the Kronecker delta function, which is equal to 1 when $w = w'$ and to 0 otherwise, then

$$\sum_{i=1}^{n} P_i(t \mid X) I(w,w_i) = C(w,t) \tag{B.11}$$

where $C(x,t)$ is again the count used in (B.8).

Taking into account the relationships (B.10) and (B.11) and the fact that, for instance, if $P_i(t,t' \mid X)$ is correctly estimated then it should represent the fraction of times the system makes the transition $t \to t'$, the following algorithm is reasonable.

The Reestimation Algorithm

1. Guess at probabilities $q(t' \mid t)$ and $r(x \mid t)$.
2. Given the data X and the probabilities $q(t' \mid t)$ and $r(x \mid t)$, compute the probabilities $P_i(t \mid X)$ and $P_i(t,t' \mid X)$.
3. Compute pseudocounts $C(w,t)$ and $C(t,t')$ by formulas (B.10) and (B.11).
4. Use computed pseudocounts to estimate new probabilities $q(t' \mid t)$ and $r(x \mid t)$ by formula (B.8).
5. Repeat from step 2.

With $P_i(t \mid X)$ and $P_i(t,t' \mid X)$ correctly computed, the reestimation algorithm actually does find the correct maximum likelihood HMM $q(t' \mid t)$ and $r(x \mid t)$ parameter values (Jelinek and Mercer, 1980). To compute $P_i(t \mid X)$ and $P_i(t,t' \mid X)$ requires in general the so-called forward-backward algorithm (described at the end of this appendix), but for most of our language models [the exceptions are (55) and (56) and (B.6) and (B.7)], the problem is much easier.

First, let us consider the basic trigram language model (9). We are after the weights q_i which represent transition probabilities, and therefore after $P_i(t,t' \mid W)$. But obviously, from (B.1),

$$\begin{aligned} P_i(s_0,t_3 \mid W) &= \frac{q_3 f(w_i \mid w_{i-2},w_{i-1})}{P(w_i \mid w_{i-2},w_{i-1})} \\ P_i(s_0,t_2 \mid W) &= \frac{q_2 f(w_i \mid w_{i-1})}{P(w_i \mid w_{i-2},w_{i-1})} \\ P_i(s_0,t_1 \mid W) &= \frac{q_1 f(w_i)}{P(w_i \mid w_{i-2},w_{i-1})} \end{aligned} \tag{B.12}$$

where $P(w_i \mid w_{i-2}, w_{i-1})$ is the obvious normalizing constant which for $i = 3$ is given by (9).

It is important to stress here that the training text W used for estimating the weights q_i must be completely distinct from the training text used to estimate the relative frequencies $f(w_3 \mid w_1, w_2), f(w_3 \mid w_2)$, and $f(w_3)$. In fact, if the two texts were identical, the result of the estimation procedure would turn out to be $q_3 = 1, q_2 = q_1 = 0$, because relative frequency is the maximum likelihood estimate of probability. All estimates aimed at smoothing* presuppose complete separation between the training texts underlying the parameters being estimated and those underlying the a priori fixed parameters of the models!

The estimation of the weights r_i for the model (30) follows exactly the pattern (B.12), so we don't need to deal with it.

Next, consider the problem of evaluating the output distributions $k(w \mid g)$ for the model (29). Here the required probabilities are given by

$$P_i(g \mid W) = \frac{k(w_i \mid g)h(g \mid w_{i-2}, w_{i-1})}{kh(w_i \mid w_{i-2}, w_{i-1})} \tag{B.13}$$

for all parts of speech g (i.e., states), where $h(g \mid w_{i-2}, w_{i-1})$ is fixed and $k(w_i \mid g)$ is the current estimate of the output probability which is being reestimated.

The estimation of the output probabilities $k(w \mid s)$ for the synonym model (8.3) follows exactly the pattern (B.13), provided the initial guess of the output distributions satisfies

$$k(w \mid s) = 0 \quad \text{if } s \text{ does not belong to the synonym set } \mathcal{S}_w \tag{B.14}$$

In general, the reestimation algorithm will converge to a local maximum of its likelihood criterion, so the initial choice of the parameters being reestimated should be made with care. However, for the simple cases treated above, L. R. Bahl and R. L. Mercer (personal communication) have proved that only one maximum exists. Therefore, initial values can be arbitrary as long as they are positive for all situations allowed by the problem, and equal to 0 for those forbidden [note (B.14)].

It is worthwhile to consider further the basic trigram model (9). We mentioned in Section 2 that the weights should depend on the number of times

---

*By smoothing we mean interpolation of components one of which is a maximum likelihood estimate of the target estimator. Thus (30) is another example of smoothing, while (52) or (55) and (56) are not.

The use of two training texts, one for deriving the basic estimators and the other for obtaining their optimal combination is called *cross validation* (Stone, 1974) because the second text validates the estimates of the first.

the conditioning bigrams w_1, w_2 have been observed.* A good way of accomplishing this is in two steps. One first estimates weights q_{22} and q_{21}, depending on the count $C(w_2)$, for the bigram probability

$$P(w_3 \mid w_2) = q_{22} f(w_3 \mid w_2) + q_{21} f(w_3) \tag{B.15}$$

and then weights q_{33} and q_{32}, depending on the count $C(w_1, w_2)$, for the trigram probability

$$P(w_3 \mid w_1, w_2) = q_{33} f(w_3 \mid w_1, w_2) + q_{32} P(w_3 \mid w_2) \tag{B.16}$$

Actually, the values q_{22} and q_{21} are estimated separately for different ranges of values of $C(w_2)$, the estimate being simply based on those bigrams w_{i-1}, w_i belonging to the training text W for which the count $C(w_{i-1})$ falls within the required range. The estimation of q_{33} and q_{32} for ranges of $C(w_1, w_2)$ is analogous.

We conclude this appendix by deriving the general formulas for $P_i(t \mid X)$ and $P_i(t, t' \mid X)$ needed in the reestimation algorithm. First of all

$$P_i(t \mid X) = \sum_{t' \text{ in } T} P_i(t, t' \mid X) \tag{B.17}$$

and it follows directly from the definition of HMMs that

$$\begin{aligned} P_i(t, t' \mid X) &= \frac{P(s_i = t, s_{i+1} = t', x_1, \ldots, x_i, x_{i+1}, \ldots, x_n)}{P(X)} \\ &= \frac{P(x_1, \ldots, x_i, s_i = t) q(t' \mid t) P(x_{i+1}, \ldots, x_n \mid s_{i+1} = t')}{P(X)} \end{aligned} \tag{B.18}$$

In Appendix A we have shown that $P(x_1, \ldots, x_i, s_i = t)$ can be computed recursively by the formula

$$\begin{aligned} &P(x_1, \ldots, x_{i-1}, x_i, s_i = t) \\ &\quad = \sum_{t' \text{ in } T} P(x_1, \ldots, x_{i-1}, s_{i-1} = t') r(x_i \mid t) q(t \mid t') \end{aligned} \tag{A.5}$$

starting with the initial condition

$$P(x_1, s_1 = t) = r(x_1 \mid t) q(t \mid s_0) \tag{A.6}$$

where s_0 is the obligatory initial state.

---

*Clearly, when the relative frequency $f(w_3 \mid w_1, w_2)$ is based on more data, it is a better estimate of $P(w_3 \mid w_1, w_2)$ and therefore its contribution (weight) q_3 to the overall estimate of $P(w_3 \mid w_1, w_2)$ should be larger. Similarly, the weight q_2 should depend at least on $C(w_2)$, the number of times the previous word had been seen.

Similarly,

$$P(x_{i+1},\ldots,x_n \mid s_{i+1} = t') = \sum_{t \text{ in } \mathcal{T}} r(x_{i+1} \mid t')q(t \mid t')P(x_{i+2},\ldots,x_n \mid s_{i+2} = t) \quad \text{(B.19)}$$

which itself can be computed recursively starting with the initial condition

$$P(x_n \mid s_n = t) = r(x_n \mid t) \quad \text{(B.20)}$$

Of course,

$$P(X) = P(x_1,\ldots,x_n) = \sum_{t \text{ in } \mathcal{T}} P(x_1,\ldots,x_n,s_n = t) \quad \text{(B.21)}$$

We have now derived the forward-backward algorithm for computing the probabilities $P_i(t,t' \mid X)$ and $P_i(t \mid X)$.

The Forward-Backward Algorithm

1. Starting with the initial condition (A.6), use formula (A.5) to compute the probabilities $P(x_1,\ldots,x_{i-1},x_i,s_i = t)$ for all t in $\mathcal{T}$ and $i = 1, 2, \ldots, n$ (the forward part of the algorithm).
2. Compute $P(X)$ from formula (B.21).
3. Starting with the initial condition (B.20), use formula (B.19) to compute the probabilities $P(x_i,\ldots,x_{n-1},x_n \mid s_i = t)$ for all t in $\mathcal{T}$ and $i = n, n-1, \ldots, 1$ (the backward back of the algorithm).
4. Compute $P_i(t,t' \mid X)$ and $P_i(t \mid X)$ from formulas (B.18) and (B.17) for all $i = 1, 2, \ldots, n$.

If we are to use the reestimation algorithm to find the parameters of the transformational or POS language models, we must decide on their initial values. A good approach to both of these models is to define a probability $k^*(t \mid w)$ that is uniformly distributed over all the synonyms (transformational model) or parts of speech (POS model) allowed to the word w. Then set initially

$$k(w \mid t) = \frac{k^*(t \mid w)f(w)}{k^*(t)} \quad \text{(B.22)}$$

where $f(w)$ is the unigram relative frequency of w and

$$k^*(t) = \sum_{w \text{ in } \mathcal{V}} k^*(t \mid w)f(w) \quad \text{(B.23)}$$

For the transformational model, the value of $h(l_3 \mid l_1, l_2)$ is not estimated; it is fixed. For the POS model, a proven procedure is to first establish a set $\mathcal{H}$ of POS trigrams $g_1g_2g_3$ which are not allowed to take place, and then have the initial $h(g_3 \mid g_1,g_2)$ be uniform over all POS's g_3 such that the trigram $g_1g_2g_3$ is not contained in $\mathcal{H}$.

Appendix C. WORD EQUIVALENCE CLASSIFICATION BASED ON MAXIMUM LIKELIHOOD AND ON MUTUAL INFORMATION

In Section 4 we discussed the desirability of dividing words into classes that might prove a better basis of language modeling than conventional parts of speech. The aim is to provide the component

$$\sum_{g_3} k(w_3 \mid g_3)\, h\,(g_3 \mid g(w_1), g(w_2)) \tag{32}$$

for an augmented trigram language model.

The *nuclear part of speech* approach is based on the assumption that a relatively small number N of identifiable words constitute examples of significantly different grammatical behavior. If one were then to classify all words into one or several of N classes, with the identified words forming class nuclei, the resulting classification might overcome the shortcomings of the natural POS model. The identification of the nuclear set is a problem which we approached heuristically by first augmenting the set of M most frequent words by $200 - M$ words that had important grammatical characteristics that were missing from the original set. Each member of the resulting set $\mathcal{N}$ of "nuclear" words was then assigned its own part of speech, and the remaining words of the vocabulary were initially allowed to belong to all 200 classes. The HMM (B.6) and (B.7) that would best model given training data W was then determined by running the F-B reestimation algorithm to find the transition and output probabilities h and k. The initial choice was

$$h(g_3 \mid g_1,g_2) = \frac{1}{200} \tag{C.1}$$

and

$$k(w \mid g) = \begin{cases} K(g)C(w) & \text{if } w \text{ is in } \mathcal{N} \text{ and } g \text{ is the assigned class of } w \\ 0 & \text{if } w \text{ is in } \mathcal{N} \text{ and } g \text{ is not the assigned class of } w \\ \dfrac{K(g)C(w)}{200} & \text{if } w \text{ is not in } \mathcal{N} \end{cases} \tag{C.2}$$

where $K(g)$ is a normalizing constant assuring that

$$\sum_{w \text{ in } \mathcal{V}} k(w \mid g) = 1$$

The directly obtained model is too complex for practical use since many of the words after training belong to a large number of classes, albeit with small probability. The simplification achieved by restricting class membership of words to the three most probable classes and running several additional iterations of the F-B reestimation algorithm does not degrade the original model noticeably. In fact, when the derived predictor is added as a fourth component to the trigram model (9), the perplexity of the resulting augmented trigram model is 20% lower than before augmentation.

The membership of the self-organized nuclear classes is intuitively satisfying. All numbers form one class, first names another, months yet another (with MAY belonging also to a class populated by auxiliaries), etc. First names share membership with MR., MRS., and MS. Plurals are not mixed with singulars, or intransitive verbs with clearly transitive ones. It is truly remarkable that this spontaneously achieved vocabulary organization is solely due to the initial choice of $\mathcal{N}$ and of the probability assignment (C.1) and (C.2), and to the technique of hidden Markov modeling!

Brown et al. (1991) originated a powerful alternate way of deriving a natural equivalence classification of words. It is based on the mutual information function introduced in Section 3. It will enable us to determine a function $g(w)$ assigning each word w of a vocabulary to one of M classes.

We take the point of view that a good assignment would be one in which the class $g(w_i)$ was a good predictor of the next word w_{i+1}. If we believe the heuristic reasoning of Section 3 that leads to the mutual information function (and it is supported by many mathematical theorems), we should then wish to select the classes so as to maximize the mutual information $I(g(w_i);w_{i+1})$ between the class $g(w_i)$ and the word w_{i+1}. Substituting into the formula

$$I(X;Y) = H(X) - H(X \mid Y)$$

$$= \sum_{y=1}^{L'} \sum_{x=1}^{L} P(y) * P(x \mid y) * \log \frac{P(x \mid y)}{P(x)} \quad (27)$$

that applies to general related events x and y, we get

$$I(g(w_i);w_{i+1})$$

$$= \sum_{w_{i+1}=1}^{L} \sum_{g(w_i)=1}^{M} P(w_{i+1})P(g(w_i) \mid w_{i+1}) \log \frac{P(g(w_i) \mid w_{i+1})}{P(g(w_i))} \quad (C.3)$$

where L is the vocabulary size and M is the number of classes.

The mathematical problem before us is to find that assignment of words into M classes that would maximize the value of (C.3). The way to proceed is to substitute counts of bigrams w_1, w_2 occurring in training text W for the probabilities of (C.3). If n is the total length of the text W, then it has $n-1$

bigrams in it. If C is the count function introduced in Section 2, we may define

$$f(w_3 \mid w_1, w_2) = \frac{C(w_1, w_2, w_3)}{C(w_1, w_2)}$$
$$f(w_3 \mid w_2) = \frac{C(w_2, w_3)}{C(w_2)}, \qquad f(w_3) = \frac{C(w_3)}{n-1} \tag{9}$$

Then the relevant mutual information function may be approximated by

$$I(g(w_1); w_2) = \frac{1}{n-1} \sum_{w_2=1}^{L} \sum_{g_1=1}^{M} C(g_1, w_2) \log \frac{C(g_1 w_2)(n-1)}{C(g_1) C(w_2)} \tag{C.4}$$

where we use the notation ($\mathcal{S}_g$ denotes the set of words w belonging to the class g)

$$C(g, w') = \sum_{w \text{ in } \mathcal{S}_g} C(w, w'), \qquad C(g) = \sum_{w \text{ in } \mathcal{S}_g} C(w) \tag{C.5}$$

Accepting the relevant approximations and noticing that class assignment does not influence the value of n or $C(w_2)$, our task is to find that assignment of words into M classes that would maximize

$$\hat{I} = \sum_{w_2=1}^{L} \sum_{g_1=1}^{M} C(g_1, w_2) \log \frac{C(g_1, w_2)}{C(g_1)} \tag{C.6}$$

One simple but nonoptimal step-by-step way is to assign each word to a different class and arrive at the M class solution by successive mergers of pairs of classes. At the beginning of the process, the relevant function has the value

$$\hat{I}_1 = \sum_{w_2=1}^{L} \sum_{w_1=1}^{L} C(w_1, w_2) \log \frac{C(w_1, w_2)}{C(w_1)} \tag{C.7}$$

If a new class should be formed by merger of the classes containing the single members w_1' and w_1'', then the resulting $\hat{I}_2$ would have the value

$$\hat{I}_1 - C(w_1', w_2) \log \frac{C(w_1', w_2)}{C(w_1')} - C(w_1'', w_2) \log \frac{C(w_1'', w_2)}{C(w_1'')}$$
$$+ \left[C(w_1', w_2) + C(w_1'', w_2)\right] \log \frac{C(w_1', w_2) + C(w_1'', w_2)}{C(w_1') + C(w_1'')} \tag{C.8}$$

Clearly those words w' and w'' should be merged into a single class for which (C.8) would be largest. Once they are found, their merger into one class will reduce the number of classes from L to $L-1$. In general, the $i-1$th merger will result in $L-i+1$ classes and some particular value $\hat{I}_i$ of the function (C.6). At that stage we will be looking for the currently existing classes g_1' and g_1'' for which

$$\hat{I}_i - C(g_1', w_2) \log \frac{C(g_1', w_2)}{C(g_1')} - C(g_1'', w_2) \log \frac{C(g_1'', w_2)}{C(g_1'')} + \left[C(g_1', w_2) + C(g_1'', w_2)\right] \log \frac{C(g_1', w_2) + C(g_1'', w_2)}{C(g_1') + C(g_1'')} \quad \text{(C.9)}$$

is the largest, and we will merge these classes once we have found them.

Although the above procedure is suboptimal, because once a merger is made it cannot be unmade, it does result in a classification leading to a language model (32) whose quality is comparable to the one based on nuclear part of speech classes arrived at by self-organization of the hidden Markov chain (see the beginning of this appendix).

In the foregoing development we did not take directly into account the fact that the desired classification was to be used in the formula (32) which predicts w_{i+1} by first predicting its class $g(w_{i+1})$ and then predicting w_{i+1} from the latter. Under these conditions, the probability $P(g(w_i) \mid w_{i+1})$ used in (C.3) will have the form

$$P(g(w_i) \mid w_{i+1}) = \frac{k(w_{i+1} \mid g(w_{i+1}))P(g(w_i), g(w_{i+1}))}{P(w_{i+1})} \quad \text{(C.10)}$$

which allows us to rewrite (C.3) as

$$I(g(w_i); w_{i+1}) = \sum_{w_{i+1}=1}^{L} \sum_{g(w_i)=1}^{M} k(w_{i+1} \mid g(w_{i+1}))P(g(w_i), g(w_{i+1})) * \log \frac{k(w_{i+1} \mid g(w_{i+1}))P(g(w_i), g(w_{i+1}))}{P(w_{i+1})P(g(w_i))}$$

Approximating probabilities by counts, the argument leading to (C.6) would then lead to the establishment of the criterion function

$$I^* = \sum_{w_2=1}^{L} \sum_{g_1=1}^{M} k(w_2 \mid g(w_2))C(g_1, g(w_2)) \log \frac{k(w_2 \mid g(w_2))C(g_1, g(w_2))}{C(g_1)} \quad \text{(C.11)}$$

But

$$k(w_2 \mid g(w_2)) = \frac{C(w_2)}{C(g(w_2))} \quad \text{(C.12)}$$

so we can rewrite (C.11) as

$$
\begin{aligned}
I^* &= \sum_{g=1}^{M} \sum_{w_2 \text{ in } S_g} \sum_{g_1=1}^{M} \frac{C(w_2)C(g_1,g)}{C(g)} \log \frac{C(w_2)C(g_1,g)}{C(g_1)C(g)} \\
&= \sum_{g=1}^{M} \sum_{w_2 \text{ in } S_g} \sum_{g_1=1}^{M} \frac{C(w_2)C(g_1,g)}{C(g)} \log \frac{C(g_1,g)}{C(g_1)C(g)} \\
&\quad + \sum_{g=1}^{M} \sum_{w_2 \text{ in } S_g} \sum_{g_1=1}^{M} \frac{C(w_2)C(g_1,g)}{C(g)} \log C(w_2) \qquad \text{(C.13)}
\end{aligned}
$$

Since the sum of $C(g_1,g)$ over all g_1 is equal to $C(g)$, then after performing its third sum, the last term of (C.13) will be equal to

$$
\sum_{g=1}^{M} \sum_{w_2 \text{ in} g} C(w_2) \log C(w_2) \qquad \text{(C.14)}
$$

Since (C.14) is independent of class assignment, it can be omitted from the criterion function. Since summing $C(w_2)$ over all words in g results in $C(g)$, then after performing its second sum, the first term on the right-hand side of (C.13) will become

$$
I\$ = \sum_{g_2=1}^{M} \sum_{g_1=1}^{M} C(g_1,g_2) \log \frac{C(g_1,g_2)}{C(g_1)C(g_2)} \qquad \text{(C.15)}
$$

Therefore, to get the final word class assignment, $I\$$ will be the appropriate criterion on which to base the above merger procedure.

The mutual information method of class assignment results in some striking word categories. Here are some examples: {January, February, March, April, June, July, August, September, October, November, December}, {can, cannot, could, might, must, shall, should, will, would} {first, second, third, last, next}, {annual, divisional, initial, international, joint, national, overall, strategic}, {appears, are, did, does, had, has, indicates, is, provides, seems, was, were}, {attention, commitment, decision, meeting, position, presentation, proposal, recommendation, trip, vehicle}.

It may be noticed that the "month" class does not contain the word may, which is found in the set {dated, early, effective, late, may, than}. This is due to the fact that our text processing ignores the distinction between upper- and lowercase letters. Similarly, the word separate belongs to {few, many, monthly, separate, several, some} and is thus treated as an adjective. But we know that it can also be a verb or even a noun. The problem simply is that the mutual information classification procedure limits a word to one class only, even when more classes would be natural. One way out of this problem is to first annotate the training text by parts of speech [using the Viterbi algorithm based on the model (B.6), (B.7) introduced in Appendix C] and then treat word–part-of-speech pairs (w,g) as elements in the classification procedure.

However, it may be that a word carries fundamentally different meanings, and thus predicts a significantly different future, even when it is associated with the same part of speech. For instance, the adjective "light" may be an antonym of "heavy" and of "dark." Or the noun "lead" may mean a metal, a wire, or an actor. On the other hand, just because a word has many senses does not mean that it is worthwhile to split it into several types and thereby both complicate the classification procedure and water down its statistical basis. In any case, the decision whether to split a given word w and its necessary annotation in text may be handled as follows.

Let $w_{1j}w_{2j}ww_{4j}w_{5j}, j = 1, 2, \ldots, J(w)$, be the context in which w appears in the training text. Suppose we suspect that w may have two aspects, s_1 and s_2 (the generalization to k aspects is straightforward). We may then take the view that the words $w_{4j}w_{5j}$ are generated from words $w_{1j}w_{2j}$ by the following mechanism:

$$P(w_{4j},w_{5j} \mid w_{1j},w_{2j}) = \sum_{k=1}^{2} P(s_k \mid w_{1j},w_{2j})P(w_{4j} \mid w_{2j},s_k)P(w_{5j} \mid s_k,w_{4j}) \tag{C.16}$$

and that we wish to find the distributions $P(s_k \mid w_1,w_2)$, $P(w_3 \mid w_1,s_k)$, and $P(w_3 \mid s_k,w_2)$ maximizing the product

$$\prod_{j=1}^{J(w)} P(w_{4j},w_{5j} \mid w_{1j},w_{2j}) \tag{C.17}$$

Since (C.16) is an HMM, the desired distributions can be found by the reestimation method presented in Appendix B. Once the distributions are determined, the word w can be annotated in each context by the value s_k, $k = 1,2$, that maximizes the product $P(s_k \mid w_{1j},w_{2j})P(w_{4j} \mid w_{2j},s_k)P(w_{5j} \mid s_k,w+j)$.

Whether a split is desirable can be decided by use of the mutual information criterion. Namely, the split is worthwhile if it increases sufficiently the value of the mutual information $I(w_4,w_5;w_1,w_2)$. But to compute $I(w_4,w_5;w_1,w_2)$ exactly is difficult, so we may make the judgment on the basis of the approximation

$$I^*(w_4,w_5;w_1,w_2) = \frac{1}{J(w)} \sum_{j=1}^{J(w)} C(w_{1j},w_{2j},w_{3j},w_{4j}) \log \frac{P(w_{4j},w_{5j} \mid w_{1j},w_{2j})}{P(w_{4j},w_{5j})} \quad \text{(C.18)}$$

where

$$P(w_{4j},w_{5j}) = \sum_{j=1}^{J(w)} \frac{P(w_{4j},w_{5j} \mid w_{1j},w_{2j})C(w_{1j},w_{2j})}{J(w)} \quad \text{(C.19)}$$

The decision to annotate w is then taken provided the value of $I^*(w_4, w_5; w_1, w_2)$ is sufficiently larger when the probabilities in (C.18) and (C.19) are evaluated by formula (C.16) than if they are given by

$$P(w_4,w_5 \mid w_1,w_2) = f(w \mid w_1,w_2)f(w_4 \mid w_2,w)f(w_5 \mid w,w_4) \quad \text{(C.20)}$$

REFERENCES

Bahl, L. R., Baker, J. K., Jelinek, F., and Mercer, R. L. (1977). Perplexity—a measure of difficulty of speech recognition tasks, 94th Meeting of the Acoustical Society of America, Miami Beach, December 15.

Bahl, L. R., Brown, P. F., deSouza, P. V., Jelinek, F., and Mercer, R. L. (1986). Using grammatical classes to obtain improved estimates of word probabilities in a speech recognition system. *IBM Tech. Disclosure Bull. 29*(6)

Bahl, L. R., De Gennaro, S. V., Gopalakrishnan, P. S., and Mercer, R. L. (1989). A fast approximate acoustic match for large vocabulary speech recognition. *Eurospeech 89, European Conference on Speech Communication and Technology*, Paris, September.

Baum, L. E. (1972). An inequality and associated maximization technique in statistical estimation of probabilistic functions of Markov processes. *Inequalities 3*(1): 1–8.

Brown, P. (1987). The acoustic modeling problem in automatic speech recognition. Ph.D. thesis, Carnegie-Mellon University, Pittsburgh.

Brown, P. F., Della Pietra, V. J., deSouza, P. V., and Mercer, R. I. (1991). Class-based *n*-gram models of natural language. *Proc. IBM Nat. Lang.* TTI., pp. 283–298.

Carroll, J. B., Davies, P., and Richman, B. (1971). *Word Frequency Book.* American Heritage, New York.

Choueka, Y., (1988). Looking for needles in a haystack or locating interesting collocational expressions in large textual databases. *Proc. RIAO '88*, Session 11, Massachusetts Institute of Technology, pp. 21–24.

Church, K. W., and Gale, W. A. (1991). Enhanced Good-Turing and Cat-Cat: Two new methods for estimating probabilities of English bigrams. *Comput. Speech Language*

DeMori, R., and Kuhn, R. (1991). A cache-based natural language model for speech recognition. *IEEE Pattern Anal. Machine Intell.*, in press.

Ferretti, M., Maltese, G., and Scarci, S. (1989). Language model and acoustic model information in probabilistic speech recognition. *IEEE Conference on Acoustics, Speech and Signal Processing*, Glasgow.

Gallager, R. L. (1968). *Information Theory and Reliable Communication*, Wiley, New York.

Jelinek, F., and Mercer, R. L. (1980). Interpolated estimation of Markov source parameters from sparse data. *Pattern Recognition in Practice*, E. S. Gelsema and L. N. Kanal (Eds.). North-Holland, Amsterdam.

Jelinek, F., and Mercer, R. L. (1985). Probability distribution estimation from sparse data. *IBM Tech. Disclosure Bull. 28*(6):2591–2594.

Jelinek, F., Mercer, R. L., and Roukos, S. (1990). Classifying words for improved statistical language models. *IEEE Conference on Acoustics, Speech and Signal Processing*, Albuquerque.

Jelinek, F., Merialdo, B., Roukos, S., and Strauss, M. (1991). A dynamic language model for speech recognition. *Fourth Darpa Speech and Natural Language*, Asilomar, California.

Katz, S. (1989). Estimation of probabilities from sparse data for the language model component of a speech recognizer. *IEEE Trans. Acoust. Speech Signal Process. ASSP-34*(3):400–401.

Markov, A. A. (1913). An example of statistical investigation in the text of "Eugene Onyegin" illustrating coupling of "tests" into "chains." *Proc. Acad. Sci. St. Petersburg, 6, VI Ser.*, *7*(3):153–162.

Merialdo, B. (1991). Using probabilistic models for tagging English text. IEEE International Conference on Acoustics, Speech and Signal Processing, Toronto, Ontario, Canada.

Shannon, C. E. (1948). A mathematical theory of communication. *Bell Syst. Tech. J. 27*:379, 623.

Sondhi, M. M., and Levinson, S. E. (1978). Computing relative redundancy to measure grammatical constraint in speech recognition tasks. *Proc. Int. Conf. Acoustics, Speech, and Signal Processing*, Tulsa, April, pp. 409–412.

Stone, M. (1974). Cross validatory choice and assessment of statistical predictions. *J. Roy. Statist. Soc. Sec. B*, pp. 111–147.

Viterbi, A. J. (1967). Error bounds for convolutional codes and an asymptotically optimal decoding algorithm. *IEEE Trans. Inform. Theory 13*(4):260–269.

22

Recent Research in Automatic Speaker Recognition

AARON E. ROSENBERG and FRANK K. SOONG AT&T Bell Laboratories, Murray Hill, New Jersey

1. INTRODUCTION

The object of speech communication is to convey messages. It follows, therefore, that the message is the most important information embedded in the speech signal. But it is not the only information. Other kinds of information embedded in the speech signal include the identity of the speaker, the language spoken, the presence and type of speech pathologies, and the physical and emotional state of the speaker. Of these other kinds of information, the identity of the speaker has received most of the attention. This is because speaker characteristics impose striking and comprehensive effects on spoken utterances and because there are many practical applications awaiting practical and effective methods of automatically extracting speaker identity information from speech signals.

The subject of this chapter is automatic speaker recognition, which we define as the automatic extraction of personal identity information from speech signal measurements. We do not consider speaker recognition by human listeners or other techniques that require subjective analyses, such as examination and comparison of speech spectrograms. In the first part of the chapter we discuss general topics and issues concerning automatic speaker recognition. The second part is devoted to discussion of more specific topics of recent interest which have led to interesting new approaches and techniques for automatic speaker recognition. The chapter concludes with a short discussion assessing the current status and possibilities for the future. This chapter is not intended to be a comprehensive review of the subject. Rather, it is intended to point to a few selected areas of relatively recent interest where we feel advances have been made in automatic speaker recognition. The reader is referred to papers by Atal (1976), Rosenberg (1976), Doddington (1985), and O'Shaughnessy (1986), for more general reviews.

2. EXTRACTING SPEAKER INFORMATION FROM THE SPEECH SIGNAL

Broadly speaking, speaker identity is correlated with physiological and behavioral characteristics of the speaker, as illustrated in Fig. 1. For example, variations in the size of vocal tract cavities produce differences in the characteristic resonances of the spectrum of speech signals. Variations in the size of vocal cords are associated with changes in the average pitch or fundamental frequency of voiced speech. Variations in the velum and size of the nasal cavities produce spectral differences in nasalized speech sounds. Atypical anatomical variations, such as the configuration of the teeth or the structure of the palate, produce atypical speech sounds such as lisps and abnormal nasality.

Behavioral correlates of speaker identity in the speech signal are more difficult to specify. "Low-level" behavioral information is associated with individual characteristics in articulating speech sounds, in characteristic transitions from sound to sound, in characteristic pitch contours, rhythm, and timing, and so forth. Generally speaking, the characteristics of speech that have to do with individual speech sounds or phones are referred to as "segmental," while those that pertain to speech phenomena involving consecutive phones are referred to "suprasegmental." Phonetic or articulatory suprasegmental "settings" individuating speakers have been identified (Laver, 1980; Nolan, 1983) which are associated with characteristic "breathy," nasal, and other voice qualities."High-level" behavioral information takes the form of characteristic choice of words and phrases and other aspects of speaking styles.

Explicit measurements of speaker characteristics are often difficult to carry out. For example, such characteristics as nasality and breathiness are difficult to measure because of the difficulty in labelling, segmenting, and measuring nasalized speech events, or in correlating and quantifying

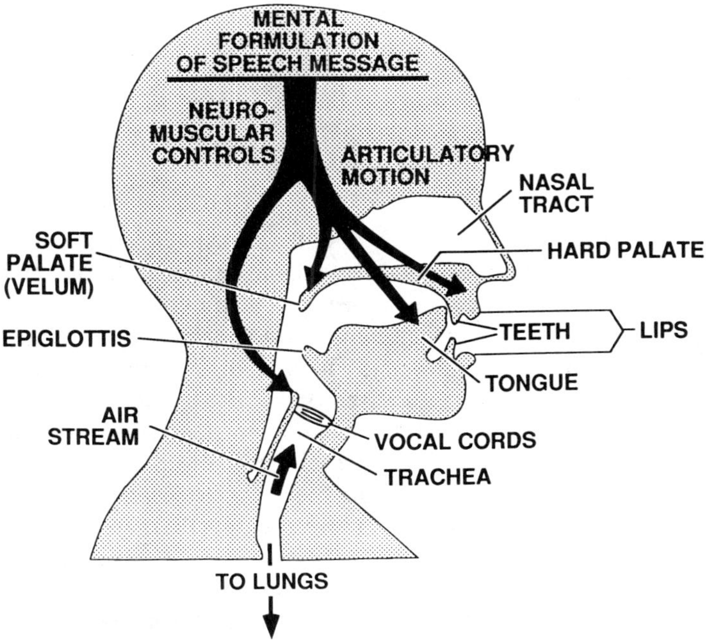

Figure 1 Simplified schematic diagram of the human vocal tract showing how speech sounds are generated. The size and shape of the various articulators differ from speaker to speaker.

breathy speech events with signal features. Voice source measurements have sometimes been mentioned as good candidates for characterizing speakers. However, other than pitch, voice source characteristics are not easy to extract from the speech signal. For example, measuring the glottal source waveform requires careful inverse filtering of the signal implying good knowledge and control of the signal recording and transmission characteristics (Rothenberg, 1973; Strube, 1974; Wong et al., 1979). Even though many voice characteristics are difficult to measure explicitly, and many others are difficult even to specify, we know that many characteristics are captured implicitly in the kinds of signal measurements that can be performed relatively easily. Such signal measurements as short-term and long-term spectral energy, overall energy, and fundamental frequency are relatively easy to obtain. Moreover, these measurements can resolve differences in speaker characteristics that often surpass human discriminability. Such analysis tools provide the means for effectively discriminating among speakers. Thus, voice characteristics, whether physiological or behavioral in origin, which can be measured automatically and easily from the speech signal are the bases for most automatic speaker recognition systems.

3. VARIABILITY

The most significant factor affecting automatic speaker recognition performance is variations in signal characteristics from trial to trial. Variations arise from differences in recording and transmission conditions, voice, and most importantly, variations generated by the speaker. Speaker-generated variations can be voluntary or involuntary. The human vocal apparatus would be useless if control could not be exercised on it. The speech signal variations induced by such control can easily change the pitch and suprasegmental timing of speech utterances from trial to trial. It is also possible for speakers to change their pronunciation, affecting spectral characteristics of individual speech sounds. One simple way to change spectral characteristics is to vary the amount of vocal effort significantly from utterance to utterance. Involuntary variations and voluntary variations are of equal concern. Speakers cannot repeat an utterance precisely the same way from trial to trial. However, it is well known that tokens of the same utterance recorded in one session are more highly correlated than tokens recorded in separate sessions (Rosenberg, 1976). In addition, there are long-term trends in voices associated with both changing speaking behavior and aging (e.g., Furui, 1974). It is important for an automatic speaker recognition system to accommodate natural and expected modifications in speech signal characteristics due to this type of variation. Speakers also produce non-speech sounds, such as breath noises and lip smacks, which must be detected and tolerated. More difficult to deal with are spectral and pitch changes arising from physical or emotional changes or from certain types of speech pathologies. Speech pathologies such as stammering or stuttering, which produce large variations in utterances from trial to trial, are especially difficult to deal with.

4. APPLICATIONS AND DECISION MODES

Practical applications suggested for automatic speaker recognition include criminal investigations, surveillance, and access control. In criminal investigations voice samples of an unknown perpetrator are compared with voice samples of suspects to determine whether or not a match can be obtained. In this kind of application it is usually not possible to exercise control over the text of the test utterance or the recording and transmission conditions. Also, it cannot be expected that speakers will be cooperative in providing voice samples. Automatic speaker recognition techniques are usually not suitable for applications such as these which are associated with large amounts of uncontrolled variability. So-called voiceprint techniques (Kersta, 1962; Bolt et al., 1973; Tosi, 1979), which include expert examination of speech spectrograms as well as expert listening, have been used in this application.

A more suitable application for automatic speaker recognition techniques is access control. In an access control application it can usually be expected that there is some control of or information about the test text and the recording and transmission conditions. Also, in this application it can be expected that the speakers are cooperative; that is, they respond to instructions and try to control variability. Moreover, in such an application, users can be prompted to provide an identity claim. When an identity claim is provided, the speaker recognition system is said to operate in the speaker verification decision mode.

Speaker recognition systems can operate in either an identification or a verification decision mode. In the identification mode a speech sample from an unknown speaker is analyzed and compared with models of known speakers. The unknown speaker is identified as the speaker whose model best matches the input speech sample. In the verification decision mode, an identity claim is made by or asserted for the unknown speaker. The unknown speaker's speech sample is compared with the model for the speaker whose identity is claimed. If the match is good enough, as indicated by passing a threshold test, the identity claim is verified. The fundamental difference between the identification and verification modes is the number of decision alternatives. In the identification mode the number of decision alternatives is equal to the size of the population, whereas in the verification mode there are two decision alternatives, accept or reject the identity claim, regardless of the size of the population. In "open set" identification a reference model for the unknown speaker may not exist. In this case, an additional decision alternative, "the unknown does not match any of the models," is required. In either the verification or identification mode an additional threshold test can be applied to determine whether the match is good enough to accept the decision or defer to a new trial. It can readily be appreciated that errors increase with the size of the population in the identification mode but, provided unknown speakers can be considered to be drawn randomly from the population at large, are independent of the population size in the verification mode (Doddington, 1985).

5. BASIC OPERATION

The basic ingredients of a speaker recognition system are shown in Fig. 2. An input utterance from an unknown speaker is analyzed to extract speaker characteristic features. The measured features are compared with prototype features obtained from known speaker models. In the identification mode, as discussed earlier, the comparison is carried out with every speaker model. In the verification mode, the comparison is carried out only with the model corresponding to the claimed identity (indicated by the dashed line). Using the matching scores provided by the comparison, an identification or verification decision is made.

Speaker recognition systems typically operate in one of two input modes, text dependent or text independent. In the text-dependent mode speakers must provide utterances of the same text for both training and recognition trials. In the text-independent mode speakers are not constrained to provide specific texts in recognition trials, although it may be required that the texts be drawn from specified vocabularies.

The "classical" approach to automatic speaker recognition in the text-dependent mode is the spectral template or "spectrogram" approach. In this approach, each speaker is represented by a sequence of feature vectors, generally short-term spectral feature vectors, analyzed for each word or phrase. This spectral template captures the spectral information displayed in a speech spectrogram of the utterance. Spectrograms are shown in Fig. 3 for three male talkers speaking a sentence-long utterance and for another male speaker repeating the same utterance three times. Since the sequence of phonetic events is the same for all of these utterances, there is an overall similarity among these spectrograms. However, the intraspeaker differences are significantly smaller than the interspeaker differences.

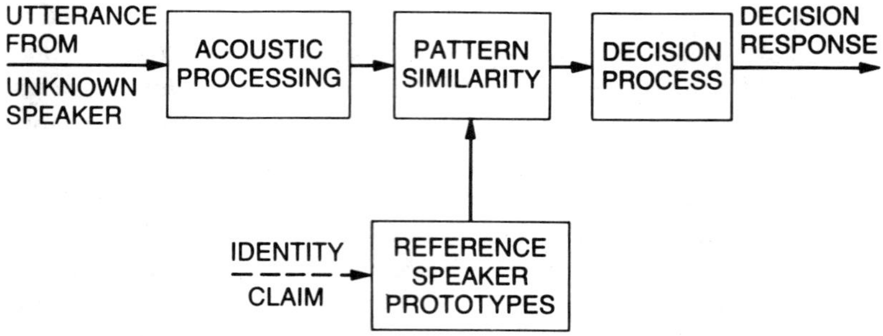

Figure 2 Block diagram of the speaker recognition process.

An important ingredient of a text-dependent speaker recognition system is a means for normalizing trial-to-trial timing variations of utterances of the same phrase by the same talker. Moderate differences in the timing of speech events, both local and overall, can be normalized by aligning the analyzed feature vector sequence of a test utterance to the template feature vector sequence using a dynamic programming time warp (DTW) algorithm (e.g., Sakoe and Chiba, 1978; Furui 1981).

In a text-independent system, the words or phrases used in recognition trials cannot generally be predicted. Therefore, it is not possible to model or match speech events at the level of words or phrases. Classical text-independent speaker recognition techniques are based on measurements for which the time dimension is collapsed. These are usually long-term sample statistics of various spectral features, e.g., the mean and variance of spectral components averaged over a series of utterances (Bricker et al., 1971; Markel et al., 1977; Markel and Davis, 1979). However, long-term spectral averages are extreme condensations of the spectral characteristics of a speaker's utterances and, as such, lack the discriminating power obtained in the sequences of short-term spectral features used as models in text-dependent systems. Moreover, recognition based on long-term spectral averages tends to be less tolerant of recording and transmission variations since, as we have noted, many of these variations are themselves associated with long-term spectral averages. More recently, text-independent speaker verification techniques based on short-duration speech events have been studied. These are reviewed in this chapter.

6. ESTABLISHMENT AND MAINTENANCE OF REFERENCE MODELS AND THRESHOLDS

Crucial to the successful operation of a speaker recognition system is the establishment and maintenance of good reference models. Maintenance is especially important for speaker verification systems used for access control, where it can be expected that user trials will take place periodically over long periods of time in which trial-to-trial variations are likely. Reference models are initially constructed for each user by analysis of training utterances collected in one or more training sessions. To cope with the problems mentioned earlier arising from variations, it is important that reference models reflect the range of speaking behavior and recording and transmission conditions to be expected in recognition trials. This requirement is difficult to fulfill because it is generally not possible to anticipate all possible variations, nor is it practical to collect large amounts of training data over a long period

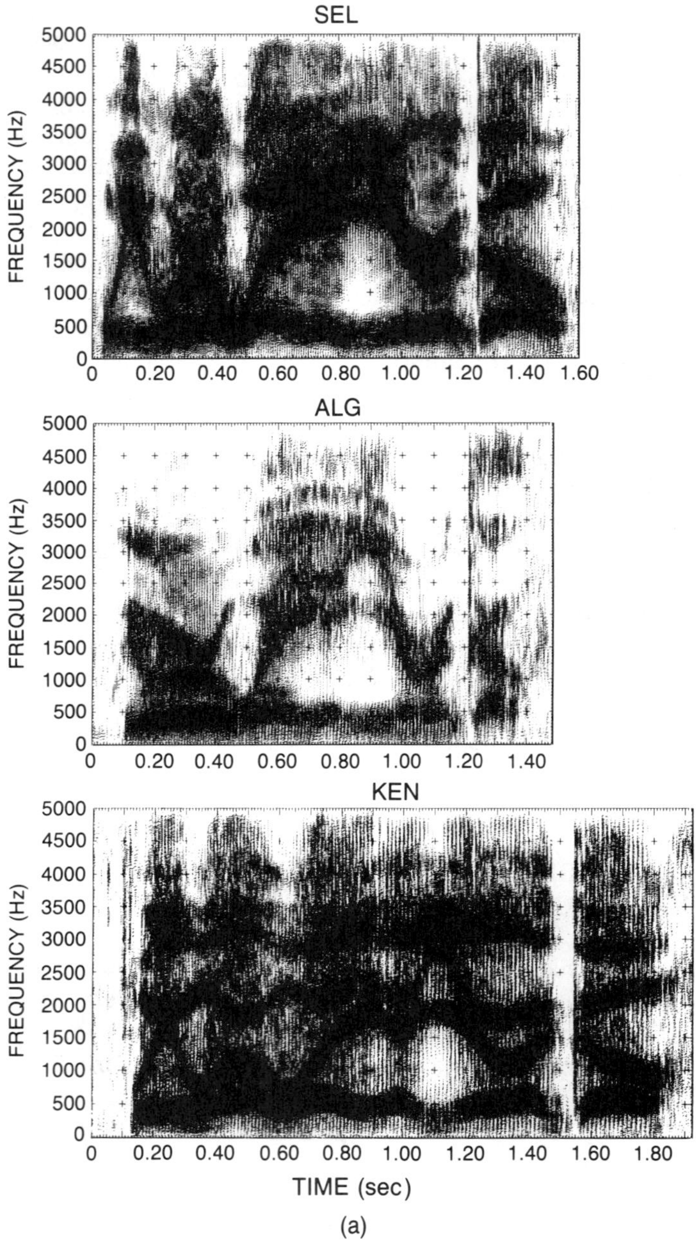

Figure 3 Wideband spectrogram of the utterance "We were away a year ago" (a) by three male speakers and (b) in three repetitions by a different male speaker.

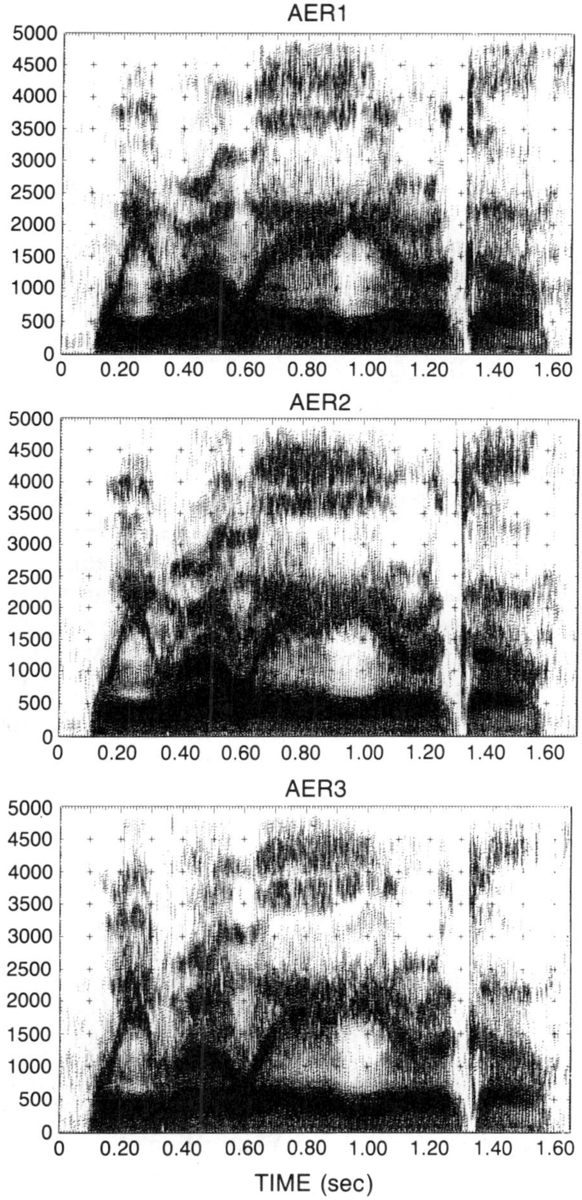
AER1
AER2
AER3
5000
4500
4000
3500
3000
2500
2000
1500
1000
500
0
0.20
0.40
0.60
0.80
1.00
1.20
1.40
1.60
TIME (sec)

(b)

of time. An effective partial solution for these limitations in a practical speaker verification system is, first, to collect and analyze relatively few samples of a new speaker's verification utterances in a controlled enrollment session. Subsequently, analyzed user utterances in successful verification trials are used to update the speaker's reference models. Adaptation of reference models carried out in this way can provide comprehensive and robust models which can accommodate reasonable trial-to-trial variations.

An associated issue is the assignment of thresholds for verification trials. Thresholds must be assigned which tolerate trial-to-trial variations and at the same time, ensure a desired level of performance. A "tight" threshold makes it difficult for impostors to be falsely accepted by the system but at the risk of falsely rejecting legitimate users (customers). Conversely, a "loose" threshold enables customers to be accepted consistently, at the risk of falsely accepting impostors. In order to set a threshold at a desired level of customer rejection and impostor acceptance, it is necessary to know the distribution of customer and impostor scores. In practice, samples of impostor and customer scores of a reasonable size to provide adequate estimates of distributions are not readily available. A satisfactory empirical procedure for setting thresholds is to assign a relatively loose initial threshold and then allow it to adapt by setting it to the average, or some other location statistic, of recent trial scores plus some margin which provides a reasonable rate of customer acceptance. For the first few verification trials the threshold may be so loose as to provide inadequate protection against impostor attempts. To prevent impostor acceptance during initial trials, they may be carried out as part of an extended enrollment.

7. DISCUSSION

In the first part of this chapter we have tried to provide some background and describe some of the issues that are associated with speaker recognition systems. In the following sections we will describe some specific recent experimental studies in speaker recognition. Topics selected for detailed discussions include short-term spectral features and associated distortion measures, speaker models based on extraction of optimal reduced sets of short-term feature vectors or segments, and channel and session-to-session variations in the spectral domain. The discussions emphasize techniques rather than performance. Performance figures are provided only to indicate the effect of varying experimental parameters. No attempt is made to provide comparative performance assessments of different systems or techniques.

8. SHORT-TERM SPECTRAL FEATURES

As noted earlier, there are many speech signal measurements that can be performed relatively easily and efficiently that capture significant voice characteristics. Short-term spectral analysis provides feature vectors that capture segmental speech phenomena over analysis windows of durations from 10 to 45 ms shifted every 5 to 15 ms. Periodicity and overall energy measured as a function of time provide segmental as well as suprasegmental speech information. In addition, measurements of the rate of change of short-term spectral features over window durations three to five times the durations of the basic analysis windows capture suprasegmental phenomena on the time scale of segment-to-segment transitions. Pitch and intensity measurements have been shown to be highly useful for speaker recognition.

In the following, we derive short-term spectral measurements based on linear predictive coding (LPC) analysis. Transitional spectral measurements are also derived in the form of linear regression fits to sequences of measured spectra. The short-time spectrum of a speech signal is modeled in LPC spectral analysis as a pth order autoregressive (AR) model driven by white Gaussian noise. This model characterizes a frame of speech samples spectrally as

$$|S(\omega)|^2 = \frac{\sigma^2}{|A(\omega)|^2} \tag{1}$$

where

$$A(\omega) = 1 + a_1 e^{-j\omega} + \cdots + a_p e^{-jp\omega} \tag{2}$$

is the frequency response of the so-called inverse filter or prediction error filter of LPC analysis and σ^2 is the corresponding LPC gain or the power of the prediction error sequence. Assuming a Gaussian distribution of speech samples, linear prediction parameters are estimated by maximizing likelihood functions (Itakura and Saito, 1968). The LPC spectral estimate, in the frequency domain, has an appealing error matching criterion, the Itakura–Saito distortion measure. The Itakura–Saito distortion measure is optimized (minimized) by matching the parametric, autoregressive model spectrum with the measured, experimental periodogram. A theoretical treatment of this topic can be found in classical references (Itakura and Saito, 1968; Gray and Markel, 1976). It is discussed only briefly here.

The Itakura–Saito distortion measure between a periodogram, or sampled spectral estimate, $|X(\omega)|^2$, and a parametric AR spectral envelope,

$|S(\omega)|^2$, is

$$d_{IS}\left(|\mathbf{X}(\omega)|^2, |\mathbf{S}(\omega)|^2\right) = \frac{1}{2\pi}\int_0^{2\pi}\left[\frac{|X(\omega)|^2}{|S(\omega)|^2} + \log\frac{|S(\omega)|^2}{|X(\omega)|^2} - 1\right] d\omega \quad (3)$$

The Itakura–Saito distortion measure is also useful for comparing two given LPC-derived model spectra. The measure is then

$$d_{IS}\left(\frac{\sigma_1^2}{|A_1|^2}, \frac{\sigma_2^2}{|A_2|^2}\right) = \frac{1}{2\pi}\int_0^{2\pi}\left(\frac{\sigma_1^2|A_2|^2}{\sigma_2^2|A_1|^2} + \log\frac{\sigma_2^2|A_1|^2}{\sigma_1^2|A_2|^2} - 1\right) d\omega \quad (4)$$

Since the spectral shape determined by the LPC prediction coefficients, A_i, is more important than the absolute loudness determined by the gain term, σ_i, the Itakura–Saito distortion needs to be modified when it is used to compare two autoregressive spectra. The relative gain between σ_1 and σ_2 should adjusted in such a way that the difference of absolute loudness has minimal effect on measuring the spectral distortion. Two different ways have been used to adjust the gain term. The first is to adjust the relative gain such that the resultant distortion d_{IS} is minimized. The resultant measure is called the gain-optimized Itakura–Saito distortion measure or the Itakura metric, d_I, given as

$$d_I = \min_{\lambda \geq 0} d_{IS}\left(\frac{\sigma_1^2}{|A_1(\omega)|^2}, \frac{\lambda\sigma_2^2}{|A_2(\omega)|^2}\right) \quad (5)$$

when the relative gain is given by λ. It can be shown that the value of λ for which d_{IS} is minimized is given by

$$\lambda = \frac{1}{2\pi}\int_0^{2\pi}\frac{\sigma_1^2}{\sigma_2^2}\, d\omega \quad (6)$$

The resultant gain-optimized d_{IS} is

$$d_I = \log\frac{1}{2\pi}\int_0^{2\pi}\frac{|A_2(\omega)|^2}{|A_1(\omega)|^2}\, d\omega \quad (7)$$

This distortion measure has a compact and simple computational form in the autocorrelation domain

$$d_I = \log\frac{\mathbf{a}_2^T\mathbf{R}_1\mathbf{a}_2}{\sigma_1^2} \quad (8)$$

commonly referred to as the log likelihood ratio (Itakura, 1975), where $\mathbf{a}_2 = [1\, a_1^2 \cdots a_p^2]$ is the prediction error filter coefficient vector of the second

LPC model, $\mathbf{R}_1$ is the Toeplitz autocorrelation matrix of the first LPC model, and T is the transpose. The second method is to set the gain terms to be equal. The resultant d_{IS} is called the gain-normalized d_{IS} or the likelihood ratio distortion measure, given as

$$d_{LR} = \frac{1}{2\pi} \int_0^{2\pi} \frac{|A_2(\omega)|^2}{|A_1(\omega)|^2} \, d\omega - 1 \tag{9}$$

The measure has a similar form in the autocorrelation domain as

$$d_{LR} = \frac{\mathbf{a}_2^T \mathbf{R}_1 \mathbf{a}_2}{\sigma_1^2} - 1 \tag{10}$$

The measure is different from the Itakura metric only in its linear rather than logarithmic form. Both measures perform about equally well in speech recognition experiments (Nocerino et al., 1985). For low distortion, the two measures are approximately equal. For larger distortion, the log likelihood ratio measure, since it is logarithmic, has a smaller, hence better controlled, dynamic range than the (linear) likelihood ratio measure. On the other hand, the likelihood ratio measure has a slight advantage over the log likelihood ratio measure because it is easier to compute the centroid of a set of vectors using this measure. This property is useful in speaker recognition based on vector quantization (Soong et al., 1987).

The current most commonly used short-term spectral measurement is LPC-derived cepstral coefficients, which are the orthogonal cosine series coefficients of the log spectrum of a given LPC inverse filter. Due to the orthogonal nature of cepstral expansion of log spectrum, a straightforward or weighted Euclidean distance is appropriate for measuring the spectral similarity between two cepstral representations. Experimental studies indicate that the LPC-derived cepstral coefficients, when appropriately weighted (Tohkura, 1987; Juang et al., 1987; Umezaki and Itakura, 1987) or rescaled on a Bark or mel scale (Davis and Mermelstein, 1980), yield better speech recognition performance than the other LPC representations. Other representations, such as log area ratios [Atal 1974] and reflection coefficients, are not as effective as some of the previously mentioned representations. More recently, line spectrum pair (LSP) coefficients (Paliwal, 1988; Liu et al., 1990) have been studied and some interesting results have been obtained.

The transitional information contained in the LPC spectrum can be represented by fitting first- or higher-order regression coefficients with respect to cepstral trajectories over a finite time window (around 50 to 100 ms). The first-order, ith Δ-cepstral coefficient, $\Delta c_i(t)$, is derived as

$$\Delta c_i(t) = \frac{\sum_{k=-K}^{K} h_k k c_i(t+k)}{\sum_{k=-K}^{K} h_k k^2} \tag{11}$$

where h_k is a window (usually symmetric) of length $2K + 1$ and c_i's are the cepstral coefficients. Higher-order Δ-cepstral coefficients can be similarly derived. These so-called Δ-cepstral coefficients are useful for characterizing a speaker (Sagayama and Itakura, 1979; Furui, 1981). When cepstral and Δ-cepstral coefficients are combined, recognition error rates are reduced by half in many speech and speaker recognition experiments (e.g., Rabiner et al., 1989; Soong and Rosenberg, 1988).

Characteristic pitch and overall intensity information can also be used to recognize speakers if it can be extracted reliably (Atal, 1972; Doddington, 1970; Lummis, 1973). Pitch information, unlike the instantaneous short-term spectral envelope, is more resistant to variations in channel or transducer characteristics. On the other hand, pitch information, as a suprasegmental feature, is easier to mimic than the short-term spectral features. Pitch information is more effective for text-dependent speaker recognition applications where pitch contours can be aligned with a prestored contour of the same text.

9. MODELS BASED ON OPTIMAL SETS OF FEATURE VECTORS AND SEGMENTS

We noted earlier that there have been several studies devoted to techniques for speaker recognition based on extracting and measuring specific phonetic events. In a speaker recognition system based on this approach, prototype measurements of speaker-specific phonetic events are made by segmenting and labeling such events over a series of training utterances. The goal of this approach is to select phonetic events which have high discriminative power over a population of speakers. Wolf (1972), for example, carried out an experiment to rate measurements of various types of speech events, segmented manually, for their speaker discriminative capability. In a test trial, target phonetic events are detected in test utterances and the measurements compared with prototype measurements for reference speakers. Such an approach can be used in either text-dependent or text-independent modes but would be especially useful in text-independent recognition systems provided a sufficient number of target events occur in test utterances. The specificity of this approach is potentially powerful and, in a sense, the opposite to the approach taken with long-term statistics which represent generalized measurements carried out over a large number of phonetic events. However, practical implementations based on this approach are lacking because of the difficulty of accurately and consistently detecting target phonetic events. (This difficulty has also inhibited the implementation of automatic speech recognition systems which explicitly segment and label phonetic events.)

Recently, new approaches to extracting and measuring salient acoustic and phonetic events for application to speaker recognition have met with more success. The bases for these approaches lie in statistical techniques for extracting and modeling reduced sets of optimally representative single feature vectors or feature vector sequences or segments. These techniques fall under the related categories of vector quantization, matrix and segment quantization, and hidden Markov modeling (HMM).

9.1. Vector Quantization

A set of short-time training feature vectors of a speaker can be used directly to represent the essential characteristics of that speaker. However, such a direct representation is not practical when the number of training vectors is large. The memory requirements for storage and computational complexity in the recognition phase are prohibitively high. Therefore, an efficient way to compress the training data has to be found. To compress the original data to a small set of representative points, a so-called vector quantization (VQ) codebook consisting of a small number of highly representative feature vectors can be used as an efficient means of characterizing speaker-specific features (Helms, 1981; Li and Wrench, 1983; Soong et al., 1987; Shikano 1985).

To illustrate this point more clearly, we use the conceptual diagram shown in Fig. 4. In the figure a two-dimensional Euclidean space is used to represent short-term spectral features. First, the training utterances from two speakers are analyzed and used to train the VQ codebooks. After the partitioning, two codebooks are generated,each with four entries. The VQ codebook entries, denoted as circles and triangles, are the centroids of the corresponding partitions of each speaker's feature space. However, with the multipartitioning of each individual speaker's feature space, the remainder of the centroids are still well separated. Even with some possible overlap, the two VQ codebook representations are capable of discriminating one speaker from the other.

Following is more formal outline for constructing a VQ codebook. Given a set of I training data vectors $\{\mathbf{x}_1, \mathbf{x}_2, \ldots, \mathbf{x}_I\}$ in a particular feature space $\mathcal{S}$, we want to find a partitioning of the space $\mathcal{S}$ into M nonoverlapping regions, or cells S_1, S_2, ..., S_M. That is $S_i \cap S_j = \varnothing$, $i \neq j$, where $\varnothing$ denotes the empty set, and $\mathcal{S} = S_1 \cup S_2 \cdots \cup S_M$, where generally $I \gg M$. Each partition, S_i, is represented by a corresponding centroid vector, $\mathbf{c}_i$. The partitioning or

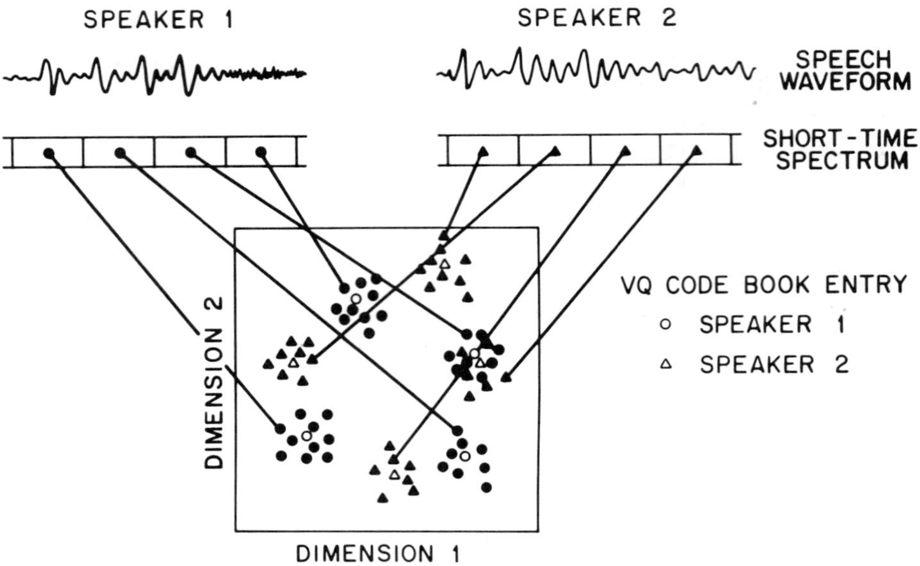

Figure 4 Conceptual diagram illustrating vector quantization codebook formation. One speaker can be discriminated from another based on the location of centroids. (From Soong et al., 1987.)

encoding is found using an iterative optimization procedure, the generalized Lloyd algorithm (Linde et al., 1980), by minimizing the average distortion between training and centroid vectors. The average distortion is

$$D = \frac{1}{I}\sum_{i=1}^{I} \min_{1\le j\le M} d(\mathbf{x}_i, \mathbf{c}_j) \tag{12}$$

where $d(\mathbf{x}_i, \mathbf{c}_j)$ is the distortion between a training vector $\mathbf{x}_i$ and a centroid vector $\mathbf{c}_j$ of the jth partition, S_j.

In the recognition process, the same distortion measure and the nearest neighbor rule are applied to an input vector sequence, $\{\mathbf{y}_1, \mathbf{y}_2, \ldots, \mathbf{y}_N\}$. Each input vector is encoded by its nearest codebook vector and the corresponding codebook distortion is accumulated. The resultant minimum average distortion

$$D = \frac{1}{N}\sum_{n=1}^{N} \min_{1\le j\le M} d(\mathbf{y}_n, \mathbf{c}_j) \tag{13}$$

is then compared with a specific threshold D_{th} to accept or to reject a speaker in verification or to choose a speaker out of a list of speakers in identification.

Variations of the training and recognition procedure exist, such as using a minimax criterion (minimizing maximum distortion) in the so-called covering algorithm (Dorsey and Bernstein, 1981), using a K-nearest neighbor rule instead of the nearest neighbor rule in computing the distortion, or using more robust statistics such as the median instead of the mean (average) in computing the total distortion. The final choice depends on specific applications.

A VQ-based speaker recognition system (Soong et al., 1987; Rosenberg and Soong, 1987) has been tested on a data base of isolated digits. The verification performance of such a system is shown in Fig. 5. Both text-independent verification performance and text-dependent performance are shown in the figure. It is generally expected that text-dependent performance should be consistently superior to text-independent performance for the same input. In text-independent operation, each input short-term spectral vector is compared with a short-term spectral model of a speaker without any memory or sequence constraint. By contrast, in text-dependent operation, the comparison is substantially constrained and therefore more precise. That is, the word (digit) prototype information constrains the sequence and subsets of codebook vectors which are compared.

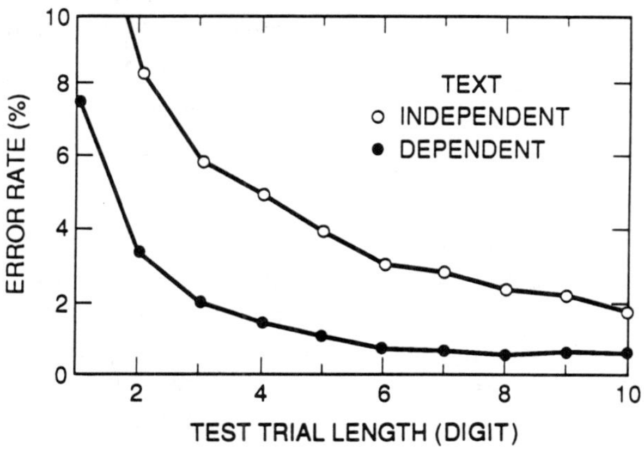

Figure 5 Speaker verification error rate versus test utterance length (digits) for text-independent and text-dependent operation. Codebook size 64, five prototypes per word. (From Rosenberg and Soong, 1987.)

9.2. Matrix Quantization

Memoryless source coding techniques such as the vector quantization technique discussed in the previous section assume that the current observation is independent of other observations in the utterance. This assumption, however, is too simplistic because it ignores the well-known observation that speech signals are highly correlated. Source coding techniques provide designs that take into account the inherent memory (interframe dependence) in the speech signals to enhance the coding efficiency. For classification purposes in recognition applications, source coding techniques which incorporate memory may have some advantages.

The concept of encoding single-frame vectors using a vector quantizer is generalized to encoding multiframe vector K-tuples using a matrix quantizer. The same generalized Lloyd algorithm can be used to train such a matrix quantizer. Given a set of I training matrices (vector K-tuples), $\{\mathbf{X}_1, \mathbf{X}_2, \ldots, \mathbf{X}_I\}$ where $\mathbf{X}_i = [\mathbf{x}_i(1), \ldots, \mathbf{x}_i(K)]$ is a matrix formed by a sequence of K vectors, $\mathbf{x}_i(n)$. The product vector space S^K can be optimally partitioned into M disjoint subspaces $\{S_1^K, \ldots, S_M^K\}$ and each subspace S_i^K is represented by its corresponding centroid C_i.

The average distortion is given by

$$D = \frac{1}{I} \sum_{i=1}^{L} \min_{1 \leq j \leq M} d(\mathbf{X}_i, \mathbf{C}_J) \tag{14}$$

where $d(\mathbf{X}_i, \mathbf{C}_j)$ is the distortion between a given matrix $\mathbf{X}_i$ and a centroid matrix $\mathbf{C}_j$. In general, the distortion between two matrices is defined as a sum of distortions between corresponding vector elements as

$$d(\mathbf{X}_i, \mathbf{C}_j) = \sum_{k=1}^{K} d(\mathbf{x}_i(k), \mathbf{c}_j(k)) \tag{15}$$

In speaker recognition, an input K-tuple matrix is compared with the matrix quantization codebook of a given speaker, the nearest codebook entry is found, and the corresponding distortion is accumulated. As in a VQ-based speaker recognizer, the average distortion is compared against a preset threshold or with the average distortions of other speakers and a verification or identification decision is then made. Experimental results (Sugiyama, 1988; Juang and Soong, 1990) indicate that matrix quantization improves speaker recognition performance over memoryless VQ. It is also observed that the performance improvement stops when the memory length goes beyond a certain range. The results (Sugiyama, 1988) indicate that performance of a matrix quantizer actually degraded when memory length exceeded 5 frames, or 75 ms, in the specific experiment where isolated digits were tested.

9.3. "Filler Template"

The advantage of a matrix quantization codebook over a vector quantization codebook representation is its characterization of the sequential nature of speech events. However, the time sequence in a K-tuple matrix is rather rigid; no "elastic" compression or expansion of the time axis is allowed. As a result, temporal variations such as the speaking rate of a speaker cannot be fully represented by using a matrix quantizer with a fixed block memory. Combining the matrix quantization codebook representation with a dynamic programming time alignment procedure provides a capability for handling the time normalization problem associated with variable speaking rates. Higgins and Wohlford (1986) proposed a segment modeling procedure for constructing a set of representative time-normalized segments, which they called "filler templates." The procedure, a combination of K-means clustering and dynamic programming time alignment, is given as follows.

Filler Template Generation

1. Using the so-called covering algorithm (Dorsey and Bernstein 1981), initial filler templates of a constant length of M frames are generated such that every M-frame training speech segment is within a preset distortion threshold of the nearest template. L such seed templates are generated.
2. A cell-occupancy histogram of the seed filler templates is obtained which provides the number of best matches for each template to training speech segments. Only the N ($N < L$) most frequently used templates are selected to form a codebook.
3. The training utterances are optimally aligned with a sequence of filler templates using a dynamic programming procedure to construct a new set of filler templates.
4. Check convergence. If convergence is not obtained, go to 3.

In recognition, each input utterance is aligned with the best template sequence such that the resultant distortion is minimized.

A speaker recognition system based on filler templates was tested on a data base of conversational speech of 11 male speakers (Higgins and Wohlford, 1986). Speech material recorded in two different sessions, separated by 1 week, was used for enrollment (training) and testing. After removing long pauses, the duration of each conversation was about 100-s. The test speech was divided into 2.5-s segments and a verification decision was made based on single or multiple segments. The results obtained using 8-frame (80-ms),"elastic" filler templates, as expected, are better than the results obtained using a memoryless vector quantization codebook or a matrix quantization codebook where a fixed sequence of vectors is used to encode speech segments. The improved performance of 8-frame filler templates over single-frame vector quantization indicates that the dynamics of speech events in 80-ms intervals is important for speaker recognition. The filler template–based system performs better than a matrix quantization–based system because dynamic time warping is used for adjusting differences in speaking rate in different sessions.

9.4. Acoustic Segment Quantization

A method proposed by Svendsen and Soong (1987) for generating a set of acoustic segments has been applied to speech recognition (Lee et al., 1988) and speaker verification (Rosenberg et al., 1990). The segmentation procedure differs from the filler template generation procedure in its initialization. Unlike the covering algorithm, the seed raw templates are generated using a constrained vector quantization procedure. The initial segmentation is done as follows.

An utterance of T frames, $\{\mathbf{x}_1, \mathbf{x}_2, \ldots, \mathbf{x}_T\}$, of speech data can be segmented into L consecutive, nonoverlapping acoustic segments by finding the best segment boundaries according to a distortion criterion and representing all frames within a segment by their centroid (average) vector. The average distortion

$$D = \frac{1}{T} \sum_{l=1}^{L} \sum_{t=t_l}^{t_{l+1}-1} d(\mathbf{x}_t, \mathbf{c}_l) \tag{16}$$

is minimized by finding the best boundary frames, t_l's, where $\mathbf{c}_l$ is the centroid of the lth segment and $t_1 = 1$ and $t_{L+1} = T + 1$.

The computationally expensive search for the optimal set of segment boundaries can be efficiently implemented by a "segment-building" dynamic programming algorithm (Svendsen and Soong, 1987). In this algorithm, efficient recursive computations of distortions and centroids are performed from left to right (in time) and from low to high (in number of segments). The Lth segment solution is computed recursively based on the $(L - 1)$th segment solutions. Either the average distortion of an utterance or the maximum number of segments can be used as a stopping criterion for the initial segmentation. After a set of initial nonuniform-length raw segments is generated in this way, a segment clustering procedure, such as a modified generalized Lloyd algorithm (Lee et al., 1988), can be used for constructing a segment codebook in which single-frame centroids are used as codebook entries. A segment-to-vector distortion is used as the design criterion. More complex segment-to-segment distortion criteria can be defined for comparing two given segments of different frame lengths for which higher computational complexity is required. An objective criterion is also needed to determine the optimal length of a segment codebook entry.

A segment codebook generated in this way can be used directly in a template-based recognizer or the segments grouped in each cluster can be used to construct a set of multistate hidden Markov models representing acoustic segments for an HMM-based recognizer, as described below.

9.5. HMM

Hidden Markov modeling has been successfully used in various speech recognition applications (e.g., Lee, 1990; Chow et al., 1987; Bahl et al., 1989). HMMs have the same advantages for speaker recognition as for speech recognition. Remarkably robust models of speech events can be obtained with only small amounts of specification or information accompanying training utterances. In the HMM formulation the speech signal is considered as a sequence of Markov states representing transitions from one speech event to another. The Markov states themselves are "hidden" but are indirectly observable from the sequences of spectral feature vectors. The parameters of an HMM are the transition probabilities between the states and the probabilities of observing spectral vectors in each state.

Poritz (1982) proposed using an ergodic HMM (i.e., all possible transitions between states are allowed) for text-independent speaker recognition. An utterance is characterized as a sequence of transitions through a 5-state HMM in the acoustic feature space. The five hidden states, by association with labeled utterances were roughly grouped into five different categories, such as strong vocalization, silence, nasals and liquids, stop bursts, and fricatives. A linear predictive or a Gaussian autoregressive HMM was adopted by Poritz in his study for characterizing the output probability density function. The approach was applied to a preliminary speaker identification experiment. Tishby (1991) expanded Poritz's idea by using an 8-state ergodic autoregressive HMM represented by continuous probability density functions with two to eight mixture components per state, which has a higher spectral resolution than Poritz's model.

Naik et al. (1989) applied HMM representations to speaker verification over long-distance telephone lines. A word-based HMM model was used for a text-dependent speaker recognition experiment. The best performance achieved by the HMM using full covariance matrices was a 4.6% equal error rate, a significant improvement over the performance of a template-based DTW system which achieved an equal error rate of 6.2%.

Rosenberg et al. (1990) have studied the use of phone like units (PLUs) and acoustic segment units (ASUs) in an HMM-based speaker verification system. PLUs are based on phonetic transcriptions of utterances and ASUs are based directly on signal properties of the utterances. Each subword unit is represented by a left-to-right HMM containing two to three states. The observation probability for each state is represented by a continuous probability density function (pdf) in the form of a mixture of Gaussian densities. Thus the probability of observing the feature vector $\mathbf{o}_t$ at time t in state j is given by

$$b_j(\mathbf{o}_t) = \text{prob}(\mathbf{o}_t \mid \text{state } j) = \sum_{m=1}^{M} c_{jm} \mathbf{N}(\mathbf{o}_t; \boldsymbol{\mu}_{jm}, \mathbf{U}_{jm}) \tag{17}$$

where $\mathbf{N}$ represents a multivariate Gaussian pdf with mean vector $\boldsymbol{\mu}_{jm}$ and (diagonal) covariance matrix $\mathbf{U}_{jm}$ for the mth component of the M-component mixture and c_{jm}, $m = 1, 2, \ldots, M$, are the set of mixture weights. The maximum number of mixture components, M, varies from 1 to 3 in these experiments. In addition to the sets of c_{jm}, $\boldsymbol{\mu}_{jm}$, and $\mathbf{U}_{jm}$ for each state, the model parameters include a set of state transition probabilities a_{ij}, representing the probability of a transition from state i to state j. In a left-to-right model with transitions allowed only to adjacent states, a_{ij} is nonzero for j equal to i or $i + 1$. In addition

$$a_{ii} + a_{i,i+1} = 1 \quad \text{and} \quad \sum_{m=1}^{M} c_{jm} = 1 \tag{18}$$

are the usual probability constraints which must be satisfied. The HMM parameters are estimated with a segmental K-means algorithm (Rabiner et al., 1986).

An experimental evaluation was carried out on a data base of isolated digit utterances spoken by 100 speakers. The verification performance obtained was quite good, better than what had been obtained previously with the same data base for both VQ (Soong et al., 1987) and ergodic HMM (Tishby, 1987) representations. Also, differences in performance between the PLU and ASU representations were slight.

10. ACOUSTIC RESOLUTION AND SPEAKER RECOGNITION PERFORMANCE

Accurately modeling the acoustic space of a speaker is crucial to the success of a speaker recognition system. The acoustic space of a speaker is multimodal in general, and cannot be adequately characterized by a unimodal Gaussian probability density function which is represented by a mean and a covariance matrix. A model with higher acoustic resolution generally gives better recognition performance. For example, it was observed (Soong et al., 1987) that by using a larger VQ codebook, the acoustic space of a speaker can be characterized with better accuracy and less quantization error. The speaker identification error rate obtained by such a VQ-based system is plotted against the codebook size (rate) in Fig. 6. The error rate, as shown in the figure, decreases monotonically with the codebook size. To further illustrate the effect, the interspeaker and intraspeaker means, standard deviations, and histograms of two different codebook sizes (8 and 32) are shown in Fig. 7. Both intraspeaker statistics collected from the "true" speaker trials and interspeaker statistics collected from "false" speaker trials are shown in the figure. It is observed from the figure that the mean separation between inter- and intraspeaker distributions stays approximately the same for the two different codebook sizes whereas the standard deviations are appreciably decreased when a larger codebook is used. Also the overlapped region between the intra- and interspeaker distortion distributions is reduced with a larger codebook.

Similar performance improvement was observed by Rose and Reynolds (1990) in a probabilistic framework where a variable number of Gaussian mixture components was used to characterize the output pdf's of a speaker's space. The performance improvement, as shown in Fig. 8, is most appreciable when the mixture components are increased from 2 to 5 and from 5 to 10.

In an HMM subword-based system, the number of mixtures per state, which determines the spectral resolution of the model, and the number of states per phone, which determines the temporal resolution of the model, have been shown (Rosenberg et al., 1990) to be crucial to a speaker recognition system.

11. CHANNEL AND RECORDING SESSION EFFECTS

In the introduction we described sources of trial-to-trial variability in speaker recognition and underscored this as the most significant problem affecting the

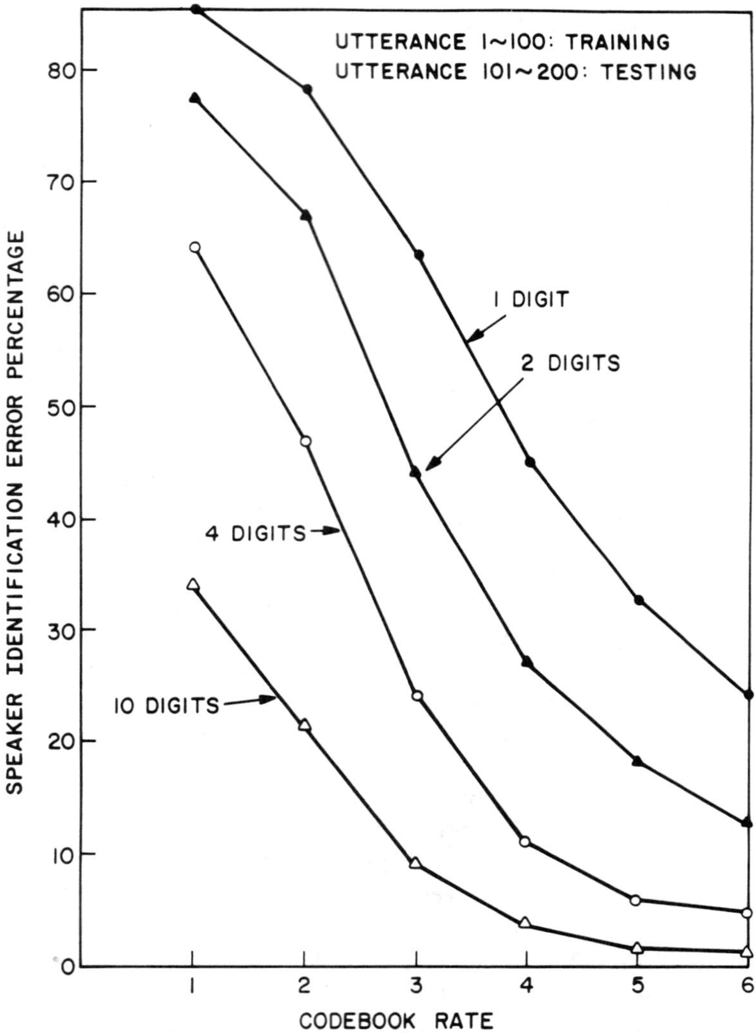

Figure 6 Speaker identification error rate versus codebook rate (bit) for test sequences of 1, 2, 4, and 10 digits. (From Soong et al., 1987.)

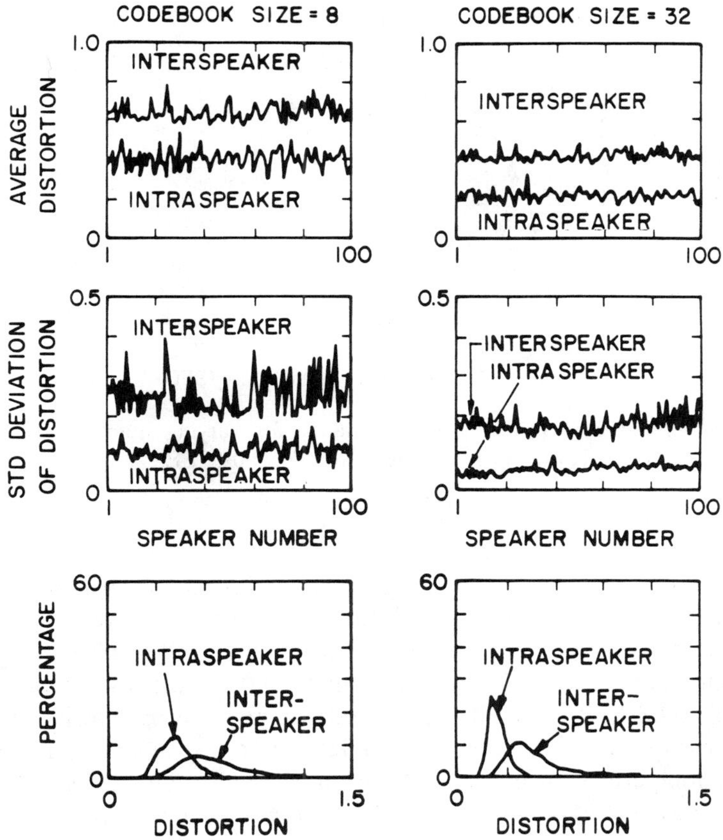

Figure 7 Effects of different codebook sizes on the mean and the standard deviation of the VQ distortions. (From Soong et al., 1987.)

operation of an automatic system. In this section we will present some ideas and techniques for dealing with some types of variabilities.

11.1. Channel Effects

The transmission channel can affect the speech signal in many different ways, Some of the most significant factors are (1) band limiting effects; (2) spectral distortion due to a nonflat frequency response in the passband; and (3) additive or nonadditive channel or environmental noise.

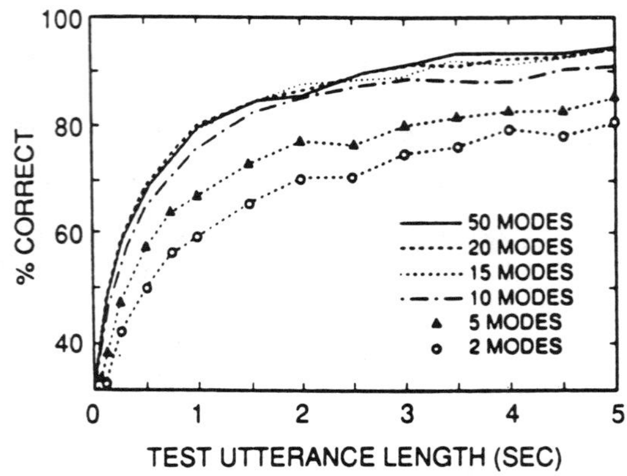

Figure 8 Speaker identification performance for varying numbers of mixture components (From Rose and Reynolds, 1990; © 1990 IEEE.)

A typical telephone channel has a passband from approximately 300 to 3300 Hz. Signal energy outside this range is attenuated. However, this nominal bandwidth can vary from one call to another and a mismatch of the passband characteristics between a training and testing session can therefore result. Gish (1990) has demonstrated that by simply prefiltering different recordings of telephone speech with a fixed filter, text-independent speaker recognition performance over telephone lines can be improved significantly. Signals thus filtered are less mismatched in their channel characteristics.

For text-dependent speaker recognition applications using sufficiently long utterances, linear channel effects (i.e., those that are additive in the log spectral domain) can be fairly well compensated by the "blind equalization" method. In this method the long-term average of the spectrum over the entire recording is subtracted, under the assumption that the channel is linear and time invariant within the whole utterance period. Unavoidably, some text-dependent and speaker-specific features are removed by this blind equalization technique. However, many experiments have shown that the degradation of performance is not appreciable, at least for sentence-long utterances (Furui, 1981). Long-term average subtraction is inappropriate for short utterances in speaker recognition applications. The long-term average of short utterances cannot resolve speaker-related utterance information and channel information. Removing that information can be disastrous to speaker recognition performance.

In addition to the blind equalization method, channel-invariant acoustic features can be useful in dealing with variable channels. For example, Soong and Rosenberg (1988) conducted an experiment in which a mild pre-emphasis was used to create a spectral difference between test and training material. They showed that time derivatives of cepstral features are resistant to linear channel mismatch between training and testing. However, speaker recognition performance was degraded significantly when only instantaneous cepstral features were used.

Other transitional features such as second- or higher-order time derivatives should also be resistant to a similar linear, time-invariant channel mismatch. Channel-invariant short-term spectral features, even though they tend to be more resistant to channel variations, bear less speaker-specific information than their channel-noninvariant counterparts. We have indicated that such features obtained by subtracting the long-term average (Furui, 1981) or derived from time derivatives like the Δ-cepstrum (Soong and Rosenberg, 1988) inevitably degrade speaker-specific information while compensating for channel variations. Experiments have shown that when there is no channel mismatch between training and testing data such features actually yield lower recognition performance. These adverse effects may be alleviated by equalizing channel-sensitive features such as instantaneous short-term spectra and then combining them with channel-invariant features such as pitch and transitional short-term spectra.

A more effective way to construct a channel-resistant speaker recognition system is to model the channel statistically. This can be achieved if enough training samples of channels to be modeled are available. Gish et al. (1985, 1986) using multivariate Gaussian probability density functions to model channels. The observed data from a speech utterance are assumed to be a sequence of N Gaussian distributed p-dimensional, statistically independent spectral vectors $\mathbf{x}_n$, $n = 1, \ldots, N$.

The log likelihood that the data are from a speaker j is given as

$$\log p_j = -\frac{N-1}{2}\log|\Sigma_j| - \frac{N}{2}\,\mathrm{tr}(\Sigma_j^{-1}\mathbf{S}) - \frac{1}{2}\log\left|\frac{\Sigma_j}{N}\right| - \frac{1}{2}(\bar{\mathbf{x}}-\mu_j)^T\left(\frac{\Sigma_j}{N}\right)^{-1}(\bar{\mathbf{x}}-\mu_j) + K \qquad (19)$$

where S is the sample covariance

$$\mathbf{S} = \frac{1}{N}\sum_{n=1}^{N}(\mathbf{x}_n - \bar{\mathbf{x}})(\mathbf{x}_n - \bar{\mathbf{x}})^T \qquad (20)$$

$\bar{\mathbf{x}}$ is the simple mean

$$\bar{\mathbf{x}} = \frac{1}{N} \sum_{n=1}^{N} \mathbf{x}_n \tag{21}$$

μ_j and Σ_j are the mean and covariance matrix of speaker j, K is a constant, and tr indicates the trace of the matrix.

A modified likelihood function can be constructed as

$$\log p_i' = (1 - \alpha)Cov + \alpha Mean \tag{22}$$

which is a linear combination of two components of the original function,

$$Cov = -\frac{n-1}{2} \log|\Sigma_j| - \frac{N}{2} \operatorname{tr}(\Sigma_j^{-1}\mathbf{S}) \tag{23}$$

$$Mean = -\frac{1}{2} \log\left|\frac{\Sigma_j}{N}\right| - \frac{1}{2}(\bar{\mathbf{x}} - \mu_j)\left(\frac{\Sigma_j}{N}\right)^{-1}(\bar{\mathbf{x}} - \mu_j) \tag{24}$$

The first term measures how well the sample covariance $\mathbf{S}$ matches Σ_j, while the second measures how well the sample mean $\bar{\mathbf{x}}$ matches μ_j. The combination factor α is a positive number used for adjusting the contributions where

$$0 \leq \alpha \leq 1 \tag{25}$$

The optimal α is usually estimated by using a subset of training data.

The channel of a recording can be modeled as a linear, time-invariant filter. The cepstral vector of an observation, $\mathbf{y}_n$, is represented as

$$\mathbf{y}_n = \mathbf{x}_n + \mathbf{c} \tag{26}$$

where $\mathbf{x}_n$ is the cepstral vector produced by a speaker and $\mathbf{c}$ is the cepstral component associated with the channel. It is assumed that the component $\mathbf{c}$ is fixed for each trial but varies from trial to trial. Accordingly, $\mathbf{c}$ is modeled as a Gaussian random vector with mean μ_j and covariance Σ_c. The observed speech data are then modeled as a combination of the speaker and channel model. The mean term μ_j in Eqs. (23) and (24) is replaced with the sum of μ_j and μ_c and the combined covariance is the sum of Σ_j/N and Σ_c. Estimates of μ_j, μ_i, Σ_j, and Σ_i are described in the papers (Gish et al., 1985, 1986). A series of experiments were carried out to evaluate the performance of the combined speaker and channel modeling procedure in a speaker identification mode. A 20-speaker data base of conversational speech recorded over long-distance telephone lines was used. Either 20 or 60 s of speech recorded over these different channels was used to estimate model parameters. Results are shown in Fig. 9. It can be seen that better performance is obtained with the modified Gaussian formulation which includes channel modeling.

More recently, experimental results have shown that performance can be improved by removing outliers in low-dimensional subsets of the data. The speaker recognition discriminator is based on the set of eigenvalues of $\mathbf{\Sigma}_j^{-1}\mathbf{S}$ where $\mathbf{\Sigma}_j$ is the covariance of the matrix of the training data (N cepstral vectors) of speaker j and $\mathbf{S}$ is the sample covariance matrix of the test data. If all the eigenvalues are equal to 1, there is a perfect match between the training and test data. Gish (1990) used

$$d_{\mathrm{ROB}}(k_1,k_2) = \sum_{i=k_1}^{k_2} |\lambda_i - 1| \tag{27}$$

as a robust distance metric between the test and training data, where the λ_i are the ordered eigenvalues and k_1 and k_2 are specified to exclude eigenvalues that are too small or too large. A significant improvement of performance was obtained by using this new metric on a long-distance telephone data base.

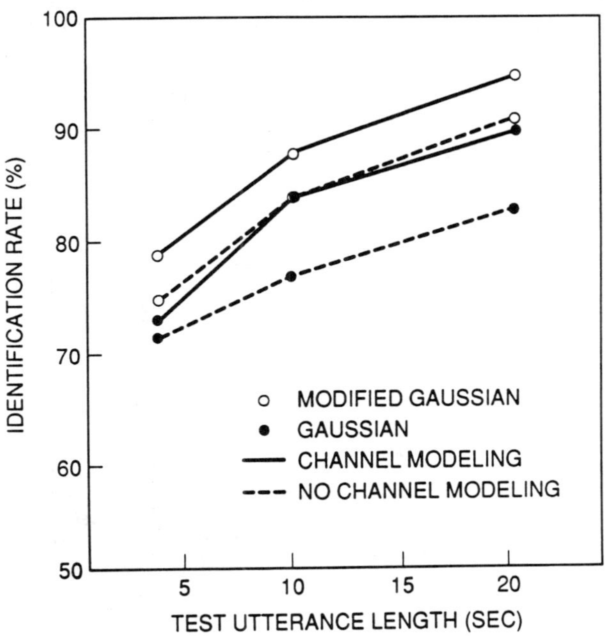

Figure 9 Speaker identification performance with/without modified Gaussian channel modeling (From Gish et al., 1986; © 1986 IEEE.)

Speaker identification accuracy improved from a very low 22% to 59% when the new robust metric was used but no eigenvalues were dropped. With appropriate selection of eigenvalues, the performance improved to 70%. A plot of the eigenvalue distributions can be seen in Fig. 10, where poorly performing models have low and high eigenvalues deviating significantly from unity. The modified metric is more robust for two reasons: first, it sets a bound on the contributions of those near-zero eigenvalues; second, extremal eigenvalues are deleted when the values of k_1 and k_2 are set appropriately. As with other robust procedure in statistics, there is a price attached to the modified metric. It was shown in the same study that performance degrades when there are no or few outliers present in the data or when there is no or little channel mismatch. An improved design of the metric may provide better modeling accuracy as well as account for the presence of outliers.

The effects of different recording sessions on speaker recognition performance have been noted by many researchers (Furui, 1974). For example, it is reported (Soong et al., 1985) that in-session recognition performance is very different from cross-session performance. In their experiments five different sessions (40 isolated digits/session) were recorded over a period of 2 months. The first 100 digits were used to train a VQ-based speaker recognizer and the remaining 100 digits were tested against trained models. As depicted in Fig. 11, the test data in session 3 performed significantly better than the test data in sessions 4 and 5 because the first half of session 3 was used in the training, hence the speaker and channel characteristics in session 3 were faithfully represented in the VQ codebook. The remaining 20 digits in

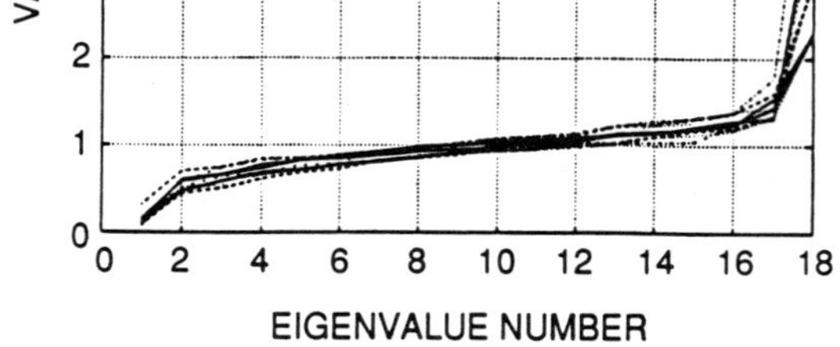

Figure 10 Relative eigenvalues for poorly performing, nonrobust models. (From Gish, 1990; © 1990, IEEE.)

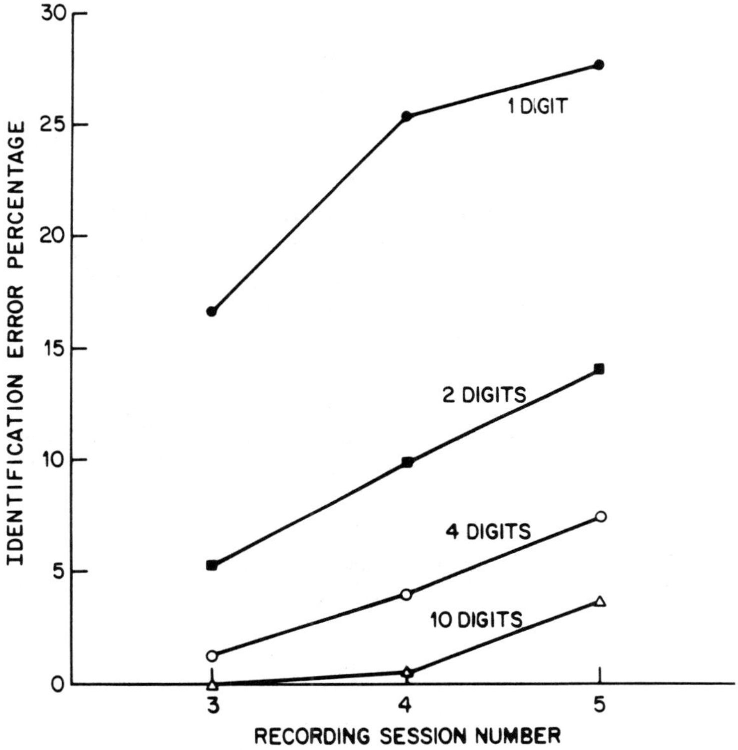

Figure 11 Speaker identification error rate (%) versus recording session number. (From Soong et al., 1985.)

the third recording session when used for testing matched quite well with the trained VQ codebook, and the performance is therefore high. On the other hand, since the data in the fourth and fifth sessions were not characterized as well as the data in the third session, the speaker recognition performance was inferior. The fifth session, which is more different from the training data due to the time span in between, suffered the most degradation in recognition performance.

12. CONCLUSION

Ideas in automatic speaker recognition have largely paralleled ideas in automatic speech recognition. In both speech and speaker recognition many studies have been carried out to find ways to measure specific events in spoken utterances. In the case of speech recognition such measurements are made to explicitly segment and label phonetic events. For speaker recognition, such measurements are made to extract and characterize specific events thought to have good speaker discriminating properties. For the most part, however, these studies have not resulted in practical recognition systems. The problem, as we noted previously, is spectral and temporal variations which make it difficult to reliably segment and label specific speech events across different utterances and speakers. However, as we also noted previously, many advances in automatic speech and speaker recognition have come about by using measurement techniques which capture speech or speaker characteristics reliably and consistently in an implicit rather than explicit manner. Examples drawn from topics discussed in this chapter include the following: DTW techniques in template-based recognizers reliably align corresponding events in speech utterances without explicitly labeling the events; short-term spectral analysis techniques are capable of making fine discriminations between speech events but are less reliable in associating the events with specific phonetic or speaker characteristic labels; statistical modeling approaches make no definite assumptions about the actual relation between models and phonetic events but are capable of making fine discriminations among different speech events and speakers.

Successful speaker recognition systems have been obtained by devising feature measurements and modeling techniques which can efficiently and reliably discriminate speakers without measuring explicit speaker characteristics or necessarily understanding the exact relationship between the measurements or models and known speaker characteristics. As an example, we have described the use of cepstral and Δ-cepstral measurements for speaker recognition. One practical reason for using a set of low-order cepstral coefficients to characterize short-term spectra is that the spectral envelope reconstructed from this truncated set of cepstral coefficients is smooth, much smoother in general than an envelope reconstructed from, say, LPC coefficients. Thus, although the spectral envelope does not focus on specific speaker characteristics, a smooth spectral envelope provides a stabler representation from one repetition to another of a particular speaker's utterances. As for Δ-cepstral coefficients, they were intended to provide information characterizing dynamic changes within a speech sound and transitions from one speech sound to another. However, the information they provide is general and does not focus on particular speech sounds.

Speaker models based on optimal reduced sets of features and segments, described in Section 9, are another example of an advance in speaker recognition capability, paralleling similar advances in speech recognition, which does not extract explicit speaker characteristics from utterances but rather more efficiently and reliably captures general speaker discriminating characteristics. The simplest example of this approach is the use of VQ codebooks to represent speakers. The VQ codebook is simply a distillation of all the training feature vectors available for a speaker. It is intended to be an optimal representation of the training vectors but the codebook vectors do not have an obvious association with explicit speaker characteristics. The matrix quantization and filler template ideas represent generalizations of this technique which account for temporal relations in vector sequences. The HMM representation can also be considered a generalization of this technique in which speech events, e.g., words, subword phonelike units, or acoustic segment units, are represented not just by characteristic sets or sequences of feature vectors but also by probabilistic models of these events calculated from feature vector "observations" in the training utterances. Here again there are no explicit representations of speaker characteristics.

Although there have been many recent advances and successes in automatic speaker recognition, there are still many problems for which good solutions remain to be found. Most of these problems arise from variability, a recurrent topic in this chapter. As we have seen, this includes speaker-generated variability and variability in channel and recording conditions. The speaker-generated variability problem can be summed up as how to model or account for speaking behavior that is not observed in training utterances. We have seen that adaptation by updating models with current test utterances represents a partial solution to this problem. However, adaptation cannot anticipate sudden or singular changes in behavior. Similarly, although many advances have been made in measuring and compensating for recording and channel variations and disturbances, such as the work described in Section 11, the problem remains a difficult one. The reasons are essentially the same as noted for speaker-generated variations. That is, it is difficult to anticipate and thereby compensate for conditions that have not been observed in training.

There is an essential difference between speech recognition and speaker recognition with regard to the variability problem. Speech recognition has advanced significantly recently with the development of robust and accurate speaker-*independent* recognizers. This has come about by using large amounts of training data from large populations of speakers. The use of HMMs has been especially powerful in characterizing these variations provided sufficient training data are available. It is even possible to model utterances recorded under varying recording and channel conditions in this way. For this application, the training data are collected under supervised conditions from a designated set of reference speakers recruited specifically to provide the amount and type of training data that are required. This kind of training cannot be duplicated in a practical speaker recognition system. In the case of speaker recognition, individual speaker models must be created and the user population is intended to be the same as the reference population. It would not be practical to require each user to exercise the system to provide sufficient training to account for large amounts of variation.

In summary, there have been significant recent advances in automatic speaker recognition techniques which parallel advances in speech recognition. These advances have not necessarily come about as an outgrowth of new or better understanding of speaker characteristics or how to extract them from the speech signal, but rather through improvements in techniques for making speaker-sensitive feature measurements and models. It can be expected that improvements will continue with this approach, but we hope that current and future understanding of speaker characteristics in the speech signal can be applied to provide even more effective speaker recognition systems. Finally, variability is a persistent problem and likely to be more serious for speaker recognition than for speech recognition.

REFERENCES

Atal, B. S. (1972). Automatic speaker recognition based on pitch contours. *J. Acoust. Soc. Am. 52*:1687–1697.

Atal, B. S. (1974). Effectiveness of linear prediction characteristics of the speech wave for automatic speaker identification and verification. *J. Acoust. Soc. Am. 55*:1304–1312.

Atal, B. (1976). Automatic recognition of speakers from their voices. *Proc. IEEE 64*: 460–475.

Bahl, L., Bakis, R., Bellegarda, J., Brown, P. F., Burshtein, D., Das, S. K., de Souza, P. V., Gopalakrishnan, P. S., Jelinek, F., Kanevsky, E., Mercer, R. L., Nadas, A. J., Nahamoo, E., and Picheny, M. A. (1989). Large vocabulary natural language continuous speech recognition. *Proc. 1989 IEEE Int. Conf. Acoustics Speech, and Signal Processing* (ICASSP89), vol. 1, pp. 465–467.

Bolt, R. H., Cooper, F. S., David, E. E., Denes, P. B., Pickett, J. M., and Steven, K. N. (1973). Speaker identification by speech spectrograms: Some further observation. *J. Acoust. Soc. Am. 54*:531–534.

Bricker, P. D., Gnanadesikan, R., Mathews, M. V., Pruzansky, S., Tukey, P. A., Wachter, K. W., and Warner, J. L. (1971). Statistical techniques for talker identification. *Bell Syst. Tech. J. 50*:1427–1454.

Chow, Y-L., Dunham, M. O., Kimball, O. A., Krasner, M. A., Kubala, G. F., Makhoul, J., Price, P. J., and Roucos, S. (1987). BYBLOS: The BBN continuous speech recognition system. *Proc. 1987 IEEE Int. Conf. Acoustics Speech, and Signal Processing* (ICASSP87), vol. 1, pp. 89–92.

Davis, S. B., and Mermelstein, P. (1980). Comparison of parametric representations for monosyllable word recognition in continuously spoken sentences. *IEEE Trans. Acoust. Speech Signal Process. ASSP-28*:357–366.

Doddington, G. R. (1979). A computer method of speaker verification. Ph.D. dissertation, University of Wisconsin.

Doddington, G. R. (1985). Speaker recognition—identifying people by their voices. *Proc. IEEE 73*:1651–1664.

Dorsey, E., and Bernstein, J. (1981). Inter-speaker comparison of LPC acoustic space using a minimax distortion criterion. *Proc. 1981 IEEE Int. Conf. Acoustics Speech, and Signal Processing* (ICASSP81), vol. 1, pp. 16–119.

Flanagan, J. L. (1972). *Speech Analysis Synthesis and Perception*. Springer-Verlag, Berlin.

Furui, S. (1974). An analysis of long-term variation of feature parameters of speech and its application to talker recognition. *Electron. Commun. Jpn. 57a*:34–42.

Furui, S. (1981). Cepstrum analysis technique for automatic speaker verification. *IEEE Trans. Acoust. Speech Signal Process. ASSP-29*:254–272.

Gish, H. (1990). Robust discrimination in automatic speaker identification. *Proc. 1990 IEEE Int. Conf. Acoustics Speech, and Signal Processing* (ICASSP90), pp. 289–292.

Gish, H., Karnofsky, K., Krasner, K., Roucos, S., Schwartz, R., and Wolf, J. (1985). Investigation of text-independent speaker identification over telephone channels. *Proc. 1985 IEEE Int. Conf. Acoustics Speech, and Signal Processing* (ICASSP85), pp. 379–382.

Gish, H., Krasner, M., Russel, W., and Wolf, J. (1986). Methods and experiments for text-independent speaker recognition over telephone channels. *Proc. 1986 IEEE Int. Conf. Acoustics Speech, and Signal Processing* (ICASSP86), vol. 1, pp. 865–868.

Gray, A. H., and Markel, J. D. (1976). Distance measure for speech recognition. *IEEE Trans. Acoust. Speech Signal Process. ASSP-24*:380–390.

Helms, E. H. (1981). Speaker recognition using linear prediction vector codebooks. Ph.D. thesis, Southern Methodist University.

Higgins, A. L., and Wohlford, R. E. (1986). A new method of text-independent speaker recognition. *Proc. 1986 IEEE Int. Conf. Acoustics Speech, and Signal Processing* (ICASSP86), vol. 2, pp. 869–872.

Itakura, F. (1975). Minimum prediction residual principle applied to speech recognition. *IEEE Trans. Acoust. Speech Signal Process. ASSP-23*:67–72.

Itakura, R., and Saito, S. (1968). An analysis-synthesis telephony based on maximum likelihood method. *Proc. Int. Congr. Acoust.*, C-5-5, Tokyo, August.

Juang, B.-H., and Soong, F. K. (1990). Speaker recognition based on source coding approaches. *Proc. 1990 IEEE Int. Conf. Acoustics Speech, and Signal Processing* (ICASSP90), vol. 1, pp. 613–616.

Juang, B.-H., Rabiner, L. R., and Wilpon, J. G. (1987). On the use of bandpass liftering in speech recognition. *IEEE Trans. Acoust. Speech Signal Process. ASSP-35*: 947–954.

Kersta, L. G. (1962). Voiceprint identification. Nature *196*:1253–1257.

Laver, J. (1980). *The Phonetic Description of Voice Quality*. Cambridge University Press, London.

Lee, C-H., Soong, F. K., and Juang, B.-H. (1988). A segment model based approach to speech recognition. *Proc. 1988 IEEE Int. Conf. Acoustics Speech, and Signal Processing* (ICASSP88), pp. 501–504.

Lee, K-F. (1989). *Automatic Speech Recognition—The Development of the SPHINX System*. Kluwer Academic Publishers, Boston.

Li, K-P., and Wrench, E. H., Jr. (1983). An approach to text-independent speaker recognition with short utterances. *Proc. 1983 IEEE Int. Conf. Acoustics Speech, and Signal Processing* (ICASSP83), vol. 2, pp. 555–558.

Linde, Y., Buzo, A., and Gray, R. M. (1980). An algorithm for vector quantization. *IEEE Trans. Commun. COM-28*:84–95.

Liu, C-S., Lin, M-T., Wang, W-J., and Wang, H-C. (1990). Study of line spectrum pair frequencies for speaker recognition. *Proc. 1990 IEEE Int. Conf. Acoustics Speech, and Signal Processing* (ICASSP90), pp. 277–280.

Lummis, R. C. (1973). Speaker verification by computer using speech intensity for temporal registration. *IEEE Trans. Audio Electroacoust. 21*:80–89.

Markel, J. D., and Davis, S. B. (1979). Text-independent speaker recognition from a large linguistically unconstrained time-spaced data bases. *IEEE Trans. Acoust. Speech Signal Process. ASSP-27*:74–82.

Markel, J. D., Oshika, B., and Gray, A. H. (1977). Long-term feature averaging for speaker recognition. *IEEE Trans. Acoust. Speech Signal Process. ASSP-25*:330–337.

Naik, J. M., Netsch, L. P., and Doddington, G. R. (1989). Speaker verification over long distance telephone lines. *Proc. 1989 IEEE Int. Conf. Acoustics Speech, and Signal Processing* (ICASSP89), vol. 1, pp. 524–527.

Nocerino, N., Soong, F. K., Rabiner, L. R., and Klatt, D. H. (1985). Comparative study of several distortion measures for speech recognition. *Speech Commun. 4*: 317–331.

Nolan, G. (1983). *The Phonetic Bases of Speaker Recognition*. Cambridge Univ. Press, London.

O'Shaughnessy, D. (1986). Speaker recognition. *IEEE ASSP Mag. 3*(4):4–17.

Paliwal, K. K. (1988). A study of line spectrum pair frequencies for speech recognition. *Proc. 1988 IEEE Int. Conf. Acoustics, Speech, and Signal Processing* (ICASSP88), pp. 485–588.

Poritz, A. B. (1982). Linear predictive hidden Markov models and the speech signal. *Proc. 1982 IEEE Int. Conf. Acoustics, Speech, and Signal Processing* (ICASSP82), vol. 2, pp. 1291–1294.

Rabiner, L. R., Wilpon, J. G., and Juang, B.-H. (1986). A segmental K-means training procedure for connected word recognition. *AT&T Tech. J. 65*:21–31.

Rabiner, L. R., Wilpon, J. G., and Soong, F. K. (1989). High performance connected digit recognition using hidden Markov models. *IEEE Trans. Acoust. Speech Signal Process. ASSP-37*:1214–1225.

Rose, R., and Reynolds, R. A. (1990). Text independent speaker identification using automatic acoustic segmentation. *Proc. 1990 IEEE Int. Conf. Acoustics Speech, and Signal Processing* (ICASSP90), pp. 293–296.

Rosenberg, A. E. (1976). Evaluation of an automatic speaker verification system over telephone lines, *Bell Syst. Tech. J. 55*:723–744.

Rosenberg, A. E. (1976). Automatic speaker verification: A review, *Proc. IEEE 64*: 475–587.

Rosenberg, A. E., and Soong, F. K. (1987). Evaluation of a vector quantization talker recognition system in text independent and text dependent modes. *Comput. Speech Language 22*:143–157.

Rosenberg, A. E., Lee, C-H., and Soong, F. K. (1990). Sub-word unit talker verification using hidden Markov models. *Proc. 1990 IEEE Int. Conf. Acoustics, Speech, and Signal Processing* (ICASSP90), pp. 269–272.

Rothenberg, M. (1973). A new inverse-filtering technique for deriving the glottal air flow wave and deriving voicing. *J. Acoust. Soc. Am. 53*:1632–1645.

Sagayama, S., and Itakura, F. (1979). On individuality in a dynamic measure of speech. *Proc. Spring Meeting Acoust. Soc. Jpn.*, pp. 589–590 (in Japanese).

Sakoe, H., and Chiba, S. (1978). Dynamic programming algorithm optimization for spoken word recognition. *IEEE Trans. Acoust. Speech Signal Process. ASSP-26*: 43-49.

Shikano, K. (1985). Text-independent speaker recognition experiments using code-books in vector quantization. *J. Acoust. Soc. Am. S1 77*:S11 (abstract).

Soong, F. K., and Rosenberg, A. E. (1988). On the use of instantaneous and transitional spectral information in speaker recognition. *IEEE Trans. Acoust. Speech Signal Process. ASSP-36*:871–879.

Soong, F. K., Rosenberg, A. E., Rabiner, L. R., and Juang, B.-H. (1987). A vector quantization approach to speaker recognition. *AT&T Tech. J. 66*:14–26.

Strube, H. W. (1974). Determination of the instant of glottal closure from the speech wave. *J. Acoust. Soc. Am. 56*:1625–1629.

Sugiyama, M. (1988). Segment based text independent speaker recognition. *Proc. Spring Meeting Acoust. Soc. Jpn.*, pp. 75–76 (in Japanese).

Svendsen, T., and Soong, F. K. (1987). On the automatic segmentation of speech signals. *Proc. 1987 IEEE Int. Conf. Acoustics Speech, and Signal Processing* (ICASSP87), vol. 1, pp. 77–80.

Tishby, N. (1991). On the application of mixture AR hidden Markov models to text independent speaker recognition. *IEEE Trans. Signal Proc. SP-39*:563–570.

Tohkura, Y. (1987). A weighted cepstral distance measure for speech recognition. *IEEE Trans. Acoust. Speech Signal Process. ASSP-35*:1414–1422.

Tosi, O. (1979). *Voice Identification: Theory and Legal Application*. University Park Press, Baltimore.

Umezaki, T., and Itakura, F. (1987). Distance measure for speech recognition based on the smoothed group delay spectrum. *Proc. 1987 IEEE Int. Conf. Acoustics Speech, and Signal Processing* (ICASSP87), pp. 1257–1260.

Voiers, W. D. (1964). Perceptual bases of speaker identity. *J. Acoust. Soc. Am. 36*: 1965–1973.

Wolf, J. J. (1972). Efficient acoustic parameters for speaker recognition. *J. Acoust. Soc. Am. 51*:2044–2055.

Wong, D. Y., Markel, J. D., and Gray, A. H. (1979). Least squares glottal inverse filtering from the acoustic speech waveform. *IEEE Trans. Acoust. Speech Signal Process. ASSP-27*:350–355.

Part IV

Text-to-Speech Synthesis

23

Overview of Text-to-Speech Systems

JONATHAN ALLEN Massachusetts Institute of Technology,
Cambridge, Massachusetts

1. INTRODUCTION

Text-to-speech (TTS) systems convert an input string of text characters (assumed to be in machine-readable form) to an output speech waveform. Their design has been motivated by a number of factors arising from the interaction of evolving computer technology and the nature of human language performance. As computers become more prevalent, the need for communication between humans and computers is increasing, and there is a tendency for computing machines to accommodate the characteristics of their human users. Speech is universal in all cultures, but many languages have no normally used orthographic form, so that speech communication is always available even when no orthographic representation is available. Although an often cited advantage of orthography is the durability and reliability of its stored forms, this is probably an accident of the history of technological evolution, and in fact the capability to store and manipulate speech on personal computers is becoming increasingly widespread. Further, orthographic interaction with computers presupposes typing and reading skills that are not universal in any language community, even when an orthographic representation of the language is available. Historically, TTS development has been strongly motivated by the needs of the handicapped, particularly the blind, to participate in mainstream society. All of these factors have been driving forces for research on TTS systems, resulting in currently available commercial systems that can produce speech of acceptable intelligibility from unrestricted input text. The nature of these systems is dictated by the characteristics of the intended applications, the constraints used in their design, and the levels of intermediate representation utilized in ways that will be described below.

Historically, TTS can be seen as one of several techniques for providing speech output from computers, involving highly abstract internal representations and transformations. While speech output from computers always aims to provide acceptable output speech, the nature of the input has varied widely depending on the intended application. The simplest way to provide speech output is simply to record the speech (in analog or digital form) and play it back on command. The design of such systems is dominated by the memory technology employed, but high-quality speech will require about 50,000 bits per second to represent digitally sampled speech. Thus, if the message set is small and not likely to be changed, this approach may be satisfactory. From the research point of view, little knowledge of speech and language is needed, and only a modest amount of signal processing is required. In order to cut down storage requirements for speech waveforms, a variety of techniques have been developed to code the waveform in a reduced number of bits (Jayant, 1984). The required encoding and decoding algorithms can easily be implemented on a single integrated circuit, and high-quality speech at 16 to 32 kilobits per second can be obtained, thus retaining good speech quality with lower memory cost. In order to reduce the memory cost further, some form of speech production model must be introduced. The source/filter model is in wide use, whereby the acoustic generation of the waveform is modeled as the excitation of a time-varying filter by an appropriate excitation function. The (supraglottal) filter has been characterized in terms of its resonances, the frequencies of which are called *formants*. These formants can be discovered by Fourier transform spectral analysis and also by the linear prediction coding (LPC) analysis process, both of which are widely used. An articulatory model can also be utilized, but these models and their control are much less well developed than the resonance-based filter models. Nevertheless, the articulatory model is fundamental to speech production and provides a natural way to represent coarticulation and other aspects of articulatory behavior. Speech production models allow the speech waveform to be generated from the model based on a small set of parameters that vary slowly with time. Formant and LPC-based models are typically controlled by a set of approximately 20 parameters which are updated every 5 or 10 ms. This technique provides further savings in memory and can still provide high-quality speech from a parameter stream of 5000 to 10,000 bits per second. Each set of parameter values is called a *frame*, and a speech utterance can be represented by a succession of such frames, which are then interpreted by the corresponding speech production model. Once it is possible to generate speech from frames, it is natural to abstract again to seek a model at a higher level which can produce frame sequences corresponding to a sequence of higher-level units. A variety of such units has been utilized, ranging from the single phonetic segment to the diphone, demisyllable, syllable, morpheme, and word. The goal has been to produce the frame sequence *by rule* from the *names* of the input units. Thus a sequence of phonetic segment symbols are converted by a *synthesis by rule* program into a corresponding sequence of frames. This new level of abstraction has exploited the notion of *target* so that the frame parameter values are considered to be moving through smoothed trajectories between target values corresponding to articulatory positions and movements. Because of the introduction of targets, and the rules which elaborate the frame parameters from them, phonetic segments (and other units) can be compactly represented in tables of target values and other numbers needed to facilitate the synthesis by rule process. Typically, less than 100 phonetic segments are utilized in a synthesis inventory, so that the tables needed for the rule synthesis do not require excessive memory. Although the synthesis interpretation process provides for smoothing between phonetic segments (or other units), these units need to be further marked for prosodic variation including intonation and stress. In most TTS systems, a *prosodic framework* provides the durations for all segments and, having thus established the temporal organization of the utterance, then describes an intonation contour in terms of the fundamental frequency by means of establishing targets of the fundamental frequency waveform together with smoothed transitions between these values. The complete input, including the sequence of segment names and the prosodic framework, provides a *narrow phonetic transcription* of the utterance, which can then be synthesized by the interpretive synthesis process. In TTS systems, this transcription must be produced by means of text analysis for the utterance to be satisfactorily synthesized.

From the above description, it is clear that speech can be produced from a wide variety of inputs. These range from stored speech itself, through encoded speech waveforms, frame parameterized speech, and segmental synthesis symbols, to complete phonetic characterizations including the prosodic frame. All of these inputs have been used in applications depending on many factors. Typically the simpler schemes are used where the number of messages and the message lengths are small and it is desirable to preserve the speaker identity. The more complex procedures provide the flexibility of abstraction to accommodate a large (potentially infinite) set of possible utterances, smooth composition of units, flexibility and consistency of output, and the possibility of several voices and dialects. The price paid for this increased flexibility, scope, and control is a loss of speech quality, although the resultant speech obtained from phonetic transcriptions is easy to understand and acceptable for many applications. Over time, the synthesis by rule procedures have improved considerably and the cost of implementing these complex algorithms has decreased substantially, thus opening up a continually increasing scope of applications. Coupled with text analysis procedures, robust low-cost TTS systems can be readily configured for these innovative uses.

It is worthwhile to observe that the various kinds of speech synthesis systems described above can be thought of as being determined by the interaction of a set of constraints which are successively introduced in order to provide for the use of increasingly abstract inputs. Four major kinds of constraints can be recognized. Certainly the *task* is a constraining force on any synthesis system. It determines the number of utterances, their length, the vocabulary size, syntactic range, and number of voices required. Additionally, the task will determine the binding time of synthesis decisions, dictated by flexibility and update considerations, and also implementation considerations of size, weight, power, interface connection, and cost. As more abstract systems are introduced, due to either the need to compress storage or the need to accommodate synthesis by rule strategies, the nature of the *human vocal apparatus* becomes an additional constraining factor. The human speech production mechanism can be modeled in a variety of ways that provide for low data rate parametric control while still providing high-quality speech output. Furthermore, these models provide a natural way to represent the smoothing action of the articulators which is essential for intelligible and natural sounding speech. When complete flexibility is required in a TTS system, the third constraint, *language structure*, is utilized. It is believed that all aspects of linguistic structure affect the speech waveform. The constraining effects of syllable organization, morpheme structure, word morphology, stress, morphophonemics, syntax, semantics, and pragmatics are all exploited to determine the narrow phonetic transcription utilized by the synthesis procedures. Finally, the fourth constraint, *technology*, has a very strong effect on both speech research and synthesis system implementation. X-rays, motion pictures, digital signal processing, and computing in all forms have been essential tools for speech research. Indeed, some aspects of speech research, such as the study of articulation, are fundamentally limited by the constraints of available technology to observe the motion and shape of the articulators. Modern integrated circuit technology has also constrained the nature of system implementations. Digital signal processing chips, together with large memories, have made it possible to construct small, low-cost TTS systems of broad applicability. Much insight can be gained in the appreciation of the various approaches to speech synthesis by keeping these four constraining factors in mind. Each constraint provides part of the determining considerations for any particular speech synthesis application. It is especially important to note that TTS systems aggressively exploit all four of these constraint domains, since they aim to provide comprehensive conversion of text to speech corresponding to a human reading aloud.

In addition to the motivating factors cited above for the development of TTS systems, other more specific attributes of applications have presented a demand for TTS capability. There is always a need to decrease *storage* of messages, and text requires less memory than the corresponding high-quality speech signal. TTS systems also solve the *composition* problem, allowing message units to be cascaded with natural and smooth interfaces in the speech representation. TTS systems provide a general means of converting text to speech so that language elements can be *reused* in many contexts, thus avoiding the storing of speech variants of a given text element corresponding to a wide range of contextual variation. Many applications require *consistent* speech output for multiple uses, so that output speech has identical characteristics over time and a broad scope of uses in many application instances. TTS systems can, of course, create *new* speech which has never been recorded and analyzed before, since they are designed to convert *unrestricted* text to speech. Because of this capability, they provide great *flexibility* in the output utterance set. Even if the vocabulary at a given time is small, it may change quickly in many applications, so that the ability to avoid having to record and analyze speech (required in speech playback schemes) is highly desirable. Lastly, several application domains require transmission of speech representations at very low bit rates, less than 500 bits per second. Analyzed text, in a form appropriate for speech synthesis by rule, fulfills this requirement and hence is exploited where speech synthesis is needed at the end of a very low bandwidth channel.

2. OVERALL TTS SYSTEM DESIGN CONSIDERATIONS

Having established the motivation for TTS systems, together with a view of the knowledge constraints that they exploit and the nature of their applications, overall considerations for TTS system design are discussed next. In the sequel, reference is made largely to TTS systems for English, although multilingual systems will also be mentioned. It is believed, however, that the overall structure of TTS systems will be similar for the languages of the world, although the detailed nature of both analysis and synthesis procedures will certainly vary considerably.

Historically, speech synthesis by rule schemes (Holmes et al., 1964) were developed before there was any means to convert text to their input. For this reason, the need to built a TTS system required a technique for converting input text (as a string of characters) to a narrow phonetic transcription suitable as input to a (phonemic) synthesis by rule system. It is important to remember that this requirement was for a *linear* phonological representation in keeping with the phonological views of the 1960s (Chomsky and Halle, 1968). Most complete contemporary TTS systems adopt this linear format, although more complex structures are now beginning to be utilized. Thus in the late 1960s, although speech synthesis by rule was still in its infancy, the analysis of text was the missing link in an overall TTS system.

The problem of converting words from their spelling to their sound representation has been called the *letter-to-sound* (L/S) problem. Its importance had been established well before the advent of speech synthesis by rule because of the interest in characterizing human ability to read aloud and the need to teach this skill to schoolchildren. In the middle 1960s, there was no comprehensive algorithmic notion as to how this conversion might be accomplished, and reading was thought to employ either the *look-say* method or the *phonic* method. In the former process, readers were thought to "look up" the pronunciation of words in their mental lexicon, whereas in the latter method L/S rules were supposed to provide the needed sound representation. Although a number of L/S regularities had been observed (Venezky, 1970), they had not been formalized and it was assumed that some sort of moving-window contextual analyzer could compute the pronunciation from the spelling. Unfortunately, no such direct conversion process has been possible, and it has been necessary to discover the underlying linguistic structure common to *both* the orthographic and speech forms of language. These two *surface* forms, of very different physical makeup, are thus linked by a common underlying abstract linguistic representation. The fact that no direct transformation between the two surface representations has been found has led to the overall TTS process being characterized as an *analysis/synthesis* task, and this two-part process is the highest or first level of structure introduced into the overall TTS transformation.

Given the partitioning of TTS conversion into analysis and synthesis components, it remains to elaborate these two parts. Just as it has not been feasible to convert text to speech in one monolithic process, the separate analysis and synthesis phases have also been provided with internal structure. This further partitioning is typical of the solution of many complex systems, the overall performance of which is realized through the interaction of a number of interacting constraint domains. Each of these domains (e.g., syntax, morphology, phonology) has a substantial internal theory, but significant overlaps between the domains are also important to the overall theory. For example, the stress of English words can be determined largely within the domain of phonology, but a few bisyllabic noun/verb ambiguities (such as "incline," "survey") require knowledge of the syntactic role of the word for the determination of the proper stress in a sentence context. Because of these overlaps, *all* of the qualitatively distinct forms of linguistic structure affect the pronunciation of text. These include the lexicon, phonology, morphology, syntax, semantics, and pragmatics. The constraining effect of each of these domains is characterized by a *model* which interrelates elements of the domains. Thus, in syntax a *grammar* characterizes the syntagmatic constraints of the language.

It is apparent, however, that not only must the analysis and synthesis processes be partitioned into these interacting constraint domains, but also there is a structural relationship within each domain whereby bigger elements are composed from a set of smaller ones. Thus a set of features can be composed to form a phoneme, morphemes can be composed to form a word, and words can be composed to form clauses. Each of these compositional processes introduces additional forms of constraint, since they characterize the internal structure within each qualitatively distinct constraint domain. It is essential to recognize that these compositional processes are not in one-to-one correspondence, so that there is often no simple "line up" of elements from one domain with those of another. Thus the plural "sheep" can be represented abstractly as having an underlying PLURAL morpheme, whereas the singular "sheep" does not, so there are no letters to represent the PLURAL morpheme. In the characterization of speech, it has increasingly been realized that features, such as nasality, span substantial epochs of the utterance and cannot be neatly confined to single phonetic segments, such as the consonants and vowels. From this perspective, language must be represented in a two-dimensional, nonlinear form. For the purposes of characterizing TTS systems, it will be necessary to elaborate models of the different constraint domains (comprising one dimension of description) while also (along the other dimension) characterizing the compositional structure of each domain, together with the cross-domain links necessary to build the entire coherent representation of the utterance. It is the belief of this author that all of the constraint domains of language, together with their internal compositional structure, contribute to the determination of the utterance waveform, and that all of these structurally modified signal perturbations are relevant to phonetic categorization and are utilized by experienced listeners as potentially relevant cues for perception (Diehl and Kluender, 1987). The surface forms of language, both text and speech, are thus seen as composed forms of elements from structured constraint domains, and it is the task of text analysis to *undo* this composed structure to reveal the constituent underlying domain structures and determine the common underlying abstract linguistic structure. Similarly, although not addressed here, it is the task of speech analysis (or speech recognition) to undo the composed structure of the speech waveform to reveal the constituent underlying domain structures common to both text and speech, and in so doing provide the basis for synthesizing the text representation of the utterance. Failure to recognize the full richness of the structure of these underlying constraint domains will lead to impoverished models of both synthesized speech from analyzed text and synthesized text from analyzed speech.

A further characteristic of the interaction between constraint domains and structural hierarchy is the nature of the *atomic* symbolic unit of speech. A number of units, ranging from the syllable to demisyllable, diad, and phonetic segment, have been utilized as the atomic unit, for a variety of reasons discussed below. The selection of this unit is important in the analysis phase since, unless these units are completely contained within lexical entries, it is necessary to be able to analyze the text down to the level of these atomic units. Thus words are typically not analyzed procedurally into syllables due to the ambiguity of this process; instead the morph (the surface correlate of the abstract morpheme) is used as the atomic unit of textual representation since analysis procedures can reveal them reliably. As will become clear, the morph is not only a convenient and practical atomic unit of analysis, it is also *essential* in order to derive the correct pronunciation because analysis in terms of morphs reveals structure that contributes to the determination of the phonetic transcription. From this discussion, it is clear that below the word level, units of *analysis* don't have to be the same as units of *synthesis*. TTS systems can be built that analyze to the morph level, but morphs can then be represented in terms of a variety of units reflecting ease of compositional processing, such as the syllable, demisyllable, diad, and phonetic segment.

3. TEXT ANALYSIS

Having characterized the overall TTS conversion process as text analysis followed by speech synthesis, we now focus on the analysis phase. At the outset, it is well to point out that text may not be the input to a system that uses TTS capability. Systems based on "speech synthesis from concept" (Young and Fallside, 1979) have been developed which have the advantage of generating a text string, the syntactic structure of which is known because it is generated by the algorithm. This advantage may also accrue to the increasing number of systems using a discourse model (Brown and Yule, 1983). These models may be based on speech acts (Austin, 1976; Searle and Vanderveken, 1985) as the basic units of discourse, permitting correct determination of emphasis and pronominal reference. Discourse focus and the relative contrast of new and old information can also be exploited in such discourse-based TTS systems to improve the speech quality. Unfortunately, systems based on conceptual or discourse model inputs are not well developed at this time, and although their use will undoubtedly become more important, no further description of analysis techniques utilizing these models, and aimed at refining the common underlying abstract linguistic description for TTS systems, can be made at this time.

Although there is variation over the TTS systems described in the literature, the following set of capabilities forms a superset of the analysis tasks undertaken in these systems.

- Format analysis
- Single word pronunciation
    - Morphological analysis and lexical retrieval
    - Letter-to-sound rules
    - Lexical stress
    - Morphophonemic rules
- Parsing
- Phrase and clause level stress
- Low-level phonetic adjustments

In the sequel we justify the need for each of these tasks, describe each of them, and show how each of them contributes in a modular way to the development of the comprehensive underlying abstract linguistic description. Corresponding to each task is a constraint domain characterized by a model. The analysis procedures explicitly establish the constraints that are operating for a particular text input, providing the context for conversion rules which incrementally build the underlying framework needed for synthesis. Each constraint model, analyzer, and rule system constitutes a module of the TTS analysis phase, and each of these modules is characterized by a well-defined set of input-output conventions. The latter makes it possible to configure subsets of the overall TTS analysis subsystem for purposes of development, experimentation, demonstration, and system application where some of the analysis has been completed at a previous time (early binding).

3.1. Format Analysis

In any practical TTS application, the input text may arise from a variety of sources utilizing many different spelling conventions. The format analysis task is to convert these input letter strings to reformatted or "normalized" letter strings which can be properly processed by the analysis phase. Unfortunately, no concise formal model of these input character strings has been developed, so that analysis consists of a number of heuristic procedures which accomplish the normalization. Most of the objects examined and manipulated in this module have no well-established linguistic significance and are instead the surface manifestations of conventions for abbreviations and printing.

The normalization analysis phase detects blank space(s), sentences, and paragraphs, and some systems require knowledge of the limits of a text line. Capital letters, both sentence-initial and otherwise, must be detected, and a number of common abbreviations must be converted to normal spelling. These include $, %, &, #, and @, as well as titles (Mr., Mrs., Ms., Dr., Prof.), abbreviations for months (e.g., Jan), and "etc." There are many increasingly used mnemonics, such as UK, USA, ATT, NTT, and MIT, which must be spelled out, as well as names such as LATEX and UNIX which the system designers may not want to submit directly to the word pronunciation modules. This usage points out the general capability of this initial normalization module to avoid inappropriate processing by later modules, by converting the input letter string into a new string that can be processed by the remainder of the system to provide the desired pronunciation.

Normalization must also process numbers including integers, numbers with a decimal point, dates, and time. It is not always certain what meaning category these strings belong to, but satisfactory algorithms are available. There is, of course, a wide variety of punctuation, and special marks are often inserted into the text stream to be later interpreted by the parsing analysis routines and the prosodic synthesis procedures, since they have a strong effect on the overall intonation and stress contour of the utterance. Procedures for analyzing monetary expressions, such as $17.95, must be provided, and depending on the degree of specialization of the TTS system, capability may also be provided for interpreting mathematical formulas and other complex structured symbols. While ignored by many TTS systems, italics, boldface, and underlining can be marked, as well as symbols in some specialized "alphabets."

The aggregate procedural capability to normalize all (unrestricted) input text constitutes the formatting module, or text analysis preprocessor. It is an essential component of all TTS systems, but arises, as noted above, from the conventions of orthography, rather than from fundamental linguistic considerations. Full descriptions of these procedures may be found in the literature (Allen et al., 1987).

3.2. Single-Word Pronunciation: Conceptual Form

This multimodule part of TTS analysis systems can be viewed as a largely self-contained set of procedures for determining the pronunciation of isolated words. Of course, as remarked earlier, these pronunciations may be later modified by phrase and sentence level context, but a great deal of constraining strength on the ultimate pronunciation is computed at the word level. Historically, many TTS systems were built from single-word systems, and there is a great deal of algorithmic and linguistic interest in these modules.

At the isolated word level, the task is to map any input character string bounded by spaces into a string of phonetic symbols. For the present discussion these phonetic symbols may represent units that span any subset of the word, ranging from single conventional phonetic segments through diads to syllables and entire words. (The criteria for selection of these units will be considered when methods for speech *synthesis* are described.) There are several ways in which the mapping might be done.

The most obvious approach to establishing word pronunciation is to store the spelling/pronunciation correspondence for *all* words. The number of English words is large, but bounded, so a word dictionary is a possibility. Furthermore, there are at least four reasons why a lexicon is *essential* in *any* approach, because of the difficulty of any rule system in coping with the following classes of words.

- *Foreign words*: Words such as **parfait** and **tortilla** are borrowed from other languages but retain the pronunciation of their language of origin. It is not reasonable to require any rule system to cope with multiple languages without some language identification process, except in the case of proper names. For this reason, these words are placed in an "exceptions" dictionary.
- *Function words*: An examination of phonetic environments over the language reveals seeming contradictions between high-frequency function words and lower-frequency content words. Consider the following contrasts:

Function words	**Content words**
of	fun, roof
the, this, that, ...	thesis, thimble, earth
have	shave, behave
was, is, has	atlas, canvas

These examples show that consonants are often voiced in function word environments when they are unvoiced in similar environments in content words, and that vowel quality shifts also take place. The function words appear very frequently in the language, but there are not very many of them, so it makes sense to treat them as "exceptions" to the rules governing the large class of content words and place them in the lexicon.

- *Internal structure*: Attempts to build comprehensive sets of letter-to-sound rules have encountered difficulty with the effect of internal word structure on pronunciation. The following examples show that internal morpheme boundaries break up letter sequences that might otherwise be recognized by a scanning algorithm as a functioning vowel or consonant cluster. Unfortunately, these boundaries are not explicitly available in the orthography and cannot be reliably found by moving-window scanning techniques customarily used by letter-to-sound conversion algorithms.

Cluster	**Word examples**
sch	*sch*ool, *sch*eme
	mi*sch*ance, di*sch*arge
aw	l*aw*, *aw*ful
	*aw*ake, *aw*ash
th	thirst, di*th*er
	ho*th*ouse
ea	r*ea*ch, t*ea*r, st*eak*
	chang*ea*ble

These (and many other) examples have shown that letter-to-sound rules do not apply to letter clusters spanning morph boundaries, so that a dictionary is needed to store the pronunciations for these cases.

- *Letter-to-sound rule errors*: No such set of rules has been devised that operates without errors, and this is probably true of any linguistic rule system. For this reason, it is inevitable that the pronunciations of words that are known to be improperly converted by the extant letter-to-sound rule system must be stored as exceptions in the lexicon.

At a fundamental conceptual level, there may appear to be little difference between table lookup and rule operation. Accessing of a table (or lexicon) can be viewed as a particular form of rule. Furthermore, several letter-to-sound rule systems imbed exceptions directly in the rules, by first checking that the context being examined is not part of an exceptional word. Thus it is possible to have a set of letter-to-sound rules without an exceptions dictionary, but, in effect, the exceptions dictionary is merely "distributed" throughout the text of the rules. With this understanding in mind, we will continue to distinguish letter-to-sound rules from the lexicon and its associated access scheme. This leads to a more coherent description of the analysis procedures and still leaves open the implementation option of confounding the lexicon and rule system in the interests of efficiency or uniformity of procedures.

Having established that a lexicon is essential, its nature must now be established. There are several ways to structure the lexicon, corresponding to the choice of the atomic unit of text analysis. Of course the lexicon may be designed to hold complete words. Although this is frequently done in practical TSS systems, there are two major disadvantages:

- New words enter the language continually, by compounding (**earthrise, cranapple**) and by filling phonological "accident gaps," such as **brillig**. Therefore no word-level dictionary can ever be "complete," and it may be necessary to update the lexicon frequently depending on the application environment.
- There are many regularly inflected word forms in English, and the average number of affixes and roots per word in English is approximately two. In printed dictionaries, this fact is recognized by the addition of "run-on entries" which indicate the pattern of inflection. Hence if all regularly inflected forms must be stored as separate entries in a TTS lexicon, the lexicon will be unnecessarily large.

The next smaller unit of text analysis below the word level is the *morph*. Morphemes are the minimum syntactic atomic units of the language and are sometimes also regarded as the minimum semantic atomic units. There is not necessarily any direct mapping between segments of a textually represented word and the underlying abstract morphemes, but when there is, the segments are called *morphs*. In practice, many words can be divided into morphs, such as **boys** → **boy** + **s**, **unacceptable** → **un** + **accept** + **able**, and **houseboat** → **house** + **boat**. These morphs are generally either roots or affixes, but analysis procedures typically require finer categories.

Morphs are desirable as dictionary entries for TTS systems for several reasons:

- They are the surface forms of the *atomic*, minimal syntactic units of the language (morphemes) and as such are the basic lexical building blocks from which words are composed.
- They are *stable* in the language, and it is rare for a new morpheme (and hence morph) to enter the language. This means that a morph lexicon will not have to be updated often.
- They have large *generative power* to make words that follow the morpheme composition rules of the language. A lexicon of N morphs can readily generate between $5N$ and $10N$ words.
- It is not necessary to store the regularly inflected words and regular compound words in the lexicon, since it is possible to represent these words as composed forms made up of (possibly mutated) morphs.
- Many language phenomena, such as the regularities of pronunciation represented by letter-to-sound rules, hold only within morph boundaries, so it is necessary to recognize these constituents.

For all these reasons, a morph lexicon is highly desirable in a TTS system. Experience with whole-word exception dictionaries has shown that they continually grow in size, approaching the number of entries contained in a comprehensive morph lexicon, which is often considerably less than 10,000. Furthermore, the advantages cited above can be realized in practice through the use of highly robust morph analysis techniques. Over large bodies of text, more than 98% of the running word tokens can be successfully analyzed into morph sequences, contributing in large measure to the quality of computed word pronunciations utilized in the containing TTS system.

A successful morph analysis of an English word in textual form produces a *morph covering* of the word, so that each letter of the word is mapped to exactly one morph in the covering sequence. Although this process is successful over most word tokens, there will always be input text words in a TTS system designed to process unrestricted English text that are not successfully "covered" with morphs. This can happen due to gaps or missing morphs in the lexicon, inability of the morph analysis procedures to "undo" some spelling mutations, stress markings that are not correctly computed by rule, or spelling mistakes in the input text. Some of these difficulties must be resolved through insertion of the word in the lexicon, but for a large class of content words, letter-to-sound rules may be successfully applied. Many investigators have compiled large sets of these rules (Hunnicutt, 1980), but the following considerations apply to their general structure regardless of the detailed nature of the rules:

- It has already been noted that letter-to-sound rules do not apply to letter strings (possible clusters) that span morph boundaries. Hence, an attempt must *always* be made to compute a complete morph covering for the word being analyzed. If a complete covering is found, the lexicon provides the required pronunciation representation. Only if the covering analysis fails should letter-to-sound analysis be attempted, since the letter-to-sound rules are less reliable than the lexically based procedures. Nevertheless, even if the letter-to-sound rules are invoked, an attempt must still be made to locate morph boundaries, due to possible clusters spanning morph boundaries. Since the morph covering procedure is assumed to have failed at this point, an attempt at "affix stripping" is made, whereby prefixes and suffixes are recognized and removed from the root word. In this way, any candidate word presented to the letter-to-sound procedures is first analyzed into a set of morphs, and the residual letter sequence (after removal of affixes) is assumed to be a monomorphemic root. The lexicon is used to provide the pronunciation of the affixes, so the letter-to-sound rules are applied only to the residual root. It must be emphasized that obtaining a complete morph covering for a word is vastly preferred (in terms of quality of pronunciation) to affix stripping (which is less reliable and cannot detect compounds), so letter-to-sound analysis should never be attempted when a morph covering is available. Nevertheless, several TTS systems, while providing an exceptions lexicon, do not provide morphemic covering analysis (and the corresponding morph lexicon), so each word is first searched for in the lexicon as a whole word and then, if not found there, affix stripping and letter-to-sound analysis are employed.
- Letter-to-sound conversion algorithms typically scan the input residual root with a moving window looking for letter clusters for which there are rules. A "longest match first" procedure is used; for example, **chr** is found before **ch**. If this is not done, longer functioning clusters may be missed.
- It has been found that rules for converting textual consonant clusters (e.g., **thr**) are more reliable than rules for converting textual vowel clusters (e.g., **ea**), so consonant clusters found in the residual root are converted first. This means that the available context for

converting consonant clusters includes both the letters and pronunciation of any affixes in the word, plus contextual letters in the residual root. For vowel clusters, however, the available context includes both the spelling and pronunciation of all affixes and consonant clusters, plus any previously converted vowel clusters. By staging the analysis in this way, all possible context is used to convert vowel clusters, and indeed all of this information is used by powerful letter-to-sound rule systems.

At this juncture, the process of converting any (unrestricted) English word from spelling representation to pronunciation has been described. Morph covering analysis (always used first when available) and letter-to-sound rules are the two complementary processes used. Nevertheless, adjustments to the pronunciation must still be made. These fall into three categories:

- The lexical stress contour of each word must be computed. Elaborate rule systems for taking into account the nature of the syllables (strong or weak) and the structure of affixation and compounding found in the word are utilized to generate stress marks (usually two levels) and modify the quality of reduced vowels. Many TTS systems do not include provision for stress analysis, but it is essential both for the determination of correct word pronunciation and for its effect on sentence-level prosody.
- When morph boundaries have been revealed, either by morph covering analysis or affix stripping, some adjustments in pronunciation may be triggered. These changes are produced by a set of morphophonemic rules, and they are needed in many words involving, for example, plural and past tense morphemes.
- Although the above two categories of pronunciation modification are essential, some TTS systems also provide a small set of low-level phonetic adjustment rules, generally representing coarticulatory smoothing phenomena, such as the flapping of alveolar stop consonants in selected environments. Together with the other means for modifying pronunciation, these rules contribute greatly to the overall intelligibility and naturalness of the resultant speech.

The entire set of procedures for mapping word letter strings to pronunciations has now been briefly mentioned. An ongoing research problem, however, is the determination of appropriate phonological and phonetic levels of representation to be used at the various stages of analysis, ranging from the lexicon through letter-to-sound, lexical stress, morphophonemic, and low-level phonetic adjustment processes. Clearly, each of these analysis phases must assume a definite range of phonological or phonetic representations in its input stream, and this range will affect the character and detailed structure of rules at each stage. In practice, most TTS systems have utilized a level of description that is near the phonetic surface. The way in which these constraint systems (e.g., the lexicon, stress rules, morphophonemic rules) interact, however, is still not well understood in linguistic theory, and the pronunciation strings stored in the lexicon and produced by the letter-to-sound rules should probably be more abstract than those currently used. The determination of these deep phonological abstractions is exceedingly difficult, since all following phonological and morphological processes build on them and must consistently and reliably produce, as a result of applying all of the constraints, a final accurate narrow phonetic transcription. As a simple example, consider the modification in pronunciation of **human** in **humanity** → **human** + **ity**. If the pronunciation of the root **human** is stored in the lexicon with its second vowel reduced, in accordance with its surface phonetic representation, then the appropriate vowel quality for the stressed second syllable in **humanity** may be difficult to compute. It seems clear that the lexicon should contain underlying forms which permit the stressed vowel quality to be computed, while also allowing reduced syllables to be formed where specified by the stress rules. For this process to work correctly, lexical phonological forms must be specified that anticipate all later structural, phonological, and phonetic analysis procedures. TTS systems provide a test bed for phonological theories concerned with these abstractions, and a great deal of research is needed to resolve these representational difficulties. Nevertheless, despite the frequent lack of elegance and fundamental abstractions in TTS lexical entries and subsequent processing rules, representational schemes have been devised for phonological and phonetic units in such systems that are consistent and effective, even if the lexicon is sometimes unnecessarily swelled with "exceptions" occasioned by the nature of the surface-oriented representations being used.

Summing up, the essential modules of a "one-word" TTS analysis system have been motivated and briefly described. The constraints due to lexical structure, stress, morphophonemics, and articulatory smoothing have been introduced, and the appropriate role of letter-to-sound rules described. Although much research is still needed to address the way in which these constraints interact, effective procedures have been developed in practical TTS systems that capture most of the constraints in a basic nonredundant way. The contents of the lexicon, however, certainly in "whole word" systems and also to some extent in morph lexicons, include words and morphs that should be represented at a more basic abstract level in order to capture all of the phonological and morphological regularity of the language. We are left with highly effective and capable analysis procedures, together with a continuing challenge to develop and utilize abstract units, along with appropriate procedures, to achieve the defining characteristic of the lexicon as a repository for only that which is idiosyncratic in the phonology of the language. It is interesting to note that achievement of such a goal would remove morph pronunciations from the lexicon when they are predictable by rule, even though the spelling representation of the morphs would have to be retained for morph covering analysis purposes.

3.3. Single-Word System: Implementation Considerations

Having discussed the constraints utilized to determine single-word pronunciation, we now turn to some implementation considerations. TTS systems are implemented in many different ways, but it is important to have some view of the formalisms utilized and the algorithmic techniques applied to the text analysis tasks. Much of this discussion is based on one well-known system (Allen et al., 1987), and the reader should consult this reference for additional detail.

We turn first to morph covering analysis. Five major constraining influences can be identified:

- Morphs are classified into prefixes (P), roots (R), derivational suffixes (D), inflectional suffixes (I), and absolutes (A), which can only combine with an inflectional suffix). Their order is constrained so that among the first four classes, they must appear in the order listed, unless absent, and at least one root must be present. Thus "reenlistments" contains five morphs, classed as PPRDI. A context-free morph sequence grammar represents these constraints.
- The basic morph analysis procedure consists of a longest match first recursive procedure. The recursion is needed so that shorter morphs can be examined when longer ones do not succeed, and the bias to examining the longer morphs first ensures that exceptions will always be found first if they are in the lexicon.
- Derivational and inflectional suffixes are also marked as to whether they start with a vowel (vocalic) or not. If they are vocalic, recognition of them in the analysis process triggers a spelling change investigation, since consonant doubling (**cutting** → **cut** + **ing**), dropping of a final silent "e" (**changing** → **change** + **ing**) and several other mutations to the spelling of the root must be "undone." It is also necessary to mark roots as to whether they *must* undergo a spelling change when com-

posed with a vocalic suffix (**scarred** → **scar** + **ed**), *may* undergo change (**changeable** → **change** + **able**, **changing** → **change** + **ing**), or *must not* undergo change (**alloying** → **alloy** + **ing**). The proper treatment of these spelling changes augments the basic recursive match procedure.

- If the recursive matching procedure is allowed to run all the way to completion (even if one successful covering has already been found), then in general multiple coverings for a given word will be computed. A simple set of rules can then select among them, generally preferring analyses with affixation to those with compounding. Thus **formally** → **form** + **al** + **ly** is preferred to **formally** → **form** + **ally**. The constraints represented by these selections, along with the morph sequence grammar, can be combined in a finite state machine which can always terminate an analysis when it can be predicted that completion of the current analysis cannot possibly produce an acceptable morph covering.
- Some morphs are strong forms in that they map onto two or more underlying morphemes. For the above analysis to work correctly, these morphs must have their underlying morphemic constituency marked in the lexicon. Thus **teething** can be analyzed into either **teeth** + **ing** or **teethe** + **ing**. The former covering is rejected since **teeth** has morpheme sequence RI and the simpler morpheme sequence of the latter analysis (RI) is preferred to the more complex former one (RII). Thus the selection rules discussed above operate at the underlying morphemic level.

Morph covering analysis requires, of course, a lexicon of morphs. This can be constructed by "priming" a nascent lexicon with affixes and function words that are known to be necessary and then interactively analyzing a large corpus (e.g., Kucera and Francis, 1967), sorted in order of increasing word length, to build up the lexicon (Hunnicutt, 1976). Each tentative analysis must be examined for acceptability on grounds of correct pronunciation and identification of correct functioning morpheme and morph sequences. Furthermore, when utilizing any corpus of sampled text, polymorphemic words will be encountered where one or more of the constituent morphs have not been previously encountered. This will require "back-insertion" of the required morphs in the lexicon. These processes are tedious, and although much of the work can be done automatically, there is still the need for human consideration of each analysis. Furthermore, any change to the morph covering analysis procedures will require, in general, a complete reexamination of the test corpus. In addition to providing each lexical entry with the morph spelling, morph type code, spelling change code, and pronunciation code, parts-of-speech information is also included to facilitate the determination of the parts-of-speech code for each word, which is required by later parsing.

Letter-to-sound rules, as described earlier, require affix stripping as a first step to reveal possible morph boundaries. Since letter-to-sound rules must not be applied until morph covering analysis has failed, an elegant way to perform the affix stripping is to include it as a second step in the morph matching procedures. In practice, however, it is simpler to perform this task separately. Furthermore, some letter-to-sound rules recognize "high-frequency endings" (such as **-ace** in **lace** and **menace**) which do not have morph status, but which do lead to reliable pronunciations. Thus it is common for letter-to-sound rules to recognize any letter string that provides a reliable sound correspondence, irrespective of its linguistic status. Since the affix-stripping process is not as reliable as the morph covering analysis process, additional constraining power is added to confirm the correctness of an affix being stripped from the word. A small categorial grammar can be written which describes the syntactic role of an affix in relation to the affix or residual root on its left. As an example of such grammar rules, the letter-to-sound rules may strip off three affixes from **dictatorship** to form **dict + ate + or + ship**. The sequence of affixes is shown to be syntactically compatible by the categorial relations that **ship** makes nouns and attaches to nouns, **or** makes nouns but attaches to verbs, and **ate** makes verbs. Hence the affix string can be parsed by these simple category relationships to confirm a satisfactory analysis. This check has the added advantage that the part of speech of the analyzed word is also computed. These techniques can also reject a possible affix analysis. Thus in the tentative analysis **astonishing → aston + ish + ing, ing** must attach to a noun or verb on its left, but **ish** produces adjectives, so the analysis is rejected, and **ish** is made part of the residual root. A large set of rules (several hundred) is necessary to provide good coverage and may include several ordered sequences of rules. It has been found convenient to express the rules using standard phonological rule notation, so that the overall set of rules can then be easily compiled or interpreted. An information theoretic approach to the discovery of letter-to-sound rules (Lucassen and Mercer, 1984) has been developed which automatically determines the optimal questions to be asked to construct a decision tree for the correct pronunciation. This approach is attractive in that it utilizes the statistical information in letter-to-sound correspondences in an optimal way, but without knowledge of morph boundaries, pronunciation mistakes are still made. This outcome illustrates once again that scanning window conversion processes, even when coupled with powerful statistical techniques, cannot compensate for lack of information on the internal structure of words. Finally, we mention the important case of proper name pronunciation. By examination of three-letter sequences, it has been possible (Church, 1985) to recognize the language of origin of the name and then apply the appropriate letter-to-sound rules for that language. This two-step procedure is very effective and adds a much-needed capability to TTS systems for many applications.

Lexical stress is a large topic which has received (and continues to receive) much emphasis in the literature (Chomsky and Halle, 1968; Halle and Keyser, 1971). In a "stress-timed" language such as English, proper stress determination is essential. For example, the correct pronunciation of the words **systematic** and **systematize** can be derived by rule, despite the substantial difference in stress contour and vowel quality. Indeed, the intellectual contribution of the rules of English stress to word-level phonology is one of the great achievements of modern linguistics. Even the "main stress rule" has many cases, and we cannot hope to convey the richness of the overall stress rule system here. For TTS systems, it has been important to extend the original rules to include the effect of affixes, and both cyclic and noncyclic rules are combined into one overall system (Allen et al., 1987). The interaction of stress and morphology in a coherent lexical phonology has been studied (Mohanan, 1986), examining ways in which the previously separate constraint domains of stress and morphology can be interleaved and unified.

The implementation of morphophonemic rules is straightforward. Environments spanning morph boundaries are examined for the appropriate context. By far the most important rules to implement are the PLURAL and PAST rules, which have exactly the same form. They are essentially voicing agreement rules, with the additional provision of breaking up a sequence of two consonants with the same (alveolar) place of articulation with a short vowel. Rules for palatalization are also frequently employed, as in the smoothing between **-ate** and **-ion** in **determination**. Morphophonemics, together with lexical stress, serve to adjust the word-level pronunciation in a way dictated by the morph sequence. No TTS system can afford to ignore either of them.

Finally, low-level phonetic adjustments are made by simple searches for relevant contexts. They generally apply to environments where some coarticulatory smoothing is needed, or where adjacent words present adjacent segments that must be smoothed. At the single-word level, examples include the flapping of alveolar stops, but most application environments occur between words and are covered below.

3.4. Syntactic Analysis

At this juncture in the overall text analysis process, a narrow phonetic transcription for each word of the text has been computed. Merely concatenating these words together, however, will not produce a satisfactory phonetic characterization for the ensuing speech synthesis. There are several reasons why further analysis is necessary:

- The pronunciation of some individual words is affected by the syntactic role of these words. There are approximately 30 bisyllabic "noun/verb ambiguities" in English, such as **survey**, **incline**, and **refuse**, as well as some "noun/adjective ambiguities," such as **invalid.** Clearly a syntactic analysis is needed in order to select the correct pronunciation. Differences in tense must also be resolved, as in **read.**
- Although it is not difficult to determine, the type of sentence (declarative, imperative, or question) must be known to determine the correct prosody. In the case of questions, it is important to discriminate between "yes/no" questions (**Is Harry tall?**) and "Wh" questions (**Where is the train station?**), each of which has its own distinctive prosody.
- Phrases and clauses must be recognized in order to compute an appropriate temporal framework. The distribution of pauses is important, as is the position of "head" in phrasal constituents. Segments are lengthened before pauses, and in general, there is some lengthening toward the end of syntactic constituents at all levels. Pauses and durations must be correctly established in order to facilitate human perceptual processing. The distribution of pauses and lengthening helps the listener to organize the utterance into manageable and semantically coherent chunks, and without such cues, sentence level perception becomes very difficult.
- Just as the syntactic structure of sentences must be determined to facilitate establishment of the temporal framework, it is also needed to compute the fundamental frequency contour of the sentence. Taken together, duration and fundamental frequency constitute the main prosodic correlates that facilitate human perception of speech. Both are correlates of stress, which is also determined, in part, by the syntactic structure of the sentence. Thus a detailed syntactic analysis is essential for the establishment of the prosodic framework, which in turn is necessary for facile perception.
- Reduced constructions, particularly reduced relative clauses, must be identified in order to facilitate the computation of correct prosodic cues for these structures. For example, the sentence **The player kicked the ball kicked him** must be recognized as a reduced version of the longer sentence **The player who was kicked the ball kicked him** or **The player kicked the ball which was kicked to him.** This sentence illustrates the highly prevalent nature of syntactic ambiguity in En-

glish. The classic examples, such as **They are flying planes** and **He saw the man in the park with the telescope**, have ambiguous readings, and in general, conjunctions and prepositional phrases lead to ambiguity which usually cannot be resolved at a syntactic level within the sentence. It is interesting to note that in the case of some ambiguous constructions, there does not seem to be much phonetic variation between the alternate readings. But in other cases, there is substantial change between the versions. Thus **They must not be married** has two readings, depending on whether the stress is on **must** or **not**, and **She slapped him on the face and then she hit the man** refers to either one man or two distinct men, which can be indicated by the relative stress on **hit** and **man**. From examples like these, it is clearly important to recognize ambiguous readings, even though this may be difficult or impossible within a sentence context. In some extended TTS systems, such as "synthesis from concept," where the deep underlying syntactic structure is known, the intended interpretation may be available and hence the corresponding syntactic structure will be known. In any case, the syntactic analysis capability of a TTS system can produce only one syntactic structure, since only one utterance will be produced. Care must be taken to design the syntactic analysis procedures so that acceptable parses are produced as frequently as possible for the intended application domain.

In order to perform a syntactic analysis (parse) of a sentence, it is first necessary to determine the parts-of-speech (POS) codes for the individual words of the sentence. For English, less than 100 parts of speech are usually sufficient, although attribute-value feature markings are also used as an adjunct. Too few parts of speech lead to parsing errors and an excessively coarse analysis, while attempts to introduce several hundred parts of speech have generally not improved the quality of the syntactic analysis. Small classes for the function words are useful, and in many systems the single word **to** is accorded its own syntactic POS class due to its importance in the parsing process.

Once the syntactic POS categories have been chosen, a set of POS must be determined for each word. Of course, for the function words and the short, highly frequent content words, the POS set can be stored directly in the lexicon. The remaining words usually function as nouns, verbs, adjectives, or adverbs. One of the greatest difficulties arises from the fact that in English every common noun can also function as a verb. Also the final suffix **-s** can serve to indicate either a plural noun or a third person present tense singular verb, further exacerbating the POS determination.

At the single-word level, however, the *set* of all possible functioning POS for a word is determined. It has already been seen that morph covering analysis, or letter-to-sound affix stripping, can provide important POS information. An algorithm for utilizing this information has been developed (Allen et al., 1987) which effectively utilizes these morphemic syntactic constraints. Thus, while **-s** is syntactically ambiguous, **−er** + **−s** as the rightmost two morphs of a word unambiguously denote a plural noun. Numerous other schemes have been utilized to determine the POS set. A small window surrounding the word in question has been utilized when no affixes are recognized, and statistical techniques based on trigram (word) frequencies have been utilized in a similar way to provide a local contextual analysis. Statistical modeling can be very effective, but training is needed on the texts of the intended application. Thus **light table** is always a compound in a set of laser patent texts, rather than functioning as an attributive construction. Since these techniques utilize a small local window of words around the word being analyzed, it is natural to view this POS determination process as a form of local parsing, which could be incorporated directly into the overall sentence level parsing scheme. While this approach might provide an overall coherence and unity to the parsing procedure, in practice the POS determination process is performed separately. Finally, we note that attribute-value pairs of features are often used with some of the POS (such as singular/plural number for nouns), but this alternative to more elaborated POS categories is just a notational variant used for implementation convenience.

There has been a great deal of study of parsing procedures. It is important, however, to keep in mind the *purpose* of the parsing, since many different goals have been addressed by parsing research. For use in TTS systems, a parser does not need to provide "deep structure" but should show all phrases and clauses in each sentence. Only one parse can be used, so extra parsings in the case of ambiguity are useless. Also, TTS systems require parsing of *unrestricted* text, with no limits on the vocabulary or range of syntactic constructions. The parsing process should "fail gracefully" so that if a parse fails at the top sentence or clause level, at least the previously detected phrases will be available. For these reasons, parsers which emphasize "coverage" rather than linguistic sophistication have been preferable for TTS systems. Simple grammatical formalisms, such as augmented context-free grammars, have proved useful since they are well understood and efficient parsing procedures for them exist.

A complete clause-level analysis of a sentence is always preferable, but it may be excessively costly in some application domains. Experience has shown that bottom-up noun and verb phrase parsing is fast (although sometimes in error, due to noun/verb ambiguities and reduced relative clauses) but that ensuing clause-level parsing is far more expensive computationally. Furthermore, ambiguity must be dealt with at the clause level, or avoided due to use of some pragmatic bias such as locality of binding. For this reason, many TTS systems use phrase-level parsing only, possibly augmented with simple clause-level tests. A set of pragmatic observations motivates the overall parsing strategy. In English, there will generally be many more noun phrases than verb phrases. The start of noun phrases, often characterized by determiner sequences, is easier to detect than the start of verb phrases, which often start with a verb itself (which can often function as a noun as well). On the other hand, nominal compounds can be very long and hard to detect, so that detecting the end of a noun phrase is often difficult. All of these observations imply a global strategy of scanning for noun phrases first and extending them to maximal length consistent with the rules of the grammar. Once the noun phrases have been found, verb phrases are sought, and simple clause-level tests may be applied such as satisfying the need for at least one main verb in each sentence.

With these comments in mind, parsing for TTS often becomes largely a matter of searching first for noun phrases and then for verb phrases. Grammars for these phrasal constituents have been written (Allen et al., 1987), and they exploit the strongest syntagmatic constraints of English. Determiner sequences in noun phrases and auxiliary sequences in verb phrases undoubtedly exhibit the strongest syntagmatic relations in English, so it is useful to rely on them as a central component of the parsing strategy. Parsers for these phrase grammars are compact and fast, although they do occasionally classify a functioning verb as a noun due to the global strategy bias toward extending noun phrases. Accurate POS determination obviously contributes strongly to parser performance, thus favoring the inclusion of a morph covering analysis procedure in TTS systems.

At a purely syntactic level, many of the problems of parsing are likely to remain. Compound analysis, prepositional phrase attachment, ambiguous reduced relative clauses, preposition/particle confusions, and the analysis of conjoined constructions present insurmountable difficulties unless the parsing process can be complemented by use of semantic and pragmatic constraints, but these are currently beyond the scope of TTS systems. Building on robust phase-level analysis, contemporary implementations seek to provide efficient pragmatically biased clause-level analysis where these tests are reliable, thereby providing the best possible structure for the generation of the prosodic framework.

3.5. Phrase-Level Stress, and Phonetic Adjustments

The parsing process described above is assumed to produce a phrase-level bracketing of the sentence being processed. This information is sufficient to apply phrase-level stress rules using the nuclear stress and compound stress rules. These are simple rules (Chomsky and Halle, 1968) which apply recursively. They require, however, that the syntactic analysis recognize compounds and right-branching attributive structures. The essential notion is that, of the two adjacent constituents in a right-branching structure, the right constituent retains main stress, whereas if the structure is left branching (as in the case of a compound), then main stress is retained on the left constituent. Successive application of the nuclear stress rule implies that in a right-branching structure, stress will start at the left on a secondary level and go down until the "head" is reached, which will get main stress. The compound stress rule provides a more "up-down" stress contour, and of course, if the syntactic analysis can provide a reliable bracketing, then the two rules may be used in an intermixed fashion. Application of these rules may imply many levels of stress, but more than three levels is unusual in a practical TTS system, so that low-stress levels are gathered into an equivalence class. For example, **psychological** has main stress on the third syllable, but in an attributive construction, such as **psychological reality**, *all* syllables of **psychological** may be held at a low level.

In addition to syntactically based stress placement, emphasis may be placed on important words in the sentence. When the speaker wishes to emphasize his or her attitude toward the truth value of a proposition, modal elements of the language can be used. For example, in English, sentential adverbs (e.g., **surely**), past tense modal auxiliaries (e.g., **might**), and negative elements (e.g., **not**) are used for this purpose (O'Shaughnessy and Allen, 1983), and some TTS systems mark such lexical items for appropriate inflection in the prosodic synthesis routines. Focus-shifting transformations, the most frequent of which is the Passive transformation (in English), also have a strong effect on the stress of phrases and words. For example, in a simple right-branching declarative sentence such as **John bought the books**, main stress will be on the last word of the sentence. But when passivized, the sentence becomes **The books were bought by John**, and **John** gets the main stress. Applying agent deletion at this stage produces **The books were bought**, with main stress on the last word. In all three of these cases, main stress falls on the last word, but the emphasized word varies. With other focus-shifting transformations, a variety of constituents can be stressed or reduced, providing salience to the focused items. Although many of these effects have been observed, they are not typically employed in TTS systems due to the difficulty of detecting the transformations and the complex prosodic contours associated with them. The distinction between new and old information can also affect emphasis. In general, old information, even if it is contained in syntactic constituents that would normally receive stress, is deemphasized. This is often observed in discourse environments, where speakers do not emphasize that which has already been introduced. A correct treatment of these phenomena would require an elaborate syntactic analysis together with determination of pronominal reference, and is currently beyond the scope of contemporary TTS systems. Nevertheless, some systems have simply kept track of the n (say 50) most recent words, and deemphasized a content word if it also occurred in this set. Lastly, we mention that an English speaker can emphasize *any* word of a sentence. In most TTS systems, the information on which such emphasis would be based is not available, so these emphases are not provided.

At the sentence and phrase level, a number of phonetic adjustments need to be made to reflect coarticulation between words. These are similar to phonetic adjustments within words and arise for the same reasons. For example, alveolar flapping can occur within a word, as in **ladder**, but also between words, as in **sat about**. Glottalized dental stops arise at word boundaries, such as **that one**, and in general, glottalization can occur between words when there are vowels on either side of the boundary and the vowel tongue height difference is small. Clearly, some hiatus gesture is needed for this situation when the vowels are identical (**he eats**), but when they are clearly distinct, as in **two apples**, there is no need for this junctural marking. Two plosives can interact, as in **Pat caught cold**, removing the explosion of the first stop of the sequence. Finally, we mention that the pronunciation of the vowel in **the** varies with the following vowel. It is schwa before a consonant, but is articulated as a high front vowel before another vowel, thus minimizing the need for a hiatus gesture. The collection of phonetic adjustments such as those described here is part of the means of providing for articulatory smoothing, both within and between words. These changes are essential for good-quality synthetic speech, since otherwise the speech will sound choppy and overarticulated.

4. SPEECH SYNTHESIS

In the previous section, we described the utilization of a number of constraint domains to convert the input text string to a narrow phonetic description. Most TTS systems assume a linear representation for this description, in accordance with the phonological views of the 1960s and 1970s. More recently, a variety of forms of "nonlinear" phonology have been introduced, which argue for the use of considerably richer multidimensional geometric structures.

While these configurations are likely to be useful for TTS systems, they are not yet in use and must await further development since they imply large changes in overall TTS strategy as well as detailed algorithms.

In the earlier discussion of text analysis, the most important unit of analysis was the morph, representing the minimum syntactic atomic unit of the language. We emphasized that although the detection of morph boundaries is essential for the computation of word pronunciations, the selection of units for synthesis is an entirely separate matter. That is, even though the pronunciation is represented by a string of phonetic segments with stress and junctural markings, this sequence can be mapped onto a number of other classes of units, depending on the requirements of the synthesis scheme. Before discussing the various approaches to synthesis, then, it is important to consider the desiderata for these synthesis units.

Any TTS system forces composition of a small number of basic units to form larger utterances. There are clear trade-offs involving the *size* of the units. The larger the unit, the more elements are needed in the unit set. Thus there can be as few as 40 phonemes but at least 1000 diphones and an even larger number of syllables. Attempts to record and play back words have not been successful, largely due to the large and changing number of words and the need to make contextual adjustments. This fact illustrates that the choice and representation of units must allow sufficient degrees of freedom to *control* all significant aspects of the waveform. Furthermore, it must be possible to *analyze* speech in terms of these units and their parameters. Recently, there has been much interest in automatic extraction and training of units, since the establishment of a unit set can be tedious and time-consuming. The difficulty of analyzing speech to establish articulatory gestures as units is very great, since these are underlying entities not readily apparent in the speech waveform. Thus even though articulation is a fundamental way to characterize speech production, it implies great difficulty in building up a repertoire of articulatory gestures. It must be possible to *compose* an utterance by interpretive processes on the units. This requirement highlights the tension between internal and external structure and cohesion among the units. The quest for units has led to the selection of some structures, such as the diphone, which minimize the need for compositional adjustments. Thus although the number of phonemes is small, they are difficult to compose, whereas the number of diphones is more than an order of magnitude larger, but diphones are selected for their easy composition properties. These considerations show that the units should have a *natural relation* to the speech production model used in the synthesis system. Source/filter models and articulatory models emphasize aspects of speech production that strongly imply the nature of the units selected. It is further desirable that the units should provide *insightful* relevance to the research literature. For this reason, units based on both formants and LPC have enjoyed popularity, whereas there is much less guidance from the literature for the selection of articulatory units.

A number of important implementation considerations affect the selection of units. They can either be used in compiled static form or be processed interpretatively. Compiled forms imply less run-time complexity in synthesis, but also less flexibility for change. Thus the larger units, such as diphones and syllables, may bind many important features internally within the unit, but they also deny accessibility to these features, such as durations and vowel quality, which are unavailable to the compositional process. Of course, the shorter units, such as phonemes, require knowledge of these adjustments in order to synthesize speech, whereas the larger units do not require the specific manipulation of many aspects of speech production that may not be understood comprehensively. As research understanding grows, there is a tendency for late binding and flexibility to favor smaller units, but at the same time implementation technology may favor large unit sets, since the cost of storing these inventories is rapidly decreasing.

In discussions of speech synthesis, the notion of unit is usually thought of as referring to the sequence of *phonetic segments* in the utterance. But larger units for the specification of prosodic variables are also important, although not nearly as well formulated or understood as those at the phonetic segment level. Nevertheless, it is important to realize that in general there is a hierarchy of units of varying *scope*. Furthermore, there is substantial overlap of many units, in accordance with the notions of autosegmental phonology (Goldsmith, 1990). There is an increasing tendency to modularize understanding at a number of levels, together with an interactive structure between these levels, involving both morphology and phonology (Durand, 1990).

The notion of *target* has been important in characterizing units and has served as a bridge between abstractions and physical properties. There has been a tendency for the targets themselves to become complex objects exhibiting composition, scope, and internal cohesion. For example, fundamental frequency contours must be characterized at both the phrase and sentence level and in terms of "segmental" effects. It is thus seen that the term "target," originally motivated by the characterization of formant trajectories, is increasingly synonymous with "abstract representational structure." In fact, targets do not need to be thought of as rigid structures, since they can incorporate contextually determined cue trade-offs. This means that units must be able to support the integrated variability of cue complexes that are known to be important to perception. Whereas this capability is difficult to provide with acoustically based units designed for source/filter models of speech production, it is a natural consequence of the selection of articulatory units. This is indeed the major motivation for articulatory synthesis, which has had a long history despite the lack of empirical basis and fundamental motor control theory needed for such models. A measure of the growth of this approach, however, is the recent proposal to treat articulatory gestures as basic phonological units, and the emerging class of synthesis systems built on this perspective (Browman and Goldstein, 1989). Emphasis on provision of naturally coherent cue complexes indicates increasing concern with the ability of synthesis units to support speech production that is perceptually very close to human speech, rather than merely trying to provide the necessary parameters to imitate human speech. While such "copy synthesis" is undoubtedly useful and important, the need to relate the critical choice of speech synthesis units to perceptual requirements is becoming increasingly realized. Simply put, much of contemporary speech synthesis relies on the invariant utilization of a few strongly sufficient cues in a quest to exceed some threshold of intelligibility. To realize both intelligible and natural-sounding speech consistently, however, a much more flexible approach is needed. Whether this requirement can be met by acoustic modeling or by articulatory means will become apparent as studies of synthesis and perception become increasingly intermixed.

With this discussion of synthesis units in mind, we can now elaborate the entire speech synthesis process. The nature of the units used will pervade this discussion. For the most part, practical synthesis systems have been based on use of the source/filter production model, where the spectral information is represented either via formants or LPC. Units of various size, ranging from the phoneme through the diphone, demisyllable, and syllable, have been used with this model, and indeed the nature of the targets used, the means afforded for compositional smoothing, and the way in which perceptually relevant cue complexes can be specified with these unit sets are directly related to the structure of source/filter model. As indicated above, articulatory synthesis not only attempts to back up the speech production chain to a more fundamental level but also one which is more difficult to characterize. While the more traditional synthesis units can be used to drive an articulatory model (Coker, 1973), the current trend is to develop an inventory of articulatory gestures directly, and hence *phonologically* specify the utterance in terms of these gestures, which would then excite an appropriate articulatory model. In this way, explicit reference to *acoustic* properties is avoided altogether, the transformation from gestures to the speech waveform being handled entirely and implicitly within the confines of the articulatory model. Thus the entire speech synthesis process may be initiated from a remarkably varied set of units, each with its corresponding speech production model. Nevertheless, the overall synthesis process can be logically built up through consideration of the following constraint domains:

- Temporal framework
- Unit instantiation
  Fundamental frequency contour units
  Spectral synthesis units (target instantiation and elaboration, with contextual smoothing)
- Model interpretation (determination of model control parameters and their interpretation to produce the output speech waveform)

The view expressed by this sequence is somewhat novel in that it makes a strong break between the two traditional prosodic domains, timing and fundamental frequency. Knowledge of the unit sequence is needed to set up the temporal framework, but what is really generated is an *event stream* to which the units get attached (instantiated). These units comprise both fundamental frequency contours and the more traditional segmental synthesis units. We group them together because, while they represent different classes of constraint in acoustically based systems, they both represent parameterizations of the synthesis model triggered by events in the temporal framework and smoothed by the interpretive instantiation process. In the articulatory framework, there appears to be no justification, in fact, to distinguish gestures governing fundamental frequency from those that govern cavity shapes and hence indirectly spectral properties. In this way, we aim to provide generality and insight into the nature of all synthesis processes.

4.1. Temporal Framework

We assume that a previous text analysis (or other means) has provided a detailed phonetic transcription of the utterance. Differences will arise in systems as to the "narrowness' of this specification, and in some systems it may be appropriate to provide a phonological description where the detailed phonetic modifications are accommodated by aspects of the synthesis model. Thus we will not be concerned with maintaining a rigid distinction between phonetic and phonological specifications, since the nature of needed input derives from the character of the units and speech production model. In order to fix the sequence of units in time, a temporal framework must be erected. In acoustically based systems, this process consists of determining the durations of segments or larger units (e.g., the syllable), which are then concatenated, whereas in articulatory synthesis the event times for the initiation of articulatory gestures are computed. Thus for acoustically based systems, both *when* and for how *long* each segment or unit must be instantiated must be found, for both spectral units and fundamental frequency contour units. In the articulatory world of description, however, gestures are launched and assume their own course in time except as influenced by other gestures. In either case, we can unify the description by considering the temporal framework to consist of a sequence of events, which may or may not have durations associated with them.

Modeling the temporal framework raises a number of difficult issues centered on the notions of rate and inherent duration and the arbitrary way in which durations are measured from acoustic data. The use of segment durations derives from the implicit assumption that phonologically derived segments have a physically distinct reality, but at the acoustic level they overlap and interfere due to coarticulation. For this reason it is important to utilize consistent measuring conventions, sometimes implemented in an algorithm, but the definition of boundaries is always arbitrary. As long as the synthesis-by-rule algorithms follow the same convention, however, there is little difficulty in practice. The notions of *rate* and *inherent duration*, however, do not arise directly from physical measurements and take on meaning only in an overall model for utterance timing. In a TTS system, it would be desirable to have a single control for speaking rate, but this would imply the need to control all aspects of the timing framework in order to achieve the desired rate while maintaining both intelligibility and naturalness. Despite substantial study of temporal phenomena, no comprehensive model exists with rate as an explicit parameter. Indeed, prosodic modeling is a major challenge for TTS systems, and improvements in this area will require improved text analysis, including parsing, to have the most impact on speech quality.

Over a wide range of speaking rates, it has been observed that measured segments have a *minimum duration* (Gaitenby, 1965). For this reason, segmental durations have often been modeled as a fixed minimum duration plus a variable part determined by a number of factors (Klatt, 1979; Umeda, 1975, 1977). Pauses are inserted before each sentence-internal main clause and at boundaries delimited by a syntactic comma. There are a number of lengthening effects at syntactic boundaries. The vowel or syllabic consonant just before a pause is lengthened, as well as any consonants following the vowel nucleus. Such lengthening has also been observed at the end of phrases and clauses, leading to a view of recursive nested prejunctural lengthening. Many segments are shortened in a variety of environments. Syllabic shortening in nonphrase-final and nonword-final positions is observed, and noninitial consonants are also shortened. There is significant shortening in polysyllabic words, as well as in consonant clusters (O'Shaughnessy, 1974). Unstressed segments are reduced in length and also have a shorter minimum duration. An important perceptual cue is provided by the length of vowels before consonants, voiced consonants being associated with longer preceding vowels than unvoiced consonants. Another segmental effect is the lengthening of vowels preceded by a voiceless plosive. Finally, we mention that substantial lengthening is provided for emphasized vowels, particularly those arising from semantic modality contexts. This collection of phenomena has been codified in simple formulas adequate for intelligible speech. Data for this purpose should not be averaged across speakers, as there is much individual variation. A major need now is to collect large amounts of durational data to train more sophisticated models that reflect natural durational changes as a result of these many factors. It must be emphasized that shortening is not a simple linear compression over the segments involved, but is a much more complicated process that is yet to be modeled either accurately or insightfully over a broad set of linguistic environments.

4.2. Unit Instantiation

Once the temporal framework has been established, synthesis units can be attached to it as events. These units are of two types. Phonemic segments, diphones, demisyllables, and syllables have been used as synthesis units. These are used to characterize the spectral consequences of both excitation and cavity shape. The factors affecting choice of these units have been discussed above. In addition, during voiced intervals, a fundamental frequency contour is specified as a set of unit trajectories. Alignment between the spectrally oriented units, the fundamental frequency units, and the temporal framework is obviously essential, and slight misalignments can lead to substantial degradations in utterance quality.

4.2.1. Fundamental Frequency Contour Units

The characterization of fundamental frequency contours has been difficult due to the multiple functional loads carried by this parameter. As a result, no unified theory exists that combines all factors known to affect the contour over a broad range of linguistic contexts. At the sentence level, fundamental frequency (F_0) indicates a declaration or a question, and for the latter, whether it is a "yes/no" question or a "wh" question. The archetype contour for declarative sentences rises quickly on the first stressed syllable, and then declines slowly throughout the utterance until the last stressed syllable, when it falls rapidly to a low level. Clearly the temporal extent of the sentence must be known before this overall contour can be established, since otherwise the extent of the initial F_0 rise and the declination slope cannot be determined.

Furthermore, although many investigators have attempted to fit a "declination line" to the contour, such a line can only have status by virtue of a theory in which it plays a part in determining the actual detailed F_0 contour. For "yes/no" questions, the contour rises quickly on the first stressed syllable but tends to stay high on a nearly level plateau before rising slightly on the final syllable. In the case of "wh" questions, the declination after the initial rise tends to fall more quickly than for a declarative sentence, rising slightly on the final stressed syllable, then falling to the final low level associated with declaratives.

When there are multiple clauses in a sentence, the archetype contours, as described above, are joined by a "fall-rise" continuation rise indicating that although a clause is ending, the end of the entire sentence has not yet been reached. Note that the initial rapid rise in F_0, the declination, the final rise for questions, the final fall for declaratives, and the continuation rise are all F_0 gestures, or partial contours that can be thought of as inserted on the temporal framework, and composed to form the total contour.

The smooth fall of the declination line is perturbed by a number of influences. Within phrasal units, F_0 rises on the first stressed syllable, then declines, and falls back to the main declination line at the last stressed syllable of the phrase. These phrasal rises are often thought of as the "hat pattern" ('t Hart and Cohen, 1973). In addition to these raised plateaus, individual stressed syllables are indicated by short F_0 perturbations. The way in which the F_0 contour changes at syntactic boundaries and stress points has been characterized by a set of smoothed steps and impulses (Öhman, 1967; Maeda, 1974), each change corresponding to an event in the temporal framework. The magnitude of the steps and the impulses can be thought of as targets for the F_0 contour, although they may not be achieved due to the smoothing. An alternate way of characterizing the F_0 trajectory is to first establish an upper and a lower declination line, with the understanding that F_0 perturbations take place between these declining lines, except for the final fall. The difference (in Hz) between the two lines diminishes monotonically with time, so that F_0 perturbations are more pronounced on an absolute scale earlier in the utterance than later. Targets are then established for salient syntactic and stress events, and the target peaks are then connected by canonical catenary transition contours (Pierrehumbert, 1981). From this discussion, it is clear that all approaches to F_0 modeling have employed targets, smoothing, and canonical contour segments or gestures, from which the overall contour is generated.

In addition to the influences on F_0 described above, there are a number of segmental effects (Umeda, 1981). When an unvoiced consonant (particularly a plosive) precedes a vowel, there is a downward transient of F_0 into the succeeding vowel. On the other hand, when the preceding consonant is voiced, the F_0 transient rises into the following vowel. Positive correlation of F_0 level with tongue height has also been reported, but no consistent pattern has been found in fluent speech. The location in time of the F_0 peak in vowels is perceptually significant. When the peak occurs early in the vowel, it is perceived as stressed, but when the peak is late in the vowel, the F_0 pattern may be perceived as a continuation rise.

Apart from linguistic stress, the part of speech of a word can contribute to strong F_0 perturbations. These often arise due to the speaker indicating his or her attitude toward the truth value of the utterance proposition. These "modality effects" (O'Shaughnessy and Allen, 1983) often arise from the expression of positive belief, doubt, or negation and can be predicted lexically. Thus sentential adverbs are strongly marked, as well as negative elements, interrogative adjectives, quantifiers, and stressable (past tense) modals. In addition, there are strong effects on F_0 due to the presence of new versus old information and focus-shifting transformations. When syntactic entities refer to old information, they are destressed and receive a correspondingly low F_0 contour. Due to this destressing, other elements of the utterance may be perceived contrastively, with large F_0 excursions. Furthermore, transformations that serve to focus part of the sentence, such as topicalization, place extra F_0 emphasis on these elements. Although these effects have been noted in read speech, they have not been codified in F_0 contour algorithms due to the lack of ability to detect the focus-shifting transformations. The presence of these emphases, however, would undoubtedly enhance the liveliness and naturalness of the synthetic speech.

A number of additional influences on F_0 have been noted. For female voices, F_0 values are approximately 1.7 times the male values, but a simple linear expansion does not give a realistic contour. Thus, F_0 is a cue to gender, and it is also correlated with head size, psychological state, and speaker attitude. For use in TTS systems, however, these influences are not utilized, apart from the adjustment in F_0 for female voices. On the other hand, some systems allow manual contrastive marking of stress for emphasis and even the introduction of custom F_0 contours for some application needs.

For the most part, F_0 contour algorithms in TTS systems have been derived from multifactor curve fitting rather than any comprehensive theory (O'Shaughnessy, 1977). A great deal of additional study is needed of F_0 contours in multiple environments, as well as the development of an insightful framework for the expression of F_0 events. The fact that F_0 is used by speakers for so many purposes makes it especially hard to isolate effects, and the difference between laboratory conditions and fluent speech has led to a tension between the need for experimental control and the desire for realistically obtained data. Inasmuch as F_0 is tightly tied to the presence of particular syntactic units, correct parsing is particularly important for the assignment of appropriate F_0 contours. It seems prudent to first build a basic model for F_0 for simple syntactic constructions read in well-controlled linguistic environments. While the resulting contours may not be as natural as desired, it is more important to build up fundamental understanding of the constraints on the F_0 contour. There are far too many factors influencing the trajectory to try to build an initial theory from unrestricted data. On the other hand, there is ample opportunity for new insightful modeling of restricted classes of utterances, and the contours desired from such models will augment the intelligibility and perceptual ease of the synthetic speech substantially.

4.2.2. Spectral Synthesis Units

We now turn to units that contribute directly to the spectral character of the utterance. This includes the nature of the excitation, both voiced and unvoiced, as well as the resonance characteristics imposed by the geometry of the oral, nasal, and pharyngeal cavities. All of these units participate in some form of speech synthesis by rule, but they vary in the nature and degree of abstraction utilized.

The least abstract form of spectral unit corresponds to some chunk of the utterance coded in terms of the formant or LPC parameters. For this purpose, phoneme-sized units are too small, since simple concatenation, followed by some sort of smoothing, cannot accommodate the extensive degree of coarticulation at this level. The next biggest unit is the *diphone*, which seeks to place the region of formant transitions within the unit, while placing the unit borders in the middle of (hopefully) stable vowel nuclei. The idea is quite old and was impractical when recorded diphones were merely butted together, but in current practice the diphones are represented in LPC and smoothed on the basis of area functions (Olive and Spickenagle, 1976; Olive, 1977). Approximately 1000 diphones are needed to adequately cover all phonetic contexts. This approach has the advantage that diphones can be automatically extracted from speech and that little knowledge of perceptually relevant cues is needed, since the representation of the utterance chunks is direct at the acoustic level. Despite the area function smoothing, mismatches at diphone boundaries can occur, and it is difficult to adjust the timing and fundamental frequency within the diphone units in a flexible way. Furthermore, for use in an unrestricted TTS system, some effort must be made to obtain consistent speech for analysis, otherwise the "speaker" may appear to change in the middle of an utterance due to change in voice quality. The outstanding attribute of this approach is the fact that synthesis systems using diphones can be constructed and modified relatively easily without the need for basic understanding of speech production. With respect to the quality of contemporary TTS systems, diphones produce a good level of intelligibility and naturalness and are easily and inexpensively utilized in the ambient technology. Following the same reasoning that led to the diphone, *demisyllable* and *syllabically* based synthesis systems have been proposed (Browman, 1980; Fujimura and Lovins, 1978). The expected advantage of these larger units is that more of the important coarticulation is expected to occur internal to the units, but there is still a need to adjust the units at their borders. Of course, as the size of the units increases, their number does also, and as many as 10,000 syllables may be needed for English, although the number of demisyllables is comparable to the number of diphones. The choice between diphones and demisyllables is based on finding the best place to establish unit boundaries. At present, although no direct comparison of these two approaches has been made, diphones are much more widely used in practice. Concatenation of diphones or demisyllables may not appear to constitute speech "synthesis by rule," but it can be regarded as utilizing a simple selection and smoothing control structure, together with some limited control of duration and F_0, and hence the use of rules is minimal. These units do not utilize any abstraction beside those of signal processing and do not require any commitment to theories or models of speech production. Thus they do not advance the fundamental understanding of either the articulatory or acoustic theory of speech production, but they do strongly contribute to the establishment of good-quality practical TTS systems.

Formant-based speech synthesis by rule has an extensive history, and several robust systems have been developed. For this approach, a phoneme-sized unit is used, and approximately 40 of these suffice. They strongly utilize the accumulated understanding of speech production regarded as the extensively coarticulated elaboration of an allophone sequence. Much of the basic research for this point of view can be found in Lehiste (1967), Mattingly (1968), and Klatt (1987), and we do not propose to review this extensive literature here. The most important abstraction introduced for this approach is the notion of *target*. For each phonetic segment, target values for the individual formants are stored in a table. A parameter generation algorithm then operates on the target sequence to provide the spectral *frames* required by the synthesis model at regular intervals, usually either 5 or 10 ms. A notion of dominance has been introduced (Holmes et al., 1964) between adjacent segments which determines the relative influence of each of two neighboring segments on the formant transitions between them. A variety of smoothing functions have been applied to the parameter contours, and the needed table data have been determined from extensive analysis of CV, VC, CVC, and VCV utterances. By far the most difficult segments to synthesize have been the stop consonants. Here, proper management of the set of perceptually relevant cues, including the burst spectrum, formant transitions, and voice onset time (among many others, see Lisker, 1986), has led to the necessity of introducing several subclasses of segment category. For example, the generation of proper burst spectra has required specification of the following vowel as front, back unrounded, or back rounded. Since every potential acoustic cue has been shown to have some perceptual value (Dorman et al., 1977), there is a tension in the design of these systems between trying to maintain generality and abstraction and providing a great amount of contextually determined variation at a low acoustic level. The result has been that, to a substantial extent (directly correlated with the level of abstraction and inversely correlated with the fineness of acoustic detail utilized), formant synthesis-by-rule systems tend to use a few of the strong cues over a broad range of contexts, when in natural speech there would be a more varied and integrated set of cues. This represents a trade-off between the efficiency of abstraction and the storage of coordinated phonetic detail, to the extent that it is known. With time, a great many enhancements have been added to these systems. Careful manipulation of vowel nasalization, voice onset time, target undershoot, vowel-vowel coarticulation across intervening consonants, and many other phenomena has incrementally improved the quality of formant-based synthesis-by-rule through the progressive accommodation of an increasingly wide range of phonetic detail. Thus, while the abstraction represented by the notion of target and transition seeks to provide some theoretical basis for such synthesis, in fact the added phonetic detail is agglutinated to the basic model, without revising its nature or providing richer abstractions. Nevertheless, it is a tribute to the designers of these systems that the quality of synthetic speech produced is very good in terms of intelligibility and somewhat less good in terms of naturalness, and this approach is certainly competitive with the diphone approach while encapsulating a level of explicit phonetic knowledge in a procedurally coordinated way.

There is some trade-off between the refinement of the narrow phonetic transcription at the symbolic level in the text analysis phase and the adjustment of parameters within the spectral synthesis units during the speech synthesis phase. A wide variety of coarticulatory smoothing is essential to achieving good-quality synthetic speech, but there is no clear guidance as to where these constraints should be stated in the overall TTS system. Many of these adjustments have to do with the position of a segment, both in a word (initial, medial, or final) and at other junctural points. Many of these important cues are discussed by Lehiste (1959, 1964) and Klatt (1987). It has sometimes been thought that these modifications arise due to "sloppy speech," but in fact they have been found to be important cues to the perception of fluent speech since they reflect the nested syntagmatic units of the utterance. That is, far from being a nuisance, these refinements to the acoustic characterization of the utterance are of great perceptual utility to the listener, and significantly improve the intelligibility of the speech (Church, 1988). For example, while we have emphasized the importance of discovering the morphemes of textual words, it is correspondingly important to mark these boundaries phonetically as an aid to perception of the syntactic constituents of words. Thus, the first two segments of **scat** are pronounced differently than the corresponding fricative-stop sequence in **discourteous**, due to the intervening morph boundary. Many such junctural cues abound in fluent speech, and their utilization in formant-based synthesis, whether symbolically or parametrically, is required for high quality.

Although not yet used in practical systems, articulatory synthesis provides a fundamental means for speech production with many valuable properties. Early articulatory synthesizers directly manipulated the cross-sectional area function of a vocal tract model, but current systems permit the modification of a limited set of parameters that control the positions of the major articulators, including the lips, jaw, tongue body, tongue tip, velum, and hyoid bone (Rubin et al., 1981). Such systems have been intended for perceptual research, but they have also been incorporated into TTS systems (Coker, 1976). Articulatory synthesis is attractive since it includes dynamic constraints of the articulators and implies a control strategy for the basic parameters. Whether or not the control is based on articulatory variables directly or in part on auditory feedback is debatable, it seems that the articulatory level of specifying speech production is more basic and logically prior to the acoustic level of description. Certainly, acoustic cues that may appear to be distinct at the acoustic level are often "integrated" or unified by articulation, and the dynamics of movement can provide a natural explanation for seemingly spurious acoustic events, such as the production of epenthetic stops which arise due to articulatory overlap. An interesting area of investigation is the study of articulatory "look ahead" for unconstrained articulators. For example, when a bilabial closure takes place, the tongue is free to move to a position appropriate to the next sounds in sequence. Velum lowering in anticipation of an upcoming nasal is also well known. Understanding of these movements is extremely valuable in modeling coarticulation and is of value to formant-based synthesis as well as articulatory synthesis.

Laryngeal modeling is also important for synthesis, since it reveals a variety of forms of voiced excitation and has a direct effect on the spectral representation of speech. Successful modeling of vocal fold vibrations has been coupled to supralaryngeal synthesis to provide high-quality synthesized speech, but not by rule (Flanagan and Ishizaka, 1976, 1978; Flanagan et al.,

1975). Particularly at the beginning and end of voiced intervals, vocal cord motion is very complex, giving rise to a variety of spectral effects. For this reason, it may be too simple to characterize voicing in terms of a single scalar number, F_0, for purposes of synthesis. More complex models for voiced excitation will undoubtedly be incorporated into TTS systems. Furthermore, natural means of generating mixed excitation, as in voiced fricatives and devoicing in vowels, can probably be better achieved in the articulatory domain than by purely acoustic means. This is particularly true given the structure of formant synthesizers, where mixed excitation is sometimes introduced in unrealistic ways. The fact that fricatives sometimes sound separate from the rest of the synthetic speech probably arises in part from such approximations.

Research in articulatory synthesis, difficult as it is, continues to contribute to the basic understanding of speech production. Articulatory "gestures" are not only being proposed as basic phonological units (Browman and Goldstein, 1989), but are also being used as the input for an articulatory synthesizer. These results are valuable not only for articulatory synthesis, which can be expected to evolve over a long time, but also for the understanding provided, which can be directly exploited by acoustic formant synthesis. Thus articulatory studies are of increasing utility for TTS systems even though articulatory models are not employed directly.

4.3. Model Interpretation

Once the synthesis units (both fundamental frequency and spectral) have been attached to the temporal framework, and their control parameters smoothed, a fully elaborated set of control frame values can be obtained to drive the synthesis model. Thus, for formant synthesis, smoothing of the targets (Holmes, 1988) produces approximately 20 control values for each frame, and these values are used for a brief interval of 5 or 10 ms, which is short enough to capture the perceptually significant dynamics of the vocal tract configuration. Similarly, for articulatory synthesis, positions of the key articulators must be converted to cross-sectional area frames. This is usually done by first transforming from the key articulator positions to sagittal cross dimensions, which are then converted (using tabular data) to cross-sectional areas. The resulting area function is then approximated by a sequence of uniform tube sections, which change at the frame rate.

The frames thus contain the parametric data that control the synthesizer for the short intervals of time between frame updates (5 to 10 ms). It then remains to exercise the synthesizer to produce an output speech waveform. In the case of formant synthesis, the synthesizer is made up of resonances connected either in cascade or in parallel (or both) which are excited by a sequence of glottal pulses (the shape of which can sometimes be manipulated) and a fricative noise source. The selection of excitation source, as well as mixed excitation, is parametrically controlled. The design of these synthesizer models has received a lot of attention, with the goal of matching spectral detail (Allen et al., 1987; Holmes, 1983), and it has been shown that when the parameters for such models are controlled manually, speech that is indistinguishable from natural speech can be produced (Holmes, 1973). Thus any weakness in speech synthesis by rule arises from inappropriate parameters rather than the synthesizer model itself. In contemporary technology, the synthesizer is implemented as a sampled digital filter system, requiring the parameter values contained in each frame to be converted to coefficient values for the difference equations used in the synthesizer. These coefficients must then be loaded into the digital system at the frame rate, and the sampled data system then produces speech waveform samples at the system sampling rate, which is usually 10 kHz. Of course, the output samples must be smoothed by an appropriate low-pass filter to produce the output analog waveform.

For systems using concatenated stored parametric units, such as diphones, the synthesis process is similar. These systems typically represent speech in terms of an LPC frame sequence. This frame sequence, obtained from storage, is smoothed at the unit (diphone) boundary so as to provide smooth changes in area function. The resulting set of smoothed frames is then converted to appropriate coefficients for a digital LPC model, and these are loaded into the digital synthesizer at the frame rate. The synthesizer then produces output speech samples, which are smoothed to form the analog waveform, as with formant synthesis.

In the case of articulatory synthesis, an acoustic transfer function is computed from the uniform-tube-section representation. (In some approaches, eigenvalues [resonant frequencies] of the tube model are computed and used as formant frequencies in a formant synthesizer, as described above.) The acoustic waveform corresponding to glottal excitation can either be supplied directly or computed from a laryngeal model. Fricative excitation can be provided by insertion of a noise source just anterior to the tube section with minimum cross-sectional area. The acoustic output of the vocal tract is then obtained by exciting the acoustic transfer function, which is implemented as a digital filter.

In all synthesis schemes described here, the synthesizer model is typically implemented on a single digital signal processing integrated circuit. Contemporary circuits are more than adequate to provide the real-time digital signal processing capability needed for these synthesizers. In fact, the entire speech synthesis process, starting from the narrow phonetic transcription (produced by the text analysis system) and ending at the final output analog speech waveform, can be readily accommodated on such a circuit together with memory for tables.

5. RULE FRAMEWORKS AND LANGUAGES

In the past, most TTS systems have been built using ad hoc approaches to the representation of rules and data, as well as the way in which the several constraint domains interact. It is not uncommon for systems to be implemented in low-level programming languages with modest use of structured programming concepts in order to achieve flexibility and efficiency. As TTS systems have come to be developed in substantial number, however, the need for an overall formalism and language particularly suited to the needs of text analysis and speech synthesis has been felt. Such capability lends clarity to the structure and content of the system, makes modification easy, increases accessibility of TTS capability to a wider group of interested researchers, undoubtedly increases the productivity of TTS system developers, and encourages a theoretical framework for expressing the constraints involved in the overall conversion process. The introduction of such formalisms, even if not utilized computationally, can be very useful. Perhaps the best-known example of rule formalism for linguistic constraints is the set of notational conventions used for the expression of phonological rules (Chomsky and Halle, 1968). This format permits the expression of a linear sequence of elements, varying numbers of elements, disjunctive choice of elements, and arbitrary context. Rules formulated in this way are virtually impossible to understand when expressed in a natural language, to say nothing of additional constraints introducing cyclicity and ordering. Without doubt, this rule format has enhanced research and communication of results in phonology and illustrates the well-known fact that notation can often have either a positive or a negative influence on the advancement of science. There is, however, a negative side to such formalisms, since they tend to constrain the representation of rules and at times force these rules into unnatural frameworks that may even lead to incorrect results. The evolution of "post SPE" (Sound Pattern of English; Chomsky and Halle, 1968) phonology has illustrated these difficulties. A great many new forms of phonological representation have been introduced, their common heritage being a rejection of the *linear* representation of rules utilized in SPE. Contemporary phonology has introduced a much richer set of structures, and argued for their necessity in order to describe language phenomena insightfully and parsimoniously. Although it is thus easy to criticize the use of linear formalisms for phonology, it is undoubtedly true that this representation "took a stand" on the requisite form of rules and stimulated a large amount of research. When research results are not presented in clear formal terms, they do not become effectively codified, and differences in results are harder to detect, as are contradictions to established results. Thus the search for an adequate phonological formalism is part of the ongoing research process, and it can be expected to evolve, parts of it being discarded from time to time in favor of new forms. The formalism sharpens and focuses research issues and avoids much confusion in research claims. Of course, it has also given rise to the occasional finding that two allegedly conflicting results are finally seen as "notational variants" of each other, but the understanding of such equivalences is in itself useful.

Building on the experience in phonology, a number of researchers have developed special-purpose programming languages for the formulation of TTS rules. These languages permit the introduction of abstract classes of groups of language elements, such as phonemes, in a way that provides generality and conciseness for the rule system (Carlson and Granström, 1975, 1976). These rules can either be interpreted or compiled into conventional programming languages, providing both flexibility and ease of implementation. Similar experiences with other rule systems (Hertz, 1982) have led to the introduction of a new data structure (Hertz et al., 1985) more in spirit with the recent developments of nonlinear phonology. This structure, called DELTA, recognizes the need to characterize multiple constraint domains simultaneously and also to show the "alignment" or interaction between these domains. With DELTA, it is possible to define any sequential *stream* of tokens corresponding to a constraint domain, such as a textual character string, the names of words, and their constituent morphs, syllables, and phonemes. These token streams are then *aligned* at various points to show how the streams are related. A language has been introduced to manipulate this structure, and a very wide class of rules can be readily accommodated in this way. As experience builds up with this approach, in a variety of application domains including both acoustic and articulatory synthesis, new insights and understandings can be expected with respect to the needed formalisms for TTS and other language-related systems. The continuing evolution of such languages and data structures is to be encouraged, since they provide the best framework for pinpointing the strengths and weaknesses of current theories. It is no accident that most work on multilingual TTS systems has taken place in laboratories where there was an emphasis on rule languages and data structures (Carlson and Granström, 1986; Hertz et al., 1985). The ability to configure a new rule system quickly has led to TTS systems for many languages, including English (both American and British), Swedish, Norwegian, Spanish, French, German, Italian, Japanese, and Hebrew. Undoubtedly many more such systems will be readily assembled. Comparison of the resulting rule systems can be expected to lead to new understanding of universal capability needed in rule and data structures. Furthermore, these systems, being expressed in a common formalism for all languages, can be readily implemented in portable technology, providing easily switched multilingual capability for an increasingly interactive world.

6. PERCEPTUAL EVALUATION OF TEXT-TO-SPEECH SYSTEMS

It was not until the late 1970s that a concerted effort was made to assess the perceptual quality of speech produced by TTS systems (Pisoni and Hunnicutt, 1980). At that time, a survey of methods available for this purpose revealed a surprising lack of techniques, and an almost total concern with segmental intelligibility rather than other measures such as naturalness and comprehension. Many of the existing tests had been developed to assess perception in noise and other forms of degraded speech and were used with isolated words and nonsense sentences. It is still difficult to devise tests which are both sensitive and meaningful, so direct testing of a system for its intended application is always preferable. Nevertheless, a substantial amount of comparative testing of TTS systems has been done (Pisoni et al., 1985; Klatt, 1987), with most emphasis on intelligibility and comprehension. Since these references provide a comprehensive discussion, we will not review these results in detail here, but will provide an overview of the main findings and their implications.

Intelligibility has been tested using the modified rhyme test, nonsense words, anomalous sentences, and speech in noise. Tests of CVC intelligibility with the modified rhyme test show error rates of 3 to 10%, compared to 1% for natural speech. This level of performance appears to be adequate for many applications. What is of more interest, however, is the distribution of errors over the class of phonemic segments. The data are not uniformly available, but for one system (Allen et al., 1987), errors arose for initial stops and affricates, but mostly for final fricatives and particularly nasals. Both of these latter two classes have accordingly received considerable attention, resulting in lowered error rates. Unfortunately, testing is often an afterthought in the development of many systems, not just those for TTS. Much better would be an ongoing series of tests designed to reveal weaknesses in the system and hence provide a more effective distribution of effort toward the improvement of the system.

Comprehension from TTS-synthesized speech compared with visual reading has also been tested (Pisoni et al., 1985), showing that the ability to answer multiple-choice questions concerning a "read" paragraph is comparable between visual reading and TTS listening and that TTS speech is even somewhat more comprehensible than natural speech. To what extent these tests measure memory performance as opposed to speech or visual perceptual performance is not known, but the fact that TTS systems can be used in many systems with good comprehension is encouraging. Personal experience listening to a disabled "speaker" using a TTS system for a 1-hour lecture has shown that the speech was essentially perfectly comprehensible for a broad audience in a large lecture theater.

Naturalness is an elusive subjective attribute of speech which attempts to characterize the degree to which synthetic speech approximates the characteristics of natural human speech. Although there are applications where it is desirable that the speech be recognized as *non*human, a general goal of TTS systems has been to improve naturalness. Techniques exist for the comparison of two systems producing synthetic speech from the same text, so that even though no clear definition of naturalness is available, consistent comparative judgments are possible. While these tests provide a desirable selection criterion, little insight as to the characteristics of synthetic speech responsible for naturalness has been obtained.

7. FUTURE DIRECTIONS

The fact that TTS systems are commercially available suggests that they are adequate for many applications. Although this is true, they can hardly be regarded as having reached optimal performance, and much basic research remains to be done.

In the text analysis phase, there is a need for more abstract underlying forms and for the principled interaction of morphological and phonological processes, as contemplated by lexical phonology. Syntactic analysis must be improved to provide a better basis for prosodic frameworks. The characterization of speech acts and discourse in a form that can be exploited by TTS systems will be increasingly needed as interactive human-machine dialogs become more prevalent.

In the speech synthesis phase, many challenging problems must be solved in order to improve the output speech quality. The dynamic selection of synthesis units of varying size may provide better speech quality with a more robust compositional process. A great deal of detailed phonetic knowledge is needed to characterize a large variety of utterance contexts. Junctural cues that are related to the syntactic structure of the utterance must be discovered and implemented, and the deterministic selection of cue subsets for the maintenance of phonemic contrasts is needed. Articulatory synthesis, coupled with new forms of phonology where articulatory gestures are basic, holds the promise of unifying a great deal of phonetic detail that seems otherwise unrelated. The difficulty of devising articulatory synthesis strategies is large, but great benefit from this approach is expected, despite the long time and large effort needed to accumulate the needed data and build useful synthesis systems. These articulatory studies will benefit not only the characterization of spectral sequences but also the proper representation of vocal tract excitation including frication, laryngeal function, and the combination of these two sources. Such basic studies of speech production will also help to address the problems of providing multiple voices for synthesis. One can anticipate a time when new voices are readily accommodated in TTS systems, affording a degree of personalization not currently contemplated.

Although there are clearly many distinct research tasks needed to improve both text analysis and speech synthesis, a number of general observations can also be made. The study of the constraint domains relevant to speech production has only just begun, and the *representation* of these constraints, together with their interaction, over a large corpus of utterances must be provided. This means that the speech production process must be *modeled* at all levels. Most current models have been developed for sequences of spectral frames, but comprehensive models for the temporal framework and fundamental frequency are also needed. These models must be formulated in computationally tractable ways, so that their parameters can be automatically trained without the need for human interaction and judgment. As these formal models become more sophisticated, they will motivate new rule formalisms and data structures, which will in turn form the basis for improved TTS programming languages.

Just as speech production must be modeled using appropriate representations, the nature of human speech perception must be related to integrated cue ensembles, so that synthesis not only aims to mimic human speech production but also benefits from an understanding of the ways in which perceptual distinctions can be made. New forms of nonlinear phonology, which seek to explain the regularities of these contrasts, are becoming increasingly concerned with representations, such as articulatory gestures, that form a *direct* part of the production process, so that there will be an increasing overlap of phonological formalisms with the rule and data structure formalisms used in TTS systems. These trends imply an increasing level of interaction with many domains of speech and language science, providing not only benefit to TTS research, but also motivation and an important test-bed for the results of perceptual research and phonological theory.

Research in TTS systems thus participates in a widening spectrum of research in language, speech, and hearing. These systems embody much of what is known in these disciplines and thus comprise (in part) the continuing evolution of our collective understanding in these disciplines. They provide not only a useful technology for many applications but also an intellectual focus for the integration of many studies of highly complex human language behavior. In this way, TTS systems will continue to evolve in capability and utility for practical needs, and also constitute the fabric of much of speech and language understanding. They will not be merely derivative from that knowledge but will be a powerful means for extending it. To speak of TTS research pejoratively as "incremental" or "mature," while motivated by the appreciable capability already available, is thus to miss the intellectual excitement of current research, as vigorous now as when TTS systems were in their infancy, and the appreciation of potential for very significant growth and achievement. Given the central representational and formal role of TTS systems in speech and language research, the problem of providing high-quality TTS systems is not "solved" any more than the problems of speech and language research are completely resolved. These systems will thus continue to grow; motivating, focusing, and extending basic speech and language research and satisfying an increasing need for speech output technology in an increasingly computerized society. This evolving process promises to continue its rapid growth and achievement, reminding us of what we have already understood and stimulating us to discover more.

REFERENCES

Allen, J., Hunnicutt, M. S., and Klatt, D. H. (1987). *From Text to Speech: The MITalk System*. Cambridge University Press, London

Austin, J. L. (1976). *How to Do Things with Words*. Oxford University Press, London.

Browman, C. P. (1980). Rules for demisyllable synthesis using Lingua, a language interpreter. *Proc. Int. Conf. Acoustics, Speech, and Signal Processing*, ICASSP-80, pp. 561–564.

Browman, C. P., and Goldstein, L. (1989). Articulatory gestures as phonological units. *Phonology 6*(2):201.

Brown, G., and Yule, G. (1983). *Discourse Analysis*. Cambridge University Press, London.

Carlson, R., and Granström, B. (1975). A phonetically oriented programming language for rule description of speech. *Speech Communication*, G. Fant, ed., vol. 2, 245. Almqvist and Wiksell, Uppsala.

Carlson, R., and Granström, B. (1976). A text-to-speech system based entirely on rules. *Proc. Int. Conf. Acoustics, Speech, and Signal Processing*, ICASSP-76, pp. 686–688.

Carlson, R., and Granström, B. (1986). Linguistic processing in the KTH Multilingual text-to-speech system. *Proc. Int. Conf. Acoustics, Speech, and Signal Processing*, ICASSP-86, pp. 2403–2406.

Chomsky, A. N., and Halle, M. (1968). *Sound Pattern of English*. Harper & Row, New York.

Church, K. W. (1985). Stress assignment in letter-to-sound rules for speech synthesis. *Proc. 23rd Meeting Association for Computational Linguistics*, pp. 246–253.

Church, K. W. (1988). *Phrase Structure Parsing: A Method for Taking Advantage of Allophonic Constraints*. Kluwer Academic Publishers, Norwell.

Coker, C. H. (1973). Speech synthesis with a parametric articulatory model. In *Speech Synthesis*, J. L. Flanagan and L. R. Rabiner, eds. Dowden, Hutchinson, and Ross, Stroudsburg, Pa.

Coker, C. H. (1976). A model of articulatory dynamics and control. *Proc. IEEE 64*: 452.

Diehl, R. L., and Kluender, K. R. (1987). On the categorization of speech sounds. In *Categorical Perception*, S. Harnad, ed. Cambridge University Press, London.

Dorman, M. F., Studdert-Kennedy, M., and Raphael, L. J. (1977). Stop consonant recognition: Release bursts and formant transitions as functionally equivalent context-dependent cues. *Percept. Psychophys. 22*:109.

Durand, J. (1990). *Generative and Non-Linear Phonology*. Longman, London.

Flanagan, J. L., and Ishizaka, K. (1976). Automatic generation of voiceless excitation to a vocal cord–vocal tract speech synthesizer. *IEEE Trans. Acoust. Speech Signal Process. ASSP-24*:163.

Flanagan, J. L., and Ishizaka, K. (1978). Computer model to characterize the air volume displaced by the vibrating vocal cords. *J. Acoust. Soc. Am. 63*:558.

Flanagan, J. L., and Rabiner, L. R. (Eds.) (1973). *Speech Synthesis*. Dowden, Hutchinson, and Ross, Stroudsburg, Pa.

Flanagan, J. L., Ishizaka, K., and Shipley, K. L. (1975). Synthesis of speech from a dynamic model of the vocal cords and vocal tract. *Bell Syst. Tech. J. 54*:485.

Fujimura, O., and Lovins, J. (1978). Syllables as concatenative phonetic elements. In *Syllables and Segments*, A. Bell and J. B. Hopper, eds., p. 107, North Holland, New York.

Gaitenby, J. (1965). The elastic word. Status Report on Speech Research SR-2. Haskins Laboratories, New Haven.

Goldsmith, J. (1990). *Autosegmental and Metrical Phonology*. Blackwell, Oxford.

Halle, M., and Keyser, S. J. (1971). *English Stress: Its Form, Its Growth, and Its Role in Verse*. Harper &Row, New York.

't Hart, J., and Cohen, A. (1973). Intonation by rule: A perceptual quest. *J. Phonet. 1*:309.

Hertz, S. R. (1982). From text to speech with SRS. *J. Acoust. Soc. Am. 72*:1155.

Hertz, S. R., Kadin, J., and Karplus, K. J. (1985). The delta rule development system for speech synthesis from text. *Proc. IEEE 73*(11):1589.

Holmes, J. N. (1973). The influence of the glottal waveform on the naturalness of speech from a parallel formant synthesizer. *IEEE Trans. Audio Electroacoust. AU-21*:298.

Holmes, J. N. (1983). Formant synthesizers: Cascade or parallel. *Speech Commun.* 2:251.

Holmes, J. N. (1988). *Speech Synthesis and Recognition.* Van Nostrand Reinhold (UK), Wokingham.

Holmes, J. N., Mattingly, I. G., and Shearme, J. N. (1964). Speech synthesis by rule. *Lang. Speech* 7:127.

Hunnicutt, M. S. (1976). A new morph lexicon for English. *Proc. Sixth International Conference on Computational Linguistics*, Ottawa, Canada.

Hunnicutt, M. S. (1980). Grapheme-to-phoneme rules, a review. Speech Transmission Laboratory, QPSR 2-3:38. Royal Institute of Technology, Stockholm.

Jayant, N. S. (1984). *Digital Coding of Waveforms: Principles and Applications to Speech and Video.* Prentice-Hall, Englewood Cliffs, N.J.

Klatt, D. H. (1979). Synthesis by rule of segmental durations in English sentences. In *Frontiers of Speech Communication Research*, B. Lindblom and S. Öhman, eds., pp. 287–300. Academic Press, New York.

Klatt, D. H. (1987). Review of text-to-speech conversion for English. *J. Acoust. Soc. Am. 82*(3):737.

Kucera, H., and Francis, W. N. (1967). *Computational Analysis of Present-Day American English.* Brown University Press, Providence, R. I.

Lehiste, I. (1959). An acoustic-phonetic study of internal open juncture *Suppl. Phonet.* 5:1.

Lehiste, I. (1967). *Readings in Acoustic Phonetics.* MIT Press, Cambridge, Mass.

Lucassen, J. M., and Mercer, R. L. (1984). An information theoretic approach to the automatic determination of phonemic base forms. *Proc. Int. Conf. Acoustics, Speech and Signal Processing*, ICASSP-84, pp. 42.5.1–42.5.4.

Maeda, S. (1974). A characterization of fundamental frequency contours of speech. Research Laboratory of Electronics QPR 114:193. MIT, Cambridge, Mass.

Mattingly, I. G. (1968). Synthesis-by-rule of general American English. Supplement to Status Report on Speech Research. Haskins Laboratories, New Haven.

Mohanan, K. P. (1986). *The Theory of Lexical Phonology.* Reidel, Dordrecht, The Netherlands.

Öhman, S. E. G. (1967). Word and sentence intonation: A quantitative model. QPSR 2-3:20, Speech Transmission Laboratory, Royal Institute of Technology, Stockholm.

Olive, J. P. (1977). Rule synthesis of speech from diadic units. *Proc. Int. Conf. Acoustics, Speech, and Signal Processing*, ICASSP-77, pp. 568–570.

Olive, J. P., and Spickenagle, N. (1976). Speech resynthesis from phoneme-related parameters. *J. Acoust. Soc. Am.* 59:993.

O'Shaughnessy, D. (1974). Consonant durations in clusters. *IEEE Trans. Acoust. Speech Signal Process. ASSP-22*(4):282.
O'Shaughnessy, D. (1977). Fundamental frequency by rule for a text-to-speech system. *Proc. Int. Conf. Acoustics, Speech, and Signal Processing,* ICASSP-77, pp. 571–574.
O'Shaughnessy, D., and Allen, J. (1983). Linguistic modality effects on fundamental frequency in speech. *J. Acoust. Soc. Am. 74*(4):1155.
Pierrehumbert, J. (1981). Synthesizing intonation. *J. Acoust. Soc. Am. 70*:985.
Pisoni, D. B., and Hunnicutt, M. S. (1980). Perceptual evaluation of MITalk: The MIT unrestricted text-to-speech system. *Proc. Int. Conf. Acoust. Speech Signal Process. ICASSP-80*, pp. 572–575.
Pisoni, D. B., Nusbaum, H. C., and Greene, B. G. (1985). Perception of synthetic speech by rule. *Proc. IEEE 73*:1665.
Rubin, P., Baer, T., and Mermelstein, P. (1981). An articulatory synthesizer for perceptual research. *J. Acoust. Soc. Am. 70*(2):321.
Searle, J. R., and Vanderveken, D. (1985). *Foundations of Illocutionary Logic.* Cambridge University Press, London.
Umeda, N. (1975). Vowel duration in American English. *J. Acoust. Soc. Am. 58*(2): 434.
Umeda, N. (1977). Consonant duration in American English. *J. Acoust. Soc. Am. 61*(3):846.
Umeda, N. (1981). Influence of segmental factors on fundamental frequency in fluent speech. *J. Acoust. Soc. Am. 70*:350.
Venezky, R. L. (1970). *The Structure of English Orthography.* Mouton, 's Gravenhage.
Young, S. J., and Fallside, F. (1979). Speech synthesis from concept: A method for speech output from information systems. *J. Acoust. Soc. Am. 66*(3):685.

24

Text Analysis and Word Pronunciation in Text-to-Speech Synthesis

MARK Y. LIBERMAN University of Pennsylvania, Philadelphia, Pennsylvania

KENNETH W. CHURCH AT&T Bell Laboratories, Murray Hill, New Jersey

1. INTRODUCTION: THE STRUCTURE OF A TEXT-TO-SPEECH SYSTEM

A text-to-speech (TTS) system maps a sequence of numbers representing the characters of a text into another sequence of numbers representing the samples of an acoustic waveform. It is convenient to divide this mapping into four types of processing:

- Text analysis
- Word pronunciation
- Phonetic interpretation
- Signal generation

This chapter will focus on the first two of these. *Text analysis* involves a number of aspects of text "understanding": dividing the input letter sequence into words, dividing the word sequence into sentences, assigning parts of speech to the word tokens, and parsing them into intonation phrases. The text analysis phase must interpret punctuation (for example, to decide which periods end sentences and which do not) and treat such special orthographic categories as abbreviations, acronyms, dates, fractions, and amounts of money. Many of these tasks turn out to be more difficult (and more interesting) than one might have thought. We have found that a systematic attack that makes heavy use of a representative corpus gives better results than an approach based solely on memory and intuition.

Word pronunciation is the problem of translating spelling (orthography) into an IPA-like phonological representation of the sort found in most dictionaries. High-accuracy word pronunciation is a challenging problem, especially in languages like English and Japanese, where the writing system is not phonetically transparent—as opposed to Spanish, for example, where spelling is related to pronunciation in a more regular way. We will describe algorithms for pronunciation of English words, as they occur in unrestricted text, that reduce the error rate to only a few tenths of a percent for ordinary text, about two orders of magnitude better than the word error rates of 15% or so that were common a decade ago. In this area as well, we have found that a corpus-based approach—surveying frequency-weighted word lists—is the best road. Since many "words" (as much as 5% of newswire text, for instance) may be pronounced in a way that depends on their context of use, some form of text analysis is also crucial to achieving these low error rates. Even in languages such as Spanish, where standard word pronunciation can be determined fairly accurately by simple rules, the pronunciation of some orthographic word tokens (such as numbers in dates, fractions, amounts of money, addresses) may depend on their context of use.

The result of text analysis and word pronunciation is an explicit representation of the linguistic structure of the message encoded in the original text. The *phonetic interpretation* phase of a TTS system assigns quantitative phonetic values to the various aspects of this linguistic representation: durations of phonetic segments, F_0 target values for pitch accents, and so forth. The *signal generation* phase of a TTS system then uses this detailed phonetic specification to produce time functions of the control parameters for an acoustic or articulatory speech synthesis model, which are then used to calculate the samples of the speech waveform.* We will have nothing more to say in this chapter about these two types of processing.

The nature and difficulty of the text analysis and word pronunciation problems depend very much on the language and on the genre of text. For instance, in ordinary English text it is relatively easy to divide the text into words, but in Chinese this is a very hard problem (Sproat and Shih, 1991) because the average word length is about 2.3 characters, the set of words is open ended, and boundaries between words are not indicated in the orthographic

*See Chapters 23 and 25 for more discussion of these types of processing.

system. On the other hand, the problem of Chinese word pronunciation can be almost completely solved by table lookup on characters with a simple combination rule: there are only a few thousand characters, each of which corresponds almost always to a fixed syllable-plus-tone, and the pronunciation of a multicharacter word is almost always just the concatenation of the character pronunciations with rightward stress assignment. By contrast, high-accuracy word pronunciation in English is quite hard, requiring large tables of information about the pronunciation of words and word fragments, complex rules to bring these tables to bear on particular cases, and accurate text analysis to deal with contextually variable pronunciations.

Even within a single language, different types of text can raise very different sorts of difficulties. For instance, text analysis and word pronunciation programs that do a good job in reading the newspaper may fail quite badly in dealing with some data base text fields, which are all capital letters with erratic punctuation and spacing:

```
WILLIAM F BYRNE 4025 SANDY HILL RD APT 1C W SPRINGFIELD MA 01075
VIETTA J MCSWAIN 72 1/2 KING ST  PROSPECT HTS IL 60068
01011   GEL-KAM DTP FRUIT&BERRY 3.50Z   0.4%
01552   NYSTATIN TOPICAL OINT    15GM100000U
01610   MULTIVITAMIN W/FL CHEW TAB031.5MG
01727   ANTIBIOTIC OTIC SOLN 10CC
```

To cope with such material, a system needs to apply a variety of somewhat specialized text analysis methods. In our experience, applications are never "just plain text"; each application highlights a certain set of problems and stresses (or breaks!) some of a system's algorithms. The range of problems and solutions is very large, and we do not see a clean distinction between "standard" or "general" text on the one hand and "special" text on the other. Instead, each application has features of varying degrees of generality, from those that are shared with nearly all English text to those that are entirely unique. For instance, the headers and titles of newswire text, or of electronic mail, present problems like those of the name-and-address fields and drug catalog entries shown above.

This chapter will feature the approaches that we know best, primarily those that have been developed at AT&T Bell Laboratories.* We will discuss

---

*Among those who have contributed to the text analysis and word pronunciation aspects of the AT&T Bell Laboratories text-to-speech synthesizer, in addition to those contributions explicitly referenced in the text, we would like especially to cite Adam Buchsbaum, Jill Burstein, Cecil Coker, Julia Hirschberg, Peter Pavlovcik, James Rowley, and David Schulz.

a range of genres and highlight algorithms that are adaptable, or even directly trainable from text samples.

Throughout, our practice will be to evaluate each problem in terms of its frequency in a particular type of text and to evaluate each solution in terms of its performance and its costs. Except where explicitly noted, all examples are selected from some source of on-line text, such as the Associated Press newswire, rather than invented.

2. MOTIVATIONS FOR TEXT ANALYSIS

There are two reasons for a TTS system to do text analysis. One reason is that word pronunciation sometimes depends on usage: the word **I** can be a pronoun, a letter name, or a Roman numeral; the word **wind** can rhyme with "bind" or "binned"; **Dr.** can be "doctor" or "drive"; **2/3** can be "two thirds" or "February third" or "two slash three." A second and equally important reason for text analysis arises in the phonetic interpretation phase—we should modulate the pitch, timing, amplitude of the speech, and even its spectral trajectories so as to present the text's message clearly. In other words, we want the program to read as if it were a skilled speaker who had understood the text.

Although this chapter will not discuss phonetic interpretation at any length, we wish to illustrate how pervasively the results of text analysis affect the phonetic "shape" of a spoken phrase, even in very simple structures

Table 1

	Duration		F_2		F_0		RMS	
Position	Mean	S.E.	Mean	S.E.	Mean	S.E.	Mean	S.E.
1	110	4.6	1326	14.1	158	.97	2711	78
2	110	5.3	1366	26.6	157	2.7	2446	209
3	144	5.7	1261	16.4	178	3.0	3075	115
4	92	3.4	1342	14.5	152	2.3	1808	67
5	98	5.0	1335	11.5	139	1.13	1985	115
6	131	4.7	1273	17.6	173	1.07	2316	88
7	100	3.5	1326	15.8	143	3.7	1547	121
8	115	5.0	1282	15.1	171	2.8	2565	147
9	98	3.1	1309	18.9	144	3.2	1697	87
10	152	3.7	1179	7.1	115	1.3	1225	58

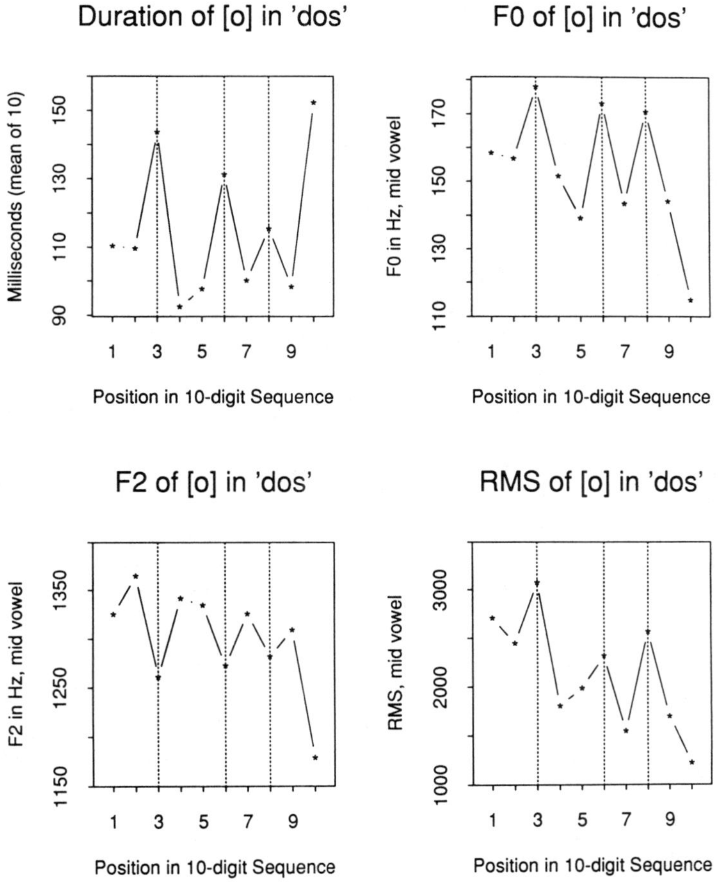

Figure 1 Phonetic shape of Spanish telephone numbers.

such as telephone numbers. We will illustrate this point briefly with some data from Spanish; we have looked at similar data from English as well as several other languages and believe that qualitatively similar results will emerge in any language.

Table 1 shows that the Spanish digit "dos" ("two") is realized very differently depending on its position within a 10-digit telephone number. In particular, the table shows the value of four phonetic factors for the vowel [o] in "dos" as a function of position within a 10-digit telephone number. Each value in Table 1 represents the mean of 10 measurements, or the standard error of that same set of 10 measurements, from a set of 100 ten-digit numbers arranged so that each digit occurs equally often in each position and each pair of digits occurs equally often spanning each pair of positions. Durations are given in milliseconds, and F_2 and F_0 values are given in hertz. The F_2, F_0, and rms amplitude measures are the mean of the central 40 ms of each vowel,* averaged over the 10 cases in each position. The mean values are also displayed in Fig. 1.

The 3 + 3 + 2 + 2 grouping of the digits is plain to see in all measures. In Fig. 1 there is a clear peak in duration, F_0, and rms energy at positions 3, 6, and 8, as indicated by the dotted lines. F_2 shows a clear valley at the same positions. As we expect from the literature on prosody, the vowel is longer at the end of each of the groups, giving a higher duration measure in position 3, 6, 8, and 10. In shorter syllables, the F_2 target for [o] is raised by greater coarticulation with the surrounding coronal consonants, so that the F_2 measure correlates negatively with duration. As is common in Spanish, nonterminal intonational phrases end in an F_0 rise, producing greater F_0 values in positions 3, 6, and 8, while the final phrase falls, giving a lower F_0 value in position 10. As expected, the amplitude shows the same pattern as F_0.

In general, this sort of phrasal shape expresses what is sometimes called the "information structure" of the text, so that to shape a phrase skillfully requires understanding the text. Since full machine understanding of unrestricted text remains a far-off goal, in a text-to-speech system we do what we can, ranking the problems according to their impact on speech quality and trying to find the approximate solutions that give the best overall results for a particular application.

3. DIVISION INTO SENTENCES

Consider the problem of dividing a text into sentences. This is a fundamental problem in text analysis—as a practical matter, it's the first step in processing most types of text. In addition, our discussion of this problem will help to exemplify the approach we advocate.

Let's assume that the text is made up of sentences to begin with. Some text isn't, at least not in the sense that we mean—a computer program, a table

*Exactly the same trends are visible if we use simply the center value, or the extreme value, or a number of other choices.

of numbers, an inventory list, a return address. Such text must be recognized, often in an application-specific way, and dealt with appropriately.

We also assume that fairly normal English punctuation applies—we accept that our algorithm will be baffled by Molly Bloom's soliloquy*—so sentence boundaries will be marked by one or more characters in the set {.!?}, possibly followed by a string from the set {',",''} indicating the end of a quotation. As Table 2 shows, the frequency of these punctuation marks depends on the type of text. We show the frequency (per million characters) of period, question mark, and exclamation point in four corpora: a collection of about 12 million words of AP newswire text from 1987, a collection of about 5 million words of *Wall Street Journal* (WSJ) text from 1989, the million-word Brown corpus (Francis and Kucera, 1982), and a collection of about half a million characters of email interchanges between puzzled users and computer system administrators, collected at the University of California, Berkeley. The relative frequencies of these punctuation marks vary across these collections of text, in a way that makes some sense—serious journalists are sober and unexclamatory, and the email exchanges are full of questions—but in general, periods far outnumber the others.

It would be easy to find sentence boundaries in text if these punctuation marks always marked the ends of sentences, but unfortunately, each of these characters has other uses. In the Brown corpus, about 90% of all periods mark the end of sentences, while in the WSJ only about 53% do. In the WSJ, there is much more talk of dollar amounts and percentages and much more use of abbreviated titles such as **Mr.**

How should we distinguish the periods that end sentences from those that don't? Even a small sample of cases suggest some obvious heuristics. We give here the ten first examples of sentence-ending periods, followed by

Table 2 Frequency (per million characters) of {. ? !}

	Wall Street Journal	Associated Press	Brown corpus	Help mail
Period	14,153	10,910	8,987	9,292
Question mark	86	76	759	2,343
Exclamation point	0	17	258	347

*The last section in James Joyce's *Ulysses*, in which a lengthy stream-of-consciousness monologue is presented without any punctuation at all.

the ten first examples of sentence-medial periods, from a single WSJ story. We have replaced newlines with spaces for clarity in display.

Ends:	ears ago, researchers reported	. The asbestos fiber, crocidol
	ecades later, researchers said	. Lorillard Inc., the unit of
	nite cigarette filters in 1956	. Although preliminary finding
	g new attention to the problem	. A Lorillard spokeswoman said
	an said, "This is an old story	. We're talking about years ag
	ng any questionable properties	. There is no asbestos in our
	o asbestos in our products now	." Neither Lorillard nor the r
	smokers of the Kent cigarettes	. "We have no useful informati
	s Dana-Farber Cancer Institute	. Dr. Talcott led a team of re
	iversity and Boston University	. The Lorillard spokeswoman sa
Not Ends:	esearchers said. Lorillard Inc	., the unit of New York-based
	t of New York-based Loews Corp	. that makes Kent cigarettes,
	ers are at risk," said James A	. Talcott of Boston's Dana-Far
	na-Farber Cancer Institute. Dr	. Talcott led a team of resear
	in 1956. From 1953 to 1955, 9	.8 billion Kent cigarettes wit
	tos-related diseases," said Dr	. Talcott. The percentage of l
	rkers at the West Groton, Mass	., paper factory appears to be
	ned by Hollingsworth & Vose Co	., was under contract with Lor
	ort those who argue that the U	.S. should regulate the class
	t those who argue that the U.S	. should regulate the class of

Not surprisingly, the basic pattern for an end-of-sentence period seems to be: a period, followed by an optional end quote, followed by one or more space characters,* followed by an optional start quote, followed by a capitalized word. In regular expression notation,

$$\{\textbf{period}\}\ \{\textbf{endquote}\}?\ \{\textbf{whitespace}\} + \{\textbf{startquote}\}?\ \{\textbf{capitalized word}\}$$

In the text shown above, both the *startquote* and *endquote* symbols are " (double quote), but they might have been " and ", or ' and ', or whatever. Overgeneralizing a bit, we might therefore amend the pattern to read

$$\{\textbf{period}\}\ \{\textbf{quote character}\} * \{\textbf{whitespace}\}$$
$$+ \{\textbf{quote character}\} * \{\textbf{capitalized word}\}$$

This will catch nearly all the sentence-ending periods—the false negatives will include cases where the following sentence begins with a nonalphabetic

*Although many style manuals prescribe the use of double spaces following an end-of-sentence period, this prescriptive rule is violated more often than it is followed, in our experience.

character such as a number, of where a typographical error omits the space or the capitalization. Such cases do occur, but fairly rarely, so the rate of false negatives in the WSJ or the AP will be quite low even without any further elaboration of the pattern. In a sample of about 248,000 sentences from the WSJ, we found only two cases of a sentence beginning with a number* and about 28 false negatives due to omitted spaces, exemplified below:

end or a rejection of multilateralism	. 1992 merely intensifies an ongoing p
red to 20.7% last year, to 8% in 1990	.Economists now predict the growth ra
rb is needed in asbestos removal work	."Some of these countries use child la
e company due to its high debt level	.They added that fiscal first quarter
iment returned to the volatile market	.The All-Ordinary Share Price Index ro

Thus our simple sentence-ending pattern seems to turn up false negatives in the WSJ at a rate of less than 2 per 10,000 sentences, with most of the problems due to typographical errors. However, the situation is not so good with respect to false positives. In the WSJ corpus mentioned above, the pattern matches 287,163 periods, of which only about 84% actually end a sentence.

In order to get rid of the 16% (44,669) false positives, we will try to find a set of patterns that (if they match a case) will reliably prevent a false sentence break. Of the false positives, 32,997 involve abbreviated titles before names, such as **Mr., Ms., Rep.** These essentially never end a sentence (although the previous sentence is a counterexample!), so that matching a disjunction of abbreviated titles will reduce the false positive rate from 16 to 4%, while adding only a negligible amount to the rate of false negatives.

Another 6687 of the false positives are initials, like the **E.** in **Robert E. Lee.** There is a slightly greater problem in this case with adding to the set of false negatives: there are 37 sentences (among more than 248,000) that end with a capital letter, like the following:

She's Delores (Anjelica Houston), his neurotic airline-hostess mistress who's tired of being Madame X.
Look at the experience of poor Ms. J.
Nobody seems to feel compelled to set some priorities, and declare that X environmental problem needs to be addressed before Y.
Of the people diagnosed with colon cancer, 21,000 with have stage Dukes' C.
The story has been passed around among friends who have chortled as much as I.

---

*There were also six sentences starting with names such as **3Com, 1st Nationwide Bank**, and so on.

> It's only a matter of time, Mr. Brooks feels, before the Japanese finally concede that their domestic jerky is as inferior as he knows it is, discard arcane restrictions on imports of muscle-meat jerky, and throw open the door to the real, he-man article from the U.S. of A.
>
> His office said that Mr. Quayle continues to hold 19,800 shares of Class A and 25,000 shares of Class B.
>
> The book, an autobiography, surely sets a record for use of the word I.
>
> The experiment was spawned when 598 workers in Boston hospitals, clinics and medical laboratories were vaccinated with Merck's Heptavax-B vaccine for hepatitis B.
>
> It is true that there were thousands of men who probably deserved the medal more than I.

So at the expense of another couple of false negatives (missed breaks) per 10,000, we can reduce the rate of false positives (wrongly inserted breaks) to 1.7%. The next largest group of false positives (2993) are caused by dual initials, as in **W. R. Grace** or **R. P. Scherer** or **U.S. Army**. In this case, the effect on overall error rate (although still an improvement) is less marked, since there are 965 cases where such dual initials end a sentence.

Continuing this process, we would examine ways to distinguish periods based on other characteristics of the words that follow them; thus **The** is not very likely to occur within a sentence, while **I** is. With a couple of days of work, we can build a heuristic sentence-division algorithm whose equal error rate on newswire text is low enough that it's hard to estimate accurately, but seems to be somewhere around a tenth of a percent.

Of course, such a method depends on its source of text behaving as expected. If, for instance, case distinctions are lost, or nearly all words are capitalized (as often in titles and headers), or the rate of typographical errors goes up, then we will have to revise the decision tree that performs the discrimination or else face greatly degraded performance. Luckily, techniques are available that make it possible for such decision trees to be constructed automatically from training data (Breiman et al., 1984).

Given a body of classification examples, each of which is described in terms of a set of features that may be either quantitative or categorical in nature, these approaches "grow" a stepwise optimal binary decision tree to perform the indicated classification. These algorithms work by choosing, at each step, the way of splitting the data that minimizes the entropy of the resulting split, that is, the split that results in the "purest" categories. To avoid overtraining, a crossvalidation technique is used that prunes the tree to an "honest" size.

Riley (1989) discussed the application of such classification tree techniques to a number of natural-language problems, including the sentence-finding task that we are discussing here. He used the following features (where *preceding word* means the word just before the period and *following word* means the word just after the space):

1. The probability that the preceding word occurs at sentence end
2. The probability that the following word occurs at sentence start
3. Length (in characters) of preceding word
4. Length of following word
5. Case of preceding word: {upper, lower, capitalized, numbers}
6. Case of following word
7. Punctuation after period (if any)
8. Abbreviation class of preceding word: month name, unit of measure, title, address name, etc.

The indicated probabilities were computed from 25 million words of AP news, using the beginning and end of paragraphs (which are unambiguously marked in the AP) rather than the beginning and end of all sentences. Applied to the Brown corpus, the resulting classification tree correctly identifies the ends of declarative sentences 99.8% of the time, so that its performance is quite similar to that of a handcrafted heuristic algorithm. The advantage of such an approach is that it can be retrained for new typographical conventions with relatively little trouble, as long as training data can be obtained. By contrast, handcrafted heuristic decision procedures are notoriously tricky to revise to fit new circumstances without introducing bugs.

4. WORD TOKEN AMBIGUITIES

After sentences have been identified, it is necessary to disambiguate words (tokens) in various ways. In particular, it is necessary to assign parts of speech, expand abbreviations and acronyms, etc.

4.1. Part-of-Speech Assignment

There are some well-known words like *wind* and *use* that are pronounced one way as a noun and another way as a verb. Fisher (1982) argued against the necessity of syntactic analysis for word pronunciation in text-to-speech, estimating that only a relatively small number of errors (0.015% by phoneme) would be introduced by simply assigning each word its most frequent pronunciation. He found that this approximation would introduce only 0.015% errors by phoneme. This error rate of 0.015% by phoneme corresponds to a 0.75% error rate by word, assuming that there are five phonemes in a word on average. Fisher estimated that a week's work on the more pressing problems with his letter-to-sound rules would net him about four times as much improvement, leading him to conclude, quite reasonably, that syntactic analysis was not a priority for his system.

We applaud (and imitate) Fisher's approach to weighing the costs and benefits of particular aspects of text analysis in speech-synthesis applications. However, now that word pronunciation is much improved, due to advances in dictionary-based approaches such as those described in Section 6, we come to a different conclusion about the value of syntactic analysis to word pronunciation. Nevertheless, it must be recognized that we are talking about a fraction of a percent by word. The more common words with generally used multiple pronunciations are listed in Table 3, along with their frequency (per million words) in a collection of 44.34 million words from the 1988 AP newswire. Our complete list contains 428 such words with a total frequency of 4131 per million.

We do not mean to suggest that such part-of-speech ambiguity is the only source of words with ambiguous pronunciations. In a few cases of noun/verb pronunciation ambiguities, there is a de-nominal verb that is pronounced

Table 3

use	319	rebel	48	discount	30	experiment	17
increase	230	finance	46	concert	30	transplant	17
close	215	estimate	46	desert	28	offset	17
record	195	conflict	45	associate	27	merchandise	16
house	150	export	43	chose	26	prospect	16
contract	143	moderate	42	import	26	ally	16
lead	131	abuse	41	transfer	26	mouth	15
live	130	contest	40	recall	25	object	15
lives	105	delegate	40	transport	24	proceeds	14
protest	94	combat	40	upset	23	perfect	14
survey	91	document	39	contrast	22	overthrow	14
project	90	wind	38	reject	22	excess	13
separate	87	permit	36	legitimate	22	advocate	13
present	80	affect	35	graduate	21	compound	12
read	72	pace	35	minute	20	implement	12
subject	68	elaborate	32	wound	20	content	12
progress	66	tear	31	frequent	19	overhaul	11
impact	65	suspect	31	entrance	18	warehouse	10
produce	59	appropriate	31	refuse	17	decrease	10
conduct	52						

like the noun, such as *contract* used to mean "make a contract" rather than "shrink," or *wind* used to mean "cause to be out of breath" rather than "twist around." And *tear* has both a nominal and a verbal use in both of its pronunciations. A sample of other cases not differentiated by part of speech is given in Table 4. Disambiguating these cases requires algorithms for word-sense selection, which we do not have space to discuss here. The overall frequency with which such cases occur is generally less that the frequency of pronunciation ambiguities correlated with part of speech.

Another important area in which the pronunciation of a word varies according to its function is the text is the case of monosyllabic "function words" that have an extremely reduced form in some cases, but may also commonly be pronounced in a full form or even be emphasized. The reduced forms of such words are called "clitics," and in a few cases the distinction between the reduced and the full form correlates with a part-of-speech distinction. For example, the word *that* is usually reduced when it is a complementizer, introducing a subordinate or relative clause, as in "I'm sure that John left" or "the one that he chose," whereas it is not normally reduced when used as a demonstrative, as in "what was that?" or "that book is on order." In other cases of uncertain reduction, such as *some*, there is not a part-of-speech difference but simply a difference in meaning or in emphasis. The reduced forms of such words are called *clitics.*

Table 4

axes	more than one axis; more than one axe
bass	music; fish
bow	ribbon; boat part or gesture
cleanly	hygienic; neatly
coop	cage; store
lead	kind of paragraph or punch; kind of metal
peaked	infirm; having a peak
primer	book; paint
row	alignment; altercation
mares	horses; moon feature
equivalence	sameness; having a similar chemical property
prayer	supplication; supplicant
sewer	garbage conduit; one who sews
tier	stratum; one who ties
stingy	miserly; tending to sting
Jesus	English pronunciation; Spanish name

These cases of uncertain reduction are much more common than the cases like *use*, *wind*, *axes*, or *bow*. The word *that* alone makes up 1.0% of the Brown corpus and 0.7% of the 1988 AP corpus. Considering the frequency of other potential clitics such as *some*, *can*, *to*, *in*, *on*, *up*, *him*, and *her*, it is easy to see how a proper treatment of clitics could affect the pronunciation of several percent of the words in a text.

We have discussed elsewhere (Church, 1988) a program that assigns part of speech with very high accuracy. The program, like others such as those in DeRose (1988) and Jelinek (1985), uses a linear time dynamic programming algorithm to find an assignment of parts of speech to words that optimizes the product of (1) lexical probabilities (probability of observing part of speech i given word i) and (2) contextual probabilities (probability of observing part of speech i given n following parts of speech). Probabilities were estimated from the tagged Brown corpus (Francis and Kucera, 1982). Performance is good enough that a growing user population has found that it meets their needs better than most alternative programs they have access to.

Some of the probabilities are very hard to estimate by direct counting. In general, the lexical probabilities are more difficult to estimate than the contextual probabilities.* We need to estimate how often each word appears with each part of speech. Unfortunately, it is not possible to estimate all of these parameters. If there were only 50,000 words and 10 parts of speech, there would be 500,000 parameters to estimate. Given that we have only 1,000,000 words of training material, it is just not possible to estimate so many parameters even if the data were nicely distributed (and they aren't). In general, no matter how much text we look at, there will always be a large tail of words that appear only a few times. In the Brown corpus, for example, 40,000 words appear five times or less. Many of these words are not rare; all of the words in Table 5 appeared just once in the Brown corpus. The number indicates the frequency (per million words) in the 1988 AP corpus.

Given that many noninfrequent words will be found just a few times in our corpus, how do we come up with accurate estimates of their lexical probabilities? For instances, the word "yawn" appears once as a noun and once as a verb. What is the probability that it can be an adjective? It is impos-

*Apparently, contextual probabilities may not be as difficult to estimate. For example, Ayuso et al. (1990) report that they were able to obtain reliable estimates of contextual probabilities from only 100,000 words of tagged training material. Furthermore, they did not find it necessary to use fancy smoothing techniques such as the Good–Turing estimate (Good, 1953): they simply added one to each of their observed counts.

Table 5

bushel	1.6	options	0.6	injuring	0.5	evangelist	0.4
cocaine	1.6	citing	0.6	investor	0.5	wounding	0.4
lawsuit	1.5	monitored	0.6	monitor	0.5	seasonally	0.4
demonstrators	1.1	funding	0.6	memo	0.5	smuggling	0.4
helicopter	1.0	guidelines	0.6	cruise	0.5	booster	0.4
leftist	1.0	ousted	0.6	layoffs	0.5	deficits	0.3
hostage	0.9	justices	0.5	kidnapping	0.4	shareholder	0.3
unidentified	0.9	referendum	0.5	rivals	0.4	advocates	0.3
worldwide	0.8	terrorists	0.5	lawsuits	0.4	trillion	0.3
caucuses	0.8	longtime	0.5	dissident	0.4	undercover	0.3
detained	0.8	riots	0.5	felony	0.4	losers	0.3
kidnapped	0.7	sentencing	0.5	detention	0.4	banning	0.3
hospitalized	0.7						

sible to say without more information. There are a number of places to try to find additional sources of information. In particular, one might turn to a collegiate dictionary.

Many other estimation problems have a big impact on the performance of a part-of-speech algorithm. Proper nouns and capitalized words are particularly problematic; some capitalized words are proper nouns and some are not. Estimates from the Brown corpus can be misleading. For example, the capitalized word "Acts" is found twice in the Brown corpus, both times as a proper noun (in a title). It would be a mistake to infer from this evidence that the word "Acts" is always a proper noun. Most proper nouns are now dealt with in a prepass that labels words as proper nouns if they are "adjacent to" other capitalized words (e.g., "White House," "State of the Union") or if they appear several times in a discourse and are always capitalized. This very simple prepass appears to work well enough that the main routine does not need a very good model of lexical probabilities for proper nouns.

4.2. Acronyms and Abbreviations

Quite a few acronyms and abbreviations are ambiguous. There may be several acronyms and abbreviations that are written the same way, such as **Dr.** for "doctor" or "drive." There may also be ordinary words that are homographic with an acronym or abbreviation, such as **Jan.**, which can be a first name or the abbreviation for the month of January. Measure-word abbreviations are systematically ambiguous between singular and plural interpretations: **1 in.** is "one inch," while **2 in.** is "two inches."

The wizard of Oz.	*is not*	The wizard of ounces.
Let me in.	*is not*	Let me inches.
OK, I agree.	*is not*	Oklahoma, I agree.
Al Gore.	*is not*	Alabama Gore.
per square ft	*is not*	per square fort
3 square ft	*is not*	3 square foot
Ft Collins	*is not*	Feet Collins
in ms.	*is not*	in Miz (*might be "in manuscript" or "in milliseconds"*)
Ms. Jones	*is not*	Millisecond Jones
3 Ms.	*is not*	3 Mississippi (*likely to be "3 milliseconds"*)
305 ST CHARLES ST WINONA MS	*is not*	305 STREET CHARLES SAINT WINONA MANUSCRIPT

We will not discuss in detail the heuristics for making these distinctions.* In general, we have taken the same approach to these problems that we took in the case of sentence identification—to develop a decision tree for each case based on the facts as seen in appropriate corpora. In some cases, the alternative outcomes belong to different parts of speech and so could be distinguished in principle by the algorithm just described. In general, however, these ambiguities are more like ambiguities of word sense, for which somewhat different techniques are needed.

4.3. Digits

There are many complexities in the analysis (and subsequent pronunciation) of bits of text that contain digits. Digit expressions can have many different semantic and pragmatic roles, and these roles impose particular (and varying) modes of pronunciation.

For instance, it would be quite confusing to read **the 1964 San Francisco 49ers** as *the one thousand nine hundred and sixty four San Francisco four nine E R S*, or **the early 20th century** as *the early two zero T H century*. A **50-50 proposition** is not a *fifty minus fifty proposition*, although if the dash is silent in **the change was −10%**, the meaning is seriously changed. To read **$5.27 billion** as *five dollars and twenty seven cents billion* is likely to leave listeners puzzled.

*We will cover the problem of pronouncing (as opposed to interpreting) acronyms in a later section.

Within this discussion, in which we analyze the textual function of digit-containing words, we will also treat the problem of how to pronounce them. Within strings of digits proper, there are quite a few different strategies of pronunciation. We can distinguish the main options as follows:

- **Serial**: each digit is read independently—**4567** is *four five six seven*.
- **Combined**: the digit string is treated as an integer, and read with all place quantifiers made explicit—*four thousand five hundred and sixty seven*.
- **Paired**: each pair of digits is translated as a **combined** unit—*forty five sixty seven*. If the length of the string is odd, one of the nonfinal digits will stand alone, usually the first one.
- **Hundreds**: for a four-digit string, or for the last four digits of a longer string—*forty five hundred (and) sixty seven*.
- **Trailing Unit**: a string that ends in one or more zeros is read in **serial** mode up to the last nonzero digit, which is followed by the appropriate unit—**9263000** becomes *nine two six three thousand*.

For strings of digits after a decimal point, the main choices are

- **Serial: 3.45** is *three point four five*.
- **Decimal fraction: 3.45** is *three and forty five one hundredths*.
- **Cents: 3.45** is *three dollars and forty five cents*.

To some extent, we can map directly from orthographic characteristics to the choice of pronunciation mode. For instance, number strings that contain commas separating groups of three digits, counting from the right, are almost always to be read in **combined** mode, regardless of the function of the number.

However, it is often true that the same digit string needs to be pronounced in different ways depending on the context. For instance, in American mailing addresses, the usual practice is to read street numbers in a **paired** fashion. Thus **18225 BURN RD CALEXICO CA 92231** becomes *one eighty two twenty five Burn Road, Calexico, California, nine two two three one*.* Four-digit dates are always read in **paired** or **hundreds** mode—to read a date in **combined** mode is disastrous. Money amounts, on the other hand, should be in **combined** mode—**1927 francs** is *one thousand nine hundred twenty seven francs*, not *nineteen twenty seven francs*, which sounds like a reference to the

*As we indicated in the discussion of Table 1, we also need to determine an appropriate grouping to strings of numbers read in serial mode.

currency of the year 1927. Such facts imply that text analysis for speech synthesis must include algorithms for recognizing categories like *date* and *money amount*.

There are several other problems involving digits that we mention and exemplify for completeness, although we do not have the space here to discuss them.

The treatment of mixed digit and alphabetic strings:

35 mm	BMW 318is	Tupelev-144	Cray-3	pre-1933	mid-1970s	the '60s
30-day	47-year-old	76er	21st	42nd	53rd	18th

The distinction among fractions, dates, rates, alternatives, and other uses of the slash, and the interaction of fractions with other signs such as %:

4/16/86	3½ years
the benchmark 11¾% bond due 2003/2007	STN 50-528/529/530
53.6 +/- 18.6 pg/ml	20/20 hindsight
extending from 1½–2' from the trunk	3/16ths in.
27-½ years	1½ to 2½ high
over $2.00/- square foot	50/1,000ths

The distinction among numerical ranges, subtraction, dates, fractions, and simple separation in the use of the dash, and the interaction of ranges with number interpretation in general:

dating from A.D. 300–900.	a 9–1 advantage
the $200–$500 million child pornography industry	from 1969–71
the estimated 15–20 million Kurds	was defeated 5–4.
traded at 6 5-16 percent	Jan. 16–24.
the $80- to $90-a-share range.	Lotus 1-2-3
call, toll-free, at 800-822-8987.	20-20 hindsight
$30–$35,000 a year	the 6–12 MeV range
an $80–100 million market	up to −10%
At 14–15 MeV of neutrons	

The use of colons with numbers in expressing times, biblical citations, ratios, and other things:

2:30 a.m.	Isaiah 40:26
a contrast ratio of 1:1000	stoichiometry of 1:1:2 for Na:K:Cl
1:100,000-scale maps	a 4.5:4.5:1 ratio
Main Estimates 1986:87	a 50:50 cost-sharing
Mon, 17 Jul 89 17:36:10	

It does no good to complain about the erratic, irregular, and sometimes downright incorrect way that texts combine these signs—a TTS system, like a reader, has no choice but to cope. We have had good results using the same approach here that we have been recommending throughout: choose a clear structure for the decision algorithm, then take an appropriate data base of texts, and use a combination of exploratory data analysis and statistical training methods to set the algorithm's parameters in detail. We do not have space to describe this exercise in detail for the text analysis problems involving digits, but is should be clear how the approach would apply.

5. PARSING

5.1. General Solutions

As pointed out in Allen et al. (1987, p. 40) the text-to-speech application places somewhat unusual demands on a parser. The parser should have a broad (though possibly superficial) coverage of unrestricted text, rather than a deep analysis of a restricted domain. In addition, it is important to tune the trade-off between type I and type II errors appropriately. In particular, it is probably more serious to insert a spurious boundary than to leave a real one out. Finally, there are severe constraints on time: the parser must run in real time. Memory is also a critical resource in many practical implementations.

> The parser for the text-to-speech system is designed to satisfy a unique set of constraints. It must be able to handle arbitrary text quickly, but does not need to derive semantic information. Many parsers attempt to build a deep structure parse from the input sentence so that semantic information may be derived for such uses as question-answering systems. The text-to-speech parser supplies a surface structure parse, providing information for algorithms which produce prosodic effects in the output speech ....
>
> It is well known that parsing systems which parse unrestricted text often produce numerous ambiguous or failed parses. Although it is always possible to choose arbitrarily among ambiguous parsings, a failed parsed is unacceptable in the text-to-speech system. When one examines ambiguous results from full sentence-level parsers, one finds ... much of the structure at the phrase level has been correctly determined. The phrase-level parser takes advantage of this reliability, producing as many phrase nodes as possible for use by the MITalk prosodic component.
>
> The phrase-level parser uses comparatively few resources and runs in real-time. This is quite unusual for parsers which handle unrestricted text, but is necessary for a text-to-speech system.

The phrase-level parser in the MITalk system produces output such as (Allen et al., 1987, p. 51):

Noun Group:	Most of the exercises
Verb Group:	are

Noun Group:	translations
Unclassified:	.
Verb Group:	There are
Noun Group::	several important changes
Prepositional Phrase:	in the way
Noun Group:	the quantifier rules
Verb Group:	will work
Prepositional Phrase:	for the remainder of the course
Unclassified:	.

We agree with Allen et al.—for synthesis applications it is currently necessary to take a very robust, conservative, and unambitious approach toward parsing. In fact, we have taken an even more conservative approach than they have. It has been our experience that a very simple pattern of the form

{function word}∗{content word}∗

will produce results that are about as good as can be expected. For example, in this case, this pattern would produce

Most of the exercises
are translations.
There are several important changes
in the way
the quantifier rules
will work
for the remainder
of the course.

Such a method works, of course, because English is largely a right-headed language, whose grammatical (function) words tend to pile up at the left edge of constituents, being what syntacticians call *specifiers*. We call such a sequence of function words followed by content words a *function-word group*, or *f-group*.

Our simple f-group parser can be easily improved to a certain extent, at the expense of slightly greater complexity. For example, a pronoun in the objective case, such as *him* or *them*, is more likely to function like a noun, ending a unit, than like a determiner, beginning one. In addition, bare tensed verb forms (such as *provoked*) are likely to start a phrase.* This last improvement is particularly helpful for breaking up long sequences of content words such as: *The United States Supreme Court decision produced angry protest marches*.

*The *-ed* suffix acts somewhat like an auxiliary verb such as *be* or *have*.

With the improvement, the f-group parser can split this sequence of nine content words into two phrases:

The United States Supreme Court decision
produced angry protest marches.

In practice, content words are usually broken up by function words. This example was actually made up; it is hard to find real examples from the Associated Press newswire with even as many as six content words in a row (see example below). The fact that long sequences of content words are usually broken up by function words also helps explain why the simple f-group parser works as well as it does.

These improvements basically work by modifying the categories in the pattern to correspond more closely to the structure of English phrases. Thus, to avoid confusion, we will refer to members of the *function word* category as *chinks* and members of the *content word* category as *chunks*. Then we put objective pronouns into the *chunk* category and tensed verb forms into the *chink* category and greedily match the pattern: (chink* chunk*)*, a process that we can now dub the *chinks 'n chunks* algorithm.* These modifications generally produce better behavior (Table 6).

These improvements depend on the existence of a robust part-of-speech algorithm like that in Church (1988). For example, *her* is ambiguous between the genitive form, which usually begins a unit, and the objective pronoun, which usually ends one, and therefore accurate tagging is a precondition for separating the cases. Likewise, tensed verb forms are not easy to distinguish from past participles (in the past form) and from nouns (in the present from). When the parser for our text-to-speech system was first designed, the tagging algorithm was not sufficiently reliable to justify treating tensed verbs as chinks, since such a heuristic would produce too many inappropriate boundaries.

Comparing the extremely simple chinks 'n chunks approach to the outcome of more complex parsers, we have found that the division into small f-group–sized units is nearly the same, while the complex parsers' decisions on grouping of larger units are wrong too often to be relied on.

Of course, we must make it clear that this simple algorithm is not very good. It can produce mistaken groupings, as in the example below:

*This nomenclature is inspired by an algorithm for generating permuted indices, due to Tukey.

Table 6

Before	After
A convicted murderer and rapist whose parole several years ago provoked public outrage has been charged with attacking a woman.	A convicted murderer and rapist whose parole several years ago provoked public outrage has been charged with attacking a woman.
I asked them if they were going home to Idaho and they said yes and anticipated one more stop before getting home	I asked them if they were going home to Idaho and they said yes and anticipated one more stop before getting home
and a Kansas state trooper helped them on Interstate 70 near the Colorado border	and a Kansas state trooper helped them on Interstate 70 near the Colorado border
Ellsberg testified Friday that a protester much like those on trial persuaded him to leak the Pentagon Papers	Ellsberg testified Friday that a protester much like those on trial persuaded him to leak the Pentagon Papers

Output	Should Be
The outspoken conspiracy theorist will not pass up a chance to speak.	The outspoken conspiracy theorist will not pass up a chance to speak.

More important, the output f-groups are too small to be very useful—what we really need to know is how the f-groups combine hierarchically to form sentence-sized units. As a result, only a rather weak amount of phonetic modulation is appropriate to mark the f-group structuring. In order to prevent long sentences from being too monotonous, we can also impose mildly alternating patterns of tonal prominence on accented words within f-groups and f-groups within phrases. Again, we would really like to know what the text's information structure is, and in the absence of effective methods to find it we must follow the medical principle of "first, do no harm."

We believe that current research in text analysis, combining sensible models of linguistic structure with techniques for learning from very large text corpora, is likely to produce techniques that work well enough to be used safely in TTS applications.

5.2. A Focused Approach: Attributive Tags

One small, but instructive, example of such techniques is provided by an algorithm to locate attributive tags, done as a summer project in our lab by Katherine Baker in 1988. Tag phrases, e.g., *a department spokesman said Friday*, are very common in certain types of text, especially in the Associated Press corpus, where every twelfth sentence contains a tag. It is useful for speech synthesis applications to be able to identify tags and separate them form the surrounding text, so that they can be assigned appropriate intonation parameters. Tags are identified with a number of heuristics using two types of cues: (1) length constraints (tags are often fairly short and the surrounding quoted material is often relatively long) and (2) verb constraints (the main verb in a tag is usually *said* or some other "verb of saying," e.g., *reported*, *announced*).

Baker developed a set of (robust) heuristics that identify tag phrases. For synthesis applications, errors of omission are not nearly as bad as errors of commission, since inappropriate use of tag intonation sounds extremely silly. In one evaluation, Baker found that her program made no egregious errors of commission and correctly identified 80% of the tags in a restricted context.

What is a tag phrase? Some typical tags are:

he said
he later added
the pope said in his homily
Horowitz replied
said store manager Anita Dickey
concluded the eight-page opinion written by Winter
one recommendation reads

said one Dutch diplomat, speaking on condition of anonymity
she told the cast
Young, 55, said in a recent interview

The main verb in a tag phrases is usually *said*, or some other verb of "saying," such as:

added, adding, adds, admits, advises, angering, answered, apologizing, asked, asks, asserted, asserts, beamed, boasted, cautions, changed, chided, chimed, chuckle, confided, countered, crowed, deadpanned, deadpans, declared, declares, decried, elaborating, exclaimed, exclaims, explained, explains, foresees, fretted, groused, growled, grumbles, grumped, implored, interjected, joked, lamented, laughed, marveled, misspoke, mused, observed, observes, praises, quipped, quoted, reads, reasoned, recalled, recalls, recounted, recounts, remarked, remembers, replied, replies, replaying, reported, responded, responds, retorted, said, say, says, shouted, shouts, shrugged, sighed, sighs, sings, smiled, stated, tells, told, translates, uttered, ventured, warns, writes, wrote, yelled

Many tag phrases contain one of these verbs of "saying," though of course there are also some tags that do not: "the statement went on," "runs an old adage," and "the translator came in."

Baker's algorithm uses a voting framework to combine the length constraints and the verb constraints. Let q be candidate quote and t be a candidate tag.* The question is: do we believe that t is, in fact, a tag? Baker's algorithm concludes "yes" if at least two of the following three conditions hold and "no" otherwise.

1. q is "long" ($> T_{\text{long}}$ tokens).
2. t is "short" ($< T_{\text{short}}$ tokens).
3. t contains a verb of "saying" (according to a short list of known verbs of "saying").

A simple parameter-fitting exercise indicated that $T_{\text{long}} = 8$ and $T_{\text{short}} = 7$.

Baker implemented this algorithm and found that it correctly identified 80% of tags following quotes in direct discourse. No errors of commission were found in this evaluation. The error rate was determined by testing on a corpus of approximately 13,000 sentences selected from the 1987 AP wire.

An examination of the omissions found that the second condition was responsible for 5% of the omissions and that the third condition was respon-

*The candidate strings must meet a number of syntactic conditions. In particular, there should be explicit quote marks around q. In addition, there are certain requirements on the distribution of commas, periods and other punctuation.

sible for 15%. Apparently, one could improve this algorithm dramatically by extending the list of known verbs of "saying." During the evaluation, it became clear that there were a number of very common verbs that should have been on the list, including *promised*, *continued*, and *noted*.

We find Baker's experience to be very encouraging. The spirit of her work is similar to the spirit of Riley's work on sentence division described earlier: the investigator selects a set of relevant indicators and a general form for the heuristic classifier and then uses training on an appropriate body of text to settle the details of the classification. We expect that similar techniques can be applied to many other problems in the area of text analysis.

5.3. Another Approach: Clause Parsing

Ejerhed (1988) has experimented with two methods of parsing sentences into "clause-like" phrases, one based on a finite-state grammar and the other based on a stochastic Markov model. Ejerhed constructed a corpus for developing and training the parsers. The corpus consisted of 136,340 words extracted from three genres (news, learned, general fiction) of the tagged Brown corpus. Ejerhed used a semiautomatic procedure to insert 27,678 "clause" boundaries. A small fragment of this corpus is given below. Note that the definition of "clause" is somewhat nonstandard. Because Ejerhed wants to use finite-state methods, she has adopted a nonrecursive notion of clause which elevates fragments such as *that the City Executive Committee* to the level of "clause" status.

[the/AT Fulton/NP-TL County/NN-TL Grand/JJ-TL Jury/NN-TL said/VBD Friday/NR]
[an/AT investigation/NN of/IN Atlanta/NP 's/$ recent/JJ primary/NN election/NN produced/VBD no/AT evidence/NN]
[that/CS any/DTI irregularities/NNS took/VBD place/NN ./.]

[the/AT jury/NN further/RBR said/VBD in/IN term-end/NN presentments/NNS]
[that/CS the/AT City/NN-TL Executive/JJ-TL Committee/NN-TL ,/,]
[which/WDT had/HVD over-all/JJ charge/NN of/IN the/AT election/NN ,/,]
[deserves/VBZ the/AT praise/NN and/CC thanks/NN of/IN the/AT City/NN-TL of/IN-TL Atlanta/NP-TL for/IN the/AT manner/NN]
[in/IN which/WDT the/AT election/NN was/BEDZ conducted/VBN ./.]

[that/AT September-October/NP term/NN jury/NN had/HVD been/BEN charged/VBN by/IN Fulton/NP-TL Superior/JJ-TL Court/NN-TL Judge/NN-TL Durwood/NP Pye/NP]

[to/TO investigate/VB reports/NNS of/IN possible/JJ irregularities/NNS in/IN the/AT hard-fought/JJ primary/NN]
[which/WDT was/BEDZ won/VBN by/IN Mayor-nominate/NN-TL Ivan/NP Allen/NP Jr./NP ./.]

Based on this corpus, Ejerhed developed a regular expression that attempted to identify "clause" boundaries from an input sequence of parts of speech. This approach was compared with a stochastic model that found the optimal well-formed bracketing of the input sequence of parts of speech subject. Each candidate bracketing is denoted by a set of indexes, identifying the location of boundaries. The candidates are scored by

$$\prod_{i \text{ in } indexes} Pr(boundary \mid pos_i, pos_{i+1}) \prod_{i \text{ not in } indexes} 1 - Pr(boundary \mid pos_i, pos_{i+1})$$

where $Pr(boundary \mid pos_i, pos_{i+1})$ is an empirically determined estimate of the probability of finding a bracket between the two parts of speech pos_i and pos_{i+1}.

Ejerhed evaluated both the regular expression and the stochastic approach and found that the stochastic method performed slightly better, a result that she did not expect when she began the experiment.

5.4. Parsing and Interpretation of Complex Nominals

By "complex nominals" we mean English noun phrases in which a head noun is preceded by one or more nouns or adjectives. Such phrases are extremely common—their frequency depends very much on style, but in most kinds of English text they are approximately as common as the word *the*, which is the commonest word in the language.

As a matter of practice, the location of the main stress in such sequences is quite variable. Normally, **[Noun Noun]** sequences are left dominant (have main stress on the left), while **[Adjective Noun]** sequences are right dominant (have main stress on the right):

Right Dominate	Left Dominate
an important **meeting**	a **staff** meeting
a red **jacket**	a **life** jacket
an empty **can**	a **garbage** can

Of course, emphasis or contrast in the context of use can change this: "I only go to **important** meetings;" "it's a garbage **pan**, not a garbage **can**." In addition, there are systematic exceptions to the stress pattern in both directions—there are normally right-dominant **[Noun Noun]** cases and left-dominant **[Adjective Noun]** cases (Table 7).

By simply counting stress patterns in read text, we find that about 75% of **[Noun Noun]** patterns are left dominant and about 90% of **[Adjective Noun]** Nominals are right dominant. Thus simple parsing and part-of-speech assignment can help a bit. To improve performance beyond this point, two broad classes of techniques can be applied.* One approach relies on an analysis of the semantic type of the elements of the nominal. Thus we expect *safety board* to be left dominant because it is an instance of a semantic pattern informally stated as **X group-whose-concern-is-X:**

anthropology department	oil industry
Weather Service	Cancer Society
fire brigade	savings-and-loan association

Alternatively, we can rely on the fact that the individual words in the phrase *safety board* tend to yield a left-dominant stress pattern in other cases—that is, the probability of left stress given that the left-hand word is *safety* times the probability of left stress given that the right-hand word is *board* is greater than the probability of right stress under the same conditions. A sample of the data on which such an evaluation depends is shown below:

Table 7

Exceptional cases: right-dominant **[Noun Noun]**			
girl swimmer	enemy invasion	quarterback sneak	student government
garage door	attic roof	summer palace	Christmas dinner
gallon jug	dollar bill	bathtub gin	corduroy suit
Exceptional cases: left-dominant [Adjective Noun]			
square knot	clean room	safe house	soft drink
black belt	red neck	crazy quilt	medical building
sick room	cold spell	straight edge	french fries

*See Sproat and Liberman (1987) and Liberman and Sproat (1991) for further discussion.

safety belt	advisory board
safety binding	bulletin board
safety council	chalk board
safety equipment	draft board
safety feature	school board

Both of these techniques have difficulty in dealing with the vast scope of actually occurring nominal elements and relations, but can achieve a considerable improvement in correct stress assignment even for unrestricted text.

A further problem arises in the assignment of stress to longer Nominals such as *government tobacco price support program*, *power generating station control room complex*, *grinding wheel truing mechanism engaging device*, and *state highway department public relations director*. Such examples are mercifully rare in all but the most clotted prose, but when they occur we need to parse them and analyze their constituent parts in order to pronounce them in a helpful way. Syntactic analysis does not help with such analysis, since a sequence of modifiers preceding a noun is in principle consistent with all possible parse trees. We have to use some combination of semantic and lexical analysis. As sketched in Liberman and Sproat (1991) and Sproat (1991), an approach founded on evidence derived from a large text corpus can achieve respectable results in such cases.

6. PRONUNCIATION OF ORDINARY (ALL-ALPHABETIC) WORDS

Typically, the conversion of spelling (orthography) to an IPA-like phonological representation is accomplished in one of two ways: (1) by looking the words up in a dictionary (with possibly some limited morphological analysis), or (2) by sounding the words out from their spelling using basic principles.

Both approaches have their advantages and disadvantages, Most speech synthesizers adopt a hybrid strategy: employing letter-to-sound rules for most words and catching the most common irregular words with a small "exceptions dictionary" of 5000 words or less. MITalk took a radical dictionary-based approach for its day. A dictionary of 10,000 morphemes (Allen et al., 1987, p. 25) covered the vast majority of the input words. Only 5% of the input words could not be handled by the *decomp* module and had to be passed to Hunnicutt's letter-to-sound rules (Allen, personal communication). The Bell Laboratories text-to-speech system, TTS, takes an even more radical dictionary-based approach; dictionary methods are used for 99.9% of the input words, and only the remaining 0.1% will be passed to *namsa*, a letter-to-sound rule system designed for surnames (Church, 1986). Now that the dictionary is the rule and not the exception, the term "exceptions dictionary" seems somewhat dated.

The main motivation for moving to a dictionary-based approach is accuracy. In general, table lookup is *much* more accurate than starting from basic principles. Dictionary-based systems make a few errors per 10,000 words. In contrast, a good letter-to-sound system such as Hunnicutt's (Allen et al., 1987, chapter 6) will make about 100 times as many errors. Self-organizing/connectionist systems such as that of Sejnowski and Rosenberg (1987) are so much worse that they report error rates by letter, not by word. For instance, Rosenberg (1988) reports that his best system achieved 92% by letter (including the spaces between words) on words outside its training set, which corresponds to a word error rate of approximately 50%.

In the early days of speech synthesis, the dictionary-based approach faced two problems: memory and coverage. The memory problem has been much alleviated with declining memory prices, though for some applications the memory requirements (approximately 0.25 megabyte) are still a concern. Coverage is a more serious issue, especially for surnames, which are generally thought to be more difficult than ordinary words that one might find in a collegiate dictionary.

Names are particularly hard because there are so many of them. A list obtained in 1987 from the Donnelly marketing organization contains 1.5 million names (covering 72 million households in the United States). In order to appreciate just how large 1.5 million is, note that it is three times larger than the number of entries in an unabridged dictionary e.g., *Merriam Webster's Third New International* (Gove et al., 1961), which is considerably larger than a collegiate dictionary. In addition, it takes a much larger list of names to achieve a certain amount of coverage, in comparison with ordinary words. For example, to cover half of the surnames in the United States requires more than 2300 names. In contrast, 50% of nonnames are drawn from a list of only 141 words. And names are thought to be less amenable to derivation techniques. Names come from many different languages; the methods of derivation are more diverse and language specific.

David Schulz and Beth Schulz (AT&T Bell Laboratories, Indian Hill Park) have recently constructed a dictionary of the 50,000 most frequent surnames in the United States, so that it is no longer necessary to use the letter-to-sound system *namsa* for these names. This greatly improves performance on a corpus of names such as the Kansas City telephone book. It is believed that *namsa* by itself produces good results 50% of the time and acceptable results about 85% of the time (Schulz, personal communications).

The dictionary itself covers 87% of Kansas City. Thus, if *namsa* is somewhere between 50% and 85% correct by frequency, the combination of the dictionary plus *namsa* should yield between 87% + (50%)(13%) = 93.5% and 87% + (85%)(13%) = 98% performance, a significant improvement over *namsa* alone.

The argument becomes considerably stronger when we consider morphology and analogical extensions to the dictionary. Names such as *Walters* and *Lucasville* can be derived from other names by very simple morphological processes. These stress-neutral processes increase the coverage of the names dictionary by 25%, as indicated in Table 8. More complicated stress shifting processes such as *Jordan* → *Jordanian* and *Washington* → *Washingtonian* have also been implemented. One might note, with some disappointment, that they do not produce great benefits in coverage. Table 8 shows that

Table 8 Methods Used for Frequent Names in Donnelly Marketing List

	Raw Counts		Percentage		Evaluation		
Method	Type	Token	Type	Token	Good	OK/?	Poor
Direct hit	40,208	40,354,563	16.08	59.34	95	3	2
Name + stress-neutral ending	42,454	16,996,558	16.98	24.99	98	6	4
Name + primary-stress ending	627	69,016	0.25	0.10	94	6	6
Name + *ity*-class ending	6,760	984,662	2.70	1.45	127	4	8
Name + *al*-class ending	2,393	341,301	0.96	0.50	90	8	6
Name + Name (compound)	16,619	1,663,444	6.65	2.45	75	3	13
Name + suffix-exchange	14,523	1,010,813	5.81	1.49	101	5	3
Rhyme with name	29,924	1,579,499	11.97	2.32	84	8	4
Partial letter-to-sound	13,796	500,923	5.52	0.74	71	8	5
Prefix	1,230	117,857	0.49	0.17			
Combinations of above	43,874	3,023,149	17.55	4.44			
All dictionary-based methods	212,408	66,641,785	84.96	97.99			
Remainder (to be handled by *namsa*)	37,592	1,364,118	15.04	2.01			
Totals	250,000	68,005,903	100.00	100.00			

stress shifting morphology (primary-stress endings, suffix-exchange,* *ity*-class endings, *al*-class endings) contributes considerably less than stress-neutral morphology.

Sometimes surprisingly simple processes provide the greatest benefits. The rhyme analogy method is one such case. The pronunciation of an unknown name such as *Plotsky* is determined by analogy with *Trotsky*, which happens to be in the names dictionary. The pronunciation of Plotsky is computed from the pronunciation of Trotsky by removing the initial /tr/ of Trotsky and replacing it with /pl/. It is remarkable just how many names can be pronounced in this way. As Table 8 shows, the rhyme method covers more names than many of the more complicated morphological processes.

There is, of course, some chance of error. For example, we wouldn't want to derive *Jose* from *hose*. It is not possible to know for sure if two words rhyme by looking at their spelling alone. The heuristic employed by the rhyme analogy method is correct about 90% of the time. Although far from perfect, this heuristic is more reliable than letter-to-sound rules. Given a choice between the rhyming heuristic and letter-to-sound rules, it is much safer to choose the rhyming heuristic.

The rhyming method uses letter-to-sound rules in a relatively safe region (the beginning of the word) and uses dictionary methods for the rest of the word. Another method, which we call *interior letter-to-sound*, also combines letter-to-sound rules and dictionary methods. This method uses letter-to-sound rules for an interior syllable such as the *-ar-* in *Daddario*. The remainder of Daddario is derived from *Addonizio*, which happens to be in the dictionary, by suffix-exchange and rhyming. Suffix exchange is used to replace the *-izio* in Addonizio with *-io*; rhyming is used to add the initial *d-*. Thus we use interior letter-to-sound as an extension of suffix exchange, to infer pronunciation for a syllable preceding a recognized stress-forcing ending. The ending tells whether the preceding syllable will be unstressed or primary stressed; and if stressed, whether the vowel will be tense or lax.

Table 8 gives the coverage for the quarter-million most frequent names in the Donnelly marketing list. The table also shows the result of an informal evaluation by a single human judge, Jill Burstein. The judge listened to almost 1000 names and graded them on a three-way scale: (1) good ("like I would have said it"), (2) OK or don't know, and (3) poor ("yuck"). This evaluation shows that compounding is considerably more risky than the other processes.

---

*We introduce the term *suffix-exchange* to refer to a process (like Aronoff's truncation operation) of substituting one affix for another (in the same class) such as *nominate* → *nominee*.

In general, surnames are very hard; the error rate for ordinary words is much smaller.

The following list gives a number of examples of the more common decomposition methods:

- stress-neutral ending: *abandons* = *abandon* + *s*; *abandoning* = *abandon* + *ing*; *abandonment* = *abandon* + *ment* (names) *Abbotts* = *Abbott* + *s*; *Abelson* = *Abel* + *son*
- primary-stress ending: *addressee* = *address* + *ee*; *abductee* = *abduct* + *ee*; *accountability* = *account* + *ability*; *activization* = *active* + *ization*; *adaptation* = *adapt* + *ation*
- *ity* class ending: *abortion* = *abort* + *ion*; *abnormality* = *abnormal* + *ity*; *academician* = *academic* + *ian*; (names) *Adamovich*: *Ambrosian*: *Anagnostakis*
- *al*-class ending: *accidental* = *accident* + *al*; *adjectival* = *adjective* + *al*; *combative* = *combat* + *ive*
- suffix-exchange: *auditoria* = *auditorium* − *um* + *a*; *collusive* = *collude* − *ude* + *usive*; *eldress* = *elder* − *er* + *ress* (names) *Agnano* = *Agnelli* − *elli* + *ano*; *Bierstade* = *Bierbaum* − *baum* + *stadt*
- prefix: *adjoin*; *cardiovascular*; *chlorofluorocarbon* (names) *O'brien*; *Macdonald*; *Distephano*
- compound: *airfield*; *anchorwoman*; *armrest* (names) *Abdulhussein*; *Baumgaertner*
- Rhyming: (names) *Aliano* (from *Califano*); *Anuszweski* (from *Januszewski*)

Table 9 shows the coverage of the various methods for words distributed over the Associated Press newswire during 1988. This corpus is very different from the Donnelly list of surnames. There is a large number of uppercase words in the AP corpus, only some of which are names. For the purposes of this chapter, a word is considered to be a name if it appears in uppercase at least 100 times more often than it appears in lowercase.

The pronunciation problem has traditionally been divided into two very separate modules: letter-to-sound rules and the exceptions dictionary. The focus has been on letter-to-sound rules, which work from first principles. In contrast, the present work resorts to letter-to-sound rules only when all alternatives have been exhausted. The most reliable inference is table lookup. Failing that, the system tries to make as safe an inference as possible from two words in the dictionary. Stress-neutral morphology is considered fairly safe; rhyming is more dangerous but far more reliable than letter-to-sound rules. Our approach breaks down the traditional barriers between letter-to-sound rules and dictionary-based methods. The rhyme method, for example, uses

Table 9 Methods Used in 1988 Associated Press Newswire

	Ordinary words (nonnames)				Capitalized words (names)			
	Raw Counts		Percentage		Raw Counts		Percentage	
Method	Type	Token	Type	Token	Type	Token	Type	Token
Direct hit	13,356	20,646,656	18.27	75.13	26,965	4,147,321	22.87	70.24
Stress-neutral	24,050	4556078	32.90	16.58	25,638	811,751	21.75	13.75
Primary-stress	679	76,222	.93	.28	433	17,128	.37	.29
ity-class	1,943	332,587	2.66	1.21	2,209	96,469	1.87	1.63
al-class	1,174	237,979	1.61	.87	1,119	67,817	.95	1.15
Suffix-exchange	497	36,763	.68	.13	2,409	23,356	2.04	.40
Rhyme					6,888	137,054	5.84	2.32
Prefix	2,907	436,839	3.98	1.59	780	12,945	.66	.22
Partial L-to-S					5,133	23,833	4.35	.40
Compound	3,718	170,877	5.09	.62	5,591	79,753	4.74	1.35
Combinations	15,151	966,033	23.46	3.51	32,705	496,026	27.76	8.40
All methods	63,475	27,460,034	89.58	99.92	97,849	5,752,566	83.01	97.43
Remainder	9,618	21,076	10.42	.06	20,032	151,951	16.99	2.57
Totals	73,093	27,481,110	100.00	100.00	117,881	5,904,517	100.00	100.00

letter-to-sound rules to pronounce the initial consonant onset and dictionary methods to pronounce the remainder.

7. PRONUNCIATION OF SPECIAL WORDS

When we try to calculate the pronunciation of words, we find that a certain fraction of them are not "ordinary" words but are special in some way. Some obvious examples are words made up of letters not in the 26 from A to Z—digits, dollar signs, and other such things. Another set of "special" words, whose pronunciation has to be treated somewhat differently, consists of acronyms and abbreviations. In newswire text, 3 to 4% of all words are special in one of these ways. Of course, other genres of text may have much larger or much smaller proportions.

In addition to having special properties from the point of view of word pronunciation, these cases often require some special interaction with the text analysis component. For example, we won't expect every date and dollar amount to be found in a dictionary of lexical probabilities for our part-of-speech analyzer—we need a pattern matcher that can recognize such things when they occur and give them an appropriate value in the input to the sequence optimizer.

Treatment of such cases becomes crucial to statistics on word pronunciation performance, once we are getting high accuracy on fully spelled alphabetic words. In the section on lexical ambiguity, we sketched some of the issues that arise in treating abbreviations and tokens containing digits. Here we will content ourselves with a brief treatment of our approach as applied to the pronunciation of acronyms.

7.1. Pronunciation of Acronyms

There are two ways to write acronyms in English—with periods, as in **I.R.S.**, and without periods, as in **IRS**. Both patterns can occur, even for the some acronym in the same style of text. In a corpus of about 12 million words of Associated Press newswire text sampled during 1987, we find the results in Table 10. Overall, there were about 62,000 acronyms without periods in the 12-million-word corpus, of about 0.5% of the words, and about 31,000 acronyms with periods. The acronyms with periods are nearly always spelled out, by concatenating their constituent letter names. Such acronyms may be canonically all caps, like those in Table 10; they may contain a mixture of cases, like **G.m.b.H.** and **S.p.A.**, or they may be all lowercase, like **a.m., p.m.**.

Table 10

	With periods	No periods
I.R.S./IRS	22	428
S.E.C./SEC	0	413
L.A./LA	107	8
N.Y./NY	209	14
U.S./US	19873	95
U.S.A./USA	112	294
U.S.S.R./USSR	65	19

The periodless acronyms can be pronounced it two ways: they can be spelled out, as when **HUD*** is pronounced "aitch you dee," or they can be pronounced like an ordinary word of the same spelling, as when **HUD** is pronounced as a monosyllable rhyming with "dud." Although some acronyms, like **HUD**, are variable, most acronyms strongly prefer one pronunciation or the other. Thus we have never heard **AIDS** spelled out, but we have never heard **GOP** pronounced to rhyme with "pop." It seems rather hard to predict how any particular case will go—thus **DEC** (Digital Equipment Company) is "wordified" and pronounced like "deck," and **OPEC** (Organization of Petroleum Exporting Countries) is pronounced like "O-peck," while **NEC** (Nippon Electric Company), **SEC** (Securities and Exchange Commission), and **EEC** (European Economic Community) are all normally spelled out.

It is also sometimes hard to predict how a "wordified" acronym will actually be pronounced—**SCSI** (Small Computer Standard Interface) might have been "sexy" instead of "scuzzy," if the culture of the computer industry had been a bit different. Thus the pronunciation of acronyms requires a lexicon, both for choosing acronyms to wordify and for deciding how to pronounce them, if we are to handle it with high accuracy. A reasonable approach is to treat the spelled-out pronunciation as the default and to use a pronouncing dictionary for the cases that are normally wordified, a small sample of which is given below:

AIDS	AWACS	ESOP	NATO	PAC	SAM
ANSI	CAD/CAM	MIDI	NORAD	RICO	UNESCO
ARCO	CBEMA	NASA	NOW	SAC	UNICEF
ASCAP	CSLI	NASDAQ	OPEC	SALT	

*The acronym for Housing and Urban Development.

In AP newswire text, about 10% of acronym tokens are wordified, and a dictionary with a few hundred entries will capture the bulk of these. In some genres, such as scientific or bureaucratic prose, acronyms are coined and used more freely and mingle with coined proper names formed by other methods. Such cases form about 4% of the words in a sample of scientific abstracts, for instance, and many of the examples seem appropriate to pronounce rather than spell out, even when they are unknown to the reader. The examples below were taken from a set of 200,000 abstracts supplied by the Department of Energy:

t embodied in the laser fusion code	MEDUSA. The theory presented here
0 group constants were generated by	MINX code with the self-shielding
uranium-235) for a research reactor	(TRIGA Mark III) in Thailand. 1 t
en studied in the experiment on the	SKAT bubble chamber with freon fi
s made using this system are given,	SINTER has been designed and prod
of low-energy electron diffraction	(LEED) spot profiles. This spacin
h these problems were solved in the	MANTECH facility is illustrated.

Before leaving the topic of acronyms we should note that they can freely occur in the plural and the genitive case. These are easy to treat by rule, as long as the correct treatment of the base acronym itself is known: the final **s** or **'s** should not be spelled out: thus **CEOs** is "see ee ohs" rather than "see ee oh ess," while **PACs** is "packs." Agentive and participial forms are rarer, but do occur:

They were working with police on	IDing.
four years later became the first	DFLer elected to the U.S. Senate
CBS recently signed another	ABCer, Kathleen Sullivan, for its n
Attorney in Manhattan, in 1984	RICOed oil trader Marc Rich essenti
stice were wary of Mr. Giuliani's	RICOing of Princeton/Newport, but a
Aside from the	RICOing of Drexel, prosecutorial ab
sters had received hefty sums for	OKing the purchase of F-16 Fighting

In the foregoing discussion, we have acted as if acronyms can be unambiguously detected. In fact, there are considerable possibilities of confusion: with Roman numerals, with abbreviations, and especially in text that is capitalized for emphasis or as part of a title, or in text that is monocase (whether upper or lower).

As examples of the Roman numeral case, compare **IV drug user** with **Henry Hunter Hudnut IV**. Repeated references to "World War Aye Aye" are especially annoying. Some abbreviations are often given in all caps, especially American two-letter state postal codes such as CT, MA, and AL. such codes are typically not followed by a period—and of course a period, if present, may also be considered to be a mark of sentence punctuation rather than a signal of an abbreviation.

Some examples of capitalization in titles, reported signs, brand names, and so on are shown below:

circuit is the	AP WEEKEND ENTERTAINMENT AND ARTS package for weekend editions.
	WINDOW TINTING in autos is reviewed by the U.S.; states peer in
	DOWNEY SAVINGS & LOAN ASSOCIATION, Newport Beach, Calif., declar
	CORPORATE DOWNSIZING digs deeper.
the other day:	"COW MILK NOT DOW MILK" and "AGENT ORANGE WORKS—ASK MERRELL D
	DOWN PAYMENTS have become the biggest barrier to homeownership,

There happen to be about 8500 such words within the text portion of the 12-million-word AP corpus that we have been considering. Note that actual acronyms can be interspersed among ordinary words in these capitalized sections. In addition, text fields of everyday business data bases are often all caps, and often lack punctuation as well, like the examples we showed earlier, In order to distinguish acronyms and abbreviations in such "case-impaired" text, we need (nearly) complete lexical coverage.

7.1.1. Cost-Benefit Analysis for Acronym Pronunciation

The simplest thing to do is to spell out all words that are all capital letters, or a sequence of capital letters separated by periods. Two simple modifications will improve performance: dealing with plural and possessive forms of acronyms, and letting Roman numerals up to (say) **XVIII** preempt the spell-out convention. This approach is simple and compact. Like any other algorithm, it will make two sorts of errors: some things will not be spelled out that should be spelled out, and some things will be spelled out that should not be spelled out. The first type of errors will include ordinary words that are capitalized for other reasons—some emphasized words, words in titles, certain brand names, words in telegrams and old-fashioned computerese, and so on—as well as Roman numerals that are falsely spelled out. The second type of errors will include cases like **AIDS** and **OPEC**—about 10% of actual acronyms in AP newswire text.

Starting from this point, the first step to take is to introduce a dictionary of about 500 entries for pronouncing acronyms and to spell out every all-capitalized word that is not in this list. This fairly simple step would fix almost all type II errors at fairly small cost. Of course, in the AP corpus these errors were running at a rate of about 0.05%, or about 5 errors in 10,000 words. Now nearly all occurrences of genuine acronyms will be given an acceptable pronunciation, leaving the type I errors due to mistaken spelling out of real words that happen to be set in all caps. Again, in the AP corpus, these errors amounted to 8500 out or 12 million, or about 7 errors in 10,000 words. There are two possible ways to reduce this remaining source of error. One is to add (genre dependent) heuristics for recognizing stretches of capitalized text; the second would be to aim for almost complete lexical coverage, not only of all known acronyms but also of all words of all types, using adaptive statistical methods to disambiguate ambiguous strings. Surely our residual 7 errors in 10,000 words do not provide enough motivation for such a massive lexical investment—but if we decide that other text analysis issues motivate such a lexicon, it can be used to help distinguish spelled-out acronyms from regular words independent of case information.

8. DISCUSSION AND FUTURE DIRECTIONS

We have discussed a number of problems in text analysis and word pronunciation for text-to-speech synthesis; a similar set of problems arise in any application that attempts a linguistic analysis of unrestricted text. The literature in computational linguistics does not focus much of its attention on such problems, in part because "broad but shallow" analysis has been avoided in favor of "narrow but deep" analysis of restricted domains, and perhaps in part because some of the problems seem as much typographic as linguistic. These problems seem to require somewhat ad hoc and heuristic solutions, including large lists of special cases.

Admittedly, the solutions that we have been able to come up with do not compare with human behavior. Our solutions are not "intelligent." Our solutions do not adapt to unforeseen situations nearly as well as humans do. Humans can read text in all uppercase or all lowercase. They can even read text with uppercase and lowercase reversed, or with S P A C E S between all characters, We+can#suddenly decide-to replace%all^word*boundaries! with some.random>character?or evennone and human readers can quickly adapt. Our algorithmic solutions do not have the ability to identify new and unforeseen patterns and adapt appropriately.

Still, we have tried to outline a principled and systematic approach to dealing with such problems. We propose to avoid solutions involving an expensive and complex tangle of special-purpose, hand-built code that does not generalize easily over different kinds of text, topics, authors, and so forth.

Instead, we aim at a set of motivated heuristics, based on an appropriate selection of a general algorithmic form, combined with statistical parameter fitting to fill in the details. We have argued for the importance of exploratory data analysis (EDA) techniques applied to large text corpora in order to devise an appropriate algorithmic from, check solutions, estimate performance, and identify crucial areas for improvement. We have also shown by example how to trade off costs, such as table size and code complexity, against expected benefits in terms of performance gains, adopting a pragmatic view in order to make sure that resources are allocated in the most effective way.

This approach is very empirical. It makes heavy use of a corpus, especially for fitting (training) the parameters, estimating performance, and identifying weak points. The strength of this approach is derived from the corpus. It is very hard to design accurate text analysis procedures without a corpus of examples to draw upon—intuition and memory are unreliable guides by comparison with systematically explored data. However, the corpus can also be a source of weakness. If the corpus in not a representative sample of the kind of text that the text-to-speech program is going to be asked to read, then conclusions based on the corpus analysis may not generalize very well and the system may make serious and unnecessary errors.

Finally, now that we have brought our error rates in many areas down to small fractions of a percent, one might ask; what is an acceptable error rate? Is it OK for a particular source of error, for example, to cause mispronunciation of 0.5% of the words? of 0.05%? In fact, the answer might well be "no." An error rate of 0.5% by word is equivalent to one error per printed page. It would be equivalent to systematically mispronouncing a word that appears 5000 times per million. Only about 20 of the commonest English words have frequencies this high.

We conjecture that listeners would perceive an error type that occurred this often to be "very frequent." Some of the error rates that we have been discussing here may seem quite small and certainly represent significant progress over the techniques used just a few years ago. However, there is still room for meaningful improvement, even in the most mature algorithms, and many important problems remain almost untouched, such as the determination of phrasal emphasis.

Over the next few years, the naturalness of phonetic synthesis algorithms based on large polyphone inventories should improve to the point that they can pass, if not the Turing test, at least the Lincoln test—to fool all of the people some of the time. At the same time, text analysis algorithms are improving at a steady pace, with applications in areas like text indexing, machine-aided translation, and limited text understanding, as well as in text-to-speech conversion. Continuing progress in microelectronics makes increasingly complex processing accessible in inexpensive devices. As a result, text-to-speech synthesis will expand its niches of useful application, and with some future combination of improved quality and decreased cost it will become ubiquitous in personal computers, telecommunications systems, portable electronic reference works, and other not-yet-imagined products.

REFERENCES

Allen, J., Hunnicutt, M., and Klatt, D. (1987). *From Text to Speech: the MITalk system*. Cambridge University Press, London.

Aronoff, M. (1976). *Word Formation in Generative Grammar.* MIT Press, Cambridge, Mass.

Ayuso, D., Bobrow, R., MacLaughlin, D., Meteer, M., Ramshaw, L., Schwartz, R., and Weischedel, R. (1990). Towards understanding text with a very large vocabulary. *Proc. Darpa Workshop on Speech and Natural Language*, Hidden Valley, Pa.

Bachenko, J., Fitzpatrick, E., and Wright, C. I. (1986). The contribution of parsing to prosodic phrasing in an experimental text-to-speech system. *Proc. ACL Annual Meeting*, vol. 24, pp. 145–155.

Bernstein, J., and Nessly, L. (1981). Performance comparison of component algorithms for the phonemicization of orthography. *Proc. ACL Annual Meeting*, vol. 19, pp. 19–21.

Church, K. (1986). Stress assignment in letter to sound rules for speech synthesis. ICASSP.

DeRose, S. (1988). Grammatical category disambiguation by statistical optimization, *Comput. Linguistics 14*(1).

Ejerhed, E. (1988). Finding clauses in unrestricted text by stochastic and finitary methods. Second ACL Conference on Applied Natural Language Processing.

Fisher, W. (1982). How important is syntactic resolution of orthographic ambiguity? JASA, CC3.

Francis, W., and Kucera, H. (1982). *Frequency Analysis of English Usage*. Houghton Mifflin, Boston.

Good, I. J. (1953), The population frequencies of species and the estimation of population parameters, *Biometrika 40*:237–264.

Gove, P., et al. (1961). *Merriam Webster's Third New International Dictionary*, G. & C. Merriam Co, Springfield, Mass.

Jelinek, F. (1985). Self-organized language modeling for speech recognition. IBM Report, also available in Waibel, A., and Lee, K., eds. (1990). *Readings in Speech Recognition*. Morgan Kaufmann, San Mateo, Calif.

Kahan, S., Pavlidis, T., and Baird, H. (1987). On the Recognition of Printed Characters of any Font or Size, *IEEE Trans. Pattern Anal. Machine Intell.*, pp. 274–287.

Katz, S. (1987). Estimation of probabilities from sparse data for the language model component of a speech recognizer. *IEEE Trans. Acoust. Speech Signal Process. ASSP-35(3)*:

Leech, G., Garside, R., and Atwell, E. (1983). The automatic grammatical tagging of the LOB corpus. *ICAME News 7*:13–33.

Liberman, M., and Sproat, R. (1991). The stress and structure of modified noun phrases in English. In *Lexical Matters*, I. Sag (Ed.) University of Chicago Press, Chicago, in press.

Marcus, M. (1980). *A Theory of Syntactic Recognition for Natural Language*. MIT Press, Cambridge, Mass.

Riley, M. (1989). Some Applications of Tree-based Modelling to Speech and Language, *Proc. Darpa Workshop on Speech and Natural Language*, Cape Cod, Mass.

Rosenberg, C. (1988). Learning the connection between spelling and sound: A network model of oral reading, Cognitive Science Laboratory Report 18, Princeton University.

Sejnowski, T., and Rosenberg, C. (1987). Parallel networks that learn to pronounce English text, *Complex Syst. 1*:144–168.

Sproat, R. (1991). Stress assignment in complex nominals for English text-to-speech, *Proc. ESCA Research Workshop on Speech Synthesis*, in press.

Sproat, R., and Liberman, M. (1987). Towards treating English nominals correctly. *Proc. ACL Annual Meeting 25*:140–146.

Sproat, R., and Shih, C. (1991). A statistical method for finding word boundaries in Chinese text, *Computer Processing of Chinese and Oriental Languages*, in press.

25

Speech Synthesis for Text-to-Speech Systems

HIROKAZU SATO NTT Human Interface Laboratories, Kanagawa, Japan

1. INTRODUCTION

Speech synthesis is an integrated technology for simulating the human processes that generate speech from symbolic (or linguistic) representation of utterances to acoustic waveform. The technology has been increasingly developed in recent years from conventional speech synthesis-by-rule to text-to-speech systems, where the input to be spoken is orthographic text instead of sequences of phonetic symbols. Text-to-speech systems have increased the possible applications of speech synthesis, because sentences written in the ordinal form can be narrated without requiring complicated phonological representation. Nowadays, commercial text-to-speech systems are widely developed for many languages.

One of the issues to be addressed for text-to-speech systems is improvement of the synthesized speech quality, which is still unsatisfactory; in particular, the quality of naturalness must be improved. Synthetic speech is inferior to natural speech in voice quality, smooth articulatory movement, prosodic characteristics such as intonation and rhythm, and the presence of unnatural sounds. In this chapter, speech synthesis methods are discussed, focusing on aspects of improving synthetic speech quality. Hitherto many speech synthesis research papers have been published in Japanese. This chapter introduces several key papers and provides their basic techniques. Because of its potential, formant extraction is first addressed, and the alternating reset dual terminal analog synthesizer is explained as the most sophisticated formant-based synthesizer. The second half of the chapter describes CVC and COC synthesis methods using the linear predictive coding (LPC) technique for text-to-speech systems.

2. FORMANT EXTRACTION AND TERMINAL ANALOG SPEECH SYNTHESIZER

2.1. *Formant Extraction as Trajectories*

The terminal analog speech synthesizer is a typical well-known synthesizer for synthesis by rule. Formant frequencies, by which the synthesizer is controlled, are very convenient parameters for constructing speech synthesis systems based on rules, because the physical correspondence in speech production and the roles as cues in phoneme perception have been revealed, and, therefore, the parameters can be represented or modified by rules. Apart from synthesis using vocal tract models, formants are the only parameters that allow phone modification or generate interphoneme characteristics with rules.

Formants are trajectories varying with time and reflect the relatively slow movements of articulatory organs. From the viewpoint of speech synthesis, it is desirable that formants be continuous without abrupt changes or jumps timewise, because the continuity is needed for parameter concatenation of synthesis units, and large discontinuities cause unnatural sounds which degrade speech quality.

Formant frequencies can be precisely extracted by LPC analysis (Atal and Hanauer, 1971) or the maximum likelihood method (Itakura and Saito, 1970). These methods extract poles rather than formants. Formants are determined by selecting poles with large Q values from redundantly analyzed poles. However, the formant tracking that uses only bandwidth will often fail in trajectory extraction if the continuity of the formant frequencies is not assured. Initially introduced is a method for tracking formants using bandwidth and frequency continuity of analyzed poles, where dynamic programming is applied in the tracking calculations (Sato, 1973).

$P_{l,i}$ denotes the lth pole at time (frame) i, and $F_{l,i}$ and $B_{l,i}$ are the frequency and the bandwidth of the pole. As shown in Fig. 1, when pole $p_{m,i-1}$ and $p_{n,i-2}$ are connected and pole $p_{m,i-1}$ is connected to pole $p_{l,i}$, the cost

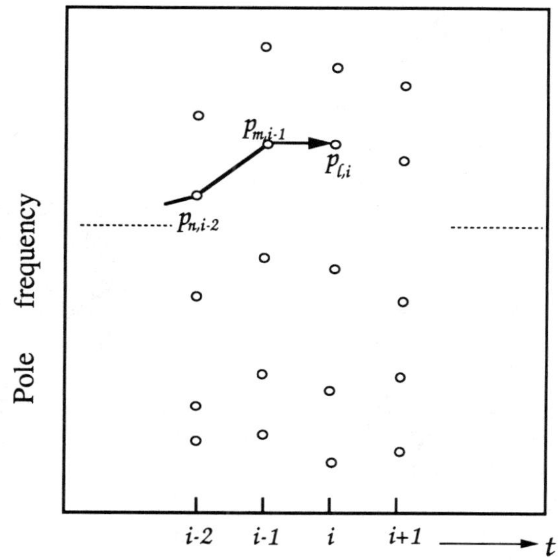

Figure 1 Formant tracking from analyzed pole series.

function $C(i,l,m,n)$ at time i is defined by

$$C(i,l,m,m) = W1 \cdot BL_{l,i} + W2 \cdot |F_{l,i} - F_{m,i-1}| + W3 \cdot |F_{l,i} - 2F_{m,i-1} + F_{n,i-2}| \quad (1)$$

where

$$BL_{l,i} = \begin{cases} 0 & (B_{l,i} < \text{BTH}) \\ \log_{10}(B_{l,i}) - 2 & (B_{l,i} \geq \text{BTH}) \end{cases} \quad (2)$$

$W1$, $W2$, and $W3$ are weighting coefficients, and BTH is the threshold of the bandwidth (200–300 Hz). Equation (1) gives the cost for smoothly tracking poles with small bandwidths. Accumulated cost S from the first to the fourth formants is minimized on the condition that formants do not cross:

$$S(i,l) = \text{Min}_m[S(i-1,m) + C(i,l,m,n)] \quad (3)$$

where

$$C(i,l,m,n) = \sum_{j=1}^{4} C(i,l_j,m_j,n_j) \quad (4)$$

$$l = (l_1,l_2,l_3,l_4) \qquad l_1 < l_2 < l_3 < l_4$$

$$m = (m_1, m_2, m_3, m_4) \qquad m_1 < m_2 < m_3 < m_4$$
$$n = (n_1, n_2, n_3, n_4) \qquad n_1 < n_2 < n_3 < n_4.$$

After all the frames are processed, optimal formant trajectories are obtained using a traceback procedure. Figure 2 shows an example of formant tracking results for the utterance.

When these formant trajectories are applied to a terminal analog speech synthesizer, the speech quality is, unfortunately, unsatisfactory. The formant frequencies by themselves do not approximate the global spectral characteristics because poles with a wide bandwidth are cast off. These poles represent source or radiation characteristics and contribute to the overall shape of the spectral envelope.

To extract local peaks (formants) from the spectrum envelope and simultaneously approximate global spectral characteristics, an A-b-S-like method using autocorrelation domain inverse filtering (ADIF) was proposed (Fushikida, 1981).

The relation between the input speech signal s_n and inverse filtered signal e_n can be represented by

$$e_n = s_n - \sum_{k=1}^{p} \alpha_n s_{n-k} \tag{5}$$

where $\{\alpha_n\}$ are linear predictive coefficients. The autocorrelation function r_i of e_n is given by

$$r_i = \sum_{n=1}^{N} e_n e_{n-i} \tag{6}$$

Substituting Eq. (5) into (6) yields Eq. (7):

$$r_i = \sum_{n=1}^{N} \left(s_n - \sum_{k=1}^{p} \alpha_n s_{n-k} \right) \left(s_{n-i} - \sum_{k=1}^{p} \alpha_n s_{n-i-k} \right) \tag{7}$$

For second-order filtering ($p = 2$), r_i is then given by

$$r_i = (1 + \alpha_1^2 + \alpha_2^2)V_i - (\alpha_1 - \alpha_1\alpha_2)V_{i+1} - \alpha_2(V_{i-2} + V_{i+2}) \tag{8}$$

where $V_i = \sum_{n=1}^{N} s_n s_{n-i}$. Second-order inverse filtering in the autocorrelation domain is shown in Fig. 3.

Moreover, the relationship for formant frequency F and its bandwidth with $\{\alpha_i\}$ is

$$\alpha_1 = r^2, \qquad \alpha_2 = 2r \cos 2\pi FT \tag{9}$$

where T is the sampling period and $r = e^{\pi BT}$.

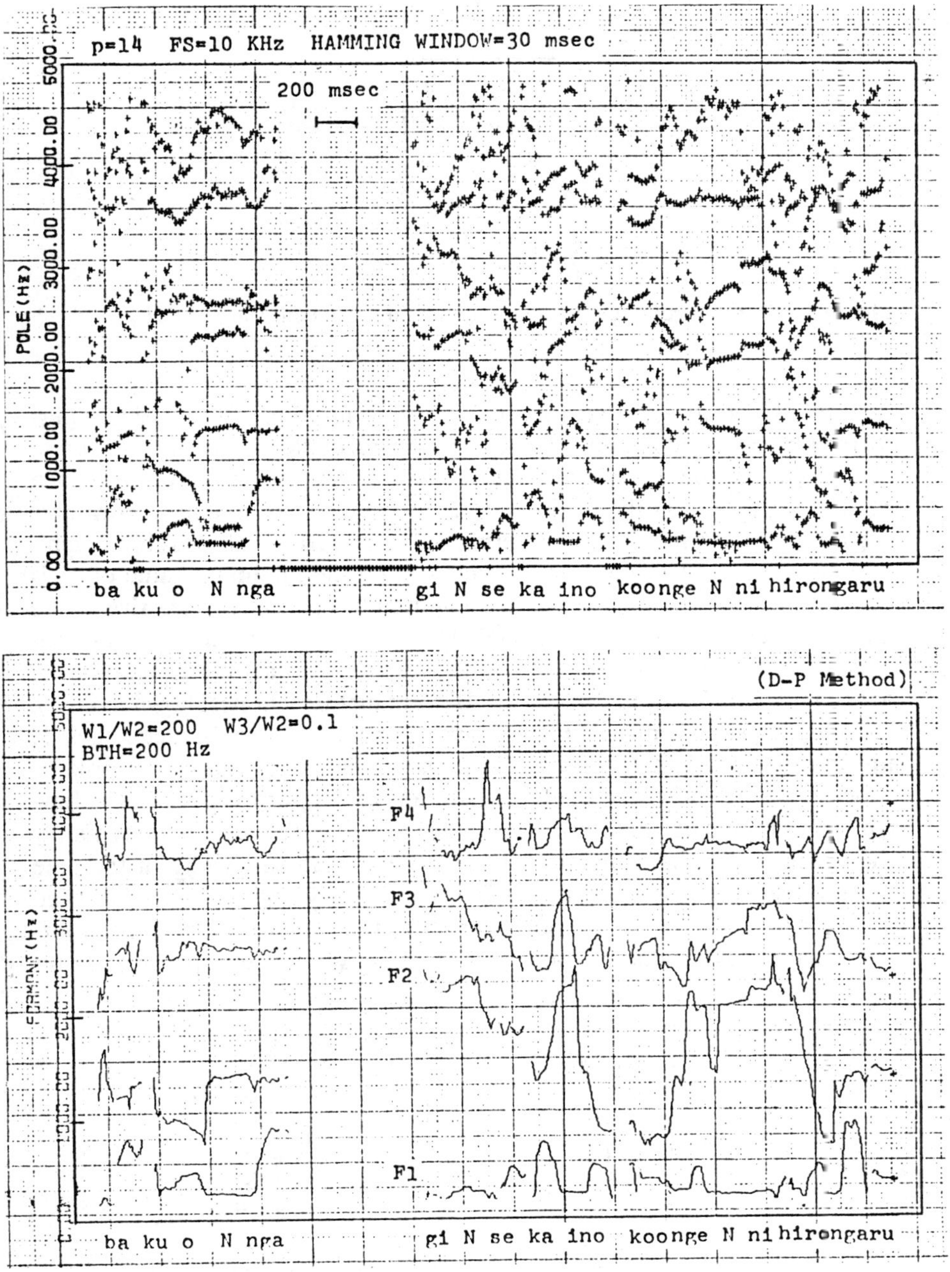

Figure 2 Pole frequencies extracted with the maximum likelihood method, and example of formant tracking.

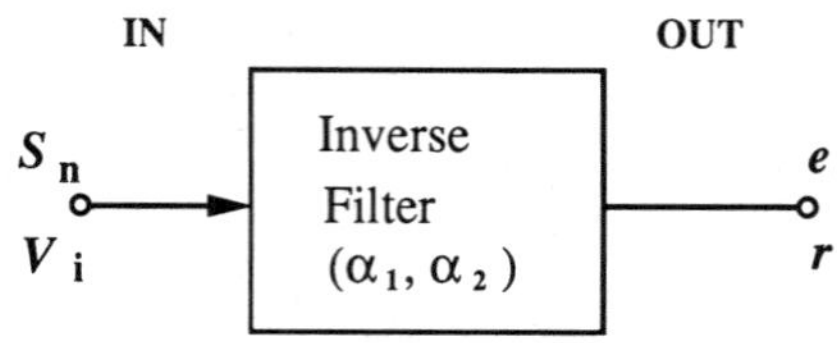

$$r = (1+\alpha_1^2+\alpha_2^2)\, v_i - (\alpha_1 - \alpha_1 \alpha_2) v_{i+1} - \alpha_2 (v_{i-2} + v_{i+2})$$

Figure 3 Second-order autocorrelation domain inverse filtering (ADIF).

To extract all essential formants, inverse filtering is successively performed by cascaded connections in the autocorrelation domain as shown in Fig. 4. The number of autocorrelation functions decreases by two with each filtering stage, and at the final stage output power r_0^k is obtained. Formants are determined when inverse filtering is performed for all possible combinations of the first to the fourth formants and the minimum value of r_0^k is detected. However, the required computation amount is excessive if the above exhaustive search is performed. Therefore, the formant extraction process is divided into several stages and proceeds from coarse to fine extraction as shown in Fig. 5. An example of ADIF formant extraction is shown in Fig. 6.

One question remains: How can the global spectrum characteristics be introduced to the ADIF method? The solution is to add a glottal pole or wide-bandwidth pole that does not have any local spectral peak in the low-frequency region of 1 kHz or less, as shown in Fig. 7. This pole reflects mainly the vocal source characteristics. The frequency and bandwidth of the pole are not fixed to constant values, but the frequency regions to be searched are restricted. As the glottal pole represents the global spectrum characteristics that change frame by frame, the resulting synthesized speech is better

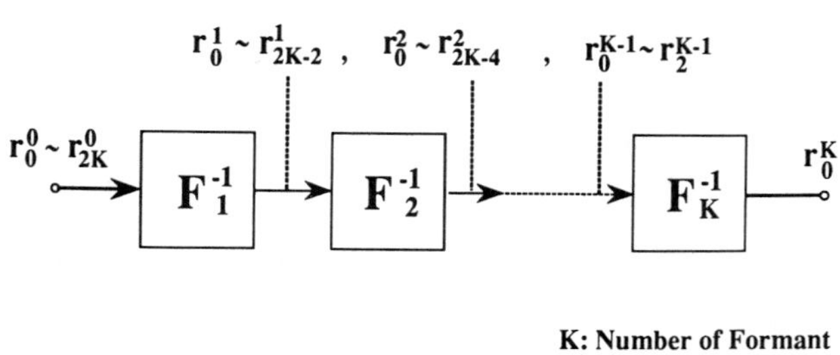

Figure 4 Cascaded inverse filtering.

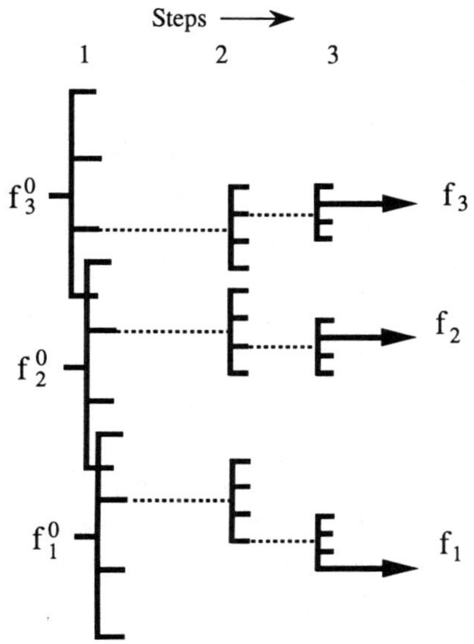

Figure 5 Schematic diagram of multistep formant tracking.

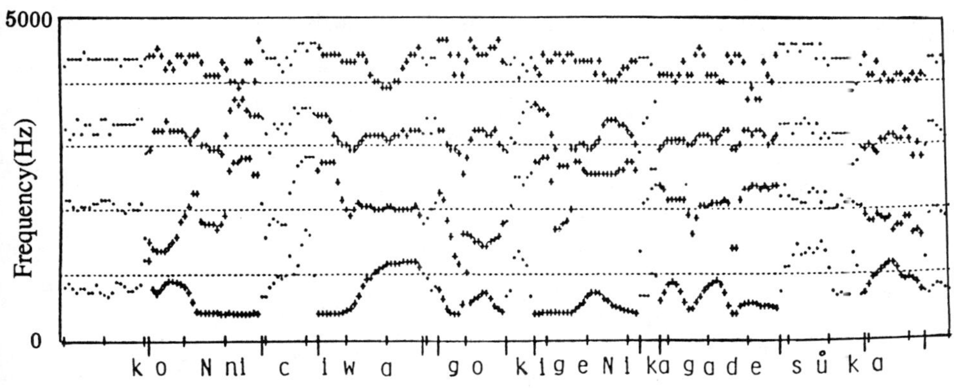

Figure 6 Example of formant extraction based on the ADIF method.

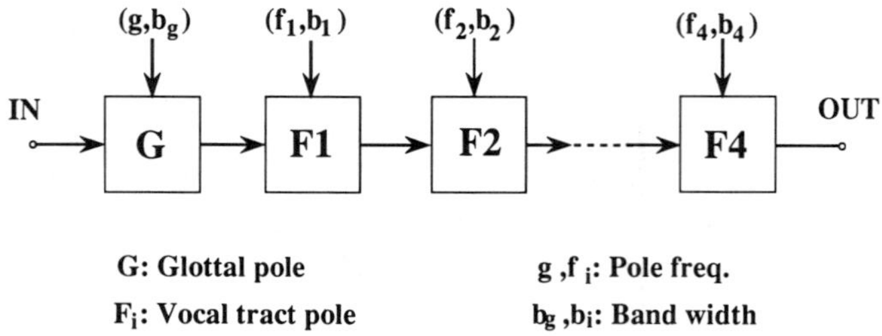

Figure 7 Glottal pole plus vocal tract pole model.

in quality than that created with poles based only on vocal tract characteristics. In pole extraction based on LPC analysis, complex-conjugate poles and real poles are observed to be scattered over the speech frequency region. These scattered poles are unusable for speech synthesis, but the extracted poles forming a smooth trajectory are useful.

2.2. Alternating Reset Dual Terminal Analog Synthesizer

Jumps in the formant frequency trajectory are well known to degrade speech quality synthesized with the terminal analog synthesizer. Hashimoto (1974)

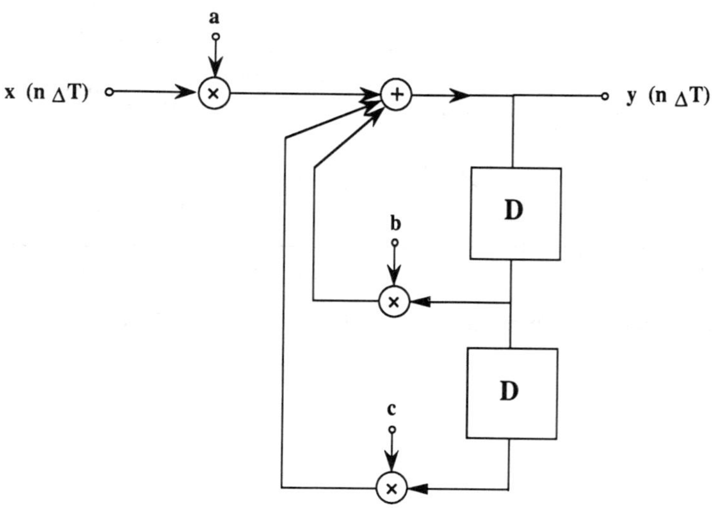

Figure 8 Second-order digital resonance circuit.

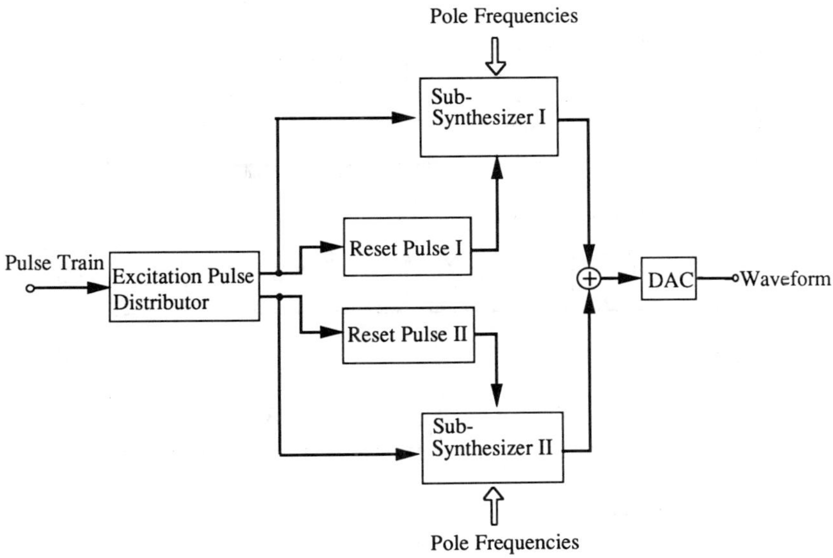

Figure 9 Block diagram of the alternating reset dual terminal analog synthesizer.

examined such instabilities in the synthesizer. He determined the conditions that caused such instabilities and proposed an improved speech synthesizer.

Figure 8 shows the second-order digital circuit that corresponds to one formant. When the memories of delay D hold the preceding value and vocal tract parameters A and B are renewed, and when the continuity of formant frequencies is broken, a large discontinuity or an oscillatory phenomenon is

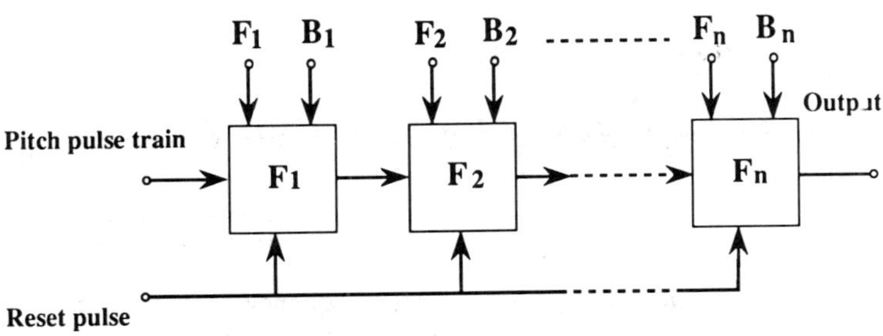

Figure 10 Block diagram of a subsynthesizer (reset-type terminal analog synthesizer).

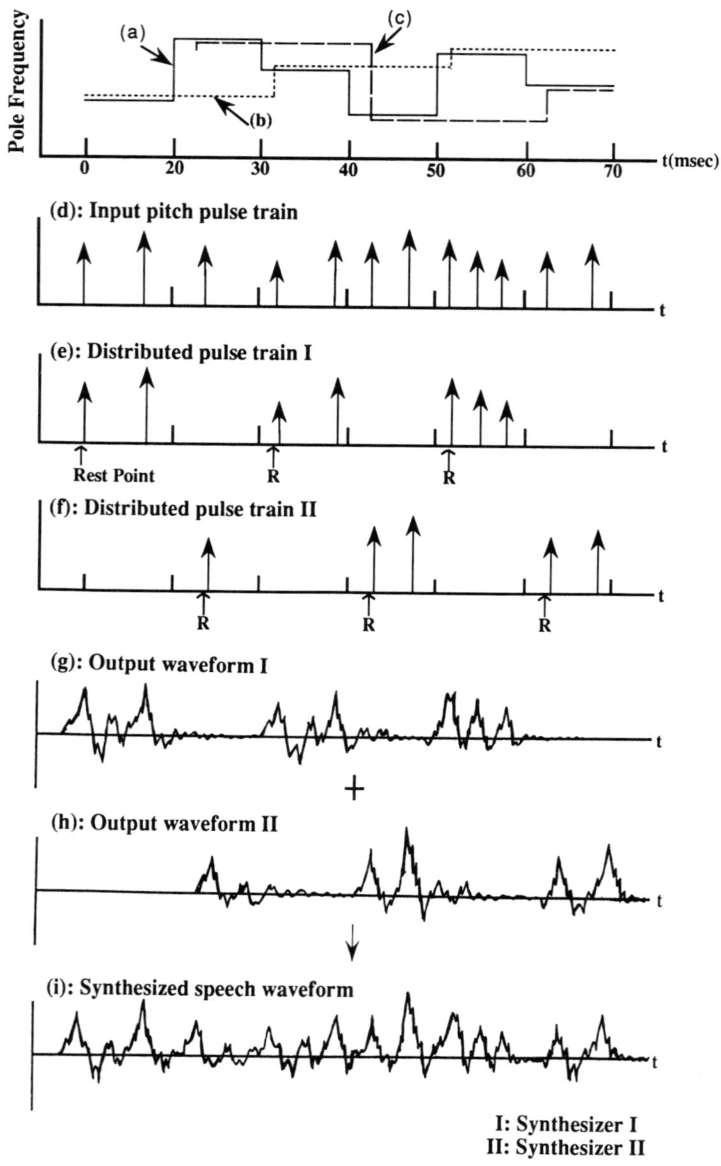

Figure 11 Dual synthesizer operation. Pole frequency control given to terminal analog synthesizer, (——); to synthesizer I, (- - - -); to subsynthesizer II, (– – – –).

then observed in the output waveforms. These abnormal phenomena can be reduced by pitch-synchronous formant control and resetting the memories to zero. However, the memory-resetting action introduces distortion of waveform truncation, especially in high-pitched voices such as those of females. Therefore, a parallel-structured cascade-connected terminal analog synthesizer was proposed. The internal memories are reset alternatingly and pitch synchronously. This synthesizer is called the alternating reset dual terminal analog synthesizer.

Block diagrams of the synthesizer and the dual subsynthesizers are shown in Figs. 9 and 10, respectively. Each subsynthesizer has a function for resetting the internal memories. Figure 11 illustrates the operation of the dual synthesizer. The control process of the synthesizer is as follows. Formant (or pole) frequencies are given to the dual synthesizer in each frame (e.g., 10 ms) as shown in Fig. 11a. Moreover, pitch pulses of the excitation source are given as shown in Fig. 11d and are distributed alternately in each frame to the subsynthesizers I and II (Fig. 11e and f). Formants are also distributed pitch synchronously (Fig. 11b and c) and each subsynthesizer is reset immediately before the first pulse in the next frame. Therefore, the waveform of the impulse response is reset after it is sufficiently damped. The resultant output waveform (Fig. 11i) is obtained by adding two waveforms (Fig. 11g and h) of the subsynthesizer outputs.

It is reported that this synthesizer is very effective, not only for high-pitched female voice synthesis but also for speech synthesis from pole sequences analyzed with the maximum likelihood method. This synthesizer does not necessarily require continuous formant trajectories. If a triple structure is developed for the synthesizer, it will be possible to reduce the distortion further by resetting.

3. UTILIZATION OF THE LPC TECHNOLOGY FOR SPEECH SYNTHESIS

Linear predictive coding was developed for speech data compression and speech transmission. It was first applied to the synthesis of arbitrary Japanese speech on the basis of vowel-consonant-vowel (VCV) units in 1974 (Sato, 1974). The LPC technique may be considered to be unsuited to speech synthesis by rule because it is difficult to compose synthesis rules using LPC parameters or to manipulate those parameters. However, LPC offers more advantages for synthesis by compiling a large number of synthesis units, like phoneme composite units or allophonic units, than synthesis based on rules to realize coarticulatory or dynamic characteristics. In LPC-based speech synthesis, rules applied for synthesis are restricted mainly to generation of suprasegmental characteristics such as pitch frequency contours. This chapter describes two speech synthesis systems using the LPC technique for text-to-speech conversion with the focus on the speech synthesis units.

3.1. CVC-Based Speech Synthesis

Hitherto, in Japanese speech synthesis, phoneme composite units such as CV or VCV have been used as speech concatenation units. The CV is the basic syllable in Japanese, and therefore, Japanese words or sentences are typically structured to alternate consonants with vowels: $C_1V_1C_2V_2 \cdots C_nV_n$. Thus, we have two triphone-like speech synthesis units, VCV and CVC. In VCV-based speech synthesis (Saito and Hashimoto, 1968; Sato, 1974), the weight was given to the consonants, and it was found that concatenation on vowels is advantageous. This is because, generally speaking, spectral characteristics are transient and change rapidly in the consonant part, while they change gently or remain constant in the vowels. However, when concatenating VCV units, artifacts are introduced to the vowels by the interpolation process. The distortion will be aurally perceptible because vowel power is much larger than consonant power.

A CVC-based speech synthesis technique was created that gave weight to natural-sounding vowel realization (Sato, 1984; Sagisaka and Sato, 1986). It was based on the consideration that achieving a more natural-sounding vowel quality makes a better impression of the overall quality of synthetic speech, so it is important to enhance the quality of synthesized vowels. By adopting CVC-type synthesis units, it is possible to avoid deterioration caused by vowel unit concatenation and to constrain distortion in consonant concatenation. Another advantage of CVC-based synthesis is that the central vowel in CVC retains the coarticulatory features in relation to the adjoining consonants. This is effective in giving synthetic speech a smooth articulatory sound.

The CVC technique introduced the idea of using waveform information to enhance the quality of synthesized speech. Most previous speech synthesis by rule techniques used simplified excitation sources: a pulse train for the voiced sounds and white noise for unvoiced sounds. The CVC speech synthesis system utilizes residual signals obtained through LPC analysis of uttered synthesis units as the excitation source signals.

For pitch control in the CVC speech synthesis technique, waveforms of one-pitch period (excitation waveform elements) should be extracted from residual signals of utterances. The question arises of how many waveform elements are needed to construct the speech synthesis system. Using one typical and universal excitation waveform may be one method, and preparing representative waveforms for each vowel may be another. In practice, however, small numbers of excitation elements do not always ensure satisfactory synthesized speech quality. In this speech synthesis scheme, each CVC has several excitation waveform elements besides the LSP spectral parameters. A CVC utterance is typically divided into five segments, namely the preceding consonant, the on-glide, the stationary vowel part, the off-glide, and the following consonant, as illustrated in Fig. 12. For each voiced speech segment, a representative one-pitch waveform element is extracted. For each unvoiced segment, the complete residual signal in the segment is stored as an excitation element. To generate the excitation waveform, these periodic waveform elements are connected repeatedly and successively according to the pitch periods with appropriate truncation and zero insertion. Figure 13 shows an example of a synthesized waveform and compares it with speech generated by the impulse train technique. The speech synthesized with CVC and excitation waveform elements is observed to be very similar to the waveforms as uttered by a human.

The CVC speech synthesis units are syllabic units attending the succeeding consonant, because Japanese syllables have an open structure like CV, in which a syllable ends in a vowel. In Japanese, other long syllables such

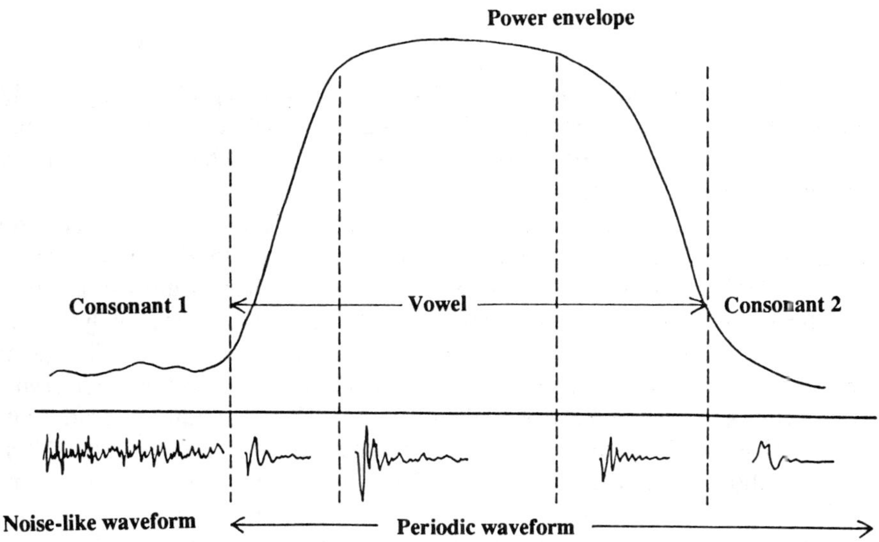

Figure 12 Extraction of excitation waveform elements from residual signal in CVC unit.

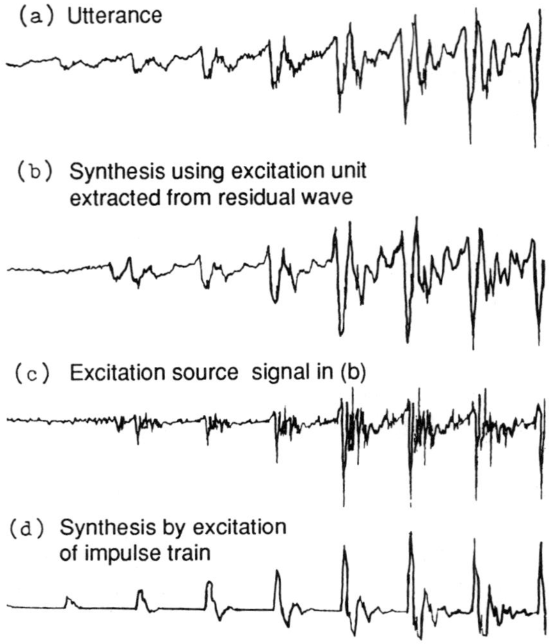

Figure 13 Natural and synthesized waveforms in /za/ of "ka*za*guruma" (pinwheel).

as CVN (N = syllabic nasal) or CVi (Vi = descending diphthong) are used. Therefore, speech synthesis units of CVNC and CViC types besides CVC were introduced in the speech synthesis system. In this chapter, these units are collectively called "CVC."

In preparing the CVC-type speech synthesis units, there is a problem; if all the possible CVCs in Japanese were adopted, the total number would exceed 6000. This number is obviously impractical, if not impossible, in a practical system. To reduce the number of CVC units, a statistical analysis was conducted using headwords in a Japanese dictionary (Sato, 1983). The most frequent concatenations (about 1100 CVCs) were extracted as the speech synthesis units and covered 90% of all the analyzed CVCs. In addition, dyad-type units such as CV, VC, and VV were also prepared to cover less frequent CVCs and diphthongs. Consequently, about 1300 synthesis units are used as shown in Table 1.

A block diagram of speech synthesis using the CVC units and the excitation waveform elements is shown in Fig. 14. When speech synthesis units are concatenated, spectrum parameters (LSP) are smoothed around boundaries

Table 1 Synthesis Units in CVC-Based Speech Synthesis

Unit	Number	Examples
CVC	795	kar, das, shon, ...
CVMC	229	kans, taik, ...
CV	135	'a, ke, Fa, pyo, shi, ...
VC	90	ab, im, ew, uz, ...
VV, VN	45	ai, ia, ue, ...
		am, an, ang, ...
Total	1294	

by averaging on the weighting of a triangular window. Generating the excitation source signal produces some waveform distortion because waveform elements are truncated to match the pitch periods. One method for reducing this distortion is to use a strongly damped waveform. Such a waveform can be reproduced by setting the phases of all original waveform elements components to zero (phase equalization). The speech synthesizer used in this system has a function that controls the frame length frame by frame. This function is very convenient when changing segmental durations.

The synthesized speech output by this system not only is intelligible but also has a realistic quality. It was confirmed by a listening evaluation

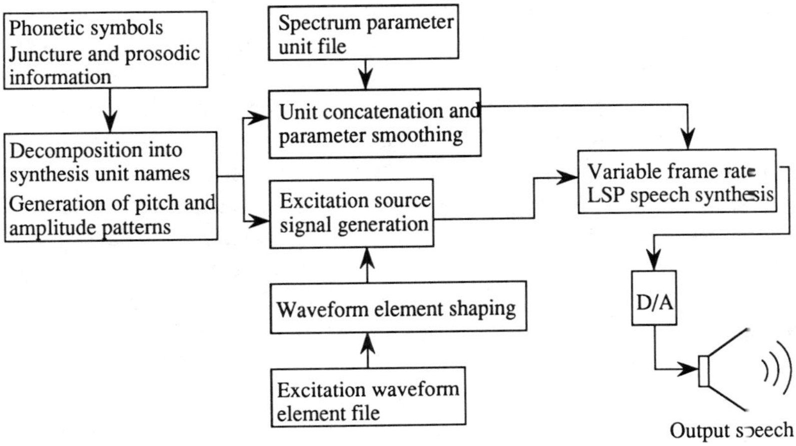

Figure 14 CVC-based speech synthesis block diagram.

test that the synthetic speech is very close to speech reproduced by the analysis/synthesis technique.

3.2. Automatic Generation of Speech Synthesis Units (COC Method)

The object of speech synthesis is to reproduce the great variety of sounds present in human speech from a limited number of speech units and synthesis rules. The realization of these sounds will lead to natural-sounding synthesized speech. Allophones, in a wide sense, have been introduced in speech synthesis to realize such phonetic variations. One of the primary factors causing the allophonic variation in utterances is the contextual or environmental effects of adjacent phonemes.

The CVC synthesis units mentioned above can be regarded for the central vowel as units including the influence of neighboring phonemes. In producing allophonic units for speech synthesis, methods based only on phonemic context, such as CVC units, may sometimes produce acoustically identical or very similar units which are redundant. On the other hand, units simply generated by a clustering of speech segments using some spectral measures will be useless for speech synthesis because they have no correspondence with environmental information to indicate where they should be used in the input text. The environmental information is used not only to distinguish phonetic features but also as information about how to construct speech. Therefore, it is expected that a great many essential phonetic segments for speech synthesis can be generated from contextual factors.

To resolve this problem, a powerful method for automatically generating synthesis units has been proposed (Nakajima and Hamada, 1988). As context-oriented clustering is used in the scheme, the procedure was called the COC method. The flow diagram of a speech synthesis system based on this method is shown in Fig. 15. The LSP technique is applied to the parametric representation of the synthesis units and speech synthesizer.

In the COC method, the phonetic context implicitly bundles a set of parameter segments and a partitioning process is performed on these implicit bundles. Figure 16 shows the simple cluster partitioning process in COC. Initially, each cluster consists of segments that have the same phonetic labels. Then each cluster is split iteratively by applying the restriction of the preceding or succeeding context. In Fig. 16, initial cluster W_1, which is a set of the segments with phonetic label /a/, is split into two clusters: W_{11}: /a/ preceding by /b/, and W_{12}: /a/ preceded by any phoneme but /b/. Each of these two clusters can be further split into two more clusters. In other words, W_{11} can be split into W_{111} (/a/ preceded by /b/ and succeeded by /s/) and W_{112} (/a/ preceded by /b/ and succeeded by any phoneme but /s/). This iterative pro-

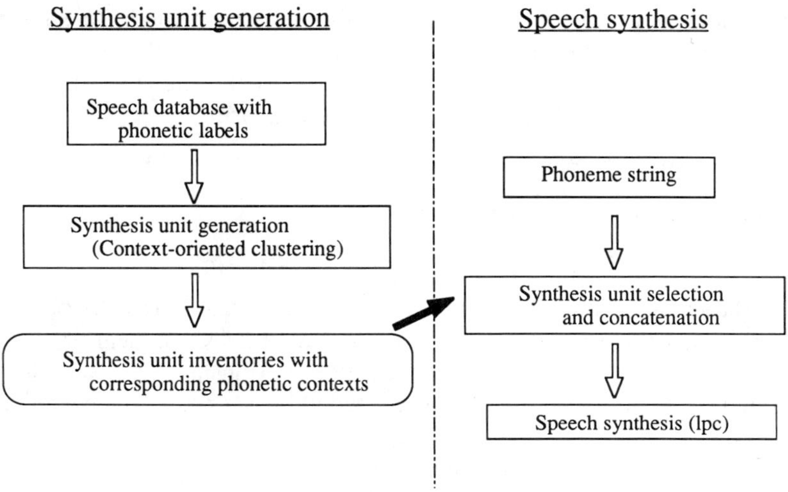

Figure 15 Block diagram of the COC speech synthesis system based on automatic synthesis unit generation.

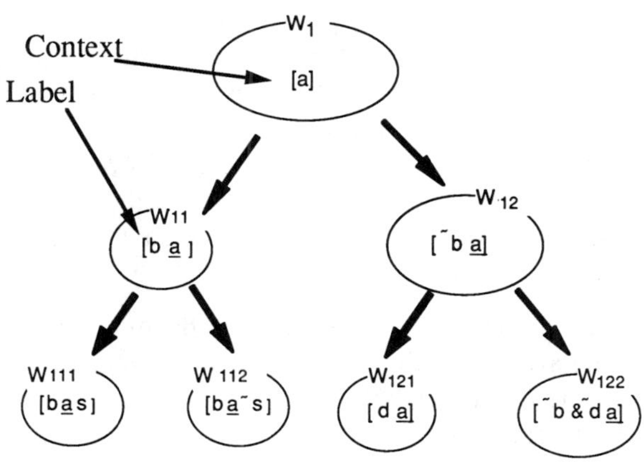

Figure 16 Example of the cluster partitioning process in context-oriented clustering.

cess is continued until some terminal conditions are satisfied. As shown in Fig. 16, each cluster has a symbolic representation, called the context label, that identifies the phoneme the cluster belongs to and the context in which the cluster is to be used. Context labels have the form:

$$[\langle \text{pre-context-list} \rangle \, \langle \textit{phonetic-label} \rangle \, \langle \text{post-context-list} \rangle]$$

The ⟨pre/post-context-list⟩ has the form

$$\langle \text{ctx-exp-1} \rangle , \langle \text{ctx-exp-2} \rangle , \ldots , \langle \text{ctx-exp-n} \rangle$$

where ⟨ctx-exp-i⟩ denotes the ith context expression and is a set-theoretic expression consisting of phonetic symbols, logical-and "&," logical-or "|," and complement "~," and "," stands for the order relationship. For example, "/a/ preceded by /b/ preceded by /u/ and succeeded by /s/ or /t/" can be expressed as "[u, b *a* s | t]."

The COC algorithm can be described in detail as follows.

1. Let each initial cluster W_i consist of the segments with the same levels, and compute each cluster evaluation value v_i^2.
2. Find a cluster W_k that gives the maximum v_k^2 and has more segments than the minimum number of segments per cluster $N_{\min}$. If no such cluster exists, terminate the procedure.
3. Find the optimum context S^* for the next partitioning step among all the neighboring contexts of W_k. S^* is the context that produces the maximum split evaluation value G.
4. Split cluster W_k into two exclusive clusters by S^*.
5. Check that the total evaluation value J is less than the threshold value J_{sth}. If this condition is not satisfied, go back to step 2. J = average of cluster evaluation values.

The cluster evaluation value is represented by an inner-cluster variance normalized by the average of the cluster evaluation values. The split evaluation value G is the difference between the cluster evaluation value before splitting and the average of those values after splitting. For splitting W_1 into W_{11} and W_{12}, G can be computed as

$$G = \frac{v_1^2 - \left(v_{11}^2 + v_{12}^2\right)}{2} \tag{10}$$

The centroid matrix of each cluster is saved as a synthesis unit together with its context label. The computation of the centroid matrix for time-varying segments is based on the procedure proposed for very low bit rate speech coding (Shiraki, 1984). Every segment is converted to a fixed dimension L by using linear interpolation or resampling techniques. In this method, L is not fixed over all clusters but is the average length of each cluster.

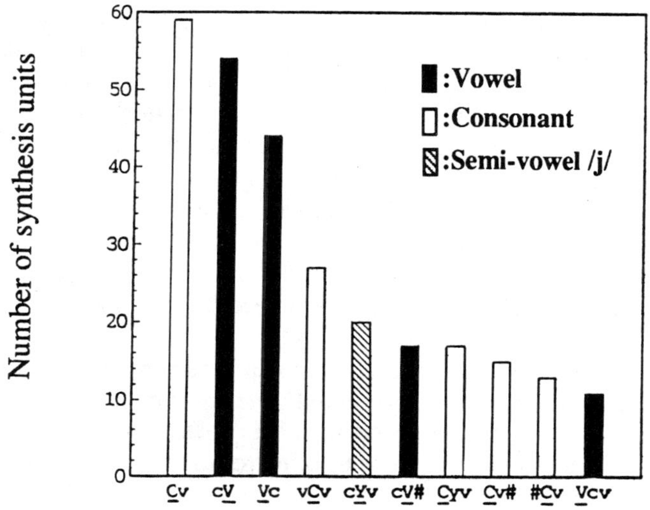

Figure 17 Frequency of synthesis units generated automatically through the COC procedure.

Applying this procedure to the synthesis unit generation of 432 words uttered by a male speaker yielded 627 units. Figure 17 shows the 10 categories with the greatest number of generated synthesis units. As seen in Fig. 17, there are many CV or CYV (Y denotes a semivowel) variations, such as *C*v, c*V*, c*V*# (# denotes a major boundary or pause) and c*Y*v. This result coincides with actual Japanese phonology because consonant-vowel and consonant-semivowel-vowel combinations are typically syllable structures in Japanese. Moreover, units that include longer contexts such as c*V*cv or v*C*vc are also generated. The CVC method is advantageous in that the synthesis units influenced by such longer contexts are automatically extracted.

A speech synthesis (text-to-speech) PC accessory board based on the CVC method has been developed (Hakoda et al., 1990). It is planned to enhance the quality of the board's output by increasing the number of synthesis units to more than 1500.

4. SUMMARY

This chapter discussed several techniques that improve synthetic speech quality. In the first half of the chapter, the methods that connect the pole extraction on the LPC analysis to terminal analog (formant) speech synthesis were described. For the terminal analog speech synthesizer, not only the local peaks of the speech spectrum should be reproduced but also the global characteristics should be approximated. Moreover, it is desirable that these characteristics, represented by poles, be extracted as continuously changing trajectories, because abrupt changes degrade the quality of synthesized speech. The importance and the method of formant extraction, including source poles to express global spectral features, were discussed. Moreover, the structure of the terminal analog speech synthesizer was presented. It consists of a parallel structure of cascade-connected synthesizers with a memory reset function.

In the second half of the chapter, two synthesis methods for text to speech using the LPC technique were presented. The production of speech synthesis units that take into account phonetic contexts was discussed. CVC speech synthesis is the method that emphasizes the reproduction of vowels with high quality. The COC method "automatically" generates speech synthesis units that can express the phonetic variation created by context.

The author expects that the techniques and method introduced here will be helpful in the further study or development of speech synthesis.

REFERENCES

Atal., B. S., and Hanauer, S. L. (1971). Speech analysis and synthesis by linear prediction of the speech wave. *J. Acoust. Soc. Am. 50*:637.

Fushikida, K. (1981). A focusing formant extraction method using autocorrelation domain inverse filtering. *Trans. Commit. Speech Res. Acoust. Soc. Jpn.*, S81-41 (in Japanese).

Hakoda, K., Nakajima, S., Hirokawa, T., and Mizuno, T. (1990). A new Japanese text-to-speech synthesizer based on COC synthesis method. *Proc. Int. Conf. Spoken Language Processing*, ICSLP-90, pp. 809–812.

Hashimoto, S. (1974). Alternating reset dual terminal analog speech synthesizer. *Proc. Spring Meet. Acoust. Soc. Jpn. 3-2-7*:367–368 (in Japanese).

Itakura, F., and Saito, S. (1970). A statistical method for estimation of speech spectral density and formant frequencies. *Electron. Commun. Jpn. 53-A*:36 (in Japanese). English version: *Speech Synthesis*, Dowden, Hutchinson & Ross, Stroudsburg, Pa., p. 293.

Nakajima, S., and Hamada, H. (1988). Automatic generation of synthesis units based on context oriented clustering. *Proc. Int. Conf. Acoustics, Speech, and Signal Processing*, ICASSP-88, pp. 133–136.

Sagisaka, Y., and Sato, H. (1986). Composite phoneme units for the synthesis of Japanese. *Speech Commun. 5*:217.

Saito, S., and Hashimoto, S. (1968). Speech synthesis system based on interphoneme transition unit. *Proc. 6th ICA*, B-5-12, p. B-195.
Sato, H. (1973). Automatic determination of formant trajectories and its application to speech synthesis system. *Trans. Commit. Speech Res. Acoust. Soc. Jpn.*, S73-05 (in Japanese).
Sato, H. (1974). Speech synthesis based on PARCOR-VCV concatenation units. *Trans. Commit. Speech Res. Acoust. Soc. Jpn.*, S74-22 (in Japanese).
Sato, H. (1983). Statistical analysis of Japanese phoneme concatenations for speech synthesis. *Trans. Commit. Speech Res. Acoust. Soc. Jpn.*, S82-77 (in Japanese).
Sato, H. (1984). Speech synthesis using CVC-concatenation units and excitation waveform elements. *Trans. Commit. Speech Res. Acoust. Soc. Jpn.*, S83-69 (in Japanese).
Shiraki, Y. (1984). Very-low-rate speech coding using time space spectrum pattern. *Trans. Commit. Speech Res. Acoust. Soc. Jpn.*, S84-06 (in Japanese).

Index

A^* search, 639
Abbreviation, 805, 827
Absolute threshold, 115
Absolutes, 758
AB test, 135
ABX test, 135
Acoustic distance, 253
Acoustic segment method, 532
Acoustic segment units (ASU), 722
Acoustic transfer function, 781
Acronym, 805, 824, 827
Adaptive bit allocation, 125
Adaptive cancellation, 332
Adaptive differential PCM (ADPCM), 89, 91, 94, 98
Adaptive learning, 536
Adaptive predictive coding, 87
Adequacy of intonation contours, 404
Advanced television systems, 101
Affix of a word, 753
Affix stripping, 755
Agglomerative clustering, 633
All-pole modeling of degraded speech, 312
Allophonic variation, 848
Alternating reset dual terminal analog synthesizer, 840
Alveolar stop consonants, 756
Ambiguities in pronouncing digits, 806
Ambiguous pronunciations, 802
Amplitude, 796
Analysis-by-synthesis, 127
  adaptive predictive coder, 159
  adaptive predictive coding, 142
  coders, 158
  predictive coding, 64
Analysis/synthesis task, 746
AP newswire, 797, 802
A posteriori probability, 515, 524
A priori probability, 524, 527
ARAEN, 362
Articulation, 362
  index, 89, 372, 407
  sound, 362

[Articulation]
  syllabic, 362
Articulators, 744
Articulatory codebook, 252
Articulatory gestures, 768
Articulatory model, 233, 245, 263, 742, 768, 770
Articulatory synthesis, 770, 779
Artificial voice signal, 375
Aspiration, 241
Associative nets, 556
Asymptotic optimality, 527
Asynchronous transfer mode (ATM), 378
Atomic syntactic units, 754
Atomic unit of speech, 748
Attributive tags, 813
Audio, 97
  conferencing, 97
  digital, 89
  quality, 99
  wideband, 85
Auditory brain stem, 456
Auditory cortex, 456
Auditory critical bands, 295
Auditory midbrain, 456
Auditory model, 101, 511
Auditory nerve, 457
  representation, 453
Auditory pathway, 456
  ascending, 456
  descending, 456
Augmented context free grammars, 764
Autoregressive (AR) model, 711
Autosegmental phonology, 769
Auxiliary function, 545
Average branching factor, 523
Average conditional entropy, 659
Average firing rate, 463
Average magnitude difference function (AMDF), 12
Average mutual information function, 662

Back propagation, 493, 498, 556, 580
  training procedure, 560
Backward adaptation, 100
Backward-adaptive, 96
  mode, 94
  system, 152
Backward masking, 132
Backward variable, 547
Bakis model, 628
Bark-scale, 444
Basilar membrane, 457
Basis functions, 332
Baum–Welch algorithm, 514, 518, 524, 680
Baum–Welch reestimation, 545, 677, 678
Bayes formula, 652
Bayes minimum error, 525, 541
Bayes' rule, 321
Bayesian framework, 515
Beam search, 493, 505, 539
  pruning, 488
Benchmarks of performance, 408
Bigram, 654, 664, 668
  models, 635
Bilinear transform, 427, 433
Bipolar electrodes, 271
Birth-death procedure, 173
Bisyllabic noun/verb ambiguities, 762
Bit allocation, 98, 101
Bit error rate, 94
Bit rate, 85
Blind equalization, 726
Blind tests, 134
Block adaptation, 69
Boltzmann machines, 556
Breathy voice, 702
Broadband continuous noise, 310
Broadcasting, 100
Brown corpus, 797, 804
  tagged, 815
Burst spectrum, 778
BYBLOS, 642

Cache language model, 674
Calculated noise threshold, 125
Cancellation condition, 346
  synthesis-dependent, 346
  synthesis-independent, 347
Canonical catenary transition contours, 774

Categorical grammar, 760
CCITT, 89, 94, 98
Ceiling effects, 397
Center clipping, 16
Central processor
  rate-place category of, 454
  temporal-nonplace category of, 454
  temporal-place category of, 454
Centroid condition, 53
Cepstral coefficients, 732
Cepstral distance measures, 257
Cepstral features, 727
Cepstral parameters, 322
Cepstral vector, 728
Cepstrum, 14, 16
  distance (CD) measure, 374
  distance, weighted, 421
Chain matrix, 237, 244
Channel model, 728
Channel noise, 92
Channel quality, 93
Characteristic frequency (CF), 459
Chinks 'n chunks algorithm, 811
Class prior model, 527
Classical pattern classification theory, 527
Classification error rate, 524
Classification tree, 801
Clause parsing, 815
Clitics, 803
Closed loop analysis, 149, 150
Closed response, 395
Coarticulation, 742, 767
Coarticulatory smoothing, 756
Cochannel distortions, 311
Cochannel problem, 311
Cochlear filtering, 114
Cochlear filters, 459
Cochlear implants, 269
COC method, 848
Codebook, 49, 210, 716
  access, 257
  adaptation, 211
  adaptive, 150, 154, 158, 210
  fixed, 150, 154, 158
  normalization, 611
  shape, 96
  speaker-dependent, 222
  speaker-independent, 222
Coded speech quality
  for ATM networks, 378
  speaker-dependency of, 376
Code-excited coder, 146, 150
Code excited linear prediction (CELP), 64, 89, 91, 92, 95, 100
Coders
  complexity of, 157
  robustness of, 157
Code vector, 49
Coding, 49
  delay, 68
  multiple stages of, 85, 93
Cognitive processes, 555
Coherence function (CF) measure, 372
Coherent neural activity, 463, 466
Commentary-quality, 98
Compact disc, 101
Complete label case, 525
Complex nominals, 816
Compound stress rules, 766
Comprehension from synthesized speech, 784
Comprehension process, 402
Comprehension tests, 406
Conditional maximum likelihood estimation, 527
Conditional probability, 524
Conjunctions, 763
Connected word recognition, 487, 489, 524
Connectionism, 556
Connectionist glue, 565
Connectionist learning, 565
Connectionist parsing, 588
Connectionist postprocessing, 585
Connectionist Viterbi training, 581
Consonant clusters, 397
Consonants, 752
Content word, 656, 752
Context free morph sequence grammar, 758
Context oriented clustering (COC), 848
Context-dependent modeling, 633
Context-free grammar, 505

Continuous speech recognition, 487, 489
Convergence in the ideal case, 336
Convergence in the nonideal case, 338
Convergence rate, 337
Cordless phone, 94
Correction rule, 526
Corrective training, 526, 534, 599, 631
Correlation
  long-term, 144
  short-term, 144
Cosh measure, 374, 425
Cost function, 248, 262
Covering algorithm, 717, 719
Creak, 5
Criminal investigations, 704
Critical band, 113
Cross-entropy, 533
Cross validation, 580, 654, 667
Cubic phase function, 176
Cubic phase interpolation, 174
Customer's loop, 329
CVC, 399
  based speech synthesis, 844

DARPA Resource Management task, 509
Data transmission, 99
Decimal point, 807
Decision rule, 524
Decision tree, 800, 806
  design, 541
Decoder, 52
Decoding hierarchy, 519, 522
Deconvolution, 299
Delay
  communication, 93, 99
  decoding, 105
  encoding, 95, 105
  in coders, 158
  noninteger, 144, 150
  processing, 85, 93
Delayed decision coding, 142
Delayed decision for excitation frames, 155
Deleted interpolation, 610
Delta-cepstral coefficients, 732
Delta cepstrum, 536, 627, 714, 727
DELTA data structure, 783
Delta modulation, adaptive, 100
Demisyllable, 743, 770
Density
  elliptically symmetric, 519
  Gaussian, 523
  heavy-tail, 520
  log-concave, 519
  mixture, 520
  multimodal, 520
  unimodal, 520
Diagnostic acceptability measure (DAM), 86, 91, 361, 405
Diagnostic rhyme test (DRT), 86, 91, 363, 394, 397
Diagonal covariance, 523
Dialog systems, 407
Dictionary, 818
  lookup, 530
Difference power, 435
Difference spectrum, 427, 435
Differential energy, 536
Differential log energy, 537
Differential spectral parameters, 599
Digital audio, 89
Digital cellular radio, 91
Digital circuit multiplication equipment (DCME), 94
Digital coding of telephone speech, 91
Digital encryption, 87
Digital signal processor, 280
Digital telephony, 86, 91, 92
Dimensionality reduction, 628
Diphone, 743, 768, 770
  concatenation of, 777
Discourse model, 749
Discriminant analysis, 534
Discriminant function, 526
Discriminant learning, 631
Discriminant methods, 599
Discrimination, 533
Distinctive features, 363
Distortion
  function, 52
  measures, 710
  spectral, 369
  spectral envelope, 369
  waveform, 369

Distribution
conditional, 534
discrete, 519, 521
estimation, 534
functions, mixture, 520
mixtures of, 519
models, nonparametric, 521
models, parametric, 521
Division into sentences, 796
Donnelly marketing list, 819, 821
Double convolution, 348
Double talking, 338
Dragon System, 509
DSI, 94
DSP processor, 523
DSP32C, 97
DTW, 732
Duration modeling, 538
Duration module, 391
Dynamically changing vocabulary, 675
Dynamically varying vocabulary, 673
Dynamic bigram distributions, 671
Dynamic programming (DP), 258, 487, 516, 644, 834
algorithms in the neural network, 493
frame synchronous, 488
matching, 487
neural network (DNN), 494
learning strategy for, 498
time warping in, 496
one pass, 488
one stage, 488
technique, 213
time alignment, 719
Dynamic range, 542
Dynamic time warping (DTW), 487, 707, 720
Dynamic trigram distributions, 671

Echo canceler
adaptive, 328, 331
line, 330
subband, 345
VLSI implementation of, 328, 353
Echo cancellation, 327
acoustic, 342
line, 343
[Echo cancellation]
subband structure for, 344
Echo path, 331
impulse response of, 332
nonlinear, 334
Echo suppressor, 330
Echoes, 99
acoustic, 328
distinct, 327
line, 328
Ensemble interval histogram (EIH), 455, 457, 463, 466, 473, 479, 482
bin, magnitude of, 464
bin, magnitude parameter of, 463
dominance property of, 464
filter bank for deriving, 455
filter characteristics for deriving, 476
"Hamming"-, 474
EIHgram, 477, 479
Electronic publishing, 101
Embedded ADPCM, 94
Embedded decoding, 530
Encoder, 52
Enrollment session, 710
Entertainment audio, 101
Entropy, 541
coding, 104, 127
Enumerative search, 636
Epoch detection, 27
Equivalence classes of histories, 653
Equivalence classification, 664, 666
Equivalent articulation loss (AEN), 362
Equivalent noise method, 360
Equivalent rectangular bandwidth (ERB), 479
Ergodicity, 657
Ergonomic, 407
Error
in pitch determination, 41
probability of, 661
rate, 99, 829
Evaluation of
intonation, 400
pauses, 400
prosody, 400
sentence accent, 400
word duration, 400

[Evaluation of]
word stress, 400
Event stream, 771
Exceptions dictionary, 818, 822
Excitation function, 742
Excitation modulation model, 87
Excitation pitch pulses, 179
Excitation source signals, 844
Excitation waveform elements, 844
Expected coverage of text, 672
Exploratory data analysis, 829
Exponential duration model, 538
Extracochlear, 270

F_0, 773
contours, 404
for female voices, 775
Fast match list of words, 660
Feedforward network, 580
Fenones, 630
Field tests, 407
Filler templates, 719
Filter bank, 115, 117, 119
hybrid, 123
in perceptual coding, 116
parameters, 322
polyphase, 122
Finite state automaton, 505
Finite-state grammar, 815
Finite state languages, 635
Finite-state vector quantization (FSVQ), 210
First-order Markov chain, 544
Fixed-length SQ, 211
Fixed-point solution, 515, 518
Flapping, 756
Focus shifting transformations, 766
Formant, 834
bandwidth, 256
extraction, 834
frequencies, 256, 834
metric, 322
synthesis, 780
tracking, 834
transitions, 778
Format analysis, 749
Formatting module, 750
Forward-adaptive system, 152
Forward and backward probabilities, 522
Forward-backward algorithm, 535, 545, 630
Forward-backward procedure, 514, 531, 546
Forward-backward training, 639
Forward induction procedure, 514, 547
Forward masking, 132
Forward spectral estimation, 95
Forward variable, 547
Frequency bin adaptive filter, 348
Frequency matching algorithm, 174
Frequency of occurrence, 540
Frequency subband solution, 349
approximation of, 349
Frequency subband structure
adaptation of, 350
extensions of the, 352
Frication, 241, 244
Fricative noise source, 781
FS1016 standard, 159
Full-duplex, 97
Function word, 656, 751, 803, 810
dependent phone models, 633
Fundamental frequency, 4, 9, 702
contour, 762, 773
Fundamental harmonic detection, 26
Fundamental period, 4

Gain normalized likelihood ratio, 713
Gaussian mixtures, 630
Generalized Lloyd algorithm, 541, 716, 718, 721
Generalized probability density function, 533
Generative power, 754
Geometric distance, 250, 253
Global optimality, 515
Glottal chink, 241
Glottal closure, instant of, 28, 38, 40
Glottal losses, 240
Glottal source, 703
Glottal waveform irregularities, 5
Golden ears, 135
Grammars
context-free, 635

[Grammars]
statistical tree-based, 635
unification, 635
Grammatical constraints, 523
Grapheme-to-phoneme conversion, 391
Greedy growing algorithm, 541
Group delay spectrum, 428
Group speciale mobile (GSM), 93

"Human" EIH, 481
"Human" filter bank, 481
Half-duplex, 97
Handicapped subjects, 407
Harmonic pattern, 9
Harpy, 642
Harvard psychoacoustic sentences, 400
Haskins sentences, 400
Hat pattern for F_0, 774
HDTV, 101
Hiatus gesture, 767
Hidden Markov model (HMM), 489, 509, 511, 513, 517, 561, 580, 599, 604, 613, 677, 715, 721, 733
based recognizer, 721
based speaker verification, 722
continuous, 521
continuous density, 630
decoding problem of, 513, 515
discrete, 630
ergodic, 721
ergodic autoregressive, 721
estimation problem of, 513
evaluation problem of, 513
higher order, 544
language model, 680
measure space, 523
probability measure, 521
semi-continuous, 521, 630
single state, 521
training, 625
training algorithm for, 518
vector quantizer, 521
Hidden units, 557
High quality, 99, 104
audio, 101
coder, 97
[High quality]
coding, 98, 100
of telephone speech, 85
of wideband audio, 85
video, 101
Huffman coder, 132
Huffman codes, 132
Huffman coding, 127
Human auditory system, 115
Human cochlea, 483
Human factors, 407
Human vocal apparatus, 744
Hybrid, 328
coder, 87, 89
coding, 92
models, 574
Hypothesis generation, 637

I-divergence discrimination information, 533
Identification
decision, 705
mode, 705
open set, 705
Identity claim, 705
Identity of speaker, 701
Incomplete data problem formulation, 532
Incomplete label case, 525, 530
Incremental learning, 565
Index assignment function, 79
Indicator function, 520
Individual information, 597
Indoor wireless communication, 94
Inflection, pattern of, 753
Information content of a source, 656
Information structure, 796, 813
Infovox, 391
Initial state probability, 512
Inner hair cells, 457
Instantaneous SNR, 319
Intelligibility, 89, 94, 361, 756, 778
of synthesized speech, 784
sentence, 362
word, 363, 401
Intensity measurement, 711
Interactive man-machine dialogs, 785
Interpolation, 522

Interval probability density, 463
Intonation, 743
  grammar, 391
Intonational phrases, 796
Intracochlear, 270
Invariant features, 598
Inverse filtering, 5, 16, 27, 836
  frequency domain, 17
IS 54 standard, 159
ISDN, 89, 94, 97
ISO, 90, 101
Isolated word modeling, 522
Isolated word recognition, 524
Isopreference contours, 361
Itakura distance, 425
Itakura metric, 712
Itakura–Saito distortion measure, 711

Just-noticeable-difference (JND), 256
Joint probability, 513
Joint segmentation and quantization, 212

Kalman filter, 246, 258
Karhunen–Loeve expansion, 170
Karhunen–Loeve transform, 118
k-Means
  algorithm, 722
  clustering, 719
    algorithm, 541
k-Nearest neighbor (KNN), 600, 717
Kullback–Liebler number, 533

Lag, 14
Lagrangian multipliers, 528
Laguerre functions, 334
Language model, 635, 651, 676
  complexity of a, 672
  perplexity, 665
  quality, 656, 660, 665
  statistical, 652
  stochastic, 635
  uniform, 635
Language structure, 744
Language weight, 636
Laryngeal cycles, 9
Laryngeal excitation cycles, 7
Laryngeal modeling, 779
Laryngeal pulses, 8, 28
Laryngeal waveform, 38
Laryngogram, 39
Laryngograph, 25, 38
Lateral inhibition, 295
LBG algorithm, 224
Leaf of tree, 60
Learning effect, 400
Learning vector organization (LVQ), 556, 560, 563
LVQ2, 563
Least mean square (LMS) algorithm, 334
Least squares (LS) algorithm, 339
Left-to-right model, 543
Letter to sound
  conversion algorithm, 752
  interior, 821
  problem, 746
  rule, 749, 755, 818, 821
    errors, 753
Level-crossing detectors, 459
Lexical decision task, 405
Lexical description, 525
Lexical phonology, 761, 785
Lexical retrieval, 749
Lexical stress, 749, 761
  contour, 756
Lexicon, 751
Likelihood function, 531
Likelihood ratio (LR), 373
  distance, 249, 257
Line spectral frequency (LSF), 74
Line spectral pairs, 713
Line spectral parameters, 209
Linear classifiers, 535
Linear code, 532
Linear gain predictor, 96
Linear prediction, 19
  residual, 14
Linear prediction coding (LPC), 742
Linear predictive coding (LPC), 91, 104, 110, 142, 309, 711, 843
  analysis, 424, 834
  cepstrum, 282
    distance, 425, 431
  coefficients, 209

[Linear predictive coding (LPC)]
derived cepstrum, 713
derived model spectra, 712
distance, 110
inverse filter, 713
model, 712
parameters, 322
spectral matching measure, 425
spectrum, 627
technique, 843
vocoder, 209
LPC-16, 403
Linear smoothing, 30
Linguistic structure, 747
List correction, 29
Listening comprehension, 406
Listening tests, 134
LMS algorithm, 343
Local parsing, 764
Log area ratio, 209, 713
Log energy, 537
Log likelihood ratio, 712
Log probability, 638
Lombard effect, 311
Long-distance telephone connection, 328
Long term average features, 727
Long term average spectrum, 726
Long-term predictor, 149
Long term sample statistics, 707
Long term spectral averages, 707
Long term statistics, 714
Longest match first procedure, 755, 758
Look-say method, 746
Loudspeaker telephony, 98
Low-delay (LD), 97
-CELP, 91, 96
coding, 100
Low-rate speech coding, 165
LS algorithm, 343
LSP, 845

Markov chain, 511, 517, 542
Masking, 112
noise, 398
threshold, 116, 123, 128
Massive parallelism, 556
Matrix quantization (MQ), 74, 210, 715, 733
codebook, 719
Matrix quantizer, 718
Maximum a posteriori estimation, 312
Maximum a posteriori probability (MAP), 516, 659
estimation with hidden Markov modeling, 317
parameter estimation, 312
versus MMSE for noise suppression, 322
Maximum likelihood, 514, 523, 526, 528, 530, 540
conditional, 527, 529, 535
criterion, 518, 654
estimation, 312, 533
method, 834
ratio measure, 425
Maximum mutual information, 527, 529, 535
estimation (MMIE), 631
method, 599
Mean opinion score (MOS), 86, 91, 92, 99, 101, 134, 135, 159, 360, 405
Mean squared error, 53
Measurement inaccuracies, 37
Median smoothing, 30
Mel-frequency spectrum, 627
Mel-scale, 444
Memory load, 405
Meta-Pi architecture, 567
Microphone arrays, 104
Minimax criterion, 717
Minimum Bayes risk, 534
Minimum classification error, 524, 534
Minimum discrimination information, 533
Minimum distortion algorithm, 225
Minimum distortion rule, 211
Minimum mean square estimation (MMSE), 313
using Gaussian models, 314
with Hidden Markov Modeling, 318
trained from speech, 314
using two states, 313
with vector quantization on filter bank data, 320
Misalignment vector, 336

Misclassification problem, 535
Misclassification rate, 524
Mispronunciation, 829
Missing-cell recovery, 379
MITalk, 391, 818
Mixed excitation, 32
source, 35
Mixing condition, 337
Mixture of Gaussian densities, 722
Mixture weights, 722
Mobile satellite communication (MSAT), 91
Modal elements of language, 766
Modality effects, 775
Model clustering and splitting, 540
Model merging, 542
Model mismatch, 526, 534
Model topology, 519
Modeling the excitation, 239
Modified discrete cosine transform (MDCT), 120
Modified rhyme test (MRT), 363, 397
Modiolar monopolar electrodes, 272
Modular incremental training, 565
Modulated noise reference unit (MNRU), 365
narrow-band, 365
wide-band, 365
Monomorphemic root, 755
Morph, 748, 753, 758
analysis, 754
covering, 754
type code, 759
Morpheme, 743, 752
Morphological analysis, 749, 818
Morphology, 744, 820
Morphophonemics, 744
Morphophonemic rules, 749, 756, 761
Most likely state sequence, 515
Multi-lingual TTS, 783
Multilayer perceptron (MLP), 556, 580
Multilingual speech input/output assessment, methodology, and standardization (SAM), 388
Multilingual systems, 745
Multimedia, 101
Multiple-choice test questions, 406
Multiple codebooks, 630
Multiple pronunciations, 633
Multiple-template method, 600
Multiplicative connections, 568
Multipulse coder, 150
Multipulse excitation, 93, 153
coder, 145
Multivox, 391
Music, 89, 99
Musical noise, 316
Mutual information, 662
average, 652

Names, 819
Naming latency, 406
Narrowband coding, 109
Nasal cavities, 702
Nasal tract, 235, 244
Naturalness, 86, 89, 360, 404, 756, 778
of synthesized speech, 785
N-best search, 646
Nearest neighbor rule, 53, 541, 716
Near-miss, 535
Near transparent coding, 86
NETgram, 586
Neural network, 246, 606, 614
dynamic (DNN), 573
for noise suppression, 322
full-word models of, 571
hidden control, 574
hidden layer of, 494
input layer of, 494
learning algorithms of, 556
linked predictive (LPNN), 581
model, 488
phonemic level, 558
predictive, 574
scaling of, 565
shift invariant, 560
time-delay (TDNN), 494, 561
Neural prediction model (NPM), 502, 574
learning procedure for, 504
Neural tuning curves, 459
n-Gram, 667
model, 665
modeling, 665
Noise-canceling systems, 104
Noise masking, 105

Noise shaping, 96, 104, 110, 147
Noise suppression
microphone selection for, 310
using sine wave analysis-synthesis, 321
with pattern matching, 309
Noise threshold, 101, 104
Non-linear classification, 559
Nonlinear distortion, 27
Non-linear modeling, 556
Nonlinear phonology, 767
Nonlinear predictors, 581
Nonuniform step size, 343
Normalized LMS (NLMS) algorithm, 338, 351
Noun phrases, 765
Nuclear stress, 766
Numerical scaling, 522
Numerical singularity, 522

Objective measurement, 368
Objective quality
assessment of coded speech, 368
evaluation, 407
measures, 368
Observation density, 543
Observation distribution, 519
form of the, 526
Observation probability measures, 513
Observation space, 521
Offline methods, 608
Online methods, 608
Open loop analysis, 149
Open response, 395
Opinion equivalent Q, 365
Opinion rating, 360
Optimal decoding, 523
Optimal nonlinear interpolation, 76
Optimal state sequence, 516
Optimization, 245, 247, 250, 263
Orthographic form, 741
Overall energy, 703, 711, 714
Overlapping block transform (OBT), 120

Packetized speech, 94
Packet network, 94
Paired comparison, 361, 404
Palatalization, 761
Palate, 702
Parallel distributed processing, 556
Parameter estimation, 523
Parameter smoothing, 522
Parsing, 749, 760, 809, 817
networks, 589
PARSNIP, 586
Part of speech (POS), 663, 668, 678, 760, 804
ambiguity, 802
assignment, 817
codes, 763
nuclear (NPOS), 664
Partial observation sequence, 547
Partition, 52
Passive transformation, 766
Pattern-matching paradigm, 515
Perceptual coders, 109
Perceptual coding, 115, 123
Perceptual quality of synthesized speech, 784
Perceptual transform coders, 131
Perplexity, 523, 652, 658, 665
Phase equalization, 847
Phase interpolation, 176
Phase unwrapping, 174
Phone-like units (PLU), 722, 733
Phonematized dictionary, 392
Phoneme, 770
distance, 397
intelligibility, 394
Phonemic synthesis, 746
Phones, 702
Phonetic balance (PB), 363
Phonetic context, 848
Phonetic interpretation, 791, 792
Phonetic segment, 743
Phonetic transcription, 744
narrow, 743
Phonetically-balanced sentences, 400
Phonic method, 746
Phonological formalism, 783
Phonological representation, linear, 746
Phonological rules, 782
Phonological units, 780
Phonotactic constraints, 396

Phonotopic maps, 560
Phrasal constituents, 765
Phrase level parsing, 765
Piece-wise linear mapping, 223
Pitch, 3, 536, 702, 714
adaptive Hamming window, 169
analysis, fractional-sample accuracy, 91
contour, 42, 702
smoothing, 29
defining, 8
determination, 3
determination algorithm (PDA), 6
artificial neural networks, 25
autocorrelation, 12, 16
center clipping, 12
comb filter, 14
evaluation of, 36, 41
harmonic pattern matching, 14
maximum likelihood, 14
short-term analysis, 11
time-domain, 11, 23
determination errors, 30
difference limen for, 10
errors, 36
frequency, 270
markers, 24, 40
measurement, 711
perception, functional models of, 14
period, 7
place, 270
virtual, 8
Polymorphemic words, 759
Postfiltering, 94, 156
Postprocessing, 29
Post-processor duration model, 540
Pragmatics, 744
Prediction
adaptive, 142
backward, 152
backward-adaptive, 148
forward adaptive, 148
Prediction gain, 96
ratio, 56
Predictive coding, 141
Predictor
long-term, 153, 155, 156
nonlinear, 76
short-term, 152, 156
Preemphasis, 727
Preference equivalent Q, 365
Preference score, 361
Prepositional phrases, 763
Principal component analysis, 537
Probabilistic descent algorithm, 536
Probabilistic modeling, 511
Probabilistic vector mapping, 319
Probability space, 517
Processing time, 405
Program broadcasting, 97
Progressive transmission, 60
Pronominal reference, 767
Pronunciation, 751, 818, 822, 824
Chinese word, 793
code, 759
modules, 750
of names, 393
word, 792
Proper nouns, 805
Prosodic cues, 762
Prosodic parser, 391
Prosody, 536, 762
Pruning, 641
hypotheses, 638
Pseudo-Gray coding, 79
Psychoacoustics, 112, 116, 130
Pulse code modulation (PCM), 87, 91, 94, 99, 101
vector (VPCM), 56
NICAM-, 100
Punctuation, 750, 791, 797

QMF filters, 117, 122, 129, 137
generalized, 117
Quality
assessment of coded speech, 357
broadcast, 358
communication, 86, 358
listening, 359
network, 86, 89, 93
of speech analysis/synthesis, 381
overall, 405
sending, 359
speech, 358
studio, 136
subjective, 360
synthetic, 86, 358

[Quality]
toll, 109, 358
transmission, 359
transparent, 136
Quantization, 209
Quantization noise, 100, 104
just-noticeable, 101
Quantization procedures, 153
Quantizer, 51
rate of, 51
resolution of, 51
Quantizing distortion unit (QDU), 367

Radial basis functions, 560
Rate-distortion performance, 220
Rate of convergence, 535
Real-time performance, 523
Recognition difficulty, 660
Recognition error rate, 526
Recurrent connections, 563
Recurrent network, 561, 581, 586
Recursive adaptation, 69
Recursive least-squares (RLS)
algorithm, 340, 343
implementation, 351
Recursive matching procedure, 759
Reduced syllables, 757
Redundancy control, 101
Reflection coefficients, 713
Regular pulse coder, 150
Regular pulse excitation, 93, 153
coder, 146
Relative frame energy, 317
Repeated encoding, 99
Residual coding, 62
Response times, 405
Reynolds number, 242
Rhyme, 821
test, 363
Rhythm, 702
Risk minimization, 536
Rms amplitude, 795
Roman numerals, 826
Root of a word, 753
Root-power sum distance (RPS), 422
Round window membrane, 270

Satellite communication, 328
Scala tympani, 269
Scalar quantization, 209
Scaling algorithm, 522
Search reduction, 638
Search space reduction, 637
Secure telephony, 91
Segment codebook, 721
Segment quantization (SQ), 74, 210, 715
speaker adaptation in, 222
temporarily overlapping, 216
Segment vocoder, 220
Segmental characteristics, 702
Segmental durations, 772
Segmental evaluation, 394
Segmental k-means algorithm, 518, 532, 626
Segmental speech phenomena, 711
Self-organized clustering, 665
Semantic differential scales, 404
Semantic processing, 544
Semantic word classes, 661
Semantics, 744
Semi-Markov chain, 538
Semi-Markov model, 540, 543
Sentence-division algorithm, 800
Sentence verification task, 401
Set-partition design, 541
Short-term memory (STM), 406
Short-time Fourier transform, 3[illegible]1
SIFT algorithm, 295
Signal bandwidth, 89
Signal energy, 536
Signal generation, 791
Signal storage, 105
Signal-to-noise ratio (SNR), 101, 371
segmental (SNR_{seg}), 371
equivalent, 403
SNR_q, 371
Simulated annealing, 54
Simulated auditory nerve activity, 460
Sine wave
amplitude, 168
coding, 193
envelope estimation, 182
analysis window, 170
cepstral model, 181, 193
cepstral transform coding, 196

[Sine wave]
coders, performance of low rate, 204
excitation, 167
phase model, 179
frame fill, 201
frequency, 169
harmonic frequency model, 189
harmonic samples, 169
harmonic zero phase model, 204
onset time, 179
parameter estimation, 168
peak matching, 173
performance results, 177
phase, 168
phase model performance, 183
phase unwrapping, 176
pitch estimation, 189
positive slope overload, 197
postfiltering, 199
spectral warping, 194
synthesis, 173
voicing-dependent phase model, 185
voicing detection, 191
warping function, 194
Sinus cavities, 235
Sinusoidal components, birth and death of, 173
Sinusoidal model, 165
Sinusoidal representation, 168
Smoothed group delay spectrum distance (SGDS), 422, 428, 440
Source coding, 110
techniques, 718
Source model, 523
Source/filter model, 742, 768
Source/filter production models, 770
Source-tract interaction, nonlinear, 231
Spanish digits, 795
Speaker adaptation, 211, 245, 608
Speaker cluster selection method, 609
Speaker-dependent recognition, 597
Speaker differences
inter, 706
intra, 706
Speaker identification, 717, 728, 730
Speaker identity, 702, 743
Speaker-independent recognition, 488, 494, 598
Speaker models, 706, 732
Speaker normalization, 608
Speaker recognition, 89, 706, 711, 713, 718, 723, 727, 729, 732
automatic, 701, 704
by human listeners, 701
reference models for, 707
text dependent, 706, 714, 722, 726
text independent, 706, 714, 726
Speaker recognizability, 86
Speaker verification, 710, 717, 720, 722
text dependent, 717
text independent, 706
thresholds for, 710
Speaking rate, 404
Speaking styles, 702
Spectral distance (SD) measure, 372
Spectral distances, 212
Spectral estimation, 95
Spectral flattening, 16
Spectral matching measures, 419
Spectral parameters, 209
Spectral segment, 211
Spectral subtraction, 311
Spectral synthesis units, 776
Spectral template, 706
Spectrogram, 706
Spectrogram reading, 604
Spectrum distance measures, 419
Speech acts, 749
Speech coding, 110, 127
very low bit rate, 209
Speech decoder log probability, 661
Speech enhancement, 309
unit, 311
Speech pathologies, 704
Speech processing hearing prostheses, 42
Speech production
model, 233, 742
coding based on, 245
physiological models of, 231
Speech quality
measurement of, 86
of text-to-speech synthesis-by-rule, 387
Speech recognition, 419

[Speech recognition]
  fundamental equation of, 634
Speech spectrograms, 701, 705
Speech synthesis, 833
  by rule, 746
  from concept, 748
  model interpretation for, 771
  systems, 744
  temporal framework for, 771
  unit instantiation for, 771
Speech technology evaluation, 388
Speech transmission index, 407
SPHINX system, 605, 610, 642
SPICOS, 642, 645
SPLIT method, 601
Spreading function, 114, 115
Stack algorithm, 539
Stack decoding, 638
Stammering, 704
State duration models, 538
State duration probabilities, 538
State-optimized likelihood, 518, 532
State sequence, 512
State transition probability matrix, 511, 519
Stationary stochastic process, 517
Statistical independence, 542
Stereo coding, 132
Stereo separation, 104
Stereo wideband channels, 99
Stochastic convergence, 536
Stochastic gradient algorithm, 334
Stochastic process, 513
Stop consonants, 778
Storage media, 101
Store-and-forward, 97
Stress, 743, 816
  analysis, 756
  assignment, 818
  clause level, 749
  linguistic, 775
  patterns, 817
  phrase level, 749
  rule system, 761
  timed language, 761
Stuttering, 704
Subband ADPCM, 129, 136
Subband coder, 98
Subband coding, 100, 101, 104
Subjective assessment by listeners, 338
Subjective evaluation, 134
Subjective quality assessment of coded speech, 363
Substitution/insertion/deletion errors, 397
Sub-word unit, 521, 525, 540
Suffixes
  derivational, 758
  inflectional, 758
Supervised adaptation, 612
Supervised learning, 527
Supervised methods, 608
Supraglottal filter, 742
Suprasegmental characteristics, 702
Suprasegmental features, 714
Suprasegmental speech phenomena, 711
Surnames, 819
Switched-adaptive interframe vector prediction (SIVP), 74
Syllable, 743, 770
Synchronous satellite, 331
Synchrony, 463
Synonym language model, 680
Synonym language modeling, 677
Synonym sets, 676, 679
Syntactic ambiguity, 762
Syntactic analysis, 761, 801
Syntactic processing, 544
Syntactic word classes, 661
Syntagmatic constraints, 747
Syntagmatic relations, 765
Syntax control, 488
  by finite state automaton (FSA), 489
Syntax, 744
Synthesis, 743
  allophone-based, 390
  articulatory-based, 390
  by rule, 743, 833
  diphone-based, 390
  units, 768, 770, 773
Synthesizer, 834
  control parameters for, 233, 245
Synthetic phase residuals, 187

Tag phrases, 813
TASI, 94

TDNN-LR-DP, 578
Teaching aids for foreign languages, 42
Teleconferencing, 85, 98, 104
Telephone channel, 399
Telephone connection via satellite, 331
Template, 719
  based recognizers, 732
Temporal flow model, 561
Temporal nonplace approach, 463
Temporally dynamic networks, 560
Temporally static networks, 559
Terminal analog speech synthesizer, 834
Text analysis, 748, 791, 794, 796
  preprocessor, 750
Text corpora, 829
Text-dependent methods, 608
Text-independent methods, 608
Text preprocessing, 391
Text-to-speech (TTS), 741
  system, 791, 818, 833
    design, 745
    single word, 751
  unrestricted, 745
Threshold test, 705
Time synchrony models, 322
Time-varying bit allocation, 91
Time-varying excitation model, 87
Time-varying signal, 518
Timing-synchrony analysis, 455, 473
Tonotopic organization, 275
Training, 514
  aids for the deaf, 42
  sequence, 514
  utterances, 707
Transform coders, 118
Transform coding, 100, 101, 104, 119
Transform gain, 117
Transition probability, 511, 543
Transitional features, 727
Transitional information, 713
Transmission bit error, 368
Transmission errors, 85
Transparent coding, 101
Transversal filter, 334
Trellis, 522
Trigram, 654, 663, 669
  cache, 669
[Trigram]
  language model, 654, 669, 674, 677, 680
  model, 635, 665
  probabilities, 656, 668, 675, 678
  word frequencies, 764
Triphone, 632
  generalized, 633
  modeling, cross-word, 632
Two echo path models, 339
Two-mass model of glottis, 239

Unigram, 654, 663, 670
  cache, 669
  models, 635
  probability distribution, 670
Unit instantiation, 773
Units for speech synthesis, 391
Unrestricted text, 809, 828
Unsupervised adaptation, 616
Unsupervised methods, 608
Unvoiced-to-voiced errors, 36

Variable bit rate, 156
Variable-length SQ, 211
Variational method, 528
Varying discourse domains, 674
VCV, 843
VCV, VC, and CV words, 396
Vector excitation coding (VXC), 64
Vector excited coder, 146
Vector predictive coding, 63
Vector quantization (VQ), 49, 210, 488, 493, 505, 627, 713, 715, 718, 720
  based preprocessor, 601
  based speaker identification, 723
  based speaker recognition, 717
  based speaker recognizer, 730
  codebook, 715, 719, 723, 730, 733
  entropy-constrained (ECVQ), 55
  gain adaptive, 56
  gain-shape (GSVQ), 58, 67
  high resolution, 55
  learning, 627
  multistage (MSVQ), 59
  for nonlinear prediction, 76

[Vector quantization (VQ)]
redundancy-free error protection codes for, 79
structurally constrained, 58
supervised, 627
tree-structured (TSVQ), 60
versus HMM for noise suppression, 322
Vector quantizer, 51, 718
design, 541
Vector sum excitation, 67
Vector sum excited linear prediction (VSELP), 67
Velum, 233, 702
Verb of saying, 813
Verb phrases, 765
Verification decision, 705
Verification mode, 705
Viterbi algorithm, 489, 516, 531, 540, 548
Viterbi beam search, 642, 644
Viterbi net, 583
Viterbi training, 518
Vocabulary selection, 672
Vocal apparatus, 231
Vocal cords, 702
Vocal fold vibration, 4
Vocal fry, 5, 24, 37
Vocal tract, 233, 702
curvature of, 236
input impedance of, 238
input reflectance, 244
losses, 237
variable, 263
nominal length of, 251
transfer function, 238, 244
wall vibration, 237
wave motion in, 235
Vocoder, 87
systems, 42
Vocoding, 91
Voice
individuality, 597
mail, 97
onset time, 778
pathology, 42
source, 703
analysis, 3
[Voice]
fluctuations of the, 37
Voiced-to-unvoiced error, 33, 36
Voiceless excitation, 32
Voiceprint, 705
Voicing agreement rules, 761
Voicing, degree of, 32
Voicing determination, 3
algorithm (VDA), 32
evaluation of, 36, 41
pattern recognition, 32, 35
threshold analysis, 32
errors, 36, 41
Voicing errors, 37, 41
Voicing transition frequency, 187
VOTRAX, 404
Vowel quality, 752

Wall Street Journal, 797
Waveform coder, 87
Wavelet theory, 137
Weighted centroid, 79
Weighted likelihood ratio (WLR) measure, 373, 421, 431
Weighted nearest neighbor, 79
Weighted spectral distortion (WSD) measure, 372
Weighted squared error, 53
Wideband audio, 98
Wideband coding, 109
Wideband speech, 85, 97, 101
Wiener filter, 312, 317
Window function, 119
Window length of interval histogram, 463
Wireless access, 94
channel, 91
Word gating paradigm, 403
Word *n*-grams, 652
Word predictability, 402
Word pronunciation, 791
Word-sense selection, 803
Word token ambiguities, 801

Zero-crossing analysis, 27
Zero-crossing detector, 291